THE GROWING FUNGUS

THE GROWING FUNGUS

Edited by

Neil A.R. Gow

Department of Molecular and Cell Biology
University of Aberdeen
Aberdeen
UK

and

Geoffrey M. Gadd

Department of Biological Sciences
University of Dundee
Dundee
UK

CHAPMAN & HALL

London · Glasgow · Weinheim · New York · Tokyo · Melbourne · Madras

Published by Chapman & Hall, 2–6 Boundary Row, London SE1 8HN, UK

Chapman & Hall, 2–6 Boundary Row, London SE1 8HN, UK

Blackie Academic & Professional, Wester Cleddens Road, Bishopbriggs, Glasgow G64 2NZ, UK

Chapman & Hall GmbH, Pappelallee 3, 69469 Weinheim, Germany

Chapman & Hall USA, One Penn Plaza, 41st Floor, New York NY 10119, USA

Chapman & Hall Japan, ITP-Japan, Hirakawacho Kyowa Building, 3F, 2-2-1 Hirakawa-cho, Chiyoda-ku, Tokyo 102, Japan

Chapman & Hall Australia, Thomas Nelson Australia, 102 Dodds Street, South Melbourne, Victoria 3205, Australia

Chapman & Hall India, R. Seshadri, 32 Second Main Road, CIT East, Madras 600 035, India

First edition 1995
Reprinted 1995

©1995 Neil A.R. Gow and Geoffrey M. Gadd

Typeset in 9/11 Palatino by Photoprint, Torquay, Devon
Printed in Great Britain at the Alden Press, Oxford

ISBN 0 412 46600 7

A catalogue record for this book is available from the British Library

Library of Congress Catalog Card Number: 94-72015

∞ Printed on permanent acid-free text paper, manufactured in accordance with ANSI/NISO Z39.48–1992 and ANSI/NISO Z39.48–1984

To our families –

Ann and Cameron (the Gows), Julia, Katy and Richard (the Gadds)

– our own fun-guys!

CONTENTS

CONTRIBUTORS

T.H. ADAMS
Department of Biology,
Texas A & M University,
College Station,
Texas 77843, USA

A.M. AINSWORTH
School of Biological Sciences,
University of Bath,
Claverton Down,
Bath BA2 7AY, UK

D.B. ARCHER
Institute of Food Research,
Norwich Research Park,
Colney,
Norwich NR4 7UA, UK

G.W. BEAKES
Department of Biological and Nutritional Sciences,
Agriculture Building,
The University of Newcastle upon Tyne,
NE1 7RU, UK

M.J. CARLILE
Department of Biology,
Imperial College at Silwood Park,
Ascot, Berkshire SL5 7PY, UK

A.J. CLUTTERBUCK
Institute of Genetics,
Church St,
University of Glasgow,
Glasgow G11 5JS, UK.

G.M. GADD
Department of Biological Sciences,
University of Dundee,
Dundee DD1 4HN, UK

A. GARRILL
School of Biological Sciences,
Flinders University,
GPO Box 2100 Adelaide,
South Australia 5001, Australia

G.W. GOODAY
Department of Molecular and Cell Biology,
Marischal College,
University of Aberdeen,
Aberdeen AB9 1AS, UK

N.A.R. GOW
Department of Molecular and Cell Biology,
Marischal College,
University of Aberdeen,
Aberdeen AB9 1AS, UK

G.S. GRIFFITH
School of Biological Sciences,
University of Bath,
Claverton Down,
Bath BA2 7AY, UK

I.B. HEATH
Department of Biology,
York University,
4700 Keele St,
North York, Ontario,
Canada M3J 1P3

P. MARKHAM
Microbiology Department,
King's College London,
Campden Hill Road,
London W8 7AH, UK

D. MOORE
Microbiology Research Group,
School of Biological Sciences,
Stopford Building,
The University of Manchester
Manchester, M13 9PT, UK

J.I. PROSSER
Department of Molecular and Cell Biology,
Marischal College,
University of Aberdeen,
Aberdeen AB9 1AS, UK

A.D.M. RAYNER
School of Biological Sciences,
University of Bath,
Claverton Down,
Bath BA2 7AY, UK

A. VAN LAERE
Katholieke Universiteit Leuven,
Laboratory for Developmental Biology,
Botany Institute,
Kardinaal Mercierlaan 92,
B 3001 Heverlee–Leuven, Belgium

C. STABEN
T.H. Morgan School of Biological Sciences,
101 Morgan Building,
University of Kentucky,
Lexington, Kentucky 40506-0225, USA

D.A. WOOD
Horticulture Research International,
Worthing Road,
Littlehampton,
West Sussex BN17 6LP, UK

PREFACE

This book is about the growth and differentiation of filamentous fungi. The impetus for this work stems from our perception that the coverage of this highly diverse and important group of organisms has been neglected in recent years, despite many significant advances in our understanding of the underlying mechanisms of growth. This situation contrasts with the treatment of *Saccharomyces cerevisiae*, for example, which because of its ideal properties for genetic analyses, has established itself as **the** model eukaryote for the analysis of the cell cycle, and basic studies of biochemical and genetic regulation. This book does not deal with the detailed growth physiology of *S. cerevisiae*, but the authors have attempted to show how studies on this organism have aided our understanding of filamentous fungi.

In attempting to put together a series of chapters that is focused yet comprehensive we have had to make certain choices over the subject matter. Rather arbitrarily we have chosen not to cover slime moulds, which like yeast have been well championed over the years. For the most part we have, however, included oomycetes, which are not true fungi but rather algae in disguise. Oomycetous fungi were included because they exhibit a mycelial growth habit, with hyphae that resemble fungal hyphae remarkably closely. We have not dealt directly with applied aspects of the growth of these organisms in relation to their symbiotic and pathogenic associations with plants and animals or in biotechnological or industrial contexts. Furthermore, the ecology of filamentous fungi has been summarized in several other recent texts. We hope this book is comprehensive in examining the basic processes underlying the growth and differentiation of fungi and that it provides the reader with adequate source references for further information.

It is estimated conservatively that there are more than 1.5 million species of fungi – more than five times the number of vascular plants and second only in diversity to the insects. The extreme diversity of form in the fungi has always been a source of inspiration for mycologists. This book is concerned mainly with those systems that have been well characterized from the biochemical, physiological or genetic points of view. Although it has not been possible to illustrate the breadth of structures and forms of fungi, it is hoped that the principles learnt from the model systems described here can be applied widely to the understanding of how other less well studied fungi grow, divide and sporulate. The taxonomy and classification of fungi remains an underdeveloped aspect of biology. We hope that this text illustrates what is interesting and unique about this group of organisms and stimulates interest in their evolution.

The authors of the chapters were asked to provide a broad review of the general subject area in which they had specialized research interests. We thank them for their efforts in bringing together a wide range of material and for responding quickly to suggested revisions. We would also like to thank the many people who provided original micrographs and figures for the various chapters. Finally we thank Yvonne Knox and Julia Gadd for their help in checking and assembling the chapters in the final stages.

Neil A.R. Gow and Geoffrey M. Gadd

PART ONE

THE GROWING FUNGUS

THE SUCCESS OF THE HYPHA AND MYCELIUM

1

M.J. Carlile
Department of Biology, Imperial College at Silwood Park, Ascot, Berks, UK

1.1 INTRODUCTION

Most fungi, through most of their life cycle, consist of hyphae, cylindrical cells that increase in length by growth at one end (Chapters 13–15). A typical fungal life cycle can be thought of as starting when a spore, a cell that has perhaps just arrived after dispersal through the air, or has been lying dormant awaiting favourable conditions, germinates on a suitable substratum. The germ tube that emerges from the spore is a slender hypha. It grows and branches, and the branches in turn branch, to form a radiating system of hyphae known as the mycelium (Chapters 14–15; Jennings and Rayner, 1984). In a Petri dish in the laboratory the mycelium forms a circular colony on the surface of the agar, the colony increasing in diameter at a constant rate and also growing down into the agar. In nature, less uniform conditions give less uniform mycelial growth, but here too the mycelium spreads on surfaces and penetrates the substratum. The nutrients available in the substratum, whether in the laboratory or in nature, are absorbed and support growth. As exhaustion of nutrients approaches, the production of spores, either for dispersal, a period of dormancy, or both, is initiated. There is an immense diversity in how this is achieved, but the spores are commonly borne on hyphae, or on elaborate fruit bodies constructed of hyphae, that rise above the substratum. There are some fungi that are not hyphal, and some other organisms that are, but hypha and mycelium are associated with the growth and form of fungi to a greater extent than with any other major group of organisms.

1.2 THE FILAMENTOUS FUNGI AS MODULAR ORGANISMS

1.2.1 Unitary organisms and modular organisms

Multicellular organisms are of two kinds, unitary organisms and modular organisms. The concept of a unitary organism is easy to understand, since we are unitary organisms. Our form is determinate; we have four limbs, two eyes and a structure that, apart from sexual differences and accidents, is similar to that of all other human beings. Size variation is limited, with the biggest healthy adults only a few times heavier than the smallest, even when members of different races are compared. Growth is accompanied by a change in form, at first rapid and then more gradual, as the fertilized egg develops into the adult. Development is complete and the ability to produce offspring attained in about 15 years, vigour and fertility decline after about 40 years, and even under the most favourable conditions, death occurs after a few more decades. There is no doubt about what constitutes an individual, and apart from identical twins, individuals differ genetically. Most animals are unitary organisms, and statements similar to those above could be made for animals as different as rats and chickens, frogs and fruit flies. Most of our ideas on the structure, physiology, behaviour and individuality of organisms, on competition and cooperation, on resource capture and on population ecology and genetics, have resulted from the study of unitary organisms.

Many of these ideas, however, have proved unhelpful in the study of plant populations. This

The Growing Fungus. Edited by Neil A.R. Gow and Geoffrey M. Gadd. Published in 1994 by Chapman & Hall, London. ISBN 0 412 46600 7

Table 1.1 Typical features of unitary and modular organisms

Unitary organisms	*Modular organisms*
Growth accompanied by change in form	Growth by the iteration of modular units
Growth determinate, ending when development is complete	Growth open ended, continuing as long as conditions are favourable
Adults vary little in size	Size can vary greatly
The individual clearly recognizable	The individual may be difficult to recognize; the genetic individual (genet) may consist of many physiological individuals (separated ramets)
Usually actively motile	Usually non-motile
Resources usually reached through motility	Resources usually reached through growth
Response to environment as an integrated whole	Localized response to environment
Local damage can be serious, often fatal	Local damage unimportant
Reproductive potential reaches maximum early in adult life	Reproductive potential can increase indefinitely
Clonal reproduction unusual	Clonal reproduction common
Each individual usually genetically different	Members of a clone genetically identical
A defined life span, ending in senescence and death	Senescence and death may be local only, with loss of unwanted modules

led John Harper to formulate the concept of the modular organism (Harper *et al.*, 1986; Begon *et al.*, 1990). Modular organisms grow by the repeated iteration of modules, usually to yield a branching pattern. In most plants a hierarchy of modules can be recognized, for instance branches, branchlets and twigs down to the basic module of a leaf, bud and associated stem (node and internode). Some modular plants, such as trees, grow vertically but others may also extend laterally. In the creeping buttercup or the strawberry plant, for example, horizontal shoots (stolons or runners) produce roots at the nodes and then leaves, forming a module that can survive and flourish even if the internode is severed or decays. Hence a single genetic individual or **genet** can consist of many modules or **ramets** capable of existence as individuals. Modular organisms of this type are referred to as clonal organisms and are abundant. For example, about half of the native perennial plants in the intensively studied flora of Britain are of this type, and so are many non-motile animals worldwide, of which corals are perhaps the best known. Modular organisms contrast with unitary organisms in many ways, which are summarized in Table 1.1 and considered further in later subsections.

The concept of the modular organism is clearly applicable to the fungal mycelium. The basic module from which the mycelium is built is the hyphal growth unit, a hyphal tip with an associated length of hypha (Trinci, 1984; Chapter 14). Some of the features that make mycelial fungi successful are those normally to be expected in modular organisms. Such features will now be considered, before going on to the special attributes of fungal hyphae and mycelia. Other assessments of the fungi as modular organisms are those of Trinci and Cutter (1986) with respect to mycelial structure and growth and of Andrews (1991) in an ecological context.

1.2.2 The exploitation of resources by modular organisms

Some organisms, in order to obtain food, have to look for or pursue other organisms. Only highly integrated unitary organisms can do this effectively. If, however, a supply of nutrients adequate for growth can be obtained at a single site, then an organism can be static in its growth phase, dispensing with the expenditure of energy and thus resources on motility. The way in which modular construction can facilitate the exploitation of a static resource by a non-motile organism will now be considered.

A unit that is consuming a resource creates around itself a resource depletion zone (Harper, 1985; Begon *et al.*, 1990). This applies, for example, to a leaf absorbing light, a root taking up minerals, a protozoan that ingests bacteria or a hypha assimilating organic compounds. An effective solution to the problem, for a static organism, is to grow away from the area depleted of resources and at the same time to produce further modules, located so that their resource-depletion zones overlap as little as possible yet fully exploit the available resources. The way in which this is done accounts for the varied branching patterns of trees, roots and fungal mycelium. The branching pattern in a young colony of *Phycomyces blakesleeanus* and in two young colonies that have encountered each other is illustrated by Harper (1985). No instance was found in the young colonies of one hypha growing over another, the hyphal branches ceasing to grow just before they would have made contact with another hypha. The entire area occupied by the colonies could be divided into polygons, centred on hyphal tips and mostly of similar area, suggesting an efficient exploitation of resources. A different type of analysis was applied to young colonies of *Thamnidium elegans* (Gull, 1975) and *Candida albicans* (Gow and Gooday, 1982), one that showed that the branching pattern was efficient, in giving access to an area, with economy in the total length of hyphae (Chapter 14).

Ecologists studying laterally spreading clonal plants have distinguished between guerilla and phalanx growth forms and strategies (Lovett Doust and Lovett Doust, 1982; Harper, 1985; Begon *et al.*, 1990). A typical guerilla, such as the creeping buttercup, has long stolons that are short-lived and which link nodes at which rooting occurs. As would be expected of a guerilla, it exploits favourable situations, fails to survive severe competition, disappears from some areas and appears in others. The phalanx growth form is named after the infantry units of Alexander the Great, advancing relentlessly in close formation with parallel pikes. Examples of the phalanx growth form are the tussock grasses and bamboos, which with repeatedly branching rhizomes advance slowly but decisively, occupying space and denying it to competitors. Most clonal species have intermediate positions on a continuum from extreme guerilla to extreme phalanx. Within a species, races from different habitats can differ in their position in the spectrum, having evolved in a way appropriate to their habitat. Finally, within a race, internode length and branch frequency can vary with the opportunities available.

The concept of guerilla and phalanx growth forms and strategies can be applied to fungal mycelia. Sugars are readily utilized by almost all bacteria and fungi, but for that reason a localized supply may not remain for long. A guerilla growth form with a long hyphal growth unit and infrequent branching is hence appropriate for fungi that are primary colonists of substrates rich in sugar, enabling them to get there first. Fungi able to utilize cellulose or other refractory substrates have access to a resource that may be massive and long lasting. A phalanx growth form with profusely branching hyphae could facilitate the colonization of such a substrate, and the production of the high local concentrations of extracellular enzymes that may be needed to convert polysaccharides into soluble sugars that can be taken up. It could also permit the production of high local concentrations of antibiotics that inhibit the growth of those bacteria and fungi which, as secondary invaders, may utilize the soluble sugars without having contributed to their release. The phalanx growth form is common in Ascomycetes and Deuteromycetes that utilize polysaccharides. The size of hyphal growth units and branch frequency have been much studied in relation to the mechanics of fungal growth, but more information on branching in relation to natural habitats and substrates would be of interest. Whether isolates of the same species from different habitats show differences in growth form seems not to have been studied, although great differences in hyphal growth unit length and branch frequency have been found in laboratory mutants of *Neurospora crassa* (Trinci, 1973). Within a strain, a rich medium that gives an increased specific growth rate will reduce the length of the hyphal growth unit and hence increase branch frequency, as has been shown for several species (Trinci, 1984). The control of branch frequency by local conditions is shown when a sparsely branching mycelium, spreading from a food base across water agar, passes onto a nutrient rich agar and branches profusely. A single mycelium can thus adopt guerilla or phalanx strategies as required, although the characters of the species impose limits on such adaptation.

1.2.3   Modularity and the mitigation of predation and damage

A unitary organism is readily destroyed by physical injury or the attentions of a predator, but modular organisms can survive a great deal of both. Grasses thrive on grazing and mowing, and every gardener is aware of the ability of clonal weeds to survive, and even be dispersed by, mechanical attempts at control. Fungi, being modular organisms, are well able to survive physical damage and predation. In the laboratory a few seconds in a Waring blender is a good way to convert a mycelium into numerous propagules for inoculating a fermenter. In nature it is probable that the break-up of substrates such as soil and plant materials by animal and human activity will break up and disperse mycelia. Many invertebrates, such as mites, feed on fungi, and it is likely that the benefit arising from their dispersal of spores outweighs any harm done by the consumption of vegetative modules. Some species are known to thrive on the grazing by the tropical Attine ants and other insects that cultivate fungi or encourage their growth (Cooke, 1977).

1.2.4   The survival and dispersal of modular organisms

Unitary organisms reach their maximum physiological capacity for reproduction soon after maturity. Thereafter the probability that offspring will be produced declines with time, because of the high mortality experienced by unitary organisms in nature. There will therefore be strong selection for early maturity and for vigour and reproductive success in the young, and very little for maintaining fitness in the chronologically old. Hence mutations that enhance reproductive performance in the young will be favoured even if they have later deleterious effects. This is thought to be the basis for the evolution of the senescence and limited life span that occurs in all unitary organisms.

There was no need to define 'reproduction' with respect to unitary organisms, but what constitutes reproduction in modular organisms is debatable. An appropriate concept, applicable also to unitary organisms, is that of the production of new individuals that lack physical continuity with a parent. This can occur in modular organisms by

fragmentation and by the asexual or sexual production of seeds or spores. The ability, in modular organisms, to reproduce in such ways does not, as in unitary organisms, reach an inevitable maximum, but increases as module number increases. Hence natural selection will not necessarily favour the health of the young at the expense of the chronologically old, so the evolution of senescence is not inevitable. It may occur, as for example in annual plants, which convert all their resources into dormant seeds before the favourable season ends. However, in many other modular organisms, especially clonal species, senescence and death is local, limited to modules that have outlived their usefulness. As a result clones of some plants may cover many hectares and be hundreds of years old.

The same is true of some fungal mycelia. One clone of *Armillaria bulbosa*, spreading through the soil of a Canadian forest and infecting tree roots, occupies 15 hectares and is thought to be 1500 years old (Smith *et al.*, 1992), and clones of other *Armillaria* species of comparable size and age are known (Brasier, 1992). The circular fairy rings produced by many grassland Basidiomycetes are also evidence for mycelial growth that, starting from a point, has advanced through the soil for decades or even centuries (Buller, 1922; Ramsbottom, 1953; Burnett and Evans, 1966). During this mycelial growth the fruit bodies of the *Armillaria* and the fairy ring fungi will have released enormous numbers of basidiospores, a high proportion of which will have been deposited locally but, as indicated by genetic uniformity, without establishing colonies. Mycelial hyphae are hence far more effective than spores in colonizing the habitats occupied by these fungi. The continued production of fruit bodies in these species, however, indicates the importance of the occasional success of basidiospores in colonizing new sites. Long-continued mycelial growth, and the dormant survival of mycelium or mycelial structures such as sclerotia, are also crucial for the success of the many soil fungi that do not produce spores, such as *Sclerotium* and other Agonomycetes (Mycelia Sterilia). The spread of such fungi to distant sites must be through the dispersal of sclerotia or mycelium in soil particles or in fragments of plant material.

Whereas mycelial growth of indefinite duration is an asset in fungi that live in effectively continuous habitats, it may not be so in fungi that colonize habitats of limited volume. In animal droppings, for example, growth has to come to an end and new sites be reached by spore dispersal. In fungi living in such habitats, selection to postpone senescence indefinitely will be unlikely (Chapter 2). It is hence of interest that one of the few well-established examples of clonal senescence in fungi is in a coprophilous fungus, *Podospora anserina* (Esser, 1991).

1.3 THE POTENTIALITIES OF THE HYPHA

1.3.1 Germ tube behaviour

The germ tube is the slender hypha that emerges from a germinating spore. Nutrient reserves within a spore are limited, so a germ tube must quickly reach an environment having nutrients and otherwise favourable for growth. Accomplishment of this task is facilitated by a remarkable range of sensory responses that determine the direction of growth of the germ tube. Fungi differ in their environmental requirements, so there are differences between species in the nature of these responses, known as tropisms. Germ tubes may grow away from each other (Robinson, 1980), towards amino acids (Manavathu and Thomas, 1985), towards volatile compounds emitted by substrates (Carlile and Matthews, 1988), towards (Robinson, 1973) or away from (Carlile and Tew, 1988) oxygen, away from light (Carlile, 1970) and probably towards water (Wynn, 1981). The germ tubes of fungi that infect plants have to overcome formidable defences, which they do by a variety of routes (Agrios, 1988). Some enter via stomata and others penetrate the cuticle directly. The sensory systems of such germ tubes are varied and sophisticated (Wynn, 1981; Hoch and Staples, 1987). They can include thigmotropism or contact guidance, in which germ tube growth is guided to a suitable site for penetration by microscopic surface features of the plant cuticle. Having reached a suitable site, the germ tube tip may swell into an **appressorium** adhering firmly to the plant surface. Invasion is often by a slender hypha, the infection peg, which grows from the appressorium and penetrates the plant cuticle by both mechanical pressure and enzymic attack. The versatility of the fungal hypha is displayed to the full in plant infection, and also in the infection of insects (Evans, 1988), which involves the penetration of a tough chitin–protein cuticle.

1.3.2 The mature hypha: growth over, through and away from substrata

Fungi live on organic materials. Such materials may be in the form of the dead or living bodies of plants or of animals, especially insects, of fragments of organisms, including fallen leaves, and of wastes, such as animal droppings or plant exudates. A great deal of organic material is in soil, a medium which also contains mineral particles and liquid and gas phases. Hence, in order to reach nutrients, hyphae must be able to spread over surfaces, penetrate substrata, both loose and dense, and cross gaps to reach other substrata. Fungi differ in their habitats, and so their hyphae differ in their capabilities, but the hyphae of most fungi will be able to perform the above activities to varying extents.

The older parts of a fungal hypha adhere to the substratum. If this were not so, cell elongation would be much less effective in advancing the hyphal tip. Growth is limited to the hyphal tip – intercalary growth would lead to hyphal buckling and damage, through friction between the more advanced parts of the hypha and substratum, and through encounter between the tip and obstacles. Protoplasmic streaming brings materials to the hyphal tip. This enables metabolism over a considerable length of hypha, the hyphal growth unit (Chapter 14), to support rapid advance of the hyphal tip, a highly competitive feature. It also enables the hypha to traverse regions devoid of nutrients to reach new substrates. The older parts of the hyphal wall are rigid with considerable mechanical strength. This permits high turgor pressures to be developed, and considerable hydrostatic pressure to be exerted at the plastic and extensible hyphal tip. This pressure will assist the hypha in forcing its way through such substrata as soils and plant tissues. A tubular cell with growth confined to a tapering apex is clearly an effective device for penetrating substrata. It is seen not only in the fungal hypha but in the pollen tubes and root hairs of higher plants. The former penetrate the female tissues (stigma and style) of flowers to effect fertilization, the latter grow through the soil, enormously increasing the absorptive surface of roots.

The ability of hyphae to grow away from substrata, enabling them to cross a gas phase to reach new sites for growth, is evident in those fungi which produce copious aerial mycelium on agar media. Specialized stout hyphae, stolons, able to cross very wide gaps, occur in the Zygomycete genera *Rhizopus* and *Absidia*.

Information on the guidance of vegetative hyphae, other than germ tubes, is sparse; mature hyphae, which are almost always accompanied by other hyphae and by branches, are less suitable than germ tubes for the study of tropisms. There is evidence for negative autotropism, the avoidance of other hyphae, in mature vegetative hyphae (Hutchinson *et al.*, 1980; Trinci, 1984) as well as in germ tubes. There is scepticism (Gooday, 1975) about the occurrence of tropic responses by vegetative hyphae to exogenous factors, other than positive chemotropism to amino acids in aquatic Oomycetes, substances that are not effective in bringing about tropism in other fungi. Most fungi are, however, terrestrial, living in environments in which liquid continuity is interrupted by gas phases, hindering the diffusion of substances that are not volatile. It may hence be more rewarding for fungi to respond to – and investigators to seek – chemotropic factors that are volatile as well as soluble (Carlile and Matthews, 1988; Carlile and Tew, 1988).

1.4   LOWER FUNGI AND HIGHER FUNGI

Two fungal groups, the Ascomycetes and Basidiomycetes, bear sexually produced spores in complex fruit bodies visible to the naked eye. Such fruit bodies, suggesting an evolutionarily advanced condition, have led to these two groups being termed the higher fungi, along with a third group, the Deuteromycetes. The latter are fungi that are related to the Ascomycetes, or less commonly the Basidiomycetes, but have lost the sexual process and hence lack fruit bodies. The hyphae of higher fungi have numerous cross walls or septa (Chapter 3; Gull, 1978). Hyphal fusion (Buller, 1931, 1933) is also common, short side branches from hyphae fusing with corresponding side branches from adjacent hyphae and establishing protoplasmic continuity. A radiating system of hyphae is thus converted into a three-dimensional network (Gregory, 1984). These features led Gregory to describe the mycelium of higher fungi as septate-reticulate. There are several groups of fungi, among which the Oomycetes and Zygomycetes are the best known, which do not produce large and complex fruit bodies. Because of this they have been regarded as evolutionarily primitive and termed the lower fungi. Septa and hyphal fusions are relatively uncommon in the mycelium of lower fungi. Gregory (1984) termed such mycelia as filamentous-coenocytic. The term is perhaps unfortunate as the hyphae of higher fungi too are filamentous and also coenocytic – there are usually several nuclei in the compartments between septa, and the compartments are usually linked by septal pores. Here the mycelium of the lower fungi will be described as **simple** in contrast to the **septate-reticulate** mycelium of higher fungi. The association between large and complex fruit bodies and the septate-reticulate mycelium suggests that such a mycelium is needed for the production of large fruit bodies; this possibility will be considered further (page 12).

1.5   THE SIMPLE MYCELIUM

In some lower fungi mycelial development is very limited. The Oomycetes and Zygomycetes, however, both produce extensive mycelia. Many members of the Mucorales, the best studied Zygomycete order, are saprotrophic sugar fungi, unable to attack more refractory carbon sources. They have rapidly extending hyphae which facilitate the swift colonization of newly available substrates. As hyphae advance, protoplasm moves into the younger parts of the mycelium, vacating the older hyphae. These empty hyphae are then cut off by septa, which will also develop to seal off any damaged hyphae. The Oomycetes include aquatic fungi and also important plant pathogens, and like the Zygomycetes produce extensive mycelia. Molecular and ultrastructural evidence, however, now makes it clear that the relationship between the Oomycetes and other fungi is very remote (Cavalier-Smith, 1987). Indeed, it is probable that we and other animals are more closely related to the true fungi than are either to the Oomycetes, termed by Cavalier-Smith the pseudofungi. Hence in the evolution of the Oomycetes there has been an independent origin of hypha, mycelium, sporulation and a life style closely simulating those of the true fungi.

Extensive septum production occurs in the growing mycelium of some Zygomycete orders (Alexopoulos and Mims, 1979) and has even been observed in young hyphae of *Mucor hiemalis* and *M. ramannianus* (Fiddy and Trinci, 1977). Fusion between vegetative hyphae occurs in the Zygomycetes *Mortierella* (Griffin and Perrin, 1960) and *Syncephalis* (Gregory, 1984) and in the Oomycete *Phytophthora* (Shaw, 1983). The lower fungi are thus fully capable of hyphal septation and fusion. It would seem that for their life style a simple mycelium is normally adequate, and septum production and hyphal anastomosis a waste of resources and commonly lacking.

1.6 THE ADDITIONAL CAPABILITIES OF THE SEPTATE-RETICULATE MYCELIUM

1.6.1 Damage limitation

The hyphae of fungi are liable to be damaged in a variety of ways, such as by grazing by mites and other Arthropods and by osmotic rupture through sudden changes in external water potential. It is likely that such damage can be more effectively limited and remedied in the higher than in the lower fungi. In the mycelium of higher fungi cross-walls develop some distance behind the extending hyphal tip. These septa have pores of various types linking the adjacent hyphal compartments (Gull, 1978). In *Endomyces geotrichum* (anamorphic state, *Geotrichum candidum*), and probably in other members of the order Endomycetales, the pores are numerous and minute, but in most higher fungi they are large enough to permit rapid streaming of protoplasm and even the movement of nuclei between compartments. The mycelium of higher fungi is hence, like that of lower fungi, functionally coenocytic. In the lower fungi a damaged region can be sealed off by the development of a septum, but may spread for a long time, as much as 20 minutes, before this occurs (Buller, 1933). In higher fungi, however, the spread of damage from one compartment to others can be prevented by the prompt blocking of septal pores. This can occur either through the occlusion of pores by protoplasmic inclusions such as crystals or by the rapid synthesis of pore plugs (Trinci and Collinge, 1974; Chapter 5). In addition, the readi-

ness with which hyphal fusion occurs in higher fungi means that protoplasmic flow need not be interrupted for long. Hyphae can grow from compartments adjacent to a damaged one, even through the damaged compartment, and fuse, re-establishing protoplasmic continuity. Alternatively protoplasmic streaming can be diverted through the mycelial network to bypass the damaged area (Buller, 1933).

1.6.2 Dikaryon formation and maintenance

In most organisms fusion between two haploid cells that differ in mating type (plasmogamy) is followed by fusion between their nuclei (karyogamy), to give a diploid cell. The haploid cells involved are usually ones that have undergone a commitment to the sexual process, and in most fungi the mating is soon followed by the formation of spores. In Basidiomycetes, however, cell fusion follows an encounter between vegetative hyphae that differ in mating type. The hyphae may both be from established mycelia, or one of them may have arisen from a spore that has recently germinated on or near an established mycelium. The hyphal fusion results in a hyphal compartment with two haploid nuclei of different mating types. Instead of nuclear fusion, nuclear division follows, nuclei of each mating type migrating through septal pores into compartments that lack that mating type. The result is the conversion of mycelia that are monokaryons (i.e. with haploid nuclei of one mating type) into a dikaryon, with hyphal compartments that contain two haploid nuclei, one of one mating type and one of the other. When a mycelium has become a dikaryon, the dikaryotic state is maintained in hyphae advancing at the colony margin by means of clamp connections (Figure 1.1). Hence, as a result of hyphal fusion and nuclear migration, a possibly already extensive monokaryotic mycelium can be converted into a dikaryon and, as further growth occurs, be maintained in that state (Casselton, 1978). Fruit bodies, in which nuclear fusion, meiosis, genetic recombination and spore production occur, can then be initiated at appropriate seasons and optimal sites on the mycelium. The latter advantages would only be pronounced in mycelia that are extensive and long lived. Those of many Basidiomycetes are both, some covering large areas and surviving for decades (page 6).

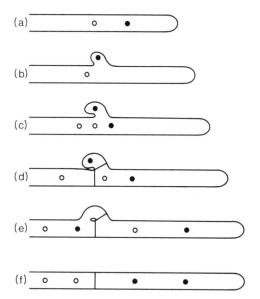

Figure 1.1 Diagram illustrating how the dikaryotic state is maintained by means of clamp connections. (a) A dikaryotic hyphal tip, with two nuclei of different mating types. (b) The development of a backwardly growing side branch, into which one of the nuclei has moved. (c) Synchronous division of the two nuclei has occurred, and one of the daughter nuclei from the side branch nucleus has moved back into the main hypha. (d) A septum has developed between the daughter nuclei derived from the nucleus that remained in the main hypha, and a second septum cutting off the side branch. Septal pores are not indicated. (e) The tip of the side branch has fused with the main hypha, giving the clamp connection typical of the Basidiomycete dikaryon, and the side branch nucleus has moved into the main hypha. A new compartment with two nuclei of differing mating type has thus been formed, with the apical compartment remaining dikaryotic. (f) An illustration of the probable consequence of nuclear division in a dikaryotic hyphal tip as illustrated in (a) but without clamp connection development, resulting in hyphal compartments reverting to the monokaryotic state.

1.6.3 Eelworm traps

Eelworms (nematodes) are very abundant in the soil, with numbers ranging from 1 to 20 million m^{-2}, and are attacked by a variety of nematode-destroying fungi (Barron, 1977). Some of these fungi infect when their spores, ingested by eelworms, germinate. Others, especially members of the higher fungi, trap the eelworms, after which they are invaded by hyphae. The traps in some species consist of specialized hyphae ending in

adhesive knobs, and in others three-dimensional adhesive hyphal networks are produced. There are also species with hyphal rings (non-constricting rings) in which fast-moving eelworms get jammed, and others with constricting rings. These, when an eelworm touches their inner surface, inflate in about 0.1 s, trapping the eelworm.

1.6.4 The construction of multihyphal organs and tissues

Multihyphal organs visible to the naked eye are rare in the lower fungi. The only well-known examples are in a Zygomycete order, the Endogonales, in which many species produce sporocarps, clusters of zygospores surrounded by hyphae and ranging in diameter from 1 to 20 mm. (Alexopoulos and Mims, 1979; Webster, 1980). In the higher fungi, however, multihyphal organs are widespread, and are often composed of several distinct tissues. They can have a role in vegetative growth, in crossing inhospitable terrain to colonize a new site, in dormant survival and in both sexual and asexual reproduction. Examples of each role follow.

(a) Lichen thalli

Lichens (Hale, 1983; Hawksworth and Hill, 1984) bear some resemblance to small plants such as mosses and liverworts in size, and often in form and colour. They are not, however, plants but an intimate symbiotic association between a fungus and a photosynthetic organism, either an alga or a blue–green bacterium. Each lichen species has a different fungal component, an organism not known in nature except as part of the lichen. The photosynthetic organisms belong to relatively few species, and some of them are also common in the free-living state. Lichens are now classified as fungi, the vast majority of species being assigned to Ascomycete families, although a few are Basidiomycetes. The vegetative growth (thallus) may be crustose, firmly encrusting rock or bark, foliose or leaf-like (latin, *folium*, a leaf), or fruticose (latin, *fruticosus*, like a bush). The thallus consists of hyphae, often organized into several distinct tissues, and of cells of the photosynthetic partner, often confined to an upper or outer layer of the thallus where light intensities will be adequate for photosynthesis. Hyphae may also be organized into strands that attach to or penetrate the substratum, and pores that facilitate gas exchange between thallus and atmosphere. Multihyphal structures associated with asexual and sexual reproduction are also formed, the latter permitting the assignment of lichens to various Ascomycete orders.

(b) Mycelial strands and rhizomorphs

Mycelial strands (also termed mycelial cords) and rhizomorphs (Watkinson, 1979; Webster, 1980; Thompson, 1984) are widespread in Basidiomycetes that produce large fruit bodies, and also occur in a few Ascomycetes, again in species that produce large fruit bodies, such as *Xylaria polymorpha*. Mycelial strands develop behind an advancing mycelial front by the growth of hyphae alongside and in contact with an existing hypha, to give a highly organized and differentiated structure consisting of hundreds or thousands of parallel hyphae. The most intensively studied mycelial strands are those of *Serpula lacrymans*, the notorious dry rot fungus. Reaching up to 5 mm in diameter, these strands provide efficient conduits for transporting water and nutrients to sustain the advance of a mycelial front across, for example, many metres of brickwork, to reach, colonize and rot more timber. The term 'rhizomorph' means 'having the form of a root', and the most intensively studied rhizomorphs, those of the honey fungus *Armillaria mellea*, simulate plant roots closely (Watkinson, 1979). There is a 'meristem', a region of cell division, about 25 μm behind the tip, which extends backwards as a cylinder corresponding to the cambium of roots. In front of the 'meristem' is a region corresponding to a root cap, protecting the meristem from damage by friction against soil particles as the rhizomorph advances. Considerable hyphal differentiation occurs behind the region of rapid growth. Rhizomorphs, advancing from a food base into non-nutrient media, can grow over ten times as fast as can normal mycelial hyphae under similar conditions. In nature they are able to extend many metres through the soil from diseased roots to attack hitherto healthy trees and shrubs.

(c) Sclerotia

Many higher fungi form sclerotia (Willetts, 1978), multihyphal structures capable of prolonged survival. They result from the repeated branching and interweaving of hyphae to form a compact mass which, depending on species, commonly ranges in size from about 1 mm to a few cm. Sclerotia can, however, be much larger. Those produced by the Australian *Polyporus melittae* reach 15 kg and, as 'Blackfellow's Bread', are eaten by Aborigines. Many sclerotia show hyphal differentiation, with the outer layers of cells thick walled and pigmented to form a tough rind. Sclerotia may be formed in response to adverse conditions or as a normal step in the life cycle. The sclerotia of *Claviceps purpurea*, the ergot of rye, are formed at the end of the host's growing season, fall to the ground and, having survived winter, germinate to give fruit bodies which carry perithecia from which ascospores are dispersed. The sclerotia of the onion pathogen, *Sclerotium cepivorum*, can survive in soil for decades in the absence of their host. Then, on stimulation by volatile compounds emitted by onion roots, they germinate to give a mycelium that infects the host. Prolonged survival in adverse conditions is possible with very small structures, such as some ascospores and basidiospores and the still smaller endospores of bacteria. Multicellular sclerotia, however, can carry sufficient nutrient reserves to allow massive spore production and dispersal, or attack on resistant host plants when favourable conditions return.

(d) Coremia (synnemata)

In the higher fungi asexual sporulation commonly occurs on conidiophores, erect hyphae which may be branched. Spores may, however, be formed on, or in, more complex multihyphal structures. An example of such a structure is the coremium (plural, coremia) or synnema (plural, synnemata), which occurs in *Penicillium* and some other Deuteromycetes and Ascomycetes (Watkinson, 1979). In many species of *Penicillium* there is a tendency for conidiophores to adhere to each other to form tufts. Where this tendency is highly developed, the result is a coremium, a structure that is much stouter than an individual conidiophore and which can hence rise much higher above the substratum, to give enhanced opportunities for spore dispersal. The most spectacular coremia in the genus are those of *Penicillium claviforme* (Carlile *et al.*, 1961) and *Penicillium isariforme* (Carlile *et al.*, 1962a) which illustrate the versatility of even relatively simple multihyphal systems. The coremia of *P. claviforme* are determinate, with a sequence of developmental steps ending in sporulation, whereas those of *P. isariforme* are indeterminate, with growth continuing at the apex while sporulation occurs lower down the coremium. Parallel orientation of hyphae in the coremium is brought about by the phototropism of the constituent hyphae, and numerous hyphal fusions help to maintain coherent growth. Such hyphal fusions are much less common in the rudimentary coremia of some other species, such as *Penicillium italicum* (Carlile *et al.*, 1962b).

(e) Fruit bodies

Sexual sporulation in higher fungi usually occurs in or on fruit bodies. These multihyphal structures protect the developing spores, facilitate spore dispersal, or both, and display a great diversity in size and form (Ramsbottom, 1953; Alexopoulos and Mims, 1979; Webster, 1980). In many Ascomycetes the ascospores are actively discharged from asci, which are borne on dish- or cup-shaped apothecia or within flask-shaped perithecia. In others the asci occur within roughly spherical cleistothecia, and the ascospores are released by the lysis of the asci and the rupture of the cleistothecia. Perithecia may develop on larger multihyphal structures, stromata, that are several centimetres high. The development of perithecia on stromata may be preceded by a phase in which conidia are formed on the stromatal surface. Truffles are underground stromata containing cleistothecia. They are relished and sought by mammals ranging in size from rodents to gourmets, although the former contribute more to the scattering of spores. Fruit bodies are even larger and more varied in the Basidiomycetes (Chapter 20). The fruit bodies of the cultivated mushroom, *Agaricus bisporus*, are typical of those of a large order, the Agaricales. The role of the fruit body in these fungi is to bear, at some distance above the ground, an extensive downward-facing area of basidia, the cells from which basidiospores are discharged and carried away by air currents. Tissue formation in the fruit bodies of one Agaric, the Ink Cap *Coprinus cinereus*, is considered in detail in Chapter 20. In contrast to the fleshy and relatively short-lived fruit bodies of the Agarics, those of another order, the Aphyllophorales, which includes the bracket fungi of trees, are often leathery or woody in texture and may discharge spores for months or even years. Their tough fruit bodies may be made of up to three distinct types of hyphae, generative hyphae that bear basidia, very thick-walled skeletal hyphae, and much branched binding hyphae that weave themselves between the other hyphae (Webster, 1980). A wide variety of fruit bodies and spore-dispersal mechanisms occur in the Gasteromycetes, the best-known representatives of which are the Puffballs. The evolution of large fruit bodies has permitted the release of prodigious numbers of spores (Ingold, 1971). For example, it is estimated that the Giant Puffball *Calvatia gigantea* contains about 7×10^{12} spores, and that the large bracket fungus *Ganoderma applanatum* discharges about 3×10^{10} spores per day for about 6 months, to liberate a similar total. Even a small and short-lived Agaric, the Ink Cap *Coprinus comatus*, discharges about 9×10^{9} spores in its two to three days of activity.

(f) Hyphal fusion and septation in relation to multihyphal organs

As a fungal colony spreads, protoplasm is withdrawn from the older hyphae. In the lower fungi this results in the break-up of the colony into many physiologically isolated ramets (Figure 1.2). The protoplasm and nutrient reserves in such a ramet are an adequate resource for the development of the simple reproductive structures that occur in the lower fungi, but not for the production of sclerotia and the multihyphal organs commonly involved in sexual and to a lesser extent asexual sporulation in the higher fungi. In higher fungi, however, hyphal fusions can link the hyphae within a clone giving a developing fruit body or other multihyphal organ access to the resources of an extensive mycelial network. As already indicated (page 11), hyphal fusions appear to have a role in maintaining coherent growth in coremia. They are also frequent in the development of mycelial strands, sclerotia and fruit bodies. Whereas hyphal fusions can allow the passage of nuclei and cytoplasm between hyphae, septum formation can limit such movement, even within a hypha. Gull (1978) pointed out that septa and septal pores were most elaborate in fungi that formed complex fruit bodies, and suggested that

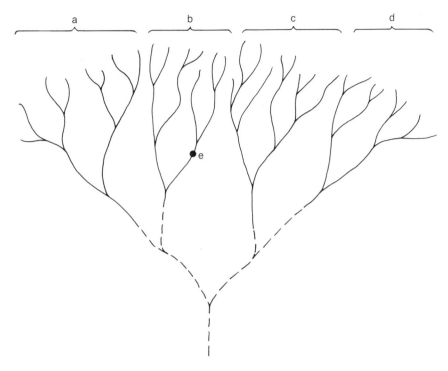

Figure 1.2 Diagrammatic representation of a simple mycelium colonizing a substratum. Protoplasm has been withdrawn from the older hyphae, indicated by broken lines. The hyphal systems a–d have hence become physiologically separate although still belonging to the same genetic individual (genet). If the development of a sporophore is initiated at, for example, site e, materials for its construction can be drawn only from hyphal system b, imposing limitations on the size of the sporophore. In a septate-reticulate mycelium, however, fusion between hyphae within a clone could link mycelial systems a–d, enabling resources to be withdrawn from all of them to support the construction of a large fruit body at e.

septa had a role in the delimitation of cell types in tissue differentiation (Chapters 5 and 20).

1.7   AN ALTERNATIVE TO THE HYPHA – THE YEAST CELL

1.7.1   The disadvantages of being a hypha

The hypha, as already discussed, is an effective device for advancing over surfaces and especially for penetrating substrata. It is, however, expensive in terms of nutrient requirements. For every 1 mm that a hypha advances, 1 mm of cylindrical hyphal wall has to be synthesized. Ultimately protoplasm is drained from older hyphae as the mycelium continues to advance or sporulation occurs, abandoning a considerable amount of carbon and nitrogen in the form of chitin and other hyphal wall components. Not only is hyphal growth costly in resources, but the hyphal apex is a vulnerable structure. This results from the advantageous feature of a high growth rate combined with the necessity of confining hyphal extension to the tapering apex (Chapter 13). Thus a strain of *Neurospora crassa* has hyphae that advance at 38 μm min^{-1} with a hyphal extension zone of only 29 μm (Steele and Trinci, 1975), requiring that in the extension zone a length of wall material equal in length to the entire extension zone has to be synthesized in less than a minute. Even for a slow-growing fungus, *Penicillium digitatum*, the corresponding time is under 5 minutes. Hence, although the wall in the mature hypha can be thick and tough, the extension zone is a site of rapid structural change and is thin and plastic. It is therefore vulnerable to mechanical damage and, if there is a sudden increase in ambient water potential, to osmotic rupture as water flows into the cell (Robertson and Rizvi, 1968; Bartnicki-Garcia and Lippman, 1972).

1.7.2 The success of the yeast cell

The form that has the smallest surface area per unit volume is the sphere. Hence the greatest economy in wall material is achieved with a spherical cell. Although few yeasts are precisely spherical, most are spheroids with the long axis not much longer than the minor axes, and are hence economical in wall materials. A yeast cell replicates by producing a bud that becomes a daughter cell. Since the mother cell survives to produce further daughter cells, wall material is not wasted. As well as being economical in resources, yeast cells are very resistant to osmotic rupture. The potentially vulnerable site, corresponding to a hyphal tip, is the bud that gives rise to a daughter cell. The bud, however, typically takes about an hour to expand to the size of the mother cell, so that the low rate of wall extension permits a different and much more resilient structure than that at a hyphal tip. Furthermore when the bud wall is thinnest, at the beginning of bud formation, the radius of the bud is small. Stress in the wall of a pressure vessel is directly proportional to the radius of the vessel, so with a yeast bud is minimal at the potentially vulnerable stage.

The yeast form is hence advantageous when there is no need to advance across a surface or penetrate a substratum, and when there is a risk of rapid changes in ambient water potential. Many fungal species from a wide range of taxonomic groups are able, in response to appropriate environmental stimuli, to switch from hyphal extension to yeast-like budding – the phenomenon of yeast–hyphal dimorphism (Chapter 19). About 1% of known fungal species are regarded as yeasts. These are the fungi that are incapable of hyphal growth or that live predominantly in the yeast phase. Such 'true yeasts' live mainly on the surfaces of plants (Phaff and Starmer, 1987), where they do not have to penetrate substrata, can be spread by dew or rain splash, and are exposed to violent changes in ambient water potential, as occurs when sugary exudates are diluted by sudden rain.

Opportunistic pathogens of man, such as *Blastomyces dermatitidis* (anamorph *Ajellomyces dermatitidis*), usually produce hyphae when living saprotrophically in soil or in animal droppings, but assume a yeast form when they infect man (Kwon-Chung and Bennett, 1992). Probably yeast cells are transported more efficiently in the vascular and lymphatic systems of warm-blooded animals than are hyphae. In *Candida albicans* both the hyphal and yeast forms appear to be important in disease, with some organs being colonized most effectively by the yeast form and some by hyphae (Ryley and Ryley, 1990; Cutler, 1991). Fungi attacking the skin (dermatophytes) have a tough structure to invade, and are usually mycelial. Many yeasts have a commensal relationship with insects, often living in their gut (Phaff and Starmer, 1987), but insect pathogens, which have to penetrate the tough chitinous exoskeleton, are usually mycelial (Evans, 1988; McCoy *et al.*, 1988).

1.8 THE SUCCESS OF THE FUNGI

As has been indicated in the preceding pages, the hypha and mycelium are versatile devices. The most ingenious man-made devices, however, may not be a success in the market, and in nature there are remarkable structures that are possessed merely by a few rather rare species. The real test for the success of hypha and mycelium is whether the principal mycelial organisms, the fungi, are a success. The success of a group of organisms can be assessed in various ways. Here three criteria will be used. First, how many species are there? The number of known species in each taxonomic group and in the fungi as a whole is given in the 7th edition of *Ainsworth & Bisby's Dictionary of the Fungi* (Hawksworth *et al.*, 1983). The total for fungi is about 64 000. However, hundreds of new species are discovered every year, and the numbers so far known are likely to be only a small fraction of those that exist. Hawksworth (1991) regards 1.5 million as a conservative estimate, omitting species that inhabit insects. Many insects are hosts for parasitic, mutualistic or commensal fungi, and these fungi often have a limited host range. Since insects include far more species than occur in the rest of the animal kingdom or among plants, it is probable that there are very large numbers of undescribed fungal species among insects. It is difficult to estimate how many, but Hawksworth (1991) regards 1.5 million as possible, thus suggesting a total of 3 million fungal species, a number greater than for any biological group except the insects. A further criterion of success is that of the scale of activity – do fungi constitute an important or even indispensable part of the biosphere? Finally, have fungi survived for a long time and are they likely to continue to do so? These latter criteria, and a further consideration of species numbers, are dealt with below.

1.8.1 The success of the lower fungi

As indicated earlier (page 8), two groups of lower fungi, the Oomycetes and Zygomycetes, produce extensive mycelia. Since hyphal fusion and septation are rare in these fungi, their degree of success is indicative of that of the simple mycelium. Only about 600 Oomycete species are known, but the Oomycetes include many successful plant parasites, including what is probably in economic terms the most important of all genera of plant pathogens, *Phytophthora*. One species, the causal agent of potato blight, *Phytophthora infestans*, influenced history by causing a terrible famine in Ireland and massive emigration to the USA. The Zygomycetes are also unimpressive in numbers of species, about 700 being known, but include some very common saprotrophs (Mucorales) and insect pathogens (Entomophthorales). Another Zygomycete order, the Endogonales, includes *Glomus* and other genera responsible for the formation of vesicular-arbuscular mycorrhiza. This mutualistic relationship results in improved phosphorus uptake and enhanced growth of plants in grassland and in the tropics (Read, 1991). It has existed at least since the Triassic period. Fungi that may have had a similar relationship with their hosts, and been essential in facilitating the colonization of land by plants, have been found in the earliest known fossil plants, from the Silurian period (Hawksworth, 1991). The lower fungi, although at present low in the number of species, have had a long history and remain important as parasites, mutualistic symbionts and saprotrophs.

1.8.2 The success of the higher fungi

The higher fungi, with their septate-reticulate mycelium, include far more species than do the lower fungi. About 29 000 Ascomycete species are known, of which nearly half occur as lichens. Deuteromycetes, most of which are likely to have arisen from Ascomycetes by loss of the teleo-

morphic phase, number about 17 000. About 16 000 Basidiomycetes are known, over half of which – about 8000 Hymenomycetes (Agaricales and Aphyllophorales) and 1000 Gasteromycetes – produce macroscopic fruit bodies. The important plant parasites, Rusts and Smuts, number about 6000 and 1000 species respectively.

The higher fungi also fulfil an important ecological role. There is, through photosynthesis, an annual biomass production of about 100×10^9 tons. Higher fungi, many of which can degrade cellulose, have a major role in breaking down this biomass and returning carbon dioxide to the atmosphere and minerals to the soil. About 20×10^9 tons of the biomass are in the form of lignin, attack on which is accomplished almost wholly by Basidiomycetes – without the Basidiomycetes the globe's carbon dioxide would become lignin, and life end under accumulated wood. Basidiomycetes are also responsible for the formation of ectomycorrhizas, which in the Cretaceous period enabled forests to spread from the tropics into temperate regions and poor soils (Hawksworth, 1991). Ectomycorrhizas are highly effective in the uptake of combined nitrogen and remain important for the survival of forests on soils in which nitrogen availability limits growth (Read, 1991). Finally, lichens are able to grow in environments too poor in nutrients and too harsh to support the growth of higher plants, including the Arctic tundra. Some lichens even grow within the outer translucent layers of porous stones in the cold deserts of Antarctica (Friedmann, 1982), where they account for a substantial biomass and are perhaps the longest-lived organisms on earth (Prince, 1992).

1.8.3 The future of the fungi

The fungi have been a success in the past, but what of their future? It is estimated (Hawksworth, 1991) that the current assault on the tropical forests could lead to the disappearance of about 400 000 fungal species, far more than have so far been described, in the next 25 years. In temperate lands pollution and changes in grassland and forest management are leading to a decline in the numbers of many fungi, especially mycorrhizal species (Arnolds, 1991) and lichens (Richardson, 1991). The prospects for future fungal biodiversity seem bleak. However, although the number of fungal species may diminish, fungal biomass may not. The fungal hypha remains an effective device for the penetration and utilization of plant tissues, dead or alive, and extensive areas of genetically uniform plantations and crops provide an ideal environment for the explosive spread of plant pathogenic fungi well adapted for attack on such hosts.

There is a further route by which some fungi will survive through, rather than in spite of, human activities – domestication. There are many animals and plants which are now rare or extinct in nature but which had attributes that led to their domestication and to the survival of their descendants. These descendants – compare for example the dog and its ancestor the wolf – may now occur in far greater numbers and diversity than their wild ancestors. Fungi too have attributes that have led and will continue to lead to their domestication and a major role in biotechnology. These attributes include the ability to infiltrate solid substrates with hyphae and to secrete extracellular enzymes, especially those able to break down macromolecules. These attributes are displayed by the cultivated mushroom *Agaricus bisporus* (current yields about one million tons per year) and Shiitake, *Lentinus edodes*, cultivated by about 200 000 Japanese farmers. The former attacks the cellulose and lignin in horse manure and straw composts and the latter the same components in wooden stakes. Similar features are shown by the strains of *Aspergillus oryzae* and other moulds used in traditional oriental fermentations (Hesseltine, 1983), still vital for the nutrition of millions of people. Some of these strains attack the cellulose and proteins of indigestible soybeans as a step in the formation of palatable products (e.g. miso, soy sauce) and others convert the starch in rice into sugar for fermentation to saké and other alcoholic beverages. The modern fermentation industry, using pure cultures and giant fermenters, produces a wide range of enzymes, organic acids, antibiotics and other products from fungi. Attempts to produce food (single cell protein) in this way from bacteria and yeasts have been expensive failures, but success has been achieved with a fungus, *Fusarium graminearum*, marketed as Quorn. The success of this product is in part due to the constituent hyphae giving a meat-like texture (Trinci, 1991). With the advent of gene cloning,

microorganisms are being used to produce an increasing range of pharmacologically active mammalian peptides and proteins, hitherto obtained in small yields and at great expense from animal sources. At first the bacterium *Escherichia coli* and then the yeast *Saccharomyces cerevisiae* were used for gene expression. However, these species tend to retain proteins within the cell, making recovery and purification difficult, whereas filamentous fungi – notable producers of extracellular enzymes – have an exceptional capacity for protein export. Moulds such as *Aspergillus nidulans* are hence being increasingly used as hosts for the expression of heterologous genes, the extent of current interest in this area being demonstrated by the publication of three books in two years (Leong and Berka, 1991; Peberdy *et al.*, 1991; Kinghorn and Turner, 1992). Environmental concerns about the effects of chemical pesticides are leading to increasing interest in the biological control of pests, with fungi among the agents being considered (Burge, 1988; Whipps and Lumsden, 1990). One approach is the manipulation of environmental conditions to encourage the growth of fungi or other microbes antagonistic to the pest. Alternatively, fungal spores or other microbial preparations may be applied by means of the techniques used with chemical pesticides. Fungal preparations have been used or are being considered for the control of insect pests (McCoy *et al.*, 1988), weeds (TeBeest, 1991), plant pathogenic fungi (Whipps and Lumsden, 1990) and eelworms that infect plants (Stirling, 1991).

1.9 CONCLUSIONS

The success of the fungi is in part due to their modular construction, appropriate for non-motile organisms with an absorptive heterotrophic nutrition. It is due also to the nature of the module, the hypha, a superb device for the penetration of such substrata as plant tissues and insect cuticles. In general terms, the morphology and physiology of mycelial growth are now well understood, due to intensive work on a few species in pure culture (Chapters 13–15). Comparative studies on the rates of hyphal growth and the frequency of branching in species and strains with varied habitat preferences and under different conditions relevant to natural situations would be of interest. In addition to pure culture studies, more work on mycelia in the natural environment is needed. Some important work has been done where mycelial strand formation makes individual mycelia easy to recognize (e.g. Thompson, 1984). In the past, when such features were lacking, recognition of a clone required laborious isolations and genetic analysis, as was done in a study on the fairy rings of *Marasmius oreades* (Burnett and Evans, 1966). However, advances in DNA technology now permit the recognition of a clone from nucleic acid isolated from very small samples, a development with enormous potential for the study of fungal mycelia in the field. Such methods are already being applied, as in a study on a 1500-year-old clone of *Armillaria bulbosa* (Smith *et al.*, 1992). Field studies are now feasible that could answer many questions, for example how long it usually is before a monokaryotic Basidiomycete mycelium becomes dikaryotic, and how well monokaryotic mycelia grow and survive compared with dikaryotic. Studies on long-lived mycelia in the field, such as old fairy rings and forest mycelia, are now not only feasible but urgent, since their habitats, such as primary forest and grassland throughout the world, and secondary but ancient woodland and grassland in long settled areas, are disappearing rapidly.

REFERENCES

Agrios, G.N. (1988) *Plant Pathology*, 3rd edn, Academic Press, London.

Alexopoulos, C.J. and Mims, C.W. (1979) *Introductory Mycology*, 3rd edn, Wiley, Chichester.

Andrews, J.H. (1991) *Comparative Ecology of Microorganisms and Macroorganisms*, Springer, New York.

Arnolds, E. (1991) Mycologists and nature conservation, in *Frontiers in Mycology*, (ed.D.L. Hawksworth), CAB International, Wallingford, pp. 243–64.

Barron, G.L. (1977) *The Nematode-Destroying Fungi*, Canadian Biological Publications, Guelph, Ontario.

Bartnicki-Garcia, S. and Lippman, E. (1972) The bursting tendency of hyphal tips of fungi: presumptive evidence for a delicate balance between wall synthesis and wall lysis in apical growth. *Journal of General Microbiology*, 73, 487–500.

Begon, M., Harper, J.L. and Townsend, C.R. (1990) *Ecology: Individuals, Populations and Communities*, 2nd edn, Blackwell, Oxford.

Brasier, C.M. (1992) A champion thallus. *Nature*, 356, 382–3.

Buller, A.H.R. (1922) *Researches on Fungi*, vol. 2, Longmans Green, London.

Buller, A.H.R. (1931) *Researches on Fungi*, vol. 4, Longmans Green, London.

Buller, A.H.R. (1933) *Researches on Fungi*, vol. 5, Longmans Green, London.

Burge, M.N. (ed.) (1988) *Fungi in Biological Control Systems*, Manchester University Press, Manchester.

Burnett, J.H. and Evans, E.J. (1966) Genetical homogeneity and the stability of the mating-type factors of 'fairy rings' of *Marasmius oreades*. *Nature*, 210, 1368–9.

Carlile, M.J. (1970) The photoresponses of fungi, in *Photobiology of Microorganisms*, (ed. P. Halldal), Wiley, Chichester, pp. 309–44.

Carlile, M.J. and Matthews, S.L. (1988) Chemotropism of germ-tubes of *Chaetomium globosum*. *Transactions of the British Mycological Society*, 90, 643–44.

Carlile, M.J. and Tew, P.M. (1988) Chemotropism of germ-tubes of *Phytophthora citricola*. *Transactions of the British Mycological Society*, 90, 644–46.

Carlile, M.J., Lewis, B.G., Mordue, E.M. and Northover, J. (1961) The development of coremia. I. *Penicillium claviforme*. *Transactions of the British Mycological Society*, 44, 129–33.

Carlile, M.J., Dickens, J.S.W., Mordue, E.M. and Schipper, M.A.A. (1962a) The development of coremia. II. *Penicillium isariiforme*. *Transactions of the British Mycological Society*, 45, 457–61.

Carlile, M.J., Dickens, J.S.W. and Schipper, M.A.A. (1962b) The development of coremia. III. *Penicillium clavigerum*, with some observations on *P. expansum* and *P. italicum*. *Transactions of the British Mycological Society*, 45, 462–64.

Casselton, L.A. (1978) Dikaryon formation in higher basidiomycetes, in *The Filamentous Fungi, vol. 3, Developmental Mycology*, (eds J.E. Smith and D.R. Berry), Arnold, London, pp. 275–97.

Cavalier-Smith, T. (1987) The origin of fungi and pseudofungi, in *Evolutionary Biology of the Fungi, Twelfth Symposium of the British Mycological Society*, (eds A.D.M. Rayner, C.M. Brasier and D. Moore), Cambridge University Press, Cambridge, pp. 339–53.

Cooke, R.C. (1977) *Fungi, Man and his Environment*, Longmans, London.

Cutler, J.E. (1991) Putative virulence factors of *Candida albicans*. *Annual Review of Microbiology*, 45, 187–218.

Esser, K. (1991) Molecular aspects of ageing: facts and perspectives, in *Frontiers of Mycology*, (ed. D.L. Hawksworth), C.A.B International, Wallingford, pp. 3–25.

Evans, H.C. (1988) Coevolution of entomogenous fungi and their insect hosts, in *Coevolution of Fungi with Plants and Animals*, (eds K.A. Pirozynski and D.L. Hawksworth), Academic Press, London, pp. 149–71.

Fiddy, C. and Trinci, A.P.J. (1977) Septation in mycelia of *Mucor hiemalis* and *Mucor ramannianus*. *Transactions of the British Mycological Society*, 68, 118–20.

Friedmann, E.I. (1982) Endolithic microorganisms in the Antarctic cold desert. *Science*, 215, 1045–53.

Gooday, G.W. (1975) Chemotaxis and chemotropism in fungi and algae, in *Primitive Sensory and Communications Systems: the Taxes and Tropisms of Microorganisms and Cells*, (ed. M.J. Carlile), Academic Press, London, pp. 155–204.

Gow, N.A.R. and Gooday, G.W. (1982) Growth kinetics and morphology of the filamentous form of *Candida albicans*. *Journal of General Microbiology*, **128**, 2187–94.

Gregory, P.H. (1984) The fungal mycelium: an historical perspective. *Transactions of the British Mycological Society*, **82**, 1–11.

Griffin, D.M. and Perrin, H.N. (1960) Anastomosis in the Phycomycetes. *Nature*, **187**, 1039–40.

Gull, K. (1975) Mycelial branch patterns in *Thamnidium elegans*. *Transactions of the British Mycological Society*, **64**, 321–4.

Gull, K. (1978) Form and function of septa in filamentous fungi, in *The Filamentous Fungi, vol. 3, Developmental Mycology*, (eds J.E. Smith and D.R. Berry), Arnold, London, pp. 78–93.

Hale, M.E. (1983) *The Biology of Lichens*, 3rd edn, Arnold, London.

Harper, J.L. (1985) Modules, branches and the capture of resources, in *Population Biology and Evolution of Clonal Organisms*, (eds J.B.C. Jackson, L.W. Buss and R.E. Cook), Yale University Press, New Haven, pp. 1–33.

Harper, J.L., Rosen, R.B. and White, J. (eds) (1986) The growth and form of modular organisms. *Philosophical Transactions of the Royal Society of London, series B*, **313**, 1–250.

Hawksworth, D.L. (1991) The fungal dimension of biodiversity: magnitude, significance and conservation. *Mycological Research*, **95**, 641–55.

Hawksworth, D.L. and Hill, D.J. (1984) *The Lichen-forming Fungi*, Blackie, Glasgow.

Hawksworth, D.L., Sutton, B.C. and Ainsworth, G.C. (1983) *Ainsworth & Bisby's Dictionary of the Fungi*, 7th edn, Commonwealth Mycological Institute, Kew.

Hesseltine, C.W. (1983) Microbiology of oriental fermented foods. *Annual Review of Microbiology*, **37**, 575–601.

Hoch, H.C. and Staples, R.C. (1987) Structural and chemical changes among the rust fungi during appressorium development. *Annual Review of Phytopathology*, **25**, 231–7.

Hutchinson, S.A., Sharma, P., Clarke, K.R. and Macdonald, I. (1980) Control of hyphal orientation in colonies of *Mucor hiemalis*. *Transactions of the British Mycological Society*, **75**, 177–91.

Ingold, C.T. (1971) *Fungal Spores, their Liberation and Dispersal*, Clarendon Press, Oxford.

Jennings, D.H. and Rayner, A.D.M. (eds) (1984) *The Ecology and Physiology of the Fungal Mycelium, Eighth Symposium of the British Mycological Society*, Cambridge University Press, Cambridge.

Kinghorn, J.R. and Turner, G. (eds) (1992) *Applied Molecular Genetics of Filamentous Fungi*, Marcel Dekker, New York.

Kwon-Chung, K.J. and Bennett, J.E. (1992) *Medical Mycology*, Lea & Febiger, Philadelphia.

Leong, S.A. and Berka, R.M. (eds) (1991) *Molecular*

Industrial Mycology: Systems and Applications for Filamentous Fungi, Marcel Dekker, New York.

Lovett Doust, L. and Lovett Doust, J. (1982) The battle strategies of plants. *New Scientist*, **95**, 81–4.

Manavathu, E.K. and Thomas, D. des S. (1985) Chemotropism of *Achlya ambisexualis* to methionine and methionyl compounds. *Journal of General Microbiology*, **101**, 65–70.

McCoy, C.W., Samson, R.A. and Boucias, D.G. (1988) Entomogenous fungi, in *Handbook of Natural Pesticides, vol. 5, Microbial Insecticides, part A, Entomogenous Protozoa and Fungi*, (eds C.M. Ignoffo and N.B. Mandava), CRC Press, Boca Raton, FL, pp. 151–236.

Peberdy, J.F., Caten, C.E., Ogden, J.E. and Bennett, J.W. (eds) (1991) *Applied Molecular Genetics of Fungi, Eighteenth Symposium of the British Mycological Society*, Cambridge University Press, Cambridge.

Phaff, H.J. and Starmer, W.T. (1987) Yeasts associated with plants, insects and soil, in *The Yeasts*, 2nd edn, *vol. 1, Biology of Yeasts* (eds A.H. Rose and J.S. Harrison), Academic Press, London, pp. 123–80.

Prince, R.C. (1992) The Methusalah factor: age in cryptoendolithic communities. *Trends in Ecology and Evolution*, **7**, 211.

Ramsbottom, J. (1953) *Mushrooms and Toadstools*, Collins, London.

Read, D.J. (1991) Mycorrhizas in ecosystems, in *Frontiers in Mycology*, (ed. D.L. Hawksworth), CAB International, Wallingford, pp. 101–30.

Richardson, D.H.S. (1991) Lichens and man, in *Frontiers in Mycology*, (ed. D.L. Hawksworth), CAB International, Wallingford, pp. 187–210.

Robertson, N.F. and Rizvi, S.R.H. (1968) Some observations on the water relations of the hyphae of *Neurospora crassa*. *Annals of Botany*, **32**, 279–91.

Robinson, P.M. (1973) Oxygen – positive chemotropic factor for fungi? *New Phytologist*, **72**, 1349–56.

Robinson, P.M. (1980) Autotropism in germinating arthrospores of *Geotrichum candidum*. *Transactions of the British Mycological Society*, **75**, 151–3.

Ryley, J.F. and Ryley, N.G. (1990) *Candida albicans* – do mycelia matter? *Journal of Medical and Veterinary Mycology*, **28**, 225–39.

Shaw, D.S. (1983) The cytology and genetics of *Phytophthora*, in *Phytophthora: its Biology, Taxonomy, Ecology and Pathology*, (eds D.C. Erwin, S. Bartnicki-Garcia and P.H. Tsao), The American Phytopathological Society, St Paul, Mn, pp. 81–94.

Smith, M.L., Bruhn, J.N. and Anderson, J.B. (1992) The fungus *Armillaria bulbosa* is among the largest and oldest living organisms. *Nature*, **356**, 428–31.

Steele, G.C. and Trinci, A.P.J. (1975) The extension rate of mycelial hyphae. *New Phytologist*, **75**, 583–7.

Stirling, G.R. (1991) *Biological Control of Plant Parasitic Nematodes: Progress, Problems and Prospects*. CAB International, Wallingford.

TeBeest, D.O. (ed.) (1991) *Microbial Control of Weeds*, Chapman & Hall, New York.

Thompson, W. (1984) Distribution, development and

functioning of mycelial cord systems of decomposer Basidiomycetes of the deciduous woodland floor, in *The Ecology and Physiology of the Fungal Mycelium, Eighth Symposium of the British Mycological Society*, (eds D.H. Jennings and A.D.M. Rayner), Cambridge University Press, Cambridge, pp. 185–214.

Trinci, A.P.J. (1973) The hyphal growth unit of wild type and spreading colonial mutants of *Neurospora crassa*. *Archiv für Mikrobiologie*, **91**, 127–36.

Trinci, A.P.J. (1984) Regulation of hyphal branching and hyphal orientation, in *The Ecology and Physiology of the Fungal Mycelium, Eighth Symposium of the British Mycological Society*, (eds D.H. Jennings and A.D.M. Rayner), Cambridge University Press, Cambridge, pp. 23–52.

Trinci, A.P.J. (1991) Mycoprotein: a twenty year overnight success story. *Mycological Research*, **96**, 1–13.

Trinci, A.P.J. and Collinge, A.J. (1974) Occlusion of the septal pores of damaged hyphae of *Neurospora crassa* by hexagonal crystals. *Protoplasma*, **80**, 57–67.

Trinci, A.P.J. and Cutter, E.G. (1986) Growth and form in lower plants and the occurrence of meristems. *Philosphical Transactions of the Royal Society of London, series B*, **313**, 95–113.

Watkinson, S.C. (1979) Growth of rhizomorphs, mycelial strands, coremia and sclerotia, in *Fungal Walls and Hyphal Growth, Second Symposium of the British Mycological Society*, (eds J.H. Burnett and A.P.J. Trinci), Cambridge University Press, Cambridge, pp. 93–113.

Webster, J. (1980) *Introduction to Fungi*, 2nd edn, Cambridge University Press, Cambridge.

Whipps, J.M. and Lumsden, R.D. (eds) (1990) *Biotechnology of Fungi for Improving Plant Growth, Sixteenth Symposium of the British Mycological Society*, Cambridge University Press, Cambridge.

Willetts, H.J. (1978) Sclerotium formation, in *The Filamentous Fungi, vol. 3, Developmental Mycology*, (eds J.E. Smith and D.R. Berry), Arnold, London, pp. 197–213.

Wynn, W.K. (1981) Tropic and taxic responses of pathogens to plants. *Annual Review of Plant Pathology*, **19**, 237–55.

MYCELIAL INTERCONNECTEDNESS

2

A.D.M. Rayner, G.S. Griffith and A.M. Ainsworth
School of Biological Sciences, University of Bath, Bath, UK

2.1 INDETERMINACY AND INTERCONNECTEDNESS

Fungal mycelia are capable of growth and persistence over extended, potentially unlimited periods (see also Carlile, Chapter 1). In this respect they contrast with determinate body forms, such as those of many animals and unicellular organisms, which have pre-set limits in space and time. Moreover, as different parts of a mycelial boundary expand, they remain physically linked. Events occurring within the boundary cannot therefore be totally independent, notwithstanding that there may be considerable variation in the degree of autonomy of different parts of the system. Equally importantly, all local environments that come within range of the boundary of a mycelium cannot be treated as entirely separate domains.

Such interconnectedness is likely to be of fundamental importance to the organization and ecological roles of mycelial systems. Yet, both experimentally and conceptually, it has all but been ignored – perhaps largely because it introduces 'undesirable' unpredictability and a need to address systemic issues. On the other hand, there has been a palpable desire to treat mycelia ecologically, physiologically, genetically and evolutionarily as if they were fully particulate entities with readily calculable and hence predictable behaviour. This desire has been manifest in attempts to quantify mycelial distribution by counting and weighing, and the experimental cultivation of single, well-characterized strains under uniform conditions. Also, concepts derived from studies of organisms with determinate development have often been applied too readily to fungi. Even recent attempts to describe mycelia as 'modular' organisms (Carlile, Chapter 1) represent only a partial recognition of the true nature of indeterminacy, and a clinging to the notion of absolute

quantifiability in the form of discrete hyphal growth units.

There is a danger if such approaches are used alone that the resulting understanding of mycelial dynamics may at best be limited and approximate, and at worst be sterile and distorted. The purpose of this chapter is therefore to expand upon some of the themes introduced by Carlile (Chapter 1) in order to develop an alternative, but complementary approach to understanding mycelial patterns. This approach will be based on exploration of the wider and deeper implications of mycelial interconnectedness, drawing attention to those observable properties of mycelia that cannot readily be explained simply by concepts based on unicellular biology.

2.2 ALTERNATIVE VIEWPOINTS – INSIDE-OUT OR OUTSIDE-IN?

Thinking about interconnectedness enforces consideration of relationships between events within the mycelium. However, in most recent thinking the mycelium has been regarded from outside, as a contrivance which optimizes resource capture, with its most fundamental dynamic properties – apical extension, branching and anastomosis of hyphae – all being interpreted as means to this end. The principal questions addressed have therefore concerned how these properties, as adaptive mechanisms, are brought into play at appropriate times and places. As a result, considerable emphasis has been placed on the role of appropriate responses to external environmental gradients in generating mycelial patterns (cf. Prosser, 1991, 1993). Following the same kind of argument, however, a river system might be described as an adaptation to environmental topography such as to achieve maximum drainage to the sea.

The Growing Fungus. Edited by Neil A.R. Gow and Geoffrey M. Gadd. Published in 1994 by Chapman & Hall, London. ISBN 0 412 46600 7

By contrast, from an inside-out viewpoint, such properties as polarized extension, branching and anastomosis are not mechanisms *per se*, but processes found in all kinds of dynamic systems, living and non-living, where expansive trends are counteracted by constraints. Although gene action will govern the operation of these processes, their fundamental origin is not to be sought therein, but rather in the way energy becomes channelled along paths of least resistance through variably bounded systems. In these terms, a branching pattern reflects how a system distributes captured energy back to its environment as an intrinsic consequence of automatic processes, rather than how it responds to its environment in a calculated manner. The emphasis therefore shifts from what is **functionally adaptive** to what is **organizationally impelled** (cf. Gould and Lewontin, 1979). Notwithstanding that the former may well coincide – or be made to coincide – with the latter, this shift engenders a radical change in the thinking used to explain mycelial patterns, with the focus on the dynamic relationship between uptake and throughput of nutrients and water.

2.3   INTEGRATION AND DIFFERENTIATION

As long as the rate of uptake into an expanding system (which depends on the amount and/or activity of absorptive surface) is well below the throughput (carrying) capacity (determined by total resistance to flow), the system will exhibit exponential growth. A doubling in the absorptive surface will result in doubling of the uptake rate and therefore of throughput to the expanding boundary. When uptake becomes equal to the carrying capacity, the system will expand linearly (i.e. with equal generations or no net acceleration of growth). These special cases correspond to the behaviour described by Trinci (1978) of young, 'undifferentiated' mycelia and of apical compartments in the peripheral growth zone of mature mycelia (Prosser, Chapter 9).

However, when uptake exceeds throughput capacity, the system will be prone either to leak or oscillate or to increase its distributive surface by subdividing (differentiating), i.e. to branch. The frequency of branching will depend on the excess of uptake over throughput capacity to existing release points on the boundary. The angle of branching to the main axis will depend on whether

the region of maximum boundary deformation is symmetrical or asymmetrical about the point of origin.

When branches coalesce or anastomose (integrate), proliferation will be reduced as the carrying capacity of the system as a whole is greatly increased. This change is equivalent to replacing the 'in series' resistances of a radial system by the 'in parallel' resistances of a network. Symmetry is effectively restored, as energy within the system is cycled rather than distributed to expansive sites on the boundary. Unless the cycle can be broken, this could have the effect of limiting increase in diameter of the system under conditions of continuous supply of nutrients, a situation that has been described in fungal mycelia (Gottlieb, 1971). On the other hand, further rounds of symmetry-breaking, differentiation and integration, would allow the system to progress through a series of organizational levels evident externally as 'phase shifts'. The latter would involve fundamental changes in operational scale, but with each stage repeating the same basic generative patterns.

A computer-simulated developmental pattern based on an uptake-throughput model is illustrated in Figure 2.1. Although only preliminary, it already exhibits many of the dynamic properties characteristic of mycelia. Processes leading to formation of such patterns are initiated when a system breaks spherical symmetry to produce an elongating structure, as when a spore germinates. This event may be preceded by a variable period of isotropic expansion or swelling, as internal pressure builds with increasing uptake to a threshold value when it becomes directed towards one or more irregularities which act as stress foci in the boundary. Thereafter, expansion becomes polarized, i.e. oriented along particular axes, to an extent which depends on the degree to which uptake is translated into a hydraulic or vectorial form of throughput, thrust (Rayner *et al.*, 1994).

It is important to note that thrust, due to the maintenance of **internal** pressure at an expanding boundary, such as the hyphal tips of a mycelium, is not equivalent to mass flow. The latter occurs along a tubular system as the result of an osmotic gradient and consequent water potential difference across its axial boundaries. Similarly it may not be legitimate to equate thrust with turgor, a non-vectorial term more appropriately applicable to stationary systems (cf. Lockhart, 1965), and

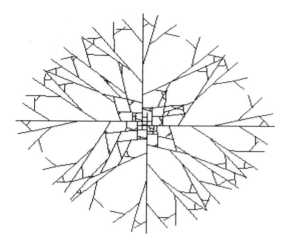

Figure 2.1 Computer simulation based on an uptake-throughput model of mycelial development in a plane. The model calculates the total uptake of the system and converts this into expansion at individual hyphal tips. As throughput increases, hyphae branch less frequently, branch angles become more acute and hyphae insulate more rapidly. Hyphal density is regulated in part by hyphal anastomosis, which occurs wherever a hyphal tip contacts another hypha, and has the effect of reducing branch proliferation. The protoplasm is assumed to be continuous, accounting for the system's coherence.

certainly not with a turgor gradient. The latter term is a misnomer, due to the equalization of pressure within a continuous system (e.g. Money, 1990; cf. Jennings, 1984, 1987).

The higher the thrust, for a given rate of uptake, the more polarized, that is the less frequently and more acutely branched, will be the pattern. A measure of the degree of polarity is the fractal dimension, which approaches two in a plane and three in a volume, the more laterally branched (i.e. less polarized or line-like) the system is (e.g. Ritz and Crawford, 1990; Prosser, Chapter 9). Correspondingly, the amount of thrust in an elongated system is a function of the ratio between the resistance of its lateral boundary to deformation and leakage relative to that of its axial boundary. In other words, thrust is dependent on how well sealed off or 'insulated' the lateral boundary is from its surroundings.

Whereas the sealing off of lateral boundaries in this way provides for efficient distribution, it also limits further net uptake, and so the rate of sealing must not exceed the rate of boundary expansion, if

the system is not to stagnate. It follows that while still enlarging, the system will be in a pattern-determining state of dynamic balance between assimilative and distributive processes, with at least some part of its boundary in open communication with its surroundings.

Of equal importance to the degree of communication across the system's boundaries in determining thrust, is the degree of communication within the system itself due to isolating or partitioning processes (e.g. septation, localized cell death) that affect the pattern of pressure distribution. Here it may be noted that although the local significance of pressure as a means of driving hyphal extension has been widely appreciated, and incorporated into explanatory models (Bartnicki-Garcia, 1973; Wessels, 1986), its systemic significance as a means of communication seems to have received little attention. Nonetheless, it may help to explain examples of apparently very rapid long-range 'signalling' (see below).

Pressure within an uncompartmented system, such as a fully coenocytic mycelium, brings all parts into virtually instantaneous communication. The presence of septa still allows the system to operate in this way as a protoplasmic continuum, but will impede throughput. The latter fact may account for the correlation between septation and anastomosis (e.g. Boddy and Rayner, 1983; Ainsworth and Rayner, 1991), a process which will tend to restore and/or enhance throughput capacity, but be prone to limit extension (see above). It may also relate to the correlation between septation and branching (e.g. Trinci, 1978), although it may be that septa are laid across sites where throughput capacity has been reached rather than vice versa.

If adjacent system components become discontinuous, e.g. by septal sealing with Woronin bodies or by other means (e.g. Todd and Aylmore, 1985; Markham, Chapter 5), a competitive disequilibrium will be possible whereby the more metabolically active (actively transporting) component will drain resources from the less active one. This will have the effect of reversing the normal direction of thrust from that in a protoplasmic continuum, so enabling redistribution and breaking any limitations imposed by anastomosis. Moreover, the potential for competitive disequilibria, and consequent 'self-parasitism' to arise in the absence of communication channels may be of considerable relevance to the evolution of multicellular structures of all kinds.

2.4   VERSATILITY AND DEGENERACY

It should be apparent from the above discussion that by varying the resistance to deformation and penetration of its external boundary (i.e. its 'insulation') and the partitioning of its interior, an indeterminate system is capable of generating variable thrust and consequently diverse organizational patterns. Such versatility may be evident from the earliest stages of mycelial development. Whereas the process of spore germination commonly results in symmetry-breaking to establish the indeterminate mycelial phase, many fungi under appropriate circumstances can exhibit determinate development at this stage. This results in such phenomena as mycelial–yeast dimorphism (Gow Chapter 19), microcycle conidiation (Anderson and Smith, 1971) and multiple secondary spore types (Ingold, 1984; Webster, 1987).

Once mycelial development is under way, branches begin to emerge, and may do so with varying frequency and at varying angles, representing different degrees of commitment to radial (explorative) or tangential (exploitative/consolidative) growth. The higher the ratio between the radial and tangential growth processes, the lower is the fractal dimension of the structure, the higher the thrust, and the greater its polarity. Changes in branching pattern, when coordinated (e.g. by anastomosis), give rise to changes in the organization and extension rate of the mycelial margin. Where these changes occur abruptly, they result in 'phase shifts' (see above) that have been termed slow-dense/fast-effuse transitions (Rayner and Coates, 1987), and are often manifested as sectoring or 'point growth' phenomena as the mycelium changes its operational scale (Coggins *et al.*, 1980; Figure 2.2).

Also the emerging hyphal tubes may or may not, depending partly on organism and partly on circumstance, be divided by septa into uninucleate, binucleate or multinucleate compartments. In the basidiomycetes *Phlebia rufa* and *Phlebia radiata*, a marginal zone of rapidly extending, sparsely branched, non-anastomosed, coenocytic hyphae is established, followed by a phase of densely branched, septate, anastomosed hyphae (Boddy and Rayner, 1983). These fungi

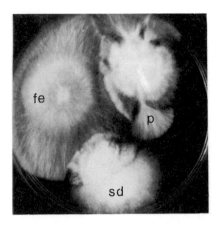

Figure 2.2   Outgrowth of mycelia from fruit body tissue of *Hypholoma fasciculare* onto 2% malt agar showing slow-dense (sd) and fast-effuse (fe) colony forms, with the former switching to the latter by 'point growth' (p).

therefore combine the 'guerilla' and 'phalanx' characteristics described by Carlile (Chapter 1), driving a short-lived exploratory phase ahead of a persistent exploitative phase. In *Phanerochaete magnoliae*, basidiospore-germination results in a rapidly extending, sparsely branched (fractal dimension close to 1) coenocytic mycelium (Figure 2.3). When the latter reaches about 1.5 cm diameter, septa, lateral branches and anastomoses appear in rapid succession, resulting in a partial network within which extensive autolysis of non-interconnected hyphae occurs as hymenial surfaces form and discharge their spores (Ainsworth and Rayner, 1991).

The formation of hymenia relates to another important feature of hyphae, the capacity to develop either diffusely or in compact associations that enable the mycelium to undergo phase shifts. Hyphal aggregation may fulfil migratory or connective roles in the case of rhizomorphs and mycelial cords which, being corporate structures, are often capable of far more rapid extension than individual hyphae, whilst exhibiting parallel patterns of branching and anastomosis (Thompson and Rayner, 1983; Rayner *et al.*, 1985; Dowson *et al.*, 1988a). Protective roles of hyphal aggregation are found in sclerotia and pseudosclerotia, and reproductive roles in stromata and fruit bodies.

Such corporate roles depend in their turn on the ability of mycelial systems to shift between assimilative and non-assimilative states. Generally, this

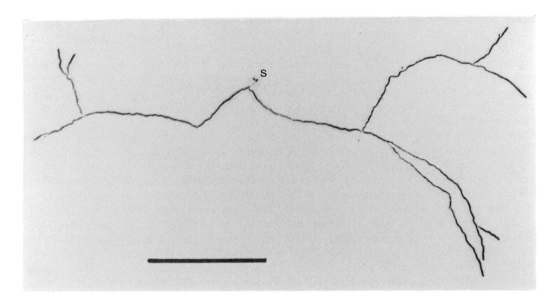

Figure 2.3 Coenocytic mycelium of *Phanerochaete magnoliae* originating from a basidiospore (s). Branches have only formed terminally, resulting in a structure with a very low fractal dimension. Scale bar, 1 mm.

ability enables mycelia to penetrate into or across physicochemically hostile regions, and in some cases to ameliorate conditions for superseding exploitative phases. This has considerable practical as well as ecological relevance. A good example is provided by certain fungi causing root disease, which advance over the root surface as a superficial mycelium before producing exploitative mycelium within the root tissues (Garrett, 1970; Stenlid and Rayner, 1989).

It is also arguable that without the ability to alternate exploratory and exploitative phases, mycelial margins would be unable, as they do, to maintain a constant rate of extension in more than one dimension without becoming progressively more sparse. The latter would follow from the inability of lateral branches to infill between leader hyphae that are already at maximum extension rate.

An illustration of how the interdependence between exploration and exploitation may work in practice is provided by the leaf-litter-decomposing basidiomycete, *Clitocybe nebularis* (Dowson *et al.*, 1989a). The mycelium of this fungus consists of a 'fairy ring' with exploratory mycelial cords at its outer margin behind which is a zone of diffuse, exploitative mycelium in which bleaching of the

leaf litter occurs. If the exploratory cords enter barren domain greater than the width of the annulus, the cords cease extension, the exploitative phase is not produced and a persistent gap forms in the ring. Maintenance of exploration therefore allows the colony to expand its boundaries, but can only be sustained by exploitation over relatively small distances.

The formation of gaps and the presence of a degenerative trailing edge to the enlarging mycelium suggests that the latter depends not only on a dynamic relationship between exploration and exploitation, but on establishment of powerful, self-sustaining, source-to-sink fluxes from redundant to expanding regions. This possibility is substantiated by the observation that segments cut experimentally from the annulus extend with conserved rates and polarity if transplanted inside or outside the ring, but degenerate if re-oriented before being re-inserted within the annulus (Dowson *et al.*, 1989a).

Such behaviour is not restricted to fairy ring fungi. Another example is provided by the wood-decomposing basidiomycete, *Peniophora lycii*, which exhibits 'polar growth' along twigs of ash (*Fraxinus excelsior*) such that it cannot be isolated from locations through which its growth front has already passed (Griffith and Boddy, 1991). More generally, redistributive processes are vital to the functioning of mycelial foraging systems of all kinds (see below) as well as to the emergence of sporophores (e.g. Watkinson, 1977; Sietsma and Wessels, 1979; Ruiters and Wessels, 1989).

Redistribution therefore clearly provides a functional role for degenerative processes in indeterminate mycelial systems. These processes can also have a protective effect in the sealing-off of a mycelial boundary against a hostile biotic or abiotic environment, and, as will be amplified below, are themselves an inevitable consequence of insulating mechanisms. Degeneration, which in some guises has been referred to as senescence (e.g. Kück, 1989) is therefore both a product and vital component of mycelial versatility, whilst always engendering the potential for self-destruction if uncontrolled. To coin a phrase, death is a way of life in indeterminate systems, but is not without risks!

2.5 ENERGY CAPTURE, DISTRIBUTION AND CONSERVATION

If the degree of insulation between a mycelium and its environment can be altered to be minimal when external supplies are maximal, and maximal when these supplies are depleted, then an efficient balance would automatically develop between processes of energy capture, distribution and conservation. Where this balance is struck would depend further on the spatiotemporal frequency and scale of environmental fluctuations, defining in turn the system's niche and ecological strategy (cf. Cooke and Rayner, 1984).

The question of energy-efficiency *per se* is an important aspect of physiological and ecological considerations of mycelial functioning, since discontinuities in resource supply are an ever-present feature of the natural lives of many filamentous fungi. Temporal discontinuities necessitate the formation of highly insulated, enclosed (hence 'conservative') survival structures, such as sclerotia and constitutively dormant spores ('memnospores' according to the terminology of Gregory, 1966). Spatial discontinuities necessitate formation of distributive structures, either in the form of dispersal spores ('xenospores' in Gregory's terminology), or assemblages of migratory hyphae. Migratory hyphae exhibit 'foraging strategies'

between spatially separated resource depots. They are characteristic of fungi growing in a wide range of habitats, e.g. bark-inhabiting fungi that grow between medullary rays (Rayner *et al.*, 1981), mycorrhizal fungi that spread between the roots of neighbouring plants (Read, 1984, 1992), and decomposer fungi that form links between pieces of woody material on the forest floor (Thompson, 1984).

The study of foraging strategies provides a highly effective way of exploring interconnectedness and its relation to heterogeneity. It clearly demonstrates the nature of circumstance-driven development and the consequent departure in the behaviour of mycelial systems from that which would be expected if they only operated as additive assemblages of discrete hyphal growth units.

A foraging experiment basically consists of placing a mycelial inoculum amidst an array of sources of organic carbon to which it can only have access by forming connections across non-nutritive domain. The fungus is thereby enforced to separate out and alternate its assimilative and non-assimilative phases, allowing the properties and interrelationships of these phases to be investigated.

The simplest experimental design consists of a pair of nutritive sites, one of which is colonized and serves as the inoculum, the other of which is uncolonized and serves as 'bait' (Dowson *et al.*, 1986). More complex designs can consist of elaborate matrices or lattices that can be perturbed once the experiment is in progress (Dowson *et al.*, 1989b). The experiments may make use of purely artificial media and materials (Figure 2.4), or they may be made seminatural (Dowson *et al.*, 1986, 1988b, 1989b) or even be conducted entirely in the field (Dowson *et al.*, 1988c,d).

An example of an experiment made under seminatural conditions is illustrated in Figure 2.5. This involved a wood-decomposing basidiomycete able to grow out into unsterile soil in the form of mycelial cords. The central issue with respect to efficient resource allocation in such organisms is whether to broadcast exploratory mycelium sparsely but over a long range or more densely over a shorter range. Generally, long-range foraging is most appropriate when resources are widely scattered, but large, whereas short-range foraging is more effective amidst smaller but more frequent

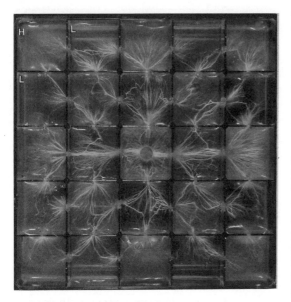

Figure 2.4 Networking of mycelium of the basidiomycete, *Coprinus picaceus*, grown in a matrix of compartments alternately containing high (H) and low (L) nutrient media (2% malt agar and tap water agar). Note the formation of interconnecting mycelial cords predominantly on the low nutrient media, and the alignment of consolidated hyphal aggregates with localized communication channels cut between compartments. (Courtesy of Louise Owen and Erica Bower.)

resources, and also ensures that the fungus will remain within the vicinity of a large resource.

Short-range foraging patterns, as exemplified in Figure 2.5, are strongly redistributional, having many features in common with the behaviour of fairy ring fungi described previously (Dowson *et al.*, 1989b). Contact with a bait by a dense, initially radially symmetrical explorative mycelium results in virtually immediate cessation of extension of the margin. This is followed by thickening of connective mycelial cords and regression of all non-connective mycelium as renewed, unilaterally directed outgrowth occurs from the bait.

Long-range foraging is more individualistic, with sparsely distributed, clearly separate foraging units extending rapidly and independently from an inoculum source, and not all ceasing extension when a successful contact is made. Rhizomorphic species of *Armillaria* are probably the longest range foragers, as is reflected by the enormous domains, measurable in hectares, that individual genets of

these fungi can occupy (Smith *et al.*, 1992; Carlile, Chapter 1).

Even though the foraging patterns exhibited by particular species may tend to be towards one end or other of the long-range–short-range spectrum, suggesting genetic adaptation to resource distribution, they do vary according to circumstances. A precisely similar situation occurs in army ants (Rayner and Franks, 1987; Franks *et al.*, 1991).

Of particular importance to the foraging patterns is the resource availability in inoculum and baits. A superior availability in baits compared with inocula has the effect of enhancing the degree of redistribution (Dowson *et al.*, 1986). Mineral nutrients have also been shown to be distributed preferentially to baits with superior resource availability (Wells and Boddy, 1990), and a superior availability in inocula leads to an increase in the fractal dimension of mycelial outgrowth (Bolton and Boddy, 1993). The latter increase can lead in its turn to enhancement of lateral anastomoses and thereby to greater persistence and distributive capacity of initially explorative mycelium.

The production of persistent networks as a consequence of anastomosis adds the ability to forage in time to the ability to forage in space. This is because any resource that comes into contact with the network, for example when it falls from a tree canopy, can be quickly colonized. Networks therefore conserve their energy until the opportunity for assimilation arises, rather than actively locating resources. Several mycelial cord-forming wood-decomposing fungi illustrate this strategy, e.g. *Phallus impudicus* and *Tricholomopsis platyphylla*, and correspondingly seem to have a higher threshold for production of actively explorative systems (Dowson *et al.*, 1988b). In tropical forests, rhizomorphic fungi that entrap falling litter in tree canopies have been likened by Hedger (1990) to 'filter feeders'.

The formation of redistributive networks between the living roots of different plants, both of the same and of different species, is a feature of certain mycorrhizal fungi. The resultant opening of communication channels between plants has fundamental implications in terrestrial ecosystems. It may enable adult plants to 'nurse' seedlings until they have developed sufficiently to become established in their own right (Read, 1984), reduce competition between plant species and enhance overall community uptake of soil

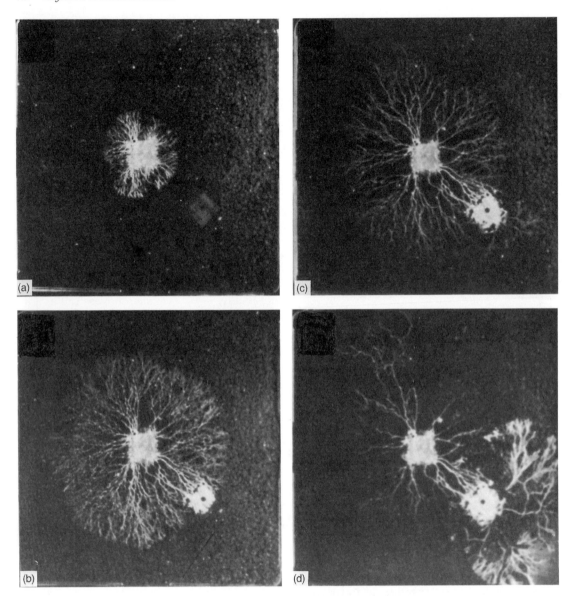

Figure 2.5 (a–d) Stages in short-range foraging by a mycelial cord system of *Hypholoma fasciculare*, which has been introduced on a beechwood block inoculum centrally into a tray containing non-sterile soil and allowed to grow towards an uncolonized 'bait' wood block placed 4 cm away. Note initially symmetrical outgrowth until contact with the bait followed by cessation of marginal extension and regression of non-anastomosed mycelium as renewed outgrowth occurs from the bait. (From Dowson *et al.*, 1989b.)

nutrients (Perry *et al.*, 1989). On the other hand, it also facilitates piracy, the possibility of tapping into the network and drawing off its resources while not contributing anything in return. Some plants clearly do just this (Harley and Smith,

1983). An example is the Yellow Bird's nest, *Monotropa hypopitys*, which lacks chlorophyll and draws its energy from tree roots via ectomycorrhizal mycelium. Some fungi can also piratize the network, e.g. *Cordyceps capitata* which parasitizes ectomycorrhizal truffles. A slightly different situation occurs with orchids which, in effect, parasitize parasitic and decomposer fungi by forming mycorrhiza with them.

2.6 SELF-REGULATION

To state that a mycelium adjusts to circumstances by altering its internal partitioning and the resistance of its boundary to penetration and deformation, so varying its pattern of pressure distribution, does not explain how it does so. To be able to provide such an explanation requires knowledge of the actual materials involved, how they interact, and how their production and interactions are affected by environmental factors.

The most studied of these issues concerns the way in which the deformability of hyphal walls can be varied, but even here many uncertainties remain – sufficient for there to be two major alternative hypotheses to account for the apical extension of hyphae (see also Chapters 3 and 13). The 'balanced lysis' model of Bartnicki-Garcia (1973) suggests that the apical wall is rendered plastic by the action of lytic enzymes. These continually loosen the microfibrillar framework – usually consisting of chitin – that lines the inner part of the hyphal wall and gives it its tensile strength, at the same time that this framework is being assembled by synthetic enzymes. The 'steady state' model of Wessels (1986) suggests that the apical wall is synthesized as a viscoelastic material that becomes progressively rigidified by means of chemical cross-linking and hydrogen bonding.

Both these models are strongly non-linear in that they envisage a dynamic balance between counteractive associative and dissociative processes regulating the extensibility and rigidity of the wall. So long as this balance is sustained, the hypha will maintain a constant diameter and extension rate, appropriate to the amount of thrust in the system. If it is not sustained, then the tip may become more plastic, increase in diameter and in extreme cases burst, or it may rigidify more quickly, leading to narrowing or total inextensibility, thereby reducing the throughput capacity. All these features regularly occur in fungal colonies.

Whereas components affecting the strength properties of hyphal walls have received much attention, components affecting penetrability seem to have been neglected. Of particular interest are water-resistant materials that coat, impregnate or line hyphal walls. Such materials would be expected to have considerable influence on interconversions between assimilative and non-assimilative states and mycelia emerging from a nutrient source are often visibly hydrophobic, e.g. in certain mycorrhizal fungi (Unestam, 1991). Current candidates for such a role include two main categories of compounds.

The first category consists of a group of water-repellent, cysteine-rich polypeptides, known as 'hydrophobins' (Wessels, 1991, 1992; see Chapter 3). These polypeptides probably correspond with the 'rodlets' which coat the surface of hyphae and spores of a wide range of fungi (e.g. Cole *et al.*, 1979) and appear to contain strongly conserved sequences of amino acids in both ascomycetes and basidiomycetes (Stringer *et al.*, 1991). In *Schizophyllum commune* the spatiotemporal distribution of hydrophobins coincides with the behaviour that might be expected of insulating compounds. They are bound to the walls of emergent (non-assimilative) aerial hyphae and fruit body initials, but released into the medium from submerged hyphae (Wessels, 1991, 1992).

With regard to the consumption of nitrogen, however, the use of polypeptides as insulators is very costly. The resultant requirement to have access to sufficient nitrogen would thereby limit the developmental competence of a mycelium to produce non-assimilative phases until it had generated adequate assimilative biomass. Moreover, competition for nitrogen between alternative non-assimilative phases would account for numerous examples of reciprocal relationships between these phases, and for the likely role of nitrogen-source depletion in fruit body initiation (Leatham, 1985). By the same token, growth in habitats with low nitrogen availability is liable to impose considerable developmental constraints on the organism.

An alternative, less nitrogen-costly means of insulation could be provided by aromatic and terpenoid products from the acetate, polyketide and shikimate pathways (notably absent from animals with determinate development, but widespread in plants and fungi). Many such compounds may be capable of conversion into free radicals when acted upon by phenol-oxidizing enzymes (laccases, tyrosinases and peroxidases) leading in turn, perhaps as influenced by redox potential, both to chemical cross-linking (polymerization) and depolymerization. Such processes acting outside the plasmalemma would affect the insulating properties of hyphal walls, for example by chemical cross-linking and quinone-tanning (including melanization), as well as the degradation of external substrates such as lignocellulose. Inside the plasmalemma they would instigate cell death. Phenol-oxidizing activity would also be expected to be substrate-induced (e.g. Ainsworth and Rayner, 1991) and to affect cysteine-rich polypeptides.

It is therefore of interest that changes in phenoloxidase activity have been clearly linked with a variety of developmental landmarks, including the initiation of fruit bodies, sclerotia and rhizomorphs (e.g. Willetts, 1978; Leatham and Stahmann, 1981; Ross, 1985; Worrall *et al.*, 1986). Moreover, free radical production is implicated in the 'senescence' of *Podospora pauciseta* (Frese and Stahl, 1990) and the autolysis of *Peniophora lycii* (A.D.M. Rayner and G.S. Griffith, unpublished; see above). Evidence is also accumulating for a key role for phenoloxidizing activity in determining the outcome of intra- and interspecific mycelial interactions (see below).

Mycelia do seem, therefore, to possess the material wherewithal to bring about autodegeneration and to vary the insulating properties of their boundaries. As has already been implied, it would make both logistic and ergonomic sense for insulation to be minimized when rates of uptake at and resource supply to hyphal boundaries are greatest and vice versa. In other words, there should be a direct feedback between external resource availability and the boundary properties affecting uptake and distribution processes. The proliferation of branches under local conditions allowing high rates of assimilation, and switchover to explorative, redistributive or conservative patterns at low local rates of assimilation would then follow automatically.

Such a switchover would be accomplished by a change from assimilative or 'inductive' metabolic pathways involved in the generation of ATP and its utilization in active transport to non-assimilative or 'transductive' pathways involved in sealing off boundaries and initiating autodegeneration. These pathways in their turn appear to have much in common, if not to be identical with those which have previously been described as 'primary' and 'secondary' metabolism. The functional significance

of the latter has been a cause of endless speculation in the face of the widespread microbiologist's view that all there is to life is growth (Bushell, 1989a). However, the fact that many of its products are hydrophobic and will interact with phenoloxidizing enzymes suggests that it could have an important role in insulation.

It is well known that a major cue for the initiation of secondary metabolism is a disruption in resource supply, whether by nutrient limitation or by other means, associated with a local reduction in 'growth rate'. It has been suggested that such a disruption will result in a drop in energy charge and associated increase in concentration of the second messenger, cAMP. By the same argument, catabolite repression explains the inoperation of secondary metabolism at high assimilative rates (cf. Bushell, 1989a, b). However, this may be only one aspect of the relation between energy turnover and metabolic regulation. Any accumulation of ADP, as may be influenced by changes in cytosolic Ca^{2+}, for example, will have positive feedback effects on glycolysis. This will in turn lead to the accumulation of those intermediates of primary metabolism that are the starting points for secondary metabolism. There may then be further feedback effects if secondary products, such as phenolics, inhibit ATP production, for example through uncoupling of oxidative phosphorylation and free radical formation in the presence of phenol-oxidizing enzymes. Evidence for a reciprocal relationship between oxidative phosphorylation and phenoloxidase activity in mycelial fungi has been reported by Lyr (1958, 1963).

The above discussion highlights the likely significance of switches to secondary metabolism in the regulation of indeterminate development and, in view of its pivotal significance in energy generation, of the mitochondrion as their executive. These possibilities are further illuminated by the responses shown by mycelial systems when they encounter neighbours.

2.7 NON-SELF RELATIONSHIPS

Encounters between neighbouring mycelia belonging either to the same or to different species occur frequently in natural populations and communities of fungi inhabiting relatively undisturbed, stress-free habitats (Cooke and Rayner, 1984; Rayner *et al.*, 1987a). With interest in the real lives of fungi as mixed rather than pure cultures gaining ground, such encounters are also becoming increasingly frequent in the laboratory, where changes in morphogenetic and distributional pattern induced when mycelia are inoculated opposite one another in Petri dishes are monitored.

These encounters, involving, as they do, challenges between dynamically unstable systems, provide a powerful means both of illustrating and investigating mycelial interconnectedness. Moreover, their outcomes can be equally challenging to interpret (Rayner, 1991a); they can be both complex and unpredictable, belying the traditional tendency to try to classify them in absolute, 'plus or minus' terms. For example, in a Petri dish pairing on 3% malt agar between *Coriolus versicolor* and *Peniophora lycii* which was replicated as exactly as possible 20 times, three fundamentally differing interaction patterns were observed: in two of these *C. versicolor* replaced *P. lycii*, in the other, *P. lycii* replaced *C. versicolor* (G.S. Griffith and A.D.M. Rayner, unpublished). Such variability is characteristic of non-linear systems in which small changes in initial conditions can be amplified by feedback into radical changes in output pattern (e.g. Gleick, 1988).

2.7.1 Interspecific encounters

Where the neighbour is of an unrelated species, there is normally little likelihood of hyphal fusion leading to significant protoplasmic continuity between opposing colonies (e.g. Gregory, 1984). Under these circumstances, the outcome of interaction seems to depend first on the sensitivity of individual hyphae to protoplasmic degeneration at long or short range from opposing hyphae, and second on the corporate capacity of mycelia to produce non-assimilative phases.

Hyphal degeneration at long range is generally attributed to diffusible or volatile factors or antibiotics. It commonly results in mutual inhibition between colonies, suggesting that there may be feedback between the induction of degenerative processes in one colony and the production of factors eliciting these processes by the other (Rayner and Webber, 1984). Degeneration following contact or very close proximity is described as hyphal interference, and is usually thought to be particularly characteristic of interactions involving basidiomycetes (Ikediugwu and Webster, 1970).

These degenerative responses appear likely to have much in common, in their biochemistry, genetic regulation, mode of induction and biological consequences, to those exhibited by plants as a result of microbial infection (Rayner, 1986). They can both protect the mycelium from parasitism and enhance its susceptibility to invasion, depending on circumstances.

The latter point is illustrated by three wood-inhabiting basidiomycete species, *Lenzites betulina*, *Pseudotrametes gibbosa* and *Phanerochaete magnoliae* which, both in culture and in nature, are able selectively to take over domain formerly occupied by other species. *L. betulina* and *P. gibbosa* respectively replace *Coriolus* and *Bjerkandera* species by means of invasive, mycoparasitic mycelial fronts which, unlike other fungi, elicit no immediate degenerative response by the host species (Rayner *et al.*, 1987b). *P. magnoliae* exhibits a 'guerrilla-strategy' (cf. Carlile, Chapter 1; Lovett-Doust, 1981). It produces rapidly extending, sparsely distributed hyphae (cf. Figure 2.3), penetrating deep into the interior of *Datronia mollis* colonies and initiating widespread hyphal interference reactions (both in itself and *D. mollis*), and then emerges, phoenix-like, as sole survivor (Ainsworth and Rayner, 1991).

Phase shifts, involving the emergence of assemblages of presumably non-assimilative hyphae that either accumulate at or invade across mycelial interfaces are critical to the outcome of interactions involving many basidiomycetes and some ascomycetes inhabiting durable substrata such as decaying wood (Figure 2.6). In order of their fractal dimension, these assemblages take the form of stationary 'barrages', which resist invasion, and invasive mycelial fronts, mycelial fans and rhizomorphic structures. Marked redistributional effects, evident as a reduction in density of mycelium distal to the interaction interface, accompany these phase shifts, and render the colony more susceptible to invasion if the interfacial region is broached. There may also be evidence of consolidation of 'lines of communication' connecting the distal parts of the colony to emergence points at the interface. Mycelial patterns in invasive phases range across a continuum from diffuse fronts to rhizomorphic, all of which can sometimes be expressed in a single interaction. These phases exhibit strong directionality, suggesting that they are driven forward from the assimilative

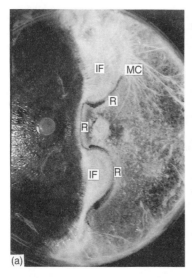
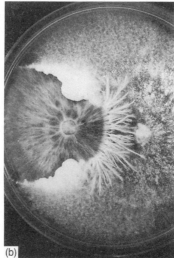

Figure 2.6 a–c Patterns of mycelial emergence at interfaces between different species of wood-inhabiting basidiomycetes paired in 9 cm Petri dishes containing 3% malt agar and incubated in darkness at 20 °C. (a) *Phanerochaete velutina* (left) against *Phlebia radiata*. Periodically extending invasion fronts (IF) produced by *P. velutina* have halted where *P. radiata* has produced ridges (R) of salmon pink aerial mycelium, but these ridges have subsequently been broached by penetrative mycelial cords (MC). (b) *Hypholoma fasciculare* (left) against *Coriolus versicolor*. *H. fasciculare* has produced invasive mycelial cords at the centre of the interface, but is being invaded by dense aerial mycelial masses produced by *C. versicolor* at the periphery of the interface. (c) *Hypholoma fasciculare* (left) against *Phlebia rufa*. *H. fasciculare* has invaded *P. rufa* both in the form of dendritically branched mycelial cords (bottom) and as more diffuse bands showing complex oscillations in hyphal density. (From Rayner *et al.*, 1994.)

mycelium, and they often exhibit simple to complex oscillations in hyphal density.

All these features point to the fluid-like or non-linear dynamic properties of the emergent phases (Rayner *et al.*, 1994). In such terms, the effect of an encounter with a neighbour may be interpreted as imposing a resistance to axial deformation (radial extension), so increasing the ratio between uptake and throughput capacity of the system. This sets the stage for successive rounds of branching and anastomosis leading to phase shifts and oscillations (cf. Schaffer, 1987).

Production of inter- and intracellular pigments commonly accompanies these events, implicating changes in phenol-oxidizing activity as a contributor to the interaction responses; these changes have been confirmed both qualitatively and quantitatively (White and Boddy, 1992; G.S. Griffith and A.D.M. Rayner, unpublished). Moreover, there often appears to be a reciprocal relationship between the activity of tyrosinase and laccase enzymes, associated respectively with the promotion and suppression of formation of aerial mycelium.

Marked differences have been detected in the release of hydrophobic metabolites in culture extracts from mycelia at various stages of development in pure and interactive cultures (G.S. Griffith, A.D.M. Rayner and H.G. Wildman, unpublished). Moreover, exposure of monocultures to the uncoupling agent, 2,4-dinitrophenol (DNP) had very similar effects to interactions both on metabolite profiles (generally suppressing release of phenolic/quinonic compounds) and on development (Figure 2.7). By contrast, incorporation of a phenoloxidase inhibitor in the medium promoted release of metabolites. These findings require further substantiation, but do point to a relationship between mycelial pattern, hydrophobic extracellular chemistry and phenol-oxidizing activity, as predicted by the general self-regulatory model described earlier. As will now be described, a similar relationship is evident from studies of intraspecific mycelial interactions.

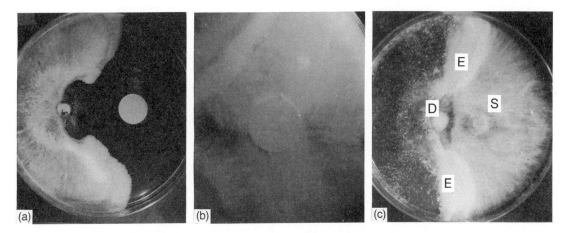

Figure 2.7 Effects of interspecific interaction and antibiotic assay discs loaded with 125×10^{-5} g 2,4 – dinitrophenol on mycelial emergence patterns in *Coriolus versicolor*. (a) *C. versicolor* 2 weeks after inoculation on 3% malt agar opposite an assay disc. The degenerative central portion of the colony is flanked by peripheral zones of emergent mycelium. (b) Part of the same plate as in (a) after further incubation, viewed from below. Mycelium, with bands of variable density, has spread out from the peripheral zones, covering the assay disc. (c) *C. versicolor* (left) paired against *Peniophora lycii*, showing central degenerative zone (D), peripheral zones of emergent mycelium (E), and mycelium (S) spreading out from the peripheral zones across the opposing mycelium.

2.7.2 Intraspecific encounters

Where a neighbouring mycelium belongs to the same or a closely related species, then, at least in higher fungi (ascomycetes and basidiomycetes), the processes that lead up to and follow hyphal anastomosis are regulated at a hierarchical series of decision points (Figure 2.8). The following discussion concerns only those processes that operate between sexually undifferentiated structures.

The primary option concerns whether hyphae can be brought into sufficiently intimate contact to allow fusion processes to be initiated. Failure at this stage will obviate all possibility of genetic exchange, as has been noted between 'anastomosis groups' (equivalent to biological species) of *Thanatephorus cucumeris* (*Rhizoctonia solani*) (e.g. Anderson, 1984). Repulsion, such as is observed between growing tips of main hyphae, even within the same colony (e.g. Ainsworth and Rayner, 1986), and long-range inhibition and/or initiation of degeneration are obvious causes of such failure. In the absence of such preventive effects, contact may occur by chance, or may actually be promoted.

Promotive processes may involve induction of hyphal tip formation at a distance and/or curvature or 'homing' towards recipient sites. They are probably due to chemical signalling and reflect evolutionary affinity between organisms based either on their relatedness or involvement in mycoparasitic associations (Kemp, 1977; Fries, 1981; Rayner, 1986). In the basidiomycete, *Phanerochaete velutina*, long-range homing over up to 250 µm and leading to tip-to-side fusions has been observed to be common, as has curvature of longer 'donor' to shorter 'recipient' branches in the formation of 'H-bridges' between adjacent main hyphae (Ainsworth and Rayner, 1986). This led to the suggestion that incipient branch initiation sites in the lateral wall of recipient hyphae instigated and directed the process. However, in *Stereum* spp., long-range homing was much less marked, and the shorter rather than longer branches exhibited most curvature during H-bridge formation, indicating a reversal of roles from those in *P. velutina* (Ainsworth and Rayner, 1989).

Once contact is established, subsequent development depends on whether protoplasmic continuity is achieved through the opening up of a fusion pore in adjacent hyphal walls. If the pore does not open, there may be a hyphal interference reaction (see above) in which one or both hyphae undergo protoplasmic degeneration, there may be

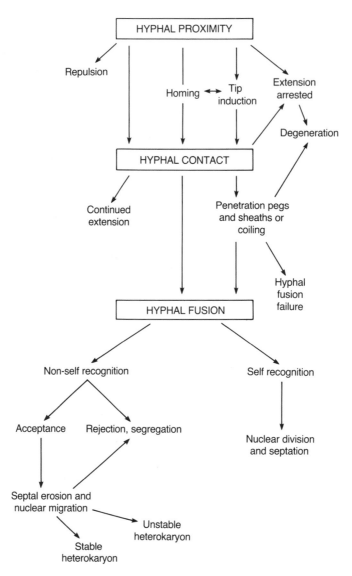

Figure 2.8 Flow diagram illustrating decision points governing processes leading up to and following hyphal anastomosis in higher fungi. (Adapted from Ainsworth and Rayner, 1989.)

renewed growth away from the contacted hypha, or a mycoparasitic reaction may ensue. The latter may involve actual penetration or encoiling of the contacted hypha, and corresponds with the disequilibrium that occurs between adjacent cells in the absence of a communication channel (see above). There have been several reports of intraspecific and even of 'self' hyphal parasitism (Nguyen and Niederpruem, 1984; Yokoyama and

Ogoshi, 1986; Rayner and Coates, 1987; Ainsworth and Rayner, 1989), and it may have an important role in the redistribution of resources from hyphae incapable of anastomosis.

Events following establishment of protoplasmic continuity depend on whether there is sufficient genetic difference between the participants for 'recognition' of 'non-self'. Self-fusions are characterized by an absence of protoplasmic degeneration or of septal erosion and nuclear migration. However, in basidiomycetes with strictly monokaryotic or dikaryotic hyphal compartments, they are characteristically followed by 'nuclear replacement reactions' in which nuclei in recipient compartments are destroyed and become replaced by the mitotic daughters of nuclei in donor compartments (Aylmore and Todd, 1984; Todd and Aylmore, 1985). These reactions do not normally seem to occur in basidiomycetes with multinucleate compartments (Ainsworth and Rayner, 1986, 1989).

Self-fusions occur not only between individual hyphae, but also between multihyphal aggregates, such as mycelial cords (Dowson *et al.*, 1988a). In such cases they can be associated with marked redistributional effects (Figure 2.9).

Responses to non-self fusions are of two, opposite types: rejection and acceptance. Rejection is a consequence of somatic incompatibility and involves the initiation of degenerative processes, probably associated with production and oxidation of phenolic compounds, often giving rise to coloured products (e.g. Hiorth, 1965; Li, 1981). It is usually localized to the fusion compartments themselves, so sealing off the participating mycelial systems from one another and obviating any further genetic or physiological communion. These processes may occur very rapidly, and be akin in expression to those associated with hyphal interference, as in *Phanerochaete velutina* (Ainsworth and Rayner, 1986; Aylmore and Todd, 1986). They can also occur between multihyphal aggregates (Dowson *et al.*, 1988a).

Non-self acceptance reactions occur between outcrossing homokaryons of basidiomycetes (see Clutterbuck, Chapter 11), and involve the initiation of processes, notably the erosion of dolipore septa and proliferation of lateral anastomoses, giving access to invasive nuclei. The rate and extent of septal erosion will determine whether nuclei predominantly invade down radial routes through pre-existing hyphae, or whether they are channelled tangentially, through anastomoses, and outwardly, through hyphae proliferated after the initial mycelial encounter.

Given access, the resulting relationship between immigrant and resident nuclei and cytoplasm may or may not be stable. If stable, then a vigorous, independently growing, secondary mycelial phase, or mating-type heterokaryon will emerge in a

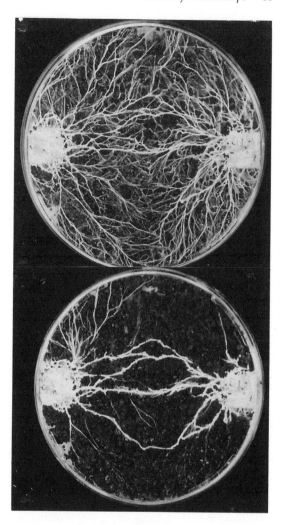

Figure 2.9 Early (top) and late (bottom) stages of interaction between heterokaryotic mycelial cord systems of *Phanerochaete velutina* grown from beechwood inoculum blocks in 14 cm diam. Petri dishes containing non-sterile soil. Anastomosis, followed by regression of all but a few, thickened, connective cords between systems of like genotype.

pattern which reflects the access routes. The developmental properties of this phase commonly differ from those of progenitor homokaryons: it often has a more fast-effuse morphology (see above), is more able to exhibit phase shifts and rejects genetically unlike heterokaryons and homokaryons. Its emergence is commonly associated with the suppression of extension of the progenitors, indicating that the latter become redistributive, e.g. in *Coniophora puteana* (Ainsworth and Rayner, 1990).

The emergence of a secondary mycelium involves the incorporation, following numerous separate fusion events, of populations of two types of nuclei within common cytoplasm. Moreover, since mitochondria do not appear to migrate over long range (e.g. Casselton and Economou, 1985), different 'genomic symbioses' (Lederberg, 1952; Ainsworth *et al.*, 1990a; Rayner, 1991c) may be set up on either side of an interaction interface. The process of associating two types of nuclei with mitochondria from different parents is therefore re-iterated many times during basidiomycete mating, within systems that are already finely balanced between explorative, exploitative and degenerative modes. The potential for unstable relationships is therefore great, raising the issue of how it can be avoided in a compatible mating.

One approach to this issue is to realize that unstable outcomes of non-self access can and do occur, and then to investigate their basis. Evidence for such instability comes from four main kinds of observation.

First, it is common for both non-self access and rejection to be evident in the same interaction. In basidiomycetes, this may often apply when there is not full complementarity of mating factors, for example in what has been described as the 'bow-tie' interaction in *Stereum hirsutum* (Coates and Rayner, 1985). However, it can even be found where there is such complementarity, where rejection limits the domain occupied by secondary mycelium. Moreover, in certain ascomycetes emergence of a secondary mycelial phase occurs at rejection interfaces, in a manner paralleling basidiomycetes, but does not remain stable on subculture (Sharland and Rayner, 1986, 1989). Such observations indicate a reciprocal relationship between access and rejection, both of which are activated by the same trigger; non-self difference (Rayner *et al.*, 1984).

Second, the outcome of nuclear non-self access need not be the same on both sides of the interaction interface and may be complex. In pairings between homokaryons of *Stereum complicatum* from the USA and *Stereum hirsutum* from Russia, reciprocal nuclear exchange resulted in a uniform secondary mycelium, resembling its progenitor,

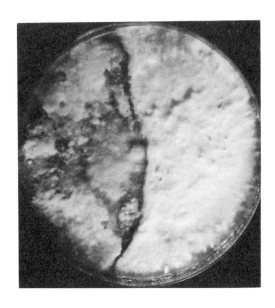

Figure 2.10 Pairing between a homokaryotic strain of *Stereum complicatum* from USA (right) and a homokaryotic strain of *S. hirsutum* from what at the time was the USSR. Following reciprocal exchange of nuclear, but not mitochondrial DNA, the USA strain has retained a uniform morphology, like that of its progenitor. The USSR strain has broken down into numerous subdomains of variable morphology and separated by pigmented zones. (From Ainsworth *et al.*, 1992)

on the USA side. However, the Russian side was divided into numerous subdomains, some of which contained phenotypes resembling *S. complicatum* and others containing *S. hirsutum*-like phenotypes (Figure 2.10). There was some molecular evidence both for limited re-organization, and suppression of expression of *S. hirsutum* nuclear genomes during these interactions (Ainsworth *et al.*, 1992).

Third, degenerative development can occur throughout the accessed region; in an interaction between English and Australian strains of *S. hirsutum*, such degeneration was associated with the outgrowth of crystalline aggregates of the sesquiterpene, (+)-torreyol from the Australian side (Ainsworth *et al.*, 1990b).

Finally, non-self access can result in the replacement of resident nuclear genomes by invasive nuclear genomes: for example, this has been found to occur in certain outcrossing strains of

S. hirsutum when paired with non-outcrossing strains (Ainsworth *et al.*, 1990a).

A general interpretation of these instabilities may lie in the possibility of disparity between the relationships formed by different nuclear genomes with resident mitochondria (Rayner, 1991a; Rayner and Ross, 1991). This would be in accord with the pivotal role played by mitochondria in the metabolic processes regulating indeterminate development. Whatever their explanation, these observations of instability as a consequence of non-self access have obvious implications for fungal speciation. They also interconnect the developmental regulation of individual mycelial systems with the interactions of genomic organelles that bring diversity in natural fungal populations and communities.

2.8 CONCLUSIONS

By operating as feedback-regulated hydrodynamic systems, fungal mycelia can vary the balance between processes of exploration and assimilation, conservation and redistribution of resources and so respond efficiently to changes in local circumstances. How this balance is varied both determines, and is determined by ecological niche and depends on the specification of mechanisms that open and seal boundaries within, between and around hyphae. These mechanisms are in turn both influenced by and have influence over the relationships between populations of genomic organelles within variably partitioned, variably anastomosed hyphal systems.

Fungal mycelia therefore epitomize, at cellular and subcellular levels, organizational principles that are familiar in distributive systems of all kinds, including human and animal societies. Fundamentally, these principles involve the counteraction of expansive (dissociative) and constraining (associative) processes to produce richly heterogeneous patterns. By growing fungi in homogeneous regimes, both the origins and significance of heterogeneity have become obscured. On the other hand, use of heterogeneous regimes in which mycelia have the freedom to interconnect otherwise discrete microenvironments, (e.g. Figure 2.4) provides prospects for ordering, analysing, and making more effective use of fungal indeterminacy.

REFERENCES

Ainsworth, A.M. and Rayner, A.D.M. (1986) Responses of living hyphae associated with self and non-self fusions in the basidiomycete, *Phanerochaete velutina*. *Journal of General Microbiology*, **132**, 191–201.

Ainsworth, A.M. and Rayner, A.D.M. (1989) Hyphal and mycelial responses associated with genetic exchange within and between species of the basidiomycete genus *Stereum*. *Journal of General Microbiology*, **135**, 1643–59.

Ainsworth, A.M. and Rayner, A.D.M. (1990) Mycelial interactions and outcrossing in the *Coniophora puteana* complex. *Mycological Research*, **94**, 627–34.

Ainsworth, A.M. and Rayner, A.D.M. (1991) Ontogenetic stages from coenocyte to basidiome and their relation to phenoloxidase activity and colonization processes in *Phanerochaete magnoliae*. *Mycological Research*, **95**, 1414–22.

Ainsworth, A.M., Rayner, A.D.M., Broxholme, S.J. and Beching, J.R. (1990a) Occurrence of unilateral genetic transfer and genomic replacement between strains of *Stereum hirsutum* from non-outcrossing and outcrossing populations. *New Phytologist*, **115**, 119–28.

Ainsworth, A.M., Rayner, A.D.M., Broxholme, S.J. *et al.* (1990b) Production and properties of the sesquiterpene, (+)-torreyol, in degenerative mycelial interactions between strains of *Stereum*. *Mycological Research*, **94**, 799–809.

Ainsworth, A.M., Beeching, J.R., Broxholme, S.J. *et al.* (1992) Complex outcome of reciprocal exchange of nuclear DNA between two members of the basidiomycete genus *Stereum*. *Journal of General Microbiology*, **138**, 1147–1157.

Anderson, J.G. and Smith, J.E. (1971) The production of conidiophores and conidia by newly germinated conidia of *Aspergillus niger* (microcycle conidiation). *Journal of General Microbiology*, **69**, 185–97.

Anderson, N.A. (1984) Variation and heterokaryosis in *Rhizoctonia solani*, in *The Ecology and Physiology of the Fungal Mycelium*, (eds D.H. Jennings and A.D.M. Rayner), Cambridge University Press, Cambridge, pp. 367–82.

Aylmore, R.C. and Todd, N.K. (1984) Hyphal fusion in *Coriolus versicolor*, in *The Ecology and Physiology of the Fungal Mycelium*, (eds D.H. Jennings and A.D.M. Rayner), Cambridge University Press, Cambridge, pp. 103–25.

Aylmore, R.C. and Todd, N.K. (1986) Cytology of non-self hyphal fusions and somatic incompatibility in *Phanerochaete velutina*. *Journal of General Microbiology*, **132**, 581–591.

Bartnicki-Garcia, S. (1973) Fundamental aspects of hyphal morphogenesis. *Symposia of the Society for General Microbiology*, **23**, 245–67.

Boddy, L. and Rayner, A.D.M. (1983) Mycelial interactions, morphogenesis and ecology of *Phlebia radiata* and *Phlebia rufa* in oak. *Transactions of the British Mycological Society*, **80**, 437–8.

Bolton, R.G. and Boddy, L. (1993) Characterisation of the spatial aspects of foraging mycelial cord systems using fractal geometry. *Mycological Research*, **97**, 762–8.

Bushell, M.E. (1989a) Biowars in the bioreactor. *New Scientist*, **124**, 42–5.

Bushell, M.E. (1989b) The process physiology of secondary metabolite production. *Symposia of the Society for General Microbiology*, **44**, 95–120.

Casselton, L.A. and Economou, E. (1985) Dikaryon formation, in *Developmental Biology of the Higher Fungi*, (eds D. Moore, L.A. Casselton, D.A. Wood and J.C. Frankland), Cambridge University Press, Cambridge, pp. 213–29.

Coates, D. and Rayner, A.D.M. (1985) Genetic control and variation in expression of the 'bow-tie' reaction between homokaryons of *Stereum hirsutum*. *Transactions of the British Mycological Society*, **84**, 191–205.

Coggins, C.R., Hornung, U., Jennings, D.H. and Veltkamp, C.J. (1980) The phenomenon of 'point growth', and its relation to flushing and strand formation in the mycelium of *Serpula lacrimans*. *Transactions of the British Mycological Society*, **75**, 69–76.

Cole, G.T., Sekiya, M., Kasai, R. *et al.* (1979) Surface ultrastructure and chemical composition of the cell walls of conidial fungi. *Experimental Mycology*, **3**, 132–56.

Cooke, R.C. and Rayner, A.D.M. (1984) *Ecology of Saprotrophic Fungi*, Longman, London and New York.

Dowson, C.G., Rayner, A.D.M. and Boddy, L. (1986) Outgrowth patterns of mycelial-cord-forming basidiomycetes from and between woody resource units in soil. *Journal of General Microbiology*, **132**, 203–11.

Dowson, C.G., Rayner, A.D.M. and Boddy, L. (1988a) The form and outcome of mycelial interactions involving cord-forming decomposer basidiomycetes in homogeneous and heterogeneous environments. *New Phytologist*, **109**, 423–32.

Dowson, C.G., Rayner, A.D.M. and Boddy, L. (1988b) Foraging patterns of *Phallus impudicus*, *Phanerochaete laevis* and *Steccherinum fimbriatum* between discontinuous resource units in soil. *FEMS Microbiology Ecology*, **53**, 291–8.

Dowson, C.G., Rayner, A.D.M. and Boddy, L. (1988c) Inoculation of mycelial cord-forming basidiomycetes into woodland soil and litter. I. Initial establishment. *New Phytologist*, **109**, 335–41.

Dowson, C.G., Rayner, A.D.M. and Boddy, L. (1988d) Inoculation of mycelial cord-forming basidiomycetes into woodland soil and litter. II. Resource capture and persistence. *New Phytologist*, **109**, 343–9.

Dowson, C.G., Rayner, A.D.M. and Boddy, L. (1989a) Spatial dynamics and interactions of the woodland fairy ring fungus, *Clitocybe nebularis*. *New Phytologist*, **111**, 699–705.

Dowson, C.G., Springham, P., Rayner, A.D.M. and Boddy, L. (1989b) Resource relationships of foraging mycelial systems of *Phanerochaete velutina* and *Hypholoma fasciculare* in soil. *New Phytologist*, **111**, 501–509.

Franks, N.R., Gomez, N., Goss, S. and Deneubourg, J.L. (1991) The blind leading the blind in army ant raid patterns: testing a model of self-organisation (Hymenoptera: Formicidae). *Journal of Insect Behaviour*, **4**, 583–606.

Frese, D. and Stahl, U. (1990) Ageing in *Podospora anserina* – a consequence of alternative respiration?, in *Fourth International Mycological Congress IMC4 Abstracts*, (eds A. Reisinger and A. Bresinsky), University of Regensburg, p. 184.

Fries, N. (1981) Recognition reactions between basidiopores and hyphae in *Leccinum*. *Transactions of the British Mycological Society*, **77**, 9–14.

Garrett, S.D. (1970) *Pathogenic Root-Infecting Fungi*, Cambridge University Press, Cambridge.

Gleick, J. (1988) *Chaos*, Heinemann, London.

Gottlieb, D. (1971) Limited growth in fungi. *Mycologia*, **63**, 619–29.

Gould, S.J. and Lewontin, R.C. (1979) The spandrels of San Marco and the Panglossian paradigm: a critique of the adaptionist programme. *Proceedings of the Royal Society, London, Series B*, **205**, 581–98.

Gregory, P.H. (1966) The fungus spore: what it is and what it does, in *The Fungus Spore*, (ed. M.F. Madelin), Butterworth, London, pp. 1–13.

Gregory, P.H. (1984) The fungal mycelium: an historical perspective. *Transactions of the British Mycological Society*, **82**, 1–11.

Griffith, G.S. and Boddy, L. (1991) Fungal decomposition of attached angiosperm twigs. IV. Effect of water potential on interactions between fungi on agar and in wood. *New Phytologist*, **117**, 633–41.

Harley, J.L. and Smith, S.E. (1983) *Mycorrhizal Symbiosis*. Academic Press, London.

Hedger, J. (1990) Fungi in the tropical forest canopy. *The Mycologist*, **4**, 200–2.

Hiorth, J. (1965) The phenoloxidase and peroxidase activities of two culture types of *Phellinus tremulae* (Bond.) Bond. & Boriss. *Meddelelser Norske Skogforsöksvesen*, **20**, 249–72.

Ingold, C.T. (1984) Patterns of ballistospore germination in *Tilletiopsis*, *Auricularia* and *Tulasnella*. *Transactions of the British Mycological Society*, **83**, 583–91.

Ikediugwu, F.E.O. and Webster, J. (1970) Antagonism between *Coprinus heptemerus* and other coprophilous fungi. *Transactions of the British Mycological Society*, **54**, 181–204.

Jennings, D.H. (1984) Water flow through mycelia, in *The Ecology and Physiology of the Fungal Mycelium*, (eds D.H. Jennings and A.D.M. Rayner), Cambridge University Press, Cambridge, pp. 143–64.

Jennings, D.H. (1987) Translocation of solutes in fungi. *Biological Reviews*, **62**, 215–43.

Kemp, R.F.O. (1977) Oidial homing and the taxonomy and speciation of basidiomycetes with special reference to the genus *Coprinus*, in *The Species Concept in Hymenomycetes*, (ed. H. Clémençon), J. Cramer, Vaduz, pp. 259–76.

Kück, U. (1989) Mitochondrial DNA rearrangements in *Podospora anserina*. *Experimental Mycology*, **13**, 111–20.

Leatham, G.F. (1985) Growth and development of *Lentinus edodes* on a chemically defined medium, in *Developmental Biology of Higher Fungi*, (eds D. Moore, L.A. Casselton, D.A. Wood and J.C. Frankland), Cambridge University Press, Cambridge, pp. 403–27.

Leatham, G.F. and Stahmann, M.A. (1981) Studies on the laccase of *Lentinus edodes*: specificity, localization and association with developing fruit bodies. *Journal of General Microbiology*, **125**, 147–57.

Lederberg, J. (1952) Cell genetics and hereditary symbiosis. *Physiological Reviews*, **32**, 403–30.

Li, C.Y. (1981) Phenoloxidase and peroxidase activities in zone lines of *Phellinus weirii*. *Mycologia*, **73**, 811–21.

Lockhart, J.A. (1965) Cell extension, in *Plant Biochemistry*, (eds J. Bonner and J.E. Varner), Academic Press, New York, pp. 826–49.

Lovett- Doust, L. (1981) Population dynamics and local specialization in a clonal perennial (*Ranunculus repens*). I. The dynamics of ramets in contrasting habitats. *Journal of Ecology*, **69**, 743–55.

Lyr, H. (1958) Die Induktion der Laccase-Bildung bei *Collybia velutipes* Curt. *Archiv für Mikrobiologie*, **28**, 310–324.

Lyr, H. (1963) Enzymatisches Detoxifikation chlorierter Phenole. *Phytopathologie Zeitschrift*, **38**, 342–54.

Money, N.P. (1990) Measurement of hyphal turgor. *Experimental Mycology*, **14**, 416–25.

Nguyen, T.T. and Niederpruem, D.J. (1984) Hyphal interactions in *Schizophyllum commune*: the di-mon mating, in *The Ecology and Physiology of the Fungal Mycelium*, (eds D.H. Jennings and A.D.M. Rayner), Cambridge University Press, Cambridge pp. 73–102.

Perry, D.A., Margolis, H., Choquette, C. *et al.* (1989) Ectomycorrhizal mediation of competition between coniferous tree species. *New Phytologist*, **112**, 501–11.

Prosser, J.I. (1991) Mathematical modelling of vegetative growth of filamentous fungi, in *Handbook of Applied Biology*, vol. 1, (eds D.H. Arora, B. Rai, K.G. Mukerji and G.R. Knudsen), Marcel Dekker, New York, pp. 591–623.

Prosser, J.I. (1993) Growth kinetics of mycelial colonies and aggregates of ascomycetes. *Mycological Research*, **97**, 513–28.

Rayner, A.D.M. (1986) Mycelial interactions – genetic aspects, in *Natural Antimicrobial Systems*, (eds G.W. Gould, M.E. Rhodes Roberts A.K. Charnley, *et al.*, Bath University Press, pp. 277–296.

Rayner, A.D.M. (1991a) The challenge of the individualistic mycelium. *Mycologia*, **83**, 48–71.

Rayner, A.D.M. (1991b) Conflicting flows – the dynamics of mycelial territoriality. *McIlvainea*, **10**, 24–35.

Rayner, A.D.M. (1991c) The phytopathological significance of mycelial individualism. *Annual Review of Phytopathology*, **29**, 305–23.

Rayner, A.D.M. and Coates, D. (1987) Regulation of mycelial organisation and responses, in *Evolutionary Biology of the Fungi*, (eds A.D.M. Rayner, C.M. Brasier

and D. Moore), Cambridge University Press, Cambridge, pp. 115–36.

Rayner, A.D.M. and Franks, N.R. (1987) Evolutionary and ecological parallels between ants and fungi. *Trends in Ecology and Evolution*, **2**, 127–33.

Rayner, A.D.M. and Ross, I.K. (1991) Sexual politics in the cell. *New Scientist*, **129**, 30–3.

Rayner, A.D.M. and Webber, J. (1984) Interspecific mycelial interactions – an overview, in *The Ecology and Physiology of the Fungal Mycelium*, (eds D.H. Jennings and A.D.M. Rayner), Cambridge University Press, Cambridge, pp. 383–417.

Rayner, A.D.M., Bevercombe, G.P., Brown, T.C. and Robinson, A. (1981) Fungal growth in a lattice: a tentative explanation for the shape of diamond-cankers in sycamore. *New Phytologist*, **87**, 383–93.

Rayner, A.D.M., Coates, D., Ainsworth, A.M. *et al.* (1984) The biological consequences of the individualistic mycelium, in *The Ecology and Physiology of the Fungal Mycelium*, (eds D.H. Jennings and A.D.M. Rayner), Cambridge University Press, Cambridge, pp. 509–40.

Rayner, A.D.M., Powell, K.A., Thompson, W. and Jennings, D.H. (1985) Morphogenesis of vegetative organs, in *Developmental Biology of Higher Fungi*, (eds D. Moore, L.A. Casselton, D.A. Wood and J.C. Frankland), Cambridge University Press, Cambridge, pp. 249–79.

Rayner, A.D.M., Boddy, L. and Dowson, C.G. (1987a) Genetic interactions and developmental versatility during establishment of decomposer basidiomycetes in wood and tree litter. *Symposia of the Society for General Microbiology*, **41**, 83–123.

Rayner, A.D.M., Boddy, L. and Dowson, C.G. (1987b) Temporary parasitism of *Coriolus* spp. by *Lenzites betulina*: a strategy for domain capture in wood decay fungi. *FEMS Microbiology Ecology*, **45**, 53–8.

Rayner, A.D.M., Griffith, G.S. and Wildman, H.G. (1994) Differential insulation and the generation of mycelial patterns, in *Shape and Form in Plants and Fungi*, (ed. D.S. Ingram), Academic Press. London, pp. 293–312.

Read, D.J. (1984) The structure and function of the vegetative mycelium of mycorrhizal roots, in *The Ecology and Physiology of the Fungal Mycelium*, (eds D.H. Jennings and A.D.M. Rayner), Cambridge University Press, Cambridge, pp. 215–40.

Read, D.J. (1992) The mycorrhizal fungal community with special reference to nutrient mobilization, in *The Fungal Community*, 2nd edn, (eds G.C. Carroll and D.T. Wicklow), Marcel Dekker, New York, pp. 631–52.

Ritz, K. and Crawford, J. (1990) Quantification of the fractal nature of colonies of *Trichoderma viride*. *Mycological Research*, **94**, 1138–1141.

Ross, I.K. (1985) Determination of the initial steps in differentiation in *Coprinus congregatus*, in *Developmental Biology of Higher Fungi*, (eds D. Moore, L.A.

Casselton, D.A. Wood and J.C. Frankland), Cambridge University Press, Cambridge, pp. 353–73.

Ruiters, M.H.J. and Wessels, J.G.H. (1989) *In situ* localization of specific RNAs in whole fruiting colonies of *Schizophyllum commune*. *Journal of General Microbiology*, **135**, 1747–54.

Schaffer, W.M. (1987) Chaos in ecology and epidemiology, in *Chaos in Biological Systems*, (eds H. Degn, A.V. Holden and L.F. Olsen), Plenum Press, New York and London, pp. 233–48.

Sharland, P.R. and Rayner, A.D.M. (1986) Mycelial interactions in *Daldinia concentrica*. *Transactions of the British Mycological Society*, **86**, 643–50.

Sharland, P.R. and Rayner, A.D.M. (1989) Mycelial interactions in outcrossing populations of *Hypoxylon*. *Mycological Research*, **93**, 187–98.

Sietsma, J.H. and Wessels, J.G.H. (1979) Evidence for covalent linkages between chitin and β-glucan in a fungal wall. *Journal of General Microbiology*, **114**, 99–108.

Smith, M.L., Bruhn, J.N. and Anderson, J.B. (1992) The fungus *Armillaria bulbosa* is among the largest and oldest living organisms. *Nature*, **356**, 428–31.

Stenlid, J. and Rayner, A.D.M. (1989) Environmental and endogenous controls of developmental pathways: variation and its significance in the forest pathogen, *Heterobasidion annosum*. *New Phytologist*, **113**, 245–58.

Stringer, M.A., Dean, R.A., Sewall, T.C. and Timberlake, W.E. (1991) *Rodletless*, a new *Aspergillus* developmental mutant induced by directed gene inactivation. *Genes and Development*, **5**, 1161–71.

Thompson, W. (1984) Distribution, development and functioning of mycelial cord systems of decomposer basidiomycetes of the deciduous woodland floor, in *The Ecology and Physiology of the Fungal Mycelium*, (eds D.H. Jennings and A.D.M. Rayner), Cambridge University Press, Cambridge, pp. 185–214.

Thompson, W. and Rayner, A.D.M. (1983) Extent, development and functioning of mycelial cord systems in soil. *Transactions of the British Mycological Society*, **81**, 333–45.

Todd, N.K. and Aylmore, R.C. (1985) Cytology of hyphal interactions and reactions in *Schizophyllum commune*, in *Developmental Biology of Higher Fungi*, (eds D. Moore, L.A. Casselton, D.A. Wood and J.C. Frankland), Cambridge University Press, Cambridge, pp. 231–48.

Trinci, A.P.J. (1978) The duplication cycle and vegetative development in moulds, in *The Filamentous Fungi*, vol. 3, (eds J.E. Smith and D.R. Berry), Arnold, London, pp. 132–63.

Unestam, T. (1991) Water repellency, mat formation, and leaf-stimulated growth of some ectomycorrhizal fungi. *Mycorrhiza*, **1**, 13–20.

Watkinson, S.C. (1977) Effect of amino acids on coremium development in *Penicillium claviforme*. *Journal of General Microbiology*, **101**, 269–75.

Webster, J. (1987) Convergent evolution and the functional significance of spore shape in aquatic and semi-aquatic fungi, in *Evolutionary Biology of the Fungi*, (eds A.D.M. Rayner, C.M. Brasier and D. Moore), Cambridge University Press, Cambridge, pp. 191–201.

Wells, J.M. and Boddy, L. (1990) Wood decay, and phosphorus and fungal biomass allocation, in mycelial cord systems. *New Phytologist*, **116**, 285–95.

Wessels, J.G.H. (1986) Cell wall synthesis in apical hyphal growth. *International Review of Cytology*, **104**, 37–79.

Wessels, J.G.H. (1991) Fungal growth and development: a molecular perspective, in *Frontiers in Mycology*, (ed. D.L. Hawksworth), CAB International, Kew, Surrey, pp. 27–48.

Wessels, J.G.H. (1992) Gene expression during fruiting in *Schizophyllum commune*. *Mycological Research*, **96**, 609–20.

White, N.A. and Boddy, L. (1992) Extracellular enzyme localization during interspecific fungal interactions. *FEMS Microbiology Letters*, **98**, 75–80.

Willetts, H.J. (1978) Sclerotium formation, in *The Filamentous Fungi*, vol. 3 (eds J.E. Smith and D.R. Berry), Arnold, London, pp. 197–213.

Worrall, J.J., Chet, I. and Hüttermann, A. (1986) Association of rhizomorph formation with laccase activity in *Armillaria* spp. *Journal of General Microbiology*, **132**, 2527–33.

Yokoyama, K. and Ogoshi, A. (1986) Studies on hyphal anastomosis of *Rhizoctonia solani*. IV. Observation of imperfect fusion by light and electron microscopy. *Transactions of the Mycological Society of Japan*, **27**, 399–413.

PART TWO
THE ARCHITECTURE OF FUNGAL CELLS

CELL WALLS

G.W. Gooday

Department of Molecular and Cell Biology, Marischal College, University of Aberdeen, Aberdeen, UK

3.1 INTRODUCTION

Fungal walls have commanded much attention as they are the major cellular features that distinguish fungi from other organisms. The shape of the fungal cell is the shape of its wall. The mechanical strength of their walls enables fungi to assume a variety of forms, such as penetrative, ramifying hyphae, proliferating yeast cells and spores of many shapes and sizes. The chemical and physical make-up of the wall gives protection to the protoplast from a range of environmental stresses. Its physical strength gives protection against osmotic bursting. Specific components, especially of spore walls, give protection against damage from ultraviolet radiation, enzymic lysis, organic solvents, toxic chemicals and desiccation. Some secreted materials may be considered as periplasmic, as they are secreted from the plasma membrane but do not move through the wall to the outside. These include some lytic enzymes involved in breakdown of nutrients into molecules that can be transported through the plasma membrane. An example is invertase, which hydrolyses sucrose to give glucose and fructose. Some other lytic enzymes, such as cellulases, pass through the wall into the medium (cf. Chapter 7). This process probably occurs chiefly at the hyphal apex where the wall may be most porous. The extent of permeability of the wall is unclear. Extrapolating from results with yeast cells, some reports suggest that lateral hyphal walls are relatively impermeable, only allowing ready passage, in or out, of molecules with molecular masses up to about 700. Other studies suggest much greater permeabilities, for example of dextran with molecular mass up to 70 000 Da which is equivalent to a globular protein of molecular mass 400 000 Da (De Nobel *et al.*, 1989), but these authors caution that other factors such as glycosylation and charge are important in the movement of molecules through the wall.

The outer surface of the wall is the primary interface between the cell and its environment. It is the site of antigens and of agglutinins involved in mating and in adhesion to substrates, hosts or other cells. Some polysaccharides of the wall matrix, such as α(1–3)-glucans and water soluble β(1–3), β(1–6)-glucans can act as food reserves. When cultures are starved, these glucans may be hydrolysed to glucose and re-utilized.

3.2 CELL WALL CONSTITUENTS: OCCURRENCE AND PROPERTIES

The fungi are polyphyletic and analyses of their walls have shown a wide range of polymers, chiefly polysaccharides (Bartnicki-Garcia, 1968; Wessels and Sietsma, 1981; Table 3.1). In electron micrographs walls usually appear to be composed of layers. In some cases it is possible to assign particular components to particular layers. For example in spore walls distinct layers are often deposited in turn during development. In hyphal and yeast cells, however, the layers are best thought of as merging into one another to form one structure. Nevertheless, generalizations can be made (Figure 3.1). The shape-determining, more fibrillar polysaccharides, chitin, chitosan, chitin–glucan complexes and cellulose, make up the inner layer of the wall. These are embedded in more gel-like matrix polymers such as glucans and glycoproteins which extend outwards to make up the outer layer of the wall. That this basic make-up is composed of different polymers in different taxonomic groups of fungi suggests that convergent evolution has occurred to provide a range of chemical solutions for a common physical

The Growing Fungus. Edited by Neil A.R. Gow and Geoffrey M. Gadd. Published in 1994 by Chapman & Hall, London. ISBN 0 412 46600 7

Table 3.1   Major polymers occurring in fungal cell walls

Taxonomic group	More fibrous polymers	More gel-like polymers
Basidiomycetes	Chitin β-(1–3),β-(1–6)-Glucan	Xylomannoproteins α-(1–3)-Glucan
Ascomycetes	Chitin β-(1–3),β-(1–6)-Glucan	Galactomannoproteins α-(1–3)-Glucan
Zygomycetes	Chitin Chitosan	Polyglucuronic acid Glucuronomannoproteins Polyphosphate
Chytridiomycetes	Chitin Glucan	Glucan
Hyphochytridiomycetes	Chitin Cellulose	Glucan?
Oomycetes	β-(1–3),β-(1–6)-Glucan Cellulose	Glucan

Adapted from Bartnicki-Garcia (1968) and Wessels and Sietsma (1981).

problem, namely the construction of a wall by a growing hypha or yeast cell.

3.2.1   Chitin and chitosan

Chitin is the β(1-4)-linked polymer of *N*-acetylglucosamine and chitosan is the β(1-4)-linked polymer of glucosamine. Chitosan is formed by progressive enzymic deacetylation of chitin and it is likely that there is great variety in degree of acetylation in these molecules in nature. Chitin is abundant throughout the natural world. It occurs as a structural polysaccharide in most invertebrates and in many protists and is probably a universal component of fungal walls. Fungi that were thought to lack chitin, such as many Oomycetes, e.g. *Saprolegnia* species and the fission yeasts, *Schizosaccharomyces* species, have been shown to have small amounts. The molecular conformation of chitin gives it great mechanical strength. Individual polysaccharide chains are hydrogen bonded between adjacent sugar units, giving them rigidity. In α-chitin, the form found in fungal walls, individual chains are aggregated into microfibrils, with hydrogen bonds holding adjacent chains together in an antiparallel fashion, i.e. the chains run in opposite directions. The final result is a very strong rigid structure (Gooday, 1979). The chitin microfibrils give shape to the cell, as enzymic and/or chemical removal of other wall components gives chitinous cell 'ghosts' which retain the original shape of the cell. The chitin

microfibrils are in the innermost part of the wall. In electron micrographs of chitin preparations the microfibrils appear in a range of forms, such as short stubby ones from yeast walls and long interwoven ones in many hyphal walls (Gow and Gooday, 1983). In most cases they have no preferred orientation and appear as a random network (Burnett, 1979). In the apices of germ tubes, however, they may be predominantly longitudinal. In the walls of cells in the stipes of the mushroom *Coprinus cinereus* microfibrils are arranged in a shallow helix (Kamada *et al.*, 1991a). This gives these walls great lateral strength, so that the turgor pressure of each cell is contained by the stress-bearing fabric of its wall. This results in the multicellular structure having great strength, enabling the mushroom to exert vertical force as it pushes up through substrates. Septa typically are rich in chitin (Figure 3.1) and have microfibrils arranged in a tangential fashion. The occurrence of chitin is demonstrated by the septa staining with Calcofluor White, a fluorescent brightener, and with fluorescein-labelled wheat germ agglutinin, a plant lectin, both of which have high affinities for chitin, and by their dense labelling in autoradiographs of growing hyphae fed with tritiated *N*-acetylglucosamine. Chitin does not occur by itself and its interactions with other components are of importance in the make-up of the wall. In particular, covalent cross-links involving peptides are formed between chitin and other wall components, notably glucans (Wessels, 1986, 1990; see below).

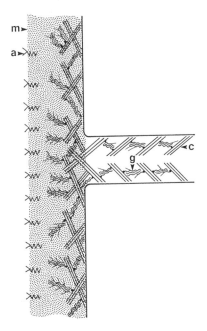

Figure 3.1 Diagram of distribution of major components of the lateral wall and septum of a typical fungus. The wall and septum are bordered on their inner faces by the plasma membrane (continuous line). The more fibrous components, chitin (c) and β(1–3), β(1–6) glucan (g), are predominantly in the inner layer of the wall and are the major components of the septum. The more gel-like mannoproteins (m) form a matrix throughout the wall and are predominant in the outer layer. Antigenic glycoproteins, agglutinins and adhesins (a) are on the outer face of the wall, exposed to the environment.

Chitosan occurs as a major component in walls of Zygomycetes, but probably is a minor wall component in many other groups of fungi. It occurs as a distinct layer in the ascospore wall of *Saccharomyces cerevisiae* (Briza *et al.*, 1990). It is unusual as a natural polymer as it is a polycation with strong adsorptive properties. In walls of Zygomycetes it is accompanied by anionic polymers rich in glucuronic acid and probably by polyphosphate.

3.2.2 Glucans

β-Linked glucans are major constituents of most fungal walls. Walls of all members of Ascomycetes and Basidiomycetes contain β(1–3)-glucans with branches of one or more β(1–6)-linked glucose residues. These branched β(1–3)-glucans are

accompanied by varying amounts of glucans with β(1–6) links only, β(1–3) links only or alternating β(1–3) and β(1–6) links (Wessels, 1986, 1990). These branched glucans can form gels of interconnected triple helices. Covalent links with chitin microfibrils give them a major role in the structure of the mature wall. In *Schizophyllum commune*, hyphae may also become covered with a mucilage of β(1–3), β(1–6)-glucan which may build up to cover the colony with a gelatinous mass or gelatinize the entire culture medium (Wessels and Sietsma, 1979). Also occurring are microcrystalline α(1–3)-glucans which can comprise a major component of the outer matrix of the wall. α(1–3)-Glucans from different fungi also contain various amounts of α(1–4) links. In the Oomycetes, a cellulosic polymer with β(1–4) links provides the structural role analogous to that of chitin in other fungi (Fevre, 1979). It occurs together with β(1–3), β(1–6)-glucans and may be covalently linked to them (Wessels and Sietsma, 1981).

3.2.3 Glycoproteins and proteins

Glycoproteins, especially mannoproteins, galacto-mannoproteins and xylomannoproteins, are major constituents of the matrix of many fungal walls. They may have various amounts of mannosyl-6-phosphoryl linkages. Glycoproteins containing galactosamine and/or N-acetylgalactosamine have also been reported from a range of fungal walls. In pathogenic fungi, such as *Candida albicans* and *Aspergillus fumigatus*, they are important surface antigens. Most work on their structure has been with *S.cerevisiae*, where the major mannoprotein is a large molecule, containing about 95% carbohydrate linked to protein in two ways. There are short mannose chains O-linked to serine or threonine residues, and long α(1–6)-mannose chains, with α(1–2) and α(1–3) side chains, some with phosphomannose, linked to two β(1–4)-N-acetylglucosamine units which are N-linked to asparagine. Van Rinsum *et al.* (1991) have characterized glucomannoproteins from cell walls of *S.cerevisiae*. Upon initial extraction from the walls, the carbohydrate part of these molecules contained N-acetylglucosamine, mannose and glucose in a molar ratio of 1:53:4. Removal of N-linked and O-linked chains followed by alkaline hydrolysis gave a glucose-rich product with a molar ratio of 1:17:18. Van Rinsum *et al.* (1991) suggest that these glucose-rich chains are responsible for linkages between mannoproteins and glucans in the cell wall. Adjacent mannoprotein molecules may be cross-linked by disulphide bridges.

The hydrophobins are a group of closely related extremely hydrophobic proteins that have been characterized from *Schizophyllum commune* (Wessels, 1992). The most hydrophobic hydrophobin (pSc3) is mainly found in the walls of aerial hyphae, where it probably is a component of the outer wall layer consisting of arrays of parallel rodlets. Such rodlet arrays occur on aerial hyphae of other fungi, and on the surfaces of hydrophobic conidia such as those of *Neurospora crassa* and *Aspergillus nidulans*. Other hydrophobins, pScl and pSc4, are involved in fruit body formation in *S.commune*, and may have a role in binding hyphae together. Hydrophobins represent 6–8% of all proteins synthesized by developing cultures of *S.commune*. Wessels (1992) suggests that when hyphae are submerged the hydrophobins are excreted to the medium from the hyphal tips. Then when hyphae emerge from the medium into the air the hydrophobins polymerize on their surfaces by hydrophobic interactions to form the rodlet layer (Figure 3.2). This layer presumably gives the aerial hyphae properties such as increased resistance to desiccation and perhaps increased strength required for growth in the air.

The outermost of the layers of ascospore walls of *S.cerevisiae* is a cross-linked insoluble polymer of D, L dityrosine, glycine, alanine, glutamic acid and a trace of glycine (Briza *et al.*, 1990). This layer is closely associated and possibly covalently linked to the second outer layer that consists of chitosan. Wild-type spores are resistant to adverse environmental conditions, such as lytic enzymes and organic solvents. Mutant spores lacking the dityrosine-rich layer are much less resistant, suggesting that this outer layer confers much of the resistance.

3.2.4   Other wall polymers

As well as chitin and chitosan, walls of Zygomycetes such as *Mucor* species contain polysaccharides rich in glucuronic acid (Bartnicki-Garcia, 1968). Datema *et al.* (1977) have characterized a heteroglycuronan from walls of *Mucor mucedo* which contains residues of fucose, mannose, galactose and glucuronic acid in the molar ratio of 5:1:1:6. This polyanion is

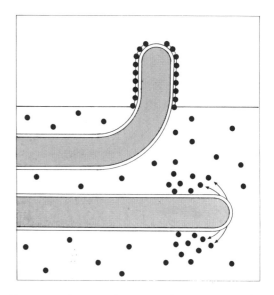

Figure 3.2   Diagram of secretion of hydrophobins (●) at hyphal apices. When the hypha is submerged (lower one), the hydrophobins diffuse into the medium; when it emerges into the air (upper one), they aggregate to form a hydrophobic coat on the hyphal surface. (From Wessels, 1992).

held insoluble in the wall by ionic bonding with the polycationic chitosan. It can be rendered water soluble by extraction with salt solutions of high ionic strength or by degradation of chitosan by treatment with nitrous acid. Datema *et al.* (1977) also found polyphosphate associated with their wall preparations from *M.mucedo*, but caution that this may have been a cytoplasmic component that bound to the walls during preparation.

Melanins are dark brown to black pigments that accumulate in distinct layers of walls of spores, vegetative hyphae or fruit bodies of many fungi (Rast *et al.*, 1981; Wheeler and Bell, 1988). They are polymers of phenolic metabolites, such as tyrosine, dihydroxynaphthalenes, catechol, catecholamines and γ-glutaminyl-3, 4-dihydroxybenzene. They can confer resistance to lysis by enzymes. For example, walls of melanin-deficient strains of a range of fungi are more readily digested than those of pigmented strains by lytic enzymes such as chitinases. Another aromatic polymer produced by fungi is sporopollenin. This is an exceptionally resilient molecule, being very resistant to chemical and physical attack. It is the characteristic component of the walls of pollen grains and is present

Table 3.2 Some characteristics of chitin synthase preparations from different fungi

Fungus	$K_{0.5}{}^{a}$ (μM) UDP-GlcNAc	K_i for inhibition^{a} (μM)			MIC in vivo^{a} (μM)	
		UDP	Polyoxin	Nikkomycin	Polyoxin	Nikkomycin
Basidiomycetes						
Coprinus cinereus	900	500	3	–	0.5	–
Ascomycetes						
Neurospora crassa	2100	–	1.4	2	190	–
Deuteromycetes						
Candida albicans	2000	2000	1.2	–	20	1
Zygomycetes						
Mucor rouxii	500	400	0.6	–	100	0.5
Chytridiomycetes						
Allomyces macrogynus	1200	–	–	–	25	0.01
Oomycetes						
Saprolegnia monoica	–	–	20	–	–	–

Compiled from Gooday and Trinci (1980), Gooday (1990), Bulone *et al.* (1992).
^{a} Values for kinetic parameters vary with different preparations, different conditions and different preparations of antibiotics. $K_{0.5}$ values are for substrate concentrations giving half maximal activity.

in some fungal spore walls such as those of zygospores of *Mucor mucedo* and ascospores of *Neurospora crassa* (Gooday, 1981). Sporopollenin, melanin and related polymers (such as the green pigments of conidia of *Penicillium* and *Trichoderma* species), provide mechanical strength to fungal walls and will protect fungal cells from a range of environmental stresses such as ultraviolet light, solar radiation, desiccation and enzymic lysis (Wheeler and Bell, 1988).

3.3 CELL WALL BIOSYNTHESIS

Cell wall biosynthesis takes place in three sites: cytoplasm, plasma membrane and the wall itself. Structural polymers such as chitin, β(1–3)- and β(1–4)-linked glucans are synthesized vectorially at the plasma membrane, by transmembrane synthases, accepting nucleotide sugar precursors from the cytosol and feeding the polymerized chains into the wall (cf. Chapter 4). Matrix polymers such as mannoproteins are synthesized in the cytoplasmic secretory pathway of endoplasmic reticulum through Golgi vesicles to secretory vesicles. Wall assembly, involving activities such as covalent cross-linking of polymers and modifications such as deacetylation of chitin, takes place in the wall itself.

3.3.1 Chitin metabolism

Chitin is synthesized by intrinsic plasma membrane proteins, the chitin synthases. These enzymes catalyse glycosidic bond formation from the nucleotide sugar substrate, uridine diphospho-*N*-acetylglucosamine:

$$2\text{UDP-GlcNAc} + (\text{GlcNAc})_n \rightarrow (\text{GlcNAc})_{n+2} + 2\text{UDP}$$

Where investigated in detail, chitin synthases are activated by *N*-acetylglucosamine and its oligomers. In the case of *C. cinereus*, chitin synthase is allosterically activated, both by UDP-GlcNAc and by *N*-acetylglucosamine and its oligomers (Gooday, 1977, 1979). As discussed in Chapter 4, chitin synthases are clearly associated with the plasma membrane lipid bilayer and a phospholipid environment is required for their activity. Divalent cations such as magnesium, manganese or cobalt are required for activity. The UDP product is a competitive inhibitor (Gooday, 1979), as are the antibiotics polyoxin and nikkomycin (see below).

Chitin synthase activities have been characterized from a wide range of fungi. In all cases the enzymes have similar properties (Table 3.2). It is probable that in most cases there is more than one type of chitin synthase in these preparations but nevertheless generalizations can be made. Thus their K_m values for UDP-GlcNAc are about 1 mM, the approximate concentration in the cell (Gooday, 1979). Most preparations are, at least to some extent, zymogenic, i.e. produced as proenzymes requiring activation by specific proteases. This proteolytic activation presumably plays a role in the temporal and spatial regulation of the enzyme,

Table 3.3 Some characteristics of chitin synthases of *Saccharomyces cerevisiae*

Chitin synthase	K_m^a (μM) UDPGlcNAc	K_i^a (μM) for inhibition	Gene	% Chitin loss in mutant	Zymogenicity	Proposed role
I	500	3(P),0.25(N)	CHS1	< 10	Yes	Cell wall repair
II	250	22(P), 6(N)	CHS2	< 10	Yes	Primary septum
III	800	1.5 (P)	CSD2[b]	> 90	No	Growth, mating, sporulation

Compiled from Orlean (1987), Cabib (1991), Bulawa (1992, 1993).
[a] Values for kinetic parameters vary with different preparations, different conditions and different preparations of antibiotics; (P) polyoxin, (N) nikkomycin.
[b] Requires other gene products for activity; allelic with *CAL1, CAL4, DIT101*.

by locally activating it in the membrane when and where its activity is required. As well as being in zymogenic and active forms in the plasma membrane, zymogenic chitin synthase also occurs in fungal cells in chitosomes, which are membrane-bound microvesicles about 70 nm in diameter (Bartnicki-Garcia *et al.*, 1979; Kamada *et al.*, 1991b; cf. Chapter 5). After purification by differential centrifugation, chitosomes can be activated by treatment with proteolytic enzymes, and then produce chitin microfibrils if incubated with UDP-GlcNAc.

Chitin synthesis in *S. cerevisiae* involves at least three chitin synthases, I, II and III (Table 3.3). These three enzymes have different properties. Chitin synthases I and II are zymogenic. Chitin synthase III does not require proteolysis for activity. Chitin synthase I is at least an order of magnitude more sensitive to inhibition by 0.5 M-NaCl, polyoxin D and nikkomycins X and Z than chitin synthase II (Sburlati and Cabib, 1986; Cabib, 1991). Chitin synthase I has a pH optimum at about 6.5, chitin synthase II at 7.5–8.0. Chitin synthase II is inhibited by cobalt ions, whereas chitin synthase II is stimulated (Sburlati and Cabib, 1986). Chitin synthases I and II have similar K_m values for UDP-GlcNAc, of about 0.8–0.9 mM.

Chitin synthase I zymogen is the most abundant of the three enzymes in the yeast cell. Disruption of its gene did not affect chitin synthesis during normal growth (Cabib *et al.*, 1992). The only apparent deficiency of these mutants was a tendency to leak cell contents from the septum following release of the daughter cells when growing in media of low pH. As described below, daughter cell release is the result of chitinase activity. Cabib *et al.* (1992) suggest that the leakage of cell contents is the result of overaction of chitinase, favoured by acidic pH, which in wild-type cells would be corrected by the synthesis of

chitin via chitin synthase I. Thus chitin synthase I is a 'repair enzyme'. Chitin synthase II is responsible for chitin synthesis in the primary septa, as mutants of *CHS2* did not form them, and showed clumpy growth with aberrant cell shapes and sizes (Silverman *et al.*, 1988).

The structural genes for chitin synthases I and II are well characterized, but the situation with chitin synthase III is more complex, as more than one gene appears to be involved in its activity (Bulawa, 1993). Several mutants are deficient in chitin synthase III. These include *Cal1* and *Cal4*, resistant to Calcofluor White, a fluorescent brightener which binds to nascent chitin and prevents its orderly deposition (Valdivieso *et al.*, 1991; Shaw *et al.*, 1991) and *CSD2* (chitin synthesis deficient) which is allelic to *CAL1* and *CAL4* (Bulawa, 1992). Chitosan, formed by deacetylation of chitin, is a major component of spore walls of *S. cerevisiae*, where it occurs together with a dityrosine polymer (Briza *et al.*, 1990). *CSD2* mutants produce no chitosan or dityrosine layers in their spore walls. The mutant *dit101*, identified by lack of dityrosine in its spore walls, has proved to be allelic to *CSD2* (Bulawa, 1992). Mutants of *CAL1* produce thin chitinous septa but no chitinous ring and no chitin in their cell walls (Shaw *et al.*, 1991). Assessment of mutants deficient in chitin synthase III activity shows that this enzyme is responsible for synthesis of chitin during growth, mating and sporulation (Bulawa, 1992, 1993). A mating yeast cell, called a shmoo, has an elevated chitin content in the walls of its conjugation tube. Expression of *CSD2* is required for synthesis of most of this chitin, but there appears to be no increase in activity of chitin synthase III (Orlean, 1987). The observed increase in chitin synthesis thus appears to be the result of increased flux through the chitin synthesis pathway, with no change in enzyme activity.

The amino acid sequences predicted from the

genes for chitin synthases I and II are substantially similar for their carboxyl terminal two-thirds, whereas the amino terminal one-third segments are different (Silverman, 1989). This suggests that the catalytic sites are in the similar regions, whereas the unrelated regions may be involved in regulation or localization of the two enzymes. The similar regions have four potential membrane-spanning domains, consistent with them being intrinsic membrane proteins. The two genes are unlinked, but their similarity suggests that they have arisen from duplication of an ancestral gene.

Mutants in any one of these three genes are not lethal. The cell has sufficient back up activity from the other enzymes to be able to grow. A triple mutant, however, was inviable (Shaw *et al.*, 1991). It could be rescued by a plasmid containing the *CHS2* gene under the control of a *gal1* promoter. Transfer of the mutant from galactose to glucose as carbon source resulted in cell division arrest followed by cell death.

Genes have been identified from *C. albicans* that encode chitin synthases. Two proteins from genes designated *CHS1* and *CHS2* (Chen-Wu *et al.*, 1992) are zymogenic, that from *CHS1* having a pH optimum of 6.5. Northern analyses of mRNA expression showed a marked difference between the two genes, with very much higher levels of *CHS2* message during hyphal outgrowth at 40°C than during yeast growth at 30°C. A chitin synthase gene, designated *CHS1*, has been identified in *Neurospora crassa* that has homologies with *CHS1* and *CHS2* of *S. cerevisiae* (Yarden and Yanofsky, 1991). Disruption of this gene led to a sparsely growing mutant strain with much lower chitin synthase activity than wild-type, with abnormal hyphal swellings, but with septa that stained with Calcofluor White as strongly as those of wild-type hyphae. Mutant colonies were much more sensitive to inhibition by nikkomycin Z than wild-type colonies.

Using polymerase chain reactions with oligo-meric DNA primers designed from conserved sequences in *S. cerevisiae* *CHS1* and *CHS2* and *C. albicans* *CHS1* genes, 32 genes for chitin synthases have been identified from a further 13 species of fungi (Bowen *et al.*, 1992). Their sequences fall into three classes, except for *S. cerevisiae* *CHS1* which is separate.

Chitin synthases are very specifically and potently inhibited by two related families of antibiotics, the

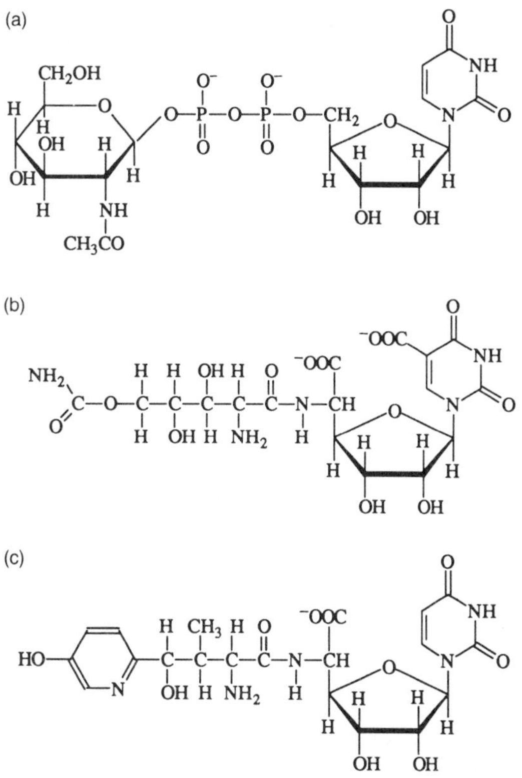

Figure 3.3 Structures of the substrate of chitin synthase, UDP-GlcNAc (a), and its competitive inhibitors, polyoxin (b) and nikkomycin (c).

polyoxins and nikkomycins (Table 3.2), produced by soil-dwelling *Streptomyces* species (Gooday, 1990). These are competitive inhibitors, being nucleoside peptides that are structural analogues of UDP-GlcNAc (Figure 13.3). They are nucleosides with peptide side chains and are transported into cells by peptide permeases. K_1 values for their inhibitory potency for different chitin synthase preparations are about 1 μM, but different fungi show great differences in susceptibilities. These differences can be attributed to differences in peptide uptake systems in different fungi. Their effects on susceptible fungi demonstrate the importance of chitin synthesis to the cell. When treated with polyoxin or nikkomycin, hyphal tips rapidly swell and burst and dividing yeast cells form 'exploding pairs', with disrupted separation of daughter from mother cells (Gooday, 1990; Zhu and Gooday, 1992). Polyoxins are used as

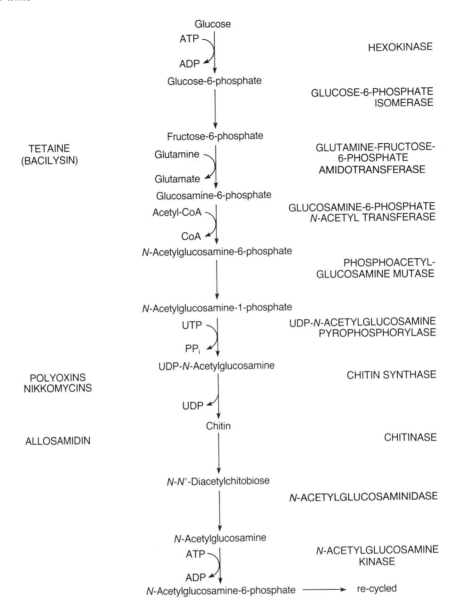

Figure 3.4 Major pathways of metabolism of chitin, with enzymes involved (RHS), and antibiotics specifically inhibiting these (LHS). In addition there are synthetic analogues of N-acetylglucosamine which are specific inhibitors of N-acetylglucosaminidase.

agricultural fungicides, but resistance rapidly occurs by the treated fungi acquiring altered peptide uptake systems which no longer transport the antibiotics into the cell.

UDP-N-acetylglucosamine is the substrate for chitin synthases, and also for enzymes adding N-acetylglucosamine to glycoproteins. It is biosynthesized from glucose by a pathway of six enzymes, all of which have been characterized from fungi (Figure 3.4). This pathway is regulated in part by feedback control of the glucosamine -6-phosphate synthase (L-glutamine: fructose-6-

phosphate amidotransferase) by the level of UDP-GlcNAc. Borgia (1992) characterized chitin-deficient temperature-sensitive mutants of *Aspergillus nidulans, orlA* (osmotically remedial lysis) and *tsE*. He showed that these two genes are necessary for production of L-glutamine: fructose-6-phosphate amidotransferase which is the key enzyme on the pathway to UDP-*N*-acetylglucosamine, the precursor of chitin and glycoproteins. Growth of the mutants was remedied by osmotic stabilizers and by *N*-acetylglucosamine. L-Glutamine:fructose-6-phosphate amidotransferase is inhibited by the antibiotic tetaine (synonyms: bacilysin, bacillin). This dipeptide is taken up and hydrolysed in the cell to yield L-alanine and anticapsin, an analogue of glutamine, which binds irreversibly at the active site of the enzyme (Milewski *et al.*, 1986). This results in inhibition of synthesis of chitin and glycoproteins.

Also occurring in fungal cells are three further enzymes involved in chitin metabolism; chitinase, *N*-acetylglucosaminidase and *N*-acetylglucosamine kinase (Figure 3.4). Together they provide a pathway for recycling of chitin (e.g. during autolysis) or the digestion and utilization of exogenous chitin (e.g. by soil saprophytes, insect pathogens or mycoparasites). Chitinases, however, may play a more fundamental role during hyphal morphogenesis (Gooday *et al.*, 1992). They are always present in an active form in exponentially growing cells, and probably act to regulate the plasticity of expanding areas of walls during branching and perhaps also during apical hyphal growth (Bartnicki-Garcia, 1973; Rast *et al.*, 1991; Figure 3.4). A clear role for chitinase is during budding of yeast cells. During budding, a ring of chitin forms at the site of bud emergence which grows inwards to form the chitin-rich primary septum on the mother cell wall when budding is completed (Cabib, 1987; Cabib *et al.*, 1982, 1992). Release of the daughter cell involves the lytic action of chitinase. Cells of *S.cerevisiae* in which the chitinase gene has been disrupted accumulate as clumps, as daughter cells are unable to separate from mother cells (Kuranda and Robbins, 1991). Such clumps also accumulate when *S.cerevisiae* or yeast cells of *C.albicans* are grown in the presence of allosamidin or demethylallosamidin, antibiotics that specifically inhibit chitinase (Sakuda *et al.*, 1990; Gooday *et al.*, 1992).

After synthesis, chitin chains of the growing

hyphal walls undergo important changes, the occurrence and extent of which differ in different fungi, giving different properties to the walls. Hydrogen bonding between individual chains leads to their crystallization to form microfibrils. Formation of covalent links between chitin and glucans is of particular importance in the rigidification of the wall (Wessels, 1986, 1990). Enzymes involved in cross-linking chitin with other wall components have not been identified, but all polysaccharidases, such as chitinase, can act as transglycosidases, particularly in conditions of low water activity and so may play a role.

A further modification of chitin is its progressive deacetylation to give chitosan, the β(1–4)-linked polymer of glucosamine. This process is probably quite widespread, but is especially prominent in the walls of members of the Zygomycotina, such as *Mucor rouxii*, in which chitosan is a major wall polymer. Davis and Bartnicki-Garcia (1984) demonstrate how chitin deacetylase acts in consort with chitin synthase to give chitosan, with deacetylation of chitin occurring as it is synthesized.

3.3.2 Glucan metabolism

Fungal wall β(1–3)-glucans are biosynthesized via the nucleotide sugar, UDP-glucose. The glucan synthases responsible, like chitin synthases, are intrinsic proteins of the plasma membrane. There they accept the substrate from the cytosol and feed the growing glucan chains into the wall (Shematek *et al.*, 1980; Szaniszlo *et al.*, 1985; Jabri *et al.*, 1991):

$$\text{UDP-Glc} + (\text{Glc})_n \rightarrow (\text{Glc})_{n+1} + \text{UDP}$$

For enzyme preparations from a range of fungi activity is stimulated by very low concentrations of guanosine triphosphate (GTP) (Szaniszlo *et al.*, 1985). The enzyme preparations can be separated into two components, a membrane-bound one and a soluble fraction which has properties of a GTP-binding regulatory protein (Cabib *et al.*, 1988). Both are required for activity. Preparations of membranes from *Saprolegnia monoica*, when provided with UDP-Glc, elaborate polymers containing varying amounts of β(1–4) and β(1–3) links, with low substrate concentrations favouring (1–4) links and high concentrations favouring (1–3) links, respectively (Girard and Fevre, 1984).

β(1–3)-Glucan synthesis in fungi is inhibited by several lipopeptide antibiotics of fungal origin, including echinocandin, aculeacin and papulocandin (Debono and Gordee, 1990; Taft and Selitrennikoff, 1988). These inhibit glucan synthases in a specific but non-competitive fashion. Treatment of growing hyphae with these antibiotics results in bursting of hyphal apices (Zhu and Gooday, 1992).

β(1–6)-Glucan synthesis has been investigated in *S.cerevisiae* by studying mutant strains that are resistant to yeast killer toxin Kl, the receptor for which is β(1–6)-glucan. Such mutants define three genes, *KRE1*, *KRE5* and *KRE6*, whose components are required for the sequential assembly of β(1–6)-glucan (Meader *et al.*, 1990). The *KRE5* and *KRE6* genes code for proteins early in the pathway. The structure of the *KRE5* protein is consistent with its activity in the lumen of the endoplasmic reticulum. Meader *et al.* (1990) suggest that linear chains of β(1–6)-glucan are made in the endoplasmic reticulum, probably with UDP-glucose as substrate, and are released to the wall via the secretory pathway. The *KRE1* gene codes for a secreted protein involved in the assembly of the β(1–6)-glucan in the wall.

The vectorial synthesis of β(1–3)-glucan chains allows only linear molecules to be made and thus any β(1–6) branches must be added in the wall. This process is part of the maturation of the cell wall and occurs in the apical dome of the growing hypha (section 3.3.5). Hartland *et al.* (1991) have characterized a protein from culture medium of *Candida albicans* that they suggest is an enzyme involved in this branching in the wall. The protein has no glucanase activity, but has a unique glucanosyl transferase activity. When laminaripentaose (β(1–3)-linked Glc_5) was incubated with the protein, it disappeared, to be replaced by Glc_2 and higher oligomers of Glc_8 to Glc_{11}. The Glc_2 originated from the reducing end of the substrate. Nuclear magnetic resonance spectroscopy showed that the octomer, Glc_8, contained a new β(1–6) glycosidic linkage. Other proteins in fungal walls with no known function may also prove to have such branching and cross-linking activities.

There is circumstantial evidence for the involvement of glucanases in hyphal growth and branching. Evidence for the role of cellulase in hyphal morphogenesis comes chiefly from studies with species of *Achlya* and *Saprolegnia*. Male hyphae of *Achlya* species respond to antheridiol, a sterol pheromone produced by female hyphae, by formation of antheridial lateral branches. This response is accompanied by an increase in production and secretion of cellulase (Mullins, 1979). Electron microscopic studies of male hyphae of *Achlya ambisexualis* responding to antheridiol show aggregations of vesicles at the sites of branching, with clear indications of their exocytosis into the wall (Mullins and Ellis, 1974; Mullins, 1979). Fevre (1979) has studied the involvement of cellulases and β(1–3)-glucanases in hyphal growth of *Saprolegnia monoica*. He reported that their highest activities were in the youngest areas of colonies and subcellular fractionation showed them to be associated with membranous fractions corresponding to dictyosomes and apical vesicles. Evidence for the involvement of β(1–3)-glucanases in apical growth was provided by Kritzman *et al.* (1978), who used immunofluorescent antibody staining with a polyclonal antibody prepared with a purified fraction of β(1–3)-glucanase from *Sclerotinium rolfsii*. The antibody stained hyphal tips, clamp connections, new septa and branch points. Specificity of staining was shown by lack of reaction with a range of other fungi, and inhibition of reaction by diethylpyrocarbonate, which inhibited enzyme activity. Glucanase activity is involved in the transient process of septal dissolution that occurs in monokaryotic hyphae of Basidiomycetes such as *Schizophyllum commune* to allow nuclear migration during dikaryotization (Wessels and Sietsma, 1979). Glucanases of *S.commune* are also involved in re-utilization of some of the secreted mucilagenous β(1–3), β(1–6) glucan that can occur during carbon starvation (Wessels and Sietsma, 1979). In *Aspergillus nidulans*, α(1–3) glucan from the wall is re-utilized during cleistothecial production via hydrolysis by glucanases and uptake of the resultant glucose (Zonneveld, 1974).

3.3.3 Glycoprotein metabolism

Wall glycoproteins are biosynthesized in the secretory pathway: endoplasmic reticulum → Golgi bodies → secretory vesicles → release at the plasma membrane (cf. Chapter 5). The transmembrane stages of this biosynthesis involve sugar precursors linked to the polyprenol dolichol, the 'lipid intermediates' (Lehle, 1981; Cabib *et al.*, 1988). In the *O*-linked chains, the first mannose unit is linked to the protein via the precursor dolichol-phosphomannose, in the endoplasmic reticulum. The other mannose units are added via the nucleotide sugar guanosine diphosphomannose, GDP-Man, in the Golgi bodies. The N-linked chains are assembled by a more complex scheme, giving the lipid intermediate dolichol-diphospho-$(GlcNAc)_2$-Man_9-Glc_3 which is N-linked to asparagine in the protein in the endoplasmic reticulum, with the release of the terminal four sugar units, Man-Glc_3. The outer chain of many mannose units is added by several linkage-specific mannosyl transferases, with GDP-Man as substrate, in the Golgi bodies. Some secreted enzymes, notably invertase, acid phosphatase and chitinase, are also mannoproteins, synthesized and secreted in a similar fashion (Kuranda and Robbins, 1991; cf. Chapters 5 and 7). In hyphal apices, some of the vesicles in the Spitzenkörper contain carbohydrate-rich material, almost certainly including wall mannoproteins and secreted enzymes (cf. Chapter 5). The involvement of autolytic enzymes in processing mannoproteins as well as all other major wall polysaccharides was suggested by studies with *Aspergillus nidulans*, in which autolysis of mannans was highest in newly formed walls (Rosenberger, 1979). Mannoprotein biosynthesis is inhibited along with chitin synthesis by tetaine which after hydrolysis to anticapsin inhibits the synthesis of UDP-GlcNAc (Milewski *et al.*, 1986), and by tunicamycin, an antibiotic which inhibits synthesis of the lipid intermediates.

3.3.4 Metabolism of other components

In the Zygomycetes, such as *M.rouxii*, the biosynthesis of the glucuronans and glucuronomannoproteins is via the nucleotide sugar, UDP-glucuronic acid (Dow *et al.*, 1981). This is a membrane-bound enzyme activity, probably active in Golgi bodies so that the polymers are released to the wall via secretory vesicles. There is evidence for hydrolysis of glucuronans during growth of *M.rouxii*, with oligosaccharides being released to the medium where they are rapidly degraded (Dow and Villa, 1980).

3.3.5 Dynamics of the cell wall

In most fungal cells, wall synthesis is highly polarized. Thus autoradiographic studies of the incorporation of radioactive N-acetylglucosamine and glucose into growing hyphal walls have shown that nearly all synthesis of chitin and glucan occurs in the apical 1 μm of the hyphae (Gooday, 1971). Staebell and Soll (1985) have investigated the patterns of wall growth of hyphae and yeast cells of *Candida albicans* by attaching polylysine-coated microbeads to cell surfaces at different times in their growth. They conclude that during mycelial growth, at least 90% of expansion is due to a small, highly active apical zone which continues in activity during growth. In contrast, during about the first two-thirds of bud growth a very small highly active apical zone accounts for about 70% of surface expansion, the remaining 30% being a general, non-localized expansion. When the bud reaches about two-thirds its final surface area, the apical zone shuts down and subsequent expansion is completed by a general mechanism (Chapters 13 and 19).

Nascent walls, for example at hyphal tips, are plastic. In the apical dome they progressively mature to become the rigid lateral walls of the hyphae (Figure 3.5). Increasing evidence implicates three processes in this maturation. One process is the cross-linking of different polymers, most notably the formation of covalent bonds between chitin and branched $\beta(1$–$3)$-glucan perhaps by peptide bridges. Evidence for this comes from solubilization of previously insoluble glucans by alkaline extraction following specific depolymerization of chitin by deacetylation and treatment with nitrous acid (Wessels, 1986, 1990). Surarit *et al.* (1988) suggest the presence of a glycosidic linkage between position 6 of N-acetylglucosamine in chitin and position 1 of glucose in $\beta(1$–$6)$-glucan in the wall of *Candida albicans*. The evidence for this comes from examination of products following selective enzymic digestion and partial acid hydrolysis of alkali-insoluble glucan isolated from regenerating protoplasts and from intact cells. Another process is crystallization of chitin microfibrils by an increase in hydrogen bonding. Evidence for this comes from greater susceptibility of microfibrils in growing hyphal apices to disruption and to dissolution by chitinase than those in non-growing hyphae (Vermeulen and Wessels, 1984). Also occurring is the addition of new materials leading to thickening of microfibrils and accumulation of matrix materials overlying the fibrillar skeleton. Evidence for these processes comes from electron microscopy of microfibrils, from the appearance of native and extracted hyphal surfaces and from differential staining of hyphae, showing that

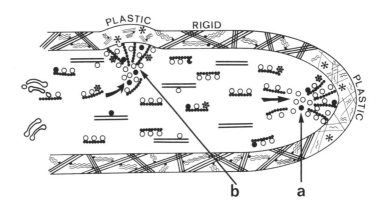

Figure 3.5 Model of wall development during apical hyphal growth and branching of a typical fungus. Microvesicles of a variety of types are produced by the Golgi bodies and transported to the apex (a) and the new branch site (b) via cytoskeletal elements: microfilaments (beaded lines) and microtubules (parallel lines). Membrane-bound enzymes, including chitin and β(1–3)-glucan synthases, are incorporated into the apical plasma membrane by fusion of microvesicles (solid circles). These enzymes act to produce nascent wall fibrillar components: chitin (straight lines) and glucans (wavy lines). More gel-like matrix materials such as α-glucans and mannoproteins are released to the wall by exocytosis of microvesicles (open circles). In the apical dome the chitin and β(1–3)-glucan become more crystalline by hydrogen bonding and become progressively cross-linked by formation of covalent bonds (black dots) as the wall matures. Branch formation occurs by localized weakening of the wall via release of lytic enzymes (stars). Turgor pressure pushes out this weakened wall, and the branch site becomes a new tip (cf. Chapter 4). It is likely that wall lytic enzymes are also secreted at the apex, but their role in apical wall growth is unclear.

material appears on the subapical hyphal surface and masks underlying material that was exposed in the apical dome (Burnett, 1979; Gooday and Trinci, 1980; Wessels and Sietsma, 1981).

3.4   CELL–CELL FUSIONS

In ascomycetous and basidiomycetous yeasts, fusions between mating cells involve interactions between complementary mating-type-specific surface agglutinins (Gooday, 1992). In *S. cerevisiae*, mating is between pairs of cells of *MATa* and *MATα* mating types. Vegetative cells develop into gametes, called shmoos, under the influence of complementary peptide pheromones, α-factor and a-factor. Development of shmoos involves a range of developmental changes, for example an increase in chitin content of the cell walls, but a key response is production of agglutinins by *MATa* and *MATα* cells. The induction of *MATa* agglutinin by α-factor is a particularly sensitive response. The K_{50} value (for 50% response) is 6 × 10^{-11} M α-factor, a concentration which is much lower than those required for arrest of cell division and formation of conjugation tubes (Moore, 1983).

When complementary competent cells were mixed, *MATa–MATα* cell pairs formed rapidly, but most cells formed large aggregates in about one hour (Yanagishima and Yoshida, 1981). When cells of one mating type were marked by staining with a vital dye and the composition of the aggregates was scored the two cell types were seen in equal numbers. The complementary agglutinins have been characterized as glycoproteins which form a molecular complex with each other. Transmission electron micrographs of cells responding to pheromones show fibrous material on the outer surfaces of the conjugation tubes. This material presumably is the sexual agglutinin. Fluorescent antibodies to the agglutinins stain the conjugation tubes. Other species of ascomycetous yeasts have similar mating systems. The pheromones have some cross-reactivity between species, but the agglutinins that they induce are strictly species-specific.

Some basidiomycetous yeasts have similar mating systems. Conjugation between two yeast cells of compatible mating types *ab* and *AB* of *Tremella mesenterica*, the 'yellow brain fungus', is regulated by production of complementary pheromones, tremerogens *a* and *A*, respectively. In response to the complementary pheromone, a yeast cell produces a conjugation tube, the surface of which is covered with a proteinaceous agglutinin (Miyakawa *et al.*, 1987). When pheromone-induced gametes of *ab* and *AB* were mixed, there was about 70% agglutination in one hour, to give clumps of 80–100 cells. When vegetative cells were mixed, agglutination was delayed for several hours until conjugation tubes were formed. The sexual nature of this agglutination was confirmed by prestaining the cell wall chitin of cells of one of the mating types with Calcofluor White, and mixing with unstained cells of opposite mating type, in ratios from 1:9 to 9:1. In all cases the agglutination complexes contained nearly equal numbers of cells of the two mating types. Pretreatment of pheromone-induced gametes with proteolytic enzymes abolished or very much reduced the agglutination. Polypeptides were released to the medium by treatment of cells of both mating types with thermolysin. These polypeptides inhibited agglutination and so Miyakawa *et al.* (1987) suggested that they were the products of proteolysis of specific agglutinins which retained their specific binding properties.

Conjugation in the anther smut fungus of the Caryophyllaceae, *Ustilago violacea*, is between two yeast-like cells, sporidia, of opposite mating type, a_1 and a_2. Cells may mate if they are in contact, or at distances up to 20 μm apart. Day and Garber (1988) have described fimbriae on the cell walls of the sporidia and suggest that they are involved in conjugation. Sporidia carried more than 200 fimbriae, showing by shadow-cast electron microscopy as fibrils of 0.5–>10 μm long with a diameter of 6–7 nm. Cells could be defimbriated by sonication or other physical treatments, and the fimbriae regenerated at a rate of 1–2 μm h^{-1}. Observations of conjugating cells suggested that the fimbriae connected between mating cells and guided the growth of the conjugation tube. Cells lacking fimbriae could not conjugate. Fimbriae were digestible by proteases and were characterized as being composed of a 74 kDa protein that could assemble spontaneously. Similar fibrils were observed on the surfaces of a range of fungal cells and so such fimbriae may be quite widespread and may have other roles as well as in mating of *U. violacea*.

After fusion of cell walls, during both mating and vegetative anastomoses, there is localized wall lysis to allow contact of the two plasma membranes (cf. Chapter 4). As yet very little is known of this process, which presumably involves directed release of wall lytic enzymes or localized activation of their zymogens.

Less specific adhesion is shown during the phenomenon of flocculation, i.e. cells clumping together but not fusing. This is an important property to brewers as its occurrence determines the state of yeast cells at the end of a fermentation. Different strains of *S. cerevisiae* show different degrees of flocculation. A major factor involved is the ionic cross-bridging of surface mannoproteins of adjacent cells by divalent cations such as Ca^{2+}.

3.5 CELL WALLS IN PATHOGENESIS AND SYMBIOSIS

As their walls are the primary site of interaction between fungal cells and their environment, it is not surprising that wall metabolism plays a major role in the establishment and development of pathogenic and symbiotic relationships between fungi and other organisms.

Cell wall components are involved in adhesion of pathogenic and symbiotic fungi to their hosts. Cells of the human pathogen *Candida albicans* adhere to human epithelial cells. Cells of less virulent species of *Candida* show less adherence. There is now considerable evidence for a mannoprotein adhesion on the surface of cells of *C. albicans* which binds to fucose-containing glycoproteins on the surfaces of epithelial cells (Douglas, 1987). The extent of adherence of the fungal cells depends on their growth medium. For example, yeast cells grown in medium containing 500 mм-galactose were up to ten times more adherent than cells grown in 50 mм-glucose. The more adherent cells had a fibrillar surface layer (Douglas, 1987). Yeast cells in clinical specimens from patients with *Candida* infections have similar fibrous surfaces. The hyphal growth form of *C. albicans* is generally considered to be more pathogenic than the yeast growth form. In culture, yeast cells of *C. albicans* do not normally aggregate but extensive aggregation accompanies the induction of hyphal growth. This indicates the occurrence of cell surface changes during the yeast to hyphal transition. By manipulating levels of divalent cations and by treatment with protein-disrupting agents, Holmes *et al.* (1992) conclude that this aggregation is mediated by divalent cation cross-bridging between opposing anionic sites and proteins in a synergistic fashion. Immunofluorescent studies have shown the development of germ tube-specific antigens on the surfaces of the hyphae and germ tubes develop the property of binding albumin, fibrinogen and transferrin from human plasma (Page and Odds, 1988). These proteins bound with high avidity to germ tubes formed by *C.albicans* but not to yeast cells of *C.albicans* or of other species of *Candida*. This binding of plasma proteins did not appear to be related directly to virulence, as a strain of fungus with much reduced lethality to mice showed the greatest binding of any isolate tested (Chapter 19).

Tunlid *et al.* (1992) review the range of attachment mechanisms of nematophagous fungi to their host nematodes. *Arthrobotrys oligospora* produces hyphal traps. If a nematode comes into contact with a trap it adheres to it. The surfaces of trap hyphae and vegetative hyphae have a layer of fibrillar material. Initial trapping of a nematode induces increased secretion of this material. Analysis of the fibrils has shown that their major component behaves as a glycoprotein of molecular mass at least 100 000 Da, containing neutral sugars (75%), uronic acids (6%) and protein (19%). Further fractionation has shown the presence of a haemagglutinating glycoprotein of molecular mass about 16 000 Da. The involvement of a lectin in the fungus–nematode interaction was investigated in experiments in which trapping ability was assessed in the presence of a range of sugar solutions. Low concentrations of *N*-acetylgalactosamine (GalNAc) inhibited nematode capture, suggesting that the adhesion of a nematode to *A. oligosporus* is mediated by a GalNAc-specific lectin binding to a GalNAc-containing receptor on the nematode surface. The putative lectin has been characterized from surface polymers from trap-containing mycelium but not from vegetative mycelium as a protein of molecular mass about 20 000 Da. Infection of nematodes by *Drechmeria coniospora* is via radiating fibrils on adhesive knobs of the conidia. The conidia stick to the surface of a passing nematode, particularly around the sensory organs of its head. This specific adhesion is inhibited in the presence of *N*-acetylneuraminic acid (sialic acid) and by treatment of nematodes with sialidase. These results suggest that a sialic acid-specific lectin is a

component of the surface of the conidial knobs. The infectious stage of *Catenaria anguillulae* is a zoospore. The zoospore is chemotactically attracted to the nematode surface, where it adheres and rapidly encysts by the secretion of fibrillar material. Results obtained using reflectance infrared spectroscopy suggest the involvement of proteins on the zoospore surface in the initial adhesion.

Germ tubes of the mucoraceous mycoparasite, *Piptocephalis virginiana*, adhere to hyphae of their host fungus *Mortierella pusilla*, but not to those of a non-host species, *Mortierella candelabrum*. Manocha and Chen (1991) have characterized glycoproteins from the surface of host hyphae but not of non-host hyphae that agglutinate spores of *P. virginiana*. They suggest that these agglutinins are involved in the attachment of mycoparasite germ tubes to host hyphal walls.

Adhesion of fungal propagules to their host plant surfaces is a prerequisite for pathogenesis. For example, germlings of rust fungi are attached to the host surface by an extracellular matrix. Epstein *et al.* (1987) identify protein as an important component of the adhesive material of uredospore germlings of *Uromyces appendiculatus* because proteolytic enzymes greatly reduced adhesion. Conidia of the rice blast pathogen *Magnaporthe grisea* attach firmly by their tips to the hydrophobic rice leaf surface (Figure 3.6). Microscopy shows that the spore apex has a coating of mucilage (Hamer *et al.*, 1988). This mucilage binds concanavalin A (ConA, a plant lectin with a high affinity for mannose units) and treatment of spores with ConA antagonizes early adhesion. This suggests that the adhesin may be glycoprotein in nature. The apices of dry conidia have large periplasmic deposits of material that is presumably dehydrated adhesin. On wetting the spores, this material is rapidly released and gives a mucilaginous coating to the tips of the spores (Figure 3.6).

Zoospores of the plant pathogenic oomycete *Phytophthora cinnamomi* attach themselves to root surfaces of host plants by very rapid exocytosis of adhesive material. By immunogold labelling of thin sections, Gubler and Hardham (1988) have shown that this adhesin is stored in small peripheral vesicles, and about 85% is secreted in the first minute of encystment on the plant surface. It contains a group of high-molecular-weight

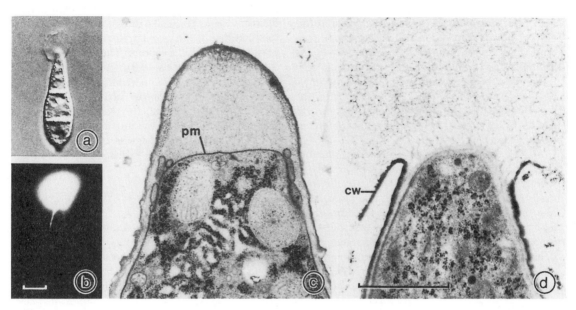

Figure 3.6 Release of mucilage from apices of conidia of the rice blast pathogen, *Magnaporthe grisea* upon wetting. Light (a and b) and electron (c and d) micrographs of conidia before (c) and after (a,b, and d) wetting. (a and b) The same fluoroscein-ConA-labelled conidium imaged with interference contrast optics and epifluorescence, respectively; scale bar, 5 μm. (c) A dry conidium apex contains a deposit outside the plasma membrane (pm). (d) When wetted, the cell wall (cw) is broken down at the apex and extracellular fibrous material is released; scale bar, 1 μm. (From Hamer *et al.*, 1988).

glycoproteins. Studies with plant lectins specific for particular sugars suggested that the adhesin is rich in *N*-acetylgalactosamine residues.

Many species of plants form mycorrhizal associations. Bonfante-Fasolo *et al.* (1987) have investigated the role of surface components of the ericoid mycorrhizal fungus *Hymenoscyphus ericae* in adhesion to plant roots. An infective strain of the fungus, but not a non-infective strain, had a coating of fibrous surface material that stained strongly with ConA. These authors suggest that this material represents glycoprotein or oligosaccharide material involved in the establishment of the contact between fungus and its host plant.

The make-up and structure of cell walls of pathogenic or symbiotic fungi may alter during interactions with the host. This occurs for example in the vesicular–arbuscular mycorrhizas, where the fungus is a zygomycete belonging to the Glomales. Hyphae of the fungus penetrate root cortical cells and develop into a range of specialized structures including arbuscules, which are highly branched intracellular structures. During the infection process, the fungal cell walls undergo characteristic modifications (Bonfante-Fasolo *et al.*, 1992). They become much thinner, from 500 nm to 20–30 nm, and their texture changes from fibrous in the intracellular hyphae to amorphous in the arbuscules. By using labelled lectins, Bonfante-Fasolo *et al.* (1992) suggest that the amorphous walls still contain chitin and glucan, but no longer in crystalline and cross-linked forms. The development of these thinner, more amorphous walls may aid interflow of nutrients and other molecules between the symbionts.

Cell wall melanins are involved in examples of both plant and animal pathogenesis. Melanin can confer great strength to a fungal wall and melanin deposition plays a key role in the development of the hydrostatic pressure which allows penetration of plant tissue by appressoria of the rice blast pathogen *M. grisea* (Howard and Ferrari, 1989). Appressoria of melanin-deficient mutants or of fungi treated with inhibitors of melanin biosynthesis were unable to penetrate leaf surfaces. Melanin biosynthesis is an important target for agricultural fungicides, for example tricyclazole. Melanin formation may be involved in virulence of the human pathogens *Cryptococcus neoformans* and *Wangiella dermatidis* as melanin-deficient strains of both fungi have reduced pathogenic effects in mice (Wheeler and Bell, 1988). In these cases the melanin may aid in protecting the fungal cells against attack by the host's chemical defence mechanisms.

As well as these examples of melanin formation, walls of human pathogenic fungi have a range of features that may be associated with pathogenesis. Wall components, particularly mannoproteins, may be major antigens. Gross wall composition and assembly may change during pathogenesis, for example in the dimorphic pathogenic fungi (Chapter 19). Commonly walls become thicker, such as in fungal cells in mycetoma grains, giving the cells greater physical and chemical protection against host defences. An important example is the basidiomycetous yeast, *Cryptococcus neoformans*. In human tissue, cells of *C.neoformans* elaborate an extensive polysaccharide capsule (Bancroft *et al.*, 1992). This is chiefly composed of an $\alpha(1-3)$ mannose backbone substituted with xylose and glucuronic acid, with varying extents of *O*-acetylation. This polysaccharide capsule is a major virulence factor for *C.neoformans*; it confers resistance to phagocytosis and can induce immunological suppression.

Plants have a large repertoire of defences against potentially pathogenic fungi. Amongst these are the 'pathogenesis related proteins', a group of proteins induced rapidly after a plant is attacked. Notable amongst these is a range of chitinases and glucanases (Boller, 1985). Some of these enzymes have immediate antifungal activity as they weaken the chitin and glucan fabrics of the wall, resulting in cell bursting. Others may have indirect effects, causing the release of specific oligosaccharides from the fungi which act as elicitors, signalling to the plant to trigger further active defence mechanisms. Elicitors include oligomers of chitin and chitosan and specific glucan fragments, such as a $\beta(1-3)$, $\beta(1-6)$-linked glucose heptamer from walls of *Phytophthora megasperma* (Ride, 1992).

Killer toxins are proteins produced by some strains of a range of yeast species that kill cells of susceptible strains. Cell wall components are the primary sites of recognition for the toxins, and different toxins recognize different cell wall receptors, binding to them as a preliminary to reaching plasma membrane receptors. Thus K1 toxin from *S. cerevisiae* interacts with $\beta(1-6)$-glucans of susceptible strains, K28 toxin interacts with $\alpha(1-3)$-mannan and *Kluyveromyces lactis* toxin interacts with chitin (Meader *et al.*, 1990; Gooday, 1993). K1 toxin then inserts into the membrane, leading to pore formation and leakage of cytoplasmic ions, whereas K28 and *K. lactis* toxins are taken up and act inside the cell.

3.6 CONCLUSIONS

Studies of cell walls were initially chiefly concerned with their compositions. Once these were established most interest turned to the biochemistry of their synthesis. These studies are continuing, in particular with the impetus from powerful molecular biological techniques. Increasingly, however, we are appreciating the dynamics of fungal walls and studies are being undertaken of the modelling and interactions of wall components during growth and development.

Its wall protects the fungal cell but is a vulnerable shield. Cell wall biosynthesis is the target of agricultural fungicides such as polyoxin which inhibits chitin synthase and the inhibitors of melanin biosynthesis, and of potential antifungal drugs such as derivatives of echinocandin which inhibit glucan synthesis. Plants and some animals have wall lytic enzymes in their repertoire of antifungal defences. Other organisms such as invertebrate grazers and soil microbes produce an array of lytic enzymes for the digestion of dead or living fungi. The production of protoplasts by the controlled action of cocktails of lytic enzymes in buffered medium of appropriate osmotic strength is an important technique. Under appropriate incubation conditions, protoplasts will regenerate cell walls. Such protoplasts can be transformed with alien DNA to produce novel organisms and different protoplasts can be fused to produce hybrid organisms. Studies of regenerating protoplasts have been of value in elucidating aspects of wall assembly.

This chapter has served to introduce the topic of cell walls but as they are so important to the life of the fungus, most successive chapters will discuss other aspects of them.

REFERENCES

Bancroft, G.J., Rockett, E.R. and Collins, H.L. (1992) Capsule synthesis and immunity to *Cryptococcus neoformans*, in *New Strategies in Fungal Disease*, (eds J.E. Bennett, R.J. Hay and P.K. Peterson), Churchill Livingstone, Edinburgh, pp. 179–91.

Bartnicki-Garcia, S. (1968) Cell wall chemistry, morphogenesis and taxonomy of fungi. *Annual Review of Microbiology*, **22**, 87–108.

Bartnicki-Garcia, S. (1973) Fundamental aspects of hyphal morphogenesis, in *Microbial Differentiation, Society for Experimental Microbiology Symposium*, vol. **23**, (eds J.M. Ashworth and J.E. Smith), Cambridge University Press, Cambridge, pp. 245–67.

Bartnicki-Garcia, S., Ruiz-Herrera, J. and Bracker, C.E. (1979) Chitosomes and chitin synthesis, in *Fungal Walls and Hyphal Growth*, (eds J.H. Burnett and A.P.J. Trinci), Cambridge University Press, Cambridge, pp. 149–68.

Boller, T. (1985) Induction of hydrolases as a defence reaction against pathogens, in *Cellular and Molecular Biology of Plant Stress*, (eds J.L. Key and T. Kosuge), Alan R. Liss, New York, pp. 247–62.

Bonfante-Fasolo, P., Peretto, S., Testa, B. and Faccio, A. (1987) Ultrastructural localization of cell surface sugar residues in ericoid mycorrhizal fungi by gold-labelled lectins. *Protoplasma*, **139**, 25–35.

Bonfante-Fasolo, P., Peretto, R. and Peretto, S. (1992) Cell surface interactions in endomycorrhizal symbiosis, in *Perspectives in Plant Cell Recognition*, (eds J.A. Callow and J.R. Green), Cambridge University Press, Cambridge, pp. 239–55.

Borgia, P. (1992) Roles of the *orlA*, *tsE*, and *bimG* genes of *Aspergillus nidulans* in chitin synthesis. *Journal of Bacteriology*, **174**, 384–9.

Bowen, A.R., Chen-Wu, J.L., Momany, M. *et al.* (1992) Classification of fungal chitin synthases. *Proceedings of the National Academy of Sciences, USA*, **89**, 519–23.

Briza, P., Ellinger, A., Winkler, G. and Breitenbach, M. (1990) Characterization of a D, L-dityrosine-containing macromolecule from yeast ascospore walls. *Journal of Biological Chemistry*, **265**, 15118–23.

Bulawa, C.E. (1992) *CSD2*, *CSD3*, and *CSD4*, genes required for chitin synthesis in *Saccharomyces cerevisiae*: the *CSD2* gene product is related to chitin synthases and to developmentally regulated proteins in *Rhizobium* species and *Xenopus laevis*. *Molecular and Cellular Biology*, **174**, 1764–76.

Bulawa, C.E. (1993) Genetics and molecular biology of chitin synthesis in fungi. *Annual Review of Microbiology*, **47**, 505–34.

Bulone, V., Chanzy, H., Gay, L. *et al.* (1992) Characterization of chitin and chitin synthase from the cellulosic cell wall fungus *Saprolegnia menoica*. *Experimental Mycology*, **16**, 8–21.

Burnett, J.H. (1979) Aspects of the structure and growth of hyphal walls, in *Fungal Walls and Hyphal Growth*, (eds J.H. Burnett and A.P.J. Trinci), Cambridge University Press, Cambridge, pp. 1–25.

Cabib, E. (1987) The synthesis and degradation of chitin. *Advances in Enzymology*, **59**, 59–101.

Cabib, E. (1991) Differential inhibition of chitin synthetases 1 and 2 from *Saccharomyces cerevisiae* by polyoxin D and nikkomycins. *Antimicrobial Agents and Chemotherapy*, **35**, 170–3.

Cabib, E., Roberts, R. and Bowers, B. (1982) Synthesis of the yeast cell wall and its regulation. *Annual Review of Biochemistry*, **51**, 763–93.

Cabib, E., Bowers, B., Sburlati, A. and Silverman, S.J. (1988) Fungal cell wall synthesis: the construction of a biological structure. *Microbiological Sciences*, **5**, 370–5.

Cabib, E., Silverman, S.J. and Shaw, J.A. (1992) Chitinase and chitin synthase I: counter balancing activities in cell separation of *Saccharomyces cerevisiae*. *Journal of General Microbiology*, **138**, 97–102.

Chen-Wu, J.L., Zwicker, J., Bowen, A.R. and Robbins, P.W. (1992) Expression of chitin synthase genes during yeast and hyphal growth phases of *Candida albicans*. *Molecular Microbiology*, **6**, 497–502.

Datema, R., van den Ende, H. and Wessels, J.G.H. (1977) The hyphal wall of *Mucor mucedo*. 1. Polyanionic polymers. *European Journal of Biochemistry*, **80**, 611–26.

Davis, L.L. and Bartnicki-Garcia, S. (1984) Chitosan synthesis in tandem action of chitin synthetase and chitin deacetylase from *Mucor rouxii*. *Biochemistry*, **23**, 1065–73.

Day, A.W. and Garber, E.B. (1988) *Ustilago violaceae*, anther smut of the Caryophyllaceae. *Advances in Plant Pathology*, **6**, 457–82.

De Nobel, J.G., Dijkers, C., Hooijberg, E. and Klis, F.M. (1989) Increased cell wall porosity in *Saccharomyces cerevisiae* after treatment with dithiothreitol or EDTA. *Journal of General Microbiology*, **135**, 2077–84.

Debono, M. and Gordee, R.S. (1990) Drug discovery: nature's approach, in *Chemotherapy of Fungal Disease*, (ed J. Ryley), Springer-Verlag, Berlin, pp. 77–109.

Douglas, L.J. (1987) Adhesion of *Candida* species to epithelial surfaces. *CRC Critical Reviews of Microbiology*, **15**, 27–35.

Dow, J.M. and Villa, V.D. (1980) Oligoglucuronide production in *Mucor rouxii*: evidence for a role of endohydrolases in hyphal extension. *Journal of Bacteriology*, **142**, 939–44.

Dow, J.M., Carreon, R.R. and Villa, V.D. (1981) Role of membranes of mycelial *Mucor rouxii* in synthesis and secretion of cell wall matrix polymers. *Journal of Bacteriology*, **145**, 272–9.

Epstein, L., Laccetti, L.B., Staples, R.C. and Hoch, H.C. (1987) Cell-substratum adhesive protein involved in surface contact responses of the bean rust fungus. *Physiological and Molecular Plant Pathology*, **30**, 373–88.

Fevre, M. (1979) Glucanases, glucan synthases and wall growth in *Saprolegnia monoica*, in *Fungal Walls and Hyphal Growth*, (eds J.H. Burnett and A.P.J.

Trinci), Cambridge University Press, Cambridge, pp. 225–63.

Girard, V. and Fevre, M. (1984) β-1-4- and β-1-3-glucan synthases are associated with the plasma-membrane of the fungus *Saprolegnia*. *Planta*, **160**, 400–6.

Gooday, G.W. (1971) An autoradiographic study of hyphal growth of some fungi. *Journal of General Microbiology*, **67**, 125–33.

Gooday, G.W. (1977) Biosynthesis of the fungal wall: mechanisms and implications. The first Fleming Lecture. *Journal of General Microbiology*, **99**, 1–11.

Gooday, G.W. (1979) Chitin synthesis and differentiation in *Coprinus cinereus*, in *Fungal Walls and Hyphal Growth*, (eds J.H. Burnett and A.P.J. Trinci), Cambridge University Press, Cambridge, pp. 203–23.

Gooday, G.W. (1981) Sporopollenin, in *The Fungal Spore: Morphogenetic Controls*, (eds G. Turian and H.R. Hohl), Academic Press, New York, pp. 307–23.

Gooday, G.W. (1990) Inhibition of chitin metabolism, in *The Biochemistry of Cell Walls and Membranes in Fungi*, (eds P.J. Kuhn, A.P.J. Trinci, M.J. Jung, M.W. Goosey and L.G. Copping), Springer, Berlin, pp. 61–79.

Gooday, G.W. (1992) The fungal surface and its role in sexual agglutination, in *Perspectives in Plant Cell Recognition*, (eds J.A. Callow and J.R. Green), Cambridge University Press, Cambridge, pp. 33–58.

Gooday, G.W. (1993) Cell surface diversity and dynamics in yeasts and filamentous fungi. *Journal of Applied Bacteriology Symposium Supplement*, **74**, 12S–20S.

Gooday, G.W. and Trinci, A.P.J. (1980) Wall structure and biosynthesis in fungi, in *The Eukaryotic Microbial Cell, Society for General Microbiology Symposium*, vol. **30**, (eds G.W. Gooday, D. Lloyd and A.P.J. Trinci), Cambridge University Press, Cambridge, pp. 207–51.

Gooday, G.W., Zhu, W-Y. and O'Donnell, R.W. (1992) What are the roles of chitinases in the growing fungus? *FEMS Microbiology Letters*, **100**, 387–92.

Gow, N.A.R. and Gooday, G.W. (1983) Ultrastructure of chitin in hyphae of *Candida albicans* and other dimorphic and mycelial fungi. *Protoplasma*, **115**, 52–8.

Gubler, F. and Hardham, A.R. (1988) Secretion of adhesive material during encystment of *Phytophthora cinnamomi* zoospores, characterized by immunogold labelling with monoclonal antibodies to components of peripheral vesicles. *Journal of Cell Science*, **90**, 225–35.

Hamer, J.E., Howard, R.J., Chumley, F.G. and Valent, B. (1988) A mechanism for surface attachment of spores of a plant pathogenic fungus. *Science*, **239**, 288–90.

Hartland, R.P., Emerson, G.W. and Sullivan, P.A. (1991) A secreted β-glucan-branching enzyme from *Candida albicans*. *Proceedings of the Royal Society of London, Series B*, **246**, 155–60.

Holmes, A.R., Cannon, R.D. and Shepherd, M.G. (1992) Mechanisms of aggregation accompanying morphogenesis in *Candida albicans*. *Oral Microbiology and Immunity*, **7**, 32–7.

Howard, R.J. and Ferrari, M.A. (1989) Role of melanin in appressorium function. *Experimental Mycology*, **13**, 403–18.

Jabri, E., Quigley, D.R., Alders, M. *et al.* (1991) 1, 3-β-D-Glucan synthesis of *Neurospora crassa*. *Current Microbiolology*, **19**, 153–61.

Kamada, T., Takemaru, T., Prosser, J.I. and Gooday, G.W. (1991a) Right and lefthanded helicity of chitin microfibrils in stipe cells of *Coprinus cinereus*. *Protoplasma*, **165**, 64–70.

Kamada, T., Bracker, C.E. and Bartnicki-Garcia, S. (1991b) Chitosomes and chitin synthetase in the asexual life cycle of *Mucor rouxii*. *Journal of General Microbiology*, **137**, 1241–52.

Kritzman, G., Chet, I. and Henis, Y. (1978) Localization of β-(1,3)-glucanase in the mycelium of *Sclerotium rolfsii*. *Journal of Bacteriology*, **134**, 470–5.

Kuranda, M.J. and Robbins, P.W. (1991) Chitinase is required for cell separation during growth of *Saccharomyces cerevisiae*. *Journal of Biological Chemistry*, **266**, 19 758–67.

Lehle, L. (1981) Biosynthesis of mannoproteins in fungi, in *Encyclopedia of Plant Physiology*, vol. **13B**, *Plant Carbohydrates II*, (eds W. Tanner and F.A. Loewus), Springer-Verlag, Berlin, pp. 459–83.

Manocha, M.S. and Chen, Y. (1991) Isolation and partial characterization of host cell surface agglutinin and its role in attachment of a biotrophic mycoparasite. *Canadian Journal of Microbiology*, **37**, 377–83.

Meader, P., Hill, K., Wagner, J. *et al.* (1990) The yeast *KRE5* gene encodes a probable endoplasmic reticulum protein required for (1-6)-β-D-glucan synthesis and normal growth. *Molecular and Cellular Biology*, **10**, 3013–9.

Milewski, S., Chmara, H. and Borowski, E. (1986) Antibiotic tetaine – a selective inhibitor of chitin and mannoprotein synthesis in *Candida albicans*. *Archives of Microbiology*, **145**, 234–40.

Miyakawa, T., Azuma, Y., Tsuchiya, E. and Fukui, S. (1987) Involvement of cell-surface proteins in sexual cell–cell interactions of *Tremella mesenterica*, a heterobasidiomycete fungus. *Journal of General Microbiology*, **133**, 439–43.

Moore, S.A. (1983) Comparison of dose–response curves for α-factor-induced cell division arrest, agglutination, and projection formation in yeast cells. *Journal of Biological Chemistry*, **258**, 13 849–56.

Mullins, J.T. (1979) A freeze-fracture study of hormone-induced branching in the fungus *Achlya*. *Tissue and Cell*, **11**, 585–95.

Mullins, J.T. and Ellis, E.A. (1974) Sexual morphogenesis in *Achlya*: ultrastructural basis for the hormone induction of antheridial hyphae. *Proceedings of the National Academy of Sciences, USA*, **71**, 1347–50.

Orlean, P. (1987) Two chitin synthases in *Saccharomyces cerevisiae*. *Journal of Biological Chemistry*, **262**, 5732–9.

Page, S. and Odds, F.C. (1988) Binding of plasma proteins to *Candida* species *in vitro*. *Journal of General Microbiology*, **34**, 2693–702.

Rast, D.M., Stussi, H., Hegnauer, H. and Nyhlen, L.E. (1981) Melanins, in *The Fungal Spore: Morphogenetic Controls*, (eds G. Turian and H.R. Hohl), Academic Press, New York, pp. 507–31.

Rast, D.M., Horsch, M., Furter, R. and Gooday, G.W. (1991) A complex chitinolytic system in exponentially growing mycelium of *Mucor rouxii*: properties and function. *Journal of General Microbiology*, **137**, 2797–810.

Ride, J.P. (1992) Recognition signals and initiation of host responses controlling basic incompatibility between fungi and plants, in *Perspectives in Plant Cell Recognition*, (eds J.A. Callow and J.R. Green), Cambridge University Press, Cambridge, pp. 213–37.

Rosenberger, R.F. (1979) Endogenous lytic enzymes and wall metabolism, in *Fungal Walls and Hyphal Growth*, (eds J.H. Burnett and A.P.J. Trinci), Cambridge University Press, Cambridge, pp. 265–77.

Sakuda, S., Nishimoto, Y., Ohi, M. *et al.* (1990) Effects of demethylallosamidin, a potent yeast chitinase inhibitor, on cell division in yeast. *Agricultural and Biological Chemistry*, **54**, 1333–5.

Sburlati, A. and Cabib, E. (1986) Chitin synthetase 2, a presumptive participant in septum formation in *Saccharomyces cerevisiae*. *Journal of Biological Chemistry*, **261**, 15147–52.

Shaw, J.A., Mol, P.C., Bowers, B. *et al..* (1991) The function of chitin synthases 2 and 3 in the *Saccharomyces cerevisiae* cell cycle. *Journal of Cell Biology*, **114**, 111–23.

Shematek, E.M., Braatz, J.A. and Cabib, E. (1980) Biosynthesis of the yeast cell wall. I. Preparation and properties of β (1–3) glucan synthetase. *Journal of Biological Chemistry*, **255**, 888–94.

Silverman, S.J. (1989) Similar and different domains of chitin synthases 1 and 2 of *S. cerevisiae*: two isoenzymes with distinct functions. *Yeast*, **5**, 459–67.

Silverman, S.J., Sburlati, A., Slater, M.L. and Cabib, E. (1988) Chitin synthase 2 is essential for septum formation and cell division in *Saccharomyces cerevisiae*. *Proceedings of the National Academy of Sciences, USA*, **85**, 4735–9.

Staebell, M. and Soll, D.R. (1985) Temporal and spatial differences in cell wall expansion during bud and mycelium formation in *Candida albicans*. *Journal of General Microbiology*, **131**, 1467–80.

Surarit, R., Gopal, P.K. and Shepherd, M.G. (1988) Evidence for a glycosidic linkage between chitin and glucan in the cell wall of *Candida albicans*, *Journal of General Microbiology*, **134**, 1723–30.

Szaniszlo, P.J., Kang, M.S. and Cabib, E. (1985) Stimulation of β(1,3) glucan synthetase of various fungi by nucleoside triphosphates. A generalized regulatory mechanism for cell wall biosynthesis. *Journal of Bacteriology*, **161**, 1188–94.

Taft, C.S. and Selitrennikoff, C.P. (1988) LY121019 inhibits *Neurospora crassa* growth and (1-3)-β-D-glucan synthase. *Journal of Antibiotics*, **41**, 697–701.

Tunlid, A., Jansson, H-B. and Nordbring-Hertz, B. (1992) Fungal attachment to nematodes. *Mycological Research*, **96**, 401–12.

Valdivieso, M.H., Mol, P.C., Shaw, J.A. *et al.* (1991) *CAL1*, a gene required for activity of chitin synthase 3 in *Saccharomyces cerevisiae*. *Journal of Cell Biology*, **114**, 101–9.

Van Rinsum, J., Klis, F.M. and van den Ende, H. (1991) Cell wall glucomannoproteins of *Saccharomyces cerevisiae mnn9*. *Yeast*, **7**, 717–26.

Vermeulen, C.A. and Wessels, J.G.H. (1984) Ultrastructural differences between wall apices of growing and non-growing hyphae of *Schizophyllum commune*. *Protoplasma*, **120**, 123–31.

Wessels, J.G.H. (1986) Cell wall synthesis in apical hyphal growth. *International Review of Cytology*, **104**, 37–79.

Wessels, J.G.H. (1990) Role of cell wall architecture in fungal tip growth generation, in *Tip Growth in Plant and Fungal Cells*, (ed. I.B. Heath), Academic Press, New York, pp. 1–29.

Wessels, J.G.H. (1992) Gene expression during fruiting in *Schizophyllum commune*. *Mycological Research*, **98**, 609–20.

Wessels, J.G.H. and Sietsma, J.H. (1979) Wall structure and growth in *Schizophyllum commune*, in *Fungal Walls and Hyphal Growth*, (eds J.H. Burnett and A.P.J. Trinci), Cambridge University Press, Cambridge, pp. 27–48.

Wessels, J.G.H. and Sietsma, J.H. (1981) Fungal cell walls: a survey, in *Encyclopedia of Plant Physiology*, vol. 13B, *Plant Carbohydrates II*, (eds W. Tanner and F.A. Loewus), Springer-Verlag, Berlin, pp. 352–415.

Wheeler, M.H. and Bell, A.A. (1988) Melanins and their importance in pathogenic fungi. *Current Topics in Medical Mycology*, **2**, 338–87.

Yanagishima, N. and Yoshida, K. (1981) Sexual interactions in *Saccharomyces cerevisiae* with special reference to the regulation of sexual agglutinability, in *Sexual Interactions in Eukaryotic Microbes* (eds D.H. O'Day and P.A. Horgen), Academic Press, New York, pp. 261–95.

Yarden, O. and Yanofsky. C. (1991) Chitin synthase 1 plays a major role in cell wall biogenesis in *Neurospora crassa. Genes and Development*, **5**, 2420–30.

Zhu, W-Y. and Gooday, G.W. (1992) Effects of nikkomycin and echinocandin on differentiated and undifferentiated mycelia of *Botrytis cinerea* and *Mucor rouxii. Mycological Research*, **96**, 371–7.

Zonneveld, B.J.M. (1974) α-1, 3 Glucan synthesis correlated with a-1, 3 glucanase synthesis, conidiation and fructification in morphological mutants of *Aspergillus nidulans. Journal of General Microbiology*, **81**, 445–51.

CELL MEMBRANE

4

G.W. Gooday

Department of Molecular and Cell Biology, Marischal College, University of Aberdeen, Aberdeen, UK

4.1 INTRODUCTION

The fungal plasma membrane is the major interface between the cell and its environment. It displays a wide range of vectorial enzymatic activities. These include those of the osmoenzymes, catalysts which couple a chemical reaction at the molecular level with the concurrent translocation of a chemical group or an ion across the membrane (Mitchell, 1977; Harold, 1991), and systems involved in vectorial biosynthesis, such as chitin synthase, which accept substrates from the cytosol and feed out macromolecular products into the wall (cf. Chapter 9). Both groups of enzymes are intrinsic membrane proteins, with hydrophobic membrane-spanning domains in their amino acid sequences. Other intrinsic membrane proteins found in the plasma membrane include those involved in signal transduction, including receptors, proteases, phospholipase C and adenylate cyclase, which respond to environmental signals and interact with G-proteins or phosphoinositol signalling systems (cf. Chapter 9). There are also peripheral membrane proteins attached to one face of the membrane but not spanning it. These include proteins involved in the G-protein signalling pathways. There is growing evidence for both transmembrane and peripheral plasma membrane proteins interacting with cytoskeletal components such as actin, anchoring specific components at particular sites, and providing the cell with positional information (cf. Chapter 6). Such proteins have been characterized from amoebae of the slime mould *Dictyostelium discoideum* (Luna and Condeelis, 1990) and Kanbe *et al.* (1989) have shown microfilament-associated granules on the plasma membrane of growing areas of *Schizosaccharomyces pombe*. Such cytoplasm–membrane interactions may play a key role in the integrity of the hyphal apex. A further class of membrane proteins are those attached covalently via phosphatidylinositol-containing glycolipids embedded in the lipid bilayer and extending out from the cell membrane. Such proteins have been detected in *Saccharomyces cerevisiae* and may prove to have important functions (Conzelmann *et al.*, 1988), including roles in sensory signal transduction (Müller and Bandlow, 1993).

The plasma membrane is approximately half protein and half lipid, by weight. The membrane contains many different types of protein. Bussey *et al.* (1979) labelled proteins of plasma membrane ghosts of *S. cerevisiae* with ^{125}I, solubilized the proteins, subjected them to two-dimensional electrophoresis and observed about 200 species. Four of the species reacted with periodic acid–Schiff reagent, identifying them as glycoproteins. Many other glycoproteins of lower abundance can be identified by lectin-binding techniques. The many different types of membrane proteins presumably interact with each other and with the membrane lipids in a dynamic fashion during fungal growth. New membrane is being made continually and added at growth points in the cell. Hyphal tip, branch points and budding sites of yeast cells are characterized by accumulations of microvesicles. Microvesicle exocytosis, which occurs by fusion of the microvesicular membrane with the plasma membrane, has two consequences. Materials contained in the microvesicle are released to the wall and microvesicular membrane is incorporated into the plasma membrane.

The electron microscopic techniques of freeze-fracturing and freeze-etching allow fine structural details of the plasma membrane to be visualized. The membrane can be cleaved between its two

The Growing Fungus. Edited by Neil A.R. Gow and Geoffrey M. Gadd. Published in 1994 by Chapman & Hall, London. ISBN 0 412 46600 7

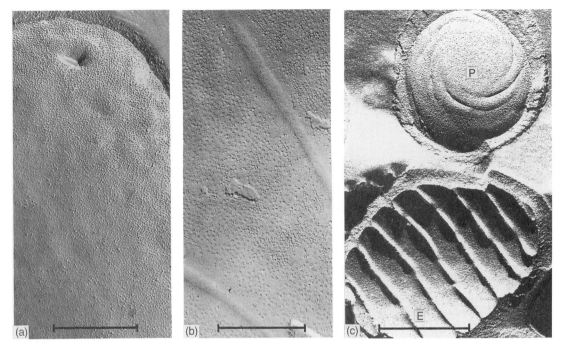

Figure 4.1 Freeze-fracture shadow-cast electron micrographs of fungal plasma membranes. (a,b) Views of P faces (plasmatic, convex fracture faces) of exponentially growing cells of the fission yeast *Schizosaccharomyces pombe*. (a) At the growing pole, showing many small globular proteinaceous particles, and a short invagination near the growing pole. Scale bar = 0.5 μm. (b) In the lateral region of the cell, showing many large proteinaceous particles with central depressions, and long invaginations free from particles. Scale bar = 0.25 μm. (c) Two cells of the budding yeast *Malassezia furfur*. In one (P) the plasma membrane shows its P face; in the other (E) it shows the E face (exoplasmatic, concave fracture face). The P face has a higher density of proteinaceous particles. Both faces show the long left-handed helical invaginations characteristic of this species. Scale bar = 1 μm. Photographs by courtesy of Dr K. Takeo.

lipid layers, and the faces revealed often show arrays of intramembranous proteinaceous particles and short straight or circular invaginations devoid of particles (Miragall *et al.*, 1986; Takeo *et al.*, 1990; Slayman *et al.*, 1990; Figure 4.1). The particles probably represent large intrinsic membrane proteins, such as ATPases. The significance of the invaginations is unclear but they may represent areas of quiescent membrane.

In order to study particular activities, preparations of purified fungal plasma membranes may be made by several different methods. These include sucrose density gradient centrifugation of cell homogenates (Serrano, 1988), stabilization of spheroplast membranes with Concanavalin A, followed by lysis, centrifugation and removal of Concanavalin A with methyl-mannoside (Smith and Scarborough, 1984; Hubbard *et al.*, 1986) and stabilization of spheroplast membranes with silica microbeads, followed by lysis and centrifugation (Aldermann and Hofer, 1984).

4.2 BIOENERGETICS AND TRANSPORT

The most abundant membrane protein in fungi is the plasma membrane proton-pumping ATPase (cf. Chapter 8). This is 5–10% of the membrane protein in *Neurospora crassa* (Slayman *et al.*, 1990) and up to 25% in *Candida albicans*. In hyphae of *N. crassa* it has been estimated that between 38 and 52% of the total ATP produced is consumed by this enzyme (Gradmann *et al.*, 1978) and to generate the observed current of 10–20 $\mu A\ cm^{-2}$ would require several thousand ATPase molecules (Slayman *et al.*, 1990). The enzyme specifically hydrolyses Mg-ATP so that one proton is normally expelled for each ADP produced. This process is electrogenic, creating both a membrane potential (internal negative) and a transmembrane pH gradient (cf. Chapter 13). Thus the enzyme supplies energy to drive the H⁺-dependent co-transport of nutrients such as glucose (Eddy, 1982; Harold, 1991) and is involved in internal pH regulation (Sanders *et al.*, 1981) and in extrusion of organic ions during fermentative growth (Sigler *et al.*, 1981). Gene disruption experiments in *S. cerevisiae* have shown the ATPase to be essential, whereas other experiments which have manipulated ATPase expression and activity have demonstrated that the ATPase controls the rate of cell growth (Portillo and Serrano, 1989). Reconstitution into liposomes to give one ATPase monomer per liposome gave full proton-pumping activity, suggesting that proton pumping is the activity of a single polypeptide (Goormaghtigh *et al.*, 1986). The amino acid sequence of the protein includes at least eight hydrophobic putative membrane-spanning domains. Figure 4.2 presents a topological model of the ATPase in the membrane, where it hydrolyses a molecule of ATP yielding a transient aspartyl phosphate intermediate. Hydrolysis of the intermediate powers changes which result in the translocation of a proton to the exterior. Both amino and carboxyl termini are hydrophilic and are on the cytoplasmic face of the membrane (Mandala and Slayman, 1989). There is evidence that the carboxyl terminus is involved in activation of the enzyme during glucose metabolism and cytoplasmic acidification, and Monk *et al.* (1991) provide direct evidence that this portion of the molecule is in the cytoplasm. Plasma membrane vesicles from *S. cerevisiae* were treated with fluorescein-labelled concanavalin A to reveal the outer surface (this lectin binds to mannose units in surface glycoproteins) and a differently labelled fluorescent antibody to reveal the carboxyl terminus. The intact vesicles proved to be of two types, staining with either one or other label according to which face of the membrane was exposed to the medium, implying that the carboxyl terminus must be on the cytoplasmic face. Vesicles prepared from a mutant strain of yeast lacking the carboxyl terminal epitope recognized by the antibody provided a key negative control.

The plasma membrane has a range of transport systems for the uptake of solutes; organic nutrients such as carbohydrates and amino acids, and inorganic cations and anions (Eddy, 1982; Prasad, 1991). Accumulation of most of these solutes is

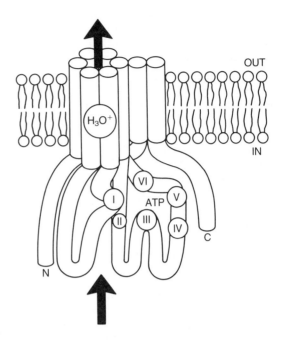

Figure 4.2 Hypothetical model of fungal proton-ATPase in the membrane, showing eight membrane-spanning domains, a proton translocating pore, the ATP binding site, and the amino and carboxyl termini in the cytoplasm. Roman numerals indicate amino acid domains conserved in all P-type ATPases ('P' signifying the mechanism of action). **N** and **C** indicate amino and carboxyl ends of the polypeptide. (Figure and information kindly supplied by Brian Monk and Ramon Serrano.)

energized by the proton gradient created by the H⁺-ATPase. There must be a range of membrane proteins mediating these transport systems, but little is yet known about them. Internal Ca²⁺ levels are kept low by continuous Ca²⁺ expulsion but sensing of environmental stimuli by receptors that control channel proteins allows calcium ions to enter the cell through these specific channels (Harold, 1991; cf. Chapter 13). Calcium influx can therefore act as both a positional and a temporal signal. Such calcium channels have been recognized in the chytrid, *Blastocladiella emersonii* (Caldwell *et al.*, 1986). Stretch-activated Ca²⁺, K⁺ and Mg²⁺ channels have been described in hyphal tip plasma membrane of the oomycete *Saprolegnia ferax* (Garrill *et al.*, 1992, 1993).

As well as the H⁺-ATPase, other transport

systems mediate the efflux of solutes. Stroobant and Scarborough (1979) have shown that efflux of calcium from *N. crassa* can be mediated by Ca^{2+}/H^+ antiports coupled to the transmembrane pH gradient generated by the plasma membrane ATPase.

Organic molecules also can be exported via membrane proteins. α-Factor, one of the two complementary sex hormones of *S. cerevisiae*, is secreted via the conventional secretory pathway (cf. Chapter 5), but **a**-factor, a very hydrophobic lipopeptide, is actively transported out of **a**-cells by a plasma membrane ATP-dependent protein with significant similarity to the mammalian multidrug-resistant P-glycoproteins (McGrath and Varshavsky, 1989).

4.3 VECTORIAL SYNTHESIS OF WALL COMPONENTS

There is good evidence that synthesis of microfibrillar wall components, such as chitin, β(1–3)- and β(1–4)-glucans, occurs only at the plasma membrane (cf. Chapters 3 and 8). Purified preparations of plasma membrane have high chitin synthase activity and this enzyme activity is used as a marker for plasma membrane during cell fractionations. Electron microscopic autoradiography shows incorporation of tritiated *N*-acetylglucosamine into insoluble material only at the plasma membrane. The enzyme is clearly associated with the lipid bilayer. Arrhenius plots show clear transition points in enzyme activity; removal of phospholipids inactivates the enzyme whereas addition of phospholipids activates and restores activity of partially inactivated preparations (Vermeulen *et al.*, 1979; Cabib *et al.*, 1983; Montgomery and Gooday, 1985; Wessels, 1986). Chitin synthase is inhibited by polyene antibiotics, which perturb fungal membranes by interacting with ergosterol (Rast and Bartnicki-Garcia, 1981), and chitin deposition is disorganized by treating fungi with antifungal azole agents, which inhibit ergosterol biosynthesis (Marichal *et al.*, 1985). There is also considerable evidence for vectorial synthesis of glucans through the plasma membrane. The active site hydrolyses substrate UDP-glucose from the cytoplasm and feeds out structural glucan chains into the wall (Shematek *et al.*, 1980; Girard and Fevre, 1984; Jabri *et al.*, 1991). Subsequent modification of wall polysaccharides may also involve membrane-bound enzymes,

such as chitin deacetylase (Davis and Bartnicki-Garcia, 1984) and chitinase (Humphreys and Gooday, 1984).

4.4 EXOCYTOSIS AND ENDOCYTOSIS

Fungal growth is characterized by fusion of microvesicles with the plasma membrane. The two consequences of this phenomenon are an increase in area of plasma membrane in the growth zone and the secretion of materials contained in the vesicles. The necessary coordination of these two processes has been investigated in *S.cerevisiae* by using the *SEC1* mutant strain, which is defective in the last step of secretion and in which large amounts of Golgi-derived vesicles accumulate (Holcomb *et al.*, 1988). The plasma membrane ATPase was used as a marker for plasma membrane assembly and acid phosphatase as a marker for secreted proteins. These two activities co-purified throughout purification procedures, including precipitation by ATPase antibody, demonstrating that a single vesicle species was involved in transport of both activities. Novick *et al.* (1988) provide a model for exocytosis in *S.cerevisiae*. They propose that the protein encoded by *SEC4* is a GTP-binding protein, which binds both to secretory vesicles and to a protein effector in the plasma membrane, and this triggers the fusion of the secretory vesicle with the plasma membrane.

To counterbalance exocytosis, part of the regulation of membrane growth may be internalization of membrane by endocytosis. This can be of two types, receptor-mediated in which a membrane-bound receptor plus ligand are internalized, and fluid-phase in which a volume of external medium perhaps containing particles is internalized. Both processes require energy and vesiculation of the plasma membrane. For most fungi the presence of a wall limits the scope of endocytosis, but both types of endocytosis occur in *S.cerevisiae* (Dulic *et al.*, 1991). Receptor-mediated endocytosis is shown experimentally by internalization of α-factor bound to its receptor during the mating response, and fluid-phase endocytosis by the uptake of the fluorescent dye lucifer yellow, and its localization in the vacuole (Riezman, 1993). As for other eukaryotic cells, transport of proteins and lipids from the plasma membrane to internal membranes in yeast is mediated by vesicles coated

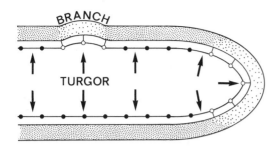

Figure 4.3 Model of activation/inhibition of stretch–responsive membrane enzymes and channels at the hyphal apex and branch site. Heavy stippling signifies rigid walls; light stippling signifies soft spots in the walls. Solid circles signify particular membrane proteins in their resting state; open circles with stalks signify their activation/inactivation by stretching of the membrane.

with polyhedral lattices of clathrin (Riezman, 1993).

4.5 POLARITY OF THE PLASMA MEMBRANE

Hyphae are polarized structures (cf. Chapter 13), and there is evidence for non-uniform distribution of activities of several of the plasma membrane proteins discussed here. In many cases these gradients in activity may represent non-uniform distribution of proteins generated by localized insertion, perhaps maintained by destruction or excision at other points. Other gradients may result from lateral movement of proteins in the membrane. Some gradients in activity may, however, represent local activation/inhibition by changes in mechanical stress on the membrane. A model for this is shown in Figure 4.3. The turgor pressure of the cytoplasm pushes the membrane uniformly out within the constraint of the wall. At a future branch point, however, the wall is locally weakened by lytic enzymes (cf. Chapter 3) and will bulge out. The underlying membrane should suffer transient mechanical stress at this location. The membrane around the apical dome of the hypha may be under similar stress. Membrane proteins in these regions that are sensitive to mechanical stress would be expected to have transiently altered activities. Examples of stress-sensitive proteins are stretch-activated ion channels in membranes of hyphal tips of *Saprolegnia ferax* (Garrill *et al.*, 1992, 1993) and in membranes of *S.cerevisiae* (Gustin *et al.*, 1988) and *Uromyces*

appendiculatus (Zhou *et al.*, 1991). Putative examples are wall synthetic enzymes, such as chitin synthase.

In all fungi examined, hyphal growth is accompanied by generation of electrical currents, usually entering near the tip and exiting subapically. These currents can be measured by a vibrating electrode in the medium adjacent to the hyphae. It is now apparent that these currents are epiphenomena and are not directly related to growth. Instead, they may be manifestations of asymmetric distributions of pump and sink activities for ions along the hyphae, implying a non-random distribution of transport proteins or their activities in the membrane (Gow, 1990; Harold, 1991; Cho *et al.*, 1991; De Silva *et al.*, 1992; see also Chapter 13).

Fungal hyphae exhibit galvanotropism. They respond to applied electric fields with oriented growth, which may be cathodotropic, anodotropic or perpendicular, depending on fungus and physiological conditions. These phenomena may reflect asymmetric re-distributions of membrane proteins and/or asymmetric responses of voltage-sensitive membrane proteins or may occur because of non-uniform perturbations of the membrane potential (Gow, 1990; Cho *et al.*, 1991; Gooday and Gow, 1994).

In the marine fungus *Dendryphiella salina* cytochemical staining shows a gradient of activity of (Na^+-K^+)ATPase, with most intense activity occurring about 200 μm behind the apex (Galpin and Jennings, 1975). Jennings (1986) has proposed that the enzyme in this fungus generates a transmembrane hyphal current involving transport of protons, potassium and sodium ions.

4.6 SIGNAL TRANSDUCTION

Responses to hormones provide well-defined examples of signal reception and transduction (cf. Chapter 9). These are probably common phenomena in fungi but very few systems have been characterized (Gooday and Adams, 1992). Of these, responses to peptide hormones have been shown to involve receptors in the plasma membrane. Haploid cells of *S.cerevisiae* are of two mating types, α and **a**. Each produces its own hormone, α-factor and **a**-factor, respectively, that acts on cells of opposite mating type, causing them to cease budding and to become gametes, called shmoos (because their characteristic pleiomorphic shapes resemble that of a cartoon character of Al Capp). α-Factor and **a**-factor are recognized by specific receptors in the plasma membranes of **a**-cells and α-cells, respectively. These receptors are integral membrane proteins that are coupled with a heterotrimeric G-protein, so that binding of α-factor or **a**-factor sets in train a sequence of events leading to formation of shmoos (Marsh *et al.*, 1991). The α-factor receptor is encoded by the *STE2* gene, which is expressed only in **a**-cells. The receptor has a typical structure for an integral membrane protein coupling with G-proteins. Seven hydrophobic domains span the membrane with the amino terminus on the surface of the cell and the carboxyl terminus inside (Burkholder and Hartwell, 1985). There are about 10 000 binding sites per cell, as determined by binding of radioactive α-factor and competition studies, with an equilibrium dissociation constant in the range of $6 \times 10^{-9} - 2 \times 10^{-8}$ M (Raths *et al.*, 1988). An immunofluorescent analysis of a *STE2*-LacZ fusion protein which retains receptor activity has shown that the receptor is located on the surface of the cell (Marsh and Herskowitz, 1988). The receptor–α-factor complex is internalized, to be replaced on the surface by new receptor, to maintain the cell in a receptive state. This process can be used to assay for endocytosis in *S.cerevisiae* (Dulic *et al.*, 1991). The **a**-factor receptor is encoded by *STE3* and is expressed only in α-cells. It also has seven membrane-spanning domains and a long hydrophilic carboxyl terminus but has little amino acid sequence identity with the α-factor receptor (Clark *et al.*, 1988). As well as these receptors, there may be other membrane-bound proteins responding to these hormones, as Hiraga *et al.* (1991) describe the inhibition of preparations of plasma membrane Ca^{2+}-ATPase from **a**-cells and **a**-cells lacking receptors (*STE2* mutants), but not from α-cells, by α-factor in a dose-dependent manner at concentrations ranging from 20 to 100 ng ml^{-1}. These concentrations would result in shmoo formation in equivalent cells. As preparations from the *STE2* mutant showed this phenomenon, Hiraga *et al.* (1991) suggest that it represents a mechanism for α-factor signalling independent from α-factor receptor. Localized inhibition of the Ca^{2+}-ATPase would result in localized influx of Ca^{2+}. This could polarize the cell at that place for the production of the conjugation tube (cf. Chapters 6,8,13). Some basidiosporogenous yeasts have similar conjugation

systems involving specific peptide hormones (Gooday and Adams, 1992). Cells of **a**-mating type of *Rhodosporidium toruloides* respond to rhodotorucine A, the lipopeptide hormone produced by cells of *A*-mating type, by forming conjugation tubes. This response is mediated by degradation of rhodotorucine *A* by a membrane-bound thiol peptidase which requires Ca^{2+} or Mn^{2+} and a phospholipid environment for activity. This 'trigger response' is accompanied by enhanced phosphorylation of membrane proteins and may represent a novel transmembrane signalling system (Miyakawa *et al.*, 1987). There is a rapid transient influx of calcium ions, peaking at about 4 min after addition of rhodotorucine *A* to receptive **a**-cells. This may result from direct inhibition of a plasma membrane Ca^{2+}-ATPase from **a**-cells but not from *A*-cells by rhodotorucine *A*, and may establish polarity of the resulting conjugation tube, as suggested for *S.cerevisiae* (Hiraga *et al.*, 1991).

In *Neurospora crassa* there is evidence for a signal-transduction pathway initiated by binding of an endogenous insulin-like molecule to a plasma membrane receptor protein. An insulin-binding protein has been characterized from *N.crassa* by Kole *et al.* (1991), who suggest that it may interact with plasma membrane protein kinases as part of a signalling mechanism involved in regulation of carbohydrate metabolism (Fawell and Lenard, 1988).

4.7 MEMBRANE FUSION

Plasmogamy, the fusion of two mating cells, involves mutual recognition at the two plasma membranes. A good example is provided by fertilizing male and female gametes of the chytrid *Allomyces macrogynus*. These are uniflagellate, actively swimming naked spores. Their fertilization was described by Hatch (1938): 'they came together smoothly, slid over each other momentarily, then fused with a rush'. Pommerville and Fuller (1976) and Pommerville (1982) have described the fine structure of this process. The cells fuse at a specific region of the plasma membrane towards the end of the cell near the flagellum. Fusion was inhibited by agents such as diphenylhydramine and chloroquine, that stabilize membranes, suggesting that membrane fluidity is required, and by trypsin, suggesting that a protein or glycoprotein surface is involved. Treatment of female gametes, but not male gametes, with cytochalasin B inhibited fusion, suggesting a role for microfilaments in the female cell. Increasing concentrations of Ca^{2+}, Mg^{2+} or Sr^{2+} decreased the time for the mating process from just over 1 min in standard conditions to half that value. This time represents the events required for gamete contact and alignment of plasma membranes and mating sites; time-lapse motion picture analysis suggested that the actual fusion process takes less than 50 ms (Pommerville, 1982). At the 'fusion interface' electron microscopy shows many small cytoplasmic bridges formed between the fusing cells, followed by a row of vesicles, then when fusion is complete the binucleate cell becomes spherical. These gametes have no walls, so this fusion represents a direct interaction between the male and female plasma membranes, presumably via complementary sexual agglutinins. In fungal cells with walls, some recognition processes during mating occur at the wall surface, but plasmogamy must involve mutual recognition by the membranes of the mating cells. In *S. cerevisiae*, two genes, *FUS1* and *FUS2*, have been identified that are involved in this process. One response to the peptide sex hormones by cells of both α- and **a**-mating type is an increase in transcription of *FUS1* by more than 50-fold (Trueheart *et al.*, 1987).

As well as during mating, fusions readily occur between vegetative hyphae within a colony or between colonies of the same or closely related species amongst the Basidiomycetes, Ascomycetes and Deuteromycetes (Gooday, 1975; Aylmore and Todd, 1984). This widespread phenomenon implies the presence of species-specific recognition molecules in the plasma membranes of these fungi, but there is no knowledge about their nature.

4.8 MEMBRANE PHOSPHOLIPIDS

The phospholipid composition of fungal plasma membranes is in general similar to that of other eukaryotic cells. The major phospholipids of the plasma membrane of *N. crassa* are phosphatidylcholine, 46%, phosphatidylethanolamine, 39%, phosphotidylinositol, 15% and phosphatidylserine <1% (Bowman *et al.*, 1987). Different membrane systems in the cell have differing phospholipid compositions, possibly regulated by specific phospholipid transfer proteins transferring lipids

through the cytosol between different membranes (Cleves *et al.*, 1991). Most work on phospholipid biosynthesis has been with *S.cerevisiae*, in particular using regulatory mutants, and with *Schizosaccharomyces pombe* and *N.crassa* (Hill *et al.*, 1990; Robson *et al.*, 1990; Figure 4.4). This work has shown that most of the phosphatidylcholine (PC) is synthesized by three successive methylations of phosphatidylethanolamine (PE), with *S*-adenosylmethionine as methyl donor. Most of the phosphatidylethanolamine is synthesized by decarboxylation of phosphatidylserine (PS). This is in contrast to the major pathway for the synthesis for both PC and PE in most other eukaryotes, the Kennedy cytidine nucleotide pathway, in which the base, choline or ethanolamine, is phosphorylated, and then reacts with cytidine triphosphate (CTP) to give CDP-choline and CDP-ethanolamine. These react with diacylglycerol (DAG) to give PC and PE, respectively. The Kennedy pathway does occur in fungi, but is less important than the methylation pathway (Robson *et al.*, 1990). This may explain the selective toxicity of the organophosphorus fungicides edifenphos and iprobenphos, which inhibit the formation of phosphatidylcholine by methylation. Phosphatidylinositol (PI) and phosphatidylserine are biosynthesized from phosphatidic acid, via CDP-diacylglycerol, which reacts with inositol and serine, respectively (Figure 4.4). This final step in the synthesis of PI is a site of action of the antibiotic validomycin (Robson *et al.*, 1990).

4.9 MEMBRANE STEROLS

Sterols are of key importance in maintaining the integrity of eukaryotic membranes. Cholesterol is the major sterol of the plasma membranes of most animals, and of the Chytridiomycotina and of the Oomycotina. Some of the pythiaceous Oomycotina, however, cannot synthesize sterols and can grow in the absence of added sterols, but their membranes have not been studied in detail (Elliott, 1977). For the majority of the fungi, ergosterol is the key membrane sterol. For the plasma membrane of *N. crassa*, Bowman *et al.* (1987) estimate that ergosterol is at least 95% of the sterol content, and sterol content is 22% of total membrane lipid. Ergosterol differs from cholesterol by being methylated at C-24 of the side chain. A consequence of having ergosterol as opposed to cholesterol in membranes is increased fluidity. The fluidity of the

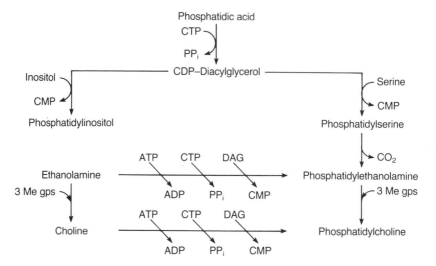

Figure 4.4   The two pathways for biosynthesis of plasma membrane phospholipids. In fungi most phosphatidyl-choline is synthesized by three successive methylations of phosphatidylethanolamine (top and RHS of Figure); whereas in most eukaryotic cells it is synthesized via the Kennedy pathway from choline, via reaction of CDP-choline with diacylglycerol (DAG) (bottom LHS of Figure).

membrane, however, is the result of its fatty acid composition as well as its sterols, so no clear inference can be made as to why ergosterol is the preferred sterol for fungi (Vanden Bossche, 1990). The presence of ergosterol provides the basis for selectivity of two types of antifungal agent, those interacting directly with the membrane and those interfering with its biosynthesis. In the first category are the polyene antibiotics, such as nystatin and amphotericin B. These antibiotics are character-ized by a carbon ring containing a hydrophobic conjugated double-bond system and a hydrophilic region. The ring is closed by lactonization. A primary mode of action of these antibiotics is to become incorporated into the fungal plasma mem-brane in a group linked hydrophobically to ergo-sterol. This produces a pore and loss of integrity of the membrane (Kerridge, 1980). In the second category are the azoles and allylamines, and their importance as antifungal agents has stimulated research on the biosynthesis of ergosterol. Enzymes of the pathway from acetate to ergosterol have been characterized, chiefly from *S. cerevisiae*. A key enzyme is squalene epoxidase, which by oxidizing the C-30 isoprene hydrocarbon squalene allows cyclization to the sterol lanosterol. The enzyme from *C. albicans* proves to be very similar to that of rat liver except that it is much more

sensitive to inhibition by the allylamine com-pounds naftifine and terbinafine and the thio-carbonate tolnaftate. This selectivity allows these compounds to be used as antifungal drugs (Ryder, 1990). The conversion of lanosterol into ergosterol is a complex process involving many enzymes. These act in the endoplasmic reticulum, except for C-24 sterol methyltransferase, which is in the mitochondrion. This enzyme uses 5-adenosyl-methionine as methyl donor, to methylate C-24 with reduction of the $\triangle24(25)$-double bond and formation of a $\triangle24(28)$-double bond. In most fungi, lanosterol is the preferred substrate, which is transported into the mitochondrion and methy-lated to give 24-methylene-dihydrolanosterol. This is transported out to the endoplasmic reticulum for further steps of removal of the methyl group at C-14, reduction of the consequent double bond and removal of both methyl groups at C-4, to give fecosterol (Figure 4.5). In *S. cerevisiae*, however, the preferred substrate for the methyltransferase is zymosterol, which has already had these trans-formations. Zymosterol is transported into the mitochondrion and methylated to give fecosterol. Fecosterol is transported out to the endoplasmic reticulum, where the final steps convert it into ergosterol: isomerization of the 8-double bond to

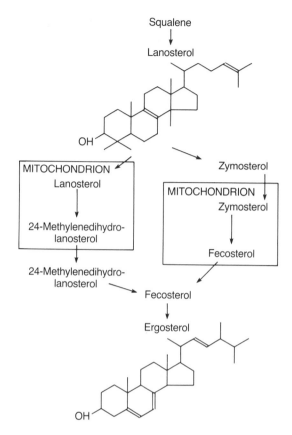

Figure 4.5 Biosynthesis of ergosterol in fungi. In most fungi (LHS) lanosterol is 24-methylated in the mitochondrion; in *S.cerevisiae* (RHS) it is zymosterol that is 24-methylated, to give fecosterol.

$\triangle 7$, introduction of $\triangle 5$- and $\triangle 22$-double bonds, and finally the reduction of the $\triangle 24(28)$-double bond. The most studied enzyme in the pathway is the cytochrome P_{450}-dependent 14-α-sterol demethylase, as this is the site of action of the azole antifungal agents. These include the *N*-substituted imidazoles and triazoles which are widely used both clinically and agriculturally (Vanden Bossche, 1990). Treatment of fungi with sublethal concentrations of these agents leads to abnormal branching and deposition of chitin which Vanden Bossche (1990) has interpreted as the results of altering fluidity of critical domains in the cell membrane by an inhibition of ergosterol biosynthesis. This disturbance in membrane structure and function will lead to growth inhibition and eventual cell death.

4.10 CONCLUSIONS

The fungal plasma membrane has a wealth of properties and activities, as befits its key role in communicating between the protoplast and the external environment. Emphasis has been placed in this chapter on features of the plasma membrane of particular relevance to fungal cells; bioenergetics, interactions with the cell wall, polarity in hyphae, and the targetting for antifungal agents. Aspects of its activities recur throughout this volume, emphasizing its importance and its being an organelle worthy of further study.

REFERENCES

Aldermann, B. and Hofer, M. (1984) Fractionation of membranes from *Metschnikowia reukaufii* protoplasts, evidence for a plasma-membrane-bound ATPase. *Journal of General Microbiology*, **130**, 711–23.

Aylmore, R.C. and Todd, N.K. (1984) Hyphal fusion in *Coriolus versicolor*, in *The Ecology and Physiology of the Fungal Mycelium, British Mycological Symposium*, vol. **8**, (eds D.H. Jennings and A.D.M. Rayner), Cambridge University Press, Cambridge, pp. 103–25.

Bowman, B.J., Borgeson, C.E. and Bowman, E.J. (1987) Composition of *Neurospora crassa* vacuolar membranes and comparison to endoplasmic reticulum, plasmalemmas and mitochondrial membranes. *Experimental Mycology*, **11**, 197–205.

Burkholder, A.C. and Hartwell, L.H. (1985) The yeast α-factor receptor: structural properties deduced from sequence of *STE2* gene. *Nucleic Acids Research*, **13**, 8463–75.

Bussey, H., Saville, M.R., Chevallier, M.R. and Rank, G.H. (1979) Yeast plasma membrane ghosts: an analysis of proteins by two-dimensional gel electrophoresis. *Biochimica et Biophysica Acta*, **553**, 185–96.

Cabib, E., Bowers, B. and Roberts, R.L. (1983) Vectorial synthesis of a polysaccharide by isolated plasma membranes. *Proceedings of the National Academy of Sciences, USA*, **80**, 3318–21.

Caldwell, J.H., Brunt, J. and Harold, F.M. (1986) Calcium-dependent anion channel in the water mold, *Blastocladiella emersonii. Journal of Membrane Biology*, **89**, 85–97.

Cho, C-W., Harold, F.M. and Schreurs, W.J.A. (1991) Electrical and ionic dimensions of apical growth in *Achlya* hyphae. *Experimental Mycology*, **15**, 34–43.

Clark, K.L., Davis, N.G., Niest, D.K. *et al* (1988) Response of yeast α-cells to **a**-factor pheromone: topology of the receptor and identification of a component of the response pathway. *Cold Spring Harbor Symposium of Quantitative Biology*, **53**, 611–20.

Cleves, A., McGee, T. and Bankaitis, V. (1991) Phospholipid transfer proteins: a biological debut. *Trends in Cell Biology*, **1**, 30–4.

Conzelmann, A., Riezman, H., Deshponds, C. and Bron, C. (1988) A major 125-kd membrane glycoprotein of *Saccharomyces cerevisiae* is attached to the lipid bilayer through an inositol-containing phospholipid. *EMBO Journal*, **7**, 2233–40.

Davis, L.L. and Bartnicki-Garcia, S. (1984) Chitosan synthesis in tandem action of chitin synthetase and chitin deacetylase from *Mucor rouxii*. *Biochemistry*, **23**, 1065–73.

De Silva, L.R., Youatt, J., Gooday, G.W. and Gow, N.A.R. (1992) Inwardly directed ionic currents of *Allomyces macrogynus* and other water moulds indicate sites of proton-driven nutrient transport but are incidental to tip growth. *Mycological Research*, **96**, 925–31.

Dulic, V., Egerton, M., Elguindi, I. *et al.* (1991) Yeast endocytosis assays. *Methods in Enzymology*, **194**, 695–710.

Eddy, A.A. (1982) Mechanisms of solute transport in selected eukaryotic microorganisms. *Advances in Microbial Physiology*, **23**, 1–78.

Elliott, C.G. (1977) Sterols in fungi. *Advances in Microbial Physiology*, **15**, 121–73.

Fawell, S.E. and Lenard, J. (1988) A specific insulin receptor and tyrosine kinase activity in the membranes of *Neurospora crassa*. *Biochemical and Biophysical Research Communications*, **155**, 59–65.

Galpin, M.F.J. and Jennings, D.H. (1975) Histochemical study of the hyphae and the distribution of adenosine triphosphatase in *Dendryphiella salina*. *Transactions of the British Mycological Society*, **65**, 477–83.

Garrill, A., Lew, R.R. and Heath, I.B. (1992) Stretch-activated Ca^{2+} and Ca^{2+}-activated K^+ channels in the hyphal tip plasma membrane of the Oomycete *Saprolegnia ferax*. *Journal of Cell Science*, **101**, 721–30.

Garrill, A., Jackson, S.L., Lew, R.R. and Heath, I.B. (1993) Ion channel activity and tip growth: tip-localized stretch-activated channels generate an essential Ca^{2+} gradient in the oomycete *Saprolegnia ferax*. *European Journal of Cell Biology*, **60**, 358–65.

Girard, V. and Fevre, M. (1984) β-1-4- and β-1-3-glucan synthases are associated with the plasma-membrane of the fungus *Saprolegnia*. *Planta*, **160**, 400–6.

Gooday, G.W. (1975) Chemotaxis and chemotropism in fungi and algae, in *Primitive Sensory and Communication Systems*, (ed. M.J. Carlile), Academic Press, London, pp. 155–204.

Gooday, G.W. and Adams, D.J. (1992) Sex hormones and fungi. *Advances in Microbial Physiology*, **34**, 69–145.

Gooday, G.W. and Gow, N.A.R. (1994) Shape determination and polarity in fungal cells, in *Shape and Form in Plants and Fungi*, (ed. D. Ingram), Academic Press, London, pp. 331–46.

Goormaghtigh, E., Chadwick, C. and Scarborough, G.A. (1986) Monomers of the *Neurospora* plasma membrane H^+-ATPase catalyze efficient proton translocation. *Journal of Biological Chemistry*, **261**, 7466–71.

Gow, N.A.R. (1990) Circulating ionic currents in microorganisms. *Advances in Microbial Physiology*, **30**, 89–123.

Gradmann, D., Hanson, U-P., Long, W.S. *et al.* (1978) Current-voltage relationships for the plasma membrane and its principle electrogenic pump in *Neurospora crassa*. *Journal of Membrane Biology*, **59**, 333–67.

Gustin, M.C., Zhou, X.L., Martinec, B. and Kung, C. (1988). A mechanico-sensitive ion channel in the yeast plasma membrane. *Science*, **242**, 762–5.

Harold, F.M. (1991) Biochemical topology: from vectorial metabolism to morphogenesis. *Bioscience Reports*, **11**, 347–85.

Hatch, W.R. (1938) Conjugation and zygote germination in *Allomyces arbuscula*. *Annals of Botany, New Series*, **2**, 583–614.

Hill, J.E., Chung, C., McGraw, P. *et al.* (1990) Synthesis and role of phospholipids in yeast membranes, in *The Biochemistry of Cell Walls and Membranes in Fungi*, (eds P.J. Kuhn, A.P.J. Trinci, M.J. Jung *et al.*), Springer, Berlin, pp. 245–60.

Hiraga, K., Tahara, H., Taguchi, N. *et al.* (1991) Inhibition of Ca^{2+}-ATPase of *Saccharomyces cerevisiae* by mating pheromone α-factor *in vitro*. *Journal of General Microbiology*, **137**, 1–4.

Holcomb, C.L., Hansen, W.J., Etcheverry, T. and Schekman, R. 1988) Secretory vesicles externalize the major plasma membrane ATPase in yeast. *Journal of Cell Biology*, **106**, 641–8.

Hubbard, M.J., Surarit, R., Sullivan, P.A. and Shepherd, M.G. (1986) The isolation of plasma membrane and characterization of plasma membrane ATPase from the yeast *Candida albicans*. *European Journal of Biochemistry*, **154**, 375–81.

Humphreys, A.M. and Gooday, G.W. (1984) Phospholipid requirement of microsomal chitin synthase from *Mucor mucedo*. *Current Microbiology*, **11**, 187–90.

Jabri, E., Quigley, D.R., Alders, M. *et al.* (1991) (1–3)-β-Glucan synthesis of *Neurospora crassa*. *Current Microbiology*, **19**, 153–61.

Jennings, D.H. (1986) Morphological plasticity in fungi, in *Plasticity in Plants, Society for Experimental Biology Symposium*, vol. 40, (eds D.H. Jennings and A.J. Trewavas), Company of Biologists, Cambridge, pp. 329–46.

Kanbe, T., Kobayashi, I. and Tanaka, K. (1989) Dynamics of cytoplasmic organelles in the cell cycle of the fission yeast *Schizosaccharomyces pombe*: three dimensional reconstruction from serial sections. *Journal of Cell Science*, **94**, 647–56.

Kerridge, D. (1980) The plasma membrane of *Candida albicans* and its role in the action of antifungal drugs, in *The Eukaryotic Microbial Cell, Society for General Microbiology Symposium*, vol. 30, (eds G.W. Gooday, D. Lloyd and A.P.J. Trinci), Cambridge University Press, Cambridge, pp. 105–28.

Kole, H.K., Muthukumar, G. and Lenard, J. (1991) Purification and properties of a membrane-bound insulin binding protein, a putative receptor, from *Neurospora crassa*. *Biochemistry*, **30**, 682–8.

Luna, E.J. and Condeelis, J.S. (1990) Actin-associated proteins in *Dictyostelium discoideum*. *Developmental Genetics*, **11**, 328–32.

Mandala, S.M. and Slayman, C.W. (1989) The amino and carboxyl termini of the Neurospora plasma membrane H^+-ATPase are cytoplasmically located. *Journal of Biological Chemistry*, **264**, 16276–81.

Marichal, P., Gorrens, J. and VandenBossche, H. (1985) The action of itraconazole and ketoconazole on growth and sterol synthesis in *Aspergillus fumigatus* and *Aspergillus niger*. *Sabouraudia: Journal of Medical and Veterinary Mycology*, **23**, 13–21.

Marsh, L. and Herskowitz, I. (1988) From membrane to nucleus: the pathway of signal transduction in yeast and its genetic control. *Cold Spring Harbor Symposium of Quantitive Biology*, **53**, 557–65.

Marsh, L., Neiman, A.M. and Herskowitz, I. (1991) Signal transduction during pheromone response in yeast. *Annual Review of Cell Biology*, **7**, 699–728.

McGrath, J.P. and Varshavsky, A. (1989) The yeast *STE6* gene encodes a homologue of the mammalian multidrug resistance P-glycoprotein. *Nature*, **340**, 400–4.

Miragall, F., Rico, H. and Sentandreu, R. (1986) Changes in the plasma membrane of regenerating protoplasts of *Candida albicans* as revealed by freeze-fracture electron microscopy. *Journal of General Microbiology*, **132**, 2845–53.

Mitchell, P. (1977) Epilogue: from energetic abstraction to biochemical mechanism, in *Microbial Energetics, Society for General Microbiology Symposium*, vol. 27, (eds B.A. Haddock and W.A. Hamilton), Cambridge University Press, Cambridge, pp. 384–423.

Miyakawa, T., Tachikawa, T., Jeong, Y.K. *et al.* (1987) Inhibition of membrane Ca^{2+}-ATPase *in vitro* by mating pheromone in *Rhodosporidium toruloides* a heterobasidiomycetous yeast. *Biochemical and Biophysical Research Communications*, **143**, 893–900.

Monk, B.C., Montesinos, C., Ferguson, C. *et al.* (1991) Immunological approaches to the transmembrane topology and conformational changes of the carboxyl-terminal regulatory domain of yeast plasma membrane H^+-ATPase. *Journal of Biological Chemistry*, **266**, 18097–103.

Montgomery, G.W.G. and Gooday, G.W. (1985) Phospholipid–enzyme interactions of chitin synthase of *Coprinus cinereus*. *Current Microbiology*, **27**, 29–33.

Müller, G. and Bandlow, W. (1993) Glucose induces lipolytic cleavage of a glycolipidic plasma membrane anchor in yeast. *Journal of Cell Biology*, **122**, 325–36.

Novick, P.J., Goud, B., Salminen, A. *et al.* (1988) Regulation of vesicular traffic by a GTP-binding protein on the cytoplasmic surface of secretory vesicles in yeast. *Cold Spring Harbor Symposium of Quantitive Biology*, **53**, 637–47.

Pommerville, J. (1982) Morphology and physiology in gamete mating and gamete fusion in the fungus *Allomyces*. *Journal of Cell Science*, **53**, 193–209.

Pommerville, J. and Fuller, M.S. (1976) The cytology of the gametes and fertilization of *Allomyces macrogynus*. *Archives of Microbiology*, **109**, 21–30.

Portillo, F. and Serrano, R. (1989) Growth control strength and active site of yeast plasma membrane ATPase studied by site-directed mutagenesis. *European Journal of Biochemistry*, **186**, 501–7.

Prasad, R. (1991) The plasma membrane of *Candida albicans*: its relevance to transport phenomenon, in *Candida albicans. Cellular and Molecular Biology* (ed. R. Prasad), Springer-Verlag, Berlin, pp. 108–27.

Rast, D.M. and Bartnicki-Garcia, S. (1981) Effects of amphotericin B, nystatin and other polyene antibiotics on chitin synthetase. *Proceedings of the National Academy of Sciences, USA*, **78**, 1233–6.

Raths, S.K., Naider, F. and Becker, J.M. (1988) Peptide analogs compete with the binding of α-factor to its receptor in *Saccharomyces cerevisiae*. *Journal of Biological Chemistry*, **263**, 17333–41.

Riezman, H. (1993) Yeast endocytosis. *Trends in Cell Biology*, **3**, 273–7.

Robson, G.D., Wiebe, M., Kuhn, P.J. and Trinci, A.P.J. (1990) Inhibitors of phospholipid biosynthesis, in *The Biochemistry of Cell Walls and Membranes in Fungi*, (eds P.J. Kuhn, A.P.J. Trinci, M.J. Jung *et al*). Springer, Berlin, pp. 261–81.

Ryder, N.S. (1990) Squalene epoxidase – enzymology and inhibition, in *The Biochemistry of Cell Walls and Membranes in Fungi*, (eds P.J. Kuhn, A.P.J. Trinci, M.J. Jung *et al*), Springer, Berlin, pp. 189–203.

Sanders, D., Hansen, U-P. and Slayman, C.L. (1981) Role of the plasma membrane proton pump in pH regulation in non-animal cells. *Proceedings of the National Academy of Sciences, USA*, **78**, 5903–7.

Serrano, R. (1988) H^+-ATPases from plasma membranes of *Saccharomyces cerevisiae* and *Avena sativa* roots: purification and reconstitution. *Methods in Enzymology*, **157**, 533–44.

Shematek, E.M., Braatz, J.A. and Cabib, E. (1980) Biosynthesis of the yeast cell wall. I. Preparation and properties of β(1-3) glucan synthetase. *Journal of Biological Chemistry*, **255**, 888–94.

Sigler, K., Kotyk, A., Knotkova, A. and Opekarova, M. (1981) Processes involved in the creation of buffering capacity and in substrate-induced proton extrusion in the yeast *Saccharomyces cerevisiae*. *Proceedings of the National Academy of Sciences USA*, **643**, 4735–9.

Slayman, C.L., Kaminsky, P. and Stetson, D. (1990) Structure and function of fungal plasma-membrane, in *The Biochemistry of Cell Walls and Membranes in Fungi*, (eds P.J. Kuhn, A.P.J. Trinci, M.J. Jung *et al*), Springer, Berlin, pp. 298–316.

Smith, R. and Scarborough, G.A. (1984) Large scale isolation of Neurospora plasma membrane H^+-ATPase. *Analytical Biochemistry*, **138**, 156–63.

Stroobant, P. and Scarborough, G.A. (1979) Active transport of calcium in Neurospora plasma membrane vesicles. *Proceedings of the National Academy of Sciences USA*, **76**, 3102–6.

Takeo, K., Sano, A., Nishimura, K. *et al.* (1990) Cytoplasmic and plasma membrane ultrastructure of *Paracoccidioides brasiliensis* yeast-phase cells as revealed by freeze-etching. *Mycological Research*, **94**, 1118–22.

Trueheart, J., Boeke, J.D. and Fink, G.R. (1987) Two genes required for cell fusion during yeast conjugation: evidence for a pheromone induced surface protein. *Molecular and Cellular Biology*, **7**, 2316–28.

Vanden Bossche, H. (1990) Importance and role of sterols in fungal membranes, in *The Biochemistry of Cell Walls and Membranes in Fungi*, (eds P.J. Kuhn, A.P.J. Trinci, M.J. Jung *et al*), Springer, Berlin, pp. 135–57.

Vermeulen, C.A., Raeven, M.B.J.M. and Wessel, J.G.H. (1979) Localization of chitin synthase activity in subcellular fractions of *Schizophyllum commune* protoplasts. *Journal of General Microbiology*, **114**, 87–97.

Wessels, J.G.H. (1986) Cell wall synthesis in apical growth in fungi. *International Review of Cytology*, **104**, 37–79.

Zhou, X.L., Stumpf, M.A., Hoch, H.C. and Kung, C. (1991) A mechanico-sensitive cation channel in membrane patches and in whole cells of *Uromyces*. *Science*, **253**, 1415–17.

ORGANELLES OF FILAMENTOUS FUNGI

P. Markham
Microbiology Department, King's College, London, UK

5.1 INTRODUCTION

Filamentous fungi are typical eukaryotes in many respects and contain a wide range of membrane-bounded subcellular compartments which are the sites of specialised functions. These are organelles as defined by the possession of a bounding membrane. Filamentous fungi contain all the major organelles with the key exception of the chloroplast, which is absent from all of these non-photosynthetic organisms. In addition, the occurrence of a structurally identifiable Golgi apparatus, with the classic dictyosome organization of stacked disc-shaped cisternae, is rare among filamentous fungi. It is common only in the Mastigomycotina, most notably in the class Oomycetes (Beckett *et al.*, 1974), a group which shows many affinities with algae and may only be tenuously related to most filamentous fungi (Beakes, 1987; Cavalier-Smith, 1987a, b). A comparison of the genomic sequences coding for small-subunit ribosomal RNA has indicated that Oomycetes, represented by *Achlya bisexualis*, are very closely related to the chrysophytes (golden-brown algae) and show far less similarity to green algae or to ascomycete fungi (Gunderson *et al.*, 1987).

Many filamentous fungi also contain more unusual organelles which are typical of these organisms and often reflect the requirements or constraints of the filamentous growth habit. There are also organelles which occur infrequently even among the fungi, perhaps in a single class or just a few genera. To this cohort of true organelles can be added a collection of other cytoplasmic inclusions of uncertain status. These may be important and consistently recognizable cytoplasmic components such as the Spitzenkörper, which cannot however be classified as a true organelle because it is not bounded by a bilayer membrane. Alternatively there are frequent reports of cytoplasmic inclusions such as multivesicular bodies, which are apparently membrane-bounded but which are not well characterized. Many such inclusions may ultimately prove to be artefacts arising from the distortion of genuine cytoplasmic components during preparation for electron microscopy. However, the frequent appearance of such ill-characterized structures does suggest that there remains a considerable amount of work to be carried out in identifying the true organelles of filamentous fungi. Some consideration will be given to these inclusions of uncertain status in order to highlight what they may indicate about subcellular organization and the potential for fruitful further investigations in filamentous fungi.

The main concern of this chapter will be to consider the structure and distribution of organelles and organelle-like components of the cytoplasm of filamentous fungi and their function as it relates to the filamentous growth habit. Detail of function and biochemistry will only be considered as it relates to mycelial organization and hyphal extension. The well-understood functioning of major organelles such as nuclei and mitochondria will not be considered.

5.2 IDENTITY, STRUCTURE AND ROLE OF FILAMENTOUS FUNGAL ORGANELLES

5.2.1 Ubiquitous organelles

(a) Nuclei

Nuclei are the most prominent of the organelles found in filamentous fungi, and in some species

The Growing Fungus. Edited by Neil A.R. Gow and Geoffrey M. Gadd. Published in 1994 by Chapman & Hall, London. ISBN 0 412 46600 7

such as *Basidiobolus ranarum* (Beckett *et al.*, 1974; Beakes, 1981) and *Erynia neoaphidis* (Butt *et al.*, 1981), are also the largest of the cytoplasmic inclusions. However, whereas in cross-section nuclei are always apparently larger than mitochondria, the latter organelles are often highly elongated and convoluted such that their total volume probably means that they are larger than nuclei in most species. Indeed, the nuclei of most filamentous fungi, with diameters generally in the range 1–3 μm, are small compared to the nuclei of most eukaryotes which are commonly 3–10 μm in diameter (Alberts *et al.*, 1989), although the nuclei of *B. ranarum* are massive by any standards and are typically 25 μm long and 10 μm wide (Beakes, 1981). The relatively small size of most fungal nuclei, however, reflects the fact that the quantity of genomic DNA that they contain is significantly smaller than the genomic DNA packaged in the nuclei of higher eukaryotes (Carlile, 1980; Chapter 12) even allowing for the fact that many of the filamentous fungi are normally haploid. The appearance of the filamentous fungal nucleus is nevertheless, very similar to that of all nuclei, the key features being the double bilayer bounding membrane perforated with nucleopores; the ribosome-free nucleoplasm apparent as moderately electron-dense material filling the bulk of the organelle; and the intensely electron-dense nucleolus. The nucleolus is the site of ribosome synthesis, occupies a significant minority of the nuclear volume and is usually approximately spherical in shape but with indistinct edges. The shape of the filamentous fungal nucleus is often approximately spherical, but particularly in the rapidly extending regions near the tips of narrow hyphae such as *Aspergillus giganteus*, *Aspergillus nidulans* and *Penicillium chrysogenum* the shape of the nuclei often becomes an oval, elongated parallel to the hyphal axis, a shape which is presumably an adaptation to the hyphal growth habit.

Nuclei are apparently free to move through the cytoplasm of many species with perforate septa, a major exception being the strictly controlled nuclear distribution observed in basidiomycetes apparently partly due to the complex dolipore septum found in that group (see below). However, even in basidiomycetes, enzymic dissolution of the dolipore septum allows the rapid migration of nuclei throughout the mycelium during specific stages of development (Moore, 1985). This movement

of nuclei in hyphae has long been observed by light microscopy (Buller, 1933) but perhaps the most dramatic image of this phenomenon is that of the nucleus actually caught in the act of passing through a narrow septal pore. Several electron micrographs of this event have been published (Hunsley and Gooday, 1974; Markham *et al.*, 1987) and the characteristic dumbbell shape adopted by the nucleus in this situation (Figure 5.1a) demonstrates the extreme flexibility of the fungal nucleus. It is therefore somewhat surprising that at least in one species, *Sordaria brevicollis*, nuclei are apparently able to act as plugs to seal septal pores and prevent cytoplasmic outflow from damaged hyphae (Collinge and Markham, 1987). This indicates that in some circumstances, nuclei can be maintained in a sufficiently rigid state to oppose the considerable force of the internal osmotic pressure of the mycelium.

(b) Mitochondria

Fungal mitochondria, the sites of oxidative respiration, are clearly recognizable as such, because they are bounded by a double bilayer membrane and contain a complex of internal membranes in common with all mitochondria. The major differences between filamentous fungal mitochondria and those of other eukaryotes is the very elongated shape they adopt in many species and the fact that their internal membranes are commonly organized as lamellae lying parallel to each other and to the long axis of the organelle. This differs from the radial arrangement of tubular cristae formed by invagination of the inner mitochondrial membrane, typical of mammalian mitochondria. Whereas mitochondria in most eukaryotes are elongated into rod-shaped compartments with rounded ends, the elongation in filamentous fungi is much more pronounced so that a single mitochondrion may extend 15 μm or more along a hypha (Beckett *et al.*, 1974). This extreme elongation appears to be an adaptation to the hyphal growth form and such mitochondria generally lie approximately parallel to the long axis of the hypha. This elongation of mitochondria appears to be most marked in older regions of mycelium. Towards the hyphal tips these organelles are often found to be much shorter rod-shaped structures of the order of

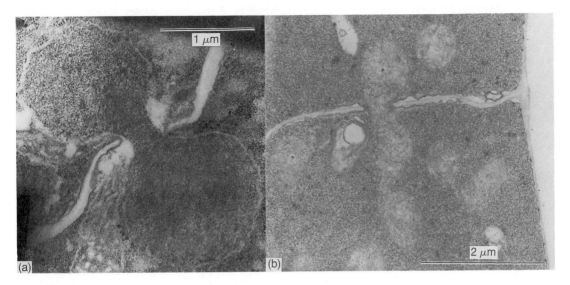

Figure 5.1 Transmission electron micrographs of organelles fixed during their passage through the septal pore in *Sordaria brevicollis*. (a) A nucleus exhibiting the typical dumbbell shape due to constriction of the central region by the narrow pore. (b) A complex elongated mitochondrion, relatively unconstricted by the pore.

1–2 μm in length (Beckett *et al.*, 1974). It is still not entirely clear to what extent filamentous fungal mitochondria might be highly branched and convoluted structures as has been found to be the case for the mitochondria of the yeast *Saccharomyces cerevisiae* (Lloyd and Turner, 1980). Certainly even a limited amount of serial sectioning of hyphae soon indicates that apparently separate mitochondria are often connected together and it seems likely that most mitochondria in hyphae are highly complex in shape, with the possible exception of those located close to the growing hyphal tip. An elegant examination, using fluorescence microscopy, of the mitochondria in the dimorphic fungus *Candida albicans* during the yeast to mycelial transition, has shown that the initial yeast cells contained giant branched mitochondria at a frequency of just one or a very few per cell (Aoki *et al.*, 1989). When germ tubes were produced, they contained elongated mitochondria which were continuous with the giant mitochondria in the yeast cells. However, as the germ tubes extended to form branched hyphae, the mitochondria became fragmented. The apical compartments and the tips of side branches contained large elongated branched mitochondria, whereas the older compartments nearer to the yeast mother cells became rapidly vacuolated and retained only small, scattered mitochondria located towards the sides of the hyphae (Aoki *et al.*, 1989). Such a distribution of large and small mitochondria is perhaps surprising as it appears to be the opposite of that reported by Beckett *et al.* (1974) for mycelial fungi in general. However, it possibly reflects a particular aspect of the organization of dimorphic fungi in general or, more likely, of this species in particular, in which subapical compartments become rapidly vacuolated and the bulk of the cytoplasm and presumably the greatest metabolic activity are restricted to the apical regions (Gow and Gooday, 1982, 1984; Aoki *et al.*, 1989).

Mitochondria are, like nuclei, capable of moving within the mycelium and can be observed doing so under the light microscope or fixed in the act of passing from one compartment to another through the septal pore in specimens examined by electron microscopy (Figure 5.1b). However, the treatment of hyphae with microtubule-binding drugs seems to inhibit mitochondrial movement, at least in certain circumstances, as demonstrated by Hoch *et al.* (1987) in *Uromyces appendiculatus*, where addition of griseofulvin stops the forward migration of mitochondria with the extending hyphal apex.

(c)   The endomembrane system

Most of the internal membranes of filamentous fungi, comprising the bounding membranes of the majority of organelles, are components of a dynamic, discontinuous structure known as the endomembrane system. The membranes of some organelles are physically connected, as in the case of the direct links between the outer nuclear envelope and the rough endoplasmic reticulum which have been observed in several species (Beckett *et al.*, 1974). The endoplasmic reticulum also appears to be intimately associated in direct physical contact (although not necessarily actually linked) to the membranes of many other organelles such as vacuoles (Beckett *et al.*, 1974). However, in most cases the endomembrane system is discontinuous such that organelles are separated within the cytoplasm, but are kept in dynamic contact with other organelles via the continuous traffic of cytoplasmic vesicles which flows between them. Thus membrane components as well as any molecules which may be carried in the lumen of these vesicles may pass through several separate organelles and ultimately out to the plasma membrane itself in the form of apical vesicles or microvesicles during the process of hyphal tip extension (see below and Chapter 13).

In many eukaryotes the most intense focus of this traffic of cytoplasmic vesicles is the Golgi apparatus, recognizable as a stack of flattened lamellae in close association and often surrounded by a cloud of vesicles in an active cell. This structure, known as a dictyosome, is rare in filamentous fungi, although it occurs in many members of the Mastigomycetes (Beckett *et al.*, 1974), but is absent from species belonging to the order Blastocladiales in the class Chytridiomycetes (Sewall *et al.*, 1989). Nevertheless, distinct Golgi equivalents have been identified in species belonging to the order Blastocladiales such as *Allomyces macrogynus*, where they consist of individual cisternae or loose associations of a few cisternae (Sewall *et al.*, 1989). This appears to be typical of the majority of fungi, in which the Golgi apparatus occurs as isolated cisternae which may be difficult to distinguish from the endoplasmic reticulum. However, in many species the Golgi equivalents can be distinguished, because they form loose but recognizable associations of small cisternae with wider lumens than the endoplasmic reticulum

(Beckett *et al.*, 1974). These associations can be irregular but are often approximately circular or semicircular and associated with localized concentrations of vesicles. Such structures are typically found close behind the most intense concentration of vesicles which occurs in filamentous fungi, the apical cluster. So the Golgi equivalents of filamentous fungi appear to be closely associated with processes which utilize large quantities of cytoplasmic vesicles as in other eukaryotes.

(d)   Vesicles

Vesicles are the smallest, but also the most numerous and one of the most important of filamentous fungal organelles. Although usually most apparent as a dense cluster in the hyphal apex, they occur in much lower concentrations throughout the cytoplasm. This reflects their vital role as transport organelles mediating a significant proportion of the cellular traffic in both structural and enzymic proteins and many other cellular components, notably lipids and polysaccharides. Localized accumulations of vesicles may occur throughout the mycelium, often associated with actual or incipient wall-formation. Thus they are often noticeable at the forming edges of septa (Figure 5.2) or at the side walls of hyphae or spores prior to branch or germ tube formation. Vesicle accumulations are also sometimes found in the angle between a mature septum and the side wall of the hypha. This may indicate a role of these cross-walls in obstructing vesicle flow along the hypha, so producing a localized concentration of vesicles as a first step towards initiation of a new branch in that region (Trinci, 1979). However, it is now clear that this is only one possible component of the complex mechanism of branch formation and regulation as discussed in Chapter 14. Occasionally streams of vesicles have been observed passing through the septal pore, presumably indicating the long-distance migration of vesicles from subapical compartments to supply the extending hyphal apex (Trinci and Collinge, 1973). More recent work has implicated microtubule components of the cytoskeleton as the possible structural determinant of such tracks along which vesicles appear to move (Howard and Aist, 1977, 1980; Howard, 1981). Hoch *et al.* (1987) have demonstrated that the microtubule inhibitor, griseofulvin, disrupts the movement of vesicles in

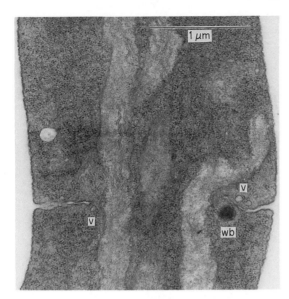

Figure 5.2 Transmission electron micrograph of longitudinal section through a hypha of *Penicillium chrysogenum*, showing an early stage of septum formation. Note the vesicles (v) and Woronin body (wb) closely associated with the incipient cross-wall. Material prepared and photographed by Dr A.J. Collinge.

the apex of *Uromyces appendiculatus*, changing the normal ordered linear translocation of vesicles to random Brownian motion (Chapter 13).

The precise origin of vesicles is still not certain, but the most likely candidates appear to be the Golgi cisternae, with which vesicles are usually found associated in significant numbers, and to a lesser extent the smooth endoplasmic reticulum. It now seems highly unlikely that the so-called multivesicular bodies, discussed below, are sources of free cytoplasmic vesicles.

There are clearly many different types of vesicles, at least in respect of the functions they perform. This is indicated by the multiplicity of specific names that have been used to describe them in an attempt to clarify some of the distinctions. These names include cytoplasmic vesicles, apical vesicles, wall vesicles, microvesicles, and chitosomes. It is not clear to what extent vesicles of different function actually differ from each other in their structural components as opposed to the functional contents of their lumen. However, there are several lines of evidence to suggest that there are real differences between the vesicles themselves.

First, it is clear that they can be significantly different in size, not just in different species, but within a single hypha. Thus most apical clusters of vesicles are made up of two size classes, large vesicles and the dramatically smaller microvesicles (Grove, 1978).

The actual sizes of vesicles vary widely from species to species, as indicated by the values quoted in Table 5.1. As demonstrated by such data, there does appear to be at least an approximate correlation between vesicle diameters and the diameters of the hyphae in which they occur. So, for example, vesicles observed in species with narrow hyphae such as *Penicillium chrysogenum* and *Aspergillus nidulans* have much smaller diameters than those of comparable vesicles observed in species with wide hyphae such as *Phycomyces blakesleeanus* and *Pythium ultimum* (Table 5.1).

There is also evidence that in some respect, presumably surface signalling molecules, vesicles originating from the same source can be recognized as different from each other by cellular sorting mechanisms. Thus vesicles from the Golgi cisternae carrying components to the vacuoles are recognized and sorted from those vesicles originating from the Golgi and carrying molecules to the plasma membrane, even when experimental means have been used to make those vesicles transport essentially the same molecules (Klionsky *et al.*, 1990).

Among the names listed above which have been used for types of vesicles, the only one which is truly specific is that of the chitosome. This name applies to the vesicles which carry the enzyme chitin synthase in its zymogenic form and is believed to be a component of the vesicle system which supplies the extending hyphal apex with the requirements for wall-assembly (Bracker *et al.*, 1976; Bartnicki-Garcia *et al.*, 1979). In terms of size these vesicles are small compared to the majority of vesicles, and are often considered to correspond to the category known as microvesicles (Bracker *et al.*, 1976; Bartnicki-Garcia *et al.*, 1979), although it is not clear whether there is a direct equivalence. Chitosomes have now been isolated, purified and extensively characterized (Bracker *et al.*, 1976; Hernandez *et al.*, 1981; Flores-Martinez *et al.*, 1990). They generally fall in the size range 40–70 nm in diameter (Bracker *et al.*, 1976; Flores-Martinez *et al.*, 1990) and are composed of approximately one-third lipid and two-thirds protein

Table 5.1   Sizes of vesicles in filamentous fungi

Source and type of vesicle	Vesicle diameters (nm)	
	Microvesicles	Large vesicles
Apical vesicles		
Oomycetes		
Pythium ultimum	40–120	200–300
Zygomycetes (typical)	≤100	200–400
Phycomyces blakesleeanus		300
Ascomycetes (typical)	≤50	c.100
Aspergillus giganteus		72±12
Aspergillus nidulans	31±6	75±12
Neurospora crassa spco9	38±3	123±19
Deuteromycetes		
Penicillium chrysogenum	29±6	68±10
Chitosomes	40–70	
(Species representative of all subdivisions of the Eumycota)		
Coated vesicles	50–80	100–180
(*Neurospora crassa, Uromyces phaseoli*)		
Uncoated vesicles	35–45	
(*Neurospora crassa, Uromyces phaseoli*)		

Data taken from Bracker *et al.* (1976); Collinge and Markham (1982); Flores-Martinez *et al.* (1990); Gooday and Trinci (1980); Grove (1978); and That *et al.* (1987).

(Hernandez *et al.*, 1981). The protein component accounts for a very small proportion of total cell protein, about 0.17% of *Mucor rouxii* yeast-cell protein, and includes chitin synthase as a major component, of which 90–96% is in the zymogenic form (Flores-Martinez *et al.*, 1990). The polypeptide composition of *M.rouxii* chitosomes has recently been shown to include a complex of molecules of different molecular mass, of which the most prominent was one with an M_r of 55 kDa (Flores-Martinez *et al.*, 1990) but it is not yet clear whether this corresponds to the chitin synthetase molecule.

A further important step in the characterization of the vesicles of filamentous fungi came with the isolation and characterization of coated vesicles in the ascomycete *Neurospora crassa* and the basidiomycete *Uromyces phaseoli* (That *et al.*, 1987). Such vesicles had previously been well-characterized from mammalian systems where they had been described as vesicles surrounded by a protein basket in the form of a network of hexagons and pentagons (Kanaseki and Kadota, 1969). The principal protein in the network surrounding such coated vesicles has been characterized as a fibrous protein with a molecular mass of 180 kDa and has been named clathrin (Pearse, 1976). That *et al.* (1987) found structurally very similar coated vesicles in *N. crassa* and *U. phaseoli*, with a 180 kDa

protein as the major component of the coat, presumably clathrin, together with a number of minor proteins of lower molecular mass. Interestingly, in view of the typical observation of two size classes of vesicle in the apices of most fungal hyphae (Grove, 1978), two size classes of coated vesicles were isolated (Table 5.1). One class had diameters in the range 100–180 nm, but the most frequently occurring coated vesicles had diameters in the range 50–80 nm (That *et al.*, 1987), very similar to the size range of microvesicles and chitosomes. During the characterization of these coated vesicles, other vesicles without coats, usually in the size range 35–45 nm in diameter, were also observed (That *et al.*, 1987). It was not clear whether these genuinely represented cytoplasmic vesicles which were normally not coated, or whether their coats had merely been lost during the isolation and processing procedures (That *et al.*, 1987). However, a very interesting observation made by examination of thin-sections of *N. crassa* hyphae, was that coated vesicles were frequently observed associated with Golgi cisternae in hyphal regions 2–3 µm behind the tip, but were rarely observed in the extreme apex, suggesting that uncoating of such vesicles might be a necessary event prior to fusion with the plasma membrane in the case of such apical vesicles (That *et al.*, 1987). It remains possible, however, that the vesicles without coats observed during the study were genuine representatives of a non-coated class of vesicles which might correspond to at least some of the apical vesicles, particularly as their small size corresponds to the lower end of the size range for chitosomes. Significantly, no prominent 180 kDa polypeptide was reported in the recent protein analysis of chitosomes, although care must be exercised in the interpretation of this information as the occurrence of faint bands above 100 kDa were indicated (Flores-Martinez *et al.*, 1990), and a coat could have been lost during the preparation procedures. In addition, Howard (1981) observed vesicles with a hexagonal profile, possibly representing coated vesicles, in the apices of *Fusarium acuminatum* hyphae prepared for electron-microscopy using freeze-substitution.

(e) Vacuoles

Probably because their appearance under the electron microscope is often that of apparently mainly empty space within hyphae, particularly in older regions, vacuoles have often been considered to be of little significance in filamentous fungi. Indeed the fact that older regions of the mycelium often become highly vacuolated has suggested that vacuoles merely result from the withdrawal of useful functional cytoplasm leaving empty non-functional space. This view appears further substantiated by the extent to which the highly active tip regions of hyphae are packed with cytoplasm and most organelles and vacuoles are not normally found to be present. To some extent this view might be correct and in certain species such as *Basidiobolus ranarum*, *Candida albicans* and *Erynia neoaphidis* the mycelium becomes highly vacuolated because the organism adopts a growth habit which effectively involves a small quantity of active biomass migrating along a network of tunnels formed by the hyphal wall, leaving much of the rest of the mycelium empty and non-functional (Webster, 1980; Gow and Gooday, 1982, 1984; Gray *et al.*, 1991). The situation is very different in the active hyphal regions of most species, where, between the fully packed hyphal tip and the dead or dying mycelial centre, there is a significant region of active biomass which contains the true membrane-bounded organelle which is the fungal vacuole. In recent years it has become apparent that this organelle, far from being an empty space or a mere storage region for water and some types of molecules, is highly complex and carries out a multiplicity of important functions. Indeed it is likely that what is recognized by its electron-microscopic appearance as a single organelle, may actually consist of a series of functionally distinct subclasses of organelle to compartmentalize the wide variety of functions performed. One subclass might therefore be largely inactive membrane-bounded regions in old hyphae. One review has described in detail many of the aspects of vacuole composition, biogenesis and functioning (Klionsky *et al.*, 1990), and although it is based to a large extent on information derived from the yeast *Saccharomyces cerevisiae* it discusses sufficient independent evidence from *Neurospora crassa* to make it clear that the vacuoles of filamentous fungi are very similar in function to those of the yeast. The principal functions so far ascribed to this organelle are: metabolite storage; a key role in cytosolic ion and pH homeostasis; and as the principal fungal equivalent of the mammalian lysosome, containing a wide range of hydrolytic enzymes in an acidic environment (Klionsky *et al.*, 1990). The possibility that vacuoles could play a physical role in the movement of cytoplasm and organelles towards the growing hyphal tip, a suggestion that had been put forward based on the concept that small vacuoles in this region could expand to squeeze the cytoplasm forwards, is now considered to be very unlikely (McKerracher and Heath, 1987).

The biochemistry of vacuole function is discussed in great detail by Klionsky *et al.* (1990) and will not be reiterated here. However, it is worth noting that although the vacuole functions described are numerous and extremely important, no vacuolar function essential for growth has yet been identified. The vacuole is biochemically characterized by the presence of α-mannosidase, the classic marker enzyme which is located in the vacuolar membrane. In fungi such as *Neurospora crassa*, the vacuolar compartment has a pH of approximately 6.0, maintained by a membrane ATPase which pumps H^+ into the vacuole, generating a membrane potential of about 25–40 mV in *N. crassa* (Klionsky *et al.*, 1990). It is this electro-chemical potential which allows this organelle to drive amino acid and ion transport mechanisms across the vacuolar membrane to achieve accumulation and storage. Such transport concentrates basic amino acids, notably arginine, making the vacuole an important site of nitrogen reserves. Acidic and neutral amino acids are not accumulated to any significant extent (Klionsky *et al.*, 1990). Certain key ions are similarly accumulated, including Ca^{2+}, Mg^{2+}, Zn^{2+}, and Fe^{2+} among others, whereas K^+ and Na^+ are usually only present in low concentrations in the vacuole. The vacuole also removes potentially toxic ions such as Sr^{2+}, Co^{2+} and Pb^{2+} from the cytosol. The vacuole is also the major site of storage for inorganic phosphorus which it accumulates and stores in the form of polyphosphate molecules with a typical size of 3–45 monomeric units in *N. crassa* (Klionsky *et al.*, 1990; see also Chapter 8, section 8.3 and Figure 8.4).

The vacuole is apparently a very dynamic structure, at least in *Saccharomyces cerevisiae*, and can exhibit very rapid changes in its morphology in wild-type strains. In mutant strains or in the presence of microtubule-inhibitors it can exhibit dramatic fragmentation into smaller vacuoles, vesicles and accumulations of unusual membranous inclusions (Klionsky *et al.*, 1990).

The origin of vacuoles is still not clear. There is some indication of a role for the endoplasmic reticulum, which has been observed completely enclosing small, apparently forming vacuoles in young hyphae, suggesting direct involvement in the process of vacuole biogenesis (Beckett *et al.*, 1974). In addition, vesicles from Golgi cisternae appear to contribute to increasing vacuole size, at least in *S. cerevisiae* as do endocytotic vesicles from the plasma membrane, emphasizing the lysosome-like aspect of some of the vacuolar functions (Klionsky *et al.*, 1990).

(f) Microbodies

If ignorance and misconception had been the major features of our understanding of vacuoles until recent years, the situation in respect of the group of organelles given the all-embracing name of microbodies could at best be described as one of confusion. The problem with this group of organelles arose largely because they were defined on the basis of a rather vague structural description. Microbodies are moderately electron-dense, spherical or ovoid cytoplasmic inclusions of diameter 0.1–2.0 μm (Beckett *et al.*, 1974) bounded by a single bilayer membrane, and sometimes containing an electron dense inclusion. Because there are

several particles within the cytoplasm of filamentous fungi that correspond to this description, the term microbody has variously been used to refer to peroxisomes, glyoxysomes, hydrogenosomes, lysosomes and Woronin bodies. In recent years two useful reviews (Maxwell *et al.*, 1977; Carson and Cooney, 1990) have served to clarify the position in respect of this group of organelles and it is now considered that the term microbody refers only to peroxisomes and glyoxysomes, which are distinguished from each other on the basis of the enzymic activities which they contain. More than 30 separate enzyme activities have now been detected in fungal microbodies, of which just over 20 have been located in filamentous fungi (Carson and Cooney, 1990). The marker enzyme for microbodies is generally considered to be catalase, and the subgroup called peroxisomes is characterized by the additional presence of flavin-linked oxidase, whereas glyoxysomes contain enzymes of the glyoxylate cycle, although in most cases they contain only some of these enzymes (Maxwell *et al.*, 1977). The occurrence of microbodies appears to be universal throughout the fungi, but the number, shape and function of these organelles is highly variable. Even within a single species, the age, developmental stage and nature of the carbon and energy source dramatically affects the microbody complement (Carson and Cooney, 1990). The number of microbodies within mycelia generally seems to increase when the carbon source is not glucose. This has been noted particularly in necrotrophic plant-pathogenic fungi where the microbodies are believed to play an important role in mobilizing the complex carbon-containing polymers of the plant host cell walls (Maxwell *et al.*, 1977). Microbodies are also very common in fungal spores, where they are often found in close association with lipid inclusions and it is thought that these organelles play a key role in mobilizing lipid-reserves during spore germination (Maxwell *et al.*, 1977). Conversely, it is also thought that these organelles may be involved in lipid biosynthesis for storage during sporulation, together with other roles in spore formation such as melanogenesis (Maxwell *et al.*, 1977). It is clear that the wide complement of different enzyme activities assigned to microbodies in different fungi means that potentially they may be involved in a wide range of degradative and biosynthetic activities and this is discussed more fully by Carson and Cooney (1990).

5.2.2 Organelles of restricted distribution amongst fungi

(a) Woronin bodies

The small cytoplasmic particles given the name Woronin bodies by Buller (1933), are perhaps the most important of the organelles found only in the filamentous fungi. They do not occur outside this group of organisms, nor are they present in all the subdivisions of the fungi. But, because they are characteristic of the ascomycetes and of the ascomycete-related deuteromycetes, they occur in the majority of species (Markham and Collinge, 1987). An important exception to this is that they apparently do not occur in yeasts, even in asco-mycete or deuteromycete yeasts. Interestingly this is true not only of permanent yeasts such as *Saccharomyces cerevisiae* but also of the yeast forms of dimorphic species such as *Blastomyces dermatitidis* and *Histoplasma capsulatum* (Garrison *et al.*, 1970), *Phialophora dermatitidis* (Oujezdsky *et al.*, 1973), *Paracoccidioides brasiliensis* (Carbonell, 1969) and *Sporotrichum schenckii* (Lane and Garrison, 1970), even though Woronin bodies were present in the hyphal form of the same cultures and were consistently found at the first transitional stage from the yeast cell towards the mycelial form (Oujezdsky *et al.*, 1973). This observation illustrates the importance of these organelles in that they are clearly a specific adaptation to the mycelial growth habit, and their taxonomic distribution indicates that they are restricted to hyphae with septa perforated by a simple pore. They are apparently not present in filamentous basidio-mycetes with their elaborate dolipore septal-complex (see below) or in the lower fungi with imperforate septa or aseptate hyphae. The nature and function of these organelles have been extensively considered (Markham and Collinge, 1987; Markham *et al.*, 1987) and only the key features will be reiterated here. The specificity of these organelles for hyphae with simple pored septa is emphasized by the fact that their cytoplasmic location is highly characteristic. They are overwhelmingly found close to the septal plate and near to the septal pore (Figure 5.3). Their appearance by electron microscopy is also very characteristic

as they are approximately spherical with a diameter in the range 0.1–0.75 μm (Markham *et al.*, 1987) and are usually intensely electron-dense, a feature which is often emphasized because of a narrow ribosome-free zone which appears to surround them, giving them a halo of electron-light cyto-plasm. They are clearly true organelles because they are bounded by a single bilayer membrane (Figure 5.3). It is now fairly clear that the major component of Woronin bodies is protein (McKeen, 1971; Mason and Crosse, 1975; Head, 1987) and this appears to be largely present in the form of a crystalline matrix as indicated by the regular lattice structure of the centre of the organelle which has been reported in several studies (Brenner and Carroll, 1968; Markham *et al.*, 1987). However, although preliminary studies have shown that none of the enzymic activities typical of microbodies are found associated with enriched fractions of Woronin bodies (Head *et al.*, 1989), the nature of the protein content has only been characterized to the extent that it appears to consist of a single type of molecule of molecular mass 29 kDa (Head, 1987). This means that there is currently no suitable marker enzyme or anti-body label for this organelle. However, Keller *et al.* (1991) have reported that the microbody-specific anti-SKL antibody, which recognizes the serine-lysine-leucine-COOH terminal tripeptide of the peroxisomal targeting signal, will bind to the hexagonal crystals of *Neurospora crassa*. This indi-cates a possible labelling approach in view of the apparent functional relationship between Woronin bodies and hexagonal crystals (see below). Whilst more specific investigations in living hyphae await such a label, our understanding of the likely origin and functions of Woronin bodies rests largely on electron-microscopic studies. Whereas most Woronin bodies are located near the septal plate and apparently take up that position from an early stage of septal formation (Figure 5.2) (Markham and Collinge, 1987), they appear to be synthesized in hyphal tips. This is supported by the obser-vation in several species of Woronin bodies very close to the hyphal apex (McClure *et al.*, 1968; Collinge and Markham, 1982), even within the apical vesicle-cluster itself. This had been thought to indicate a possible role for these organelles in plugging burst hyphal tips, but a recent study has demonstrated that this is not the case as Woronin bodies are too small to plug even the relatively

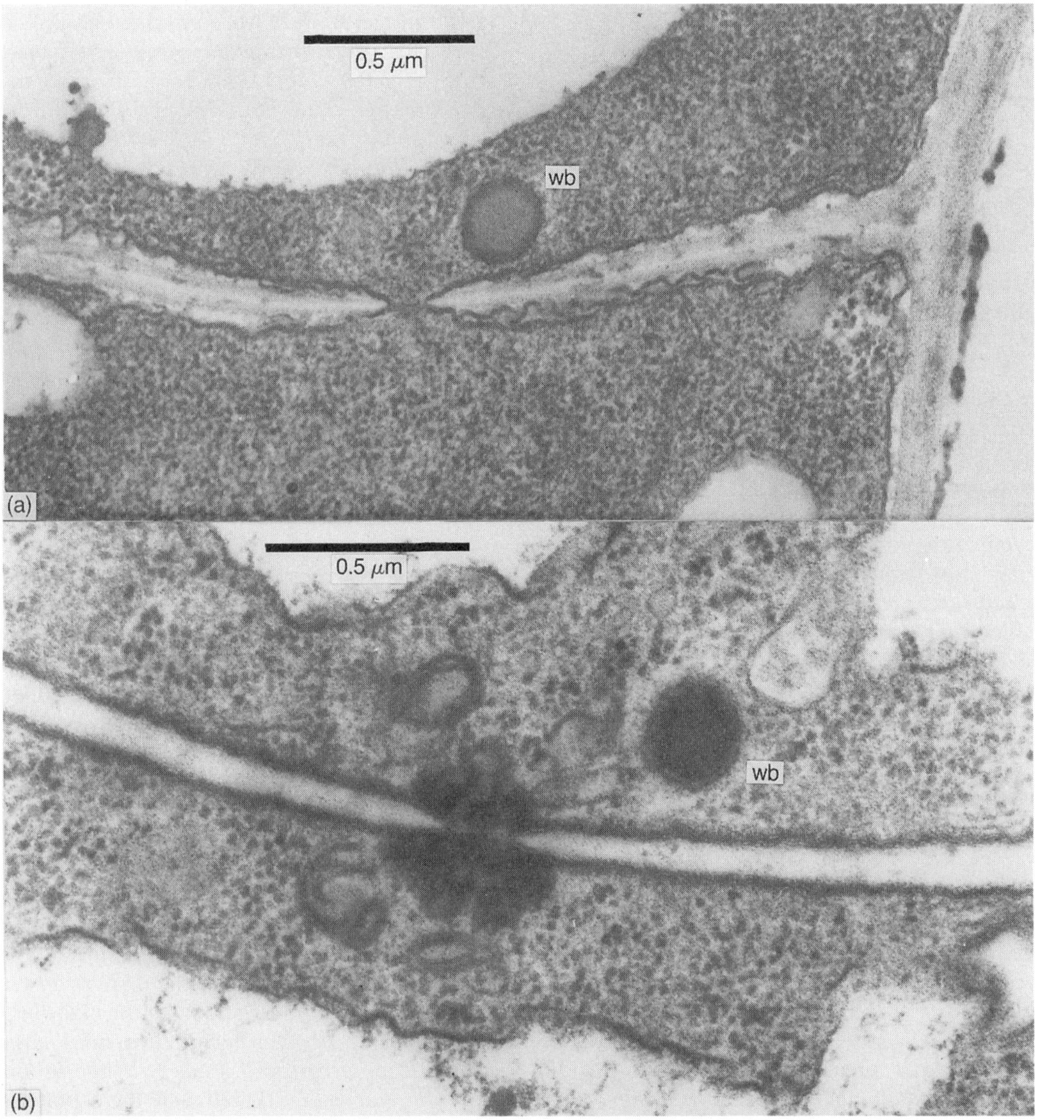

Figure 5.3 Transmission electron micrographs of cross-sections through the septal region in (a) *Aspergillus giganteus*, showing an open septal pore and a nearby electron-dense membrane-bounded Woronin body (wb); and (b) *Penicillium chrysogenum*, showing a septal pore plugged by an irregular electron-dense deposition plug, with nearby a completely normal electron-dense, membrane-bounded Woronin body (wb). Both specimens prepared and photographed by Dr A.J. Collinge.

small hole which arises in the apex of *Penicillium chrysogenum* hyphae burst by flooding with distilled water and these organelles are carried out of the apex with the extruded cytoplasm (Collinge and Markham, 1992). The presence of Woronin bodies in the hyphal apex appears to relate to the probable synthesis of these organelles within membranous structures located in the apical compartment as described by Brenner and Carroll (1968). The principal function of Woronin bodies

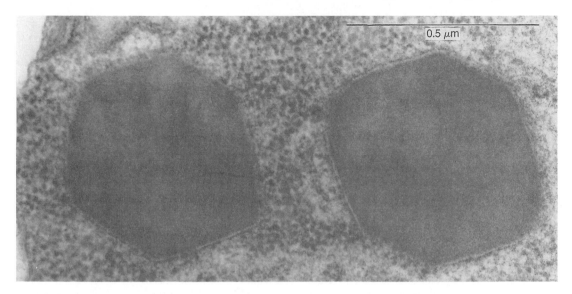

0.5 μm

Figure 5.4 Transmission electron-micrograph of two hexagonal crystals in *Sordaria brevicollis*. Note the clearly visible bounding membrane, the classic hexagonal shape of one crystal and some rounded edges of the other crystal, suggesting the existence of more than six edges in some orientations.

therefore still appears to be that which was suspected from soon after their discovery, of plugging septal pores to prevent movement of cytoplasm and organelles between compartments. It has often been suggested that this may indicate a role in controlling spatial organization of the mycelium (McKeen, 1971) but there is little evidence to support this (Markham *et al.*, 1987). However, a quantitative study of the response of *Penicillium chrysogenum* hyphae to severe damage did support the hypothesis that Woronin body plugging of septal pores is an emergency response to hyphal rupture in order to prevent excessive loss of cytoplasm and mycelial death (Collinge and Markham, 1985). Controlled and progressive sealing of septal pores during the course of normal mycelial growth is likely to be due to the formation of deposition plugs by *de novo* deposition of electron-dense material into the pore, rather than by the use of these preformed organelles. Pores plugged in such a way are often observed in undamaged hyphae, even when Woronin bodies are clearly visible nearby (Figure 5.3b).

(b) Hexagonal crystals

As their name implies, hexagonal crystals are often found to have a regular hexagonal shape in

cross-section when visualized under the electron microscope (Figure 5.4), but as shown in Figure 5.4, this is often only the most clearly recognized form. Many variants on this basic cross-section are observed, including highly elongated rhomboidal or even rectangular forms in which the number of sides is less than six, or profiles such as that seen in Figure 5.4, where some edges are rounded to suggest more than six sides and a more complex overall shape than might first be imagined. There is no clear indication as to the three-dimensional shape of these particles. Head (1987) has made some suggestions based on information from both cross-sections of intact hyphae and from observation of isolated hexagonal crystals of *Neurospora crassa*. It seems most likely that the full particle has an approximately icosahedral shape which provides the various profiles which have been observed, depending on the plane in which it happens to be sectioned, although there is no evidence that these inclusions necessarily have a consistent three-dimensional shape. The consistent features which identify these particles, however, are their extreme electron-density, the fact that they are angular, possessing at least some straight edges, and the fact that they are bounded by a single membrane. In addition it is usual to observe an internal lattice structure to the matrix, which reinforces the view that these particles are indeed crystalline. Hexagonal crystals are often found near septa in species which contain them and as a consequence of this location and the characteristics they share with Woronin bodies, they have been considered to be a specific structural subclass of Woronin bodies capable of performing the same function of septal pore plugging (Trinci and Collinge, 1974; Markham and Collinge, 1987; Markham *et al.*, 1987). This has certainly been shown to be true in *Neurospora crassa* (Trinci and Collinge, 1974). However, in a close relative of *N. crassa*, the pyrenomycete *Sordaria brevicollis*, a recent study has demonstrated that hexagonal crystals do not apparently take part in pore plugging in damaged hyphae, this function apparently being fulfilled by nuclei (Collinge and Markham, 1987). This might appear to cast doubt on the hypothesis that hexagonal crystals are simply a specific form of Woronin bodies, however the distribution of these particles seems to be unusual in *S. brevicollis*, where they are not usually closely associated with the septal plate and in particular are generally far removed from the septal pore (Collinge and Markham, 1987). This type of positioning in *S. brevicollis* could therefore explain their apparent failure to fulfil what is generally considered to be the prime function of Woronin bodies. However, it does bring into question the whole idea that the primary function of such organelles is septal pore sealing. Clearly in *S. brevicollis* the fact that they are present without performing this function indicates that such particles have an additional, possibly more important, function. It has been suggested that another function of Woronin bodies and hexagonal crystals might be as sites of protein storage (Markham and Collinge, 1987). This in turn argues that the septal pore plugging function might be fortuitous rather than of primary importance. In most species that possess these organelles this function appears to have gained importance as those species have evolved, but in others such as *S. brevicollis*, other cellular inclusions such as nuclei may be performing the pore-plugging function in the same fortuitous manner. One piece of evidence which indicates the importance of rapid pore plugging in damaged hyphae is the fact that basidiomycetes which do not have preformed plugging organelles like Woronin bodies achieve the same type of rapid septal

sealing by *de novo* deposition of electron-dense material into the pore channel (Aylmore *et al.*, 1984) suggesting the need for some specific, rather than *ad hoc*, response.

The species distribution of hexagonal crystals is apparently very limited. They have specifically been identified in *Whetzelinia sclerotiorum* (Hoch and Maxwell, 1974) and *Geotrichum candidum* (Cole and Samson, 1979) in addition to the species named above. Particles identified as hexagonal inclusions have also been reported from the *Aspergillus glaucus* group of species (Mason and Crosse, 1975) and these also appear to be genuine hexagonal crystals because Woronin bodies were recorded to be absent from the same species. This observation strengthens the hypothesis that hexagonal crystals are a subclass of Woronin bodies because their occurrence appears to be mutually exclusive. The few species in which both appear to occur may be due to the misidentification of other types of crystalline inclusion (see below) as genuine hexagonal crystals.

(c) Hydrogenosomes

The hydrogenosome is one of the most recently characterized organelles, partly because it appears to occur in only a few obligately anaerobic organisms. The term was first used for microbody-like inclusions in the anaerobic trichomonad protozoan, *Tritrichomonas foetus* by Lindmark and Müller (1973). In protozoa, these organelles have been morphologically characterized as approximately spherical or ovoid membrane bounded vesicles with a diameter in the range 0.5–2.0 μm, containing a granular matrix which sometimes has a denser core (Müller, 1980). However, it is the biochemical characteristics of these organelles which most specifically identify them, key features being the presence of the enzymes hydrogenase and pyruvate:ferredoxin oxidoreductase, and the production of molecular hydrogen. Hydrogenosomes specifically occur in protozoa from which mitochondria are apparently absent (Müller, 1980).

Hydrogenosomes have only recently been found to be present in a very limited number of fungal species. These fungi are themselves very unusual because they are obligately anaerobic, occurring in the rumen of herbivores, and can only tenuously be considered filamentous fungi because they produce rhizoids not hyphae, and apparently belong to the class chytridiomycetes (Theodorou *et al.*, 1992). Hydrogenosomes have been observed in the zoospores and rhizoids of such fungi (Theodorou *et al.*, 1992) and were first specifically identified in *Neocallimastix patriciarum* (Yarlett *et al.*, 1986). The greatest accumulation of hydrogenosomes in *N. patriciarum* appears to be close to the cellular apparatus which drives the flagella of the zoospores (Yarlett *et al.*, 1986) and it seems likely that they are involved in energy generation, as has been proposed for protozoal hydrogenosomes (Müller, 1980). Indeed Müller (1980) has suggested that these organelles are the anaerobic equivalent of mitochondria, having been derived by endosymbiosis of a clostridial-type bacterium. Consistent with this hypothesis, there are some indications that protozoal hydrogenosomes may be bounded by a double bilayer membrane and could contain a circular DNA molecule (Müller, 1980). Keller *et al.* (1991) have shown that hydrogenosomes of the protozoan *Trichomonas vaginalis* do not appear to be at all closely related to microbodies, confirming the view that their apparent structural similarity is deceptive and that hydrogenosomes are an entirely different type of organelle.

(d) Rumposomes and gamma-particles

Species belonging to the subdivision Mastigomycotina are characterized by having a motile stage in their life cycle which is a swimming spore, the zoospore, that moves by means of one or more flagella. Inevitably such a unique stage in the life cycle gives rise to dramatic differences in cellular organization not found elsewhere in the fungi and means that such species contain specific organelles unique to zoospores. Although these clearly do not relate to the filamentous stage of growth it is worth mentioning two of the most characteristic, the rumposome and the gamma-particle. These are both strange-shaped membrane-bounded inclusions in zoospores, for which the function is still uncertain. The gamma-particle has been identified in *Blastocladiella emersonii*, as a membrane-bounded organelle approximately 0.5 μm in diameter with a distinctive electron-dense elongated cup-shaped proteinaceous inclusion and an electron-light central zone (Hohn *et al.*, 1984). The investigation of these

organelles has largely been carried out by Cantino and his group (Cantino and Mills, 1979) who suggested that the gamma-particles were repositories of chitin synthase which became mobilized from the proteinaceous inclusions during the encystment stage of zoospore development as part of the wall synthesis mechanism. The suggested scheme involved the extensive production and transport of 80 nm diameter microvesicles from the gamma-particles to the plasma membrane (Cantino and Mills, 1979). However, characterization of the gamma-particle proteins has indicated that this is unlikely to be the case as they do not appear to include a protein with chitin synthase activity (Hohn *et al.*, 1984) and this leaves the function of these organelles somewhat of a mystery. Their function is clearly specific to the zoospore stage in the life cycle as these particles appear during sporulation and disappear during zoospore germination, and the appearance and disappearance of their major protein constituents correlates well with this change (Hohn *et al.*, 1984).

Rumposomes (Fuller, 1966) are very aptly named in view of their typical lateral location at the posterior end of the zoospores of members of the genus *Monoblepharella* (Fuller, 1966; Beckett *et al.*, 1974; Webster, 1980). They are curious elongated structures which are considered to consist of a network of interconnected tubules (Webster, 1980). The classic appearance is of a series of parallel columnar units which lie close to the zoospore membrane and are approximately hexagonal in cross-section. These hexagonal components are connected to one or two membranous lamellae lying at right-angles to the hexagonal structures on the cytoplasmic side of the array (Fuller, 1966). The membranous lamellae are filled with electron-dense material and connected to both the nuclear membrane and to a dictyosome-like structure in which the flagellar rootlet is anchored (Fuller, 1966). Rumposomes are zoospore-specific organelles, and although their function is still unclear, it seems likely that thay are involved in flagellar activity in view of their proximity and connection to the rootlet of that structure. Burnett (1976) has suggested they may equivalent to the complex of structures known as side bodies in species of *Allomyces* and *Blastocladiella* (Beckett *et al.*, 1974).

5.2.3   Organelle-like inclusions of uncertain status

(a)   Lomasomes

Membranous structures called lomasomes have been recognized in electron micrographs of many species of fungi (Moore and McAlear, 1961; Beckett *et al.*, 1974). They consist of ill-defined accumulations of membranous vesicles, tubules or lamellae apparently located between the plasma membrane and the hyphal wall, although in many sections the direct link to the plasma membrane is not evident. This may be because some lomasomes are genuinely free in the cytoplasm, but it is more likely that the connection to the plasma membrane simply lies out of the plane of the section. The fact that lomasomes occur in this position, and also in areas of intense biosynthetic activity such as the hyphal apex, in the angle between the septum and the hyphal wall, or the edges of developing septa, has suggested that they might be genuine organelles with a function related to wall or septal polymer synthesis (Beckett *et al.*, 1974). However, there is no real evidence to support this hypothesis and it seems more likely that these structures are simply accumulations of membranous material, either as a genuine hyphal component, perhaps indicating a localized excess of membrane supplied by vesicles over the actual requirement for incorporation into the plasma membrane (Beckett *et al.*, 1974), or possibly as an artefact. Conventional methods of preparation for electron microscopy could lead to localized infolding of the plasma membrane in much the same way that similar membranous structures in bacteria called mesosomes now appear to be artefacts which are not found in cells prepared for the electron microscope using cryofixation and freeze-substitution rather than more conventional methods (Ebersold *et al.*, 1981). Certainly Howard (1981) has demonstrated that the fungal plasma membrane is much smoother in appearance and more closely adpressed to the hyphal wall in specimens prepared for electron microscopy using freeze-substitution rather than conventional fixation procedures. The quantity of membranous material observed in fungal lomasomes perhaps supports the idea of a genuine localized accumulation rather than a mere artefactual origin, but with the greater quantity of plasma membrane available in a filamentous fungus an artefact of

preparation could become more extensive than in bacterial cells. It is clear that this visually striking component of filamentous fungi needs to be investigated in more detail to determine whether it is of any real significance.

(b)   Multivesicular bodies

A very similar type of cytoplasmic inclusion to the lomasomes is a structure called a multivesicular body. It is recognized as a membrane-bounded sac with membranous structures within it, which in cross-section look like an accumulation of cytoplasmic vesicles. This appearance often suggests that these structures might be one source of vesicles, and as they often occur just below the apical vesicle cluster it is tempting to suspect that they might represent a kind of vesicle factory. Indeed, they have been suggested as one possible origin of chitosomes (Grove, 1978; Bartnicki-Garcia *et al.*, 1979). However, as in the case of lomasomes there is no real evidence that these structures are indeed genuine organelles with a specific function, or even that their appearance in section really indicates their true structure. It is quite plausible, for example, that what appears to be a collection of vesicles within these bodies might actually merely be transverse sections through an assembly of tubular membranous structures. In *Penicillium chrysogenum* where such multivesicular bodies are often seen, the vacuoles frequently have tubular structures within them, which if sectioned transversely would give a very similar appearance to that of multivesicular bodies (A.J. Collinge and P. Markham, unpublished observations). Thus these structures might merely be projections from vacuoles, or, near the hyphal apex, an early stage in vacuole formation. Certainly here again there is an obvious need for further investigation. It is possible that these structures too are artefacts of preparation for electron microscopy, but this seems less likely than in the case of lomasomes, particularly as Howard (1981) observed them to be present in material prepared by freeze-substitution. However, there is no real evidence that they represent a distinct functional organelle and are more likely to represent a particular structural aspect of one of the better-characterized organelles.

(c) Crystalline inclusions

Most filamentous fungi contain cytoplasmic particles of a highly crystalline nature and these are simply given the name crystalline inclusions for convenience. This group of inclusions is certainly not a closely related set of cytoplasmic components and in most cases they probably do not represent true organelles. Some are not membrane-bounded, some which do have a limiting membrane probably have no function beyond the simple task of acting as a localized store of a particular metabolite. These inclusions are perhaps most noticeable in basidiomycete fungi, particularly in differentiated structures such as stipes of fruit bodies (Blayney and Marchant, 1977), and mycelial strands, cords and rhizomorphs (Wood *et al.*, 1985), although they also frequently occur in the vegetative hyphae of *Armillaria mellea* (Beckett *et al.*, 1974). This apparent high frequency in basidiomycetes may simply reflect the fact that in other fungal groups, structures of similar appearance are specifically recognized as the organelles called Woronin bodies or hexagonal crystals, previously described. It may well be, as discussed above, that crystalline inclusions were the evolutionary progenitors of Woronin bodies and hexagonal crystals which have acquired an apparently more important additional function since the evolutionary divergence of the group of species with dolipore septa from those with simple-pored septa.

(d) The Spitzenkörper

The Spitzenkörper, unlike lomasomes and multivesicular bodies, is quite clearly a genuine functional component of the hyphal cytoplasm. It is not, however, a genuine organelle according to the strict definition, because it is not bounded by a membrane. This structure is not apparent in all fungi, being recognized only in the so-called higher fungi belonging to the ascomycetes, deuteromycetes and basidiomycetes. It is located in the extreme apex of growing hyphae and is visually apparent under the light microscope as an approximately spherical phase-dark region and in ultrathin sections viewed under the electron microscope as an approximately circular region from which all or at least the largest size class of apical vesicles are excluded (Figure 5.5). In basidiomycetes in particular it is usually a totally vesicle-free region which often has an electron-dense core. This clearly represents an organized region of the cytoplasm with a specific function and its role in hyphal extension is discussed elsewhere in this volume (Chapter 13).

5.3 DISTRIBUTION OF ORGANELLES AND SPATIAL ORGANIZATION WITHIN MYCELIA

So far, the organelles of the filamentous fungi have been considered individually and essentially in isolation. However, it is clear that in an active and rapidly growing mycelium, efficient interaction between organelles will be vital for optimal functioning. To achieve such efficient interactions, a highly polarized and elongated organism such as the filamentous fungus has to implement a high degree of spatial organization of its cytoplasmic contents to bring about the most effective distribution of organelles within the mycelium. Although this is true throughout the entire mycelium, it becomes most apparent and essential at the hyphal tips, the sites of the most intense metabolic and physiological activities and the region of the most rapid change in the mycelium. Here the apical cluster is identifiable as a highly organized region of quite different ultrastructure from the rest of the hypha. In addition the walls dividing hyphae into compartments, the septa, form the focus of another specific accumulation of organelles and associated structures. These form the septal complex which in many species is another highly organized accumulation of specialized organelles associated together to carry out key functions in the growing, differentiating and senescing mycelium. Superimposed on this readily recognizable spatial organization of hyphae is the often more subtle, but nevertheless equally vital, distribution of major organelles, of which the nuclei are usually the most important and obvious examples.

5.3.1 The apical cluster

That the hyphal apex did not merely contain a random accumulation of apical vesicles, but was a highly organized region, was made clear by Girbardt's (1969) illustration of the apical region of *Polystictus versicolor*. The degree of complexity and organization manifested by the hyphal tips of a wide range of fungal species was comprehensively considered by Grove (1978). However, it was not until the advent of the dramatic improvement

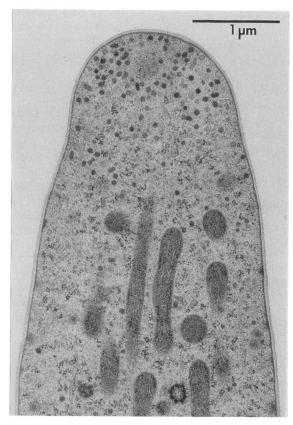

Figure 5.5 Transmission electron micrograph of an approximately median section through the apex of a hypha of the rice-blast fungus, *Magnaporthe grisea*. The Spitzenkörper is apparent as an approximately circular region just below the tip, with a filamentous core which is surrounded by apical vesicles. The micrograph also shows the cytoplasmic microtubules, peripherally located filasomes, mitochondria and fenestrated elements of smooth endomembrane cisternae, which are characteristic of the hyphal apex. The material was prepared using the process of freeze-substitution and is reproduced by courtesy of Richard J. Howard and Timothy M. Bourett (unpublished).

of preservation achieved by the use of freeze-substitution as a method for preparing hyphal apices for electron microscopy (Howard and Aist, 1979; Howard, 1981) that the truly dramatic nature of the structure, multiplicity and interactions of the various organelles and other elements of the apical cluster became fully apparent. It is immediately obvious from such studies that the degree of organization inherent in this region makes the use of the term apical cluster inadequate, implying as it does a relatively fortuitous association of cytoplasmic inclusions, and apical complex seems a more appropriate term.

The most numerous components of the apical complex are the apical vesicles, which, in *Fusarium acuminatum* prepared using freeze-substitution techniques (Howard and Aist, 1979; Howard, 1981), fall into two size classes, large spherical vesicles (70–90 nm diam.), and microvesicles (about 30 nm diam.), which were often found to be hexagonal in cross-section, suggesting that they may be coated vesicles. These vesicles appeared to originate from two distinct classes of Golgi cisternae with lumens which differed in width approximately in proportion to the size of vesicles which they produced (Howard, 1981). In addition, the large vesicles could be subdivided into two distinct types on the basis of their electron density, indicating the involvement of at least three different types of vesicle in hyphal tip extension

(Howard and Aist, 1979). The central component of the apical complex was found to be a region composed of an approximately spherical mass of vesicles, within which was a lumen containing microvesicles and microfilaments, the whole of this central component being permeated and apparently organized by a network of filaments. It is this region which was considered to correspond to the Spitzenkörper (Howard, 1981) (Figure 5.5), thus making this a more complex and rather different type of structure from that envisaged earlier. This perhaps explains why, in burst tips of *Penicillium chrysogenum*, it has been observed that, whilst there was extensive leakage of cytoplasm and some extrusion of organelles from a hole in the hyphal wall, a significant accumulation of vesicles remained within the tip (Collinge and Markham, 1992). This would be consistent with a complex of vesicles held together by microfilaments in the Spitzenkörper, but also, presumably, anchored to the plasma membrane or hyphal wall by other cytoskeletal filaments. This would also be consistent with the observations that the Spitzenkörper apparently moves as a single entity within growing hyphae (Howard, 1981) (Chapter 13).

It is perhaps surprising that such a dynamic and, for each individual component, such a short-lived region of the hypha should be so organized. However, this probably reflects the enormously critical nature of the process underway at the apex, where much of the characteristic structural and functional elements of the hypha are being constructed. Our ever-increasing understanding of this region of the hypha simply serves to emphasize the validity and perceptiveness of Robertson's famous observation that 'The key to the fungal hypha lies in the apex' (Robertson, 1965).

5.3.2 The septal complex

The septal region of the hypha has been recognized as an organized complex for many years, particularly in the case of the most structurally elaborate form, the dolipore septum of basidiomycetes. This appreciation of the interaction and organization of various structures associated with cross-walls probably stems from the distinct appearance of many of the relevant components and the apparent stability of the septal complex

once formed. Here is not the frenetic site of activity which characterizes the ever-extending region of hyphal-tip synthesis, but apparently a fixed solid structural component of the mycelium. Indeed, one of the suggestions for the likely role of septa in mycelia is as a structural support to maintain the rigidity of the hypha (Gull, 1978). Gull (1978) has reviewed the various forms of cross-wall and their associated structures which occur in fungi and has considered the possible functions of the septal complex. He favoured a more significant and active role for the septal complex, that of allowing differentiation to occur. What is clear, is that the septal complex is not merely a fixed structural component of the hypha but, at least at key moments in mycelial development, fulfils an important and active role in growth physiology, whether it be in emergencies provoked by severe stress or trauma, or in the more programmed process of differentiation or ageing. To carry out these functions the associated organelles and cytoplasmic inclusions must interact in an organized manner with the structural cross-wall. Some of these functions have already been described above, particularly those of the Woronin bodies and hexagonal crystals involved in plugging septal pores of ascomycetes and deuteromycetes. The equivalent emergency response to damage in basidiomycetes does not involve movement of preformed organelles, but it is not clear whether the pore-plugging material originates from the cytoplasm or from one of the associated organelles in the dolipore complex.

The dolipore septum (Figure 5.6) is the most elaborate septal complex found in any of the fungi, partly because the cross-wall itself is significantly different from most other septa, by having a swollen edge around the pore to give a barrel-shaped central region and a relatively narrow pore in most species, but also because it has structurally the most dramatic associated organelles, notably the parenthesomes. Moore (1985) has provided a comprehensive survey of the wide variation of dolipore septa and their associated structures. Here it is specifically the major and unique organelles associated with the dolipore septum which are of particular interest. Parenthesomes, as their name implies, bracket each side of the swellings and pore of the dolipore septum with an overarching cup or saucer-shaped structure (Figure 5.6) made up of what is believed to be

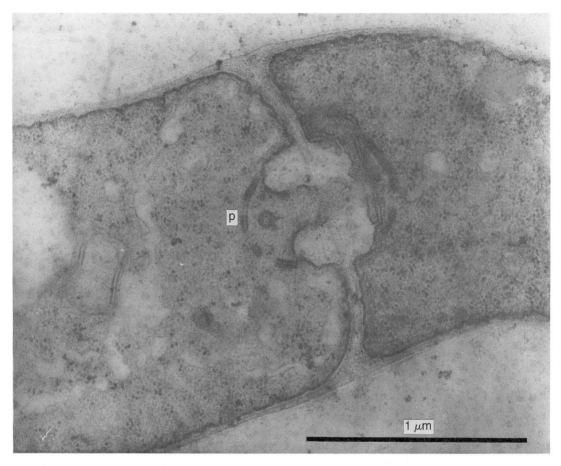

Figure 5.6 Transmission electron micrograph of an approximately median longitudinal section through a hypha of *Lepiota procera*, showing the dolipore septum. Note the electron-light swellings at the edge of the septal pore and the semicircular parenthesomes (p), perforated by pores. Material prepared and photographed by Dr O. Steele-Mortimer.

modified endoplasmic reticulum. Indeed, the parenthesomes are usually directly connected to the sheets of endoplasmic reticulum which lie alongside the septal plate and often continue along the hypha, close to and parallel with the hyphal wall (Moore, 1985), although this connection may degenerate in older hyphae (Moore, 1985). The parenthesome lumen is wider than that of the associated endoplasmic reticulum, and contains an electron-dense central layer which is often absent from the central zone of imperforate parenthesomes making that region thinner than the surrounding areas (Moore, 1985). The internal structure of the parenthesome is not clear,

although Moore (1985) has considered several possible interpretations of the electron microscopic appearance. It is, however, apparent that the parenthesomes themselves are quite stable and resilient structures, because they have been identified lying free in degenerating material associated with the discharge of spores by *Sphaerobolus stellatus* (Moore, 1985), and remain clearly distinguishable, formed into a circular structure, in the cytoplasm of damaged hyphae of *Coriolus versicolor*, after they have been detached from the plugged and remodelled dolipore (Aylmore *et al.*, 1984). Parenthesomes may or may not be perforated by large or small holes depending on the species examined, and in a limited number of species belonging to the Tremellales, they occur as vesiculate structures, made up of large, isolated, membrane-bounded vesicles apparently anchored to the dolipore complex by microfibrils (Moore, 1985). The function of the parenthesomes is still unknown, although it is tempting to suspect that they perform some kind of protective or selective function by regulating the composition of the cytoplasm immediately adjacent to the septal swellings and pore. Equally, however, they could be the site of a specific function such as the production of the electron-dense material which plugs dolipore channels in damaged hyphae (Aylmore *et al.*, 1984).

The ability to permit differentiation to occur, or to ensure that dead and dying regions of hyphae are isolated from the growing mycelium requires the septal complex to achieve pore-plugging in most species. This appears to involve Woronin bodies in many species where the developmental process is rapid, such as in asexual spore production (Cole, 1986). In contrast, the construction of highly complex seals which are believed to act like sieves to permit the selective passage of cytoplasmic components during the formation of some sexual spores such as in the asci of *Sordaria humana*, have been described (Beckett, 1981). However, in most instances the type of seal appears to be similar to that used for emergency plugging of basidiomycete septa (Aylmore *et al.*, 1984), some form of electron-dense material apparently deposited *de novo* from the cytoplasm to form what has been termed a deposition plug (Markham and Collinge, 1987; Markham *et al.*, 1987). There has been some suggestion that such plugs are produced by degeneration of Woronin bodies, at least in older hyphal regions (Beckett *et al.*, 1974). However, there is no evidence to support this suggestion other than a very plausible but merely circumstantial set of micrographs purporting to show a sequence of such degenerating organelles and the concurrent development of the plug (Beckett *et al.*, 1974). The fact that fully normal Woronin bodies are often observed very near pores sealed by a deposition plug (Figure 5.3b) and the plugging of basidiomycete septa by *de novo* deposition (Aylmore *et al.*, 1984), suggests that this is the more likely mechanism for such plug formation.

5.3.3 Nuclei

That all organelles are subject to some degree of spatial control over their distribution within the mycelium is most clearly illustrated by considering the case of nuclei. It is apparent that there is an approximate relationship between the amount of DNA and the associated volume of cytoplasm in all fungi, as exemplified in most instances by the approximately consistent number of nuclei found in each hyphal compartment, or the regular distribution of nuclei throughout the hyphae of germlings (Trinci, 1979). In *Aspergillus nidulans*, Clutterbuck (1969) has demonstrated that vegetative hyphae have half the number of diploid nuclei per compartment when compared to haploid mycelia. Most specifically it is also demonstrated by the direct relationship between ploidy and volume of the uninucleate conidiospores of *Aspergillus nidulans* (Clutterbuck, 1969) and similar volume/ploidy relationships in the asexual spores of other species (Typas and Heale, 1977; Jackson and Heale, 1985).

This organization of nuclear distribution is most apparent and most rigorous in basidiomycetes where monokaryotic forms with uninucleate compartments and dikaryotic forms with binucleate compartments are usually found (Trinci, 1979). These states are strictly maintained despite the cytoplasmic continuity which exists along the hyphae via pores in the dolipore septa. A comprehensive analysis of nuclear migration and distribution has been made in the cases of anastomoses between mycelia of *Schizophyllum commune* (Nguyen and Niedpruem, 1984; Todd and Aylmore, 1985) and *Coriolus versicolor* (Aylmore and Todd, 1984). These studies have emphasized how carefully nuclear distribution is maintained even when, as in the case of the so-called di-mon fusion between compatible dikaryon and monokaryotic mycelia of *S. commune*, septal dissolution and nuclear migration initially lead to the creation of a limited number of compartments with unusual numbers of nuclei ranging from none at all up to accumulations of 5–10 nuclei in a single compartment (Nguyen and Niederpruem, 1984). Significantly, however, the apical compartments are rapidly restored to the binucleate state by reducing the numbers of nuclei contained. This sometimes occurs by unequal distribution to leave subapical compartments containing excess nuclei, but more often involves stranding excess nuclei in pseudo-clamps, which are clamp connections which begin to form but never fuse back into the subapical compartment to release the nuclei they contain (Nguyen and Niederpruem, 1984). In the case of fusions between genetically identical monokaryons of *C. versicolor*, genetically identical dikaryons of *C. versicolor* or *S. commune*, and genetically different dikaryons of *C. versicolor* (Aylmore and Todd, 1984; Todd and Aylmore, 1985) a very limited and highly consistent phenomenon called the 'replacement reaction' occurs. In such a situation the 'donor' mycelium fuses with the 'recipient' mycelium when a hyphal tip of the donor fuses with the side wall of a recipient hypha. This temporarily creates a compartment containing twice the normal number of nuclei (although this number of nuclei would not be out of proportion to the volume of cytoplasm in the compartment). The normal nuclear number is rapidly restored, however, because the 'donor' nuclei migrate to close by the point of fusion and the recipient nuclei degenerate. This is then followed by a division of the 'donor' nuclei so that the correct DNA to biomass ratio is restored and a septation and clamp-connection formation event then divides the large compartment into two of normal size containing the standard number of nuclei. Such a consistent and highly regulated response to fusion even of genetically identical individuals emphasizes the importance and the precise control of nuclear distribution at least in basidiomycetes. Septal dissolution and nuclear migration, which is usually considered to be the 'standard' response to fusion, was only observed in fusions between compatible but genetically different monokaryons. Even then it was observed as a relatively rare occurrence and it was often merely the limited 'replacement reaction' which took place even between monokaryons genetically capable of forming a true dikaryon (Aylmore and Todd, 1984).

A more universal feature of the control of nuclear distribution in filamentous fungi is their exclusion from the extension zone in the hyphal apex. All large organelles are excluded from this region, but whereas mitochondria approach very close to the edge of the apical complex (Howard, 1981), nuclei are kept at significantly greater distances than all other organelles. For example, in *Neurospora crassa* the first nucleus occurs at between 10 and 33 μm from the hyphal apex in most strains (Collinge and Trinci, 1974) and in *Alternaria solani* the distance from the apex to the first nucleus varies between 25 and 70 μm (King and Alexander, 1969). The function and nature of such an exclusion mechanism are not known, but whilst logic would suggest that the purpose is to avoid disruption of the hyphal extension process by the presence of large organelles, the mechanism is clearly not by simple physical exclusion, at least in the case of nuclei held so far from the region of vesicle accumulation.

5.4 MOVEMENT OF ORGANELLES WITHIN MYCELIA

It has long been known that organelles could exhibit significant movement within fungal hyphae and that this was not merely due to their being carried along by the bulk flow of cytoplasm moving to the extending tip, because movement of organelles away from the tip, and movement of different organelles in opposite directions within the same hyphae had been observed frequently (Buller, 1933). The dramatic speed at which some organelles can move within hyphae is demonstrated by the report that during the formation of a dikaryon in *Coriolus versicolor*, the nuclei migrate at 25–35 μm min^{-1}, although this is clearly an exceptional instance, because similar nuclei move at a rate of 1–2.5 μm min^{-1} in the same species during the replacement reaction described above (Aylmore and Todd, 1984). Ross (1976) has recorded possibly the fastest of all examples of nuclear movement, which in *Coprinus congregatus* reached speeds of 11 μm s^{-1}.

The mechanism by which organellar movement is achieved is still not fully understood, although a much clearer picture has emerged in recent years. It is apparent that organelles do not possess their own means of independent motility, but are moved by a cytoplasmic mechanism based on the cytoskeleton and involving microtubules, actin microfilaments and energy supplied by ATP (McKerracher and Heath, 1987). It is also clear that the sophistication of the system provides for the movement of different organelles by different mechanisms as demonstrated by Oakley and Rinehart (1985). They found that the nuclei of *Aspergillus nidulans* were moved by a mechanism dependent on microtubules, whereas mitochondria moved by a separate mechanism which was either independent of microtubules, or was at least insensitive to the microtubule inhibitor benomyl and was not disrupted by mutations affecting β-tubulin synthesis. This topic of organelle movement provides one of the major challenges of modern mycology and is considered in greater detail in Chapters 6 and 13.

5.5 CONCLUSIONS

The last decade has seen a major change in the means by which fungal organelles have been studied and consequently in the sophistication of our understanding of these cytoplasmic components. Until recent years our knowledge of the structure and function of fungal organelles depended to a very great extent on conventional light and electron microscopy and only limited biochemical analysis in most cases. The advent of the new techniques in microscopy, notably freeze-substitution preparation for electron microscopy and image-enhancement and the confocal systems in light microscopy have dramatically improved the accuracy of our knowledge of organelle structure and their appearance and behaviour in living hyphae. However, it is perhaps the increasingly sophisticated use of the techniques of the biochemist and cell biologist which have been even more important in bringing such great strides in our understanding of organelles in the last decade and are certain to continue to do so ever more rapidly in the years ahead. Subcellular fractionation and isolation of enriched and purified preparations of organelles from fungal mycelia have been employed to great effect in recent years to allow effective biochemical analysis of their structural and functional components. Perhaps most importantly this has permitted the preparation of *in vitro* systems in which intact organelles are capable of carrying out what appears to be their *in vivo* function, so that it can be dissected and analysed in detail. Analysis of chitosomes is the prime example of such an approach, but similar work is likely to be highly effective in producing information about other organellar systems such as the fungal Golgi equivalents. However, the ultimate aim must be to localize and study intact organelles functioning within the living mycelium and here use of a combination of all of the techniques will be essential. Isolation and biochemical analysis of organelles has also provided another important opportunity to further this aim, by making it possible to find suitable marker molecules for organelles in living mycelia, whether they be enzymic activities, or structural components against which antibody probes can be raised, or the ability to synthesize specific nucleic acid probes in the case of organelles with their own genomes.

REFERENCES

Alberts, B., Bray, D., Lewis, J. *et al.* (1989) *Molecular Biology of the Cell*, 2nd edn, Garland New York.

Aoki, S., Ito-Kuwa, S., Nakamura, Y. and Masuhara, T. (1989) Mitochondrial behaviour during the yeast-hypha transition of *Candida albicans*. *Microbios*, **60**, 79–86.

Aylmore, R.C. and Todd, N.K. (1984) Hyphal fusion in *Coriolus versicolor*, in *The Ecology and Physiology of the Fungal Mycelium*, (eds D.H. Jennings and A.D.M. Rayner), Cambridge University Press, Cambridge, pp. 103–25.

Aylmore, R.C., Wakley, G.E. and Todd, N.K. (1984) Septal sealing in the basidiomycete *Coriolus versicolor*. *Journal of General Microbiology*, **130**, 2975–82.

Bartnicki-Garcia, S., Ruiz-Herrera, J. and Bracker, C. (1979) Chitosomes and chitin synthesis, in *Fungal Walls and Hyphal Growth*, (eds J.H. Burnett and A.P.J. Trinci), Cambridge University Press, Cambridge, pp. 149–68.

Beakes, G.W. (1981) Ultrastructure of the phycomycete nucleus, in *The Fungal Nucleus*, (eds K. Gull and S.G. Oliver), Cambridge University Press, Cambridge, pp. 1–35.

Beakes, G.W. (1987) Oomycete phylogeny: ultrastructural perspectives, in *Evolutionary Biology of the Fungi*, (eds A.D.M. Rayner, C.M. Brasier and D. Moore), Cambridge University Press, Cambridge, pp. 405–21.

Beckett, A. (1981) The ultrastructure of septal pores and associated structures in the ascogenous hyphae and asci of *Sordaria humana*. *Protoplasma*, **107**, 127–47.

Beckett, A., Heath, I.B. and McLaughlin, D.J. (1974) *An Atlas of Fungal Ultrastructure*, Longman, London.

Blayney, P.G. and Marchant, R. (1977) Glycogen and protein inclusions in elongating stipes of *Coprinus cinereus*. *Journal of General Microbiology*, **98**, 467–76.

Bracker, C., Ruiz-Herrera, J. and Bartnicki-Garcia, S. (1976) Structure and transformation of chitin synthetase particles (chitosomes) during microfibrils synthesis *in vitro*. *Proceedings of the National Academy of Sciences, USA*, **73**, 4570–4.

Brenner, D.M. and Carroll, G.C. (1968) Fine-structure correlates of growth in hyphae of *Ascodesmis sphaerospora*. *Journal of Bacteriology*, **95**, 658–71.

Buller, A.H.R. (1933) *Researches in Fungi*, vol. V, Hafner, New York.

Burnett, J.H. (1976) *Fundamentals of Mycology*, 2nd edn, Edward Arnold, London.

Butt, T.M., Beckett, A. and Wilding, N. (1981) Protoplasts in the *in vivo* life cycle of *Erynia neoaphidis*. *Journal of General Microbiology*, **127**, 417–21.

Cantino, E.C. and Mills, G.L. (1979) The gamma particle in *Blastocladiella emersonii*: what is it?, in *Viruses and Plasmids in Fungi*, (ed. P. Lemke), Marcel Dekker, New York, pp. 441–84.

Carbonell, L.M. (1969) Ultrastructure of dimorphic transformation in *Paracoccidioides brasiliensis*. *Journal of Bacteriology*, **100**, 1076–2.

Carlile, M.J. (1980) From prokaryote to eukaryote: gains and losses, in *The Eukaryotic Microbial Cell*, (eds G.W. Gooday, D. Lloyd and A.P.J. Trinci), Cambridge University Press, Cambridge University Press, Cambridge, pp. 1–40.

Carson, D.B. and Cooney, J.J. (1990) Microbodies in fungi: a review. *Journal of Industrial Microbiology*, **6**, 1–18.

Cavalier-Smith, T. (1987a) The origin of fungi and pseudofungi, in *Evolutionary Biology of the Fungi*, (eds A.D.M. Rayner, C.M. Brasier and D. Moore), Cambridge University Press, Cambridge, pp. 339–53.

Cavalier-Smith, T. (1987b) The kingdom Chromista: origin and systematics, in *Progress in Phycological Research*, Vol. 4 (eds F.E. Round and D.J. Chapman), Biopress, Bristol, pp. 309–47.

Clutterbuck, A.J. (1969) Cell volume per nucleus in haploid and diploid strains of *Aspergillus nidulans*. *Journal of General Microbiology*, **55**, 291–9.

Cole, G.T. (1986) Models of cell differentiation in conidial fungi. *Microbiological Reviews*, **50**, 95–132.

Cole, G.T. and Samson, R.A. (1979) *Patterns of Development in Conidial Fungi*. Pitman, London.

Collinge, A.J. and Markham, P. (1982) Hyphal tip ultrastructure of *Aspergillus nidulans* and *Aspergillus giganteus* and possible implications of Woronin bodies close to the hyphal apex of the latter species. *Protoplasma*, **113**, 209–13.

Collinge, A.J. and Markham, P. (1985) Woronin bodies rapidly plug septal pores of severed *Penicillium chrysogenum* hyphae. *Experimental Mycology*, **9**, 80–5.

Collinge, A.J. and Markham, P. (1987) Nuclei plug septal pores in severed hyphae of *Sordaria brevicollis*. *FEMS Microbiology Letters*. **44**, 85–90.

Collinge, A.J. and Markham, P. (1992) Ultrastructure of hyphal tip bursting in *Penicillium chrysogenum*. *FEMS Microbiology Letters*, **91**, 49–54.

Collinge, A.J. and Trinci, A.P.J. (1974) Hyphal tips of wild type and spreading colonial mutants of *Neurospora crassa*. *Archiv für Mikrobiologie*, **99**, 353–68.

Ebersold, H.R., Cordier, J–L. and Luthy, P. (1981) Bacterial mesosomes: method dependent artifacts. *Archives of Microbiology*, **130**, 19–22.

Flores-Martinez, A., Lopez-Romero, E., Martinez, J.P. *et al.*, (1990) Protein composition of purified chitosomes of *Mucor rouxii*. *Experimental Mycology*, **14**, 160–8.

Fuller, M.S. (1966) Structure of the uniflagellate zoospores of aquatic phycomycetes, in *The Fungus Spore*, (ed M.F. Madelin), Butterworths, London, pp. 67–92.

Garrison, R.G., Lane, J.W. and Field, M.F. (1970) Ultrastructural changes during the yeastlike to mycelial-phase conversion of *Blastomyces dermatitidis* and *Histoplasma capsulatum*. *Journal of Bacteriology*, **101**, 628–35.

Girbardt, M. (1969) Die Ultrastruktur der Apikalregion von Pilzhyphen. *Protoplasma*, **67**, 413–41.

Gooday, G.W. and Trinci, A.P.J. (1980) Wall structure and biosynthesis in fungi, in *The Eukaryotic Microbial Cell*, (eds. G.W. Gooday, D. Lloyd and A.P.J. Trinci), Cambridge University Press, Cambridge, pp. 207–51.

Gow, N.A.R. and Gooday, G.W. (1982) Vacuolation, branch production and linear growth of germ tubes of *Candida albicans*. *Journal of General Microbiology*, **128**, 2195–8.

Gow, N.A.R. and Gooday, G.W. (1984) A model for the germ tube formation and mycelial growth form of *Candida albicans*. *Sabouraudia*, **22**, 137–43.

Gray, S.N., Wilding, N. and Markham, P. (1991) In vitro germination of single conidia of the aphid-pathogenic fungus *Erynia neoaphidis* and phenotypic variation among sibling strains. *FEMS Microbiology Letters*, **79**, 273–8.

Grove, S.N. (1978) The cytology of hyphal tip growth, in *The Filamentous Fungi* vol. III, *Developmental Mycology*, (eds J.E. Smith and D.R. Berry), Edward Arnold, London, pp. 28–50.

Gull, K. (1978) Form and function of septa in filamentous fungi, in *The Filamentous Fungi*, vol. III, *Developmental Mycology*, (eds J.E. Smith and D.R. Berry), Edward Arnold, London, pp. 78–93.

Gunderson, J.H., Elwood, H., Ingold, A. *et al.*, (1987) Phylogenetic relationships between chlorophytes, chrysophytes, and oomycetes. *Proceedings of the National Academy of Sciences, USA*, **84**, 5823–7.

Head, J.B. (1987) The isolation and characterization of hexagonal crystals of *Neurospora crassa* and Woronin bodies of *Penicillium chrysogenum*, PhD thesis, University of London.

Head, J.B., Markham, P. and Poole, R.K. (1989) Woronin bodies from *Penicillium chrysogenum*: isolation and characterization by analytical subcellular fractionation. *Experimental Mycology*, **13**, 203–11.

Hernandez, J., Lopez-Romero, E., Cerbon, J. and Ruiz-Herrera, J (1981) Lipid analysis of chitosomes, chitin-synthesizing microvesicles from *Mucor rouxii*. *Experimental Mycology*, **5**, 349–56.

Hoch, H.C. and Maxwell, D.P. (1974) Proteinaceous hexagonal inclusions in hyphae of *Whetzelinia sclerotiorum* and *Neurospora crassa*. *Canadian Journal of Microbiology*, **20**, 1029–35.

Hoch, H.C., Tucker, B.E. and Staples, R.C. (1987) An intact microtubule cytoskeleton is necessary for mediation of the signal for cell differentiation in *Uromyces*. *European Journal of Cell Biology*, **45**, 209–18.

Hohn, T.M., Lovett, J.S. and Bracker, C.E. (1984) Characterization of the major proteins in gamma particles, cytoplasmic organelles in *Blastocladiella emersonii* zoospores. *Journal of Bacteriology*, **158**, 253–63.

Howard, R.J. (1981) Ultrastructural analysis of hyphal tip cell growth in fungi: Spitzenkorper, cytoskeleton and endomembranes after freeze-substitution. *Journal of Cell Science*, **48**, 89–103.

Howard, R.J. and Aist, J.R. (1977) Effects of MBC on hyphal tip organization, growth, and mitosis of *Fusarium acuminatum*, and their antagonism by D_2O. *Protoplasma*, **92**, 195–210.

Howard, R.J. and Aist, J.R. (1979) Hyphal tip cell ultrastructure of the fungus *Fusarium*: improved preservation by freeze-substitution. *Journal of Ultrastructure Research*, **66**, 224–34.

Howard, R.J. and Aist, J.R. (1980) Cytoplasmic microtubules and fungal morphogenesis: ultrastructural effects of methyl benzimidazole-2-ylcarbamate determined by freeze-substitution of hyphal tips. *Journal of Cell Biology*, **87**, 55–64.

Hunsley, D. and Gooday, G.W. (1974) The structure and development of septa in *Neurospora crassa*. *Protoplasma*, **82**, 125–46.

Jackson, C.W. and Heale, J.B. (1985) Relationship between DNA content and spore volume in sixteen isolates of *Verticillium lecannii* and two new diploids of *V. dahliae* (= *V. dahliae* var *longisporum* Stark). *Journal of General Microbiology*, **131**, 3229–36.

Kanaseki, T., and Kadota, K. (1969) The 'vesicle in a basket'. A morphological study of the coated vesicle isolated from the nerve endings of the guinea pig brain, with special reference to the mechanism of movements. *Journal of Cell Biology*, **42**, 202–20.

Keller, G-A., Krisans, S., Gould, S.J. *et al.* (1991) Evolutionary conservation of a microbody targeting signal that targets proteins to peroxisomes, glyoxysomes, and glycosomes. *Journal of Cell Biology*, **114**, 893–904.

King, S.B. and Alexander, L.J. (1969) Nuclear behavior, septation and hyphal growth of *Alternaria solani*. *Journal of General Microbiology*, **67**, 125–33.

Klionsky, D.J., Herman, P.J. and Emr, S.D. (1990) The fungal vacuole: composition, function, and biogenesis. *Microbiological Reviews*, **54**, 266–92.

Lane, J.W. and Garrison, R.G. (1970) Electron microscopy of the yeast to mycelial phase conversion of *Sporotrichum schenckii*. *Canadian Journal of Microbiology*, **16**, 747–9.

Lindmark, D.G. and Müller, M. (1973) Hydrogenosome, a cytoplasmic organelle of the anaerobic flagellate, *Tritrichomonas foetus*, and its role in pyruvate metabolism. *Journal of Biological Chemistry*, **248**, 7724–8.

Lloyd, D. and Turner, G. (1980) Structure, function, biogenesis and genetics of mitochondria, in *The Eukaryotic Microbial Cell*, (eds, G.W. Gooday, D. Lloyd

and A.P.J. Trinci), Cambridge University Press, Cambridge, pp. 143–80.

Markham, P. and Collinge, A.J. (1987) Woronin bodies of filamentous fungi. *FEMS Microbiology Reviews*, **46**, 1–11.

Markham, P. Collinge, A.J., Head, J.B. and Poole, R.K. (1987) Is the spatial organization of fungal hyphae maintained and regulated by Woronin bodies? in *Spatial Organization in Eukaryotic Microbes* (eds R.K. Poole and A.P.J. Trinci), IRL Press, Oxford, pp. 79–99.

Mason, P.J. and Crosse, R. (1975) Crystalline inclusions in hyphae of the glaucus group of Aspergilli. *Transactions of the British Mycological Society*, **65**, 129–34.

Maxwell, D.P., Armentrout, V.N. and Graves, L.B. (1977) Microbodies in plant pathogenic fungi. *Annual Review of Phytopathology*, **15**, 119–34.

McClure, W.K., Park, D. and Robinson, P.M. (1968) Apical organization in the somatic hyphae of fungi. *Journal of General Microbiology*, **50**, 177–82.

McKeen, W.E. (1971) Woronin bodies in *Erysiphe graminis* DC. *Canadian Journal of Microbiology*, **17**, 1557–60.

McKerracher, L.J. and Heath, I.B. (1987) Cytoplasmic migration and intracellular organelle movements during tip growth of fungal hyphae. *Experimental Mycology*, **11**, 79–100.

Moore, R.T. (1985) The challenge of the dolipore/parenthesome septum, in *Developmental Biology of Higher Fungi*, (eds D. Moore, L.A. Casselton, D.A. Wood and J.C. Frankland), Cambridge University Press, Cambridge, pp. 175–212.

Moore, R.T. and McAlear, J.M. (1961) Fine structure of mycota, 5. Lomasomes – previously uncharacterised hyphal structures. *Mycologia*, **53**, 194–200.

Müller, M. (1980) The hydrogenosome, in *The Eukaryotic Microbial Cell*, (eds G.W. Gooday, D. Lloyd and A.P.J. Trinci), Cambridge University Press, Cambridge, pp. 127–42.

Nguyen, T.T. and Niederpruem, D.J. (1984) Hyphal interactions in *Schizophyllum commune*: the di-mon mating, in *The Ecology and Physiology of the Fungal Mycelium*, (eds D.H. Jennings and A.D.M. Rayner), Cambridge University Press, Cambridge, pp. 73–102.

Oakley, B.R. and Rinehart, J.E. (1985) Mitochondria and nuclei move by different mechanisms in *Aspergillus nidulans*. *Journal of Cell Biology*, **101**, 2392–7.

Oujezdsky, K.B., Grove, S.N. and Szaniszlo, P.J. (1973) Morphological and structural changes during the yeast-to-mold conversion of *Phialophora dermatitidis*. *Journal of Bacteriology*, **113**, 468–77.

Pearse, B.M.F. (1976) Clathrin: a unique protein associated with intracellular transfer of membrane by coated vesicles. *Proceedings of the National Academy of Sciences, USA*, **73**, 1255–9.

Robertson, N.F. (1965) The fungal hypha. *Transactions of the British Mycological Society*, **48**, 1–8.

Ross, I.K. (1976) Nuclear migration rates in *Coprinus congregatus*: a new record. *Mycologia*, **68**, 418–22.

Sewall, T.C., Roberson, R.W. and Pommerville, J.C. (1989) Identification and characterization of Golgi equivalents from *Allomyces macrogynus*. *Experimental Mycology*, **13**, 239–52.

That, T.C.C-T., Hoang-Van, K, Turian, G. and Hoch, H.C. (1987) Isolation and characterization of coated vesicles from filamentous fungi. *European Journal of Cell Biology*, **43**, 189–94.

Theodorou, M.K., Lowe, S.E. and Trinci, A.P.J. (1992) Anaerobic fungi and the rumen ecosystem, in *The Fungal Community. Its Organization and Role in the Ecosystem*, 2nd edn, (eds G.C. Carroll and D.T. Wicklow), Marcel Dekker, New York, pp. 43–72.

Todd, N.K. and Aylmore, R.C. (1985) Cytology of hyphal interactions and reactions in *Schizophyllum commune*, in *Developmental Biology of Higher Fungi*, (eds D. Moore, L.A. Casselton, D.A. Wood and J.C. Frankland), Cambridge University Press, Cambridge, pp. 231–48.

Trinci, A.P.J. (1979) The duplication cycle and branching in fungi, in *Fungal Walls and Hyphal Growth*, (eds J.H.

Burnett and A.P.J. Trinci), Cambridge University Press, Cambridge, pp. 319–58.

Trinci, A.P.J. and Collinge, A.J. (1973) Structure and plugging of septa in wild-type and spreading colonial mutants of *Neurospora crassa*. *Archiv für Mikrobiologie*, **91**, 355–64.

Trinci, A.P.J. and Collinge, A.J. (1974) Occlusion of the septal pores of damaged hyphae of *Neurospora crassa* by hexagonal crystals. *Protoplasma* **80**, 57–67.

Typas, M.A. and Heale, J.B. (1977) Analysis of ploidy levels in strains of *Verticillium* using a Coulter counter. *Journal of General Microbiology*, **101**, 177–80.

Webster, J. (1980) *Introduction to Fungi*, 2nd edn, Cambridge University Press, Cambridge.

Wood, D.A., Craig, G.D., Atkey, P.T. *et al.* (1985) Ultrastructural studies on the cultivation processes and growth and development of the cultivated mushroom *Agaricus bisporus*. *Food Microstructure*, **4**, 143–64.

Yarlett, N., Orpin, C.G., Munn, E.A. *et al.* (1986) Hydrogenosomes in the rumen fungus *Neocallimastix patriciarum*. *Biochemical Journal*, **236**, 729–39.

THE CYTOSKELETON

6

I.B. Heath

Department of Biology, York University, North York, Ontario, Canada

6.1. INTRODUCTION

In this chapter we shall investigate current concepts of the structure and functions of the fungal cytoskeleton, drawing on the available information from both filamentous and yeast species, there being no reason to believe that there are fundamental differences in cellular architecture between these two growth forms. The zoosporic fungi will be included, as will the Oomycetes, which, although more closely related to some algae, show the hyphal growth form and are commonly discussed in mycology courses.

Readers unfamiliar with basic cell structure and function may find it useful to consult a modern cell biology text for background information. In addition, the book on cell movements by Bray (1992) gives a good general account of the cytoskeleton and its functions and should be consulted for relevant background.

In fungi, the term cytoskeleton is used to cover both the skeletal and muscular functions implied in the analogy with mammalian skeletons. It is a skeletal system permeating the cytoplasm in that it contains stress-bearing, rather rigid, elements which can both order and shape cellular components and provide a base against which force can be applied. However, this skeletal role differs from that of animals in that the fungal cytoskeleton can be highly dynamic, assembling and dissembling in response to changing cellular needs. Fungi also differ in receiving support from both an exoskeleton (the cell wall) and turgor pressure. Interestingly, the cytoskeleton probably interacts directly with the cell wall, and stress-activated membrane channels pass ions which directly influence the cytoskeleton and modulate osmotically significant ions (Garrill *et al.*, 1992, 1993). Thus there is likely direct integration between all skeletal components of fungi. The muscular role of the cytoskeleton derives from the diverse molecules which interact with the structural parts of the system to generate force and movement.

The fungal cytoskeleton is composed of two well-known components, microtubules and actin filaments (polymers of the proteins tubulin and actin, section 6.3), and an ever-increasing array of molecules which interact with these elements to determine their sizes, organization and interactions with each other and other cellular components. While discussion of these components will form the basis for most of this chapter, there is increasing evidence that the cytoskeleton includes other structures which will also be introduced.

6.2 TECHNIQUES FOR ANALYSIS

Although a detailed critique of the analytical techniques used to study the cytoskeleton is beyond the scope of this chapter, it is important to introduce the diversity of approaches which have been employed to better appreciate both their limitations and the multidisciplinary nature of the subject.

6.2.1 Light microscopy

Direct viewing of cells with brightfield, phase contrast or Nomarski differential interference contrast (DIC) systems generally fails to reveal most parts of the fungal cytoskeleton, except for the larger assemblies such as mitotic spindles. The cytoskeleton was essentially unknown prior to electron microscopy. However, cytoskeleton dependent processes were well known and phase contrast and DIC, in combination with inhibitors (section 6.2.4), laser and ultraviolet microbeam surgery (e.g. Aist and Berns, 1981; McKerracher and Heath, 1986a, c, 1987; Aist *et al.*, 1991; Jackson

The Growing Fungus. Edited by Neil A.R. Gow and Geoffrey M. Gadd. Published in 1994 by Chapman & Hall, London. ISBN 0 412 46600 7

and Heath, 1993a) and optical tweezers (Berns *et al.*, 1992) are the prime techniques for their study in living cells. The techniques are even more useful when, after viewing the cell alive, it is then prepared for electron microscopy, a process pioneered by Girbardt (1965) and since used to good effect by Aist and Berns (1981), McKerracher and Heath (1986a), Bourett and McLaughlin (1986), Kaminskyj *et al.* (1989) and others. The introduction of electronic contrast enhancement has extended the resolution and detectability of these techniques *in vivo* even further (Allen, 1985). The observation of organelle motility in association with cytoskeletal elements *in vitro* is now also routine (Vale *et al.*, 1985). The confocal microscope, with its very thin depth of focus, is of less value with thin cells such as fungal hyphae than thicker tissues, but has been used to good effect in some situations (e.g. Kwon *et al.*, 1991) and can expect further use,

especially in multihyphal tissues such as basidiocarps etc.

In addition to direct viewing of living cells, light microscopy of fluorescently labelled probes (such as antibodies for diverse molecules (e.g. Figures 6.15, 6.35–6.37) phalloidin for actin filaments (e.g. Figures 6.6, 6.7, 6.24–6.34, 6.39–6.43) and ion selective dyes for diverse ions) with their high specificity and detectability have also been invaluable in the study of fungal cytoskeletons. These probes are most commonly used on fixed cells, a procedure which must be used with care because fixation causes artefactual translocations (Kaminskyj *et al.*, 1992). However, they can also be used in living cells following their introduction via electroporation (Figure 6.24 and Jackson and Heath, 1990a), microinjection (Read *et al.*, 1992) or other more specific procedures (Garrill *et al.*, 1993). When using probes in living cells, it is essential to

---

Figure 6.1 Interphase SPB of the ascomycete, *Schizosaccharomyces octosporus*. The dark structure lying in the depression in the nuclear envelope is the main SPB and the arrows point to the adjacent material in the nucleoplasm which may correspond to an intranuclear SPB responsible for the formation of the spindle microtubules. ×85 000. (From Heath, 1978.)

Figure 6.2 Metaphase SPB of *Schizosaccharomyces octosporus*. The nuclear envelope is open at the arrows so that the extranuclear SPB is now in contact with the spindle microtubules. ×130 000. (From Heath, 1978.)

Figure 6.3 Median longitudinal section of an interphase SPB of the basidiomycete, *Boletus rubinellus*, showing the characteristic globular elements interconnected by a bar, all appressed to the intact nuclear envelope. ×100 000. (Reproduced from McLaughlin (1971), with permission of the Rockefeller University Press.)

Figure 6.4 Centriole (c) lying in a depression of the intact nuclear envelope of an interphase nucleus of the oomycete, *Saprolegnia ferax*. In this organism the kinetochore microtubules persist throughout the nuclear cycle and are visible in the nucleoplasm (e.g. arrows). ×114 000.

Figure 6.5 Interphase SPB of the rust fungus, *Uromyces phaseoli*, showing the characteristic interconnected double discs adjacent to the intact nuclear envelope and the bundle of filamentous material in the adjoining nucleoplasm (arrows). ×38 600. (From Hoch and Staples, 1983b.)

Figures 6.6 and 6.7 Interphase nuclei (which occur in pairs in the germ tubes at this stage of development) of *Uromyces phaseoli* stained with rhodamine-labelled phalloidin to show that the intranuclear filamentous material seen in Figure 6.5 stains strongly and is thus likely to be actin. Note the characteristic unstained centre in the stained structure seen in face view in Figure 6.7 (arrow). ×1400 (6.6) and 2300 (6.7). (From Hoch and Staples, 1983b.)

Figure 6.8 Early mitotic spindle of the zygomycete, *Zygorhynchus molleri*, showing the extranuclear SPBs (large arrows), non-kinetochore microtubules extending between the spindle poles and the short kinetochore microtubules which are already at the poles of the spindle (small arrows). ×128 300. (From Heath and Rethoret, 1982.)

Figure 6.9 Very early stage in mitotic spindle formation in *Saprolegnia ferax*, showing the separating centrioles (c), associated cytoplasmic microtubules (m) and kinetochore (k) and non-kinetochore (nk) spindle microtubules. The intact nuclear envelope (ne) is poorly contrasted in this freeze-substituted preparation. ×103 700. (From Heath *et al.*, 1984.)

Figure 6.10 Telophase mitotic spindle of the chytrid, *Catenaria anguillulae*, showing the separated chromosomes in the incipient daughter nuclei (ch), and the remains of the spindle (small arrows) in the interzone region, which is being constricted from the daughter nuclei and will eventually be discarded into the cytoplasm. ×27 200. (Reprinted from Ichida and Fuller (1968) by permission of the New York Botanical Garden.)

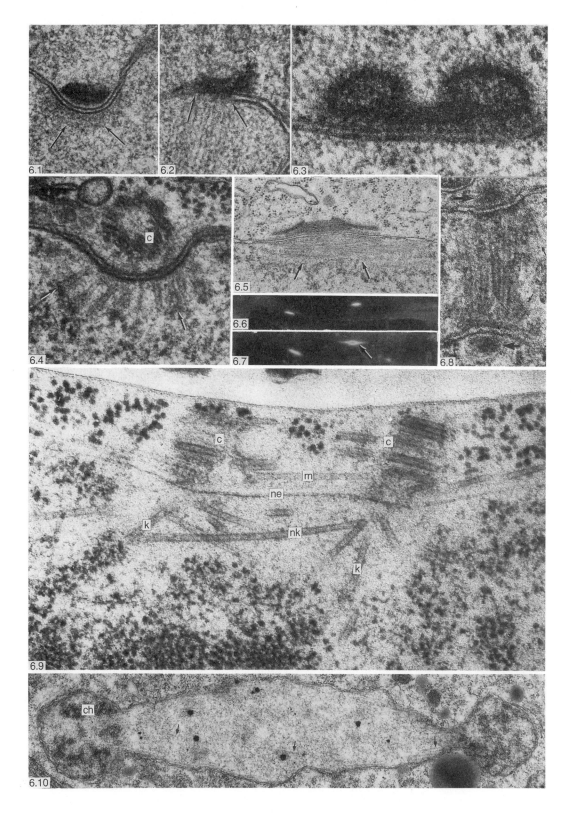

6.1

6.2

6.3

6.4 c

6.5

6.6

6.7

6.8

6.9 c c m ne k nk k

6.10 ch

demonstrate normal behaviour because they may disrupt cellular control and metabolism.

6.2.2   Electron microscopy

The routine availability of thin section-based transmission electron microscopy (TEM) was instrumental in the discovery and analysis of the cytoskeleton. Its main components were unknown prior to TEM and the technique continues to play a pivotal role in explaining how and what the system does. Serial section-based reconstructions, though laborious, provide high resolution 3-dimensional information (e.g. Heath *et al.*, 1984; Kanbe *et al.*, 1989) and immunocytochemistry (e.g. Bourett and Howard, 1992) and X-ray spectroscopy add compositional information. Because TEM primarily relies on the use of fixed material, fixation artefacts are always of concern but the introduction of freeze substitution (FS) techniques to fungal cell biology (Howard and Aist, 1979; Howard and O'Donnell, 1987) provided a dramatic improvement. However, it is clear that FS does not yet reveal all cytoskeletal components (Heath and Kaminskyj, 1989) and, conversely, conventional fixations can be as good as FS in at least some aspects of cytoskeleton preservation (Heath *et al.*, 1984).

TEM is not limited to sectioned material. Novel, and possibly more complete, views of the cytoskeleton have been obtained from extracted whole mounts (McKerracher and Heath, 1986b, 1987; Wittenburg *et al.*, 1987), but the technique has not been widely explored in fungi. Scanning electron microscopy has played no significant role in the analysis of the fungal cytoskeleton and this is unlikely to change in the near future.

6.2.3   Genetics

Some fungi have such advantages for both classical and molecular genetics that they have become major tools for cytoskeletal study. Altered sensitivity to microtubule inhibitors proved a useful screen for tubulin mutations (Morris, 1986) and restoration of function to tubulin and actin mutants has helped identify new tubulins (Oakley and Oakley, 1989), microtubule-associated proteins (MAPs) and actin-binding proteins (ABPs) (Huffaker *et al.*, 1987). To some extent the analysis of actin and tubulin mutations has tended to confirm what was already clear from earlier light and electron microscopy, but the work on new tubulins, MAPs and ABPs has shown the innovative power of the techniques. However, the genetic approaches are placing new demands on microscopical techniques for phenotypic analysis and they too require considerable care in interpretation in order to differentiate between direct, primary, versus indirect, secondary effects of the manipulations.

6.2.4   Inhibitors

Given the universal presence of the cytoskeleton in eukaryotic organisms, it is not surprising that evolution has produced natural products which selectively disrupt parts of the cytoskeleton of other organisms that they may attack, or defend themselves against. These products can then be used to investigate the functions of the cytoskeleton. They include colchicine, vinblastine, vincristine, podophyllin and maytansine which bind to tubulin and disrupt microtubules (section 6.3.1), griseofulvin which seems to disrupt microtubules by binding to a specific MAP (Roobol *et al.*, 1977), cytochalasins which disrupt actin filaments, in part, by binding to the actin monomers (Cooper, 1987), the phallotoxins, such as phalloidin, which disrupt actin functions by hyperstabilizing actin filaments (Cooper, 1987) and taxol which behaves similarly for microtubules (Dustin, 1984). In addition to these natural products, some empirically discovered pesticides have similar activities. For example, benomyl (or, more accurately, its benzimidazole ingredient most commonly known as MBC) owes its fungicidal activity to its colchicine-like effect, as do the herbicide isopropylphenylcarbamate (IPC) and the benzimidazole helminthicides such as mebendazole and the experimental anticancer drug, nocodazole (Dustin, 1984). All of these compounds differ in their potency against different species which allows them to be used as pesticides. These differences are due, at least in part, to minor variations in the target proteins (Jung *et al.*, 1992), so that care must be exercised in the choice of compound used. Furthermore, the specificity of all of them should be considered uncertain, even at the micromolar concentrations commonly used. For example, MBC and others can induce Ca^{2+} efflux from mitochondria, thereby increasing cytoplasmic free Ca^{2+} levels with unknown effects (Hertel *et al.*, 1980).

In addition to the chemical cytoskeletal inhibitors, a number of physical agents such as low temperature and high pressure can be useful (Dustin, 1984) and D_2O has been effective in hyperstabilizing fungal microtubules (Howard and Aist, 1977). When using any inhibitor, it is most desirable to demonstrate the occurrence of the expected effect, even if it is not possible to show the absence of other effects. Although the use of inhibitors is not without problems, they have provided much useful information.

6.2.5 Biochemistry

Routine biochemical techniques, such as protein separation and sequencing, activity assays, inhibitor binding assays, polymerization condition assays, organelle isolation etc., have played a similar role in cytoskeletal studies to that in other areas but three techniques deserve special mention. Actin and tubulin affinity columns have been developed and shown to be effective for isolating ABPs and MAPs, (whose cellular relevance emerged only later) (Drubin *et al.*, 1988; Liu and Bretscher, 1989a; Barnes *et al.*, 1992). Another productive approach has been the isolation of semi-intact nuclei, either from interphase, thus producing nuclear skeletons which are part of the cellular skeletal system (Cardenas *et al.*, 1990), or from mitosis, thus producing preparations enriched in MAPs (Pillus and Solomon, 1986). The third approach has been the isolation of mitotic spindles which can be induced to undergo somewhat normal elongation (section 6.5.1) *in vitro* (Masuda *et al.*, 1990). All of these approaches have opened new avenues to investigate the cytoskeleton.

6.3   COMPOSITION OF THE CYTOSKELETON

The cytoskeleton is minimally composed of two major polymers, microtubules and actin filaments, an ever-increasing list of MAPs and ABPs and a number of less well known elements. This ensemble currently totals some 60–70 proteins, but these probably represent only the tip of the iceberg. The better studied *Chlamydomonas* flagellar axoneme alone is estimated to contain over 200 proteins (Bray, 1992). However, even the currently known ensemble provides a useful insight into the way the system works.

6.3.1   Microtubules

Microtubules are the best known part of the cytoskeleton because they are stable to electron microscopy techniques and easily identified, both morphologically and immunocytochemically. There is also a very useful background of biochemical information derived from studies of the microtubule-rich nerve cells, thus facilitating their characterization in fungi.

Microtubules are composed of 13 protofilaments laterally connected into a 20–24 nm diameter rather inflexible cylinder. Each protofilament is a heteropolymer of ~110 kDa tubulin dimers, each containing closely related α and β peptides which are the products of separate genes. The extensive detailed information available for the organization of the dimers in the microtubule wall in non-fungal cells (Bray, 1992) is probably also applicable to fungal microtubules, but is beyond the scope of this review and is not essential to our current level of understanding of microtubule functions. However, the asymmetry of the dimers and their consistent orientation in the polymers (microtubules) undoubtedly contribute to the observed polarization of microtubules which have functionally important '+' and '−' ends (section 6.5). Microtubules also contain γ-tubulin which was first discovered in a fungus (Oakley and Oakley, 1989). γ-Tubulin is encoded by a separate gene and differs substantially from α and β tubulins. It does not appear to copolymerize with the α–β dimers, it only associates with the ends of microtubules at their organizing centres (section 6.4.3). α-Tubulins undergo reversible post-translational tyrosylation on their carboxy terminal ends and acetylation of a specific lysine (Bray, 1992). The proportions of these forms of α-tubulin vary in microtubules of differing stability in different regions of cells, thus indicating post-translational control of the microtubule arrays (section 6.4.2).

In the cell, microtubules occur both singly and in bundles (Figures 6.15–6.22). The bundles are held together by specific MAPs and undoubtedly are structures of greater mechanical strength than single microtubules. In fungi, the vast majority of microtubules are oriented parallel to the long axis of both hyphae (Figure 6.15) and budding cells.

The length of microtubules in cells is very variable, ranging from about 0.1 μm in the cytoplasm of some hyphae to 10–20 μm in flagellar axonemes, zoospore roots (Figures 6.35, 6.38 and 6.39) and other hyphae. The length of cytoplasmic microtubules is often not known with accuracy because they are seldom traced in serial sections in electron microscopy and immunocytochemical preparations cannot usually resolve the difference between long microtubules and bundled (possibly artefactually) short ones. However, the variability does not appear to be as great as found in actin where functionally important oligomers of only a few subunits probably exist (Pollard and Cooper, 1986).

6.3.2 Microtubule-associated proteins (MAPs)

The definition of a MAP is rather loose since although any molecule which shows a functional relationship with a microtubule is essentially included, determination of its function is not easy. In practice, *in vitro* interactions (e.g. binding to a tubulin affinity column), *in situ* co-localizations and correction of tubulin gene defects are all defining characteristics. MAPs are collectively vital to the cell because they influence initiation of polymerization, degree of polymerization (i.e. length), stability, spatial organization and both static and dynamic interactions with other cellular constituents. They are, therefore, at least partially responsible for both the organization and functions of microtubules in the cell.

Organizational MAPs include a wide range of proteins varying from about 30 to 250 kDa. None have been studied in fungi but there is no reason to believe that fungi lack MAPs homologous to those known in detail from animal cells and reviewed by Olmsted (1986). The MIP (Microtubule Interacting Protein) gene products and the SPB (section 6.4.3) and kinetochore (section 6.5.1) associated proteins listed in Table 6.1 are the best candidates for fungal organizational MAPs. The *Aspergillus* γ-tubulin, which has an organizational role, was originally isolated as a MIP gene (Weil *et al.*, 1986; Oakley and Oakley, 1989).

The best known functional MAPs are the mechanochemical translocators such as dyneins, kinesins and dynamin (Collins, 1991; Goldstein, 1991; Vallee, 1991). These proteins translocate vesicles and other membranes relative to microtubules, and microtubules relative to each other or other cytoskeletal elements, usually showing a preference for '+' or '−' end-directed movement. However, the translocatory role for dynamin is still speculative (Collins, 1991) and there is, at present, no evidence for dynamin in fungi. In contrast, fungi do contain both kinesins and dynein (Table 6.1). The flagellar axonemes of fungal zoospores appear to be identical to those of other eukaryotes, so that it is almost certain that they contain dynein, the universal eukaryotic flagellar motor (Bray, 1992) and recently a gene encoding a dynein-like protein has been identified in the non-flagellate *Saccharomyces* (Eshel *et al.*, 1992). Furthermore, dynactin is a form of actin which is distantly related to other actins and is so named because it activates cytoplasmic dyneins in

Figure 6.11 Telophase nucleus of *Saprolegnia ferax*, showing the persistent nucleolus (nu), centrioles at the spindle pole (c, the centrioles at the other pole are out of the plane of section) and, contrary to the pattern in Figure 6.10, the way in which the nucleus is undergoing median constriction such that all of the nucleoplasm will be incorporated into the daughter nuclei. ×20 966. (From Heath and Greenwood, 1968.)

Figure 6.12 Metaphase nucleus of the rust fungus, *Uromyces vignae*, showing the central bundle of non-kinetochore microtubules, peripheral chromosomes (ch) attached to their kinetochore microtubules (arrows), persistent nucleolus (nu) and plaque-like SPBs (s) set into the persistent nuclear envelope. ×49 700. (From Heath and Heath, 1976.)

Figure 6.13 Anaphase nucleus of *Catenaria anguillulae*, showing the chromosomes (ch) approaching the spindle poles while attached to their kinetochore microtubules (arrows), the bundled non-kinetochore microtubules (brackets) and one of the polar centrioles (c) adjacent to the persistent nuclear envelope. ×32 650. (Reprinted from Ichida and Fuller (1968) by permission of the New York Botanical Garden.)

Figure 6.14 One pole of an anaphase nucleus of the ascomycete, *Nectria haematococca*, showing the abundant astral microtubules (arrows) emanating from the SPB (s) which lies at the pole of the intranuclear spindle (sp). Inset shows one of the filaments which associate with the astral microtubules (between small arrow heads). ×27 500 and inset ×50 000. (Reproduced from Aist and Bayles (1991a), with permission.)

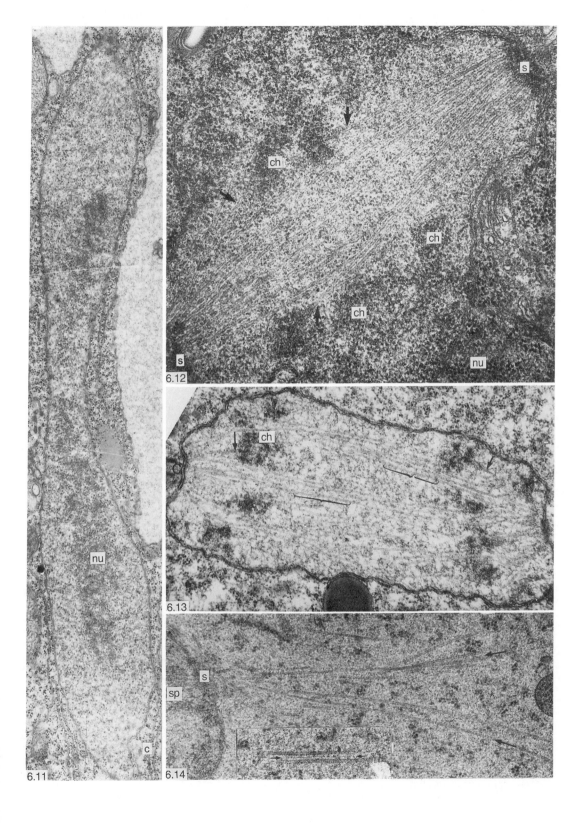

6.11

6.12

6.13

6.14

nerve cells (Gill *et al.*, 1991). The *ACT2* genes of both *Saccharomyces* (Schwob and Martin, 1992) and *Schizosaccharomyces* (Lees-Miller *et al.*, 1992) encode proteins said to be dynactin (Goldstein and Vale, 1992).

6.3.3 Actin

Like tubulins, actin is also a protein capable of polymerization. However, the ~42 kDa actin monomers are typically identical, thus forming a homopolymer. Unlike tubulin, there is only one post-translational modification known and it occurs to all monomers prior to polymerization. Two linear chains of actin monomers helically entwine to form a ~7 nm diameter, rather flexuous filament known as a microfilament or F-actin. These filaments are polarized, showing pointed and barbed ends when decorated with subfragments

Figure 6.15   A germ tube of *Uromyces phaseoli* extending from the spore (s) and stained for microtubules. Microtubules extend through the germ tube, predominantly parallel to its long axis, and reach close to the apex. ×1000. Dr H.C. Hoch, unpublished.

Figure 6.16   Interaction between a transversely sectioned microtubule and the outer membrane of a mitochondrion (m) in a hypha of *Saprolegnia ferax*, showing the characteristic spacing indicative of a functionally significant interaction. ×136 000.

Figure 6.17   Reconstructions of the microtubules associated with two *Pleurotus ostreatus* nuclei which were migrating in the direction of the arrows at the time of fixation. They extend from the SPBs (small outlines on the surface of the nuclei), predominantly parallel to the direction of migration, but in approximately equal numbers both ahead of and behind that direction. (From Kaminskyj *et al.* 1989.)

Figure 6.18   Close and even lateral spacing between a cytoplasmic microtubule and the nuclear envelope (ne) in a hypha of *Saprolegnia ferax*. Such images are common and indicate the existence of some mechanochemical linkage between the two structures. ×113 300.

Figure 6.19   Transverse section of a hypha of *Uromcyes vignae*, showing clustered microtubules, some of which (arrows) are associated with the surfaces of mitochondria (m) and transversely sectioned filaments (e.g. encircled). ×89 000. (From Heath and Heath, 1978.)

Figure 6.20   Typical interaction between a cytoplasmic microtubule and a mitochondrion (m) in a hypha of *Saprolegnia ferax*. ×68 000.

Figures 6.21 and 6.22   Cytoplasmic microtubules associated with the nuclear envelopes of nuclei (n) which were likely to have been migrating at the time of fixation in hyphae of *Uromyces vignae*. Note the fine filaments laterally associated with the microtubules (arrows). ×100 000. (From Heath and Heath, 1978).

Figure 6.23.   A hypha of *Basidiobolus magnus*, showing cytoplasmic contraction induced by an ultraviolet microbeam irradiation at the site indicated by the bar in (a). The degree of contraction is seen by comparing (a) with (h), the former being prior to irradiation and the latter 2.2 min after the end of the irradiation. (b)–(g) show the movement of cytoplasm following the irradiation at 15 s intervals. (a) and (h) ×210, (b)–(g) ×830. (From McKerracher and Heath, 1986c.)

Figure 6.24   A living and growing hypha of *Saprolegnia ferax* stained with rhodamine labelled phalloidin to reveal the population of actin which permeates the cytoplasm of this species. ×1200. (From Jackson and Heath, 1990a.)

Figure 6.25   DIC (a) and corresponding fluorescent image (b) of a hypha of *Saprolegnia ferax* in which the cytoplasm contracted during fixation. The contracted cytoplasm is rich in actin, which is stained with rhodamine-labelled phalloidin in (b). ×1124. (From Heath, 1990.)

Figures 6.26 to 6.28   Developing branches of *Saprolegnia ferax*, stained with rhodamine-labelled phalloidin to show the net of actin filaments which enclose the hyphal tips (but located in the cytoplasm, adjacent to the plasma membrane) and radiate back into the parent hyphae (to the left in each case). ×2000. (From Heath, 1987.)

Figures 6.29 to 6.33   Cells of *Neozygites* sp., stained for actin with rhodamine-labelled phalloidin at different stages in their cell cycle. The growing cell (Figure 6.29) shows higher concentrations of actin plaques at its tips. As the cells approach division, an equatorial band of actin filaments assembles around the periphery of the cell (Figures 6.30 and 6.31; in this species the cells divide at diverse lengths), and these filaments shorten to a tight disk prior to the initiation of septum formation (Figure 6.32). When the septum becomes visible in the DIC images (Figure 6.33b) and begins its centripetal growth, the actin is present as a concentration of plaques similar to those at the growing tips of the cells (Figure 6.33a). ×900. (From Butt and Heath, 1988.)

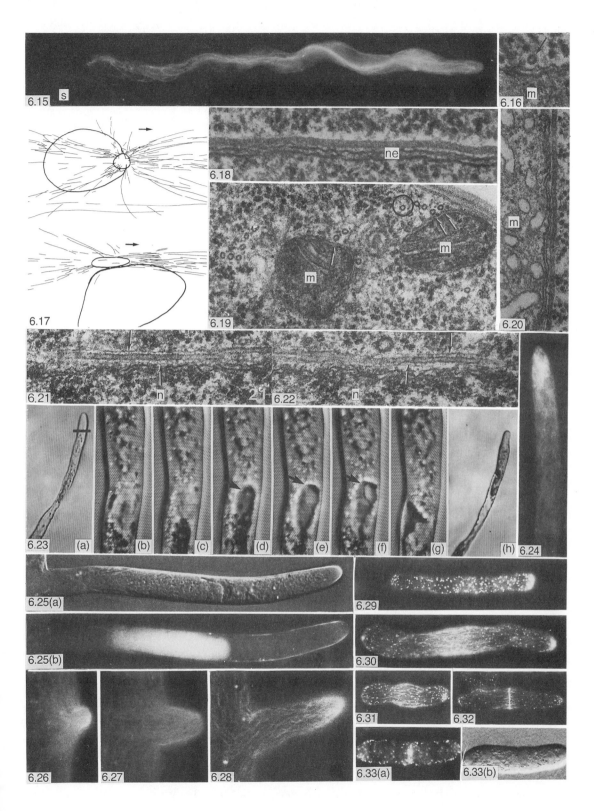

6.15 s

6.16 m

6.17

6.18 ne

6.19 m m

6.20 m

6.21 n 2 6.22 n

6.23 (a) (b) (c) (d) (e) (f) (g) (h) 6.24

6.25(a)

6.25(b)

6.26 6.27 6.28

6.29

6.30

6.31 6.32

6.33(a) 6.33(b)

Table 6.1 Fungal MAPs

Name of protein or gene	Likely function or characteristic of protein	Genus	Reference
KAR3	Kinesin related	Saccharomyces	Meluh and Rose (1990)
SMY1	Kinesin related	Saccharomyces	Lillie and Brown (1992)
KIP1=CIN9, KIP2	Kinesin related	Saccharomyces	Roof et al. (1992); Hoyt et al. (1992)
KSL2=CIN8	Kinesin related	Saccharomyces	Roof et al. (1992); Hoyt et al. (1992)
54 kDa, 66 kDa, 112 kDa, 190 kDa	Kinesin related	Saccharomyces	Roof et al. (1992); Hoyt et al. (1992)
CIN1, CIN2, CIN4	Mutations disrupt microtubules	Saccharomyces	Hoyt et al. (1990)
~25 diverse, mostly unidentified proteins	Specifically bind to tubulin affinity columns	Saccharomyces	Barnes et al. (1992)
~5 unidentified proteins	Copurify with spindle microtubules	Saccharomyces	Pillus and Solomon (1986)
80 kDa SPB associated protein	Associates with spindle microtubules	Saccharomyces	Rout and Kilmartin (1990)
64 kDa kinetochore associated protein	Has a MAP1B-like microtubule binding domain	Saccharomyces	Middleton et al. (1992)
BIK1	Colocalizes with, and mutations disrupt, microtubules, especially cytoplasmic	Saccharomyces	Berlin et al. (1990)
Dynein	Cytoplasmic dynein	Saccharomyces	Eshel et al. (1993)
ACT2 (dynactin?)	Dynein activation, associates with microtubules in other cells	Saccharomyces Schizosaccharomyces	Schwob and Martin (1992) Lees-Miller et al. (1992)
CUT7	Kinesin related	Schizosaccharomyces	Hagan and Yanagida (1990), (1992)
NUC2	Mitotic spindle component	Schizosaccharomyces	Hirano et al. (1988)
BIMA	Mitotic spindle component	Aspergillus	O'Donnell et al. (1991)
MIPB	Mutations repress β-tubulin mutations	Aspergillus	Weil et al. (1986)
BIMC	Kinesin related	Aspergillus	Enos and Morris (1990)
MIPA, MIPB	Mutations repress β-tubulin mutations	Coprinus	Kamada et al. (1990)

of myosin (section 6.3.4). F-actin is highly labile to electron microscopy preparative procedures so that it is not faithfully preserved in sections. Furthermore, immunocytochemistry probably fails to detect potentially functionally significant low concentrations of actin in the presence of nearby larger arrays. Consequently the distribution of F-actin in cells is less clear than that of microtubules. However, F-actin can be bundled (by ABPs) into extensive cables many micrometres long and possessing greater rigidity than the single filaments. Equally, *in vitro* and almost certainly *in vivo* too, very short oligomers of actin, perhaps only nanometres long, can be interconnected by ABPs to form complex 3-dimensional gels (Bray, 1992). These gels, and the cables, can be attached to the plasmalemma and thereby to the extracellular environment (Bray, 1992). In fungi, the most common and conspicuous concentration of actin is seen as plaques associated with the plasmalemma, especially in growth regions such as hyphal tips (Figures 6.29–6.33), but the cytoplasm is also often traversed by cables of varying size which are predominantly aligned parallel to the long axis of the cell (e.g. Runeberg *et al.*, 1986; Salo *et al.*, 1989 and reviewed in Heath, 1990). The cytoplasm also appears to be permeated by a diffuse meshwork of fine F-actin (Figure 6.24 and Jackson and Heath, 1990a, 1993c). Cellular actin is probably best thought of as a meshwork permeating most of the cytoplasm and reinforced by bundled cables and perhaps also microtubules. The role of the ACT2 type of actin referred to above (section 6.3.2) in this ensemble is unclear,

Table 6.2 Fungal ABPs[a]

Name of protein or gene	Likely function or characteristic of protein	Reference
Tropomyosin	Ca^{2+} regulation of actin–myosin interaction	Liu and Bretscher (1989b), (1992)
Myosin II	Myosin II heavy chain	VanTuinen *et al.* (1986); Watts *et al.* (1987)
MYO 2	Myosin type intermediate between I and II	Johnston *et al.* (1991)
CAP 1	Nucleates actin polymerization and mediates membrane attachment, α subunit	Amatruda *et al.* (1992)
CAP 2	" ", β subunit	Amatruda and Cooper (1992)
Profilin	Actin monomer sequestration and polymerization regulation	Haarer *et al.* (1990)
Cofilin	Actin monomer sequestration and polymerization regulation	Moon *et al.* (1993)
Calmodulin	Ca^{2+} regulation of actin functions, colocalizes with actin	Brockerhoff and Davis (1992)
Fimbrin(=SAC 6)	Actin filament bundling	Drubin *et al.* (1988); Adams *et al.* (1989); Adams *et al.* (1991)
SAC 1	Mutations repress actin mutations	Cleves *et al.* (1989), Novick *et al.* (1989)
SAC 2	Mutations repress actin mutations	Novick *et al.* (1989)
SAC 3	Mutations repress actin mutations	Novick *et al.* (1989)
SAC 4	Mutations repress actin mutations	Novick *et al.* (1989)
SAC 5	Mutations repress actin mutations	Novick *et al.* (1989)
SAC 7	Mutations repress actin mutations	Dunn and Shortle (1990)
ABP 1	Influences actin organization	Drubin *et al.* (1988)
RAH 3	Suppresses osmotic sensitivity in actin mutants	Chowdhury *et al.* (1992)
CDC 28	Protein colocalizes with actin	Wittenburg *et al.* (1987)
BEM 1	Has actin binding domain and mutations disrupt actin	Chenevert *et al.* (1992)
ACT 2	Sequences similar to actin, may copolymerize	Schwob and Martin (1992); Lees-Miller *et al.* (1992)

[a] All data are for *Saccharomyces cerevisiae* except ACT 2 reported by Lees-Miller *et al.* (1992) in *Schizosaccharomyces* and myosin II from *Neurospora* by VanTuinen *et al.* (1986).

but its existence does indicate the potential for as yet unexplored greater complexity in the actin system.

6.3.4 Actin binding proteins (ABPs)

As with microtubules, so the organization and function of the actin arrays in cells are determined by a large number of ABPs, many of which have been found in fungi (Table 6.2). The organizational ABPs are the largest group, which are too numerous to discuss in detail here but are reviewed by Pollard and Cooper (1986). They include proteins ranging from about 15 to 250 kDa which will (a) bind monomeric actin and thus alter the monomer–polymer equilibrium, (b) cap or fragment F-actin, thus regulating their length, (c) cross-link F-actin of various lengths into bundles of parallel filaments or more amorphous gels, (d) stiffen individual actin filaments, (e) link F-actin to assorted

membranes or other membrane-associated proteins. With this diversity of ABPs, many of which can be regulated by specific ions (e.g. Ca^{2+}), kinases and phosphatases, it is difficult to predict the form of an actin array because ABPs work simultaneously and each can affect the actions of the others (see Chapter 13).

The functional ABPs are restricted to diverse myosins, the best known of which is myosin II, the type characteristic of skeletal muscle. Myosin II forms a ~400 kDa dimer with a tail and two mechanochemical force generating head domains which interact with F-actin to produce ATP-powered mutual sliding. In muscle, many of these dimers associate tail to tail to form the bipolar thick filaments which pull oppositely oriented actin filaments together, and thus contract the muscle (Bray, 1992). Myosin II occurs in *Saccharomyces* (encoded by the *MYO I* gene) (Watts *et al.*, 1987) and *Neurospora* (VanTuinen *et al.*, 1986) and probably other fungi as well. Myosin I occurs as a ~130 kDa monomer which has a similar head region to myosin II but lacks the filament-forming tail (Hammer, 1991). Its truncated tail region interacts with actin, membranes and calmodulin in various species, thus indicating its likely function as a translocator of organelles and cross-linker of actin gels (Hammer, 1991). Myosin I is widespread in diverse cells (Hammer, 1991) and probably occurs in fungi. Also found in *Saccharomyces* is the *MYO 2* gene product which encodes a ~180 kDa myosin (Johnston *et al.*, 1991). This myosin appears to be intermediate between myosins I and II in having a short tail which mediates dimerization but not filament formation, thus potentially enhancing its capacity to contract actin gels. It seems likely that there is a spectrum of myosins with diverse sizes and characteristics which will extend the range of force-generating arrangements in which actin can participate.

6.3.5 Intermediate filaments

Animal cells contain a diverse array of filaments intermediate in size between microtubules and F-actin, at ~10 nm. These appear to be primarily skeletal (Bray, 1992), but, with the exception of the highly specialized and doubtfully homologous bud neck ring filaments of the yeasts (section 6.3.6), they are not widely reported in fungal cytoskeletons. In animal cells they are characteristically highly resistant to fixation damage, thus their rarity in the fungal ultrastructural literature is probably an accurate indication of their absence. However, a protein which is immunologically related to intermediate filaments has been described forming extensive bundles of filaments in *Neurospora* (Rosa *et al.*, 1990a, b; Alvarez *et al.*, 1991) and a protein capable of forming intermediate filaments *in vitro* has been extensively characterized from *Saccharomyces* (McConnell and Yaffee, 1993), so there is a possibility that intermediate filaments do contribute to fungal cytoskeletons. Their apparent lesser abundance may be due to the contributions of the cell wall and turgor pressure rendering them less important than in animal cells.

6.3.6 Other filaments and possible skeletal elements

There is steadily increasing evidence for the existence of fungal cytoskeletal elements in addition to actin and microtubule-based systems. For example, Roberts (1987) reviewed evidence for 2–5 nm diameter 'fine filaments' in diverse eukaryotes. He did not mention any fungi, but the zoosporic fungi contain flagellar roots, some (but not all, Figure 6.45) of which are 'fine filaments' comparable to centrin found in algae (Lechtreck and Melkonian, 1991). Heath and Heath (1978), Hoch and Staples (1983a) and McKerracher and Heath (1985) have also reported 'fine filaments' associated with microtubules in hyphae (Figures 6.19, 6.21 and 6.22). Although not characterized, clearly fungi contain 'fine filaments' which are probably heterogeneous.

Better characterized are the 12 nm diameter filaments which form a ring in the neck of budding ascomycetous yeasts (Byers and Goetsch, 1976; Heath *et al.*, 1982a). Although their gene has been identified (*CDC3*, Kim *et al.*, 1991), their function, though most likely skeletal, has not. Other morphologically characterized filamentous elements with apparent skeletal functions include the 'microscala' (McLaughlin, 1990) or 'symplechosomes' (Bauer and Oberwinkler, 1991) which link endoplasmic reticulum and mitochondria in some basidiomycetes, and the decorations which link the surfaces of oomycete mitochondria (Bracker and Grove, 1971; Heath, 1976).

Another indication of the existence of yet undescribed cytoskeletal elements is the possibility of isolating morphologically robust nuclear and cytoplasmic matrices (Clayton *et al.*, 1983; Wittenberg *et al.*, 1987; Cardenas *et al.*, 1990). These contain unidentified proteins which do not appear to correspond to known MAPs or ABPs (e.g. Clayton *et al.*, 1983; Cardenas *et al.* 1990). Clearly, future thinking about fungal cytoskeletons should not be restricted to microtubules and F-actin and their directly associated proteins.

6.4 REGULATION OF THE CYTOSKELETON

There are four levels of cytoskeletal regulation. At the genetic level there is diversity in the number of genes encoding skeletal protein isoforms, their differential transcription and the translation of their RNA products. The next level, more specific to polymeric proteins, includes post-translational processing and the dynamic interactions with co-polymerizing proteins. The third level is perhaps unique to the cytoskeleton and involves identifiable organizing centres. Finally, all of these processes are influenced by the ionic composition of the local cellular environment.

6.4.1   Transcription and translation

Unlike vascular plants (Morejohn and Fosket, 1986; Hussey *et al.* 1991) and animals (Sullivan, 1988) which have evolved multiple actin and tubulin genes, the fungi are characterized by 1–3 α- and β-tubulins, one γ-tubulin and single actin genes (Weatherbee and Morris, 1984; Fidel *et al.*, 1988; Oakley and Oakley, 1989; Alfa and Hyams, 1990; Cameron *et al.*, 1990; Drubin, 1990; Dudler, 1990; Bhattacharya *et al.*, 1991; Horio *et al.*, 1991). However, the oomycete, *Phytophthora infestans*, apparently contains several actin genes (Unkles *et al.*, 1991) so that the generalization of single actin genes may not hold up in the light of further surveys. The significance of the number of genes is not clear but may relate to functional specialization. For example, the *Aspergillus* β3-tubulin gene (also referred to as *TUBC*) appears to have some specificity for conidiogenesis, although its product can substitute for other β-tubulins in hyphae and they can substitute for it in conidiogenesis (Weatherbee *et al.*, 1985; May, 1989). A similar ability for substitution appears to occur in *Schizosaccharomyces* where the α2-tubulin gene is dispensible (Hiraoka *et al.*, 1984). However, γ-tubulin does appear to be essential to microtubule nucleation (Oakley and Oakley, 1989; Oakley *et al.*, 1990). Tubulin, but presumably not actin, gene products can also compensate for each other. For example, γ-tubulin was discovered because it could correct the phenotype of other tubulin mutations (Weil *et al.*, 1986; Oakley and Oakley, 1989) and similar compensation occurs between α and β tubulins in *Aspergillus* (Oakley *et al.*, 1987), *Saccharomyces* (Schatz *et al.*, 1988) and *Coprinus* (Kamada *et al.*, 1990). The difference in the number of tubulin versus actin genes correlates with more post-translational modifications of tubulins which might suggest differing strategies for regulation, i.e. microtubules utilize variations in monomers whereas actin relies on ABPs. However, MAPs appear to be more numerous than ABPs when sought by similar techniques in one species (25 MAPS, Barnes *et al.*, 1992 versus three ABPs, Drubin *et al.*, 1988) which is contrary to this suggestion. The significance of evolutionary pressures leading to this situation are thus obscure.

Transcriptional regulation of cytoskeletal genes in fungi has yet to show any pattern, but there are interesting observations. For example, in *Aspergillus*, there is developmental transcriptional regulation of the β3-tubulin gene (Weatherbee *et al.*, 1985) and likewise for the single β-tubulin gene of *Achlya* (Cameron *et al.*, 1990). In animal cells the cytoplasmic level of tubulin dimers modulates tubulin transcription (Cleveland *et al.*, 1981) and a similar situation occurs in *Saccharomyces* where the level of β-tubulin seems to regulate transcription of the α-tubulin gene (Burke *et al.*, 1989). The degree of regulation, however, may not be tight because a 50% reduction in the level of tubulin in the cell shows no phenotypic effect (Katz *et al.*, 1990). Further complexity is seen in *Schizosaccharomyces* where the two α-tubulin and one β-tubulin genes are all transcribed, yet there are equal quantities of α and β tubulin in the cytoplasm (Adachi *et al.*, 1986).

Translational regulation is no more comprehensively understood than transcription, but it clearly occurs. For example, in *Achlya*, a twofold increase in message correlated with a sixfold increase in β-tubulin (Cameron *et al.*, 1990) and in *Physarum*, a new isoform of α-tubulin was synthesized from preexisting message, presumably by new translation and post-translational modification (Green and Dove, 1984).

Although a unified picture of the details of transcriptional and translational regulation is unclear, it is clear that cellular levels of at least tubulin must be regulated quite accurately because the forced overproduction of β-tubulin (but not α-tubulin) is lethal in *Saccharomyces* (Weinstein and Solomon, 1990).

6.4.2   Dynamics and associated proteins

The most basic level of control is modulation of the size of the monomer pool (the αβ tubulin dimer or the actin monomer) which is influenced by their rates of synthesis and degradation. However, there are more complex controls operating. *Saccharomyces* independently monitors and selectively degrades α versus β tubulin such that in cells containing excess copies of transcribed α or β genes the monomer levels are kept optimal (Burke *et al.*, 1989). At another level, *Blastocladiella* is able to differentiate between cytoplasmic versus axonemal derived subunits and use one and not the other for construction of new microtubules (Abe and Lovett, 1982). There is no simple explanation to how this is controlled, but there is an aspect of the polymerization process which has important functional consequences. Polymerization requires nucleotide triphosphate hydrolysis (GTP for microtubules and ATP for actin), the consequence of which is a behaviour known as dynamic instability (Bray, 1992). At each end of the polymer, periods of relatively slow and decelerating polymerization alternate with rapid, catastrophic depolymerization. These are randomly variable processes influenced by monomer concentration and other factors, the net result of which (elongation or shortening) depends on the relative extent of each. However, both F-actin and microtubules are polarized (section 6.3) and the net effect of dynamic instability differs at each end. At any subunit concentration, polymerization is always more dominant at the '+' or barbed end so that at a monomer–polymer equilibrium at which there is no net change in polymer length, there can be flux of subunits along the polymer, a process known as treadmilling (Bray, 1992). The extent of dynamic instability and treadmilling *in vitro* varies with ATP or GTP levels, ionic environment (especially Ca^{2+}), post-translational modifications of the monomers, MAPs and ABPs, which indicates the likely complexity of control in the cell. Dynamic instability has been shown in living animal cells, where it was found to vary dramatically in extent between different cells and different regions of a single cell (Schulze and Kirschner, 1988). None of these behaviours has been shown in any fungi, but there is no reason to doubt that fungi differ in this respect from other cells. Consequently, one must consider that microtubules and F-actin can vary from stable polymers to highly dynamic structures depending on diverse cellular factors. The functional significance of treadmilling is especially interesting because it could translocate an organelle attached to a portion of an anchored polymer or the polymer relative to a substrate.

All of the non-motility related MAPs and ABPs in Tables 6.1 and 6.2 are likely to modulate polymerization in the cell, but none have been shown to do so in fungi. In contrast, these molecules do influence the spatial arrangement of the polymers. For example, there are changes in actin arrays attributable to alterations in *CAP 2* (Amatruda *et al.*, 1990), *BEM 1* (Chenevert *et al.*, 1992), cofilin (Moon *et al.*, 1993), *MYO 2* (Johnston *et al.*, 1991), fimbrin (Drubin *et al.*, 1988), calmodulin (Brockerhoff and Davis, 1992), tropomyosin (Liu and Bretscher, 1989b, 1992) and *CAP 1* (Amatruda *et al.*, 1992). Furthermore, profilin (Haarer *et al.*, 1990) and *USO 1* (Nakajima *et al.*, 1991) influence the organization of both actin and microtubules. Surprisingly, changes in myosin II alter the organization of microtubules, but not actin in *Dictyostelium* (Fukui *et al.*, 1990). However, one must be careful in extrapolating from a gene disruption phenotype to a conclusion of a direct cytoskeleton effect because Adams *et al.* (1990) have shown that *CDC42* and *CDC43* gene products alter actin organization, yet neither seems to have any direct relationship to the cytoskeleton. Similarly, osmotic stress changes actin organization (Chowdhury *et al.*, 1992), as can many other insults to a cell (e.g. heat shock (Pekkala *et al.* 1984), detergents (Heath, 1987, 1988) and fixation (Kaminskyj *et al.*, 1992), yet such are unlikely to be directly involved in normal cytoskeletal organization. There appears to be a general trend that the organization of the actin system is determined by more ABPs (Bray, 1992) than the tubulin system is by MAPs (Olmsted, 1986). The evolution of microtubule organizing centres (MTOCs) and the apparent absence of the equivalents for actin arrays may explain this situation, as discussed in the next section.

6.4.3 Organizing centres

Actin filaments often concentrate in cells and may appear to radiate from specific structures such as focal contacts in animal cells (Bray, 1992). These concentrations could indicate some form of organizing centres but there is no evidence for such structures in any cell, including fungi. In contrast, there is excellent evidence for the existence of microtubule organizing centres (MTOCs) in all fungi. Although microtubules can be polymerized *in vitro* in the absence of MTOCs, it is clear that the major, and possibly all, cellular microtubule arrays are polymerized from MTOCs.

Fungal MTOCs have been referred to by a large

number of names and acronyms (e.g. centrosome, spindle plaque, KCE, NAO, NAB, SPB, reviewed in Heath, 1981a). Currently spindle pole body (SPB) is most widely used, which is perhaps unfortunate because the structure spends much more of the cell cycle associated with non-dividing nuclei lacking spindles. However, there is merit in adopting the dominant term, which will be used here.

Early electron microscopy clearly showed that many of the cytoplasmic microtubules, especially the astral ones formed at mitosis and meiosis, and most spindle microtubules, emanated from nucleus-associated SPBs (Heath, 1981a). Isolated SPBs nucleate attached microtubules *in vitro* and appear to control the number, but not length formed (Hyams and Borisy, 1978). Number control by the SPBs is also indicated by the approximate correlation between SPB size and number of microtubules formed in different species (Heath, 1981a) or even within a single species (Aist and Bayles, 1991a, b), but the situation is more complex because the concentration of added tubulin influenced both numbers and length of polymers formed on isolated SPBs from *Physarum* (Roobol *et al.*, 1982).

Fungal SPBs show considerable diversity of form, which may have phylogenetic significance (Heath, 1981a), but most are discoidal (Figures 6.1, 6.2, 6.5, 6.12 and 6.14) or globular (Figures 6.3 and 6.8) and associated with the nuclear envelope. This association is also variable. In some species (e.g. *Saccharomyces*) a single SPB is inserted into a close fitting pore in the envelope (Figure 6.2) and is bifacial, forming microtubules on both its cytoplasmic and nucleoplasmic faces. In others, a globular SPB is located outside the envelope during interphase, where it may nucleate cytoplasmic microtubules, and the envelope opens to permit it to nucleate the spindle microtubules. The opening may be transient, resealing for the majority of mitosis, or more long lived, remaining open throughout mitosis. Yet other species appear to maintain two SPBs, lying on either side of a continuously intact nuclear envelope. Details of these behaviours are reviewed by Heath (1980a, 1981a, 1986).

SPB microtubule nucleating behaviour and morphology are both coupled to the nuclear cycle. Nucleation variation is most easily seen with respect to the mitotic and meiotic spindle microtubules,

most of which are only formed at the time of division. However, it should be remembered that in many fungi the kinetochore microtubules (section 6.5.1) persist throughout the nuclear cycle (Figure 6.4 and Heath, 1980b; Heath and Rethoret, 1981, and references therein). Regulation of spindle microtubules is a very precise and complex process because different subsets undergo concurrent elongation and shortening. For example, during anaphase the kinetochore microtubules shorten and the non-kinetochore ones elongate (Heath *et al.*, 1984; Aist and Bayles, 1991b). Since both types are attached to the same SPBs during these processes, it is likely that the latter plays some role in the regulatory process, but there is no direct evidence for such a role. Similarly, there is fine control of the astral microtubules during nuclear division, typically with a major increase in numbers and lengths at the onset of spindle elongation (anaphase B) (Heath and Heath, 1976; Aist and Berns, 1981; Bourett and McLaughlin, 1986; Aist and Bayles, 1991a). The molecules responsible for regulation of these changes are unknown, but we are beginning to see some clues. For example, in *Schizosaccharomyces*, the protein recognized by the MPM2 antibody only associates with the SPBs at mitosis (Masuda *et al.*, 1992). Its absence during interphase correlates with the inability of the SPBs to nucleate microtubules from exogenous tubulin (Masuda *et al.*, 1992). The same antibody also only labels G2 and mitotic SPBs in *Aspergillus* (Engle *et al.*, 1988) and shows developmentally regulated changes in *Sordaria* asci (Thompson-Coffe and Zickler, 1992). However, since this antibody specifically recognizes a phosphorylated epitope of the protein, its apparent changes in distribution during the nuclear cycle may be due to changes in phosphorylation rather than location of the protein. In a different vein, Rout and Kilmartin (1990) have identified three proteins in *Saccharomyces* SPBs. These are apparently unrelated to the MPM2 protein, show different locations on the SPBs, and one of them changes its location from inside to outside when cytoplasmic microtubules are formed. Mirabito and Morris (1993) have also identified another SPB localized protein which appears to be essential for mitosis in *Aspergillus*. The consistent localization of γ-tubulin to SPBs (Horio *et al.*, 1991; Masuda *et al.*, 1992; Oakley *et al.*, 1990) and the reduction in both numbers and lengths of associated microtubules in γ-tubulin mutants (Oakley *et al.*, 1990) both indicate its essential role in SPB-based microtubule nucleation. However, in fungi, there is no indication that γ-tubulin changes during the nuclear cycle, thus it is less likely to regulate changes in microtubule nucleating activity.

SPB morphology undergoes considerable changes during the nuclear cycle, consistent with its own duplication cycle. The basic duplication cycle involves splitting of the single discoidal or globular structure into a double disc or globule, with the formation of an interconnecting bridge (Figures 6.3 and 6.5 and Heath, 1981a). The timing of this process seems to vary systematically in the different groups of fungi. In the ascomycetes, chytridiomycetes, oomycetes and zygomycetes the interphase SPBs are single structures which duplicate at the onset of mitosis, whereas in the basidiomycetes the interphase structures are double, duplicating at the end of mitosis via an unusual intermediate (reviewed in Heath, 1981a; Heath and Rethoret, 1982; Murrin *et al.*, 1988). It is possible that the SPB duplication time is linked to the timing of DNA synthesis (Heath, 1981a; Byers and Goetsch, 1975), a concept which is supported by SPB behaviour during meiosis when SPB duplication and DNA synthesis are both suppressed between meiosis I and II, yielding small, single SPBs on the haploid nuclei (Tanaka and Heath, 1984 and references therein). This cycle is completed at karyogamy when nuclear fusion is preceded by SPB fusion (Ashton and Moens, 1982; Byers and Goetsch, 1975).

The SPB duplication cycle and its possible linkage to the DNA synthetic cycle suggest the possibility that the SPBs are self-replicating organelles with their own DNA. This concept, applied to the analogous centrioles and centrosomes of other cells, is an old and much debated one (Wheatley, 1982). Despite a recent claim for a centriole-specific genome in *Chlamydomonas* (Hall *et al.*, 1989), the current consensus is that MTOCs, including SPBs, do not contain DNA (e.g. Johnson and Rosenbaum, 1990). What does control their replication is still unclear, but again we are beginning to see some interesting indications. For example, replication appears to require calmodulin (Sun *et al.*, 1992), β-tubulin (Kanbe *et al.*, 1990), and the products of the *CDC 31* (Baum *et al.*, 1986), *KAR 1* (Rose and Fink, 1987) and *MPS 1* and *2* genes (Winey *et al.*, 1991) in yeasts. The latter are

especially interesting because mutations in *MPS 1* completely block duplication, whereas those in *MPS 2* permit the formation of a morphologically abnormal, but duplicated, SPB. The newly formed disc is unable to form spindle microtubules, but retains the ability to form cytoplasmic microtubules (Winey *et al.*, 1991).

Although most work has focused on the nucleus associated SPBs, there is also evidence for cytoplasmic MTOCs. For example, microtubule repolymerization in *Uromyces* germ tubes recovering from microtubule depolymerization begins in the tip, a region lacking nuclei (Hoch and Staples, 1985). More directly, both *Schizosaccharomyces* (Hagan and Hyams, 1988) and *Candida* (Barton and Gull, 1988) contain cytoplasmic foci from which cytoplasmic microtubules form. Rather surprisingly, whereas *S. pombe* (Hagan and Hyams, 1988) does contain such foci, *S. japonicus* (Alfa and Hyams, 1990) does not, suggesting remarkable variability within a single genus. It is also likely that the microtubules of the flagellar axonemes of fungal zoospores are polymerized by cytoplasmic MTOCs, even though they are attached to basal centrioles (which are MTOCs), because in *Chlamydomonas*, axonemal growth occurs at the tip, not base, of the flagellum (Johnson and Rosenbaum, 1992). The morphology of the putative cytoplasmic MTOCs has not been described in any fungi, although they do appear to contain γ-tubulin (Horio *et al.*, 1991), thus their characteristics and true identity are unclear. It is possible that the apparent foci represent no more than a region of cytoplasm regulated to be especially favourable to microtubule polymerization. This issue is further complicated by the concept of microtubule export from MTOCs (Gunning, 1980). According to this idea, microtubules are polymerized while attached to, and under the control of, MTOCs, then detached and transported to other regions of the cell. This scenario may explain the observations of so called 'free microtubules' (i.e. lacking connections to any detectable MTOC) which are common in both fungal cytoplasm and mitotic spindles (Heath, 1974; Heath and Kaminskyj, 1989; Aist and Bayles, 1991a). This concept, the cytoplasmic MTOC concept and the regulated cytoplasmic region model all raise the important, but totally unresolved, issue of what structures or mechanisms regulate and position these agents, which in turn regulate the microtubules.

6.4.4 Ions

The polymerization, properties and interactions of cytoskeletal components are all influenced by the concentrations of diverse ions, such as H^+, Ca^{2+}, Mg^{2+} etc. Our understanding of these effects is largely derived from *in vitro* experiments (Bray, 1992), but modulating the ionic composition of the cytoplasm *in vivo* by various means also alters the cytoskeleton (Bray, 1992). Since fungi contain such Ca^{2+} binding, cytoskeleton modulating, proteins as calmodulin (Suryanarayana *et al.*, 1985; Brocker-hof and Davis, 1992) and tropomyosin (Liu and Bretscher, 1989a,b, 1992), ionic regulation is certain. However, direct observations of these phenomena in fungi are rare. Causing Ca^{2+} influx or efflux in various ways elicits cytoplasmic contractions or relaxations or organelle movements which are likely to be due to cytoskeletal activity (Jackson and Heath, 1992; McKerracher and Heath, 1987; Kaminskyj *et al.*, 1992) and the well-known Ca^{2+} influences on hyphal growth are also likely to be cytoskeleton mediated (reviewed in Jackson and Heath, 1993b). The tip-high gradients of free cytoplasmic Ca^{2+} seen in fungi (Jackson and Heath, 1993b; Garrill *et al*, 1992, 1993) and other tip growing cells (e.g. Rathore *et al.*, 1991) and the tip-localized influx of Ca^{2+} in root hairs (Schiefelbein *et al.*, 1992) are of sufficient magnitude to influence the cytoskeleton and thus indicate the need to consider the role of ions in cytoskeleton regulation. The demonstration of coordinate regulation of Ca^{2+} and H^+ in tip growing plant cells (Felle, 1988) shows the need to also consider more than a single ion. It is also noteworthy that osmotic stress (which must alter diverse cytoplasmic ion concentrations) induces actin rearrangements in *Saccharomyces* (Chowdhury *et al.*, 1992). The role of ions in cytoskeletal organization is likely to be a major one, but one which may be difficult to investigate because of under-appreciated technical problems (Youatt, 1993). Additional studies on the behaviour of ions are reviewed in Chapters 13 and 9.

6.5 FUNCTIONS OF THE CYTOSKELETON

The cytoskeleton is arguably involved in almost all facets of fungal cell biology and many of its roles are interconnected. However, it is possible to separate these roles to some extent for ease of handling. One of the most important, and best understood, roles is in the generation of the forces necessary for chromosome movement and nuclear division during mitosis and meiosis.

6.5.1 Mitosis and meiosis

Nuclear division in the fungi involves several force-requiring processes which form a functional continuum *in vivo*, but which can be investigated separately. These include: (a) positioning of the nuclei in the correct place in the cell (especially obvious in the basidiomycetes); (b) separation and migration of the duplicated SPBs, so that they lie at the poles of the spindle; (c) movement of the chromosomes to the spindle, at or near its equator; (d) separation of the chromosomes along the spindle, to its poles (refered to as anaphase A); (e) elongation of the spindle (refered to as anaphase B); (f) separation of the nucleus into two nuclei (karyokinesis) and (g) separation of the daughter nuclei. Both (a) and (g) are really special aspects of organelle motility and will be considered in the following section, along with the movements which bring the haploid nuclei together prior to karyogamy (section 6.5.2).

Mitosis is always preceded by duplication of the SPBs (Figures 6.3 and 6.5 and section 6.4.3), each of which becomes the MTOC for each half of the bipolar spindle, and the astral microtubules which radiate away from the poles of the spindle, into the cytoplasm. There is taxon specific variation in the way this migration coordinates with spindle formation (Heath, 1980a, 1986). SPBs can migrate to the potential poles of the spindle prior to any sign of spindle formation, two half spindles can form side by side and then reorientate during migration or a bipolar spindle can form directly between the migrating SPBs (Figure 6.9 and Heath, 1980a). In a number of EM studies it was clear that in all types of migration, a few astral microtubules radiated from the moving SPBs and that new cytoplasmic microtubules developed between them (Figure 6.9). Migration could be blocked by microtubule depolymerization (Kunkel and Hadrich, 1977; Jacobs *et al.*, 1988), thus suggesting a number of possible roles for the microtubules in the migration (reviewed in Heath, 1981b). These earlier suggestions have now been augmented by the demonstration that a kinesin-related protein is needed for migration in *Saccharomyces* (Meluh and Rose, 1990), *Aspergillus* (Enos and Morris, 1990) and *Schizosaccharomyces* (Hagan and Yanagida, 1990, 1992). Presumably inter-microtubule or microtubule-membrane sliding interactions are involved. An interesting, possibly related, observation is that in many fungi when SPB migration begins there is a conspicuous 'clear zone' in the nucleoplasm, adjacent to the region of the nuclear envelope which is associated with the SPBs (and therefore may migrate with them). In *Uromyces* this zone is rich in actin (Figures 6.5–6.7 and Hoch and Staples 1983a, b), which may also have a role in the migration process (however, see below for other possible roles for this actin).

Movement and attachment of chromosomes to the mitotic spindle have received little attention in fungi. In other organisms mitotic prophase movements apparently initially involve attachment to the nuclear envelope and later they are dependent on the microtubules which are attached to the kinetochores (kinetochore microtubules). It is debatable whether kinetochore microtubules are polymerized from the kinetochores to the spindle poles or the kinetochores capture microtubules emanating from the poles (Brinkley *et al*, 1989), but in at least some fungi (e.g. *Saprolegnia* and *Saccharomyces*) this is somewhat academic because the chromosomes remain attached to the poles, via their kinetochores, throughout the mitotic nuclear cycle (Figures 6.4, 6.8 and 6.9 and Moens and Rapport, 1971; Byers and Goetsch, 1975; Heath and Rethoret, 1980, 1981). This observation is supported by the demonstration of SPB-related staining by antibodies specific to kinetochore proteins during interphase in *Saccharomyces* (Goh and Kilmartin, 1993; Jiang *et al.*, 1993) and the likely interphase localization of kinetochore DNA to the SPB in *Schizosaccharomyces* (Uzawa and Yanagida, 1992). This most unusual behaviour may relate to the reported non-random segregation of chromosomes in mitosis in *Aspergillus* (Rosenberger and Kessel, 1968) and *Saccharomyces* (Williamson and Fennell, 1980), especially since the kinetochore microtubules do disappear during meiotic prophase when pairing and recombination occur in *Saprolegnia* (Tanaka and Heath, 1984). The extent of the persistence of the kinetochore microtubules during interphase in other genera is unclear, but it does not occur in many fungi where the 'normal' pattern of behaviour produces interphase nuclei lacking any intranuclear microtubules. However, even in these genera there may be anomalous behaviour because the intranuclear 'clear zone' adjacent to the interphase SPBs, referred to in the previous paragraph, appears to act as an attachment site for interphase chromosomes and it is in this zone that the kinetochores first appear (Murrin *et al.*, 1988, and references therein). Thus, even when the kinetochore microtubules do not persist during interphase, the chromosomes may remain anchored to the region of the nuclear envelope adjacent to the SPBs. Irrespective of the origin and mode of formation of the kinetochore microtubules, the result is that when prophase is finished the chromosomes are attached to spindle poles via their kinetochore microtubules.

Metaphase in fungi is typically rather abnormal compared to the model illustrated in most introductory texts. In the majority of fungi, the characteristic plate-like array of chromosomes at the equator of the spindle is lacking. Instead the chromosomes, or at least their kinetochores, are spread along most of the length of the spindle (Figure 6.12 and Aist and Williams, 1972; Heath, 1974; Heath and Heath, 1976). This dispersal probably represents a dynamic situation in which the chromosomes are oscillating back and forth along the spindle (Heath, 1978). During these oscillations the chromosomes are attached to both spindle poles via their kinetochore microtubules so that any oscillations must involve synchronous lengthening and shortening of the microtubules as in the prometaphase movements of other cells (McIntosh and Pfarr, 1991). The absence of the metaphase plate probably occurs simply because anaphase A begins before the attainment of a metaphase equatorial equilibrium, a not unlikely situation given that fungal mitoses can be complete in less than 5 min (Aist and Williams, 1972).

Both the likely metaphase oscillations and the anaphase A movements are almost certainly mediated by the kinetochore microtubules (Figures 6.9, 6.12 and 6.13) as in other mitotic systems (McIntosh and Pfarr, 1991). The only consistent indication of any unusual behaviour in the fungi is the very low number of kinetochore microtubules per kinetochore. Most fungi have only one (Figures 6.8, 6.9, 6.12 and 6.13), at most five (Heath, 1980a), whereas 50 or more is common in plants and animals. This situation probably reflects the small size of fungal chromosomes which need few microtubules to move them (Heath, 1981a). The molecular motor for anaphase

A in fungi is as obscure as in other organisms (McIntosh and Pfarr, 1991). Kinesin-like proteins localize to the mitotic spindle (Roof *et al.*, 1992; Hoyt *et al.*, 1992; Hagan and Yanagida, 1992) and their disruption blocks mitosis, but they seem to be more important in spindle formation (Hagan and Yanagida, 1990; Roof *et al.*, 1992) and anaphase B (Enos and Morris, 1990) than anaphase A. A β-tubulin mutant blocks at mitosis with apparently normal spindles, probably due to hyperstable kinetochore microtubules (Oakley and Morris, 1981), which suggests the intuitively obvious, that kinetochore microtubules must depolymerize to permit anaphase A. A dynein-like protein has been reported in *Saccharomyces*, although there is no reason to believe that it plays a role in anaphase A, since mitosis continues when the gene is disrupted (Eshel *et al.*, 1993). The accumulation of actin found in the region of spindle formation in *Uromyces* (Figures 6.5–6.7 and Hoch and Staples, 1983a,b, see above) could become incorporated into the spindle and then play a role in anaphase A, as suggested by Forer (1988), but there is no direct evidence for this speculation and actin is reportedly lacking in other fungal spindles (Heath, 1978; Butt and Heath, 1988). At present the motor for anaphase A in fungi, and other organisms, remains enigmatic.

In contrast to the uncertainty with anaphase A, the identification of the motor (or motors) for anaphase B is more advanced, although the story is beginning to resemble the 'Pushmi-pullyu' (Lofting, 1953). In an elegant series of experiments, Aist and colleagues (Aist and Berns, 1981; Aist *et al.*, 1991; Aist and Bayles, 1991a,b) have shown that, in *Fusarium* (= *Nectria*), the asters (composed, at least in part, of astral microtubules, Figure 6.14) pull on the spindle poles and that the non-kinetochore microtubules of the spindle resist this pull and thereby regulate the rate of spindle elongation. Complementary evidence has also been obtained for *Saprolegnia* (Heath *et al.*, 1984) and the commonly observed anaphase B specific proliferation of astral microtubules in many fungi (reviewed in Heath, 1981b; Bourett and McLaughlin, 1986; Aist and Bayles, 1991a) suggests that the model described above reflects a widespread phenomenon among the fungi. However, some spindles lack astral microtubules at anaphase B (Hagan and Hyams, 1988), mutations in β-tubulin can cause a selective loss of cytoplasmic micro-tubules that does not prevent spindle elongation (Sullivan and Huffaker, 1992) and some spindles form curves or even sinusoidal bends as if they were elongating more than the poles were separating (Tanaka and Kanbe, 1986; Aist and Bayles, 1991d) all of which indicate that it is the spindle pushing rather than the asters pulling that is important during anaphase B. The resolution of the issue will probably show that both processes occur in most species and that some species or some conditions place more emphasis on pushing whereas others favour pulling. Direct evidence for this view is the demonstration of both pushing and pulling in *Fusarium* (Aist *et al.*, 1991; Aist and Bayles, 1991d). There is no contradiction between the concepts that the non-kinetochore micro-tubules are responsible for both pushing and regulating the rate of elongation. In any force-generating mechanochemical system the rate of elongation will be determined by the rate of cross-bridge cycling, within limits, independently of externally applied tensile or compressive forces, as previously discussed (Heath *et al.*, 1984).

The nature of the molecular motors operating during anaphase B is not yet certain but kinesin-like molecules seem good candidates for the spindle located component. Kinesin-like proteins localize to spindles (Roof *et al.*, 1992; Hoyt *et al.*, 1992; Hagan and Yanagida, 1992) and mutations in kinesin-like genes block spindle elongation (Enos and Morris, 1990) or spindle formation (Enos and Morris, 1990; Hagan and Yanagida, 1990). However, the story may be more complex because the characteristics of reactivation of spindle elongation in permeabilized *Schizosaccharomyces* cells point to a dynein-like motor (Masuda *et al.*, 1990) and dynein does appear to occur in *Saccharomyces* (Eshel *et al.*, 1993). Furthermore, the later stages of elongation in *Saccharomyces* involve a single pole to pole microtubule which evidently cannot be involved in any intermicrotubule sliding (King *et al.*, 1982) and mutations in both calmodulin (Davis, 1992) and a nuclear matrix protein encoding gene (Hirano *et al.*, 1988) also block spindle elongation. An interaction between non-kinetochore micro-tubules and the nuclear matrix has also been indicated on structural grounds (Heath *et al.*, 1984). The putative astral motor is even less clear. It was originally suggested that an interaction between the microtubules and the plasmalemma was involved (Aist and Berns, 1981) but detailed analysis (Aist and Bayles, 1991a) subsequently indicated that this was unlikely (see also section 6.5.2). Evidence now favours an interaction between the microtubules and an unidentified filamentous component of the cytoskeleton (Figure 6.14 and Aist and Bayles, 1991a) as previously discussed for interphase nuclear motility (McKerracher and Heath, 1986a, 1987; Kaminskyj et al., 1989; section 6.5.2). This component may be actin since both cytoplasmic microtubules and actin are apparently needed for spindle orientation (likely involving asters) (Palmer et al., 1992). Aist and Bayles (1991a) reported filaments associated with astral microtubules (Figure 6.14) but these appeared larger than actin filaments. It was also noted that cytochalasin had no effect on mitosis, but because no changes were shown in the actin arrays in the cells following this treatment the results are not conclusive and the larger size of the filaments could be due to stabilizing ABPs. Present evidence clearly indicates that the asters play a role in anaphase B, as well as nuclear motility (section 6.5.2), but the molecular motors remain obscure.

The process of karyokinesis is of fundamental significance, especially in fungi where the nuclear envelope, nucleolus and nuclear matrix remain intact throughout mitosis in most species (Heath, 1980a). The shape changes preceding and accompanying division of the nuclei (Figures 6.10, 6.11 and 6.13 and Heath, 1980a) must require force-generating systems because the nuclear matrix and envelope have considerable mechanical integrity (e.g. Hirano et al., 1988; Cardenas et al., 1990) and the diversity of these changes (Figures 6.10, 6.11, and 6.13) indicates either diverse systems or diverse regulation. Presently there are no data to indicate what those systems may be.

6.5.2 Organelle motility

Movement of organelles in fungi can be divided into two categories, positioning and motility (McKerracher and Heath, 1987). In the former, organelles are maintained in some non-random order within a cell. This may be static, as in the pattern of organelles within a zoospore, or dynamic, as in the zonation of organelles which hold position with respect to a growing hyphal tip, while migrating forward relative to the lateral walls of the hypha. In the latter, organelles often show additional rapid movements, independent

of the cytoplasm and adjacent organelles. It is unclear whether these two categories utilize different mechanisms, or differential regulation of a common mechanism. The latter seems likely, and for this reason the two phenomena will be discussed as one. The organelles involved include nuclei, mitochondria, vesicles responsible for secretion of cell wall material (which will be referred to as wall vesicles (Chapters 3 and 13)), vacuoles, assorted specialized structures contained in zoospores, and unidentified highly refractile spherical vesicles common in many hyphae (which will be referred to as sphaerosomes). It is most likely, but unproven, that all membrane-bound structures, including endoplasmic reticulum, can show positioning and motility.

There is evidence for a role for both microtubules and F-actin in organelle motility, but it is still unclear whether either can function independently of the other, or if they usually or sometimes work in concert. There are data to support all three possibilities and the burgeoning collection of identified fungal MAPs and ABPs (Tables 6.1 and 6.2) are helping to resolve the questions and identify new possibilities.

A role for microtubules is supported by morphological, inhibitor, microdissection and genetic studies. For example, interphase nuclei moving through hyphae or germ tubes (Raudaskoski and Koltin, 1973; Heath, 1974; Heath and Heath, 1978; Hoch and Staples, 1983a,b; McKerracher and Heath, 1985), into yeast cell buds (Byers and Goetsch, 1975; Heath et al., 1982a; Barton and Gull, 1988; Jacobs et al., 1988), into basidiospores (Nakai and Ushiyama, 1978; Lingle et al., 1992), along fission yeast cells (Hagan and Hyams, 1988; Kanbe et al., 1989), separating at the end of mitosis (Heath et al., 1984; Aist and Bayles, 1991a), moving together prior to karyogamy (Byers and Goetsch, 1975; Ashton and Moens, 1982; Hirata and Tanaka, 1982; Goates and Hoffman, 1986; Hasek et al., 1987) and spacing prior to zoosporogenesis (Hyde and Hardham, 1992, 1993) all do so with an extensive array of variously associated microtubules (Figures 6.15–6.22 and 6.35). Disrupting these microtubules with antimicrotubular drugs (Heath, 1982; Heath et al., 1982b; Herr and Heath, 1982; Jacobs et al., 1988; That et al., 1988; Hyde and Hardham, 1993), tubulin mutations (Oakley and Morris, 1980; Oakley and Rinehart, 1985; Oakley et al., 1987, 1990; Kamada et al., 1989; Palmer et al.,

1992) or ultraviolet microbeams (McKerracher and Heath, 1986a) disrupts the movements. Similarly, mitochondria are associated with microtubules (Figures 6.16–6.19 and 6.20 and Heath and Heath, 1978) and their positions can be disrupted with antimicrotubule agents (Howard and Aist, 1977; Herr and Heath, 1982; Hyde and Hardham, 1993). The distribution of wall vesicles is also sensitive to disruption of microtubules (Howard and Aist, 1980).

Kinesin-like molecules may be the mechano-chemical effector of microtubule based movements. Karyogamy and 'shmoo' (a type of tip growth involved in *Saccharomyces* mating and involving directed migration of wall vesicles) formation are blocked by mutations in a kinesin-like gene (Meluh and Rose, 1990). However, similar mutants in *Aspergillus* still permitted nuclear migration, even though mitosis was blocked (Enos and Morris, 1990) and mutations in the kinesin-like *KIP1* and *CIN8* genes of *Saccharomyces* did not block nuclear migration into the bud (Hoyt *et al.*, 1992). These differences may indicate the existence of other, as yet undetected, kinesins or that cytoplasmic dynein (Eshel *et al.*, 1993) may be involved especially since disruption of a dynein-like gene in *Saccharomyces* disrupted nuclear movements (Eshel *et al.*, 1993). These alternatives could also explain observations of disrupted motility in the presence of morphologically normal microtubule populations in the *NUD* mutants of *Aspergillus* (Meyer *et al.*, 1988; Osmani *et al.*, 1990). However, all of the known microtubule-based motors show a strong bias toward movement in a single direction relative to microtubule polarity (Bray, 1992, however see Sellers and Kachar (1990) for evidence that actin can be both 'pulled' and 'pushed' by myosin), yet many of the migrating nuclei listed above have microtubules both preceding and trailing them (Figure 6.17). It is unlikely that either half of these microtubules (the 'pushers' or the 'pullers') are redundant or that two different motors function in concert, which suggests that motility is based on a more complex system than, for example, axonal transport (Bray, 1992). Irrespective of the possible molecular motors involved, the way in which microtubules function in organelle movements is unclear. Movement associated microtubules are often too short to function in a track like manner (Heath *et al.*, 1982a; McKerracher and Heath, 1985, 1987)

and simple physics shows that in order to move an object such as an organelle, the necessary force must be applied to a static structure, otherwise the surrounding medium, not the organelle, moves. The latter is especially significant in the case of a large organelle such as a nucleus where the adjacent cytoplasm must be sufficiently yielding to permit passage of the organelle, yet contain elements sufficiently rigid against which to generate the force for movement. There is no answer to this paradox at present, but the observation of differentiated skeletal and motor related microtubules in axons (Miller *et al.*, 1987) may provide a clue for the fungi.

The evidence for a role for actin in organelle motility includes accumulations of secretory vesicles in actin or ABP mutants (Novick and Botstein, 1985; Drubin *et al.*, 1988; Cleves *et al.*, 1989; Liu and Bretscher, 1992), inhibition of nuclear motility in similar mutants (Watts *et al.*, 1987; Haarer *et al.*, 1990; Palmer *et al.*, 1992), the association of actin like filaments with the motility associated microtubules (Figures 6.14–6.21 and 6.22 and Heath and Heath, 1978; Heath *et al.*, 1982a; Hoch and Staples, 1983a, 1985; McKerracher and Heath, 1985), actin associated with the membranes of likely motile structures known as filasomes (Bourett and Howard, 1991; Roberson, 1992), Ca^{2+} regulation of organelle motility (McKerracher and Heath, 1986c) and the concordance between wall vesicle motility and actin arrays in hyphal tips (Heath and Kaminskyj, 1989). Shepherd *et al.* (1989) have also shown both actin and microtubules associated with the surfaces of a complex and highly dynamic vacuolar system in hyphae of *Pisolithus tinctorius*. Actin is also apparently involved in organelle motility in pollen tubes (Heslop-Harrison and Heslop-Harrison, 1989; Tang *et al.*, 1989) and other cells (Wessels and Soll, 1990; Zot *et al.*, 1992; Doberstein and Pollard, 1992; Kuznetsov *et al.*, 1992). Also consistent with a role for something other than microtubules (actin?) are the observations of normal mitochondrial motility in cells with tubulin mutation or microtubule inhibitor blocked nuclear motility (Herr and Heath, 1982; Oakley and Rinehart, 1985), nuclear motility in the absence of nucleus associated microtubules (Kaminskyj *et al.*, 1989), apparently normal tip or bud growth (and therefore normal wall vesicle motility) in microtubule-deficient cells (Howard and Aist, 1977, 1980; Herr and Heath, 1982;

Oakley and Rinehart, 1985; Hoch *et al.*, 1987; Jacobs *et al.*, 1988; Yokoyama *et al.*, 1990) and the lack of correlation between the effects of microtubule inhibitors on tip growth and microtubules (Temperli *et al.*, 1991). This 'something' may not necessarily be actin as discussed by Aist and Bayles (1991a, c) who showed unidentified filaments, larger than actin and insensitive to cytochalasin, associated with astral microtubules (Figure 6.4), and thus both nuclear and aster-associated vesicle motility. The complexity of the motility systems is further demonstrated by the observation that mutation of the *MDM2* gene (encoding a fatty acid desaturase) disrupts normal mitochondrial distribution to *Saccharomyces* buds (Stewart and Yaffe, 1991) and disruption of a gene encoding an intermediate filament-like protein in the same species disrupts both mitochondrial distribution and spindle orientation (McConnell and Yaffe, 1993).

At present it is clear that there is no simple consensus on the mechanisms by which the cytoskeleton participates in organelle motility, but cooperative interactions involving all components, possibly with varying proportions of different molecules in different situations, merit serious consideration.

6.5.3 Cytoplasmic motility

The cytoplasm in growing hyphae typically migrates forward through the hypha such that it moves relative to the lateral walls while maintaining its attachment to the tip. Reinhardt (1892) recognized this over 100 years ago and likened fungi to tube-dwelling amoebae, thereby invoking images of amoeboid movement in the hyphae. Fungal zoospores can also show amoeboid movement (e.g. Figure 6.42 and Wubah *et al.*, 1991). As with amoebae, it is clear that the organelle movements and positionings discussed above occur independently of, but within, this cytoplasm and therefore interactions and relationships between the molecular determinants of the two systems are likely. The evidence supports this speculation.

Amoeboid movement involves appropriately localized, Ca^{2+} regulated, actin–myosin based contractions and relaxations and cytoskeleton–plasmalemma attachments (Bray, 1992). Fungal cytoplasm is permeated by a diffuse network of actin filaments (Figures 6.24 and 6.41 and

Wittenburg *et al.*, 1987; Jackson and Heath, 1990a, 1993c) as well as bundles of filaments (e.g. Gull, 1975; Hoch and Staples, 1983b; Adams and Pringle, 1984; Runeberg *et al.*, 1986; Salo *et al.*, 1989), and can be induced by diverse means to undergo reversible, Ca^{2+}-dependent contractions and relaxations (Figures 6.23 and 6.25 and McKerracher and Heath, 1986c, 1987; Jackson and Heath, 1992; Kaminskyj *et al.*, 1992). The contractions can occur independently of organelle positioning (Figure 6.23 and McKerracher and Heath, 1986c), are typically unidirectional, toward the hyphal tip, and result in the formation of clumps of cytoplasm highly enriched in actin (Figure 6.25). Likewise, the cytoplasm of zoospores also contains a diffuse network of actin (Figure 6.41), although its behaviour has not yet been studied (Heath and Harold, 1992).

There is evidence for tight interactions between the actin cytoskeleton and the cell wall (via the plasmalemma), tighter at the hyphal apex (Heath, 1987; Kanbe *et al.*, 1989) and it has been suggested that the peripheral actin plaques commonly concentrated in hyphal and bud tips (Figures 6.29–6.32) represent the equivalent of focal contacts (Adams and Pringle, 1984) where the actin is attached to the plasmalemma. However, Hoch and Staples (1983b), Bourett and Howard (1991) and Roberson (1992) present an alternative interpretation of these plaques so that their true identity remains unclear.

The above observations show that fungi contain the essential elements of an amoebal contraction system and that the cytoplasm can contract, but whether this plays a role in normal growth remains to be seen.

6.5.4 Tip growth and morphogenesis

In plants and fungi with cell walls, it is traditionally thought that morphogenesis results from the interaction between turgor pressure and differential extensibility of the cell wall. Cell wall extensibility is the product of wall synthesis and the formation of interpolymer cross-linkages which impart tensile strength to the wall (Wessels, 1990 and Chapters 3 and 15). Cell wall synthesis involves the correctly localized delivery of subunits and enzymes, at least partially in wall vesicles (Chapters 3, 4 and 13), and the localization and activation of the enzymes responsible for chitin or cellulose fibril synthesis. We have seen that the cytoskeleton is responsible for wall vesicle transport (Chapters 3 and 13) and it has been suggested that F-actin arrays play a direct role in determining the distribution of plasmalemma bound enzymes (reviewed in Heath, 1990). However, Picton and Steer (1982) suggested that actin also plays a direct role in regulating the shape of tip-growing cells by controlling the extensibility of the tip. It is unclear whether this role would be in support of a contribution by the wall itself, or as the sole stress-bearing element in the tip, prior to the wall developing cross links and strength of its own. This point is hard to resolve, but a contribution by actin seems increasingly likely.

The concentration of actin in hyphal and bud tips (Figures 6.29–6.32), especially in the form of the peripheral apical net seen in the oomycetes (Figures 6.26–6.28 and Heath, 1987, 1990; Temperli *et al.*, 1990) is well placed for a morphogenic role and diverse experimental perturbations of the actin arrays in hyphae lead to distorted morphogenesis and, or, changes in growth rate (e.g. Betina *et al.*, 1972; Allen *et al.*, 1980; Grove and Sweigard, 1980; Tucker *et al.*, 1986; Jackson and Heath, 1990b, 1993a). Consistent with these effects are the observations of abnormal (usually enlarged) cell shape in mutants or deletions of the *CAP2* (Amatruda *et al.*, 1990), profilin (Haarer *et al.*, 1990), tropomyosin (Liu and Bretscher, 1992) and fimbrin (Adams *et al.*, 1991) ABP genes and increased osmotic sensitivity in actin mutants (Novick and Botstein, 1985; Chowdhury *et al.*, 1992). Essential to a role in morphogenesis is the need for some stress-bearing linkage to the cell wall, evidence for which was discussed in the previous section. If F-actin is important in morphogenesis, one might expect Ca^{2+} regulation of the actin ensemble, and thus growth. There is a substantial body of evidence for such regulation in tip-growing cells (Harold and Caldwell, 1990; Jackson and Heath, 1993b; see also Chapters 13 and 19).

In contrast to the above evidence for a morphogenic role for actin, there is little reason to believe that microtubules play any direct role in the process. Virtually all microtubule perturbations either have no effect on tip or bud shape, or have effects that can be explained by general cellular malaise.

Tip and bud morphology are not the only morphogenic processes in fungi. The enormous diversity of spore types produced by variations in hyphal expansion (e.g. conidia, basidiospores) and the endospores such as ascospores, which are delineated by directed expansion of SPB associated membranous cisternae in which cell wall material is formed (Moens and Rapport, 1971), all require precise regulation of cell wall expansion and, or, wall synthesis. There is no evidence for a cytoskeletal role in any of these types of development but they represent areas of investigation which will surely be most rewarding.

6.5.5 Cytokinesis and septation

The primary mode of fungal cytokinesis is the concomitant centripetal constriction of cytoplasm and plasmalemma and the deposition of cell wall material to form a septum. The septa so formed vary in their functions (see Chapters 1, 3 and 5) and those in hyphae often do not effect complete cytokinesis. The only other type of cytokinesis occurs during zoosporogenesis in the flagellate groups and is very different, consequently it will be discussed in section 6.5.6.

The most commonly reported involvement of the cytoskeleton in septum formation is the development of a ring of circumferentially arranged F-actin which closes centripetally as the septum grows inward (Girbardt, 1979; Hoch and Howard, 1980; Marks and Hyams, 1985; Kanbe *et al.*, 1989). This ring is located around the edge of the ingrowing septum in such a way as to suggest that it constricts and pulls the attached plasmalemma with it, in much the same way as the contractile ring effects cytokinesis in animal cells (Bray, 1992). However, this model is still strictly speculative. Consistent with a contractile role is the observation that normal cytokinesis in the neck of budding yeasts also requires myosin II (Watts *et al.*, 1985, 1987), but apparently not myosin I (Johnston *et al.*, 1991). Other observations suggest that septum associated F-actin functions in other ways. For example, in *Neozygites*, the formation of the ring is preceded by a band of filaments which are oriented parallel to the long axis of the cell, adjacent to the plasmalemma (Figures 6.30–6.33 and Butt and Heath, 1988). This band shortens to the site of the septum (Figure 6.32) and transforms to the typical ring prior to the initiation of septum synthesis and has been suggested to function in moving plasmalemma-associated molecules (e.g. cell wall synthetic enzymes) to the septum (Butt and Heath, 1988). Supporting evidence for such a role comes from the observation that myosin I is required for normal wall synthesis localization to the neck region of budding yeasts (Rodriguez and Paterson, 1990), even though it is not required for septation itself (Johnston *et al.*, 1991). The ring may also be involved in the transport of vesicles to the developing septum since vesicles have been seen closely associated with the ring (Kanbe *et al.*, 1989). Uniquely in the budding yeasts, cytoskeletal involvement in septation is complicated by the 10 nm diameter filaments (section 6.3.6) and their

associated *CDC3* and 12 gene products (Amatruda and Cooper, 1992). There may also be calmodulin control of septation since this protein accumulates to sites of septa (Brockerhoff and Davis, 1992).

6.5.6 Zoospores and sporogenesis

As might be expected, the cytoskeleton plays a major role in developing and maintaining the form of the wall-less fungal zoospores. This role of microtubules has been discussed in some detail previously (Heath *et al.*, 1982b) so that a brief review of the salient, intuitively obvious, points is sufficient. The propulsive structures, the flagella,

---

Figure 6.34 Developing sporangium of *Saprolegnia ferax*, stained with rhodamine-labelled phalloidin to show the sheets of actin which outline the cleaving zoospores. ×1358. (From Heath and Harold, 1992.)

Figure 6.35 Developing sporangium of *Phytophthora cinnamomi*, at a stage comparable to that seen in Figure 6.34, showing the fluorescently stained microtubules (arrows) radiating around each of the nuclei (dark pear-shaped structures) which are evenly spaced around the sporangium. ×1150. (Reproduced from Hyde and Hardham (1992), with permission.)

Figure 6.36 Zoospores in a sporangium of *Phytophthora cinnamomi*, treated with a DNA stain to show the nuclei (n) and the mitochondrial nucleoids which are clustered around the periphery of the spores. ×465. (Reproduced from Hyde and Hardham (1993), with permission.)

Figure 6.37 Developing sporangium of *Phytophthora cinnamomi* in which cytoplasmic cleavage has been disrupted with an antimicrotubule drug, yet the fluorescently stained vesicles have assumed a non-random distribution along the cleavage vacuoles ×575. (Reproduced from Hyde and Hardham (1993), with permission.)

Figure 6.38 Cartoons of the zoospores of *Allomyces macrogynus* (a), *Oedogoniomyces* sp. (b), *Rhizidiomyces* sp. (c) and *Saprolegnia ferax* (primary,d and secondary,e) showing the diversity of microtubule root systems and their interactions with nuclei (n) and the clustered ribosomes (stippled areas around nuclei in a and b). All approximately × 2500. (From Heath *et al.* 1982b.)

Figure 6.39 Fluorescently stained microtubule roots of a *Phytophthora cinnamomi* zoospore showing a mirror image pattern of that seen in Figure 6.38e. The flagella are arrowed. ×1650. (Reproduced from Hardham (1987), with permission.)

Figures 6.40 and 6.41 Secondary zoospores of *Saprolegnia ferax* (*cf.* Figure 6.38e) stained with rhodamine-labelled phalloidin to show actin. In Figure 6.40, actin is associated with the top of the contractile vacuole (arrows in the fluorescent and corresponding DIC images) and the posterior groove (extending down from the vacuole). In Figure 6.41, the actin surrounds the median optical section of the vacuole (arrows), also extends down the groove to the right, and permeates the rest of the cytoplasm. ×1125. (From Heath and Harold, 1992.)

Figure 6.42 Zoospore of *Neocallimastix patriciarum* stained with rhodamine-labelled phalloidin to show the abundance of actin in the amoeboid projection on the anterior of the cell as seen in the DIC image. ×1125. (From Li and Heath, 1994.)

Figure 6.43 Developing sporangium of *Neocallimastix patriciarum* at approximately the same stage of development and similarly treated to that seen in Figure 6.34. The cleavage planes are outlined with actin and the spherical nuclei are also surrounded by a shell of actin. ×1125. (From Li and Heath, 1994.)

Figure 6.44 Developing sporangium of *Phytophthora cinnamomi* at a similar stage of development to those seen in Figure 6.34, 6.35, and 6.43, showing the cleavage vacuoles (arrows), some of which contain flagella (f), surrounding the uninucleate masses of cytoplasm. ×6300. (Reproduced from Hyde *et al.* (1991), with permission.)

Figure 6.45 Zoospore of *Monoblepharella* sp. showing a face view of the most unusual discoidal root structure which surrounds the kinetosome (k) in this, and related, species. ×80 000. (From Fuller and Reichle, 1968.)

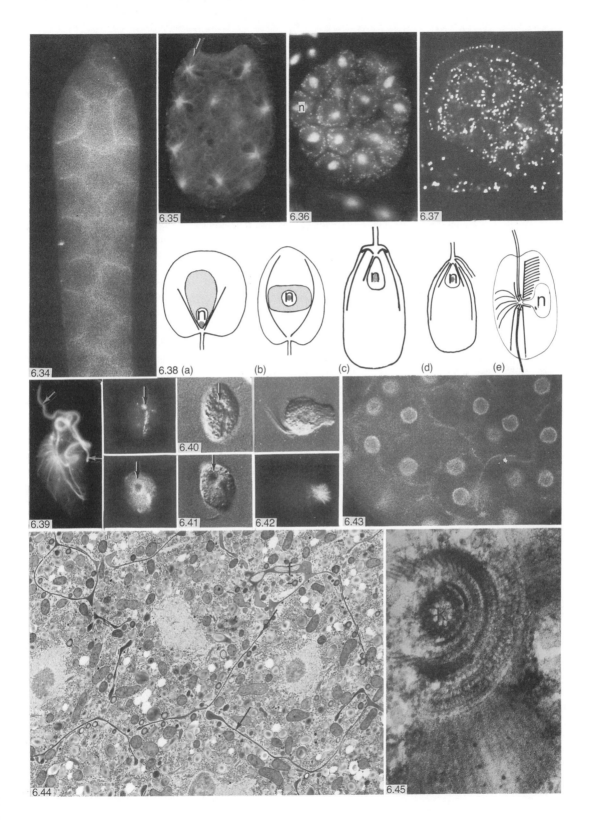

6.34

6.35

6.36

6.37

6.38 (a)　　(b)　　(c)　　(d)　　(e)

6.39

6.40

6.41

6.42

6.43

6.44

6.45

are like those of most other eukaryotes in containing a microtubule-based axoneme which generates the force for movement (Bray, 1992). Two noteworthy points are the presence of mastigonemes on the surface of anterior flagella which act as thrust reversers to make the flagellum into a pulling, rather than a pushing, motor, and the ability of multiple whiplash flagella on some chytridiomycete spores to work in synchrony to behave as a single, large, propulsive structure (Wubah *et al.*, 1991).

As in all flagellate cells, the actively beating flagella are anchored to the cytoplasm, and some organelles, by flagellar root systems which also give the zoospores characteristic shapes. There is extensive diversity in the root systems (Figures 6.38 and 6.39). This diversity has phylogenetic significance and has been used as an indicator of relationships among the fungi. It also appears to have functional significance because it is possible to detect correlations between the complexity, and apparent strength, of the roots and the swimming ability of the spores (Heath *et al.*, 1982a). The roots are composed in large part of microtubules, which connect, directly or indirectly, with the basal bodies at the bases of the flagella. These microtubules may be single or assembled into assorted groups (Figures 6.38 and 6.39), but the key question is how they interact with the cytoplasm or organelles. Clearly they will achieve little if they are not firmly attached to the spore. These attachments have not been characterized at the molecular level, but they appear to be very tight (Havercroft *et al.*, 1981) and involve lateral interactions with the membranes of nuclei (Figure 6.38) and special aggregations of organelles known as rumposomes and microbody–lipid complexes (e.g. Lange and Olson, 1979; Dorward and Powell, 1982). It has been suggested that some of these interactions are active and Ca^{2+} regulated, so that the angle of the basal body, and thus the flagella, to the surface of the spore can be altered, to effect a change in the direction of swimming (Powell, 1983). This is a most attractive idea which deserves more investigation.

Not all of the microtubule roots interact with organelles, some appear to interact only with the cytoplasm. This interaction may involve the network of actin which permeates at least some zoospores (Figure 6.41 and Heath and Harold, 1992). There is *in vitro* evidence for the potential

for such interactions (Griffith and Pollard, 1978, 1982). ATP regulated interactions between F-actin and microtubule roots have been shown in *Physarum* (Uyeda and Furuya, 1987) and at least some microtubular roots are associated with actin (Figures 6.40 and 6.41 and Heath and Harold, 1992).

Fungal zoospores contain assorted fibrous roots (e.g. Figure 6.45) which appear to differ from actin in that they typically show cross striations. Although none of these have been characterized in fungi, it seems likely that at least some of them are related to the better-known contractile and noncontractile fibres found in algal cells (Lechtreck and Melkonian, 1991).

Many fungal zoospores are known to show amoeboid movements (e.g. Wubah *et al.*, 1991) and it is most likely that these are generated by the F-actin permeating the cytoplasm (Figure 6.42 and Li and Heath, 1994) as is the case in amoeboid movement (Bray, 1992). They also contain osmoregulatory water expulsion vacuoles which are surrounded by actin (Figures 6.40 and 6.41 and Heath and Harold, 1992) and therefore probably merit their original name of contractile vacuoles. Myosin has been reported associated with similar vacuoles in other cells (Zhu and Clarke, 1992; Baines and Korn, 1990). The cytoplasmic F-actin may also play a role in determining the nonrandom organization of the organelles found in zoospores and not associated with microtubules (Holloway and Heath, 1977b).

The production of highly differentiated, uninucleate zoospores involves complex rearrangements in the cytoplasm of the sporangia and there is good evidence that the cytoskeleton also plays critical roles in these processes. Prior to cytokinesis, the nuclei and other organelles must be arranged so that subsequent cleavage produces the correct complement of organelles in each spore (Figures 6.36 and 6.37). Because the microtubular roots form a radial array around each nucleus prior to cleavage (Figure 6.35) it was suggested that they played a role in positioning the nuclei (Heath and Greenwood, 1971). This suggestion was supported by observations of abnormal nuclear positioning following application of antimicrotubule drugs (Heath *et al.*, 1982a) and recently the correlated loss of microtubules and nuclear positioning was elegantly shown by immunofluorescence (Hyde and Hardham, 1993). Microtubules also appear to be involved in organelle redistributions during cleavage since these are disturbed by antimicrotubule drugs (Hyde and Hardham, 1993), but the incomplete nature of this disruption (Figure 6.37), and the lack of effect of anti-actin drugs, indicate that other factors may be involved.

Cytokinesis itself includes the alignment and directed expansion of special cleavage vacuoles (Figure 6.44). The sensitivity of these processes to both antimicrotubule and anti-actin drugs suggests that both microtubules and actin are engaged in their regulation (Heath and Harold, 1992; Hyde and Hardham, 1993). During cleavage, these vacuoles are outlined by F-actin arrays (Figures 6.34 and 6.43), indicating a direct role for the actin in expanding or directing the vacuoles (Heath and Harold, 1992; Li and Heath, 1994). The involvement of both actin and microtubules in the zoosporogenesis processes is another example of the interaction of these two major cytoskeletal elements in cellular activities.

6.5.7 Enzyme and mRNA organization

So far we have seen that the cytoskeleton controls the organization of many of the more macroscopic aspects of the cell, at the organelle and membrane level. However, evidence is accumulating that shows that even individual, physiologically important, molecules are also organized by the cytoskeleton. In animal cells, the poly-A regions of mRNA appear to be bound to actin filaments (Taneja *et al.*, 1992), and assorted enzymes (Karkhoff-Schweizer and Knull, 1987) and small heat-shock proteins (Leicht *et al.*, 1986) are also likely to be cytoskeleton linked. In *Saccharomyces*, the product of the *CDC28* gene, a protein kinase which initiates the cell cycle, is associated with an actin rich cytoplasmic matrix (Wittenburg *et al.*, 1987). Similarly, glyceraldehyde-3-phosphate dehydrogenase, mitochondrial citrate synthase and an enzyme involved in aromatic amino acid synthesis all bind to tubulin affinity columns (Barnes *et al.*, 1992). It is premature to conclude that these observations are not attributable to technical artefacts, but they do indicate that the cytoskeleton should be considered as having a potential role in the organization of individual molecules in the cell.

6.6 CONCLUSIONS

It is abundantly clear that the cytoskeleton plays a pivotal role in all aspects of the growth and development of the fungi. The combination of structural, biochemical and genetic approaches has provided us with a rich account of many aspects of the mechanisms and controlling factors involved in cytoskeletal function. However, it is equally clear that we are nowhere near being able to provide even a superficial explanation of the way the entire system is assembled and regulated in the cell. The complexity of the task ahead is perhaps best indicated by reconsidering the fact that even the flagellar axoneme, which can be isolated and induced to show normal beating patterns *in vitro*, and is evidently highly specialized to a well-defined task, is still far from fully understood. This is not surprising since its activities are apparently encoded by about 200 genes. The less specialized cytoskeleton, with its ability to be involved in diverse tasks simultaneously in a common cytoplasm, is unlikely to yield its secrets any more easily. Nevertheless, the diversity of its functions underlines its importance to the fungi, and the diversity of ways in which the fungi themselves influence the health, food supplies and environment of the human race underlines its importance to all of us.

REFERENCES

Abe, S.S. and Lovett, J.S. (1982) Microtubular proteins and tubulin pool changes during zoospore germination in the fungus *Blastocladiella emersonii*. *Archives of Microbiology*, **131**, 323–9.

Adachi, Y., Toda, T., Niwa, O. and Yanagida, M. (1986) Differential expressions of essential and nonessential α-tubulin genes in *Schizosaccharomyces pombe*. *Molecular and Cellular Biology*, **6**, 2168–78.

Adams, A.E.M. and Pringle, J.R. (1984) Relationship of actin and tubulin distribution to bud growth in wild-type and morphogenetic mutant *Saccharomyces cerevisiae*. *Journal of Cell Biology*, **98**, 934–45.

Adams, A.E.M., Botstein, D. and Drubin, D. (1989) A yeast actin-binding protein is encoded by *SAC6*, a gene found by suppression of an actin mutation. *Science*, **243**, 231–3.

Adams, A.E.M., Johnson, D.I., Longnecker, R.M. *et al.* (1990) *CDC42* and *CDC43*, two additional genes involved in budding and the establishment of cell polarity in the yeast *Saccharomyces cerevisiae*. *Journal of Cell Biology*, **111**, 131–42.

Adams, A.E.M., Botstein, D. and Drubin, D. G. (1991) Requirement of yeast fimbrin for actin organization and morphogenesis *in vivo*. *Nature*, **354**, 404–8.

Aist, J.R. and Bayles, C.J. (1991a) Ultrastructural basis of mitosis in the fungus *Nectria haematococca* (sexual stage of *Fusarium solani*). I. Asters. *Protoplasma*, **161**, 111–22.

Aist, J.R. and Bayles, C.J. (1991b) Ultrastructural basis of mitosis in the fungus *Nectria haematococca* (sexual stage of *Fusarium solani*). II. Spindles. *Protoplasma*, **161**, 123–36.

Aist, J.R. and Bayles, C.J. (1991c) Organelle motility within mitotic asters of the fungus *Nectria haematococca*. *European Journal of Cell Biology*, **56**, 358–63.

Aist, J.R. and Bayles, C.J. (1991d) Detection of spindle pushing forces *in vivo* during anaphase B in the fungus *Nectria haematococca*. *Cell Motility and the Cytoskeleton*, **19**, 18–24.

Aist, J.R. and Berns, M.W. (1981) Mechanism of chromosome separation during mitosis in *Fusarium* (Fungi imperfecti). New evidence from ultrastructural and laser microbeam experiments. *Journal of Cell Biology* **91**, 446–58.

Aist, J.R. and Williams, P.H. (1972) Ultrastructure and time course of mitosis in the fungus *Fusarium oxysporum*. *Journal of Cell Biology*, **55**, 368–89.

Aist, J.R., Bayles, C.J., Tao, W. and Berns, M.W. (1991) Direct experimental evidence for the existence, structural basis and function of astral forces during anaphase B *in vivo*. *Journal of Cell Science*, **100**, 279–88.

Alfa, C.E. and Hyams, J.S. (1990) Distribution of tubulin and actin through the cell division cycle of the fission yeasts *Schizosaccharomyces japonicus* var. *versitilis*: a comparison with *Schizosaccharomyces pombe*. *Journal of Cell Science*, **96**, 71–8.

Allen, E.D., Aiuto, R. and Sussman, S. (1980) Effects of cytochalasins on *Neurospora crassa*. 1. Growth and ultrastructure. *Protoplasma*, **102**, 63–75.

Allen, R.D. (1985) New observations on cell architecture and dynamics by video enhanced contrast optical microscopy. *Annual Review of Biological and Biophysical Chemistry*, **14**, 265–90.

Alvarez, M.E., Rosa, A.L., Daniotti, J.L. *et al.* (1991) Antibodies against the 59 kDa polypeptide of the *N.crassa* 8–10 nm filaments immunodetect a 59 kDa polypeptide in specialized rat epithelial cells. *Molecular and Cellular Biochemistry*, **106**, 125–31.

Amatruda, J.F. and Cooper, J.A. (1992) Purification, characterization and immunofluorescence localization of *Saccharomyces cerevisiae* capping protein. *Journal of Cell Biology*, **117**, 1067–76.

Amatruda, J.F., Cannon, J.F., Tatchell, K. *et al* (1990) Disruption of the actin cytoskeleton in yeast capping protein mutants. *Nature*, **344**, 352–4.

Amatruda, J.F., Gattermeir, D.J., Karpova, T.S. and Cooper, J.A. (1992) Effects of null mutations and overexpression of capping protein on morphogenesis, actin distribution and polarized secretion in yeast. *Journal of Cell Biology*, **119**, 1151–62.

Ashton, M.L. and Moens, P.B. (1982) Light and electron microscopy of conjugation in the yeast *Schizosaccharomyces*

octosporus. *Canadian Journal of Microbiology*, **28**, 1059–77.

Baines, I.C. and Korn, E.D. (1990) Localization of myosin 1C and myosin II in *Acanthamoeba castellanii* by indirect immunofluorescence and immunogold electron microscopy. *Journal of Cell Biology*, **111**, 1895–904.

Barnes, G., Louie, A. and Botstein, D. (1992) Yeast proteins associated with microtubules in vitro and in vivo. *Molecular Biology of the Cell*, **3**, 29–47.

Barton, R. and Gull, K. (1988) Variation in cytoplasmic microtubule organization and spindle length between the two forms of the dimorphic fungus *Candida albicans*. *Journal of Cell Science*, **91**, 211–20.

Baum, P., Furlong, C. and Byers, B. (1986) Yeast gene required for spindle pole body duplication: homology of its product with Ca^{2+}-binding proteins. *Proceedings of the National Academy of Sciences USA*, **83**, 5512–16.

Bauer, R. and Oberwinkler, F. (1991) The symplechosome: a unique cell organelle of some basidiomycestes. *Botanica Acta*, **104**, 93–7.

Berlin, V., Styles, C.A. and Fink, G.R. (1990) *BIK1*, a protein required for microtubule function during mating and mitosis in *Saccharomyces cerevisiae*, colocalized with tubulin. *Journal of Cell Biology*, **111**, 2573–86.

Berns, M.W., Aist, J.R., Wright, W.H. and Liang, H. (1992) Optical trapping in animal and fungal cells using a tunable, near-infrared titanium-sapphire laser. *Experimental Cell Research*, **198**, 375–8.

Betina, V., Micekova, D. and Nemec, P. (1972) Antimicrobial properties of cytochalasins and their alteration of fungal morphology. *Journal of General Microbiology*, **71**, 343–9.

Bhattacharya, D., Stickel, S.K. and Sogin, M.L. (1991) Molecular phylogenic analysis of actin genic regions from *Achlya bisexualis* (Oomycota) and *Costaria costata* (Chromophyta). *Journal of Molecular Evolution*, **33**, 525–36.

Bourett, T.M. and Howard, R.J. (1991) Ultrastructural immunolocalization of actin in a fungus. *Protoplasma*, **163**, 199–202.

Bourett, T.M. and Howard, R.J. (1992) Actin in penetration pegs of the fungal rice blast pathogen, *Magnaporthe grisea*. *Protoplasma*, **168**, 20–6.

Bourett, T.M. and McLaughlin, D.J. (1986) Mitosis and septum formation in the basidiomycete *Helicobasidium mompa*. *Canadian Journal of Botany*, **64**, 130–45.

Bracker, C.E. and Grove, S.N. (1971) Surface structure on outer mitochondrial membranes of *Pythium ultimum*. *Cytobiologie*, **3**, 229–39.

Bray, D. (1992) *Cell Movements*, Garland Publishing, New York.

Brinkley, B.R., Valdivia, M.M., Tousson, A. and Balczon, R.D. (1989) The kinetochore: structure and molecular organization, in *Mitosis: Molecules and Mechanisms*, (eds J.S. Hyams and B.R. Brinkley), Academic Press, San Diego, pp. 77–118.

Brockerhoff, S.E. and Davis, T.N. (1992) Calmodulin concentrates at regions of cell growth in *Saccharomyces cerevisiae*. *Journal of Cell Biology*, **118**, 619–30.

Burke, D., Gasdaska, P. and Harwell, L. (1989) Dominant effects of tubulin overexpression in *Saccharomyces cerevisiae*. *Molecular and Cellular Biology*, **9**, 1049–59.

Butt, T.M. and Heath, I.B. (1988) The changing distribution of actin and nuclear behavior during the cell cycle of the mite-pathogenic fungus *Neozygites* sp. *European Journal of Cell Biology*, **46**, 499–505.

Byers, B. and Goetsch, L. (1975) Behaviour of spindles and spindle plaques in the cell cycle and conjugation of *Saccharomyces cerevisiae*. *Journal of Bacteriology*, **124**, 511–23.

Byers, B. and Goetsch, L. (1976) A highly ordered ring of membrane-associated filaments in budding yeast. *Journal of Cell Biology*, **69**, 717–21.

Cameron, L.E., Hutsul, J.-A., Thorlacius, L. and LeJohn, H.B. (1990) Cloning and analysis of β-tubulin gene from a protist. *Journal of Biological Chemistry*, **265**, 15245–52.

Cardenas, M.E., Laroche, T. and Gasser, S.M. (1990) The composition and morphology of yeast nuclear scaffolds. *Journal of Cell Science*, **96**, 439–48.

Chenevert, J., Corrado, K., Bender, A. *et al.* (1992) A yeast gene (*BEM1*) necessary for cell polarization whose product contains two SH3 domains. *Nature*, **356**, 77–79.

Chowdhury, S., Smith, K.W. and Gustin, M.C. (1992) Osmotic stress and the yeast cytoskeleton: phenotype-specific suppression of an actin mutation. *Journal of Cell Biology*, **118**, 561–72.

Clayton, L., Pogson, C.I. and Gull, K. (1983) Ultrastructural and biochemical characterization of the cytoskeleton of *Physarum polycephalum* myxamoebae. *Protoplasma*, **118**, 181–91.

Cleveland, D.W., Lopata, M.A., Sherline, P. and Kirschner, M.W. (1981) Unpolymerized tubulin modulates the level of tubulin mRNAs. *Cell*, **25**, 537–46.

Cleves, A.E., Novick, P.J. and Bankaitis, V.A. (1989) Mutations in the *SAC1* gene suppresses defects in yeast Golgi and yeast actin function. *Journal of Cell Biology*, **109**, 2939–50.

Collins, C.A. (1991) Dynamin: a novel microtubule-associated GTPase. *Trends in Cell Biology*, **1**, 57–60.

Cooper, J.A. (1987) Effect of cytochalasin and phalloidin on actin. *Journal of Cell Biology*, **105**, 1473–8.

Davis, T.N. (1992) A temperature-sensitive calmodulin mutant loses viability during mitosis. *Journal of Cell Biology*, **118**, 607–18.

Doberstein, S.K. and Pollard, T.D. (1992) Localization and specificity of the phospholipid and actin binding sites on the tail of *Acanthamoeba* myosin IC. *Journal of Cell Biology*, **117**, 1241–50.

Dorward, D.W. and Powell, M.J. (1982) Cross-linking bridges associated with the microbody–lipid globule complex in *Chytriomyces aureus* and *Chytriomyces hyalinus*. *Protoplasma*, **112**, 181–8.

Drubin, D.G. (1990) Actin and actin-binding proteins in yeast. *Cell Motility and the Cytoskeleton*, **15**, 7–11.

Drubin, D.G., Miller, K.G. and Botstein, D. (1988) Yeast actin-binding proteins: evidence for a role in morphogenesis. *Journal of Cell Biology*, **107**, 2551–61.

Dudler, D.G. (1990) The single-copy actin gene of *Phytophthora megasperma* encodes a protein considerably diverged from any other known actin. *Plant Molecular Biology*, **14**, 415–22.

Dunn, T.M. and Shortle, D. (1990) Null alleles of *SAC7* suppress temperature-sensitive actin mutations in *Saccharomyces cerevisiae*. *Molecular and Cellular Biology*, **10**, 2308–14.

Dustin, P. (1984) *Microtubules*. Springer-Verlag, Berlin. Heidelberg. New York.

Engle, D.B., Doonan, J.H. and Morris, N.R. (1988) Cell-cycle modulation of *MPM2*-specific spindle pole body phosphorylation in *Aspergillus nidulans*. *Cell Motility and the Cytoskeleton*, **10**, 432–7.

Enos, A.P. and Morris, N.R. (1990) Mutation of a gene that encodes a kinesin-like protein blocks nuclear division in *A. nidulans*. *Cell*, **60**, 1019–27.

Eshel, D., Urrestarazu, L.A., Visser, S. *et al.* (1993) Cytoplasmic dynein is required for normal nuclear segregation in yeast. *Proceedings of the National Academy of Sciences USA*, **90**, 11172–6.

Felle, H. (1988) Cytoplasmic free calcium in *Riccia fluitans* L. and *Zea mays* L.: Interaction of Ca^{2+} and pH? *Planta*, **176**, 248–55.

Fidel, A.S., Doonan, J.H. and Morris, N.R. (1988) *Aspergillus nidulans* contains a single actin gene which has unique intron locations and encodes a γ-actin. *Gene*, **70**, 283–93.

Forer, A. (1988) Do anaphase chromosomes chew their way to the pole or are they pulled by actin? *Journal of Cell Science*, **91**, 449–53.

Fukui, Y., De Lozanne, A. and Spudich, J.A. (1990) Structure and function of the cytoskeleton of a *Dictyostelium* myosin-defective mutant. *Journal of Cell Biology*, **110**, 367–78.

Fuller, M.S. and Reichle, R.E. (1968) The fine structure of *Monoblepharella* sp. zoospores. *Canadian Journal of Botany*, **46**, 279–87.

Garrill, A., Lew, R.R. and Heath, I.B. (1992) Stretch-activated Ca^{2+} and Ca^{2+}-activated K^+ channels in the hyphal tip plasma membrane of the oomycete *Saprolegnia ferax*. *Journal of Cell Science*, **101**, 721–30.

Garrill, A., Jackson, S.L., Lew, R.R. and Heath, I.B. (1993) Ion channels and tip growth: stretch activated channels generate a Ca^{2+} gradient essential for hyphal growth. *European Journal of Cell Biology*, **60**, 358–65.

Gill, S.R., Schroer, T.A., Szilak, I. *et al.* (1991) Dynactin, a conserved, ubiquitously expressed component of an activator of vesicle motility mediated by cytoplasmic dynein. *Journal of Cell Biology*, **115**, 1639–50.

Girbardt, M. (1965) Eine Zeilschnittmethode für Pilzellen. *Mikroskopie*, **20**, 254–64.

Girbardt, M. (1979) A microfilamentous septal belt (FSB) during induction of cytokinesis in *Trametes versicolor* (L. ex. Fr.). *Experimental Mycology*, **3**, 215–28.

Goates, B.J. and Hoffman, J.A. (1986) Spindle pole body fusion in the smut fungus *Tilletia foetida*. *Canadian Journal of Botany*, **64**, 1221–3.

Goh, P.-Y. and Kilmartin, J.V. (1993) NDC10: a gene involved in chromosome segregation in *Saccharomyces cerevisiae*. *Journal of Cell Biology*, **121**, 503–12.

Goldstein, L.S.B. (1991) The kinesin superfamily: tails of functional redundancy. *Trends in Cell Biology*, **1**, 93–8.

Goldstein, L.S.B. and Vale, R.D. (1992) New cytoskeletal liaisons. *Nature*, **359**, 193–4.

Green, L.L. and Dove, W.F. (1984) Tubulin proteins and RNA during the myxamoeba-flagellate transformation of *Physarum polycephalum*. *Molecular and Cellular Biology*, **4**, 1706–11.

Griffith, L.M. and Pollard, T.D. (1978) Evidence for actin filament-microtubule interaction mediated by microtubule-associated proteins. *Journal of Cell Biology*, **78**, 958–65.

Griffith, L.M. and Pollard, T.D. (1982) The interaction of actin filaments with microtubules and microtubule-associated proteins. *Journal of Biological Chemistry*, **257**, 9143–51.

Grove, S.N. and Sweigard, J.A. (1980) Cytochalasin A inhibits spore germination and hyphal tip growth in *Gilbertella persicaria*. *Experimental Mycology*, **4**, 239–50.

Gull, K. (1975) Cytoplasmic microfilament organization in two basidiomycete fungi. *Journal of Ultrastructural Research*, **50**, 226–32.

Gunning, B.E.S. (1980) Spatial and temporal regulation of nucleating sites for arrays of cortical microtubules in root tip cells of the water fern *Azolla pinnata*. *European Journal of Cell Biology*, **23**, 53–65.

Haarer, B.K., Lillie, S.H., Adams, A.E.M. *et al.* (1990) Purification of profilin from *Saccharomyces cerevisiae* and analysis of profilin-deficient cells. *Journal of Cell Biology*, **110**, 105–14.

Hagan, I. and Hyams, J.S. (1988) The use of cell division cycle mutants to investigate the control of microtubule distribution in the fission yeast *Schizosaccharomyces pombe*. *Journal of Cell Science*, **89**, 343–58.

Hagan, I. and Yanagida, M. (1990) Novel potential mitotic motor protein encoded by the fission yeast *cut7+* gene. *Nature*, **347**, 563–6.

Hagan, I. and Yanagida, M. (1992) Kinesin-related cut7 protein associates with mitotic and meiotic spindles in fission yeast. *Nature*, **356**, 74–6.

Hall, J.L., Ramanis, Z. and Luck, D.J.L. (1989) Basal body/centriolar DNA: molecular genetic studies in *Chlamydomonas*. *Cell*, **59**, 121–32.

Hammer III, J.A. (1991) Novel myosins. *Trends in Cell Biology*, **1**, 50–6.

Hardham, A.R. (1987) Microtubules and the flagellar apparatus in zoospores and cysts of the fungus *Phytophthora cinnamomi*. *Protoplasma*, **137**, 109–24.

Harold, F.M. and Caldwell, J.H. (1990) Tips and currents: electrobiology of apical growth, in *Tip Growth in Plant and Fungal Cells.*, (ed I.B. Heath), Academic Press, San Diego, pp. 59–90.

Hasek, J., Rupes, I., Svoboda, J. and Streiblova, E. (1987) Tubulin and actin topology during zygote formation of *Saccharomyces cerevisiae*. *Journal of General Microbiology*, **133**, 3355–63.

Havercroft, J.C., Quinlan, R.A. and Gull, K. (1981) Characterisation of a microtubule organizing center from *Physarum polycephalum* myxamoebae. *Journal of Ultrastructure Research*, **74**, 313–21.

Heath, I.B. (1974) Mitosis in the fungus *Thraustotheca clavata*. *Journal of Cell Biology*, **60**, 204–20.

Heath, I.B. (1976) Ultrastructure of freshwater phycomycetes, in *Recent Advances in Aquatic Mycology*, (ed. E.B.G. Jones), Paul Elek Press, London, pp. 603–50.

Heath, I.B. (1978) Experimental studies of fungal mitotic systems: a review, in *Nuclear Division in the Fungi*, (ed. I.B. Heath), Academic Press, New York, pp. 89–176.

Heath, I.B. (1980a) Variant mitoses in lower eukaryotes: indicators of the evolution of mitosis? *International Review of Cytology*, **64**, 1–80.

Heath, I.B. (1980b) Behaviour of kinetochores during mitosis in the fungus *Saprolegnia ferax*. *Journal of Cell Biology*, **84**, 531–46.

Heath, I.B. (1981a) Nucleus associated organelles of fungi. *International Review of Cytology*, **69**, 191–221.

Heath, I.B. (1981b) Mechanisms of nuclear division in fungi, in *The Fungal Nucleus*, (eds K. Gull and S. Oliver), Cambridge University Press, Cambridge, pp. 85–112.

Heath, I.B. (1982) The effect of nocodazole on the growth and ultrastructure of the fungus *Saprolegnia ferax*. Evidence against a simple mode of action, in *Microtubules in Microorganisms*, (eds P. Capuccinelli and N.R. Norris), Marcel Dekker, New York, pp. 275–311.

Heath, I.B. (1986) Nuclear division: a marker for protist phylogeny? in *Progress in Protistology* (eds J.O. Corliss and D.J. Patterson), Biopress, Bristol, pp. 115–62.

Heath, I.B. (1987) Preservation of a labile cortical array of actin filaments in growing hyphal tips of the fungus *Saprolegnia ferax*. *European Journal of Cell Biology*, **44**, 10–16.

Heath, I.B. (1988) Evidence against a direct role for cortical actin arrays in saltatory organelle motility in hyphae of the fungus *Saprolegnia ferax*. *Journal of Cell Science*, **91**, 41–7.

Heath, I.B. (1990) The roles of actin in tip growth of fungi. *International Review of Cytology*, **123**, 95–127.

Heath, I.B. and Greenwood, A.D. (1968) Electron microscopic observations of dividing somatic nuclei in *Saprolegnia*. *Journal of General Microbiology*, **53**, 287–9.

Heath, I.B. and Greenwood, A.D. (1971) Ultrastructural observations on the kinetosomes and Golgi bodies during the asexual life cycle of *Saprolegnia*. *Zeitschrift fur Zellforsch. Mikroskopisch Anatomische*, **112**, 371–89.

Heath, I.B. and Harold, R.L. (1992) Actin has multiple roles in the formation and architecture of zoospores of the oomycetes, *Saprolegnia ferax* and *Achlya bisexualis*. *Journal of Cell Science*, **102**, 611–27.

Heath, I.B. and Heath, M.C. (1976) Ultrastructure of mitosis in the cowpea rust fungus *Uromyces phaseoli* var. vignae. *Journal of Cell Biology*, **70**, 592–607.

Heath, I.B. and Heath, M.C. (1978) Microtubules and organelle movements in the rust fungus *Uromyces phaseoli* var. vignae. *Cytobiologie*, **16**, 393–411.

Heath, I.B. and Kaminskyj, S.G.W. (1989) The organization of tip-growth related organelles and microtubules revealed by quantitative analysis of freeze-substituted oomycete hyphae. *Journal of Cell Science*, **93**, 41–52.

Heath, I.B. and Rethoret, K. (1980) Temporal analysis of the nuclear cycle by serial section electron microscopy of the fungus *Saprolegnia ferax*. *European Journal of Cell Biology*, **21**, 208–13.

Heath, I.B. and Rethoret, K. (1981) Nuclear cycle of *Saprolegnia ferax*. *Journal of Cell Science*, **49**, 353–67.

Heath, I.B. and Rethoret, K. (1982) Mitosis in the fungus *Zygorhynchus molleri*: evidence for stage specific enhancement of microtubule preservation by freeze substitution. *European Journal of Cell Biology*, **28**, 180–9.

Heath, I.B., Ashton, M.L., Rethoret, K. and Heath, M.C. (1982a) Mitosis and the phylogeny of *Taphrina*. *Canadian Journal of Botany*, **60**, 1696–725.

Heath, I.B., Heath, M.C. and Herr, F. (1982b) Motile systems in fungi, in *Prokaryotic and Eukaryotic Flagella* (eds W.B. Amos and J.G. Duckett), Cambridge University Press, Cambridge, pp. 563–88.

Heath, I.B., Rethoret, K. and Moens, P.B. (1984) The ultrastructure of mitotic spindles from conventionally fixed and freeze-substituted nuclei of the fungus *Saprolegnia*. *European Journal of Cell Biology*, **35**, 284–95.

Herr, F.B. and Heath, M.C. (1982) The effects of antimicrotubule agents on organelle positioning in the cowpea rust fungus, *Uromyces phaseoli* var. vignae. *Experimental Mycology*, **6**, 15–24.

Hertel, C., Quader, H., Robinson, D.G. and Marme, D. (1980) Anti-microtubular herbicides and fungicides affect Ca^{2+} transport in plant mitochondria. *Planta*, **149**, 336–40.

Heslop-Harrison, J. and Heslop-Harrison, Y. (1989) Conformation and movement of the vegetative nucleus of the angiosperm pollen tube: association with the actin cytoskeleton. *Journal of Cell Biology*, **93**, 299–308.

Hirano, T., Hiraoka, Y. and Yanagida, M. (1988) A temperature-sensitive mutation of the *Schizosaccharomyces pombe* gene nuc2+ that encodes a nuclear scaffold-like protein blocks spindle elongation in mitotic anaphase. *Journal of Cell Biology*, **106**, 1171–84.

Hiraoka, Y., Toda, T. and Yanagida, M. (1984) The NDA3 gene of fission yeast encodes β-tubulin: a cold-sensitive nda3 mutation reversibly blocks spindle formation and chromosome movement in mitosis. *Cell*, **39**, 349–58.

Hirata, A, and Tanaka, K. (1982) Nuclear behavior during conjugation and meiosis in the fission yeast *Schizosaccharomyces pombe*. *Journal of General and Applied Microbiology*, **28**, 263–74.

Hoch, H.C. and Howard, R.J. (1980) Ultrastructure of freeze-substituted hyphae of the basidiomycete *Laetisaria arvalis*. *Protoplasma*, **103**, 281–97.

Hoch, H.C. and Staples, R.C. (1983a) Ultrastructural organization of the non-differentiated uredospore germling of *Uromyces phaseoli* variety *typica*. *Mycologia*, **75**, 795–824.

Hoch, H.C. and Staples, R.C. (1983b) Visualization of actin *in situ* by rhodamine-conjugated phalloidin in the fungus *Uromyces phaseoli*. *European Journal of Cell Biology*, **32**, 52–8.

Hoch, H.C. and Staples, R.C. (1985) The microtubule cytoskeleton in hyphae of *Uromyces phaseoli* germlings: its relationship to the region of nucleation and to the F-actin cytoskeleton. *Protoplasma*, **124**, 112–22.

Hoch, H.C., Tucker, B.E. and Staples, R.C. (1987) An intact microtubule cytoskeleton is necessary for mediation of the signal for cell differentiation in *Uromyces*. *European Journal of Cell Biology*, **45**, 209–18.

Holloway, S.A. and Heath, I.B. (1977a) Morphogenesis and the role of microtubules in synchronous populations of *Saprolegnia* zoospores. *Experimental Mycology*, **1**, 9–19.

Holloway, S.A. and Heath, I.B. (1977b) An ultrastructural analysis of the changes in organelle arrangement and structure between the various spore types of *Saprolegnia*. *Canadian Journal of Botany*, **55**, 1328–39.

Horio, T., Uzawa, S., Jung, M.K. *et al.*, (1991) The fission yeast γ-tubulin is essential for mitosis and is localized at microtubule organizing centers. *Journal of Cell Science*, **99**, 693–700.

Howard, R.J. and Aist, J.R. (1977) Effects of MBC on hyphal tip organization, growth, and mitosis of *Fusarium acuminatum*, and their antagonism by D_2O. *Protoplasma*, **92**, 195–210.

Howard, R.J. and Aist, J.R. (1979) Hyphal tip cell ultrastructure of the fungus *Fusarium*: improved preservation by freeze-substitution. *Journal of Ultrastructure Research*, **66**, 224–34.

Howard, R.J. and Aist, J.R. (1980) Cytoplasmic microtubules and fungal morphogenesis: ultrastructural effects of methyl benzimidazole-2-yl carbamate determined by freeze substitution of hyphal tip cells. *Journal of Cell Biology*, **87**, 55–64.

Howard, R.J. and O'Donnell, K.L. (1987) Freeze substitution of fungi for cytological analysis. *Experimental Mycology*, **11**, 250–69.

Hoyt, M.A., Stearns, T. and Botstein, D. (1990) Chromosome instability mutants of *Saccharomyces cerevisiae* that are defective in microtubule-mediated processes. *Molecular and Cellular Biology* **10**, 223–34.

Hoyt, M.A., He, L., Loo, K.K. and Saunders, W.S. (1992) Two *Saccharomyces cerevisiae* kinesin-related gene products required for mitotic spindle assembly. *Journal of Cell Biology*, **118**, 109–20.

Huffaker, T.C., Hoyt, M.A. and Botstein, D. (1987) Genetic analysis of the yeast cytoskeleton. *Annual Review of Genetics*, **21**, 259–84.

Hussey, P.J., Snustad, D.P. and Silflow, C.D. (1991) Tubulin gene expression in higher plants, in *The Cytosketal Basis of Plant Growth and Form* (ed. C.W. Loyd), Academic Press, San Diego, pp. 15–27.

Hyams, J.S. and Borisy, G.G. (1978) Nucleation of microtubules *in vitro* by isolated spindle pole bodies of the yeast *Saccharomyces cerevisiae. Journal of Cell Biology,* **78**, 401–14.

Hyde, G.J. and Hardham, A.R. (1992) Confocal microscopy of microtubule arrays in cryosectioned sporangia of *Phytophthora cinnamomi. Experimental Mycology,* **16**, 207–18.

Hyde, G.J. and Hardham, A.R. (1993) Microtubules regulate the generation of polarity in zoospores of *Phytophthora cinnamomi. European Journal of Cell Biology,* **62**, 75–85.

Hyde, G.J., Lancelle, S.A., Hepler, P.K. and Hardham, A.R. (1991) Freeze substitution reveals a new model for sporangial cleavage in *Phytophthora,* a result with implications for cytokinesis in other eukaryotes. *Journal of Cell Science,* **100**, 735–46.

Ichida, A.A. and Fuller, M.S. (1968) Ultrastructure of mitosis in the aquatic fungus *Catenaria anguillulae. Mycologia,* **60**, 141–55.

Jackson, S.L. and Heath, I.B. (1989) Effects of exogenous calcium ions on tip growth, intracellular Ca^{2+} concentration, and actin arrays in hyphae of the fungus *Saprolegnia ferax. Experimental Mycology,* **13**, 1–12.

Jackson, S.L. and Heath, I.B. (1990a) Visualization of actin arrays in growing hyphae of the fungus *Saprolegnia ferax. Protoplasma,* **154**, 66–70.

Jackson, S.L. and Heath, I.B. (1990b) Evidence that actin reinforces the extensible hyphal apex of the oomycete *Saprolegnia ferax. Protoplasma,* **157**, 144–53.

Jackson, S.L. and Heath, I.B. (1992) UV microirradiations elicit Ca^{2+} dependent apex-directed cytoplasmic contractions in hyphae. *Protoplasma,* **170**, 46–52.

Jackson, S.L. and Heath, I.B. (1993a) UV microirradiation implicates F-actin in reinforcing growing hyphal tips. *Protoplasma* **175**, 67–74.

Jackson, S.L. and Heath I.B. (1993b) The roles of calcium ions in hyphal tip growth. *Microbiology Reviews,* **57**, 367–82.

Jackson, S.L. and Heath, I.B. (1993c) The dynamic behavior of cytoplasmic F-actin in growing hyphae. *Protoplasma,* **173**, 23–34.

Jacobs, W., Adams, A.E.M., Szaniszlo, P.J. and Pringle, J.R. (1988) Functions of microtubules in the *Saccharomyces cerevisiae* cell cycle. *Journal of Cell Biology,* **107**, 1409–26.

Jiang, W., Lecher, J. and Carbon, J. (1993) Isolation and characterization of a gene (*CBF2*) specifying a protein component of the budding yeast kinetochore. *Journal of Cell Biology,* **121**, 513–19.

Johnson, K.A. and Rosenbaum, J.L. (1990) The basal bodies of *Chlamydomonas reinhardtii* do not contain immunologically detectable DNA. *Cell,* **62**, 615–19.

Johnson, K.A. and Rosenbaum, J.L. (1992) Polarity of flagellar assembly in *Chlamydomonas. Journal of Cell Biology,* **119**, 1605–12.

Johnston, G.C., Prydz, K., Ryd, M. and van Deurs, B. (1991) The *Saccharomyces cerevisiae MYO2* gene encodes an essential myosin for vectorial transport of vesicles. *Journal of Cell Biology,* **113**, 539–52.

Jung, M.K., Wilder, I.B. and Oakley, B.R. (1992) Amino acid alterations in the benA (β-tubulin) gene of *Aspergillus nidulans* that confer benomyl resistance. *Cell Motility and the Cytoskeleton,* **22**, 170–4.

Kamada, T., Sumiyoshi, T. and Takemaru, T. (1989) Mutations in β-tubulin block transhyphal migration of nuclei in dikaryosis in the homobasidiomycete *Coprinus cinereus. Plant Cell Physiology,* **30**, 1073–80.

Kamada, T., Hirami, H., Sumiyoshi, T. *et al.* (1990) Extragenic suppressor mutations of a β-tubulin mutation in the basidiomycete *Coprinus cinereus*: Isolation and genetic and biochemical analyses. *Current Microbiology,* **20**, 223–8.

Kaminskyj, S.G.W., Yoon, K.S. and Heath, I.B. (1989) Cytoskeletal interactions with post-mitotic migrating nuclei in the oyster muchroom fungus, *Plerotus ostreatus*: evidence against a force-generating role for astral microtubules. *Journal of Cell Science,* **94**, 663–74.

Kaminskyj, S.G.W., Jackson, S.L. and Heath, I.B. (1992) Fixation induces differential polarized translocations of organelles in hyphae of *Saprolegnia ferax. Journal of Microscopy,* **167**, 153–68.

Kanbe, T., Kobayuashi, I. and Tanaka, K. (1989) Dynamics of cytoplasmic organelles in the cell cycle of the fission yeast *Schizosaccharomyces pombe. Journal of Cell Science,* **94**, 647–56.

Kanbe, T., Hiraoka, Y., Tanaka, K. and Yanagida, M. (1990) The transition of cells of the fission yeast β-tubulin mutant *nda3–311* as seen by freeze-substitution electron microscopy: requirement of functional tubulin for spindle pole body duplication. *Journal of Cell Science,* **96**, 197–206.

Karkhoff-Schweizer, R. and Knull, H.R. (1987) Demonstration of tubulin-glycolytic enzyme interactions using a novel electrophoretic approach. *Biochemical and Biophysical Research Communications,* **146**, 827–31.

Katz, W., Weinstein, B. and Solomon, F. (1990) Regulation of tubulin levels and microtubule assembly in *Saccharomyces cerevisiae*: consequences of altered tubulin gene copy number. *Molecular and Cellular Biology,* **10**, 5286–94.

Kim, H.B., Haarer, B.K. and Pringle, J.R. (1991) Cellular morphogenesis in the *Saccharomyces cerevisiae* cell cycle: localization of the *CDC3* gene product and the timing of events at the budding site. *Journal of Cell Biology,* **112**, 535–44.

King, S.M., Hyams, J.S. and Luba, A. (1982) Absence of microtubule sliding and an analysis of spindle formation and elongation in isolated mitotic spindles from the yeast *Saccharomyces cerevisiae. Journal of Cell Biology,* **94**, 341–9.

Kunkel, W. and Hädrich, H. (1977) Ultrastructural investigations on antimitotic activity of methylbenzimidazol-2-ylcarbamate (MBC) and its influence on replication of the nucleus-associated organelle ('centriolar plaque', 'MTOC', 'KCE') in *Aspergillus nidulans. Protoplasma,* **92**, 311–23.

Kuznetsov, S.A., Langford, G.M. and Weiss, D.G. (1992) Actin-dependent organelle movement in squid axoplasm. *Nature*, **356**, 722–5.

Kwon, Y.H., Hoch, H.C. and Staples, R.C. (1991) Cytoskeletal organization in *Uromyces* urediospore germling apices during appressorium formation. *Protoplasma*, **165**, 37–50.

Lange, L. and Olson, L.W. (1979) The uniflagellate phycomycete zoospore. *Dansk Botanisk Arkiv*, **33**, 1–95.

Lechtreck, K.-F. and Melkonian, M. (1991) An update on fibrous flagellar roots in green algae. *Protoplasma*, **164**, 38–44.

Lees-Miller, J.P., Henry, G. and Helfman, D.M. (1992) Identification of act2, an essential gene in the fission yeast *Schizosaccharomyces pombe* that encodes a protein related to actin. *Proceedings of the National Academy of Sciences, USA*, **89**, 80–3.

Leicht, B.G., Biessmann, H., Palter, K.B. and Bonner, J.J. (1986) Small heat shock proteins of *Drosophila* associate with the cytoskeleton. *Proceedings of the National Academy of Sciences USA*, **83**, 90–4.

Lillie, S.H. and Brown, S.S. (1992) Suppression of a myosin defect by a kinesin-related gene. *Nature*, **356**, 358–61.

Lingle, W.L., Clay, R.P. and Porter, D. (1992) Ultrastructural analysis of basidiosporogenesis in *Panellus stypticus*. *Canadian Journal of Botany*, **70**, 2017–27.

Li, J. and Heath, I.B. (1994) The behavior of F-actin during the zoosporic phases of the chytridiomycete gut fungi *Neocallimastix* and *Oprinomyces*. *Experimental Mycology*, **18**, 57–69.

Liu, H. and Bretscher, A. (1989a) Purification of tropomyosin from *Saccharomyces cerevisiae* and identification of related proteins in *Schizosaccharomyces* and *Physarum*. *Proceedings of the National Academy of Sciences USA*, **86**, 90–3.

Liu, H. and Bretscher, A. (1989b) Disruption of the single tropomyosin gene in yeast results in the disappearance of actin cables from the cytoskeleton. *Cell*, **57**, 233–42.

Liu, H. and Bretscher, A. (1992) Characterization of TMP1 disrupted yeast cells indicates an involvement of tropomyosin in directed vesicular transport. *Journal of Cell Biology*, **118**, 285–300.

Lofting, H. (1953) *Doctor Doolittle's Circus*, Jonathon Cape, London.

Marks, J. and Hyams, J.S. (1985) Localization of F-actin through the cell division cycle of *Schizosaccharomyces pombe*. *European Journal of Cell Biology*, **39**, 27–32.

Masuda, H., Hirano, T., Yanagida, M. and Cande, W.Z. (1990) In vitro reactivation of spindle elongation in fission yeast *nuc2* mutant cells. *Journal of Cell Biology*, **110**, 417–26.

Masuda, H., Sevic, M. and Cande, W.Z. (1992) In vitro microtubule-nucleating activity of spindle pole bodies in fission yeast *Schizosaccharomyces pombe*: cell cycle-dependent activation in *Xenopus* cell-free extracts. *Journal of Cell Biology*, **117**, 1055–66.

May, G.S. (1989) The highly divergent β-tubulins of *Aspergillus nidulans* are functionally interchangeable. *Journal of Cell Biology*, **109**, 2267–74.

McConnell, S.J. and Yaffe, M.P. (1993) Intermediate filament formation by a yeast protein essential for organelle inheritance. *Science*, **260**, 687–9.

McIntosh, J.R. and Pfarr, C.M. (1991) Mitotic motors. *Journal of Cell Biology*, **115**, 577–86.

McKerracher, L.J. and Heath, I.B. (1985) Microtubules around migrating nuclei in conventionally-fixed and freeze-substituted cells. *Protoplasma*, **125**, 162–72.

McKerracher, L.J. and Heath, I.B. (1986a) Fungal nuclear behaviour analysed by ultraviolet microbeam irradiation. *Cell Motility and the Cytoskeleton*, **6**, 35–47.

McKerracher, L.J. and Heath, I.B. (1986b) Comparison of polyethylene glycol and diethylene glycol distearate embedding methods for the preservation of fungal cytoskeletons. *Journal of Electron Microscopy Techniques*, **4**, 347–60.

McKerracher, L.J. and Heath, I.B. (1986c) Polarized cytoplasmic movement and inhibition of saltations induced by calcium-mediated effects of microbeams in fungal hyphae. *Cell Motility and the Cytoskeleton*, **6**, 136–45.

McKerracher, L.J. and Heath, I.B. (1987) Cytoplasmic migration and intracellular organelle movements during tip growth of fungal hyphae. *Experimental Mycology*, **11**, 79–100.

McLaughlin, D.J. (1971) Centrosomes and microtubules during meiosis in the mushroom *Boletus rubinellus*. *Journal of Cell Biology*, **50**, 737–45.

McLaughlin, D.J. (1990) A new cytoplasmic structure in the basidiomycete *Helicogloea*: the microscala. *Experimental Mycology*, **14**, 331–8.

Meluh, P.B. and Rose, M.D. (1990) KAR3, a kinesin-related gene required for yeast nuclear fusion. *Cell*, **60**, 1029–41.

Meyer, S.L.F., Kaminskyj, S.G.W. and Heath, I.B. (1988) Nuclear migration in a *nud* mutant of *Aspergillus nidulans* is inhibited in the presence of a quantitatively normal population of cytoplasmic microtubules. *Journal of Cell Biology*, **106**, 773–8.

Middleton, K., Jiang, W., Hyman, A. et al. (1992) Molecular dissection of the yeast centromere: a molecular motor. *Molecular Abstracts*, **3**, 282a.

Miller, R.H., Lasek, R.J. and Katz, M.J. (1987) Preferred microtubules for vesicle transport in lobster axons. *Science*, **235**, 220–2.

Mirabito, P.M. and Morris, N.R. (1993) *BIMA*, a TPR-containing protein required for mitosis, localizes to the spindle pole body in *Aspergillus nidulans*. *Journal of Cell Biology*, **120**, 959–68.

Moens, P.B. and Rapport, E. (1971) Spindles, spindle plaques, and meiosis in the yeast *Saccharomyces cerevisiae* (Hansen). *Journal of Cell Biology*, **50**, 344–61.

Moon, A.L., Janmey, P.A., Louie, K.A. and Drubin, D.G. (1993) Cofilin is an essential component of the yeast cortical cytoskeleton. *Journal of Cell Biology*, **120**, 421–36.

Morejohn, L.C. and Fosket, D.E. (1986) Tubulins from plants, fungi, and protists, in *Cell and Molecular Biology of the Cytoskeleton*, (ed. J.W. Shay), Plenum, New York, pp. 257–329.

Morris, N.R. (1986) The molecular genetics of microtubule proteins in fungi. *Experimental Mycology*, **10**, 77–82.

Murrin, F., Newcomb, W. and Heath, I.B. (1988) The ultrastructure and timing of events in the nuclear cycle of the fungus *Entomophaga aulicae*. *Journal of Cell Science*, **91**, 41–7.

Nakai, Y. and Ushiyama, R. (1978) Fine structure of shiitake, *Lentinus edodes*, VI, Cytoplasmic microtubules in relation to nuclear movement. *Canadian Journal of Botany*, **56**, 1206–11.

Nakajima, H., Hirata, A., Ogawa, Y. *et al.* (1991) A cytoskeleon-related gene, *USO1*, is required for intracellular protein transport in *Saccharomyces cerevisiae*. *Journal of Cell Biology*, **113**, 245–60.

Novick, P. and Botstein, D. (1985) Phenotypic analysis of temperature-sensitive yeast actin mutants. *Cell*, **40**, 405–16.

Novick, P., Osmond, B.C. and Botstein, D. (1989) Suppressors of yeast actin mutations. *Genetics*, **121**, 659–74.

O'Donnell, K.L., Osmani, A.H. and Morris, N.R. (1991) *bimA* encodes a member of the tetratricopeptide repeat family of proteins and is required for the completion of mitosis in *Aspergillus nidulans*. *Journal of Cell Science*, **99**, 711–20.

Oakley, B.R. and Morris, N.R. (1980) Nuclear movement is β-tubulin dependent in *Aspergillus nidulans*. *Cell*, **19**, 255–62.

Oakley, B.R. and Morris, N.R. (1981) A β–tubulin mutation in *Aspergillus nidulans* that blocks microtubule function without blocking assembly. *Cell*, **24**, 837–45.

Oakley, B.R. and Rinehart, J.E. (1985) Mitochondria and nuclei move by different mechanisms in *Aspergillus nidulans*. *Journal of Cell Biology*, **101**, 2392–7.

Oakley, B.R., Oakley, C.E. and Rinehart, J.E. (1987) Conditionally lethal tubA α-tubulin mutations in *Aspergillus nidulans*. *Molecular and General Genetics*, **208**, 135–44.

Oakley, B.R., Oakley, C.E., Yoon, Y. and Jung, M.K. (1990) γ-Tubulin is a component of the spindle pole body that is essential for microtubule function in *Aspergillus nidulans*. *Cell*, **61**, 1289–1301.

Oakley, C.E. and Oakley, B.R. (1989) Identification of γ-tubulin, a new member of the tubulin superfamily encoded by mipA gene of *Aspergillus nidulans*. *Nature*, **338**, 662–4.

Olmsted, J.B. (1986) Microtubule-associated proteins. *Annual Review of Cell Biology*, **2**, 421–57.

Osmani, A.H., Osmani, S.A. and Morris, N.R. (1990) The molecular cloning and identification of a gene product specifically required for nuclear movement in *Aspergillus nidulans*. *Journal of Cell Biology*, **111**, 543–52.

Palmer, R.E., Sullivan, D.S., Huffaker, T. and Koshland, D. (1992) Role of astral microtubules and actin in spindle orientation and migration in the budding yeast, *Saccharomyces cerevisiae*. *Journal of Cell Biology*, **119**, 583–94.

Pekkala, D., Heath, I.B. and Silver, J.C. (1984) Changes in chromatin and the phosphorylation of nuclear proteins during heat shock of *Achlya ambisexualis*. *Molecular and Cellular Biology*, **4**, 1198–1205.

Picton, J.M. and Steer, M.W. (1982) A model for the mechanism of tip extension in pollen tubes. *Journal of Theoretical Biology*, **98**, 15–20.

Pillus, L. and Solomon, F. (1986) Components of microtubular structures in *Saccharyomces cerevisiae*. *Proceedings of the National Academy of Sciences USA*, **83**, 2468–72.

Pollard, T.D. and Cooper, J.A. (1986) Actin and actin-binding proteins. A critical evaluation of mechanisms and functions. *Annual Review of Biochemistry*, **55**, 987–1035.

Powell, M.J. (1983) Localization of antimonate-mediated precipitates of cations in zoospores of *Chytridomyces hyalinus*. *Experimental Mycology*, **7**, 266–77.

Rathore, K., Cork, R.J. and Robinson, K.R. (1991) A cytoplasmic gradient of Ca^{2+} is correlated with the growth of lily pollen tubes. *Developmental Biology*, **148**, 612–19.

Raudaskoski, M. and Koltin, Y. (1973) Ultrastructural aspects of a mutant of *Schizophyllum commune* with continuous nuclear migration. *Journal of Bacteriology*, **116**, 981–8.

Read, N.D., Allan, W.T.G., Knight, H. *et al.* (1992) Imaging and measurement of cytosolic free calcium in plant and fungal cells. *Journal of Microscopy*, **166**, 57–86.

Reinhardt, M.O. (1892) Das Wachsthum der Pilzhyphen. *Jahrbuch Wissenshaften Botanik*, **23**, 479–566.

Roberson, R.W. (1992) The actin cytoskeleton in hyphal cells of *Sclerotium rolfsii*. *Mycologia*, **84**, 41–51.

Roberts, T.M. (1987) Fine (2–5 nm) filaments: new types of cytoskeletal structures. *Cell Motility and the Cytoskeleton*, **8**, 130–42.

Rodriguez, J.R. and Paterson, B.M. (1990) Yeast myosin heavy chain mutant: maintenance of the cell type specific budding pattern and normal deposition of chitin and cell wall components require an intact myosin heavy chain gene. *Cell Motility and the Cytoskeleton*, **17**, 301–8.

Roobol, A., Gull, K. and Pogson, C.I. (1977) Evidence that griseofulvin binds to a microtubule associated protein. *FEBS Letters*, **75**, 149–53.

Roobol, A., Havercroft, J.C. and Gull, K. (1982) Microtubule nucleation by the isolated microtubule-organizing center of *Physarum polycephalum* myxamoebae. *Journal of Cell Science*, **55**, 365–81.

Roof, D.M., Meluh, P.B. and Rose, M.D. (1992) Kinesin-related proteins required for assembly of the mitotic spindle. *Journal of Cell Biology*, **118**, 95–108.

Rosa, A.L., Alvarez, M.E. and Maldonado, C. (1990a) Abnormal cytoplasmic bundles of filaments in the *Neurospora crassa* snowflake colonial mutant contain P59Nc. *Experimental Mycology*, **14**, 372–80.

Rosa, A., Peralta-Soler, A. and Maccioni, J.J.F. (1990b) Purification of P59Nc and immunocytochemical studies of the 8- to 10-nm cytoplasmic filaments from *Neurospora crassa*. *Experimental Mycology*, **14**, 360–71.

Rose, M.D. and Fink, G.R. (1987) KAR1, A gene required for function of both intranuclear and extranuclear microtubules in yeast. *Cell*, **48**, 1047–60.

Rosenberger, R.F. and Kessel, M. (1968) Non-random sister chromatid segregation and nuclear migration in hyphae of *Aspergillus nidulans*. *Journal of Bacteriology*, **96**, 1208–13.

Rout, M.P. and Kilmartin, J.V. (1990) Components of the yeast spindle and spindle pole body. *Journal of Cell Biology*, **111**, 1913–28.

Runeberg, P., Raudaskoski, M. and Virtanen, I. (1986) Cytoskeletal elements in the hyphae of the homobasidiomycete *Schizophyllum commune* visualized with indirect immunofluorescence and NBD-phallacidin. *European Journal of Cell Biology*, **41**, 25–32.

Salo, V., Niini, S.S., Virtanin, I. and Raudaskoski, M. (1989) Comparative immunocytochemistry of the cytoskeleton in filamentous fungi with dikaryotic and multinucleate hyphae. *Journal of Cell Science*, **94**, 11–24.

Schatz, P.J., Solomon, F. and Botstein, D. (1988) Isolation and characterization of conditional-lethal mutations in the *TUB1* α-tubulin gene of the yeast *Saccharomyces cerevisiae*. *Genetics*, **120**, 681–95.

Schiefelbein, J.W., Shipley, A. and Rowse, P. (1992) Calcium influx at the tip of growing root-hair cells of *Arabidopsis thaliana*. *Planta* **187**, 455–9.

Schulze, E. and Kirschner, M. (1988) New features of microtubule behavior observed *in vivo*. *Nature*, **334**, 356–9.

Schwob, E. and Martin, R.P. (1992) New yeast actin-like gene required late in the cell cycle. *Nature*, **355**, 179–82.

Sellers, J.R. and Kachar, B. (1990) Polarity and velocity of sliding filaments: control of direction by actin and speed by myosin. *Science*, **259**, 406–8.

Shepherd, V.A., Orlovich, D.A. and Ashford, A.E. (1993) A dynamic continuum of pleomorphic tubules and vacuoles in growing hyphae of a fungus. *Journal of Cell Science*, **104**, 495–507.

Stewart, L.C. and Yaffe, M.P. (1991) A role for unsaturated fatty acids in mitochondrial movement and inheritance. *Journal of Cell Biology*, **115**, 1249–58.

Sullivan, D.S. and Huffaker, T.C. (1992) Astral microtubules are not required for anaphase B in *Saccharomyces cerevisiae*. *Journal of Cell Biology*, **119**, 379–88.

Sullivan, K.F. (1988) Structure and utilization of tubulin isotypes. *Annual Review of Cell Biology*, **4**, 687–716.

Sun, G.-H., Hirata, A., Ohya, Y. and Anraku, Y. (1992) Mutations in yeast calmodulin cause defects in spindle pole body functions and nuclear integrity. *Journal of Cell Biology*, **119**, 1625–40.

Suryanarayana, K., Thomas, D. d. S. and Mutus, B. (1985) Calmodulin from the water mold *Achlya ambisexualis*: isolation and characterization. *Cell Biology International Reports*, **9**, 389–400.

Tanaka, K. and Heath, I.B. (1984) The behaviour of kinetochore microtubules during meiosis in the fungus *Saprolegnia*. *Protoplasma*, **120**, 36–42.

Tanaka, K. and Kanbe, T. (1986) Mitosis in the fission yeast *Schizosaccharomyces pombe* as revealed by freeze-substitution electron microscopy. *Journal of Cell Science*, **80**, 253–68.

Taneja, K.L., Lifshitz, L.M., Fay, F.S. and Singer, R.H. (1992) Poly(A) RNA codistribution with microfilaments: evaluation by *in situ* hybridization and quantitative digital imaging microscopy. *Journal of Cell Biology*, **119**, 1245–60.

Tang, X., Lancelle, S.A. and Hepler, P.K. (1989) Fluorescence microscopic localization of actin in pollen tubes: comparison of actin antibody and phalloidin staining. *Cell Motility and the Cytoskeleton*, **12**, 216–24.

Temperli, E., Roos, U.-P. and Hohl, H.R. (1991) Germ tube growth and the microtubule cytoskeleton in *Phytophthora infestans*. Effects of antagonists of hyphal growth, microtubule inhibitors, and ionophores. *Mycological Research*, **95**, 611–17.

Temperli, E., Roos, U.P. and Hohl, H.R. (1990) Actin and tubulin cytoskeletons in germlings of the oomycete fungus *Phytophthora infestans*. *European Journal of Cell Biology*, **53**, 75–88.

That, T.C.C.-T., Rossier, C., Barja, F. *et al* (1988) Induction of multiple germ tubes in *Neurospora crassa* by antitubulin agents. *European Journal of Cell Biology*, **46**, 68–79.

Thompson-Coffe, C. and Zickler, D. (1992) Three microtubule-organizing centers are required for ascus growth and sporulation in the fungus *Sordaria macrospora*. *Cell Motility and the Cytoskeleton*, **22**, 257–73.

Tucker, B.E., Hoch, H.C. and Staples, R.C. (1986) The involvement of F-actin in *Uromyces* cell differentiation: the effects of cytochalasin E and phalloidin. *Protoplasma*, **135**, 88–101.

Unkles, S.E., Moon, R.P., Hawkins, A.R. *et al.* (1991) Actin in the oomycetous fungus *Phytophthora infestans* is the product of several genes. *Gene*, **100**, 105–112.

Uyeda, T.Q.P. and Furuya, M. (1987) ATP-induced relative movement between microfilaments and microtubules in myxomycete flagellates. *Protoplasma*, **140**, 190–2.

Uzawa, S. and Yanagida, M. (1992) Visualization of centromeric and nucleolar DNA in fission yeast by fluorescence *in situ* hybridization. *Journal of Cell Science*, **101**, 267–75.

Vale, R.D., Schnapp, B.J., Reese, T.S. and Sheetz, M.P. (1985) Organelle, bead, and microtubule translocations promoted by soluble factors from the squid giant axon. *Cell*, **40**, 559–69.

Vallee, R. (1991) Cytoplasmic dynein: advances in microtubule-based motility. *Trends in Cell Biology*, **1**, 25–9.

van Tuinen, D., Ortega Perez, R. and Turian, G. (1986) A search for myosin in elongating hyphae of *Neurospora crassa*. *Botanica Helvetica*, **96**, 299–302.

Watts, F.Z., Miller, D.M. and Orr, E. (1985) Identification of myosin heavy chain in *Saccharomyces cerevisiae*. *Nature*, **316**, 83–5.

Watts, F.Z., Shiels, G. and Orr, E. (1987) The yeast *MYO1* gene encoding a myosin-like protein required for cell division. *EMBO Journal*, **6**, 3499–505.

Weatherbee, J.A. and Morris, N.R. (1984) *Aspergillus* contains multiple tubulin genes. *Journal of Biological Chemistry*, **259**, 15452–9.

Weatherbee, J.A., May, G.S., Gambino, J. and Morris, N.R. (1985) Involvement of a particular species of beta-tubulin (beta3) in conidial development in *Aspergillus nidulans*. *Journal of Cell Biology*, **101**, 706–11.

Weil, C.F., Oakley, C.E. and Oakley, B.R. (1986) Isolation of mip (microtubule interacting protein) mutations of *Aspergillus nidulans*. *Molecular and Cellular Biology*, **6**, 2963–8.

Weinstein, B. and Solomon, F. (1990) Phenotypic consequences of tubulin overproduction in *Saccharomyces cerevisiae*: differences between alpha-tubulin and beta-tubulin. *Molecular and Cellular Biology*, **10**, 5295–304.

Wessels, D. and Soll, D.R. (1990) Myosin II heavy chain null mutant of *Dictyostelium* exhibits defective intracellular particle movement. *Journal of Cell Biology*, **111**, 1137–48.

Wessels, J.G.H. (1990) Role of cell wall architecture in fungal tip growth generation, in *Tip Growth in Plant and Fungal Cells*, (ed. I.B. Heath), Academic Press, New York, pp. 1–29.

Wheatley, D.N. (1982) *The Centriole: a Central Enigma of Cell Biology*, Elsevier, New York.

Williamson, D.H. and Fennell, D.J. (1980) Non-random assortment of sister chromatids in yeast mitosis, in *Molecular Genetics in Yeast*, (eds J.F.D. von Wettstein, M. Kielland-Brandt and A. Stenderup), Alfred Benzon Symposium, Munksgaard, Copenhagen, pp. 89–102.

Winey, M., Goetsch, L., Baum, P. and Byers, B. (1991) MPS1 and MPS2: novel yeast genes defining distinct steps of spindle pole body duplication. *Journal of Cell Biology*, **114**, 745–54.

Wittenberg, C., Richardson, S.L. and Reed, S.I. (1987) Subcellular localization of a protein kinase required for cell cycle initiation in *Saccharomyces cerevisiae*. Evidence of an association between the *CDC28* gene product and the insoluble cytoplasmic matrix. *Journal of Cell Biology*, **105**, 1527–48.

Wubah, D.A., Fuller, M.S. and Akin, D.E. (1991) *Neocallimastix*: a comparative morphological study. *Canadian Journal of Botany*, **69**, 835–43.

Yokoyama, K., Kaji, H., Nishimura, K. and Miyaji, M. (1990) The role of microfilaments and microtubules in apical growth and dimorphism of *Candida albicans*. *Journal of General Microbiology*, **136**, 1067–75.

Youatt, J. (1993) Calcium and microorganisms. *Critical Reviews in Microbiology*, **19**, 83–97.

Zhu, Q. and Clarke, M. (1992) Association of calmodulin and an unconventional myosin with the contractile vacuole complex of *Dictyostelium discoidium*. *Journal of Cell Biology*, **118**, 347–58.

Zot, H.G., Doberstein, S.K. and Pollard, T.D. (1992) Myosin-I moves actin filaments on a phospholipid substrate – implications for membrane targeting. *Journal of Cell Biology*, **116**, 367–76.

PART THREE

METABOLISM AND GENETIC
REGULATION

FUNGAL EXOENZYMES

D.B Archer[1] and D.A. Wood[2]
[1]*Institute of Food Research, Norwich Research Park, Colney, Norwich, UK* and
[2]*Horticulture Research International, Worthing Road, Littlehampton, West Sussex, UK*

7.1 INTRODUCTION

The secretion of enzymes by many species of filamentous fungi is an essential feature of their lifestyle, whether it be to support saprophytic growth or pathogenicity. Fungi respond to their habitats by controlled gene expression and secretion of particular enzymes in response to environmental triggers. Although fungal enzymes have long been exploited by man, only recently have the detailed mechanisms of gene expression and enzyme secretion been investigated. Such investigations are still in their infancy and studies have necessarily concentrated on those species which are amenable to molecular approaches. Indeed, it is likely that only a small fraction of fungi have been studied at any level and, of those that have, only a minority are receptive to the techniques of molecular biology. In this chapter we have taken a broad view by discussing the impact of ecology and physiology on exoenzyme production but concentrating, where possible, on examples which have been studied at a detailed molecular level. Thus, we also describe some of the industrial uses of fungal exoenzymes and look to the future with a discussion of the molecular basis of protein secretion and the exploitation of filamentous fungi as hosts for heterologous enzyme production.

7.2 EXOENZYME PRODUCTION IN NATURE

The best studies of the production of fungal exoenzymes in nature are where enzymes have a role in pathogenesis or in their saprophytic capacity on non-living material. Fungi are capable of the colonization of a wide range of living or dead tissues including plants, wood and paper products, leaf litter, plant residues from agriculture, soils and composts, and various living or dead animal tissues (Matcham *et al.*, 1984; Cooke and Whipps, 1993).

Laboratory studies of exoenzyme production show that it is subject to several possible regulatory mechanisms (Priest, 1984). These include enzyme induction, whereby the presence of a suitable inducing substrate that can be metabolized will increase enzyme production severalfold. For example, cellulose, or related small-molecular-weight saccharides, such as cellobiose, can act as inducers of cellulases (Eriksson *et al.*, 1990). In some cases, an inducing substance is apparently non-metabolized, as is the case for cellulase induction by sophorose in *Trichoderma* spp. (Kubicek *et al.*, 1993). One unresolved problem with induction of enzymes by polymeric insoluble substrates remains the identity of the native inducer. Even for the best studied system, production of cellulases by *T. reesei*, the true inducing compound has not been identified (Kubicek *et al.*, 1993).

Enzyme production can also be repressed, by growth on compounds which are the end products of metabolism (Priest, 1984). For cellulases, growth on compounds such as glucose serves to repress cellulase production (Kubicek *et al.*, 1993). Whether this phenomenon in fungi is equivalent to carbon catabolite repression in bacteria has not yet been established (Kubicek *et al.*, 1993). Another control acting on enzyme activity is feedback inhibition whereby an end-product of a metabolic pathway acts to regulate the activity of an enzyme early in the pathway. Fungal endoglucanase activity can be inhibited by glucose (Eveleigh, 1987).

A well studied example of enzyme production during pathogenesis is the role of enzyme production by entomopathogenic (insect invading) fungi. These fungi are capable of the invasion and colonization of a wide range of insect hosts, and

The Growing Fungus. Edited by Neil A.R. Gow and Geoffrey M. Gadd. Published in 1994 by Chapman & Hall, London. ISBN 0 412 46600 7

include species of *Metarhizium*, *Beauveria* and *Verticillium*. Invasion by the fungus has been suggested to occur by a combination of mechanical pressure and enzymic hydrolysis. The cuticle is made of chitin fibrils which are embedded in a protein matrix along with small amounts of lipids, phenols and inorganic material. In culture media where insect cuticle was used as sole carbon and nitrogen source these fungi produce a range of enzymes capable of degrading the major components of the cuticle (St Leger *et al.*, 1986). These enzymes are produced in a sequential fashion, with esterase, endoprotease, aminopeptidase and carboxypeptidase activities being produced in the early stages, followed by *N*-acetylglucosaminidase, chitinase and lipase activities. The major protease activities of *Metarhizium anisoplae* have been resolved and include a chymoelastase serine protease, referred to as PR1, a trypsin-like serine protease PR2, and a trypsin-like cysteine protease (PR4) (St Leger *et al.*, 1987a). It is suggested that PR1, which is produced early in culture is the enzyme primarily responsible for cuticle degradation (St Leger *et al.*, 1987b). This enzyme is produced by all pathogenic isolates of this fungus (St Leger *et al.*, 1987a) and it is produced by infection structures. Inhibition of PR1 delays disease symptoms and mortality. Although this type of evidence provides strong circumstantial proof of the role of particular exoenzymes recent work with gene disruption techniques examining exoenzyme function in pathogenesis has led to some surprising results.

Considerable evidence exists for the role of plant pathogenic fungal cutinases as primary determinants of invasion of plant tissue (Kolattukudy, 1985). This enzyme from *Nectria haemotococca* has been shown to be present at the invasion site, and inhibitors of serine hydrolases or anticutinase sera prevent infection. Natural isolates of the fungus with low enzyme levels have reduced virulence, and similar results were obtained with UV-induced mutants. These results led to the simple model that following spore deposition onto leaf surfaces the basal level of cutinase releases cutin monomers, which in turn act as the inducer for transcription of the cutinase gene resulting in high enzyme levels. It has recently been shown that this cutinase is not responsible for fungal pathogenicity (Stahl and Schafer, 1992). Cutinase-deficient mutants were obtained by transformation-mediated gene disruption techniques, into a virulent *Nectria*

strain. Bioassays showed that such strains showed no difference in pathogenicity. It was proposed that in the absence of other, as yet undetected cutinase genes in this fungus, one role for the enzyme is for cuticle degradation in plant debris. This role would fit the known regulatory pattern for enzyme production. It has also been shown that another role for the enzyme may be in attachment of spores to the host cuticle (Deising *et al.*, 1992). Similar genetic techniques have been used to show that endopolygalacturonase is not required for pathogenicity determination (Scott-Craig *et al.*, 1990).

Light microscope and electron microscope methods, and the use of molecular probes *in situ* are beginning to reveal the sites of fungal enzyme secretion at the cell and colony level. A considerable amount of work has used the techniques of immunogold localization to determine the sites of production of these enzymes (Evans *et al.*, 1991). At the electron microscope level such techniques can be used to examine the localization of enzymes in relation to substrate degradation. In wood, enzymes such as lignin-peroxidase, laccase, cellulase and xylanase can be localized and the information integrated with biochemical models of the mode of action of these enzymes. It has been shown by immunoblotting that when *Phanerochaete chrysosporium* was grown on wood pulp as carbon source the most abundant lignolytic enzyme produced was manganese peroxidase (Datta *et al.*, 1991). Furthermore, the most abundant isoenzyme produced when analysed by amino acid sequencing was not that most abundantly produced in cultures degrading lignin in defined medium. This indicates that there is differential gene expression for this enzyme family depending on the composition of the growth medium, and that laboratory models of enzyme production do not necessarily reflect the situation in natural substrates.

7.3 PROTEIN TARGETING IN FUNGI

As this chapter is concerned with exoenzymes, i.e. secreted enzymes, the mechanisms of protein targeting and secretion are considered briefly. Protein export in fungi can be an efficient process. Fungi, in common with other organisms, have developed sorting and protein targeting mechanisms which ensure that, while some proteins are secreted, others are directed to particular subcellular *cbh1* from *Trichoderma reesei* (Shoemaker *et al.*, 1983)

ATG TAT CGG AAG TTG GCC GTC ATC TCG GCC TTC TTG

Met Tyr Arg Lys Leu Ala Val Ile Ser Ala Phe Leu

GCC ACA GCT CGT GCT CAG TCG GCC TGC ACT CTC

Ala Thr Ala Arg Ala Gln Ser Ala Cys Thr Leu

                ^

*gla*A from Aspergillus niger (Boel *et al.*, 1984)

ATG TCG TTC CGA TCT CTA CTC GCC CTG AGC GGC

Met Ser Phe Arg Ser Leu Leu Ala Leu Ser Gly

CTC GTC TGC ACA GGG TTG GCA AAT GTC ATT TCC

Leu Val Cys Thr Gly Leu Ala Asn Val Ile Ser

                       ^

AAG CGC GCG ACC TTG GAT TCA TGG

Lys Arg Ala Thr Leu Asp Ser Trp

          ↑

^   signal peptidase

↑   pro-peptide processing

Figure 7.1 *N*-terminal regions of cellobiohydrolase I and glucoamylase.

compartments. Exported proteins contain signals that are recognized by the cell and ensure passage along the export pathway. The detailed mechanisms which operate have barely been investigated in fungi but in yeast (Reid, 1991; Cleves and Bankaitis, 1992; Rothman and Orci, 1992) protein targeting has been studied extensively. The secretory mechanisms which operate in yeast and animal cells are analogous (Rothman and Orci, 1992) and it is a reasonable assumption that the fungal secretory mechanisms will also share many features (MacKenzie *et al.*, 1993). For this reason, some strategies for beginning investigation of protein secretion in fungi will exploit the progress made with yeast, e.g. by complementing yeast mutations in the secretory pathway with fungal DNA in order to isolate fungal genes involved in protein secretion. Information on protein secretion from fungi has largely been accumulated on

particular well-secreted proteins; glucoamylase from *Aspergillus niger* (and *A. awamori*) and cellobiohydrolase from *T. reesei*. The discussion that follows will lean heavily on these two systems.

Following translation of mRNA, proteins enter the lumen of the endoplasmic reticulum and this process is mediated by a signal peptide sequence at the amino terminus of the protein (Gierasch, 1989). The signal sequences of *T. reesei* cellobiohydrolase 1 and *A. niger* glucoamylase are shown in Figure 7.1. Although both proteins are exceptionally well secreted by their respective hosts there is no amino acid sequence homology between the two signal sequences. However, both signal sequences have a positively charged *N*-terminus and a hydrophobic core, typical of known secretion signals (von Heijne, 1990). A variety of signal peptides function in fungi and are processed correctly; these include signal sequences from fungal, plant, mammalian, avian and insect sources as well as synthetic signal sequences (Jeenes *et al.*, 1991; Archer, 1994). Some signal sequences promote better secretion of the target protein than others (Jeenes *et al.*, 1991; van den Hondel *et al.*, 1991). This finding has been confirmed by fusing different signal sequences to the same target protein and measuring secreted yields of processed protein. A signal sequence is not, however, a sufficient condition for secretion of a target protein. A synthetic signal sequence was effective at promoting the secretion of glucoamylase from *A. niger* but was ineffective at promoting secretion of β-glucuronidase (Punt *et al.*, 1991). Alterations to the signal sequences, particularly adjacent to the signal processing site, can reduce secreted levels of proteins in yeast (Kikuchi and Ikehara, 1991). Signal peptidases responsible for correct cleavage of the signal peptide from the mature protein recognise structural features of the signal rather than the amino acid sequence (Müller, 1992).

The importance of correct protein folding for efficient secretion and the mechanisms involved have been discussed by Gething and Sambrook (1992). Although the available information has not been obtained with filamentous fungi, it should be recognized that the important features probably have analogies in filamentous fungi and that this knowledge will dictate approaches for investigations. Correct folding *in vivo* of polypeptides in yeast is necessary for their efficient secretion (Kikuchi and Ikehara, 1991) and it is likely that this is also true in filamentous fungi. Gething and Sambrook (1992) recognize at least two classes of proteins involved in folding of polypeptides. The first class includes enzymes which catalyse specific isomerization steps, e.g. protein disulphide isomerase (PDI) and peptidyl prolyl *cis-trans* isomerase (PPI). The second class of proteins recognizes and stabilize partially folded polypeptide intermediates; proteins in this class are collectively known as chaperones (Ellis and van der Vies, 1991).

Although many chaperones were recognized by their induction in response to stress, especially heat shock, most are constitutively expressed. The response of fungi to heat shock has received attention (e.g. LeJohn and Braithwaite, 1984; Brunt and Silver, 1991) although information on heat shock-induced chaperones in filamentous fungi is much more limited (Judelson and Michelmore, 1989). Specific information on the role of chaperones in the secretion of exoenzymes from filamentous fungi is lacking.

PDI is essential in yeast (Farquhar *et al.*, 1991; Tachikawa *et al.*, 1991) and the abundance of PDI in the ER of several different eukaryotic cell types correlates with the level of secreted protein (Gething and Sambrook, 1992). Disulphide interchange and its effects on protein secretion in yeast have been studied in an elegant series of experiments using human lysozyme and derived mutants as model (heterologous) secreted proteins. It was confirmed that intramolecular shuffling of disulphide bonds occurs *in vivo* during folding (Taniyama *et al.*, 1991; Omura *et al.*, 1991) and some cysteine residues have more important roles in folding and secretion than others. One of the mutant lysozymes (Cys$_{77}$ → Ala, Cys$_{95}$ → Ala) showed an 8-fold improved secretion over the wild-type lysozyme although the three-dimensional structures of the wild-type and mutant lysozymes were virtually identical (Inaka *et al.*, 1991). Studies *in vitro* showed that the disulphide bond (Cys$_{77}$, Cys$_{95}$) contributes to the stabilization of folded lysozymes by slowing the unfolding rate (Taniyama *et al.*, 1992). Although folding *in vitro* cannot mimic *in vivo* folding, the qualitative effects on folding *in vitro* introduced by mutation may well correlate with effects on folding *in vivo* and have an impact on secretion.

Although some relevant research on targetting has been undertaken with fungi, particularly *Neurospora crassa* (Pfanner and Neupert, 1990), the secretory process has not been investigated in great detail. This position contrasts dramatically with the extensive knowledge obtained with *Saccharomyces cerevisiae* where many mutants blocked in particular stages of secretion have been obtained and characterized (Reid, 1991; Cleves and Bankaitis, 1992) which is leading to a biochemical understanding of secretion (Rexach *et al.*, 1992).

Despite the absence of detailed information on the secretion mechanism in filamentous fungi a number of studies have sought to improve secretion of heterologous proteins by fusion of its gene to the whole or part of a gene encoding a well secreted homologous protein such as glucoamylase. It is important in these studies to distinguish effects on gene expression from protein secretion by expressing the various constructs

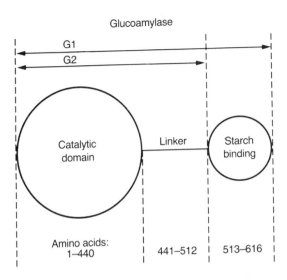

Glucoamylase

Figure 7.2 Representation (not to scale) of the domain structure of *Aspergillus niger* glucoamylase. (From Archer (1994) with permission.)

under the control of the same promoter when integrated at the same genomic locus. Production of bovine prochymosin in *A. niger* employed the *gla*A promoter in a cassette which replaced the native single copy *gla*A gene to ensure site-specific integration (van Hartingsveldt *et al.*, 1990). These results suggested that fusion of the *gla*A prepro region to prochymosin reduced secreted yields of prochymosin (relative to the *gla*A pre, i.e. signal sequence, fusion) but that a full length *gla*A – prochymosin fusion gave the highest secreted yields. A more detailed analysis of this type of approach is given elsewhere (Jeenes *et al.*, 1991; van den Hondel *et al.*, 1991). The use of full length *gla*A fusions to improve secretion has been effective not only with chymosin but also with other heterologous proteins produced in other fungal species (Ward *et al.*, 1990; Contreras *et al.*, 1991). The use of a truncated glucoamylase in gene fusions has also proven highly effective (Jeenes *et al.*, 1993).

Analysis of the structural features of glucoamylase which are responsible for the enhanced secretion has not been carried out. Detailed structural analysis of glucoamylase is only now becoming available by analysis of the domain structure (Figure 7.2) and crystal structure (Aleshin *et*

al., 1992). The α-amylase (Taka-amylase) from *A. oryzae* (Tada *et al.*, 1989, 1991) is another well-secreted amylase for which a structure is available (Matsuura *et al.*, 1984). Thus, α-amylase and glucoamylase can be studied in detail using *in vitro* mutagenesis to determine the structural features important in protein secretion. Similarly, structural information on the cellulases from *T. reesei* is accumulating (Knowles *et al.*, 1988; Rouvinen *et al.*, 1990) so that these enzymes are good candidates for study of the secretion process in *Trichoderma* spp. The importance of protein folding and conformation-dependent processing has been illustrated in the secretion of KP6 toxin from *Ustilago maydis* (Tao *et al.*, 1993). A single amino acid substitution ($Cys_{51} \rightarrow Arg$) in the α polypeptide or a double substitution ($Thr_{25} \rightarrow Pro$; $Lys_{42} \rightarrow Arg$) in the β polypeptide prevented secretion. The structural features which govern protein secretion need to be determined and this will involve a detailed analysis of folding and secretion of several target proteins and derived mutants.

Glucoamylase from *A. niger* and cellobiohydrolase I from *T. reesei* are both glycosylated and contain *N*-linked and *O*-linked oligosaccharides (Salovuori *et al.*, 1987; Svensson *et al.*, 1983, 1986). The domain structures of these two well-secreted proteins have similarities (Knowles *et al.*, 1988). The catalytic and starch-binding domains of *A. niger* glucoamylase are separated by a heavily *O*-glycosylated linker region (Figure 7.2), although the linker region is much shorter in the *A. oryzae* glucoamylase (Hata *et al.*, 1991). The linker region is known to be important in secretion of glucoamylase as a heterologous protein from *S. cerevisiae* (Evans *et al.*, 1990). Similarly, the *S. diastaticus* glucoamylase linker was shown to be important in the secretion of this glucoamylase from *S. cerevisiae* (Yamashita, 1989).

Although a role in secretion of the heavily *O*-glycosylated linkers of glucoamylase, and probably cellobiohydrolase, is likely, the role of protein glycosylation in protein secretion by filamentous fungi has barely been addressed. Introduction of an additional *N*-glycosylation site into bovine prochymosin increased the level of secreted prochymosin from *A. awamori* three-fold (Ward, 1989). Detailed analysis of glycosylation in filamentous fungal proteins is limited (Svensson *et al.*, 1983, 1986; Salovuori *et al.*, 1987; Chiba

Table 7.1 Purified fungal secreted proteases

Species	Protease	Gene (or cDNA) cloned +/−	Reference
Aspergillus awamori	Aspergillopepsin	+	Berka *et al.* (1990)
Aspergillus flavus	Metalloprotease	−	Zhu *et al.* (1990)
	Semi-alkaline protease	−	Impoolsup *et al.* (1981)
	Cysteine protease	−	Impoolsup *et al.* (1981)
Aspergillus foetidus	Aspartic protease	−	Stepanov (1985)
Aspergillus fumigatus	Alkaline protease	+	Jaton-Ogay *et al.* (1992)
Aspergillus niger	Serine carboxypeptidase	−	Dal Degan *et al.* (1992)
	Serine protease	+	Frederick *et al.* (1993)
	Semi-alkaline protease	−	Pourrat *et al.* (1988)
	Non-aspartyl acid protease	+	Takahashi *et al.* (1991)
Aspergillus oryzae	Semi-alkaline protease	+	Murakami *et al.* (1991)
	Aspartic protease	+	Berka *et al.* (1993)
	Metalloprotease	+	Tatsumi *et al.* (1991)
Candida parapilosis	Acid protease	+	De Viragh *et al.* (1993)
Endothia parasitica	Fungal rennet	−	Berkholt (1987)
Irpex lacteus	Aspartic protease	+	Kobayashi *et al.* (1989)
Metarhizium anisopliae	Serine protease	+	St Leger *et al.* (1992)
Mucor racemosus	Carboxypeptidase	−	Disanto *et al.* (1992)
Penicillium janthinellum	Penicilliopepsin	−	Hsu *et al.* (1977)
Phanerochaete chrysosporium	Acid protease	−	Datta (1992)
Russula decolorans	Aspartic protease	−	Stepanov (1985)
Rhizomucor miehei	Fungal rennet	+	Gray *et al.* (1986)
Rhizopus chinensis	Rhizopuspepsin	+	Delaney *et al.* (1987)
Rhizopus niveus	Aspartic protease	+	Horiuchi *et al.* (1988a, 1990)
Trichoderma reesei	Aspartic protease	−	Stepanov (1985)

et al., 1992) with regard to the glycan structures and the enzymology/cytology of glycosylation. Information is much more comprehensive in yeasts (Kukuruzinska *et al.*, 1987) and animal cells (Kornfeld and Kornfeld, 1985).

In yeast, proteins destined either for export or for targetting to the cell wall plasma membrane are translocated and deposited by fusion of vesicles to the plasma membrane. Fusion of vesicles to the tips of growing hyphae has been demonstrated in fungi (Grove, 1978; Hoch, 1986; Chapter 5) and this cytological evidence suggests that hyphal tips are, at least, the principal sites of protein secretion. Direct evidence for protein secretion from the growing hyphal tips of *A. niger* was recently presented by Wösten *et al.* (1991). In their system, glucoamylase, and radiolabelled proteins in general, were specifically secreted from the hyphal tips.

7.4   HOMOLOGOUS EXOENZYMES FROM FUNGI

The extensive range of enzymes secreted from fungi has already been noted. This section concentrates on those enzymes which are most well characterized, and particularly those which have been purified to a high degree or where the encoding genes have been cloned.

7.4.1   Proteases

A given species of fungus will produce a number of different proteases; a proportion of these will be secreted from the mycelia whereas others will be intracellular. There are several known families of proteases and classes of endopeptidase have been described by Barrett and Rawlings (1991). Fungi probably produce representative endoproteases in each of the classes described as well as a range of exopeptidases. Only a small proportion of fungal proteases have been described in detail and many of these studies have been driven because of the implication of proteases in pathogenesis. In addition, there is a desire to eliminate some proteolytic activity from some fungal strains used in heterologous protein production. Secreted fungal proteases which have been characterized in some detail are given in Table 7.1 (see also section 19.4.1(*b*)).

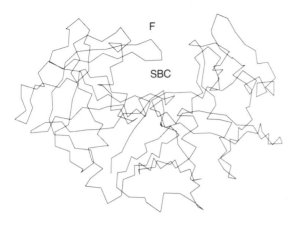

Figure 7.3 Structure of endothiapepsin, courtesy of P. Nugent and V. Dhanaraj. The flap (F) and substrate binding cleft (SBC) are labelled.

The most comprehensive studies have been with the fungal pepstatin-sensitive aspartic proteases (Thompson, 1991; Ward and Kodama, 1991). Sequence alignments of these proteases from different fungi reveal similarities, particularly around the active site aspartic acid residues (Ward and Kodama, 1991). Crystallization of the fungal pepstatin-sensitive aspartyl proteases has permitted structural analysis by X-ray crystallography. Although penicillopepsin, endothiapepsin and rhizopuspepsin have been studied by X-ray crystallography and share features with porcine pepsin and bovine chymosin (Ward and Kodama, 1991), aspergillopepsin has not yet been studied in the same detail. Figure 7.3 shows the deduced three-dimensional structure of endothiapepsin. The deep cleft containing the two active site aspartic acids separates the N- and C- terminal domains.

Interest in fungal proteases is acute although their purification often presents difficulties. However, substantial purification of some fungal proteases has been achieved in a single step by binding to the peptide antibiotics bacitracin (van Noort *et al.*, 1991) or gramicidin-S (Berka *et al.*, 1990). This approach has been successfully applied to the analysis of UV-derived protease-deficient mutants of *A. niger* and has permitted the mapping of some protease-encoding genes in the *A. niger* genome (Mattern *et al.*, 1992).

7.4.2 Amylases

Fungal amylases are used widely in starch processing. Glucoamylase (amyloglucosidase) releases glucose residues from the non-reducing end of starch by hydrolysing $\alpha(1-4)$ links. The enzyme also has a (weak) $\alpha(1-6)$ hydrolysing activity and so acts as a debranching enzyme in the enzymic saccharification of amylopectin. α-Amylase hydrolyses penultimate $\alpha(1-4)$-glycosidic links at the non-reducing end of starch to release maltose.

There is considerable information regarding glucoamylase, a direct result of the commercial uses of the enzyme (Saha and Zeikus, 1989). Various forms of glucoamylase can be distinguished (Ono *et al.*, 1988) although G1 (the full length molecule) and G2 (lacking the starch-binding domain) are the most prevalent (Svensson *et al.*, 1982). Glucoamylase genes (or cDNA genes) have been isolated and sequenced from *A. niger* (Boel *et al.*, 1984), *A. awamori* (Nunberg *et al.*, 1984; Hayashida *et al.*, 1989), *A. oryzae* (Hata *et al.*, 1991), *A. shirousami* (Shibuya *et al.*, 1990), *N. crassa* (Koh-Luar *et al.*, 1989), *Rhizopus oryzae* (Ashikari *et al.*, 1986) and *Humicola grisea* (Allison *et al.*, 1992).

The deduced amino acid sequence of the glucoamylase from *A. niger*, and the sequence from purified protein (Svensson *et al.*, 1983) is identical to that from *A. awamori* and shows 67% homology to the *A. oryzae* glucoamylase. The *A. oryzae* glucoamylase lacks most of the Ser/Thr-rich, heavily O-glycosylated, linker region which separates the catalytic and starch-binding domains of *A. niger* glucoamylase (G1) (Figure 7.2). The *A. oryzae* glucoamylase has a low affinity for raw starch and this, together with the observation that removal of the linker region from the *A. awamori* glucoamylase reduces its binding to raw starch (Hayashida *et al.*, 1989) has implicated the linker region in starch-binding. More recent results with the *A. niger* glucoamylase have shown that the linker region does not *per se* significantly increase the binding affinity for starch (Williamson *et al.*, 1992). Rather, the linker maintains a fixed distance between the catalytic and binding domains and prevents interference between the two domains which could otherwise adversely affect their activities. The domain structure of glucoamylase has been visualized by scanning tunnelling microscopy (Kramer *et al.*, 1993).

A. niger glucoamylases with molecular weights

Table 7.2 Fungal genes encoding pectin-degrading enzymes

Enzyme	Fungal species	Reference
Pectin lyase A	*Aspergillus niger*	Harmsen *et al.* (1990)
Pectin lyase B	*Aspergillus niger*	Kusters-van Someren *et al.* (1992)
Pectin lyase D	*Aspergillus niger*	Gysler *et al.* (1990)
Pectinesterase	*Aspergillus niger*	Khanh *et al.* (1990)
Polygalacturonase I	*Aspergillus niger*	Bussink *et al.* (1991b)
Polygalacturonase II	*Aspergillus niger*	Bussink *et al.* (1990, 1991a)
Polygalacturonase II	*Aspergillus tubigensis*	Bussink *et al.* (1991a)
Polygalacturonase	*Cochliobolus carbonum*	Scott-Craig *et al.* (1990)
Polygalacturonase	*Fusarium moniliforme*	Caprari *et al.* (1993)
Pectate lyase	*Aspergillus nidulans*	Dean and Timberlake (1989)
Pectate lyase	*Fusarium solani*	Gonzalez-Candelas and Kolattukudy (1992)

ranging from 46 kDa to 101 kDa have been described (Ono *et al.*, 1988). Production of this multiplicity of forms is probably a result of proteolysis. It has been suggested, however, that glucoamylases G1 and G2 are produced as a result of differential processing of the mRNA transcript from the single *glaA* gene in *A. niger*. It was proposed (Boel *et al.*, 1984) that the G2 mRNA resulted from splicing of an intervening sequence from the primary (G1) mRNA. However, the proposed splicing event could change the translation frame and give rise to G2 with a C-terminal sequence different from that known from amino acid sequencing of the purified protein (Svensson *et al.*, 1986). Comparison of this sequence with the amino acid sequence of purified G1 (Svensson *et al.*, 1983) indicates that G2 is probably derived from G1 by limited proteolysis.

Genes encoding α-amylase have been isolated from *A. oryzae* (Wirsel *et al.*, 1989; Tada *et al.*, 1989; Gines *et al.*, 1989) and *A. awamori* (Korman *et al.*, 1990). In *A. awamori* two genes (*amy*A and *amy*B) were expressed (Korman *et al.*, 1990). The sequences of the genes (including the eight introns) were identical except at the C-terminus. α-Amylase A ended Tyr-Gly, α-amylase B ended Ser-Ser-Ser (i.e. contained one more amino acid than α-amylase A). The 3′-untranslated regions shared very little homology although the 5′-untranslated regions were identical for 200 bases. The deduced amino acid sequences of the *A. awamori* α-amylases were virtually identical to the *A. oryzae* sequence. *A. oryzae* also appears to contain more than one copy of the α-amylase gene (Gines *et al.*, 1989; Wirsel *et al.*, 1989).

7.4.3　Pectin-degrading enzymes

Commercial modification and degradation of pectins has made extensive use of fungal exoenzymes, principally those derived from *A. niger* although enzymes from *Rhizopus*, *Trichoderma* and *Penicillium* spp. have also been used (Cheetham, 1985). Pectins are complex heteropolysaccharides comprising a backbone structure of poly-D-galacturonic acid interspersed with L-rhamnose residues and side chains predominated by L-, arabinose, D-galactose, D-xylose and other minor sugars. The backbone galacturonic acid residues may be modified, e.g. by *O*-acetylation and methyl-esterification. Thus, for complete degradation of pectins, a number of different enzyme activities are required (Rombouts and Pilnik, 1980) and many studies have been made of the ability of filamentous fungi to degrade pectins under various conditions (e.g. Maldonado *et al.*, 1989).

Isoenzymic forms of the various pectin-degrading enzymes may exist, depending on the fungal species and, presumably, the environmental conditions affecting enzyme production. Five distinguishable endopolygalacturonases have been purified from *A. niger* (Kester and Visser, 1990) and the genes encoding two of the enzymes (PGI and PGII) have been isolated and sequenced (Bussink *et al.*, 1990, 1991b) (Table 7.2). The deduced amino acid sequences of the two proteins showed 60% identity and differences in codon usage were observed (Bussink *et al.*, 1991b). Plant pathogenic fungi may secrete different pectinases in response to environmental conditions and it is known that some fungi, e.g. *Sclerotina sclerotiorum* (Marciano *et al.*, 1982), *Verticillium albo-atrum* (Durrands and Cooper, 1988), produce more than one polygalacturonase whereas another species, *Cochliobolus carbonum* (Scott-Craig *et al.*, 1990), appears to produce only one.

7.4.4 Cellulases

A large number of fungi produce extracellular cellulases (Eriksson *et al.*, 1990). The ability to fully solubilize crystalline cellulose is restricted to relatively few of these (Wood and Garcia-Campoyo, 1990). The distinguishing feature of this group is the production of cellobiohydrolase activity. The most detailed enzymological and genetical studies have been with the soft rot fungi such as *Trichoderma* spp., particularly *T. reesei* and *T. viride* and more recently with *Phanerochaete chrysosporium*, a white rot basidiomycete (Cullen and Kersten, 1992). These fungi produce at least three classes of extracellular activity during growth on cellulose. These are endo $\beta(1-4)$-glucanases (EG) which act to randomly cleave $\beta(1-4)$ β linkages within the cellulose molecule. These enzymes are often assayed by viscosity loss of carboxymethylcellulose (CMC), a soluble cellulose. They do not, when purified, show significant activity on crystalline cellulose. Many fungi have been shown to produce these endoglucanase activities.

The true cellulolytic fungi produce exo $\beta(1-4)$-glucanases (cellobiohydrolase, CBH) and the enzyme from *T. reesei* will remove cellobiose from the reducing ends of cellulose chains. For *P. chrysosporium* both glucose and cellulose are liberated (Cullen and Kersten, 1992). These enzymes have little or no activity on CMC. The third component is normally a $\beta(1-4)$-glucosidase activity which can hydrolyse cellobiose to glucose. Production of these enzymes by these fungi is subject to a number of regulatory controls including induction, repression and feedback inhibition (Eveleigh, 1987). A model describing an integrated scheme for full solubilization of cellulose indicates that primary attack on cellulose is by endoglucanases acting on the amorphous, less structured, regions of cellulose, followed by exo-cleavage by CBH and complete hydrolysis to cellobiose by glucosidases (Eveleigh, 1987). In addition to the hydrolytic enzymes, oxidative enzymes have been implicated in fungal cellulose degradation, e.g. cellobiose oxidase, (Eriksson *et al.*, 1990).

As well as a multiplicity of the major activity classes of cellulases it has been shown that more than one distinct protein of each class is found in culture filtrates. Thus, *Trichoderma* spp. filtrates have been shown to contain from four to eight endoglucanases, two CBHs and one or two β-glucosidases (Wood and Garcia-Campoyo, 1990). This multiplicity has been shown to be due in various fungi to genetic differences (distinct non-allelic genes), but also to differing extents of glycosylation, partial proteolysis from parent molecules, aggregation and purification artefacts. This multiplicity has led to considerable problems in assigning true substrate specificity to individual components and thus to determining the probable mode and sequence of attack for complete cellulose solubilization (Wood and Garcia-Campoyo, 1990). Substrate specificity analysis of purified components can be confounded by trace quantities of impurities with differing substrate specificity. Purified exoglucanases can apparently also act in an endo fashion (Enari and Niku Paavola, 1987).

The different activities of the cellulase complex act in a synergistic fashion to degrade cellulose (Wood and Macrae, 1979). Synergy occurs both between and within substrate specificity classes. Thus both exo-endo synergism can occur as well as endo-endo synergism. Such synergism extends to mixtures of enzymes from different fungi. The model of attack on cellulose described above includes the possibility of sequential attack but does not explain synergism. The difficulties of producing a satisfactory explanation for the substrate specificities of apparently purified cellulase components, their mode of attack, the significance of synergistic action as well as a biotechnological impetus to genetically engineer altered enzymes or to produce cellulase preparations of differing composition have led to considerable work on gene cloning, sequence analysis and structural analysis for fungal cellulases and also bacterial cellulases. Thus, for *T. reesei*, sequences of genomic and cDNA clones have been produced for *cbh1*, *cbh2*, *egl1*, *egl3* and *bgl1* (Cullen and Kersten, 1992). From such studies it has been concluded that all cellulases consist of three major regions, a catalytic core, a heavily glycosylated 'hinge' region, rich in proline, threonine and serine residues and a cellulose binding domain (CBD). In the best studied cases there are very conserved regions either at the C-terminal (for CBHI, EGI) or N-terminal region (CBHII, EGIII). CBHI and CBHII proteins have been crystallized and X-ray diffraction studies have been used to deduce that the gross domain structure of the protein has similarities to that of glucoamylase (Figure 7.2). The X-ray crystallography studies have been extended to identify the gross structure of active site regions. Thus for CBHI the catalytic core region contains a tunnel shaped active site containing four glucosyl binding sites. Comparison with similar studies of endoglucanases shows that the latter contains a more open site or 'groove', which may account for the difference in substrate specificities. In addition to X-ray studies, chemical active site modification has been used to deduce that the Glu_{126} residue is important for catalysis by CBHI. Hydrophobic cluster analysis has been used to categorize sequence data into cellulase families and also indicates sequence conservation around Glu_{126}.

7.4.5 Ligninases

Lignin is a biopolymer composed of phenyl-propanoid units. It is synthesized by condensation reactions of free radicals to give an amorphous polymer with several different carbon–carbon and carbon–oxygen linkages (Kirk and Farrell, 1987; Holzbaur *et al.*, 1991). This chemistry places certain constraints on the possible enzymic system necessary for its degradation. Such enzymic systems have to be extracellular, non-specific and non-hydrolytic (Kirk and Farrell, 1987). Lignin is localized as an amorphous polymer encrusting cellulose microfibres within plant cell walls and acts to confer mechanical strength and resistance to microbial attack.

Studies of the physiology of whole cultures of ligninolytic white fungi, including *P. chrysosporium*, show that the lignin degradation system is unusual in being expressed only in the secondary growth stage under conditions of nutrient limitation, e.g. carbon, nitrogen or sulphur, and control by oxygen partial pressure, agitation, pH and metal ion balance (Cullen and Kersten, 1992).

Two classes of ligninases have now been found. The first is referred to as lignin peroxidases (LIPs), and the second as manganese peroxidases (MnPs). The lignin peroxidases are haem proteins and belong to the general family of peroxidases. Considerable spectroscopic and kinetic evidence has accumulated indicating that the enzyme mechanism resembles that of other peroxidases

(Cullen and Kersten, 1992). The enzymes are capable of oxidizing a variety of compounds using a single mechanism, namely the one electron oxidation of the aromatic nuclei of various substrates. This produces cation radicals which degrade spontaneously via non-enzymic reactions to yield a variety of aliphatic and aromatic end products.

The lignin peroxidases (LIPs) comprise a family of isoenzymes (Holzbaur *et al.*, 1991; Cullen and Kersten, 1992) and are found only in the secondary growth stage of *P. chrysosporium* when they are the most abundant extracellular proteins. They are easily purified by concentration and ion-exchange chromatography and subsequent resolution by isoelectric focusing. All the lignin peroxidases are structurally related and are glycoproteins and haem-proteins with a molecular weight range of 38–42 kDa. They show structural similarities by peptide mapping and by cross-reactivity to polyclonal antibodies (Kirk *et al.*, 1986; Farrel *et al.*, 1989), but these antibodies only weakly cross-react with the manganese peroxidases. Structural similarities are due to the presence of a family of closely related genes and their allelic variants. Cloning of several lignin peroxidases for both cDNA and genomic DNA has been achieved (Holzbaur *et al.*, 1991; Cullen and Kersten, 1992). There are at least six related genes and their allelic variants and some of these genes are linked. The identity of some cDNA clones and genomic clones has been assigned to individual isoenzymes by amino acid sequence. The cloned ligninase genes show an overall similar structure comprising several (up to nine) small introns interspersed into the sequence. They all have glycosylation sites for both *O*- and *N*- glycosylation. Sequence analysis has revealed two conserved His residues thought to be involved in peroxidase activity, indicating that the tertiary structure resembles that of other peroxidases.

Manganese peroxidases (MnPs) also comprise a family of isoenzymes in the ligninolytic culture fluids of *P. chrysosporium*, and are also readily purified (Alic and Gold, 1991). Manganese peroxidases oxidize Mn^{2+} to Mn^{3+} using H_2O_2 as an oxidant. The Mn^{3+} species is then used to oxidize a variety of organic compounds including those found in lignin. The MnPs are also haem proteins and exist as several isoenzymes, of average molecular weight around 46kDa, and are glycosylated. The peptide maps and cross-reactivity to polyclonal antibodies

Table 7.3 Other secreted fungal enzymes

Enzyme	Species	Gene (or cDNA) cloned	Reference
Arabinanase	*Aspergillus niger*	–	Rombouts *et al.* (1988)
	Aspergillus nidulans	–	D. Ramón (personal communication)
Cutinase	*Fusarium solani*	+	Soliday *et al.* (1984)
α-Galactosidase	*Aspergillus niger*	+	den Herder *et al.* (1992)
	Aspergillus tamarii	–	Civas *et al.* (1984)
α-Glucosidase	*Aspergillus niger*	–	Kimura *et al.* (1992)
Lipase	*Aspergillus niger*	–	Sugihara *et al.* (1988)
	Humicola lanuginoas	+	Boel and Huge-Jensen *et al.* (1989)
	Rhizomucor miehei	+	Boel *et al.* (1988)
Phosphatase	*Aspergillus nidulans*	+	Caddick and Arst (1987)
	Aspergillus niger	+	MacRae *et al.* (1988)
Phytase	*Aspergillus ficuum*	+	van Gorcom *et al.* (1991)
	Aspergillus oryzae	–	Wang *et al.* (1980)
RNase	*Rhizopus niveus*	+	Horiuchi *et al.* (1988b)
	Aspergillus restrictus	+	Yang and Kenealy (1992)
Xylanase	various spp.	–/+	Smith *et al.* (1991)
			Tenkamen *et al.* (1992)
			Xue *et al.* (1992)
β-Glucanases	various spp.	–	Pitson *et al.* (1993)
Glucose oxidase	*Aspergillus niger*	+	Kriechbaum *et al.* (1989)
Invertase	*Aspergillus nidulans*	+	Berges *et al.* (1993)

indicate considerable differences to the lignin peroxidases. Both cDNA and genomic clones of manganese peroxidase have been obtained and sequenced. There are several (six) small introns interspersed in the genomic sequence, but at distinct sites to the LIP intron sites.

7.4.6 Xylanases

Xylan is the major hemicellulose component of plant cell walls. Xylans are often isolated as heteropolymers with a backbone of (1–4)-linked xylose units with variable degrees and types of substitution depending on the plant source. Xylans can also contain, when isolated, L-arabinose, D-glucuronic acid and acetylated sugars, as well as ester linked *p*-coumaric and ferulic acids. This substrate complexity is reflected in the large number of possible enzymes that are associated with depolymerization of xylan (Cullen and Kersten, 1992). For xylans these enzymes include endoxylanases, xylosidases, glucuronidases, arabinofuranosidases, and acetyl esterases (Thompson, 1993).

The occurrence of xylanases has been reported from a considerable number of fungi (Wong *et al.*, 1988). Regulation of xylanase production is controlled by induction of suitable inducing sources such as xylans or celluloses. Regulatory control can be either associated with, or independent of, the production of cellulases (Eriksson *et al.*, 1990). Their physicochemical properties and modes of action have been tabulated (Eriksson *et al.*, 1990; Bastande, 1992). They are normally low-molecular-weight enzymes (10–40 kDa) with pH optima in the acidic–neutral range.

Relatively little is known of the molecular architecture of fungal xylanase genes. Two xylanases with molecular weights of 19 and 21 kDa have been purified from *T. reesei* (Törrönen *et al.*, 1992). These also differ in their isoelectric points, 5.2 and 9.0, and account for over 90% of the culture filtrate xylanolytic activity.

7.4.7 Other secreted fungal enzymes

A large variety of enzyme activities can be detected in fungal cultures other than those described above. Some of these are listed in Table 7.3 but only those which have been extensively purified or for which the gene has been cloned. Secretion is taken to include those enzymes which are external to the cell membrane but which may be retained by the cell wall.

7.5 FUNGAL EXOENZYMES IN INDUSTRY

7.5.1 Fungal enzymes and food

The use of fungal enzymes in industry, especially in relation to food, is very well documented (e.g. Godfrey and Reichelt, 1983; Cheetham, 1985; Campbell-Platt and Cook, 1989; Lowe, 1992). The fungal exoenzymes sold as bulk commodities are used in a variety of industries (Table 7.4). Other fungal enzymes are extracted from fungal mycelia, notably glucose oxidase and penicillin amidase. In addition to the enzymes listed in Table 7.4, several thermophilic fungi (growth at ca. 40–50°C) show promise as enzyme producers (Satyanarayana *et al.*, 1992); indeed, *Thielavia terrestris* appears in Table 7.4 as an important cellulase producer. Specialist enzymes produced in lesser quantities at higher purity for research purposes are also obtained from fungi. A large list of such enzymes reported to be produced from fungi deposited in the American Type Culture Collection is given by Lowe (1992). The extensive use of glucoamylase and fungal 'rennet' merit some attention here.

Glucoamylase is produced principally from *A. niger* and *A. awamori* although other aspergilli and *Rhizopus delemar* are also used (Lowe, 1992). Glucoamylase releases β-glucose molecules from the non-reducing end of starch due to its hydrolytic activity on α(1–4) links. The enzyme also has a debranching (α(1–6) and α(1–3)) activity although this is weak in relation to the activity on α-(1–4)-linked glucose. Glucoamylase, in conjunction with the endoglucanase α-amylase, is used in the production of high glucose syrups (Lowe, 1992).

More than half the cheese produced worldwide now employs fungal acid protease (e.g. from *Rhizomucor miehei*) instead of calf chymosin (rennet) (Lowe, 1992). The use of filamentous fungi as hosts for the production of chymosin is discussed in detail in section 7.6.2. Heterologous chymosin has already claimed a large share of the market from both rennet and fungal proteases.

Fungi are used as a source of enzymes (above) but also in the production of fermented foods. These fungi are obligately aerobic and do not themselves perform the fermentation. Rather, the fungi are the source of degradative enzymes which convert the raw material into a suitable substrate for subsequent fermentation by bacteria or yeasts (Yokotsuka, 1991a, b). Alternatively, fungi may impart desirable flavours to a product such as certain cheeses (Kinderlerer, 1989; Marth and Yousef, 1991).

The role of fungal enzymes in production of fermented foods is best characterized in the production of fermented soybeans and cereals (e.g. Shoyu and Miso). A variety of fungi are exploited to initiate degradation of the raw material through their ability to secrete hydrolytic enzymes which degrade the plant tissue. The 'moulded' product is referred to in Japan as koji and this then contributes the starting point for subsequent fermentation by bacteria and/or yeasts and further processing. A variety of fungi are used although in Japan they tend to be species of *Aspergillus*, e.g. *A. oryzae*, *A. sojae* and *A. niger*. A greater variety of fungi are used in other countries and the range is extended to include *Rhizopus* spp. (e.g. *oligosporus*), *Mucor* spp., *Penicillium* spp., *Neurospora* spp. and *Fusarium* spp. The fungi used are not always thoroughly characterised although increasing attention is being paid to the strains and enzymes involved (Sakaguchi *et al.*, 1992). Understandably, concern is mainly focused on the practical production of the moulded material and levels of any mycotoxins.

The role of fungi and enzymes involved in production of Tempe has recently been reviewed in depth (Nout and Rombouts, 1990). Tempe is the generic name of fermented foods made in Indonesia from a range of raw materials. Soybeans, for example, are used to make tempe kedele. Of the moulds used in the production of tempe, *Rhizopus* spp. (*oryzae* and *oligosporus*) are the most common. *Rhizopus* spp. secrete a wide range of enzymes which, *inter alia*, facilitate the growth of the organism on the raw material and render the material suitable for use as a substrate by bacteria and yeasts which represent the normal flora of tempe. *Rhizopus oligosporus* produces pectinolytic, cellulolytic, proteolytic and lipolytic enzymes. Other activities have also been reported and, among these, is phytase (myo-inositol hexakisphosphate phosphohydrolase). Phytase is undesirable at high concentration in food and animal feed because of its capacity to complex several metals thus restricting their absorption from the gut. The phosphate in phytate is also unavailable

Table 7.4   Industrial fungal exoenzymes[a]

Enzyme	Principal species	Enzyme use
α-Amylase EC 3.2.1.1	*Aspergillus niger* *Aspergillus oryzae*	Starch processing High maltose syrup production
Amyloglucosidase (glucoamylase) EC 3.2.1.1	*Aspergillus niger* *Aspergillus oryzae*	Starch processing Glucose syrup production
Cellulase EC 3.2.1.4	*Aspergillus niger* *Trichoderma reesei* *Thielavia terrestris*	Fruit and vegetable processing Waste treatment Silage fermentations
Dextranase EC 3.2.1.11	*Penicillium* spp.	Cane sugar processing
α-Galactosidase EC 3.2.1.22	*Aspergillus niger*	Beet sugar processing
β-Galactosidase EC 3.2.1.23	*Aspergillus niger* *Aspergillus oryzae* *Penicillium emersonii*	Dairy products industry
β-Glucanase EC 3.2.1.6	*Aspergillus niger* *Aspergillus oryzae* *Penicillium emersonii*	Beer and wine production
Hemicellulase EC 3.2.1.78	*Aspergillus niger* *Aspergillus oryzae*	Fruit and vegetable processing
Lipase EC 3.1.1.3	*Aspergillus niger* *Rhizomucor miehei* *Rhizopus* spp.	Cheese ripening Waste treatment
Pectinase EC 3.2.1.15	*Aspergillus niger* *Rhizopus oryzae*	Production of wines and preserves
Acid protease EC 3.4.23.6	*Aspergillus niger* *Aspergillus oryzae* *Rhizomucor miehei* *Endothia parasitica*	Cheese production using 'microbial rennet' Meat and fish processing Detergents
Neutral protease EC 3.4.23.4	*Aspergillus oryzae*	
Tannase EC 3.1.1.20	*Aspergillus oryzae* *Aspergillus niger*	Waste treatment Beverage processing
Xylanase EC 3.2.1.32	*Aspergillus niger*	Processing of cereals, tea, coffee, cocoa, chocolate

[a] Adapted from Lowe (1992).

because of the poor capacity of man and monogastric animals to degrade it. Several species of fungi used in food fermentations produce and secrete phytase (Wang *et al.*, 1980). In this work, production of phytase by *R. oligosporus* occurred in rice flour (2% in water) but not in a synthetic medium and although some phytase was secreted most was intracellular. In contrast, higher levels of phytase were secreted by several strains of *A. oryzae* used in either miso or soy sauce manufacture. Medium composition had a profound effect on the levels of phytase recorded (Wang *et al.*, 1980) but a detailed study of the control of gene expression and phytase production has not been carried out. Phytase is, however, sufficiently important an enzyme for use in animal feeds that high level heterologous production of phytase is now a commercial reality (van Gorcom *et al.*, 1991).

The enormous range of fermented foods involving fungi has been reviewed by Yokotsuka (1991a, b). The fungi used have characteristic associated enzyme activities and, for example, the *Rhizopus* spp. used in tempe production are generally less proteolytic than the *Aspergillus* spp. used for koji. Nakadai (1985) reported some analysis of the distinguishable proteases produced by *A. sojae*. Three acid, two neutral and two alkaline proteases were identified. A more detailed survey of the proteases, and other enzymes, produced by fungi is given in section 7.4.

Fungi play an important part in cheese manufacture as sources of microbial rennet and in cheese ripening. It is cheese ripening which concerns us here because desirable flavours result from cheese maturation with appropriate moulds (Marth and Yousef, 1991). Both surface-ripened cheeses such as Camembert and Brie and the blue-veined cheeses such as Rocquefort and Stilton are ripened with *Penicillium* spp. although yeasts contribute to the development of flavour.

7.5.2 Fungal enzymes in waste treatment

Fungi exhibit sufficient enzymic versatility to be capable of degrading a variety of xenobiotics (Raj *et al.*, 1992) and organic compounds commonly found in industrial effluents (Nandan and Raisuddin, 1992). The role of fungal exoenzymes in lignocellulose degradation in industrial wastes and refuse is discussed here although the conversion of lignocellulose into fungal biomass is discussed separately below.

Several fungi contain cytochrome *P*450 which is implicated in mono-oxygenation of xenobiotic compounds. Such enzymic conversions are, however, intracellular and beyond the scope of this chapter. Similarly, the many other bioconversions catalysed by fungal enzymes (Raj *et al.*, 1992) are normally intracellular, although culture supernatants of *A. niger*, *P. cyclopodium*, *M. mucedo* and *T. reesei* were reportedly able to degrade the pesticide trichlorophan (Nandan and Raisuddin, 1992) indicating that exoenzymes were probably responsible.

Fungal degradative enzymes, in association with

Table 7.5 Mushroom production on lignocellulose

Species	Substrate
Agaricus bisporus	Composted straw
Lentinus edodes	Logs/sawdust
Volvariella volvacea	Straw/cotton waste
Pleurotus sp.	Straw/sawdust
Flammulina velutipes	Sawdust/cotton waste
Pholiota nameko	Logs/sawdust
Auricularia sp.	Logs
Tremella fuciformis	Logs

growing mycelia, are involved in the degradation of lignocellulose wastes, e.g. in pulp and paper mill wastewater, and in domestic refuse disposed to landfill. Waste degradation in landfills is brought about primarily by anaerobic processes. There, the degradative microorganisms are mainly bacteria or protozoa and, although anaerobic fungi play a part in cellulose degradation in the anaerobic environment of the bovine rumen (Orpin, 1975), the presence of anaerobic fungi in landfills, for example, has been discounted (Theodorou *et al.*, 1989). When refuse is placed in landfills there is a period of aerobic degradation during which enzymes presumed to be of fungal origin play a part in the solubilization of the pectinaceous, proteinaceous and lignocellulosic materials (Grainger *et al.*, 1984). Breakdown of lignin is an obligately aerobic process and wood-rotting fungi such as *Phanerochaete chrysosporium* have been exploited in waste treatment systems (e.g. Hammel, 1989). Many fungi also have decolourizing activities, which need not be due to exoenzymes, but which have added to their attractions for waste treatment systems (Sirianuntapiboon *et al.*, 1988).

7.5.3 Conversion of lignocellulosics into fungal biomass

The most economically successful technology for lignocellulose conversion is the production of mushrooms (Wood and Smith, 1987; Wood, 1989). About 10 species of these fungi are cultivated on a significant scale (Table 7.5). The substrates used, normally wastes from agriculture or forestry, are either inoculated directly or subjected to various microbial or physical pretreatments such as composting or pasteurization (Wood and Smith, 1987). Conversion of the substrates into fungal biomass is then mediated by the secretion of a range of hydrolytic and oxidative enzymes. Most of these fungi can efficiently degrade cellulose or hemi-celluloses. Certain of them such as *Agaricus* spp. and *Lentinus* spp. have been shown to be able to degrade lignin. *L. edodes* has been shown to be able to produce lignin peroxidase (Forrester *et al.*, 1990). It has been shown that *A. bisporus* can also degrade microbial polymers such as peptid-oglycans and glucans. Although the major role of these extracellular enzymes is nutritional it has been shown that changes in enzyme activities of laccase and endoglucanase of *A. bisporus* are also correlated with fruit body formation (Wood, 1985).

Fungal treatment of lignocelluloses has also been used to produce animal feeds (Reid, 1989; Coughlan and Amaral Collaco, 1990). Similar technology to mushroom cultivation is used with the fungus being inoculated into suitably treated materials such as straw or woodchips. A variety of fungi have been screened and *Pleurotus* spp. and other white rot basidiomycetes produce the most favourable results in terms of increase in digesti-bility (Reid, 1989). Solid substrate treatment has also been proposed for the production of paper pulps by conversion of woodchips (Eriksson *et al.*, 1990).

7.5.4 Strain selection for fungal exoenzyme producers

The objective of strain selection is to produce more of the target enzyme more cheaply. Because fungi secrete more than one enzyme the target enzyme should ideally be produced at higher absolute levels and in a higher proportion relative to the other secreted proteins. Alternatively, a commer-cial enzyme preparation may have a desired mix of activities but the precise mixture could be manipu-lated by strain selection and development.

Provided an effective screen is available, two of the most effective methods for strain improvement in regard to exoenzyme production have been mutagenesis and parasexual recombination between strains (Ball, 1984). Parasexual crosses are used because industrial strains of filamentous fungi generally lack a sexual cycle. A few examples are given by Ball (1984) of improved enzyme yields achieved either through mutagenesis or through parasexual crossing, leading to strains with altered characteristics including higher ploidy. A summary of the mutagenesis strategy aimed at improving cellulase production by *T. reesei* has been given by

Durand *et al.* (1988). The cellulolytic complex secreted by *T. reesei* has endoglucanase (EC 3.2.1.4), cellobiohydrolase (EC 3.2.1.91) and β-glucosidase (EC 3.2.1.21) activities. Improvements in the activity of the cellulase complex towards crystalline cellulose, and higher yields not subject to carbon catabolite repression or the need for cellulose induction are sought. The laborious mutagenesis and screening programme described by Durand *et al.* (1988) produced a strain with a four-fold improvement in cellulase yield and reduced susceptibility to repression by glucose and reduced requirement for cellulose as the produc-tion substrate. Also, the relative proportion of β-glucosidase was increased. Even so, the nature of the mutations generated by such approaches is obscure and optimization of yield has been achieved not only by strain development but also by cultural conditions (Allen and Roche, 1989).

Mutagenesis and screening have been success-fully applied in the optimization of yield of a number of enzymes other than the cellulases (Ball, 1984) and including the commercially very import-ant glucoamylase from *A. awamori* (Nevalainen and Palva, 1979). Mutagenesis with the intention of reducing the levels of secreted proteases has also met with some success in raising levels of target enzymes such as α-amylase (Hayashida and Teramoto, 1986) and glucoamylase (Hayashida and Flor, 1981). Hybrid strains can also be pro-duced by protoplast fusion, followed by screen-ing for desired phenotypes (Ogawa *et al.*, 1988; Ushijima *et al.*, 1990).

An alternative strategy to strain improvement is genetic manipulation, made possible by major advances in fungal molecular biology in recent years. This approach has been particularly success-ful, and necessary, with production of hetero-logous enzymes from fungi (Section 7.6.2) but has also shown promise with homologous exo-enzymes such as the cellulase complex from *T. reesei*. Several of the genes (*cbh*1, *cbh*2, *egl*1, *egl*3, *bgl*1) have now been cloned (Kubicek *et al.*, 1993). The strong *cbh*1 promoter was used to drive expression of the *egl*1 gene (Harkki *et al.*, 1991; Uusitalo *et al.*, 1991) to alter the properties of the secreted cellulase complex. Also, the *cbh*1 gene has been replaced with an inactivated gene so that strains producing elevated endoglucanase and/or no cellobiohydrolase have been constructed (Harkki *et al.*, 1991).

7.5.5 Strain improvement for lignocellulolysis and mushroom production

It has been shown that fungal pretreatment of woodchips can reduce energy use in mechanical production of paper pulp, and also enhance resulting paper strength (Eriksson *et al.*, 1990). To improve pulping performance further elimination or reduction of fungal cellulase activity is required. Various mutants of white rot fungi with loss of, or lowered levels of, cellulase have been studied (Eriksson and Kirk, 1985). Such mutants reduce energy input into pulping and enhance resulting paper strength. As with analogous mutants in other fungi the resulting strains are genetically ill-defined and exhibit pleiotropic effects for the production of other extracellular enzymes such as phenol oxidases.

Mutants in white rot fungi lacking other enzymes possibly involved in lignin degradation such as lignin peroxidase, manganese peroxidase and glucose oxidase have also been produced (e.g. Ramasamy *et al.*, 1985). In other instances various overproducing mutants have been obtained such as lignin peroxidase overproducers (Kirk *et al.*, 1986).

Despite the significance of extracellular enzyme production for edible mushroom production relatively little has been achieved in strain selection for altered enzyme levels. For the most important species *A. bisporus*, this is due to the difficulties posed by the life cycle. No stage exists where a single haploid nucleus is found in a single cell compartment (Elliott, 1985). For these reasons a model basidiomycete, *Coprinus bilanatus* has been used to produce enzyme mutants (Stephens *et al.*, 1991; Sodhi *et al.*, 1991). A range of mutants has been produced of similar phenotype to those found in lower fungi such as *Trichoderma* spp. These include overproducers, negative mutants and mutants insensitive to glucose repression of cellulase production. Some of the mutants show pleiotropic effects for enzyme production but interestingly behave as single gene recessives in segregation analysis.

7.6 EXOENZYME PRODUCTION BY RECOMBINANT FUNGI

The ability of many commercial fungi to secrete homologous enzymes at high yield has made them attractive candidates as hosts for heterologous enzyme production. A host for heterologous enzyme production should ideally be readily transformed, the recombinant strain should be stable, the recombinant enzyme should be authentic and easily extracted from the host to a good yield. Secretion of the target heterologous enzyme aids purification and, in the case of eukaryotic hosts, permits glycosylation of the enzyme. Expression and secretion of a heterologous enzyme in a fungus may be desired in order to produce commercial amounts of an enzyme, to produce sufficient amounts of an enzyme (perhaps a mutant enzyme with no natural host) for research purposes or to produce a strain of fungus with new characteristics. Each of these objectives has been met with heterologous enzyme production in fungi (Jeenes *et al.*, 1991; van den Hondel *et al.*, 1991; Archer, 1994).

7.6.1 Host strains and selectable vectors

The most successful hosts for heterologous protein secretion have been species of *Aspergillus* and *Trichoderma* although other transformable species have also been used (Jeenes *et al.*, 1991). The basis of selection for transforming DNA can be by complementation of an auxotrophic marker provided a suitable recipient host is available. The most common markers are *nia*D, *pyr*G, *arg*B and *trp*C. Alternatively, positive selection markers have been used based on drug resistance, e.g. *hph, oli*C, or a nutritional phenotype, e.g. *amd*S. A full survey of markers used and details on transformation in fungi have been given elsewhere (Fincham, 1989; Goosen *et al.*, 1992; see also section 12.3.1).

Expression vectors used in heterologous enzyme production with fungi have all been integrative. This has the advantage of offering strain stability. An autonomously replicating plasmid for use with *A. nidulans* (Gems *et al.*, 1991) may be adapted for use as an expression vector with other species. Other autonomously replicating plasmid vectors are being developed for use with a variety of fungal species and a number of species possess nuclear or mitochondrial plasmids. The isolation of telomeres and centromeres from fungi will permit the development of fungal artificial chromosomes (Goosen *et al.*, 1992) but integrative expression vectors are likely to dominate in heterologous expression studies because of demonstrated mitotic stability.

Transforming DNA can be integrated by either homologous or heterologous recombination. Also, gene replacement can occur via a double cross-over event, particularly with linear transforming DNA. For heterologous enzyme production maximized yield of the target enzyme is usually an objective and it is becoming increasingly clear that the locus within a genome can have a profound effect on the level of transcription. Thus, site-specific integration of an expression cassette to a locus of known high transcriptional activity is often desirable and this is often achieved by replacement of a known gene, e.g. *gla*A in *A. niger* (van den Hondel *et al.*, 1992) or *cbh1* in *T. reesei* (Harkki *et al.*, 1989). Integration at a known locus also avoids unwanted effects such as disruption of a gene which may result in loss of a desired character or the acquisition of a new character which may even be harmful (Chapter 12).

7.6.2 Optimizing production of heterologous enzymes

Gene expression, protein secretion and stability of the secreted enzyme are the three concerns for optimizing secreted enzyme levels. A number of different promoters have been employed to drive expression, and these include constitutive and inducible promoters (van den Hondel *et al.*, 1991; Jeenes *et al.*, 1991). Heterologous gene expression has been achieved with homologous promoters and also heterologous promoters transferred from other fungi; notably between various *Aspergillus* spp. but also between fungal genera (Jeenes *et al.*, 1991; Davis and Hynes, 1991). Although strong promoters have been sought to drive expression of heterologous genes it is recognized that measured mRNA levels are determined both by promoter activity and mRNA stability (Davies, 1991). Also, mRNA levels are not necessarily a good guide to the secreted protein levels because post-transcriptional events are important despite control at the transcriptional level. The genomic locus of the integrated expression cassette affects transcription. For example, in *T.reesei*, the prochymosin gene was expressed under the control of the *cbh1* promoter and the expression cassette was integrated specifically at the *cbh1* locus or elsewhere (Harkki *et al.*, 1989). Chymosin production was highest with the expression cassette integrated at the *cbh1* locus although the measured

chymosin mRNA and secreted chymosin protein levels were 1–2 orders of magnitude lower than the equivalent levels of both the *cbh1* mRNA and CBHI enzyme. This work suggests transcriptional limitation of chymosin production in *T. reesei*. In contrast, the importance of post-transcriptional events in the production of chymosin from *A. awamori* is more pronounced. In this work, chymosin production was driven by the *glaA* promoter and chymosin mRNA levels were similar, if slightly lower, than the glucoamylase RNA levels whereas the secreted protein level of glucoamylase was at least two orders of magnitude higher than that of chymosin (Ward, 1991). Similarly, in a separate study, integration of a *glaA* promoter–*glaA* signal sequence prochymosin cassette at the *glaA* locus (by gene replacement) resulted in a measured level of chymosin mRNA similar to that of *glaA* mRNA in the untransformed strain (van den Hondel *et al.*, 1992). With the same constructs, the level of secreted prochymosin protein was about one quarter that of secreted glucoamylase from the untransformed strain (van den Hondel *et al.*, 1992). The effects of gene copy number on production of a homologous protein (van den Hondel *et al.*, 1992) and a heterologous protein (Berka *et al.*, 1991; Dunn-Coleman *et al.*, 1991) have also been assessed and it was found that, although higher protein yields can be achieved with higher gene copy numbers, there is no strict correlation. Secreted levels of glucoamylase increased with increasing number of *glaA* genes up to about 20 (van den Hondel *et al.*, 1992). At higher gene copy numbers it was not possible to predict the secreted yield of glucoamylase and it was recognized that the sites of integration will be important and that regulation of gene expression can be complex; for example, gene-specific and regulatory proteins can be titrated out in fungi (Davis and Hynes, 1991). In species with a sexual stage, gene inactivation of duplicate genes can occur premeiotically either by MIP (methylation induced premeiotically) (Rossignol and Picard, 1991) or RIP (repeat-induced point mutation) (Selker, 1990). (See section 11.4.1.)

The use of gene fusions has been particularly rewarding in the enhancement of secreted heterologous protein yields from filamentous fungi. Typically, the gene encoding the target protein is fused downstream of the gene for a naturally well-secreted homologous protein such as glucoamylase

This approach has proven to be valuable for secretion of bovine chymosin from *A. awamori* (Ward *et al.*, 1990), human interleukin-6 from *A. nidulans* (Contreras *et al.*, 1991) and *A. niger* (van den Hondel *et al.*, 1992) and porcine phospholipase A$_2$ (Roberts *et al.*, 1992). Although chymosin autocatalytically cleaves itself from the fusion protein, it is usually necessary to introduce a processing site for release of the target protein. The *S. cerevisiae* dibasic amino acid *KEX2* site has proven to be recognized in *Aspergillus* spp. (Contreras *et al.*, 1991; van den Hondel *et al.*, 1992). Whether a fusion to the entire carrier protein is necessary for optimized secretion is not known because the reasons underlying good or poor protein secretion are not understood. Although some studies have investigated the role of different signal sequences, propeptide sequences and the *N*-terminal region of glucoamylase on the secretion of chymosin from *Aspergillus* spp. (Cullen *et al.*, 1987; van den Hondel *et al.*, 1992) a comprehensive analysis of glucoamylase fusions has not been reported. Fusion of hen egg-white lysozyme (HEWL) to truncated forms of glucoamylase is beginning to reveal details of the useful secretory features of glucoamylase as a carrier protein and secreted yields of HEWL can be much higher than when expressed without fusion to glucoamylase (Jeenes *et al.*, 1993).

Yields of secreted heterologous proteins can be dramatically reduced by proteolytic degradation. This is a more acute problem than in the production of bulk homologous enzymes because the initial yields of heterologous proteins are so much lower. The problem of proteolysis of homologous proteins has been tackled by the generation of strains deficient in proteases and this approach has also been used to increase heterologous protein yields in *Aspergillus* spp. (Berka *et al.*, 1990; van den Hondel *et al.*, 1992; Roberts *et al.*, 1992) and *T. reesei* (Uusitalo *et al.*, 1991; Nevalainen *et al.*, 1991). Whereas these approaches have mainly been pursued by resort to mutagenesis and screening for lowering of protease activity, specific deletions of the aspergillopepsin gene have also been achieved in *A. awamori* (Berka *et al.*, 1991) and *A. niger* (Mattern *et al.*, 1992). This approach can become more widely adopted as more fungal proteases are purified and their genes cloned. Some *A. niger* protease genes have now been mapped by analysis of protease-deficient mutants

(Mattern *et al.*, 1992) so control of the protease problem, particularly in *Aspergillus* spp., is now possible.

Authenticity of secreted heterologous protein is generally judged by specific enzyme activity, comigration with an authentic standard on denaturing polyacrylamide gels, *N*-terminal amino acid sequence and, sometimes, by reaction with specific antibodies. In addition, the well characterized, relatively small enzyme, HEWL was shown to have the correct fold by 2D-NMR (Archer *et al.*, 1990). Where structural analysis such as 2D-NMR is not possible the other tests provide a good guide to authenticity but before commercial use is made of the heterologous product a range of other tests must be passed before approval is granted. In the main, authenticity of heterologous proteins from fungi has been demonstrated although anomalies have been noted, e.g. incorrect processing of the *KEX2* site in *A. nidulans* (Contreras *et al.*, 1991), and authenticity of post-translational glycosylation is not usually examined in any detail. Fungi are capable of *O*- and *N*- linked glycosylation and some homologous proteins contain both *O*- and *N*-linked glycan, e.g. *T. reesei* cellobiohydrolase I (Salovuori *et al.*, 1987) and *A. niger* glucoamylase (Svensson *et al.*, 1983, 1986). Although HEWL contains exposed Ser and Thr residues none were *O*-glycosylated by *A. niger* (Archer *et al.*, 1990). Bovine chymosin expressed in *A. awamori* (Ward, 1989) and *T. reesei* (Harkki *et al.*, 1989; Nevalainen *et al.*, 1991) was not glycosylated although when expressed in *A. niger* bovine chymosin was *N*-glycosylated (van Hartingsveldt *et al.*, 1990). An extra *N*-glycosylation site introduced into bovine chymosin by *in vitro* mutagenesis was glycosylated by *A. awamori* (Ward, 1989). Some reports have indicated that heterologous proteins from fungi are hyperglycosylated in comparison to an authentic standard sample, e.g. *R. miehei* aspartyl protease expressed in *A. oryzae* (Christensen *et al.*, 1988). Hyperglycosylation can occur particularly when heterologous proteins are expressed at high yield, as in the case of the *R. miehei* aspartyl protease, expressed in *A. oryzae* (Christensen *et al.*, 1988) but not when expressed at lower yield in *M. circinelloides* (Dickinson *et al.*, 1987).

Whether authenticity of product needs to include glycosylation will depend upon the protein concerned and its intended use. Glycosylation of secreted fungal proteins remains an area requiring

Table 7.6   Commercial production of heterologous enzymes from fungi

Host	Heterologous enzyme	Reference
A. awamori	bovine chymosin	Dunn-Coleman et al. (1991)
A. niger	A. ficuum phytase	van Gorcom et al. (1991)
A. oryzae	R. miehei triglyceride lipase	Huge-Jensen et al. (1989)

further investigation because of its involvement in the secretion process and because, in some cases, correct glycosylation will be an important facet of authenticity in heterologous protein production.

In optimizing production of heterologous enzymes from fungi the value of mutagenesis and screening should not be underestimated. A combination of strategies was used to achieve commercially attractive yields of chymosin from *A. awamori* (Dunn-Coleman *et al.*, 1991). Apart from mutagenesis, multicopy integrations, glucoamylase gene fusions and a host strain devoid of aspergillopepsin were used in this work. The commercially available heterologous proteins from fungi are given in Table 7.6. Most heterologous enzymes have been produced in low yields although, apart from those listed in Table 7.6, other heterologous proteins produced in high yields (> 1 gl^{-1}) include *R. miehei* protease from *A. oryzae* (Christensen *et al.*, 1988), hen lysozyme from *A. niger* (Jeenes *et al.*, 1993) and *Streptoalloteichas hindustanus* phleomycin binding protein from *Tolypocladium geodes* (Calmels *et al.*, 1991). Other high-yielding heterologous proteins will be produced permitting commercial exploitation, and production of high value proteins will not require such high yields for commercial success. Already, recombinant antibodies are being successfully produced from *T. reesei* (Nyyssönen *et al.*, 1993).

7.7. CONCLUSIONS

Exoenzymes derived from filamentous fungi have a diversity of roles in nature, being involved in pathogenicity and degradation of many polymeric organic compounds. The versatility of fungal enzymes has long been recognized and exploited by man for use in food production and, increasingly, in a variety of other industries. Investigations of the biology of fungal enzymes and their exploitation was given further impetus by the advent of techniques for genetic manipulation of some species. The number of different fungal species for which genetic manipulation is now

feasible is ever increasing although some important species remain recalcitrant at the present time. Even so, gene cloning can be achieved despite the difficulties in producing recombinant strains. In some of the industrial strains, genetic manipulation is now routine, affording a dissection of the mechanisms for control of gene expression, protein secretion and post-translational events important in exoenzyme production (MacKenzie *et al.*, 1993). In the large majority of filamentous fungal species, however, the application of molecular approaches is not yet possible and the challenge of using such methods to unravel the role of exoenzymes in nature awaits.

REFERENCES

Aleshin, A., Golubev, A., Firsov, L.M. and Honzatko, R.B. (1992) Crystal structure of glucoamylase from *Aspergillus awamori* var. X100 to 2.2-Å resolution. *Journal of Biological Chemistry*, **267**, 19291–8.

Alic, M. and Gold, M.H. (1991) Genetics and molecular biology of the lignin degrading basidiomycete *Phanerochaete chrysosporium*, in *More Gene Manipulations in Fungi*, (ed. J.W. Bennett and L.L. Lasure), Academic Press, London, pp. 319–41.

Allen, A.L. and Roche, C.D. (1989) Effects of strain and fermentation conditions on production of cellulase by *Trichoderma reesei*. *Biotechnology and Bioengineering*, **33**, 650–6.

Allison, D.S., Rey, M.W., Berka, R.M. *et al.* (1992) Transformation of the thermophilic fungus *Humicola grisea* var. *thermoidea* and overproduction of *Humicola* glucoamylase. *Current Genetics*, **21**, 225–9.

Archer, D.B. (1994) Enzyme production by recombinant *Aspergillus*, in *Recombinant Microbes for Industrial and Agricultural Applications*, (ed. Y. Murooka), Marcel Dekker, New York, pp. 373–93.

Archer, D.B., Jeenes, D.J., MacKenzie, D.A. *et al.* (1990) Hen egg white lysozyme expressed in, and secreted from, *Aspergillus niger* is correctly processed and folded. *BioTechnology*, **8**, 741–5.

Ashikari, T., Nakamura, N., Tanaka, Y. *et al.* (1986) *Rhizopus* raw-starch-degrading glucoamylase: its cloning and expression in yeast. *Agricultural and Biological Chemistry*, **50**, 957–64.

Ball, C. (1984) Filamentous fungi, in *Genetics and Breeding of Industrial Microorganisms*, (ed. C. Ball), CRC Press, Boca Raton, FL, pp. 159–188.

Barrett, A.J. and Rawlings, N.D. (1991) Types and families of endopeptidases. *Biochemical Society Transactions*, **19**, 707–15.

Bastande, K.B. (1992) Xylan structure, microbial xylanases, and their mode of action. *World Journal of Microbiology and Biotechnology*, **8**, 353–68.

Berges, T., Boddy, L.M., Peberdy, J.F. and Barreau, C. (1993) Cloning of an invertase gene from *Aspergillus niger* by expression in *Trichoderma reesei*. *Current Genetics*, **24**, 53–9.

Berka, R.M., Ward, M., Wilson, L.J. *et al.* (1990) Molecular cloning and deletion of the aspergillopepsin A gene from *Aspergillus awamori*. *Gene*, **86**, 153–62.

Berka, R.M., Kodama, K.H., Rey, M.W. *et al.* (1991) The development of *Aspergillus niger* var. *awamori* as a host for the expression and secretion of heterologous gene products. *Biochemical Society Transactions*, **19**, 681–5.

Berka, R.M., Carmona, C.L., Hayenga, K.J. *et al.* (1993) Isolation and characterization of the *Aspergillus oryzae* gene encoding aspergillopepsin O. *Gene*, **125**, 195–8.

Berkholt, V. (1987) Amino acid sequence of endothiapepsin. *European Journal of Biochemistry*, **167**, 327–38.

Boel, E. and Huge-Jensen, B. (1989) Recombinant *Humicola* lipase and process for the production of recombinant *Humicola* lipases. European Patent Application EP 0305216.

Boel, E., Hjort, I., Svensson, B. *et al.* (1984) Glucoamylase G1 and G2 from *Aspergillus niger* are synthesized from two different but closely related mRNAs. *EMBO Journal*, **3**, 1097–102.

Boel, E., Huge-Jensen, B., Christensen, M. *et al.* (1988) *Rhizomucor miehei* triglyceride lipase is synthesized as a precursor. *Lipids*, **23**, 701–6.

Brunt, S.A. and Silver, J.C. (1991) Molecular cloning and characterization of two distinct hsp85 sequences from the steroid responsive fungus *Achyla ambisexualis*. *Current Genetics*, **19**, 383–8.

Bussink, H.J.D., Kester, H.C.M. and Visser, J. (1990) Molecular cloning, nucleotide sequence and expression of the gene encoding pre-pro-polygalacturonase II of *Aspergillus niger*. *FEBS Letters*, **273**, 127–30.

Bussink, H.J.D., Buxton, F.P. and Visser, J. (1991a) Expression and sequence comparison of the *Aspergillus niger* and *Aspergillus tubigensis* genes encoding polygalacturonase II. *Current Genetics*, **19**, 467–474.

Bussink, H.J.D., Brouwer, K.B., de Graaff, L.H. *et al.* (1991b) Identification and characterization of a second polygalacturonase gene of *Aspergillus niger*. *Current Genetics*, **20**, 301–7.

Caddick, M.X. and Arst, H.N. (1987) Structural genes for phosphatases in *Aspergillus nidulans*. *Genetics Research*, **47**, 83–91.

Calmels, T.P.G., Martin, F., Durand, H. and Tiraby, G. (1991) Proteolytic events in the processing of secreted proteins in fungi. *Journal of Biotechnology*, **17**, 51–66.

Campbell-Platt, G. and Cook, P.E. (1989) Fungi in the production of food and food ingredients. *Journal of Applied Bacteriology Symposium Supplement*, pp. 117S–131S.

Caprari, C., Richter, A., Bergmann, C. *et al.* (1993) Cloning and characterization of a gene encoding the endopolygalacturonase of *Fusarium moniliforme*. *Mycological Research*, **97**, 497–505.

Cheetham, P.S.J. (1985) The applications of enzymes in industry, in *Handbook of Enzyme Biotechnology*, (ed. A. Wiseman), Ellis Horwood, Chichester, pp. 274–379.

Chiba, Y., Yamagata, Y., Nakajima, T. and Ichishima, E. (1992) A new high-mannose type *N*-linked oligosaccharide from *Aspergillus* carboxypeptidase. *Bioscience, Biotechnology and Biochemistry*, **56**, 1371–2.

Christensen, T., Woeldike, H., Boel, E. *et al.* (1988) High level expression of recombinant genes in *Aspergillus oryzae*. *BioTechnology*, **6**, 1419–22.

Civas, A., Eberhard, R., Le Dizet, P. and Petek, F. (1984) Glycosidases induced in *Aspergillus tamarii*. Secreted α-D-galactosidase and β-D-mannanase. *Biochemical Journal*, **219**, 857–63.

Cleves, A.E. and Bankaitis, V.A. (1992) Secretory pathway function in *Saccharomyces cerevisiae*. *Advances in Microbial Physiology*, **33**, 73–144.

Contreras, R., Carrez, D., Kinghorn, J.R. *et al.* (1991) Efficient KEX2-like processing of a glucoamylase-interleukin-6 fusion protein by *Aspergillus nidulans* and secretion of mature interleukin-6. *BioTechnology*, **9**, 378–81.

Cooke, R.C. and Whipps, J.M. (1993). *Ecophysiology of Fungi*. Blackwell, Oxford.

Coughlan, M.P. and Amaral Collaco, M.T. (1990) *Advances in Biological Treatment of Lignocellulosic Materials*. Elsevier, London.

Cullen, D. and Kersten, P. (1992) Fungal enzymes for lignocellulose degradation, in *Applied Molecular Genetics of Filamentous Fungi*, (eds J.R. Kinghorn and G. Turner), Blackie, London, pp. 100–31.

Dal Degan, F., Ribadeau-Dumas, B. and Breddam, K. (1992) Purification and characterization of two serine carboxypeptidases from *Aspergillus niger* and their use in C-terminal sequencing of proteins and peptide synthesis. *Applied and Environmental Microbiology*, **58**, 2144–52.

Datta, A. (1992) Purification and characterization of a novel protease from solid substrate cultures of *Phanerochaete chrysosporium*. *Journal of Biological Chemistry*, **267**, 728–36.

Datta, A., Bettermann, A. and Kirk, T.K. (1991) Identification of a specific manganese peroxidase among ligninolytic enzymes secreted by *Phanerochaete chrysosporium* during wood decay. *Applied and Environmental Microbiology*, **57**, 1453–60.

Davies, R.W. (1991) Molecular biology of a high-level recombinant protein production system in *Aspergillus*, in *Molecular Industrial Mycology*, (eds S.A. Leong and R.M. Berka), Marcel Dekker, New York, pp. 45–81.

Davis, M.A. and Hynes, M.J. (1991) Regulatory circuits in *Aspergillus nidulans*, in *More Gene Manipulations in Fungi*, (eds J.W. Bennett and L.L. Lasure), Academic Press, San Diego, pp. 151–89.

Dean, R.A. and Timberlake, W.E. (1989) Regulation of the *Aspergillus nidulans* pectate lyase gene (*pel*A). *Plant Cell*, **1**, 275–84.

Deising, H., Nicholson, R.L., Haug, M. *et al.* (1992) Adhesion pad formation and the involvement of cutinase and esterases in the attachment of uredospores to the host cuticle. *Plant Cell*, **4**, 1101–11.

Delaney, R., Wong, N.R.S., Meng, G. *et al.* (1987) Amino acid sequence of rhizopuspepsin isozyme pI5. *Journal of Biological Chemistry*, **262**, 1461–7.

den Herder, I.F., Mateo Rosell, A.M., van Zuilen, C.M. *et al.* (1992) Cloning and expression of a member of the *Aspergillus niger* gene family encoding α-galactosidase. *Molecular and General Genetics*, **233**, 404–10.

De Viragh, P.A., Sanglard, D., Togni, G. *et al.* (1993) Cloning and sequencing of two *Candida parapilosis* genes encoding acid proteases. *Journal of General Microbiology*, **139**, 335–42.

Dickinson, L., Harboe, M., van Meeswijck, R. *et al.* (1987) Expression of active *Mucor miehei* aspartic protease in *Mucor circinelloides*. *Carlsberg Research Communications*, **52**, 243–52.

Disanto, M.E., Li, Q. and Logan, D.A. (1992) Purification and characterization of a developmentally regulated carboxypeptidase from *Mucor racemosus*. *Journal of Bacteriology*, **174**, 447–55.

Dunn-Coleman, N.S., Bloebaum, P., Berka, R.M. *et al.* (1991) Commercial levels of chymosin production by *Aspergillus*. *BioTechnology*, **9**, 976–81.

Durand, H., Clanet, M. and Tiraby, G. (1988) Genetic improvement of *Trichoderma reesei* for large scale cellulase production. *Enzyme and Microbial Technology*, **10**, 341–6.

Durrands, P.K. and Cooper, R.M. (1988) Development and analysis of pectic screening media for use in the detection of pectinase mutants. *Applied Microbiology and Biotechnology*, **28**, 463–7.

Elliott, T.J. (1985) The genetics and breeding of species of *Agaricus*, in *The Biology and Technology of the Cultivated Mushroom*, (eds P.B. Flegg, D.M. Spencer and D.A. Wood), Wiley, Chichester, pp. 111–29.

Ellis, R.J. and van der Vies, S.M. (1991) Molecular chaperones. *Annual Review of Biochemistry*, **60**, 321–47.

Enari, T.M. and Niku-Paavola, M.L. (1987) Enzymatic hydrolysis of cellulose: is the current theory of the mechanisms of hydrolysis correct? *Critical Reviews in Biotechnology*, **5**, 67–87.

Eriksson, K.E. and Kirk, T.K. (1985) Biopulping, biobleaching and treatment of kraft bleaching effluents with white rot fungi. *Comprehensive Biotechnology*, **3**, 271–94.

Eriksson, K.E.L., Blanchette, R.A. and Ander, P. (1990) *Microbial and Enzymatic Degradation of Wood and Wood Components*. Springer Verlag, Berlin.

Evans, R., Ford, C., Sierks, M. *et al.* (1990) Activity and thermal stability of genetically truncated forms of *Aspergillus* glucoamylase. *Gene*, **91**, 131–4.

Evans, C.S., Gallagher, I.M., Atkey, P.J. and Wood, D.A. (1991) Localisation of degradative enzymes in white-rot decay of lignocellulose. *Biodegradation*, **2**, 93–106.

Eveleigh, D.E. (1987) Cellulase in perspective, in *Technology in the 1990's. Utilization of Lignocellulosic Wastes* (eds B.S. Hartley, P.M.A. Broda and P.J. Senior), *Philosophical Transactions of the Royal Society of London*, pp. 321–447.

Farquhar, R., Honey, N., Murant, S.J. *et al.* (1991) Protein disulfide isomerase is essential for viability in *Saccharomyces cerevisiae*. *Gene*, **108**, 81–9.

Farrell, R.L., Murtagh, K.E., Tien, M. *et al.* (1989) Physical and enzymatic properties of lignin peroxidase isoenzymes from *Phanerochaete chrysosporium*. *Enzyme and Microbial Technology*, **11**, 322–8.

Fincham, J.R.S. (1989) Transformation in fungi. *Microbiological Reviews*, **53**, 148–70.

Forrester, I.T., Graski, A.C., Mishra, C. *et al.* (1990) Characteristics and *N*-terminal amino acid sequence of a manganese peroxidase purified from *Lentinula edodes* cultures grown on a commercial wood substrate. *Applied Microbiology and Biotechnology*, **33**, 359–65.

Frederick, G.D., Rombouts, P. and Buxton, F.P. (1993) Cloning and characterisation of *pep*C, a gene encoding a serine protease from *Aspergillus niger*. *Gene*, **125**, 57–64.

Gems, D., Johnstone, I.L. and Clutterbuck, A.J. (1991) An autonomously replicating plasmid transforms *Aspergillus nidulans* at high frequency. *Gene*, **98**, 61–7.

Gething, M-J. and Sambrook, J. (1992) Protein folding in the cell. *Nature*, **355**, 33–45.

Gierasch, L.M. (1989) Signal sequences. *Biochemistry*, **28**, 923–30.

Gines, M.J., Dove, M.J. and Seligy, V.L. (1989) *Aspergillus oryzae* has two nearly identical Taka-amylase genes, each containing eight introns. *Gene*, **79**, 107–17.

Godfrey, A. and Reichelt, J. (1983) *Industrial Enzymology. The Applications of Enzymes in Industry*. McMillan, London, UK.

Gonzalez-Candelas, L. and Kolattukudy, P.E. (1992) Isolation and analysis of a novel inducible pectate lyase gene from the phytopathogenic fungus *Fusarium solani* f. sp. *pisi* (*Nectria haematococca*, mating population VI). *Journal of Bacteriology*, **174**, 6343–9.

Goosen, T., Bos, C.J. and van den Broek, H. (1992) Transformation and gene manipulation in filamentous fungi: an overview, in *Handbook of Applied Mycology*, Vol. 4, (eds D.K. Arora, R.P. Elander and K.G. Mukerji), Marcel Dekker, New York, pp. 151–95.

Grainger, J.M., Jones, K.L., Hotten, P.M. and Rees, J.F. (1984) Estimation and control of microbial activity in landfill, in *Microbiological Methods for Environmental Biotechnology*, (eds J.M. Grainger and J.M. Lynch), Society for Applied Bacteriology Technical Series **19**, Academic Press, London, pp. 259–73.

Gray, G.L., Hayenga, K., Cullen, D. *et al.* (1986) Primary structure of *Mucor miehei* aspartyl protease: evidence for a zymogen intermediate. *Gene*, **48**, 41–53.

Grove, S.N. (1978) The cytology of hyphal tip growth, in *The Filamentous Fungi*, vol. 3, (eds J.E. Smith and D.R. Berry), Arnold, London, pp. 28–50.

Gysler, C., Harmsen, J.A.M., Kester, H.C.M. *et al.* (1990) Isolation and structure of the pectin lyase D-encoding gene from *Aspergillus niger*. *Gene*, **89**, 101–8.

Hammel, K.E. (1989) Organopollutant degradation by lignolytic fungi. *Enzyme and Microbial Technology*, **11**, 776–7.

Harkki, A., Uusitalo, J., Bailey, M. *et al.* (1989) A novel fungal expression system: secretion of active calf chymosin from the filamentous fungus *Trichoderma reesei*. *BioTechnology*, **7**, 596–603.

Harkki, A., Mäntylä, A., Penttilä, M. *et al.* (1991) Genetic engineering of *Trichoderma* to produce strains with novel cellulase profiles. *Enzyme and Microbial Technology*, **13**, 227–33.

Harmsen, J.A.M., Kusters van Someren, M.A. and Visser, J. (1990) Cloning and expression of a second *Aspergillus niger* pectin lyase gene (*pelA*): indications of a pectin lyase gene family in *Aspergillus niger*. *Current Genetics*, **18**, 161–6.

Hata, Y., Kitamoto, K., Gomi, K. *et al.* (1991). The glucoamylase cDNA from *Aspergillus oryzae*: its cloning, nucleotide sequence, and expression in *Saccharomyces cerevisiae*. *Agricultural and Biological Chemistry*, **55**, 941–9.

Hayashida, S. and Flor, P.Q. (1981) Raw starch-digestive glucoamylase productivity of protease-less mutant from *Aspergillus awamori* var. *kawachi*. *Agricultural and Biological Chemistry*, **45**, 2675–81.

Hayashida, S. and Teramoto, Y. (1986) Production and characteristics of raw starch-digesting α-amylase from a protease-negative *Aspergillus ficum* mutant. *Applied and Environmental Microbiology*, **52**, 1068–73.

Hayashida, S., Kuroda, K., Ohta, K. *et al.* (1989) Molecular cloning of the glucoamylase I gene of *Aspergillus awamori* var. *kawachi* for localization of the raw-starch affinity site. *Agricultural and Biological Chemistry*, **53**, 923–9.

Hoch, H.C. (1986) Freeze-substitution of fungi, in *Ultrastructure Techniques for Microorganisms*, (eds H.C. Aldrich and W.J. Todd), Plenum Press, New York, pp. 183–212.

Holzbaur, E.L.F., Andrawis, A. and Tien, M. (1991) Molecular biology of lignin peroxidases from *Phanerochaete chrysosporium*, in *Molecular Industrial Mycology* (eds S.A. Leong and R.M. Berka), Marcel Dekker, New York, pp. 197–223.

Horiuchi, H., Yanai, K., Okazaki, T. *et al.* (1988a) Isolation and sequencing of a genomic clone encoding aspartic proteinase of *Rhizopus niveus*. *Journal of Bacteriology*, **170**, 272–8.

Horiuchi, H., Yanai, K., Takagi, M. *et al.* (1988b) Primary structure of a base non-specific ribonuclease from *Rhizopus niveus*. *Journal of Biochemistry* **103**, 408–18.

Horiuchi, H., Ashikari, T., Amachi, T. *et al.* (1990) High-level secretion of a *Rhizopus niveus* aspartic proteinase in *Saccharomyces cerevisae*. *Agricultural and Biological Chemistry*, **54**, 1771–9.

Hsu, I-N., Delbaere, L.T.J., Jarvis, M.N.G. and Hofmann, T. (1977) Penicillopepsin from *Penicillium janthinellum* crystal structure at 2.8Å and sequence homology with porcine pepsin. *Nature*, **266**, 140–5.

Huge-Jensen, B., Andreasen, F., Christensen, T. *et al.* (1989) *Rhizomucor miehei* triglyceride lipase is processed and secreted from transformed *Aspergillus oryzae*. *Lipids*, **24**, 781–5.

Impoolsup, A., Bhumiratana, A. and Flegel, T.W. (1981) Isolation of alkaline and neutral proteases from *Aspergillus flavus* var. *columnaris*, a soy sauce koji mold. *Applied and Environmental Microbiology*, **42**, 619–28.

Inaka, K., Taniyama, K., Kikuchi, M. *et al.* (1991) The crystal structure of a mutant human lysozyme C77/95A with increased secretion efficiency in yeast. *Journal of Biological Chemistry*, **266**, 12599–603.

Jaton-Ogay, K., Suter, M., Crameri, R. *et al.* (1992) Nucleotide sequence of a genomic and a cDNA clone encoding an extracellular alkaline protease of *Aspergillus fumigatus*. *FEMS Microbiology Letters*, **92**, 163–8.

Jeenes, D.J., MacKenzie, D.A., Roberts, I.N. and Archer, D.B. (1991), Heterologous protein production by filamentous fungi. *Biotechnology and Genetic Engineering Reviews* **9**, 327–67.

Jeenes, D.J., Marczinke, B., MacKenzie, D.A. and Archer, D.B. (1993) A truncated glucoamylase gene fusion for heterologous protein secretion from *Aspergillus niger*. *FEMS Microbiology Letters*, **107**, 267–72.

Judelson, H.S. and Michelmore, R.W. (1989) Structure and expression of a gene encoding heat shock protein hsp70 from the oomycete fungus *Bremia lactucae*. *Gene*, **79**, 207–17.

Kester, H.C.M. and Visser, J. (1990) Purification and characterization of polygalacturonases produced by the hyphal fungus *Aspergillus niger*. *Biotechnology and Applied Biochemistry*, **12**, 150–60.

Khanh, N.Q., Albrecht, H., Ruttowski, E. *et al.* (1990) Nucleotide sequence and derived amino acid sequence of a pectinesterase cDNA isolated from *Aspergillus niger* strain RH5344. *Nucleic Acids Research*, **18**, 4262.

Kikuchi, M. and Ikehara, M. (1991) Conformational features of signal sequences and folding of secretory proteins in yeasts. *Trends in Biotechnology*, **9**, 208–11.

Kimura, A., Takata, M., Sakai, O. *et al.* (1992) Complete amino acid sequence of crystalline α-glucosidase from *Aspergillus niger*. *Bioscience, Biotechnology and Biochemistry* **56**, 1368–70.

Kinderlerer, J.L. (1989) Volatile metabolites of filamentous fungi and their role in food flavour. *Journal of Applied Bacteriology Symposium Supplement*, pp. 133S–144S.

Kirk, T.K. and Farrell, R.L. (1987) Enzymatic 'combustion': The microbial degradation of lignin. *Annual Review of Microbiology*, **41**, 465–505.

Kirk, T.K., Croan, S., Tien, M. *et al.* (1986) Production of multiple ligninases by *Phanerochaete chrysosporium*. Effect of selected growth conditions and use of a mutant strain. *Enzyme and Microbial Technology*, **8**, 7–32.

Knowles, J., Teeri, T.T., Lehtovaara, P. *et al.* (1988) The use of gene technology to investigate fungal cellulolytic enzymes, in *Biochemistry and Genetics of Cellulose Degradation*, (eds J.-P. Aubert, P. Beguin and J. Millet), Academic Press, London, pp. 153–169.

Kobayashi, H., Sekibata, S., Shibuya, H. *et al.* (1989) Cloning and sequence analysis of cDNA for *Irpex lacteus* aspartic proteinase. *Agricultural and Biological Chemistry*, **53**, 1927–33.

Koh-Luar, S.I., Parish, J.H., Bleasby, A.J. *et al.* (1989) Exported proteins of *Neurospora crassa*: 1-glucoamylase. *Enzyme and Microbial Technology*, **11**, 692–5.

Kolattukudy, P.G. (1985) Enzymatic penetration of the plant cuticle by fungal pathogens. *Annual Review of Phytopathology*, **23**, 223–50.

Korman, D.R., Bayliss, F.T., Barnett, C.C. *et al.* (1990) Cloning, characterization, and expression of two α-amylase genes from *Aspergillus niger* var *awamori*. *Current Genetics*, **17**, 203–12.

Kornfeld, R. and Kornfeld, S. (1985) Assembly of asparagine-linked oligosaccharides. *Annual Review of Biochemistry*, **54**, 631–44.

Kramer, G.F.H., Gunning, A.P., Morris, V.J. *et al.* (1993) Scanning tunnelling microscopy of *Aspergillus niger* glucoamylases. *Journal of the Chemical Society. Faraday Transactions*, **89**, 2595–602.

Kriechbaum, M., Heilmann, H.J., Wientjes, F.J. *et al.* (1989) Cloning and DNA sequence analysis of the glucose oxidase gene from *Aspergillus niger* NRRL-3. *FEBS Letters*, **255**, 63–6.

Kubicek, C.P., Messner, R., Gruber, F. *et al.* (1993) The *Trichoderma* cellulase regulatory puzzle: from the interior life of a secretory fungus. *Enzyme and Microbial Technology*, **15**, 90–9.

Kukuruzinska, M.A., Bergh, M.L.E. and Jackson, B.J. (1987) Protein glycosylation in yeast. *Annual Review of Biochemistry*, **56**, 915–44.

Kusters van Someren, M., Flipphi, M., de Graaff, L. *et al.* (1992) Characterization of the *Aspergillus niger pel*B gene: structure and regulation of expression. *Molecular and General Genetics*, **234**, 113–20.

LeJohn, H.B. and Braithwaite, C.E. (1984) Heat and nutritional shock-induced proteins of the fungus *Achyla* are different and under independent transcriptional control. *Canadian Journal of Biochemistry and Cell Biology*, **62**, 837–46.

Lowe, D.A. (1992) Fungal enzymes, in *Handbook of Applied Mycology*, Vol. 4, (eds D.K. Arora, R.P. Elander and K.G. Mukerji), Marcel Dekker, New York, pp. 681–706.

MacKenzie, D.A., Jeenes, D.J., Belshaw, N.J. and Archer, D.B. (1993) Regulation of secreted protein production by filamentous fungi: recent developments and perspectives. *Journal of General Microbiology*, **139**, 2295–307.

MacRae, W.D., Buxton, F.P., Sibley, S. *et al.* (1988) A phosphate-repressible phosphatase gene from *Aspergillus niger*: its cloning, sequencing and transcriptional analysis. *Gene*, **71**, 339–48.

Maldonado, M.C., Strasser de Saad, A.M. and Callieri, D. (1989) Catabolite repression of the synthesis of inducible polygalacturonase and pectinesterase by *Aspergillus niger* sp. *Current Microbiology*, **18**, 303–6.

Marciano, P., Di Lenna, P. and Magro, P. (1982) Polygalacturonase isoenzymes produced by *Sclerotinia sclerotiorum in vivo* and *in vitro*. *Physiological Plant Pathology*, **20**, 201–12.

Marth, E.H. and Yousef, A.E. (1991) Fungi and dairy products, in *Handbook of Applied Mycology*, Vol. 3 (eds D.K. Arora, K.G. Mukerji and E.H. Marth) Marcel Dekker, New York, pp. 375–414.

Matcham, S.E., Jordan, B.R. and Wood, D.A. (1984) Methods for assessment of fungal growth on solid substrates, in *Microbiological Methods for Environmental Biotechnology* (eds J.M. Grainger and J.M. Lynch), Academic Press, London, pp. 5–18.

Matsuura, Y., Kusunoki, M., Harada, W. and Kakudo, M. (1984) Structure and possible catalytic residues of Taka amylase A EC 3.2.1.1. *Journal of Biochemistry*, **95**, 697–702.

Mattern, I.E., van Noort, J.M., van den Berg, P. *et al.* (1992) Isolation and characterization of mutants of *Aspergillus niger* deficient in extracellular proteases. *Molecular and General Genetics*, **234**, 332–6.

Müller, M. (1992) Proteolysis in protein import and export: signal peptide processing in eu- and prokaryotes. *Experientia*, **48**, 118–29.

Murakami, K., Ishida, Y., Masaki, A. *et al.* (1991) Isolation and characterization of the alkaline protease gene of *Aspergillus oryzae*. *Agricultural and Biological Chemistry*, **55**, 2807–11.

Nakadai, T. (1985) The roles of enzymes produced by shoyu-koji-molds. *Journal of the Japan Soy Sauce Research Institute*, **11**, 67–79.

Nandan, R. and Raisuddin, S. (1992) Fungal degradation of industrial wastes and wastewater, in *Handbook of Applied Mycology*, (eds D.K. Arora, R.P. Elander and K.G. Mukerji), Marcel Dekker, New York, pp. 931–61.

Nevalainen, K.M., Penttilä, M.E., Harkki, A. *et al.*, (1991). The molecular biology of *Trichoderma* and its application to the expression of both homologous and heterologous genes, in *Molecular Industrial Mycology*, (eds S.A. Leong and R.M. Berka), Marcel Dekker, New York, pp. 129–48.

Nevalainen, K.M.H. and Palva, E.T. (1979) Improvement of amyloglucosidase production of *Aspergillus awamori* by mutagenic treatments. *Journal of Chemical Technology and Biotechnology*, **29**, 390–5.

Nout, M.J.R. and Rombouts, F.M. (1990) Recent developments in tempe research. *Journal of Applied Bacteriology*, **69**, 609–33.

Nunberg, J.H., Meade, J.H., Cole, G. *et al.* (1984) Molecular cloning and characterization of the glucoamylase gene of *Aspergillus awamori*. *Molecular and Cellular Biology*, **4**, 2306–15.

Nyyssönen, E., Penttilä, M., Harkki, A. *et al.*, (1993) Efficient production of antibody fragments by the filamentous fungus *Trichoderma reesei*. *BioTechnology*, **11**, 59–5.

Ogawa, K., Ohara, H. and Toyama, N. (1988) Interspecific hybridization of *Aspergillus awamori* var. *kawachi* and *Aspergillus oryzae* by protoplast fusion. *Agricultural and Biological Chemistry*, **52**, 1985–91.

Omura, F., Taniyama, Y. and Kikuchi, M. (1991) Behavior of cysteine mutants of human lysozyme in *de novo* synthesis and *in vivo* secretion. *European Journal of Biochemistry*, **198**, 477–84.

Ono, K., Shintani, K., Shigata, S. and Oka, S. (1988) Various molecular species in glucoamylase from *Aspergillus niger*. *Agricultural and Biological Chemistry*, **52**, 1689–98.

Orpin, C.G. (1975) Studies on the rumen flagellate *Neocallimastix frontalis*. *Journal of General Microbiology*, **91**, 249–62.

Pfanner, N. and Neupert, W. (1990) The mitochondrial protein import apparatus. *Annual Review of Biochemistry*, **59**, 331–53.

Pitson, S.M., Seviour, R.J. and McDougal, B.M. (1993) Noncellulolytic fungal β-glucanases: their physiology and regulation. *Enzyme and Microbial Technology*, **15**, 178–92.

Pourrat, H., Barthomeuf, C., Texier, O. and Pourrat, A. (1988) Production of semi-alkaline protease by *Aspergillus niger*. *Journal of Fermentation Technology*, **66**, 383–8.

Priest, F.G. (1984) Extracellular enzymes. *Aspects of Microbiology*, vol. 9. Van Nostrand Reinhold, Wokingham.

Punt, P.J., Zegers, N.D., Busscher, M. *et al.* (1991) Intracellular and extracellular production of proteins in *Aspergillus* under the control of expression signals of the highly expressed *Aspergillus nidulans gpdA* gene. *Journal of Biotechnology*, **17**, 19–34.

Raj, H.G., Saxena, M., Allameh, A. and Mukerji, K.G. (1992) Metabolism of foreign compounds by fungi, in *Handbook of Applied Mycology*, (eds D.K. Arora, R.P. Elander and K.G. Mukerji), Marcel Dekker, New York, pp. 881–904.

Ramasamy, K., Kelley, R.L. and Reddy, C.A. (1985) Lack of lignin degradation by glucose oxidase-negative mutants of *Phanerochaete chrysosporium*. *Biochemical Biophysical Research Communications*, **131**, 436–41.

Reid, I.D. (1989) Solid-state fermentations for biological delignification. *Enzyme and Microbial Technology*, **11**, 786–803.

Reid, G.A. (1991) Protein targeting in yeast. *Journal of General Microbiology*, **137**, 1765–73.

Rexach, M., d'Enfert, C., Wuestehube, L. and Schekman, R. (1992) Genes and proteins required for vesicular transport from the endoplasmic reticulum. *Antonie van Leeuwenhoek*, **61**, 87–92.

Roberts, I.N., Jeenes, D.J., MacKenzie, D.A. *et al.* (1992) Heterologous gene expression in *Aspergillus niger*: a glucoamylase–porcine pancreatic prophospholipase

A₂ fusion protein is secreted and correctly processed to yield mature enzyme. *Gene*, **122**, 155–61.

Rombouts, F.M. and Pilnik, W. (1980) Pectic enzymes, in *Microbial Enzymes and Bioconversions* (ed. A.H. Rose), Academic Press, London, pp. 227–82.

Rombouts, F.M., Voragen, A.G.J., Searle-van Leeuwen, M.F. *et al.* (1988) The arabinanases of *Aspergillus niger* – purification and characterization of two α-L-arabinofuranosidases and an *endo*-1,5-α-L-arabinase. *Carbohydrate Polymers*, **9**, 25–47.

Rossignol, J.-L. and Picard, M. (1991) *Ascobolus immersus* and *Podospora anserina*: sex, recombination, silencing and death, in *More Gene Manipulations in Fungi*, (eds J.W. Bennett and L.L. Lasure), Academic Press, San Diego, pp. 266–90.

Rothman, J.E. and Orci, L. (1992) Molecular dissection of the secretory pathway. *Nature*, **355**, 409–15.

Rouvinen, J., Bergfors, T., Teeri, T. *et al.* (1990) Three-dimensional structure of cellobiohydrolase II from *Trichoderma reesei*. *Science*, **249**, 380–6.

Saha, B.C. and Zeikus, J.G. (1989) Microbial glucoamylases: biochemical and biotechnological features. *Starch*, **2**, 57–64.

Sakaguchi, K., Gomi, K., Takagi, M. and Horiuchi, H. (1992) Fungal enzymes used in oriental food and beverage industries, in *Applied Molecular Genetics of Filamentous Fungi*, (eds J.R. Kinghorn and G. Turner), Blackie, Glasgow, pp. 54–99.

Salovuori, I., Makarow, M., Rauvala, H. *et al.* (1987) Low molecular weight high-mannose type glycans in a secreted protein of the filamentous fungus *Trichoderma reesei*. *BioTechnology*, **5**, 152–6.

Satyanarayana, T., Johri, B.N. and Klein, J. (1992) Biotechnological potential of thermophilic fungi, in *Handbook of Applied Mycology*, Vol. 4, (eds D.K. Arora, R.P. Elander and K.G. Mukerji), Marcel Dekker, New York, pp. 729–61.

Scott-Craig, J.S., Panaccione, D.G., Cervone, F. and Walton, J.D. (1990) Endopolygalacturonase is not required for pathogenicity of *Cochliobolus carbonum* on maize. *Plant Cell*, **2**, 1191–200.

Selker, E.U. (1990) Premeiotic instability of repeated sequences in *Neurospora crassa*. *Annual Review of Genetics*, **24**, 579–613.

Shibuya, I., Gomi, K., Iimura, Y. *et al.* (1990). Molecular cloning of the glucoamylase gene of *Aspergillus shirousami* and its expression in *Aspergillus oryzae*. *Agricultural and Biological Chemistry*, **54**, 1905–14.

Shoemaker, S., Schweickart, V., Ladner, M. *et al.* (1983) Molecular cloning of exo-cellobiohydrolase I derived from *Trichoderma reesei* strain L27. *BioTechnology*, **1**, 691–6.

Sirianuntapiboon, S., Somchai, P., Ohmomo, S. and Atthasampunna, P. (1988) Screening of filamentous fungi having the ability to decolorize molasses pigments. *Agricultural and Biological Chemistry*, **52**, 387–92.

Smith, D.C., Bhat, K.M. and Wood, T.M. (1991) Xylan-hydrolysing enzymes from thermophilic and mesophilic fungi. *World Journal of Microbiology and Biotechnology*, **7**, 475–84.

Sodhi, H.S., Elliott, T.J., Wood, D.A. and Stephens, S. (1991) Studies on enzyme mutants in *Coprinus bilanatus*. *Mushroom Science*, **13**, 131–8.

Soliday, C.L., Flurkey, W.H., Okita, T.W. and Kolattukudy, P.E. (1984) Cloning and structure determination of cDNA for cutinase, an enzyme involved in fungal penetration of plants. *Proceedings of the National Academy of Sciences of the United States of America*, **81**, 3939–43.

St Leger, R.J., Charnley, A.K. and Cooper, R.M. (1986) Cuticle-degrading enzymes of entomopathogenic fungi. Synthesis in culture on cuticle. *Journal of Invertebrate Pathology*, **48**, 85–95.

St Leger, R.J., Charnley, A.K. and Cooper, R.M. (1987a) Characterisation of cuticle degrading proteases produced by the entomopathogen *Metarhizium anisopliae*. *Archives of Biochemistry and Biophysics*, **253**, 221–32.

St Leger, R.J., Cooper, R.M. and Charnley, A.K. (1987b) Production of cuticle degrading enzymes by the entomopathogen *Metarhizium anisopliae* during infection of cuticles from *Calliphora vomitoria* and *Manduca sexta*. *Journal of General Microbiology*, **133**, 1371–82.

St Leger, R.J., Frank, D.C., Roberts, D.W. and Staples, R.C. (1992) Molecular cloning and regulatory analysis of the cuticle-degrading protease structural gene from the entomopathogenic fungus *Metarhizium anisopliae*. *European Journal of Biochemistry*, **204**, 991–1001.

Stahl, D.J. and Schafer, W. (1992) Cutinase is not required for fungal pathogenicity on pea. *Plant Cell*, **4**, 621–9.

Stepanov, V.M. (1985) Fungal aspartyl proteinases, in *Aspartyl Proteinases and Their Inhibitors*, (ed. V. Kostka), Walter de Gruyter, New York, pp. 27–40.

Stephens, S.K., Elliott, T.J. and Wood, D.A. (1991) Extracellular enzyme mutants of *Coprinus bilanatus*. *Enzyme and Microbial Technology*, **13**, 976–81.

Sugihara, A., Shimada, Y. and Tominaga, Y. (1988) Purification and characterization of *Aspergillus niger* lipase. *Agricultural and Biological Chemistry*, **52**, 1591–2.

Svensson, B., Pedersen, T.G., Svendsen, I.B. *et al.* (1982) Characterization of two forms of glucoamylase from *Aspergillus niger*. *Carlsberg Research Communications*, **47**, 55–69.

Svensson, B., Larsen, K., Svendsen, I. and Boel, E. (1983) The complete amino acid sequence of the glycoprotein glucoamylase G1, from *Aspergillus niger*. *Carlsberg Research Communications*, **48**, 529–44.

Svensson, B., Larsen, K. and Gunnarsson, A. (1986) Characterization of a glucoamylase G2 from *Aspergillus niger*. *European Journal of Biochemistry*, **154**, 497–502.

Tachikawa, H., Miura, T., Katakura, Y. and Mizunaga, T. (1991) Molecular structure of a yeast gene PDI1 encoding protein disulfide isomerase that is essential for cell growth. *Journal of Biochemistry*, **110**, 306–13.

Tada, S., Iimura, Y., Gomi, K. *et al.* (1989) Cloning and nucleotide sequence of the genomic Taka-amylase A gene of *Aspergillus oryzae*. *Agricultural and Biological Chemistry*, **53**, 593–9.

Tada, S., Gomi, K., Kitamoto, K. *et al.* (1991) Construction of a fusion gene comprising the Taka-amylase A promoter and the *Escherichia coli* β-glucuronidase gene and analysis of its expression in *Aspergillus oryzae*. *Molecular and General Genetics*, **229**, 301–6.

Takahashi, K., Inoue, H., Sakai, K. *et al.* (1991) The primary structure of *Aspergillus niger* acid proteinase A. *Journal of Biological Chemistry*, **266**, 19480–3.

Taniyama, Y., Kuroki, R., Omura, F. *et al.* (1991) Evidence for intramolecular disulfide bond shuffling in the folding of mutant human lysozyme. *Journal of Biological Chemistry*, **266**, 6456–61.

Taniyama, Y., Ogasahara, K., Yutani, K. and Kikuchi, M. (1992) Folding mechanism of mutant human lysozyme C77/95A with increased secretion efficiency in yeast. *Journal of Biological Chemistry*, **267**, 4619–24.

Tao, J., Ginzberg, I., Koltin, Y. and Bruenn, J.A. (1993) Mutants of *Ustilago maydis* KP6 toxin defective in production of one of two polypeptides from one preprotoxin. *Molecular and General Genetics*, **238**, 234–40.

Tatsumi, H., Murakami, S., Tsuji, R.F. *et al.* (1991) Cloning and expression in yeast of a cDNA clone encoding *Aspergillus oryzae* neutral protease II, a unique metalloprotease. *Molecular and General Genetics*, **228**, 97–103.

Tenkamen, M., Puls, J. and Poutanen, K. (1992) Two major xylanases of *Trichoderma reesei*. *Enzyme and Microbial Technology*, **14**, 566–74.

Theodorou, M.K., King-Spooner, C. and Beever, D.E. (1989) *Presence or Absence of Anaerobic Fungi in Landfill Refuse.* UK Department of Energy (ETSU B1246).

Thompson, S.A. (1991), Fungal aspartic proteases, in *Molecular Industrial Mycology*, (eds S.A. Leong and R.M. Berka), Marcel Dekker, New York, pp. 107–28.

Thompson, J.A. (1993) Molecular biology of xylan degradation. *FEMS Microbiology Reviews*, **104**, 65–82.

Törrönen, A., Mach, R.L., Messner, R. *et al.* (1992) The two major xylanases from *Trichoderma reesei*: characterisation of both enzymes and genes. *BioTechnology*, **10**, 1461–4.

Ushijima, S., Nakadai, T. and Uchida, K. (1990) Further evidence on the specific protoplast fusion between *Aspergillus oryzae* and *Aspergillus sojae* and subsequent haploidization, with special reference to their production of some hydrolysing enzymes. *Agricultural and Biological Chemistry*, **54**, 2393–9.

Uusitalo, J.M., Nevalainen, K.M.H., Harkki, A.M. *et al.* (1991) Enzyme production by recombinant *Trichoderma reesei* strains. *Journal of Biotechnology*, **17**, 35–50.

van den Hondel, C.A.M.J.J., Punt, P.J. and van Gorcom, R.F.M. (1991) Heterologous gene expression in filamentous fungi, in *More Gene Manipulations in Fungi*, (eds J.W. Bennett and L.L. Lasure), Academic Press, San Diego, pp. 396–428.

van den Hondel, C.A.M.J.J., Punt, P.J. and van Gorcom, R.F.M. (1992) Production of extracellular proteins by the filamentous fungus *Aspergillus*. *Antonie van Leeuwenhoek*, **61**, 153–60.

van Gorcom, R.F.M., van Hartingsveldt, W., van Parldon, P.A.B. *et al.* (1991) Cloning and expression of microbial phytase. European Patent Application 0420358A1.

van Hartingsveldt, W., van Zeijl, C.M.J., Veenstra, A.E. *et al.* (1990) Heterologous gene expression in *Aspergillus*: analysis of chymosin production in single-copy transformants of *A. niger*, in *Proceedings of the 6th International Symposium on the Genetics of Industrial Microorganisms*, Strasbourg, pp. 107–16.

van Noort, J.M., van den Berg, P. and Mattern, I.E. (1991) Visualization of proteases within a complex sample following their selective retention on immobilized bacitracin, a peptide antibiotic. *Analytical Biochemistry*, **198**, 385–90.

von Heijne, G. (1990) The signal peptide. *Journal of Membrane Biology*, **115**, 195–201.

Wang, H.L., Swain, E.W. and Hesseltine, C.W. (1980) Phytase of molds used in oriental food fermentation. *Journal of Food Science*, **45**, 1262–6.

Ward, M. (1989) Production of calf chymosin by *Aspergillus awamori*, in *Genetics and Molecular Biology of Industrial Microorganisms* (eds C.L. Hershberger, S.W. Queener and G. Hegeman), American Society for Microbiology, Washington, pp. 288–94.

Ward, M. (1991) Chymosin production in *Aspergillus*, in *Molecular Industrial Mycology*, (eds S.A. Leong and R.M. Berka), Marcel Dekker, New York, pp. 83–105.

Ward, M. and Kodama, K.H. (1991) Introduction to fungal proteinases and expression in fungal systems. *Advances in Experimental and Medical Biology*, **306**, 149–60.

Ward, M., Wilson, L.J., Kodama, K.H. *et al.* (1990) Improved production of chymosin in *Aspergillus* by expression as a glucoamylase-chymosin fusion. *Bio-Technology*, **8**, 435–40.

Williamson, G., Belshaw, N.J. and Williamson, M.P. (1992) O-Glycosylation in *Aspergillus* glucoamylase: conformation and role in binding. *Biochemical Journal*, **282**, 423–8.

Wirsel, S., Lachmund, A., Wildhardt, G. and Ruttkowski, E. (1989) Three α-amylase genes of *Aspergillus oryzae* exhibit identical intron–exon organization. *Molecular Microbiology*, **3**, 3–14.

Wong, K.K.Y., Tan, L.U.L. and Saddler, J.N. (1988) Multiplicity of β-1,4 xylanases in microorganisms: functions and applications. *Microbiological Reviews*, **52**, 305–17.

Wood, D.A. (1985) Production and roles of extracellular enzymes during morphogenesis of basidiomycete fungi, in *Developmental Biology of Higher Fungi*, (eds D. Moore, L.A. Casselton, D.A. Wood and J.C. Frankland), Cambridge University Press, Cambridge, pp. 375–87.

Wood, D.A. (1989) Mushroom biotechnology. *International Industrial Biotechnology*, **9**, 5–9.

Wood, D.A. and Smith, J.F. (1987) The cultivation of mushrooms, in *Essays in Food Microbiology*, (eds J.R. Norris and G.L. Pettipher), Wiley, Chichester, pp. 309–43.

Wood, T.M. and Garcia-Campoyo, V. (1990) Enzymology of cellulose degradation. *Biodegradation*, **1**, 147–61.

Wood, T.M. and Mcrae, S.I. (1979) Synergism between enzymes involved in the solubilisation of native cellulose. *Advances in Chemistry Series*, **181**, 181–209.

Wösten, H.A.B., Moukha, S.M., Sietsma, J.H. and Wessels, J.G.H. (1991) Localization of growth and secretion of proteins in *Aspergillus niger*. *Journal of General Microbiology*, **137**, 2017–23.

Xue, G-P., Gobius, K.S. and Orpin, C.G. (1992) A novel polysaccharide hydrolase cDNA (*cal*D) from *Neocallimastix patriciarum* encoding three multi-functional catalytic domains with high endoglucanase, cellobiohydrolase and xylanase activities. *Journal of General Microbiology*, **138**, 2397–403.

Yamashita, I. (1989) The threonine and serine-rich tract of the secretory glucoamylase can direct β-galactosidase to the cell envelope. *Agricultural and Biological Chemistry*, **53**, 483–9.

Yang, R. and Kenealy, W.R. (1992) Regulation of restrictocin production in *Aspergillus restrictus*. *Journal of General Microbiology*, **138**, 1421–7.

Yokotsuka, T. (1991a) Proteinaceous fermented foods and condiments prepared with koji molds, in *Handbook of Applied Mycology*, Vol.3, (eds D.K. Arora, K.G. Mukerji and E.H. Marth), Marcel Dekker, New York, pp. 329–73.

Yokotsuka, T. (1991b) Non-proteinaceous fermented foods and beverages produced with koji molds, in *Handbook of Applied Mycology*, Vol. 3, (eds D.K. Arora, K.G. Mukerji and E.H. Marth), Marcel Dekker, New York, pp. 293–328.

Zhu, W-S., Wojdyla, K., Donlon, K. *et al.* (1990) Extracellular proteases of *Aspergillus flavus*. *Diagnostic Microbiology and Infectious Disease*, **13**, 491–7.

TRANSPORT

A. Garrill
School of Biological Sciences, Flinders University, Adelaide, Australia

8.1 INTRODUCTION

The transport of solutes into and out of the cytoplasm across the plasma and assorted endomembranes of fungal cells is essential for their survival. These fluxes predominantly occur by means of transport proteins and are required for such processes as the uptake of nutrients, maintenance of turgor, cell expansion, development, the compartmentation of potentially cytotoxic ions and signal transduction.

The basic principles of transport follow the chemiosmotic scheme of Mitchell (1966). Primary H$^+$ pumps establish an electrochemical gradient of H$^+$ which is used to energize the uptake and concentration of solutes via carrier proteins. A third type of transport protein, ion channels (which have higher turnover rates in comparison to carriers or pumps, e.g. 10^{6-8} compared to 10^{2-4} s^{-1} (Sanders, 1990)) act as gated (or regulated) pores allowing specific ions to move down their electrochemical gradients.

There are a vast number of studies on transport in fungi and it would be impossible to cover in adequate detail all aspects in a single chapter. Therefore the major transport systems will be discussed here with particular emphasis on the most recent developments involving molecular biology and patch clamping. Both filamentous fungi and yeasts will be considered as the principles and mechanisms of transport are the same in the majority of cases. Where relevant, references to review articles are given should the reader wish to explore specific areas in greater detail.

I shall begin by discussing the transport proteins of the plasma membrane. As the outer permeability barrier of the cell the importance of this membrane in transport is perhaps unparalleled. Later discussions will concentrate on the vacuolar membrane (tonoplast) and briefly the

mitochondrial inner membrane. Due to the confines of space other endomembranes such as the endoplasmic reticulum will not be discussed, although some aspects of transport across these are considered (Chapter 9).

8.2 PLASMA MEMBRANE

8.2.1 Plasma membrane H$^+$-ATPase

Since the pioneering electrophysiological work by Slayman (1965a,b) demonstrating that hyphae of *Neurospora crassa* generate a metabolic inhibitor sensitive membrane potential of up to -250 mV across the plasma membrane, a large body of evidence has been obtained indicating the existence of an ATP-fuelled electrogenic H$^+$-pump (for review see Slayman, 1987). The pump generates a membrane potential by driving a H$^+$ current of up to 50 pA cm^{-2} out of the cell, a process energized by the hydrolysis of ATP. In addition to the membrane potential the pumping of protons also creates a pH (or concentration) gradient across the membrane and together these constitute the electrochemical gradient which may be used to energize solute uptake. The H$^+$ pumping role of the H$^+$ pump suggests an additional role in pH regulation. Measurements of both the potential and pH gradients across the plasma membrane suggest that the regulation of pH does not depend on the action of the H$^+$-ATPase *per se* although the pump is, as might be expected, an important component of the acid–base balance in the cell (Sanders *et al.*, 1981; Sanders and Slayman, 1982).

Further characteristics of the *N. crassa* pump obtained from electrophysiological work include an estimated reversal potential (E_p) of between -300 and -400 mV. From this it is possible to calculate the H$^+$ pumped to ATP hydrolysed stoichiometry (Sanders, 1988). A value of 1.36 (calculated using $E_p = -300$ mV) may suggest a

The Growing Fungus. Edited by Neil A.R. Gow and Geoffrey M. Gadd. Published in 1994 by Chapman & Hall, London. ISBN 0 412 46600 7

Table 8.1 H^+-ATPases of fungal membranes

	Plasma membrane H^+-*ATPase*	*Vacuolar* H^+-*ATPase*	*Mitochondrial* H^+-*ATPase*
ATPase type	$P\ (E_1E_2)$	$V\ (V_0V_1)$	$F\ (F_0F_1)$
Composition	Single polypeptide	At least nine polypeptides	At least ten polypeptides
Approximate M_r of subunits (for *N. crassa* enzyme)	100	(V_1) 67, 57, 48, 30, 17 (V_0) 100, 40, 20, 16	(F_1) 59, 56, 36, 15, 12 (F_0) 22, 21, 19, 16, 8
pH optima	Acidic (5.0–6.7)	Neutral (7.0–7.5)	Basic (8.0–9.0)
K_m (ATP) (mM)	0.4–4.8	0.2	0.3
Nucleotide specificity	ATP	ATP>GTP, ITP>UTP>CTP	ATP>GTP, ITP>UTP>CTP
Commonly used inhibitors	Vanadate DCCD NEM	Nitrate Bafilomycin A1 NEM	Azide

DCCD, dicyclohexylcarbodiimide.

mixed population of pumps although a value of 1 (for which $E_p = -440$ mV) cannot be discounted owing to the uncertainty of the exact value of E_p. Values of a H^+/ATP stoichiometry of 1 have been used to estimate that the H^+-ATPase accounts for about half the ATP consumption of the cell (Gradmann *et al.*, 1978).

The direct study of the pump and its mechanisms necessitates the isolation of plasma membrane vesicles and subsequent purification of the enzyme. Plasma membrane preparations which contain ATPase activity have been isolated from several species including *N. crassa* (Bowman and Slayman, 1977; Scarborough, 1977), *Schizosaccharomyces pombe* (Delhez *et al.*, 1977), *Saccharomyces cerevisiae* (Ahlers *et al.*, 1978; Serrano, 1978; Peters and Borst-Pauwels, 1979; Willsky, 1979), *Candida tropicalis* (Blasco *et al.*, 1981), *Candida albicans* (Hubbard *et al.*, 1986) and *Dendryphiella salina* (Garrill and Jennings, 1991). Purification of the enzyme may yield preparations with specific activities of 25–95 μmol min^{-1} mg^{-1} protein (Bowman and Bowman, 1986). As will be discussed below, other cellular membranes, most notably the tonoplast and mitochondrial inner membrane also contain ATPases although the respective enzymes are discernible from each other on the basis of pH optima, inhibitor sensitivity and nucleotide specificity (Table 8.1 and for review see Bowman and Bowman, 1986).

Definitive evidence that the ATPase activity is due to the H^+ pump has come from *in vitro* demonstrations of H^+ pumping. Fluorescent dyes have been used to monitor acidification of the interior of plasma membrane vesicles containing the ATPase (the vesicles are everted with respect to the situation of the intact cell and thus internal acidification is the equivalent of H^+ being pumped out of the cell) (Perlin *et al.*, 1984; Scarborough and Addison, 1984). A H^+/ATP stoichiometry of close to 1 has been estimated using these fluorescent dyes (Perlin *et al.*, 1986), resembling the values obtained by electrophysiological techniques described above.

The H^+-ATPase may be inhibited with the phosphate analogue orthovanadate and can be shown to exhibit a phosphorylated state ($E{\sim}P$) during its reaction cycle with an aspartyl residue at the site of phosphorylation (Amoury *et al.*, 1980; Dame and Scarborough, 1980; Malpartida and Serrano, 1981; Amoury and Goffeau, 1982). Further observations involving trypsin treatment of plasma membrane vesicles in the presence/absence of nucleotides or orthovanadate (Addison and Scarborough, 1982; Scarborough and Addison, 1984), P_i-HOH exchange kinetics, P_i-ATP exchange (Amoury *et al.*, 1982, 1984)) and H^+-induced changes in intrinsic fluorescence of the purified ATPase (Blanpain *et al.*, 1992) have led to the idea that the reaction mechanism involves at least two conformational states (E_1 and E_2) with a proposed reaction cycle as shown in Figure 8.1.

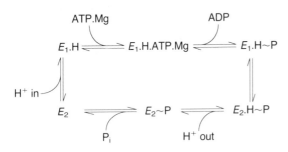

Figure 8.1 Reaction cycle of the fungal plasma membrane H⁺-ATPase. A high-energy conformation of the enzyme results from phosphorylation via MgATP. A proton binds on the internal face and a conformational change either carries the charge through the membrane or alternatively moves the membrane field around the charge releasing the proton at the external face. The protein is then dephosphorylated and undergoes a conformational change back to the original ATP binding state.

To study the active site of the enzyme Slayman and co-workers (Brooker and Slayman, 1982, 1983a,b,c; Davenport and Slayman, 1988) have used site-specific compounds such as the sulphydryl reagent *N*-ethylmaleimide (NEM) the action of which can be modified by substrate. They were able to define two sites on the enzyme, a fast site which labelled in several minutes and did not appear to be essential for ATPase activity and a slow site which required tens of minutes for labelling but could completely inactivate the enzyme. The slow site may be protected when MgATP or MgADP are bound to the enzyme with dissociation constants close to the K_m and K_i values for the nucleotides. The cysteine residues responsible for these have been identified as Cys[532] (slow site) and Cys[545] (fast site) (Pardo and Slayman, 1989). Both of these residues lie in the large central hydrophilic domain close to the presumed nucleotide binding site (see below) although neither of these residues is thought to play a direct role in binding (Pardo and Slayman, 1989).

The gene encoding the plasma membrane H⁺-ATPase (*PMA1*) has been obtained from *N. crassa* (Addison 1986; Hager *et al.*, 1986), *S. cerevisiae* (Serrano *et al.*, 1986), *S. pombe* (Ghislain *et al.*, 1987) and *C. albicans* (Monk *et al.*, 1991a). The amino acid sequence deduced from the nucleotide sequence of the cloned genes shows considerable sequence

homology between the plasma membrane H⁺-ATPases of these fungi and not surprisingly they show immunological cross-reactivity. Indeed, sequence homologies and a reaction cycle that proceeds through a covalent β-aspartyl phosphate intermediate means that the fungal H⁺-ATPases can be grouped together with the H⁺-ATPase of plants (Serrano, 1984), the Na⁺/K⁺-ATPase (Jorgensen, 1982), the H⁺/K⁺-ATPase (Sachs *et al.*, 1982) and the sarcoplasmic reticulum Ca²⁺-ATPase (Ikemoto, 1982) of animal cells and the H⁺-ATPase of *Escherichia coli*, as the P-type (or E_1E_2) ATPases.

The plasma membrane H⁺-ATPases of *N. crassa* and *S. cerevisiae* are 920 and 918 amino acids long with molecular weights of 99.9 and 99.7 kDa, respectively. These values agree with the earlier estimate of 100 kDa from SDS-PAGE preparations of the isolated enzyme (Dufour and Goffeau, 1978; Addison and Scarborough, 1981; Bowman *et al.*, 1981). The two sequences show 74% homology and of the 238 divergent residues 105 correspond to conservative substitutions of related amino acids. Higher order structures for the enzymes have been proposed based on the hydropathy profiles of the respective sequences (Chapter 4; Fig. 2). The polypeptide crosses the membrane eight times for *N. crassa* and ten times for *S. cerevisiae* with only 4% of the protein exposed to the external surface. There are three segments which extend into the cytoplasm, an *N*-terminal tail of approximately 115 amino acids and two loops of approximately 130 and 300 amino acids respectively which are thought to contain the phosphorylation and nucleotide binding sites.

A recent development has been the ability to carry out site-directed mutagenesis on specific regions of the polypeptide and begin the molecular characterization of structure-function relationships of the pump. While a functional ATPase is required for cell growth (Serrano *et al.*, 1986), a specially constructed strain of yeast has been developed (Nakamoto *et al.*, 1991). The strain SY4 contains both a wild type and mutant ATPase gene controlled by separate inducible promoters (*GAL1* and *HSE*). At the appropriate times cells are shifted from galactose medium at 23°C to glucose medium at 37°C (thereby switching from wild type gene on and mutant gene off to wild type gene off and mutant gene on). The higher temperature also inhibits the fusion of secretory vesicles with the plasma membrane as strain SY4 also carries a temperature-sensitive secretory mutation, *sec6-4* (Walworth and Novick, 1987). On the glucose medium at 37°C the cells rapidly fill with vesicles containing newly synthesized mutant ATPase which can be readily isolated.

SY4 cells have been used to study the phosphorylation site of the H$^+$-ATPase. Asp^{378} situated in the large hydrophilic segment is followed by the sequence KTGT in all P-type ATPases. These residues are thought to form a turn at the protein surface which is orientated so that it may receive the gamma-phosphoryl group of the ATP and form the β-acyl phosphate intermediate of the reaction cycle. R.K. Nakamoto, S. Verjovski-Almeida, K.E. Allen, R. Roa and C.W. Slayman (unpublished) have shown that H$^+$-ATPases in which Asp^{378} is substituted with glutamate, asparagine or serine are unable to progress normally through the biogenesis pathway and become arrested in the endoplasmic reticulum. Furthermore substitutions of Lys^{379}, Thr^{380}, Thr^{382} and Thr^{384} have shown the requirement of a positively charged residue, preferably lysine adjacent to the aspartate and three hydroxylated residues downstream of these. The arrangement within the sequence 376-CSDKTGTLT-384 is important as single insertions may lead to biosynthetic arrest of the enzyme (Rao and Slayman, 1993). It is worth noting that the Asp^{378} mutations behave as dominant negatives suggesting that the ATPase may be a functional oligomer or alternatively may interact with limiting amounts of some other accessory protein or chaperone during its biogenesis (Rao *et al.*, 1992).

Other well-conserved segments in *S. cerevisiae* include regions incorporating residues 473–477, 526–574 and 607–651 on the large central hydrophilic segment. Mutations in Asp^{534}, Asp^{560} and Asp^{638} alter the nucleotide specificity of the enzyme (Portillo and Serrano, 1988), thus implicating these regions in ATP binding as suggested by the earlier work using NEM described above. For a more detailed discussion of the structure and function of the H$^+$-ATPase the reader is referred to the reviews by Serrano (1984, 1988).

An additional gene *PMA2* is present in *S. cerevisiae* which displays 89% sequence homology with *PMA1* (Schlesser *et al.*, 1988). It has been shown that *PMA2* can functionally replace *PMA1* during normal mitotic growth although the two gene products differ in several important characteristics such as the level of expression (much lower for *PMA2*), glucose activation, pH optima, divalent cation requirement and inhibitor sensitivity (Supply *et al.*, 1993a,b). Of particular interest is the higher affinity that the PMA2 gene product displays for MgATP and Supply *et al.* (1993a) speculate that it may facilitate H$^+$-pumping under conditions of Mg$^+$ starvation.

8.2.2 Regulation of the H$^+$-ATPase

The activity of the plasma membrane H$^+$-ATPase has been shown to be regulated by glucose metabolism (Serrano, 1983) and acidification (Eraso and Gancedo, 1987). Chang and Slayman (1991) have demonstrated that glucose-mediated activation occurs by phosphorylation which is thought to occur at the carboxy-terminus and involve residues Arg^{909} and Thr^{912} (Portillo *et al.*, 1989, 1991; Monk *et al.*, 1991b). One possible scenario would be binding of the carboxy-terminus to the active site which becomes displaced upon phosphorylation in glucose-metabolizing cells, thus releasing the active site for ATP binding.

8.2.3 Solute transport

It has long been evident that fungi are able to concentrate a variety of solutes from their external medium. There are usually two components of the transport systems responsible for the uptake of these solutes which may be mediated by carriers or channels. A constitutive low affinity transporter allows facilitated diffusion (but not accumulation) when the solute is at a high external concentration. At lower concentrations a high affinity transporter which allows the thermodynamically uphill concentration of solutes becomes derepressed. The presence of two transport systems for a single solute may be energetically advantageous for the cell (Sanders, 1988). When a solute is abundant it may enter the cell via the low affinity systems which may not be energy coupled. When the external concentration of the solute is low derepression of the high affinity transporters enables continued uptake.

(a) Monovalent cations

Of the monovalent cations perhaps most is known about the transport of K$^+$. K$^+$ is required by the fungal cell for such processes as turgor maintenance, charge balancing and protein synthesis. The approximate K_m values of the respective transport systems for K^+ in *S. cerevisiae* are 20 μM for the high-affinity system and 2 mM for the low-affinity system (Rodriguez-Navarro and Ramos, 1984). These two transport processes have been shown to occur via separate proteins. These are encoded by the *TRK1* and *TRK2* genes which display significant sequence homology (Gaber *et al.*, 1988; Ko *et al.*, 1990; Ko and Gaber, 1991) and probably arose through a duplication event. *TRK1* encodes a 180 kDa, 1235 amino acid polypeptide which has been immunolocalized to the plasma membrane, contains 12 potential membrane-spanning segments and a 650 amino acid hydrophilic segment. The lower affinity transporter encoded by *TRK2* also spans the membrane 12 times but is a smaller protein of 889 amino acids with a 334 amino acid hydrophilic segment.

Cells depleted for both *TRK1* and *TRK2* may be complemented by the *KAT1* (Anderson *et al.*, 1992) or *AKT1* (Sentenac *et al.*, 1992) genes from *Arabidopsis thaliana* which are thought to encode K^+ channels. These cells are able to take up K^+ from external concentrations as low as 1 μM raising the possibility that high-affinity transport may be channel mediated, although thermodynamic considerations (e.g. Kochian and Lucas, 1993; Maathuis and Sanders, 1993) indicate that this is not possible. Furthermore, Rodriguez-Navarro *et al.* (1986) using a combination of flux and electrophysiological techniques demonstrated that the high-affinity transport system operates via H^+/K^+ symport, an idea first postulated by Boxman *et al.* (1984). The former study described high-affinity uptake of K^+ associated with stoichiometric extrusion of H^+ in K^+ starved *N. crassa* maintained in millimolar external Ca^+. The mechanism had a strong depolarizing effect on the membrane potential which indicated that two charges entered the cell with each K^+, thus implicating a 1 H^+:1 K^+ cotransport mechanism. There remains the anomaly, however, that expression of the plant K^+ channel genes enables uptake of K^+ from low K^+ medium in *trk1Δtrk2Δ* cells. Clearly this is an area in which further research is required.

Patch clamping has been used to study ion channels in the plasma membrane of several species of fungi (for review see Garrill and Davies, 1994). Two K^+ channels which under the conditions tested carried inward current and thus K^+ influx have been described in the oomycete *Saprolegnia ferax* (Garrill *et al.*, 1992, 1993; Lew *et al.*, 1992) (Figure 8.2). These channels were activated by Ca^{2+} as addition of the ionophore A23187 increased the number of channel openings (with no apparent effect on channel amplitude or open time). The channels could be blocked with quaternary ammonium. Calculated fluxes through these channels suggest that the channels play a role, albeit replaceable (Garrill *et al.*,1993), in the generation of turgor pressure (Garrill *et al.*, 1992).

The plasma membrane of *S. cerevisiae* spheroplasts has been successfully patch clamped and contains voltage-gated outward rectifying K^+ channels (responsible for K^+ efflux) (Gustin *et al.*, 1986; Bertl and Slayman, 1992; Bertl *et al.*, 1992). The channels, which were activated by positive membrane potentials, elevated cytoplasmic Ca^{2+} and increasing pH, show a high degree of selectivity for K^+ over Na^+ and can be blocked by quaternary ammonium and Ba^{2+}. They may function in steady-state turgor regulation and in balancing charge movements during proton-coupled solute uptake. Using wild type *S. cerevisiae*, Ramirez *et al.* (1989) have described K^+ channels which in addition to positive membrane potentials were also active at negative values. An important finding was that in the mutant strain *pmal-105*, in which the gene for the plasma membrane H^+-ATPase is defective, the channels opened at much lower voltages. In addition, in contrast to the wild type, the channels were activated by the intracellular addition of ATP, suggesting some relationship between the plasma membrane H^+-ATPase and channel activity. K^+ channels have also recently been described in *S. pombe* (Vacata *et al.*, 1992).

Mechanosensitive (MS) ion channels which are permeable to cations have been described in the plasma membrane of *S. cerevisiae* (Gustin *et al.*, 1988), *S. pombe* (Zhou and Kung, 1992), *Uromyces appendiculatus* (Zhou *et al.*, 1991) and *Saprolegnia ferax* (Garrill *et al.*, 1992, 1993; Levina *et al.*, 1994). Cation permeability appears to be a common feature of MS channels in walled cells (for review see Garrill *et al.*, 1994) although the channel in *S. cerevisiae* may also conduct anions (Gustin *et al.*, 1988). These channels are gated by tension which is normally applied in patch clamp experiments as

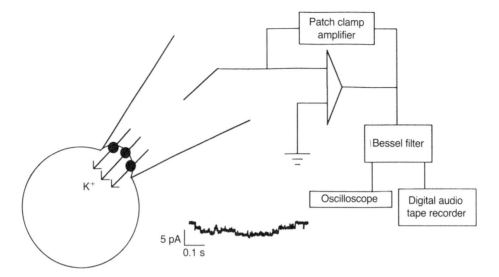

Figure 8.2 Ca^{2+}-activated K^+ channels in the plasma membrane of protoplasts from *Saprolegnia ferax*. Using micromanipulators a saline-filled glass micropipette is brought into contact with the plasma membrane of a protoplast. Application of suction promotes sealing between the glass and the membrane and isolates the membrane in the electrode tip. As the seal improves it becomes possible to resolve single channel currents which are converted into a voltage signal by the amplifier. The signal is then filtered, viewed on an oscilloscope and recorded for later analysis. Knowing the gain of the current/voltage conversion enables the calculation of the channel current amplitude from the voltage signal. Channel opening and K^+ influx is represented by the distinct downward square steps from the baseline.

suction via the patch pipette which creates a pressure gradient across the membrane. It is thought that the channels may play a role in osmoregulation and, in view of their Ca^{2+} permeability, appressoria formation (Zhou *et al.*, 1991) and tip growth (Garrill *et al.*, 1992, 1993; section 13.3.2).

Despite the fact that NH_4^+ may be utilized by fungi as a nitrogen source there have been limited studies on the mechanisms of transport. Uptake in both *N. crassa* and *S. cerevisiae* displays dual kinetics (with K_m values of 7 μM and 8.5 mM, respectively for *N. crassa*) (Slayman, 1977; Dubois and Grenson, 1979). In yeast, both of these systems may be repressed, the high-affinity system in the presence of glutamine, the low-affinity system in the presence of NH_4^+.

Divalent cations are important for growth and metabolism, but may have potentially toxic effects at high concentrations. Despite this the mechanisms responsible for their uptake remain obscure. It is clear that their uptake may be active, requires an electrochemical gradient, adequate K^+ and

phosphate, and may be inhibited with protonophores. In *S. cerevisiae* uptake is mediated by at least one transport system with a series of decreasing affinities of $Mg^{2+} > Co^{2+} > Zn^{2+} > Mn^{2+} > Ni^{2+} > Ca^{2+} > Sr^{2+}$ (Fuhrmann and Rothstein, 1968). Electroneutrality may be maintained by the efflux of two equivalents of K^+ (or in the cases of Mg^{2+} and Ca^{2+}, H^+). There is some evidence for an additional transport system, based on kinetic variability and the fact that the uptake of Mn^{2+}, Mg^{2+}, Sr^{2+} and Zn^{2+} can be driven by both H^+ and K^+ gradients whereas Co^{2+} and Ni^{2+} can only be driven by a H^+ gradient. The kinetic variability should be treated with caution, however, as it may also be due to the effect of the divalent cation on the surface potential (Borst-Pauwels, 1981; Gadd, 1993).

The transport of Ca^{2+} is a subject which has aroused much interest in view of its role as a second messenger (Jackson and Heath, 1993). The mechanisms of Ca^{2+} transport are not discussed here as they are covered by Gadd (Chapter 9) on signal transduction and Gow (Chapter 13) on tip growth and polarity.

(b) Anions

Sulphate and phosphate are the main sources of sulphur and phosphorus utilized by *N. crassa*. For both anions low affinity transport is constitutive, with K_m values of 330 μM for sulphate and 0.12 mM and 3.62 mM at pH values of 5.8 and 7.3 for phosphate (Marzluf, 1970a; Lowendorf *et al.*, 1974; Breton and Surdin-Kerjan, 1977). Transport via the high-affinity transporters is H^+ coupled with H^+/solute stoichiometries in *S. cerevisiae* of 3/1 for sulphate (with the simultaneous efflux of 1 K^+) (Roomans *et al.*, 1979) and up to 3/1 for phosphate with the efflux of 2 K^+ (or $2H^+/1K^+$ if the cells are metabolizing)) (Cockburn *et al.*, 1975; Eddy *et al.*, 1980). K_m values of 3 μM for phosphate and 5 μM for sulphate have been reported for the high-affinity transporter (Marzluf, 1970b; Lowendorf *et al.*, 1975; Burns and Beaver, 1977; Breton and Surdin-Kerjan, 1977).

The derepression of the high-affinity transport systems has been the subject of extensive study. Both systems show similarities, the structural genes *pho-4+* (formerly known as *van+* (Bowman *et al.*, 1983)) for phosphate and *cys-14* for sulphate are controlled by the *nuc-1+* and *cyc-3* gene products. These are in turn under positive control by the *preg+* and *scon-1* gene products, respectively, which may be repressed by the *nuc-2+* and *scon-2* gene products. These latter two gene products may be repressed by P_i or methionine respectively or some product derived from these (Metzenberg, 1979). Both *pho-4+* and *cyc-14* encode polypeptides which may have 12 membrane-spanning domains and a large central hydrophilic segment (Mann *et al.*, 1989; Jarai and Marzluf, 1991).

In certain fungi there may be a third phosphate transport system operating in association with Na^+. Among certain marine chytrids there is a requirement of Na^+ for phosphate uptake (Siegenthaler *et al.*, 1967; Belsky *et al.*, 1970) and Roomans and Borst-Pauwels (1977, 1979) have described a very high affinity ($K_m = 1$ μM) phosphate/Na^+ cotransport system in yeast.

In addition to NH_4^+, fungi are also able to utilize NO_3^- as a nitrogen source. NO_3^- uptake has been studied in *Candida utilis* (Eddy and Hopkins, 1985) and appears to occur via NO_3^-/H^+ symport with a stoichiometry of approximately 1/2. Electroneutrality is maintained by the efflux of 1 equivalent of K^+.

Using conventional electrophysiological techniques, Caldwell *et al.*, (1986) have described an anion channel in the water mould *Blastocladiella emersonii* permeable to chloride, nitrate and phosphate. The channel opens in response to Ca^{2+} influx and the anion efflux is thought to be responsible for the delayed phase of an action potential, the function of which is unknown. There have been as yet no patch clamp studies of anion channels in fungi apart from the mechanosensitive channel in *S. cerevisiae* described above which is permeable to both cations and anions (Gustin *et al.*, 1988).

(c) Sugar transport

In the present discussion, due to the confines of space, I will concentrate solely on the transport of the monosaccharide glucose. For detailed discussion of the transport of other monosaccharides and disaccharides the reader is referred to the recent reviews of Lagunas (1993) and Bisson *et al.* (1993). These sugars are required as carbon and energy sources.

In *N. crassa*, K_m values of 8 mM (Scarborough, 1970a) and 10 μM (Scarborough, 1970b) have been reported for the low- and high-affinity glucose transporters (designated GluI and GluII). The glucose-repressible GluII, which is able to concentrate non-metabolizable analogues of glucose by a factor of approximately 6000 (Slayman and Slayman, 1975) has been shown to operate via H^+ symport with a probable H^+/sugar stoichiometry of 1 (Slayman and Slayman, 1974).

There appears to be significant differences in the mechanisms of glucose transport in *S. cerevisiae*. As for *N. crassa* there are high ($K_m = 1$ mM) and low ($K_m = 20$ mM) affinity transporters although both appear to operate via facilitated diffusion and as a consequence yeast cells are unable to concentrate glucose. High-affinity uptake appears to require the presence of kinases although it has been argued that the transport of glucose does not involve phosphorylation and the kinase requirement may reflect a physical interaction between the kinase and the transporter (Lagunas, 1993).

Five genes (*SNF3, HXT1, HXT2, HXT3, HXT4*) encoding suspected high-affinity transporters have been identified in *S. cerevisiae* (Kruckeberg and Bisson, 1990; Lewis and Bisson, 1991; Ko *et al.*, 1993). These form part of a transporter gene

Table 8.2 Amino acid transporters of the plasma membrane of *N. crassa* (All L isomers except [a] = both D and L isomers) (modified from Horak, 1986)

System	Amino acids	Range of K_T (μM)	Gene
I (N) constitutive	Gly, Leu, Val, Phe, Trp, His, Gln, Asp	48 (Phe) – 1000 (Gly)	*mtr*
II (B) constitutive	Arg, Lys, His	1.6 (Arg) – 3500 (His)	*pmb*
III (G) constitutive	Gly, Leu[a], Val, Asn, Glu, Phe[a], Trp, Arg, Lys[a], His, Asp	1.6 (Arg) – 3400 (Asp)	*pmg*
IV	Asp, Glu	13 (Asp) 16 (Glu)	
V	Met	2.3	

superfamily which typically contains 12 membrane-spanning domains and a large hydrophilic segment. Five of the membrane-spanning domains may form amphipathic α-helices which constitute a pentagonal pore through which glucose is transported (Ko *et al.*, 1993).

(d) Amino acid transport

Amino acid transport systems of fungi can generally be grouped into those specific for one or several structurally related amino acids or a general system shared by many amino acids regardless of structure. It has been suggested by Sanders (1988) that the general amino acid permease enables the cells to scavenge for organic nitrogen whereas the more specific permeases supply amino acids for specific metabolic processes. These permeases differ from analogous systems in animal and bacterial cells in that they generally mediate only unidirectional fluxes from the external media into the cell. Uptake occurs via symport with H^+ with a usual H^+/amino acid stoichiometry of 2/1 (Horak, 1986).

The number of amino acid permeases varies between species, e.g. five in *N. crassa* (Table 8.2), two in *Aspergillus nidulans*, nine in *Penicillium chrysogenum*, nine in *Achlya* sp. and 16 in *S. cerevisiae* (Horak, 1986). I will concentrate on those of *N. crassa*; the reviews by Eddy (1982) and Horak (1986) give detailed discussions of other species. The permeases of *N. crassa* are specific for neutral and aromatic amino acids (designated N), basic amino acids (designated B), all amino acids with the exception of L-proline (designated G), acidic amino acids and L-methionine respectively. The latter two systems are active only under conditions of nitrogen and sulphur starvation whereas N, B and G are constitutive. Genetic analysis has

identified the loci responsible for the systems N, B and G which have been termed *mtr*, *pmb* and *pmg* respectively. The *mtr* gene has been cloned and sequenced (Stuart *et al.*, 1988; Koo and Stuart, 1991) and is thought to encode a 261 amino acid polypeptide with a molecular mass of 28 613 Da. Hydropathy plots suggest six potential α-helical transmembrane domains which are 20–30 amino acids long and the polypeptide contains a sequence homologous to an RNA binding motif. It is thought that the polypeptide is part of a ribonucleoprotein complex that forms the N system (Stuart, 1977).

(e) Regulation of solute transport

From estimates of the H^+/solute stoichiometries it is possible to calculate the maximum possible accumulation ratios for a given solute (Sanders, 1988). It is clear from the calculated values that these are seldom achieved and thus symport must be under regulatory control. Mechanisms responsible may occur at the levels of protein synthesis or activity via, for example, substrate inhibition, slip reactions, transinhibition and compartmentation. G amino acid permease activity may be selectively inhibited by its own substrates which kinetically may be explained by substrate inhibition (DeBusk and DeBusk, 1980). The hypothetical slip mechanisms have yet to be demonstrated but could arise due to the efflux of a solute down its concentration gradient via the non-protonated form of the symporter (Eddy, 1980, 1982). Transinhibition, in which influx of an amino acid is controlled by the concentration of the same (and other) amino acid(s) in the cytoplasm has been shown to operate for the N, B (Pall, 1971) and L-methionine (Pall and Kelly, 1971) amino acid permeases of N.

crassa. The compartmentation of solutes, specifically into the vacuole will be considered below.

8.3 TONOPLAST

8.3.1 Vacuolar H⁺-ATPase

In addition to the plasma membrane H^+-ATPase, fungal cells also contain a H^+-pumping ATPase which is located in the vacuolar membrane. There is evidence to suggest that, in *N. crassa* at least, the vacuolar H^+-ATPase, like the H^+-ATPase of the plasma membrane, is essential for growth (Bowman *et al.*, 1992). It should be noted, however, that yeast cells are able to grow without the vacuolar H^+-ATPase (Stevens, 1992). As will be shown below, the enzymes are structurally and mechanistically very distinct. Fungal vacuoles contain large amounts of basic amino acids (concentrations of up to 0.5 mM) and the vacuolar H^+-ATPase energizes their compartmentation in the vacuole. In addition it creates an acidic interior (approximately pH 6.1, Bowman *et al.*, 1992) optimizing the activity of hydrolytic enzymes such as proteases, nucleases and phosphatases in the vacuole.

The properties of the vacuolar H^+-ATPase are summarized in Table 8.1. It is a member of a group of H^+-ATPases called the V-ATPases which are present in a variety of organelles including plant vacuoles, lysosomes, chromaffin granules, clathrin-coated vesicles, Golgi and ER and has structural similarities with the F_0F_1-ATPases (see below). The enzyme can be purified from isolated vacuolar membranes by solubilization with zwitterionic detergents followed by centrifugation on a glycerol gradient (Uchida *et al.*, 1985). It displays a specific activity of 18 μmol min^{-1} mg^{-1}. The enzyme is most active around neutral pH (7.0–7.5) displays a K_m of 0.2 mM and is able to hydrolyse ATP and other nucleotides, especially GTP and ITP (Kakinuma *et al.*, 1981; Bowman and Bowman, 1982). Indeed some authors have proposed the existence of a separate GTPase enzyme (Lichko and Okorokov, 1985). Like the plasma membrane H^+-ATPase the true substrate of the enzyme is MgATP as there is an absolute requirement of Mg^{2+} for activity. The enzyme is insensitive to vanadate, which suggests that its reaction cycle does not proceed through a covalent phosphorylated intermediate, but can be inhibited with bafilomycin A1 and nitrate although the usefulness

of this latter blocker is restricted as relatively high concentrations (50–100 mM) are required (Bowman and Bowman, 1986; Bowman, E. *et al.*, 1988a).

It is possible to demonstrate the formation of an electrochemical gradient by the vacuolar H^+-ATPase *in vitro* using vacuolar vesicles and fluorescent dyes. Importantly proton pumping activity can be shown to have the same substrate specificity and affinity and inhibitor sensitivity as ATPase activity (Kakinuma *et al.*, 1981; Ohsumi and Anraku, 1981; Lichko and Okorokov, 1984; Bowman and Bowman, 1985).

The vacuolar H^+-ATPase is a large enzyme with a very characteristic structure, such that it is identifiable in electron micrographs of the tonoplast stained with 1% phosphotungstic acid (Bowman et al., 1989). The enzyme has a 'stalk and ball' appearance similar to the F_0F_1-ATPases (Figure 8.3 and below) and may be thought of as having peripheral (V_1) and membrane-associated (V_0) components (Figure 8.3). The peripheral component is thought to be composed of five different polypeptides with molecular masses of 67, 57, 48, 30 and 17 KDa which correspond to subunits A, B, C, D and E respectively. There appears to be three copies of subunits A and B in each vacuolar H^+-ATPase. A similar subunit stoichiometry of 3:3:1:1:1 has been observed with the V-ATPase of bovine-coated vesicles (Arai *et al.*, 1988). The membrane-associated portion of the vacuolar H^+-ATPase appears to be composed of polypeptides of 100, 40, 20 and 16 kDa.

The genes encoding A (vma1), B (vma2) and the 16 kDa subunit (vma3) from *N. crassa* have been isolated (Bowman, B. *et al.*, 1988; Bowman, E. *et al.*, 1988b; Sista, 1991, cited in Bowman *et al.* 1992). These genes show considerable sequence homology with those encoding subunits of the mitochondrial ATPase (discussed at length below). More recently, in yeast cells nine genes have been identified which are essential for expression of the vacuolar H^+-ATPase (Anraku *et al.*, 1992). The nucleotide sequence of the yeast gene encoding subunit A (*VMA1*) shows 73% homology with its *N. crassa* counterpart although the yeast gene contains a non-homologous insert and expresses a 119 kDa polypeptide which undergoes post-translational cleavage and splicing to give the mature 67 kDa subunit A (Kane *et al.*, 1990). In *N. crassa*, subunit A displays nucleotide protectable labelling with NEM and is thought to contain the

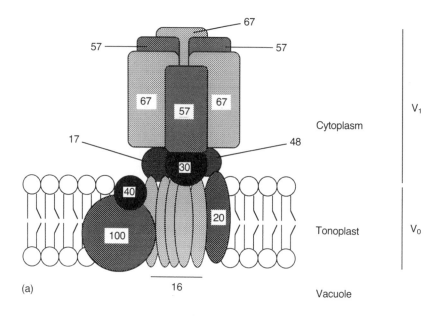

(a)

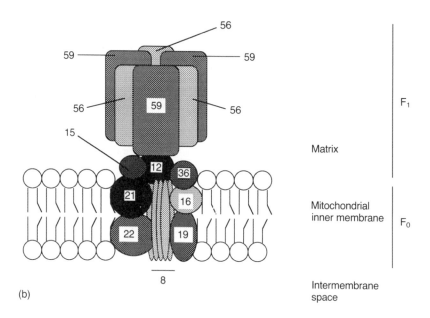

(b)

Figure 8.3 Structural models for the (a) vacuolar and (b) mitochondrial H^+-ATPases. There is significant structural homology between some subunits of the respective enzymes (see text) suggesting that they arose from a common ancestral gene. (Modified from Nelson, 1992.)

nucleotide-binding site (Bowman and Bowman, 1985). *VMA2* encoding subunit B in yeast shows 82% homology with the *N. crassa* gene. Mutants for this gene lack vacuolar H^+-ATPase activity and the ability to acidify the vacuole (Yamashiro *et al.*, 1990; Ohya *et al.*, 1991) suggesting that subunit B functions as a regulatory component. It is thought that the 16 kDa subunit may function as the proton channel (Nelson and Taiz, 1989).

8.3.2 Vacuolar pyrophosphatase

In addition to the V type H^+-ATPase the tonoplast of higher plants also contains a H^+-pumping pyrophosphatase (H^+-PPase)which utilizes inorganic pyrophosphate (PP_i) as its energy source and may, in addition to H^+, facilitate the transport of K^+ into the vacuole (Rea *et al.*, 1992; Davies *et al.*, 1992). There is at present uncertainty as to whether a similar enzyme exists in the fungal tonoplast. Lichko and Okorokov (1991) have reported H^+-translocation driven by PP_i in *Saccharomyces carlsbergenesis* although attempts to identify an enzyme homologous to the plant enzyme have not been successful (R. Eisman, E.J. Kim and P.A. Rea, unpublished, cited in Rea *et al.*, 1992). Calvert (1992) has identified a tonoplast bound enzyme in *C. albicans* with the characteristics of a PPase. However, H^+-pumping was not observed. Although the enzyme showed similarities to the yeast soluble PPase with respect to substrate and ionic requirements its pH optima and membrane location suggested the two enzymes are distinct entities.

8.3.3 Amino acid transport

It has long been recognized that large concentrations of basic amino acids are compartmentalized in the fungal vacuole. This accumulation is facilitated by H^+/amino acid antiport systems as shown by protonophore sensitive uptake of amino acids into tonoplast vesicles with H^+-ATPase activation (Ohsumi and Anraku, 1981). There may be as many as seven separate H^+/amino acid antiporters in the *S. cerevisiae* tonoplast (Sato *et al.*, 1984a) transporting arginine, arginine–lysine, histidine, phenylalanine–tryptophan, tyrosine, glutamine–asparagine and isoleucine–leucine respectively. In addition an arginine–histidine antiporter is present which utilizes the chemical potential of the histidine concentration gradient (Sato *et al.*, 1984b). There appears to be significant differences between the transport systems of different fungi, for example, the *N. crassa* tonoplast contains only a single arginine transporter thought to be a 40 kDa protein (Paek and Weiss, 1989). Under conditions of nitrogen starvation, amino acids are released from the vacuolar pool into the cytoplasm although the precise mechanisms are at present unclear.

8.3.4 Ions

The transport of cations across the tonoplast plays an important role in regulating their cytoplasmic concentrations and is thus essential for metabolic processes and the sequestration of certain potentially toxic ions. Transport is thought to occur via H^+/ion antiporters with apparent K_m values for the uptake of Ca^{2+}, Mg^{2+}, Mn^{2+}, Zn^{2+}, and P_i of 0.04, 0.3, 0.8, 0.055–0.17 and 1.5 mM respectively (Gadd, 1993). The sequestration of Ca^{2+} and maintenance of a low cytoplasmic concentration is of obvious importance in signal transduction and is covered in Chapter 9. It is thought that the uptake of Zn^{2+} and Co^{2+} into the yeast vacuole contributes to their tolerant behaviour (Gadd, 1993). Levels of polyphosphate within the vacuole appear to be influenced by the accumulation of certain ions and may therefore play a role in cation retention (Klionsky *et al.*, 1990).

Wada *et al.* (1992) have identified two Cl^- transport systems in the tonoplast which contribute to the formation of the pH gradient across the tonoplast by shunting the membrane potential generated by proton translocation. These systems differ kinetically and in their inhibitor sensitivity. A saturable component with a K_m for Cl^- of 20 mM possibly operates through a carrier system whereas a linear component which can be blocked by 4,4'-diisothiocyano-2,2'-stilbenedisulphonic acid (DIDS) may operate via channels.

The tonoplast of plant cells have proven easier to patch clamp than the plasma membrane and although reports are scarce this may also be the case for fungi (Garrill and Davies, 1994). Studies with *S. cerevisiae* tonoplast indicate a high conductance (150 pS in 100 mM-KCl) cation-selective inward rectifier channel (with opening favouring cation release from the vacuole) (Bertl and Slayman, 1990, 1992; Bertl *et al.* 1992). The channel may be activated by voltage over a feasible range of *in vivo* membrane potentials and by elevated levels of cytoplasmic Ca^{2+} (in the μM range). The sensitivity to Ca^{2+} may be lost over time but can be regained in the presence of reducing agents possibly reflecting *in vivo* protection by antioxidants or the basis of a dynamic redox regulation (Bertl and Slayman, 1990, 1992). The channel with its Ca^{2+} permeability may function in signal transduction. The fact that the channel is permeable to other cations, specifically K^+ and Na^+, may serve to stabilize the driving force for Ca^{2+} efflux by shifting the potential across the tonoplast towards the Goldmann voltage for K^+ and Na^+ (Bertl and Slayman, 1992).

8.3.5 Regulation of solute transport

There have been few studies relating to the regulation of solute transport across the tonoplast. The K_m values for the cation antiporters will clearly control transport considering their similarity to the cytosolic concentration of the various cations. The difference in the K_m values between the respective amino acid transporters of the tonoplast and plasma membrane in *S. cerevisiae* (up to 100 times higher in the tonoplast) would suggest the need for their removal from the cytosol when their concentrations become too high (Klionsky *et al.*, 1990). Regulating factors of the *S. cerevisiae* cation channel are described above.

8.4 MITOCHONDRIA

I will consider only the mitochondrial H^+-ATPase of the multitude of proteins in the mitochondrial inner membrane as it is important to compare this with the other major ATPases in the cell.

8.4.1 Mitochondrial H^+-ATPase

The inner membrane of fungal mitochondria contains the mitochondrial H^+-ATPase which, along with the H^+-ATPase from the mitochondria of other cell types and the H^+-ATPase of thylakoid membranes forms a group known as the F_0F_1-ATPases. The function of the fungal mitochondrial H^+-ATPase is the synthesis of ATP from ADP and P_i, utilizing the electrochemical gradient created by the oxidative reactions of the membrane-embedded respiratory enzymes.

Although structurally similar to the vacuolar H^+-ATPase it may be distinguished from this and the plasma membrane H^+-ATPase on the basis of inhibitor sensitivity and pH optima (Table 8.1). The F_0F_1-ATPases, like the V-ATPases, consist of separable peripheral and membrane bound components (Figure 8.3). The peripheral, or F_1 component is water soluble and consists of five subunits (designated alpha, beta, gamma, delta and epsilon) with molecular weights (in *N. crassa* (Sebald, 1977)) of 59, 56, 36, 15 and 12 kDa which are analogous to subunits A, B, C, D and E of the vacuolar H^+-ATPase. The membrane-associated, or F_0 component consists of at least five polypeptides in *N. crassa* (Sebald, 1977) of 22, 21, 19, 16 and 8 kDa. The complete enzyme consists of approximately 20 subunits as certain polypeptides are present in multiple copies.

The β subunit is thought to contain the catalytic site and the genes encoding both this polypeptide (*atp-2*), as well as the α-subunit (*atp-1*) have been isolated in both *N. crassa* and yeast (Bowman and Knock, 1992). It is thought that *atp-1*, *atp-2*, *vma-1* and *vma-2* arose from a common ancestral gene (Bowman, B. *et al.*, 1992). The 8 kDa subunit (subunit c) is analogous to the 18 kDa subunit of the vacuolar H^+-ATPase and is likewise thought to form the proton channel. There have been reports that the gene for this polypeptide has been isolated (W. Sebald, personal communication, cited in Bowman, B. *et al.*, 1992). In addition to the above, several other subunits have been characterized and their genes sequenced in yeast (e.g. Norais *et al.*, 1991).

8.5 CONCLUSIONS

Given the importance of transport systems (summarized in Figure 8.4) to a multitude of cellular processes such as those listed in section 8.1 it is obvious that an understanding of these systems is essential to aid a fuller picture of fungal biology. It is clear from the above that the study of transport systems encompassing such disciplines as electrophysiology, biochemistry, genetics, molecular biology and patch clamping has undergone significant advances in recent years. It is hoped that in the future the combination of such techniques as site-directed mutagenesis and patch clamping will enable us to probe the structure–function relationships and biophysics of the various

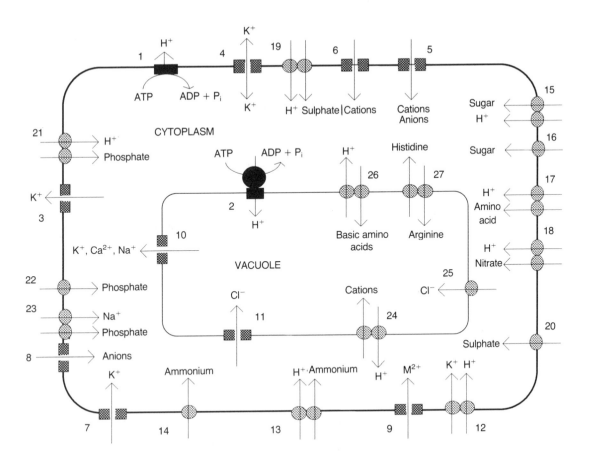

Figure 8.4 Schematic representation of transport processes operating in the fungal plasma membrane and tonoplast which are discussed in the present chapter. For actual stoichiometries see text. The low-affinity transport systems for certain solutes have been shown as carriers but may actually be channels (as indicated below). (1,2) Pumps: (1) plasma membrane H^+-ATPase (Slayman, 1977; Goffeau and Slayman, 1981; Serrano, 1984, 1988; Bowman and Bowman, 1986); (2) vacuolar H^+-ATPase (Bowman and Bowman, 1986). (3–11) Channels: (3) voltage gated outward rectifying K^+ channel (Gustin *et al.*, 1986; Bertl and Slayman, 1992; Bertl *et al.* 1992); (4) K^+ channel which activates at positive and negative voltages (Ramirez *et al.*, 1989); (5) unselective stretch-activated channel (Gustin *et al.*, 1988); (6) cation selective stretch-activated channel (Zhou *et al.*, 1991; Zhou and Kung, 1992; Garrill *et al.*, 1992, 1993); (7) Ca^{2+}-activated K^+ channel (Garrill *et al.*, 1992, 1993; Lew *et al.*, 1992); (8) anion-selective channel (Caldwell *et al.*, 1986); (9) divalent cation channel (Chapter 9); (10) tonoplast cation selective inward rectifier channel (Bertl and Slayman, 1990, 1992; Bertl *et al.*, 1992); (11) low-affinity Cl^- transporter (Wada *et al.*, 1992). (12–27) Carriers: (12) K^+/H^+ symporter (Rodriguez-Navarro *et al.*, 1986); (13) high-affinity NH_4^+ transporter (Dubois and Grenson, 1979); (14) low-affinity NH_4^+ transporter (channel mediated?) (Dubois and Grenson, 1979); (15) high-affinity sugar uptake (via H^+/sugar symport) (Scarborough, 1970b; Slayman and Slayman, 1975); (16) low-affinity sugar transporter (channel-mediated?) (Scarborough, 1970a); (17) H^+/amino acid symporter (Horak, 1986); (18) high affinity nitrate transporter (Eddy and Hopkins, 1985); (19) high-affinity sulphate transporter (Marzluf, 1970b); (20) low-affinity sulphate transporter (channel-mediated?) (Marzluf, 1970a); (21) high-affinity phosphate transporter (Lowendorf *et al.*, 1975; Burns and Beever, 1977); (22) low-affinity phosphate transporter (channel-mediated?); (23) Na^+ phosphate symporter (Siegenthaler *et al.*, 1967; Belsky *et al.*, 1970; Roomans and Borst-Pauwels, 1977, 1979); (24) H^+/cation antiporter (Gadd, 1993); (25) high affinity Cl^- transporter (Wada *et al.*, 1992); (26) H^+/amino acid antiporter (Sato *et al*, 1984a); (27) arginine/histidine antiporter (Sato *et al.*, 1984b).

transport proteins. These techniques are beginning to be applied to fungal cells and should yield major advances similar to those recently reported for plant membranes (Chasan and Schroeder, 1992).

REFERENCES

Addison, R. (1986) Primary structure of the *Neurospora* plasma membrane H$^+$-ATPase deduced from the gene sequence. Homology to Na$^+$/K$^+$, Ca^{2+} and K$^+$-ATPases. *Journal of Biological Chemistry*, **261**, 14896–901.

Addison, R. and Scarborough, G.A. (1981) Solubilisation and purification of the *Neurospora* plasma membrane H$^+$-ATPase. *Journal of Biological Chemistry*, **256**, 13165–71.

Addison, R. and Scarborough, G.A. (1982) Conformational changes of the *Neurospora* plasma membrane H$^+$-ATPase during its catalytic cycle. *Journal of Biological Chemistry*, **257**, 10421–7.

Ahlers, J., Ahr, E. and Seyfarth, A. (1978) Kinetic characterisation of the plasma membrane ATPase from *Saccharomyces cerevisiae*. *Molecular and Cellular Biochemistry*, **22**, 39–49.

Amoury, A., Foury, F. and Goffeau, A. (1980) The purified plasma membrane ATPase of the yeast *Schizosaccharomyces pombe* forms a phosphorylated intermediate. *Journal of Biological Chemistry*, **255**, 9353–7.

Amoury, A. and Goffeau, A. (1982) Characterisation of the β-aspartyl phosphate intermediate formed by the H$^+$-translocating ATPase from the yeast *Schizosaccharomyces pombe*. *Journal of Biological Chemistry*, **257**, 4723–30.

Amoury, A., Goffeau, A., McIntosh, D.B. and Boyer, P.D. (1982) Exchange of oxygen between phosphate and water catalysed by the plasma membrane ATPase from the yeast *Schizosaccharomyces pombe*. *Journal of Biological Chemistry*, **257**, 12509–16.

Amoury, A., Goffeau, A., McIntosh, D.B. and Boyer, P.D. (1984) Contribution of ^{18}O technology to the mechanism of the H$^+$-ATPase from yeast plasma membrane. *Current Topics in Cellular Regulation*, **24**, 471–83.

Anderson, J.A., Huprikar, S.S., Kochian, L.V. *et al.* (1992) Functional expression of a probable *Arabidopsis thaliana* potassium channel in *Saccharomyces cerevisiae*. *Proceedings of the National Academy of Sciences of the USA*, **89**, 3736–40.

Anraku, Y., Hirata, R., Wada, Y. and Ohya, Y. (1992) Molecular genetics of the yeast vacuolar H$^+$-ATPase. *Journal of Experimental Biology*, **172**, 67–81.

Arai, H., Terres, G. Pink, S. and Forgac, M. (1988) Topography and subunit stoichiometry of the coated vesicle proton pump. *Journal of Biological Chemistry*, **263**, 8796–802

Belsky, M.M., Goldstein, S. and Menna, M. (1970) Factors affecting phosphate uptake in the marine fungus *Dermocystidium* spp. *Journal of General Microbiology*, **62**, 399–402.

Bertl, A. and Slayman, C.L. (1990) Cation selective channels in the vacuolar membrane of *Saccharomyces*: dependence on Ca^{2+}-redox state and voltage. *Proceedings of the National Academy of Sciences of the USA*, **89**, 3736–40.

Bertl, A. and Slayman, C.L. (1992) Complex modulation of cation channels in the tonoplast and plasma membrane of *Saccharomyces cerevisiae*: single channel studies. *Journal of Experimental Biology*, **172**, 271–87.

Bertl, A., Gradmann, D. and Slayman, C.L. (1992) Calcium- and voltage-dependent ion channels in *Saccharomyces cerevisiae*. *Philosophical Transactions of the Royal Society of London, Series B*, **338**, 63–72.

Bisson, L.F., Coons, D.M., Kruckeberg, A.L. and Lewis, D.A. (1993) Yeast sugar transporters. *Critical Reviews in Biochemistry and Molecular Biology*, **28**, 259–308.

Blanpain, J.-P., Ronjat, M., Supply, P. *et al.* (1992) The yeast plasma membrane H$^+$-ATPase. An essential change of conformation triggered by H$^+$. *Journal of Biological Chemistry*, **267**, 3735–40.

Blasco, F., Chapius, J.-P. and Giordani, R. (1981) Characterisation of the plasma membrane ATPase of *Candida tropicalis*. *Biochimie*, **63**, 507–14.

Borst-Pauwels, G.W.F.H. (1981) Ion transport in yeast. *Biochimica et Biophysica Acta*, **650**, 88–127.

Bowman, B.J. and Bowman, E.J. (1986) H$^+$-ATPases from mitochondria, plasma membrane and vacuoles of fungal cells. *Journal of Membrane Biology*, **94**, 83–97.

Bowman, B.J. and Slayman, C.W. (1977) Characterisation of plasma membrane ATPase of *Neurospora crassa*. *Journal of Biological Chemistry*, **254**, 2928–34.

Bowman, B.J., Allen, K.E. and Slayman, C.W. (1983) Vanadate-resistant mutants of *Neurospora crassa* are deficient in a high affinity phosphate transport system. *Journal of Bacteriology*, **153**, 292–6.

Bowman, B.J., Allen, R., Wechser, M.A. and Bowman, E.J. (1988) Isolation of genes encoding the *Neurospora* vacuolar ATPase. Analysis of *vma-2* encoding the 57-kDa polypeptide and comparison to *vma-1*. *Journal of Biological Chemistry*, **264**, 14002–7.

Bowman, B.J., Dschida, W.J. and Bowman, E.J. (1989) The vacuolar ATPase of *Neurospora crassa* contains an F$_1$-like structure. *Journal of Biological Chemistry*, **264**, 15606–12.

Bowman, B.J., Dschida, W.J. and Bowman, E.J. (1992) Vacuolar ATPase of *Neurospora crassa*: electron microscopy, gene characterisation and gene inactivation/mutation. *Journal of Experimental Biology*, **172**, 57–66.

Bowman, E.J. and Bowman, B.J. (1982). Identification and properties of an ATPase in vacuolar membranes of *Neurospora crassa*. *Journal of Bacteriology*, **151**, 1326–37.

Bowman, E.J. and Bowman, B.J. (1985) The H$^+$-translocating ATPase in vacuolar membranes of *Neurospora crassa*, in *Biochemistry and Function of Vacuolar*

Adenosine-triphosphatase in Fungi and Plants, (ed. B.P. Marin), Springer-Verlag, Berlin, pp. 132–41.

Bowman, E.J. and Knock, T.E. (1992) Structures of the genes encoding the α and β subunits of the *Neurospora crassa* mitochondrial ATP synthase. *Gene*, **114**, 157–63.

Bowman, E.J., Bowman, B.J. and Slayman, C.W. (1981) Isolation and characterisation of plasma membranes from wild type *Neurospora crassa*. *Journal of Biological Chemistry*, **256**, 12336–42.

Bowman, E.J., Siebers, A. and Altendorf, K. (1988a) Bafilomycins: a new class of inhibitors of membrane ATPases from microorganisms, animal cells and plant cells. *Proceedings of the National Academy of Sciences of the USA*, **85**, 7972–6.

Bowman, E.J., Tenney, K. and Bowman, B.J. (1988b) Isolation of genes encoding the *Neurospora* vacuolar ATPase. Analysis of *vma-1* encoding the 67 kDa subunit reveals homology to other ATPases. *Journal of Biological Chemistry*, **263**, 13994–4001.

Boxman, A.W., Dobbelmann, J. and Borst-Pauwels, G.W.F.H. (1984) Possible energisation of K$^+$ accumulation into metabolising yeast by the proton-motive force. Binding correction to be applied in the calculation of the yeast membrane potential from tetramethylphosphonium distribution. *Biochimica et Biophysica Acta*, **772**, 51–7.

Breton, A. and Surdin-Kerjan, Y. (1977) Sulfur uptake in *Saccharomyces cerevisiae*: biochemical and genetic study. *Journal of Bacteriology*, **132**, 224–32.

Brooker, R.J. and Slayman, C.W. (1982) Inhibition of the plasma membrane [H$^+$]-ATPase of *Neurospora crassa* by N-ethylmaleimide. Protection by nucleotides. *Journal of Biological Chemistry*, **257**, 12051–5.

Brooker, R.J. and Slayman, C.W. (1983a) [^{14}C]N-ethylmaleimide labelling of the plasma membrane [H$^+$]-ATPase of *Neurospora crassa*. *Journal of Biological Chemistry*, **258**, 222–6.

Brooker, R.J. and Slayman, C.W. (1983b) Effects of Mg^{2+} ions on the plasma membrane [H$^+$]-ATPase of *Neurospora crassa*. I. Inhibition by N-ethylmaleimide and trypsin. *Journal of Biological Chemistry*, **258**, 8827–32.

Brooker, R.J. and Slayman, C.W. (1983c) Effects of Mg^{2+} ions on the plasma membrane [H$^+$]-ATPase of *Neurospora crassa*. II. Kinetic studies. *Journal of Biological Chemistry*, **258**, 8833–8.

Burns, D.J.W. and Beever, R.E. (1977) Kinetic characterisation of the two phosphate uptake systems in the fungus *Neurospora crassa*. *Journal of Bacteriology*, **132**, 511–19.

Caldwell, J.H., van Brunt, J. and Harold, F.M. (1986) Calcium-dependent anion channel in the water mold *Blastocladiella emersonii*. *Journal of Membrane Biology*, **89**, 85–97.

Calvert, C.M. (1992) Ion transport at the vacuolar membrane of *Candida albicans*. DPhil Thesis, University of York, York, UK.

Chang, A. and Slayman, C.W. (1991) Maturation of the yeast plasma membrane [H$^+$]ATPase involves phos-phorylation during intracellular transport. *Journal of Cell Biology*, **115**, 289–95.

Chasan, R. and Schroeder, J.I. (1992) Meeting report: excitation in plant membrane biology. *The Plant Cell*, **4**, 1180–8.

Cockburn, M., Earnshaw, P. and Eddy, A.A. (1975) The stoichiometry of the adsorption of protons with phosphate and L-glutamate by yeasts of the genus *Saccharomyces*. *Biochemical Journal*, **146**, 705–12.

Dame, J.R. and Scarborough, G.A. (1980) Identification of the hydrolytic moiety of the *Neurospora* plasma membrane H$^+$-ATPase and demonstration of a phosphoryl-enzyme intermediate in its catalytic mechanism. *Biochemistry*, **19**, 2931–7.

Davenport, J.W. and Slayman, C.W. (1988) The plasma membrane H$^+$-ATPase of *Neurospora crassa*. Properties of two reactive sulfhydryl groups. *Journal of Biological Chemistry*, **263**, 16007–13.

Davies, J.M., Poole, R.M., Rea, P.A. and Sanders, D. (1992) Potassium transport into plant vacuoles energized directly by a proton-pumping inorganic pyrophosphatase. *Proceedings of the National Academy of Sciences of the USA*, **89**, 11701–5.

DeBusk, R.M. and DeBusk, A.G. (1980) Physiological and regulatory properties of the general amino acid transport system of *Neurospora crassa*. *Journal of Bacteriology*, **143**, 188–97.

Delhez, J., Dufour, J.-P., Thines, D. and Goffeau, A. (1977) Comparison of the plasmamembrane bound and mitochondria bound ATPase in the yeast *Schizosaccharomyces pombe*. *European Journal of Biochemistry*, **79**, 319–28.

Dubois, E. and Grenson, M. (1979) Methylamine/ammonia uptake systems in *Saccharomyces cerevisiae*: multiplicity and regulation. *Molecular and General Genetics*, **175**, 67–76.

Dufour, J.-P. and Goffeau, A. (1978) Solubilisation by lysolecithin and purification of the plasma membrane ATPase of the yeast *Schizosaccharomyces pombe*. *Journal of Biological Chemistry*, **253**, 7026–32.

Eddy, A.A. (1980) Slip and leak models of gradient-coupled solute transport. *Biochemical Society Transactions*, **8**, 271–3.

Eddy, A.A. (1982) Mechanisms of solute transport in selected eukaryotic micro-organisms. *Advances in Microbial Physiology*, **23**, 1–78.

Eddy, A.A. and Hopkins, P.G. (1985) The putative electrogenic nitrate-proton symport of the yeast *Candida utilis*. Comparison with the systems absorbing glucose or lactate. *Biochemical Journal*, **231**, 291–7.

Eddy, A.A., Seaston, A. Gardner, D. and Hacking, C. (1980) The thermodynamic efficiency of cotransport mechanisms with special reference to proton and anion transport in yeast. *Annals of the New York Academy of Sciences, USA*, **341**, 494–508.

Eraso, P. and Gancedo, C. (1987) Activation of yeast plasma membrane H$^+$-ATPase by acid pH during growth. *FEBS Letters*, **224**, 187–92.

Fuhrmann, R. and Rothstein, A. (1968) The transport of Zn^{2+}, Co^{2+} and Ni^{2+} into yeast cells. *Biochimica et Biophysica Acta*, **163**, 325–30.

Gaber, R.F., Styles, C.A. and Fink, G.R. (1988) *TRK1* encodes a plasma membrane protein required for high-affinity potassium transport in *Saccharomyces cerevisiae*. *Molecular and Cellular Biology*, **8**, 2848–59.

Gadd, G.M. (1993) Interactions of fungi with toxic metals. *New Phytologist*, **124**, 25–60.

Garrill, A. and Davies, J.M. (1994) Patch clamping fungal membranes: a new perspective on ion transport. *Mycological Research*, **98**, 257–63.

Garrill, A. and Jennings, D.H. (1991) Isolation of a plasma membrane ATPase with H^+-ATPase-like properties from the marine fungus *Dendryphiella salina*. *Experimental Mycology*, **15**, 351–5.

Garrill, A., Lew, R.R. and Heath, I.B. (1992) Stretch-activated Ca^{2+} and Ca^{2+}-activated channels in the hyphal tip plasma membrane of the oomycete *Saprolegnia ferax*. *Journal of Cell Science*, **101**, 721–30.

Garrill, A., Jackson, S.L., Lew, R.R. and Heath, I.B. (1993) Ion channel activity and tip growth: tip localised stretch-activated channels generate an essential Ca^{2+} gradient in the oomycete *Saprolegnia ferax*. *European Journal of Cell Biology*, **60**, 358–65.

Garrill, A., Findlay, G.P. and Tyerman, S.D. (1994) Mechanosensitive ion channels, in *Membranes: Specialised Functions in Plant Cells*, (eds. M. Smallwood, J.P. Knox and D.J. Bowles), JAI Press, Greenwich, USA (in press).

Ghislain, M., Schlesser, A. and Goffeau, A. (1987) Mutation of a conserved glycine residue modifies the vanadate sensitivity of the plasma membrane H^+ATPase from *Schizosaccharomyces pombe*. *Journal of Biological Chemistry*, **262**, 17549–55.

Goffeau, A. and Slayman, C.W. (1981) The proton-translocating ATPase of the fungal plasma membrane. *Biochimica et Biophysica Acta*, **639**, 197–223.

Gradmann, D., Hansen, U.-P., Long, W.S. *et al.* (1978) Current-voltage relationships for the plasma membrane and its principal electrogenic pump in *Neurospora crassa*. I. Steady-state conditions. *Journal of Membrane Biology*, **39**, 333–67.

Gustin, M.C., Martinac, B. Saimi, Y. *et al.* (1986) Ion channels in yeast. *Science*, **233**, 1195–7.

Gustin, M.C., Zhou, X.-L., Martinac, B. and Kung, C. (1988) A mechanosensitive ion channel in the yeast plasma membrane. *Science*, **242**, 762–5.

Hager, K.M., Mandala, S.M., Davenport, J.W. *et al.* (1986) Amino acid sequence of the plasma membrane H^+-ATPase of *Neurospora crassa*: deduction from genomic and cDNA sequences. *Proceedings of the National Academy of Sciences of the USA*, **83**, 7693–7.

Horak, J. (1986) Amino acid transport in eucaryotic microorganisms. *Biochimica et Biophysica Acta*, **864**, 223–56.

Hubbard, M.J., Surarif, R., Sullivan, P.A. and Shepherd, M.G. (1986) The isolation of plasmamembrane and characterisation of plasmamembrane ATPase from the yeast *Candida albicans*. *European Journal of Biochemistry*, **154**, 375–81.

Ikemoto, N. (1982) Structure and function of the calcium pump protein of sarcoplasmic reticulum. *Annual Review of Physiology*, **44**, 297–317.

Jackson, S.L. and Heath, I.B. (1993) Roles of calcium ions in hyphal tip growth. *Microbiological Reviews*, **57**, 367–82.

Jarai, G. and Marzluf, G.A. (1991) Sulfate transport in *Neurospora crassa*: regulation, turnover, and cellular localization of the CYS-14 protein. *Biochemistry*, **30**, 4768–73.

Jorgensen, P.L. (1982) Mechanism of the Na^+, K^+ pump. Protein structure and conformations of the pure (Na^+, K^+)-ATPase. *Biochimica et Biophysica Acta*, **694**, 27–68.

Kakinuma, Y., Ohsumi, Y. and Anraku, Y. (1981) Properties of H^+-translocating adenosine triphosphatase in vacuolar membranes of *Saccharomyces cerevisiae*. *Journal of Biological Chemistry*, **256**, 10859–63.

Kane, P.M., Yamashiro, C.T., Wolczyk, D.F. *et al.* (1990) Protein splicing converts the yeast *TFP1* gene product to the 69-kD subunit of the vacuolar H^+-adenosine triphosphatase. *Science*, **250**, 651–7.

Klionsky, D.J. Herman, P.K. and Emr, S.D. (1990) The fungal vacuole: composition, function and biogenesis. *Microbiological Reviews*, **54**, 266–92.

Ko, C.H. and Gaber, R.F. (1991) *TRK1* and *TRK2* encode structurally related K^+ transporters in *Saccharomyces cerevisiae*. *Molecular and Cellular Biology*, **11**, 4266–73.

Ko, C.H., Buckley, A.M. and Gaber, R.F. (1990) *TRK2* is required for low affinity K^+ transport in *Saccharomyces cerevisiae*. *Genetics*, **125**, 305–12.

Ko, C.H., Liang, H. and Gaber, R.F. (1993) Roles of multiple glucose transporters in *Saccharomyces cerevisiae*. *Molecular and Cellular Biology*, **13**, 638–48.

Kochian, L.V. and Lucas, W.J. (1993) Can K^+ channels do it all? *The Plant Cell*, **5**, 720–1.

Koo, K. and Stuart, W.D. (1991) Sequence and structure of *mtr*, an amino acid transport gene of *Neurospora crassa*. *Genome*, **34**, 644–51.

Kruckeberg, A.L. and Bisson, L.F. (1990) The *HXT2* gene of *Saccharomyces cerevisiae* is required for high-affinity glucose transport. *Molecular and Cellular Biology*, **10**, 5903–13.

Lagunas, R. (1993) Sugar transport in *Saccharomyces cerevisiae*. *FEMS Microbiological Reviews*, **104**, 229–42.

Levina, N.N., Lew, R.R. and Heath, I.B. (1994) Cytoskeletal regulation of ion channel distribution in the tip-growing organism *Saprolegnia ferax*. *Journal of Cell Science*, **107**, 127–34.

Lew, R.R., Garrill, A., Covic, L. *et al.* (1992) Novel ion channels in the protists *Mougeotia* and *Saprolegnia*, using sub-gigaseals. *FEBS Letters*, **310**, 219–22.

Lewis, D.A. and Bisson, L.F. (1991) The *HXT1* gene product of *Saccharomyces cerevisiae* is a new member of the family of hexose transporters. *Molecular and Cellular Biology*, **11**, 3804–13.

Lichko, L.P. and Okorokov, L.A. (1984) Some properties of membrane-bound, solubilized and reconstituted into liposomes H+-ATPase of vacuoles of *Saccharomyces carlsbergensis*. *FEBS Letters*, **174**, 233–7.

Lichko, L.P. and Okorokov, L.A. (1985) What family of ATPases does the vacuolar H+-ATPase belong to? *FEBS Letters*, **187**, 349–53.

Lichko, L.P. and Okorokov, L.A. (1991) Purification and some properties of membrane-bound and soluble pyrophosphatase of yeast vacuoles. *Yeast*, **7**, 805–12.

Lowendorf, H.S., Slayman, C.L. and Slayman, C.W. (1974) Phosphate transport in *Neurospora*. Kinetic characterization of a constitutive low affinity transport system. *Biochimica et Biophysica Acta*, **373**, 369–82.

Lowendorf, H.S., Bazinet, G.F. Jr. and Slayman, C.W. (1975) Phosphate transport in *Neurospora crassa*: derepression of a high affinity transport system during phosphate starvation. *Biochimica et Biophysica Acta*, **389**, 541–9.

Maathuis, F.J.M. and Sanders, D. (1993) Energization of potassium uptake in *Arabidopsis thaliana*. *Planta*, **191**, 302–7.

Malpartida, F. and Serrano, R. (1981) Phosphorylated intermediate of the ATPase from the plasma membrane of yeast. *European Journal of Biochemistry*, **116**, 413–17.

Mann, B.J., Bowman, B.J., Grotelueschen, J. and Metzenberg, R.L. (1989) Nucleotide sequence of *pho-4+*, encoding a phosphate-repressible phosphate permease of *Neurospora crassa*. *Gene*, **83**, 281–9.

Marzluf, G.A. (1970a) Genetic and biochemical studies of distinct sulfate permease species in different developmental stages of *Neurospora crassa*. *Archives of Biochemistry and Biophysics*, **138**, 254–63.

Marzluf, G.A. (1970b) Genetic and metabolic controls for sulfate metabolism in *Neurospora crassa*: isolation and study of chromate-resistant and sulfate transport-negative mutants. *Journal of Bacteriology*, **102**, 716–21.

Metzenberg, R.L. (1979) Implications of some genetic control mechanisms in *Neurospora*. *Microbiological Reviews*, **43**, 361–83.

Mitchell, P. (1966) Chemiosmotic coupling in oxidative and photosynthetic phosphorylation. *Biological Reviews*, **41**, 445–502.

Monk, B.C., Kurtz, M.B., Marrinan, J.A. and Perlin, D.S. (1991a) Cloning and characterisation of the plasma membrane H+-ATPase from *Candida albicans*. *Journal of Bacteriology*, **173**, 6826–836.

Monk, B.C., Montesinos, C., Ferguson, C. *et al.* (1991b) Immunological approaches to the transmembrane topology and conformational changes of the carboxyl–terminal regulatory domain of yeast plasma membrane H+ATPase. *Journal of Biological Chemistry*, **266**, 18097–103.

Nakamoto, R.K., Rao, R. and Slayman, C.W. (1991) Expression of the yeast plasma membrane [H+] ATPase in secretory vesicles. A new strategy for directed mutagenesis. *Journal of Biological Chemistry*, **266**, 7940–9.

Nelson, N. (1992) Organellar proton – ATPases. *Current Opinion in Cell Biology*, **4**, 654–60.

Nelson, N. and Taiz, L. (1989) The evolution of H+-ATPases. *Trends in Biochemical Sciences*, **14**, 113–16.

Norais, N., Prome, D. and Velours, J. (1991) ATP synthase of yeast mitochondria. Characterization of subunit d and sequence analysis of the structural gene *ATP7*. *Journal of Biological Chemistry*, **266**, 16541–9.

Ohsumi, Y. and Anraku, Y. (1981) Active transport of basic amino acids driven by a proton motive force in vacuolar membrane vesicles of *Saccharomyces cerevisiae*. *Journal of Biological Chemistry*, **256**, 2079–82.

Ohya, Y., Umemoto, N., Tanida, I. *et al.* (1991) Calcium-sensitive *cls* mutants of *Saccharomyces cerevisiae* showing a Pet−phenotype are ascribable to defects of vacuolar membrane H+ATPase activity. *Journal of Biological Chemistry*, **266**, 13971–7.

Paek, Y.L. and Weiss, R.L. (1989) Identification of an arginine carrier in the vacuolar membrane of *Neurospora crassa*. *Journal of Biological Chemistry*, **264**, 7285–90.

Pall, M.J. (1971) Amino acid transport in *Neurospora crassa*. *IV*. Properties and regulation of a methionine transport system. *Biochimica et Biophysica Acta*, **223**, 201–14.

Pall, M.J. and Kelly, K.A. (1971) Specificity of trans-inhibition of amino acid transport in *Neurospora*. *Biochemical and Biophysical Research Communications*, **42**, 940–7.

Pardo, J.P. and Slayman, C.W. (1989) Cysteine 532 and cysteine 545 are the N-ethylmaleimide-reactive residues of the *Neurospora* plasma membrane H+ATPase. *Journal of Biological Chemistry*, **264**, 9373–9.

Perlin, D.S., Kasamo, K., Brooker, R.J. and Slayman, C.W. (1984) Electrogenic H+ translocation by the plasma membrane ATPase of *Neurospora*. Studies on plasma membrane vesicle and reconstituted enzyme. *Journal of Biological Chemistry*, **259**, 7884–92.

Perlin, D.S., San Francisco, M.J.D., Slayman, C.W. and Rosen, B.P. (1986) H+/ATP stoichiometry of proton pumps from *Neurospora crassa* and *Escherichia coli*. *Archives of Biochemistry and Biophysics*, **248**, 53–61.

Peters, R.H.J. and Borst-Pauwels, G.W.F.H. (1979) Properties of the plasmamembrane ATPase and mitochondrial ATPase of *Saccharomyces cerevisiae*. *Physiologia Plantarum*, **46**, 330–7.

Portillo, F. and Serrano, R. (1988) Dissection of functional domains of the yeast proton-pumping ATPase by directed mutagenesis. *EMBO Journal*, **7**, 1793–8.

Portillo, F., de Larrinoa, I.F. and Serrano, R. (1989) Deletion analysis of yeast plasma membrane H+-ATPase and identification of a regulatory domain at the carboxyl terminus. *FEBS Letters*, **247**, 381–5.

Portillo, F., Eraso, P. and Serrano, R. (1991) Analysis of the regulatory domain of yeast plasma membrane H+-ATPase by directed mutagenesis and intragenic suppression. *FEBS Letters*, **287**, 71–4.

Ramirez, J.A., Vacata, V., McKusker, J.H. *et al.* (1989) ATP-sensitive K+ channels in a plasma membrane H+-ATPase mutant of the yeast *Saccharomyces cerevisiae*.

Proceedings of the National Academy of Sciences of the USA, **86**, 7866–70.

Rao, R. and Slayman, C.W. (1993) Mutagenesis of conserved residues in the phosphorylation domain of the yeast plasma membrane H⁺-ATPase. Effects on structure and function. *Journal of Biological Chemistry,* **268**, 6708–13.

Rao, R., Nakamoto, R.K., Verjovsky-Almeida, S. and Slayman, C.W. (1992) Structure and function of the yeast plasma membrane H⁺-ATPase. *Annals of the New York Academy of Sciences,* **671**, 195–203.

Rea, P.A., Kim, Y., Sarafian, V. *et al.* (1992) Vacuolar H⁺-translocating pyrophosphatases: a new category of ion translocase. *Trends in Biochemical Sciences,* **17**, 348–53.

Rodriguez-Navarro, A. and Ramos, J. (1984) Dual systems for potassium transport in *Saccharomyces cerevisiae. Journal of Bacteriology,* **159**, 940–5.

Rodriguez-Navarro, A., Blatt, M.R. and Slayman, C.L. (1986) A potassium-proton symport in *Neurospora crassa. Journal of General Physiology,* **87**, 649–74.

Roomans, G.M. and Borst-Pauwels, G.W.F.H. (1977) Interaction of phosphate with monovalent cation uptake in yeast. *Biochimica et Biophysica Acta,* **470**, 84–91.

Roomans, G.M. and Borst–Pauwels, G.W.F.H. (1979) Interactions of cations with phosphate uptake by *Saccharomyces cerevisiae. Biochemical Journal,* **178**, 521–7.

Roomans, G.M., Kuypers, G.A.J., Theuvenet, A.P.R. and Borst-Pauwels, G.W.F.H. (1979) Kinetics of sulfate uptake by yeast. *Biochimica et Biophysica Acta,* **551**, 197–206.

Sachs, G., Wallmark, B., Saccomani, G. *et al.* (1982) The ATP-dependent component of gastric acid secretion. *Current Topics in Membranes and Transport,* **16**, 135–59.

Sanders, D. (1988) Fungi, in *Solute Transport in Plant Cells and Tissues,* (eds D.A. Baker and J.L. Hall), Longman, Harlow, UK, pp. 106–65.

Sanders, D. (1990) Kinetic modelling of plant and fungal membrane transport systems. *Annual Review of Plant Physiology and Plant Molecular Biology,* **41**, 77–107.

Sanders, D. and Slayman, C.L. (1982) Control of intracellular pH. Predominant role of oxidative metabolism, not proton transport, in the eukaryotic microorganism *Neurospora. Journal of General Physiology,* **80**, 377–402.

Sanders, D., Hansen, U.-P. and Slayman, C.L. (1981) Role of the plasma membrane proton pump in pH regulation in non-animal cells. *Proceedings of the National Academy of Sciences of the USA,* **78**, 5903–7.

Sato, T., Ohsumi, Y. and Anraku, Y. (1984a) Substrate specificities of active transport systems for amino acids in vacuolar-membrane vesicles of *Saccharomyces cerevisiae. Journal of Biological Chemistry,* **259**, 11505–8.

Sato, T., Ohsumi, Y. and Anraku, Y. (1984b) An arginine/histidine exchange transport system in vacuolar membrane vesicles of *Saccharomyces cerevisiae. Journal of Biological Chemistry,* **259**, 11509–11.

Scarborough, G.A. (1970a) Sugar transport in *Neurospora crassa. Journal of Biological Chemistry,* **245**, 1694–8.

Scarborough, G.A. (1970b) Sugar transport in *Neurospora crassa.*II. A second glucose transport system. *Journal of Biological Chemistry* **245**, 3985–7.

Scarborough, G.A. (1977) Properties of the *Neurospora crassa* plasma membrane ATPase. *Archives of Biochemistry and Biophysics,* **180**, 384–93.

Scarborough, G.A. and Addison, R. (1984) On the subunit composition of the *Neurospora* plasma membrane H⁺-ATPase. *Journal of Biological Chemistry,* **259**, 9109–14.

Schlesser, A. Ulaszewski, S., Ghislain, M. and Goffeau, A. (1988) A second transport ATPase gene in *Saccharomyces cerevisiae. Journal of Biological Chemistry,* **263**, 19480–7.

Sebald, W. (1977) Biogenesis of the mitochondrial ATPase. *Biochimica et Biophysica Acta,* **463**, 1–27.

Sentenac, H., Bonneaud, N., Minet, M. *et al.* (1992) Cloning and expression in yeast of a plant potassium ion transport system. *Science,* **256**, 663–5.

Serrano, R. (1978) Characterisation of the plasma-membrane ATPase of *Saccharomyces cerevisiae. Molecular and Cellular Biochemistry,* **22**, 51–63.

Serrano, R. (1983) In vivo glucose activation of the yeast plasma membrane ATPase. *FEBS Letters,* **156**, 11–14.

Serrano, R. (1984) *Plasmamembrane ATPase of Plants and Fungi,* CRC Press, Boca Raton, FL.

Serrano, R. (1988) Structure and function of proton translocating ATPase in plasma membranes of plants and fungi. *Biochimica et Biophysica Acta,* **947**, 1–28.

Serrano, R., Kielland-Brandt, M.C. and Fink, G.R. (1986) Yeast plasma membrane ATPase is essential for growth and has homology with (Na⁺+K⁺), K⁺- and Ca²⁺-ATPases. *Nature,* **319**, 689–93.

Siegenthaler, P.A., Belsky, M.M. and Goldstein, S. (1967) Phosphate uptake in an obligately marine fungus: a specific requirement for sodium. *Science,* **155**, 93–4.

Sista, H.S. (1991) *Characterization of a component of the proton channel of the vacuolar ATPase.* PhD thesis, University of California, Santa Cruz, CA.

Slayman, C.L. (1965a) Electrical properties of *Neurospora crassa.* Effects of external cations on the intracellular potential. *Journal of General Physiology,* **49**, 69–92.

Slayman, C.L. (1965b) Electrical properties of *Neurospora crassa.* Respiration and the intracellular potential. *Journal of General Physiology,* **49**, 93–116.

Slayman, C.L. (1977) Energetics and control of transport in *Neurospora,* in *Water Relationships in Membrane Transport in Plants and Animals* (eds A.M. Jungreis, T.K. Hodges, A. Kleinzeller and S.G. Schultz), Academic Press, New York, pp. 69–89

Slayman, C.L. (1987) The plasma membrane ATPase of *Neurospora*: a proton-pumping electroenzyme. *Journal of Bioenergetics and Biomembranes,* **19**, 1–20.

Slayman, C.L. and Slayman, C.W. (1974) Depolarisation of the plasma membrane of *Neurospora* during active transport of glucose: evidence for a proton-dependent cotransport system. *Proceedings of the National Academy of Sciences of the USA*, **71**, 1935–9.

Slayman, C.W. and Slayman, C.L. (1975) Energy coupling in the plasma membrane of *Neurospora*: ATP-dependent proton transport and proton-dependent sugar transport, in *Molecular Aspects of Membrane Phenomena*, (eds H.R. Kaback, H. Neurath, G.K. Radda, R. Schwyzer and W.R. Wiley), Springer-Verlag, Berlin pp. 233–48.

Stevens, T.H. (1992) The structure and function of the fungal V-ATPase. *Journal of Experimental Biology*, **172**, 47–55.

Stuart, W.D. (1977) New class of ribonucleic acid in *Neurospora* associated with the outer cell envelope. *Journal of Bacteriology*, **129**, 395–9.

Stuart, W.D., Koo, K. and Vollmer, S.J. (1988) Cloning of *mtr*, an amino acid transport gene of *Neurospora crassa*. *Genome*, **30**, 198–203.

Supply, P., Wach, A. and Goffeau, A. (1993a) Enzymatic properties of the PMA2 plasma membrane-bound H^+-ATPase of *Saccharomyces cerevisiae*. *Journal of Biological Chemistry*, **268**, 19753–9.

Supply, P., Wach, A. Thinès-Sempoux, D. and Goffeau, A. (1993b) Proliferation of intracellular structures upon overexpression of the PMA2 ATPase in *Saccharomyces cerevisiae*. *Journal of Biological Chemistry*, **268**, 19744–2.

Uchida, E., Ohsumi, Y. and Anraku, Y. (1985) Purification and properties of H^+-translocating, Mg^{2+}-adenosine triphosphatase from vacuolar membranes of *Saccharomyces cerevisiae*. *Journal of Biological Chemistry*, **260**, 1090–5.

Vacata, V. Hofer, M., Larsson, H.P. and Lecar, H. (1992) Ionic channels in the plasma membrane of *Schizosaccharomyces pombe*. Evidence from patch clamp measurements. *Journal of Bioenergetics and Biomembranes*, **24**, 43–53.

Wada, Y., Ohsumi, Y. and Anraku, Y. (1992) Chloride transport of yeast vacuolar membrane vesicles: a study of in vitro vacuolar acidification. *Biochimica et Biophysica Acta*, **1101**, 296–302.

Walworth, N.C. and Novick, P.J. (1987) Purification and characterization of constitutive secretory vesicles from yeast. *Journal of Cell Biology*, **105**, 163–74.

Willsky, G.R. (1979) Characterisation of the plasma-membrane Mg^{2+}-ATPase from the yeast *Saccharomyces cerevisiae*. *Journal of Biological Chemistry*, **254**, 3326–32.

Yamashiro, C.T., Kane, P.M., Wolczyk, D.F. *et al.* (1990) Role of vacuolar acidification in protein sorting and zymogen activation. *Molecular and Cellular Biology*, **10**, 3737–49.

Zhou, X.-L. and Kung, C. (1992) A mechanosensitive ion channel in *Schizosaccharomyces pombe*. *EMBO Journal*, **11**, 2869–75.

Zhou, X.-L., Stumpf, M.A., Hoch, H.C. and Kung, C. (1991) A mechanosensitive cation channel in membrane patches and in whole cells of the fungus *Uromyces*. *Science*, **253**, 1415–17.

G.M. Gadd
Department of Biological Sciences, University of Dundee, Dundee, UK

9.1 INTRODUCTION

Eukaryotic cell function is dependent on a variety of signalling mechanisms which translate external physicochemical and biochemical stimuli into specific intracellular responses involving second messengers, e.g. Ca^{2+}, cyclic AMP (cAMP), inositol lipids (Figure 9.1). Signal transduction is believed to underpin virtually all important cellular processes, including growth, differentiation and metabolism. In comparison to mammalian, and to a lesser extent plant systems, information on signal transduction in fungi is limited, particularly for filamentous

species. However, with the accelerating use of both budding and fission yeast as eukaryotic cell models, considerable progress is being made at the molecular and biochemical level and it is now possible to review general aspects of signal transduction in fungi in the light of other more extensively studied eukaryotic models.

Many signalling pathways, e.g. those involving cAMP and diacylglycerol (DAG), consist of a series of proteins, including specific receptors, GTP-binding (G) proteins, second messenger-generating enzymes, protein kinases, target and regulatory proteins (Figure 9.1). Furthermore, there is a multiplicity of interactions between different signalling systems (Nishizuka, 1992a, b). The important role of Ca^{2+} in eukaryotic cell function is now appreciated though Ca^{2+} homeostasis is maintained by complex interactions between many signalling systems including those mediated by cAMP and phosphoinositide hydrolysis. For convenience, second messengers are dealt with separately in this chapter but the dynamic interactions between different signal transduction systems should not be overlooked.

9.2 CALCIUM

Although the role of Ca^{2+} as a second messenger in fungal growth and differentiation is not as well understood as that in mammalian cell systems, it now seems clear that Ca^{2+} is highly important in fungi with direct and indirect evidence being obtained from a variety of experimental systems (Jackson and Heath, 1993). Although it is sometimes difficult to demonstrate a requirement for Ca^{2+} in fungi due to difficulties in rendering media Ca^{2+}-free and Ca^{2+}-contamination of medium components (Kovac, 1985; Youatt, 1993), calcium is found in all fungal cells examined and both excess external Ca^{2+} or the presence of substances

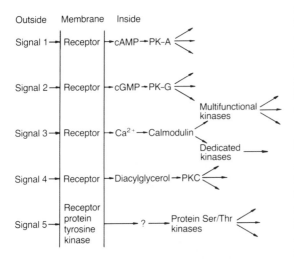

Figure 9.1 Simplified diagram of the principal signalling systems which operate in eukaryotic cells. PK-A, cAMP-dependent protein kinase; PK-G, cGMP-dependent protein kinase; PKC, protein kinase C; ?, hypothetical second messenger. Ca^{2+} may be released from internal stores by Ins1, 4,5P3; CaM-dependent kinases may be dedicated and multifunctional (adapted from Cohen, 1992).

The Growing Fungus. Edited by Neil A.R. Gow and Geoffrey M. Gadd. Published in 1994 by Chapman & Hall, London. ISBN 0 412 46600 7

Table 9.1   Examples of Ca^{2+}/Ca^{2+}-CaM-mediated effects on fungal differentiation

Organism	Process	References
Achlya bisexualis	Sporangium formation: Ca^{2+} regulates transcellular electric current	Thiel *et al.* (1988)
Candida albicans	Dimorphic transitions	Sabie and Gadd (1989) Paranjape and Datta (1990) Holmes *et al.* (1991)
Cephalosporium acremonium	Ca^{2+} important for antibiotic production	Crabbe *et al.* (1988)
Fusarium graminearum	Regulation of hyphal extension and branching	Robson *et al.* (1991a,b)
Neurospora crassa	Morphology of hyphal apex; apical dominance and extension; regulation of branching; transhyphalion currents	Ortega Perez *et al.* (1981) McGillivray and Gow (1987) Reissig and Kinney (1983) Schmid and Harold (1988) Dicker and Turian (1990)
Ophiostoma (Ceratocystis) ulmi	Regulation of dimorphism	Muthukumar and Nickerson (1984, 1985) Muthukumar *et al.* (1986, 1987) Brunton and Gadd (1991) Gadd and Brunton (1992)
Phytophthora palmivora	Maintenance of zoospore state and transition to cyst stage	Griffith *et al.* (1988)
Pythium sp.	Mediation of zoospore motility	Donaldson and Deacon (1993) Deacon and Donaldson (1993)
Rhodosporidium toruloides	Pheromone-mediated sexual reproduction	Liu *et al.* (1990)
Saccharomyces cerevisiae	Growth, proliferation, cell cycle control, cytokinesis	Iida *et al.* (1990a, b) Anraku *et al.* (1991)
Saprolegnia ferax	Hyphal growth	Jackson and Heath (1989)
Schizosaccharomyces pombe	Vegetative growth	Takeda and Yamamoto (1987)
Sporothrix schenckii	Conidial germination	Rivera-Rodriguez and Rodriguez-del Valle (1992)

which alter free Ca^{2+} concentration in the medium or inside cells, e.g. chelators and ionophores, have induced diverse effects on growth, differentiation and sporulation (Pitt and Ugalde, 1984; Ugalde *et al.*, 1990; Robson *et al.*, 1991a,b; Deacon and Donaldson, 1993) (Table 9.1).

Calcium is believed to transduce stimuli at the cell surface, which may include chemical, electrical or physical signals, into specific intracellular effects. The regulation by calcium of diverse cellular processes underlines the complexity of the situation: several Ca^{2+}-mediated events may occur simultaneously.

Although the basic features of Ca^{2+} signalling in fungi may be similar to those in mammalian systems, it seems there are some important differences, particularly regarding regulation of cytosolic free Ca^{2+} ($[Ca^{2+}]_c$). Since calcium is potentially toxic, largely because of its ability to bind phosphate groups, including those of ATP, ADP and AMP, fungal cells possess several mechanisms for maintenance of a low $[Ca^{2+}]_c$. Such a low and precisely controlled concentration is an important characteristic of a second messenger and $[Ca^{2+}]_c$ may fluctuate, e.g. from 10^{-7} to 10^{-5} M, without significantly altering the overall cellular ionic composition and modulate Ca^{2+}-sensitive cell processes by selective binding to appropriate substrates (Carafoli, 1987; Rasmussen and Rasmussen, 1990).

In order to understand the significance of Ca^{2+}

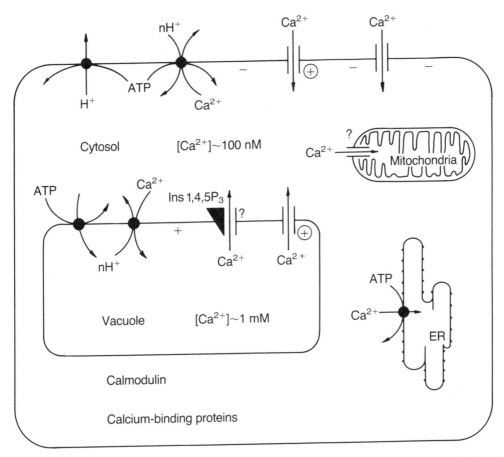

Figure 9.2 Simplified diagram of the identity and location of Ca^{2+} homeostatic systems in fungi. On the plasma membrane, systems are (left to right) primary electrogenic H^+-ATPase, Ca^{2+}-ATPase, voltage-dependent- and other types of Ca^{2+} channel. Transport systems on the vacuolar membrane are (left to right) primary electrogenic H^+-ATPase, Ca^{2+}/H^+ antiport, Ins1,4,5P$_3$-gated Ca^{2+} channel, voltage-dependent Ca^{2+} channel. A Ca^{2+} channel is shown on the mitochondrion as well as a Ca^{2+}-ATPase on the endoplasmic reticulum. Adapted from Sanders *et al.* (1990) and Johannes *et al.* (1991, 1992).

as an intracellular messenger in fungi, it is necessary first to describe the mechanisms of regulation of cytosolic Ca^{2+} *via* transport processes and intracellular compartmentation/sequestration. This will be followed by consideration of reception and transduction of the Ca^{2+} signal and the mechanisms by which Ca^{2+} may regulate diverse biochemical processes and induce specific effects. Interaction of other important second messengers, cyclic AMP and inositol lipids, with Ca^{2+} signalling processes will also be described.

9.2.1 Regulation of intracellular calcium

Transient elevation of cytosolic free Ca^{2+} ($[Ca^{2+}]_c$) is the frequent trigger for intracellular events (see later). Since Ca^{2+} is toxic at high concentrations, cells utilize a complex system of sequestrative and homeostatic mechanisms which maintain $[Ca^{2+}]_c$ at a low level (Jackson and Heath, 1993) (Figure 9.2). Because $[Ca^{2+}]_c$ is maintained at a low level, a significant change in $[Ca^{2+}]_c$, adequate to produce the stimulus, can be achieved by movement of only a small number of Ca^{2+} (Carafoli and Penniston,

1985). Since $[Ca^{2+}]_c$ is a product of fluxes from outside and within the cell, it is necessary to outline the nature of these Ca^{2+} movements, and their relative roles in Ca^{2+} signalling.

(a) Calcium transport

Fungal transport systems for ions are usually classed as carrier or channel systems. In the former, conformational changes in the transport protein may result in alternate exposure of the transport binding site(s) on either side of the membrane. Carriers include metabolically coupled and H^+-gradient driven transport systems (Sanders, 1990).

Ca^{2+} transport processes at the plasma and vacuolar membranes are important in regulation of $[Ca^{2+}]_c$; in resting fungi $[Ca^{2+}]_c$, has been recorded as 100–300 nM (Halachmi and Eilam, 1989; Miller *et al.*, 1990). A transmembrane electrochemical Ca^{2+} gradient ($\Delta\tilde{\mu}_{Ca^{2+}}$) is necessary for signal transduction (Miller *et al.*, 1990). Intracellular Ca^{2+} may be compartmentalized in various membrane-bound organelles and/or bound to calcium-binding proteins, e.g. calmodulin. This is essential for the avoidance of Ca^{2+} toxicity and maintenance of $[Ca^{2+}]_c$ at a low level.

Calcium influx

Ca^{2+} enters cells *via* ligand- or voltage-gated channels. These are membrane proteins which, in the open position, allow passive flux of Ca^{2+} down the electrochemical gradient (Pietrobon *et al.*, 1990). Ca^{2+} influx in fungi is metabolism-dependent in that it proceeds in the presence of a metabolic substrate (Boutry *et al.*, 1977; Jones and Gadd, 1990; Eilam and Othman, 1990). However, Ca^{2+} influx may be difficult to distinguish because of energy-dependent vacuolar Ca^{2+} accumulation in fungi (Eilam *et al.*, 1985). Ca^{2+} influx in *S. cerevisiae* has been measured without interference, and it was suggested that Ca^{2+} influx was mediated by gated Ca^{2+}-channels sensitive to the membrane potential ($\Delta\psi$) (Eilam and Chernichovsky, 1987; Eilam and Othman, 1990). However, changes in $\Delta\psi$ only account partly for glucose-stimulated Ca^{2+} influx, with intracellular acidification also being involved (Eilam *et al.*, 1990). Protoplasts of the plant pathogen *Uromyces appendiculatus* possess a mechanosensitive ion channel which can pass a variety of cations, including Ca^{2+}. It is suggested that such channels may transduce membrane

stresses into an influx of ions, including Ca^{2+}, that may trigger differentiation (Zhou *et al.*, 1991). In the Oomycete *Saprolegnia ferax*, the concentration of stretch-activated Ca^{2+} channels at the hyphal tip supports a direct role in the tip growth process (Garrill *et al.*, 1992, 1993) (Chapters 8 and 13).

Calcium efflux

Earlier evidence demonstrated the existence of a Ca^{2+}/H^+ antiport on the plasma membrane, the requisite electrochemical proton gradient ($\Delta\tilde{\mu}_{H^+}$) being generated by the plasma membrane H^+-ATPase (Eilam, 1982; Slayman *et al.*, 1990). However, it is now proposed that Ca^{2+} efflux is *via* a H^+/Ca^{2+}-ATPase, with a stoichiometric ratio of at least $2H^+/Ca^{2+}$ in *N. crassa* (Miller *et al.*, 1990). Although final confirmation of the existence of a fungal H^+/Ca^{2+}-ATPase depends on purification and reconstitution studies, similarities in the mode of Ca^{2+} efflux from animal, plant and fungal cells, and a low K_m for ATP ($< 100\ \mu M$), may indicate that eukaryotes share a common class of enzyme (Miller *et al.*, 1990).

(b) Organellar compartmentation

In eukaryotic cells, a variety of membrane-bound compartments may be involved in Ca^{2+} compartmentation, e.g. mitochondria, endoplasmic reticulum, cytoplasmic vesicles and vacuoles.

Vacuoles

Fungal vacuoles appear to be the major storage organelles for Ca^{2+} (Ohsumi *et al.*, 1988; Kitamoto *et al.*, 1988, Miller *et al.*, 1990) and several other divalent cations, e.g. Mn^{2+}, Zn^{2+} (Okorokov, 1985; White and Gadd, 1987; Jones and Gadd, 1990; Klionsky *et al.*, 1990; Halachmi and Eilam, 1993). Calcium transport across the vacuolar membrane (tonoplast) occurs *via* a Ca^{2+}/nH^+ antiport system driven by a proton motive force generated by the vacuolar membrane H^+-ATPase (Klionsky *et al.*, 1990; Anraku *et al.*, 1991). The vacuolar H^+-ATPase utilizes the energy arising from ATP hydrolysis to pump H^+ into the vacuole and a pH gradient of 0.5–1.5 units may be generated across the tonoplast; such activity is also important in regulation of cytosolic pH (Klionsky *et al.*, 1990; Jones and Gadd, 1990). Gradients of pH and electrical potential difference across the fungal tonoplast may be regulated by interactions of a membrane potential-dependent cation channel dependent on Ca^{2+} (Bertyl and Slayman, 1990), chloride transport systems (Wada *et al.*, 1986, 1987), and the vacuolar H^+-ATPase (Anraku *et al.*, 1989). Vacuolar compartmentation of Ca^{2+} enables growth in high Ca^{2+} concentrations, e.g. > 100 mM-$CaCl_2$ (Ohya *et al.*, 1986a) with most vacuolarized Ca^{2+} being stored as Ca-polyphosphates (Ohsumi *et al.*, 1988). Thus, the vacuolar H^+-ATPase and Ca^{2+}/nH^+ antiporter provide a regulatory system for maintenance of $[Ca^{2+}]_c$ homeostasis under widely differing external Ca^{2+} concentrations (Anraku *et al.*, 1991). Further evidence for the importance of vacuoles in Ca^{2+} homeostasis was obtained using Ca^{2+}-sensitive mutants of *S. cerevisiae*. These have defects in the vacuolar H^+-ATPase and vacuolar Ca^{2+} uptake which result in an elevated $[Ca^{2+}]_c$ and serious metabolic lesions (Ohya *et al.*, 1991).

Although the existence of a Ca^{2+} accumulation mechanism and an intravacuolar Ca^{2+}-buffering system now appears well established in fungi and yeasts, precise identification of the Ca^{2+} release mechanism from the vacuole has received little attention. It seems likely that such a Ca^{2+} release mechanism would comprise Ca^{2+} channels which mediate release of Ca^{2+} into the cytosol (Corzo and Sanders, 1992) as found in a variety of animal and plant cell types (Pietrobon *et al.*, 1990; Johannes *et al.*, 1991). In plants, it appears that a voltage-sensitive as well as an inositol 1, 4, 5-trisphosphate ($Ins1,4,5P_3$)-mediated Ca^{2+} channel exists on the vacuolar membrane (Alexandre and Lassalles, 1992; Johannes *et al.*, 1992). The propagation of this (and other) Ca^{2+} signals will depend on cytosolic properties since the distribution of cytosolic Ca^{2+} is probably non-uniform. A rise in $[Ca^{2+}]$ in a particular region may be propagated by diffusion and triggering of Ca^{2+} release at further points (Berridge and Irvine, 1989; Tsunoda, 1991). Cornelius and Nakashima (1987) and Cornelius *et al.* (1989) presented evidence for an $Ins1,4,5P_3$-mediated Ca^{2+} channel in the vacuolar membrane of *N. crassa* but work on other fungi as well as Ca^{2+} signal propagation in the fungal cytosol is still lacking. The existence of, and involvement, of a vacuolar H^+-translocating inorganic pyrophosphatase (H^+-PPase) in ionic homeostasis and compartmentation has not been established (Rea *et al.*, 1992) although such a pump has been described in *S. carlsbergensis* (Okorokov, 1985; Lichko and Okorokov, 1991).

Mitochondria

Although mitochondria have been proposed as being of some importance in fungal Ca^{2+} homeostasis (Pitt and Ugalde, 1984), there is little evidence for this, and it has been suggested that fungal mitochondria show little affinity for Ca^{2+} (Carafoli *et al.*, 1970). It should be noted that interest in mammalian mitochondrial Ca^{2+} homeostasis has now declined and despite intensive research, the nature of proposed Ca^{2+} transport (channel or carrier) and efflux is still poorly understood (Pietrobon *et al.*, 1990). It appears that mitochondrial Ca^{2+} transport is important for intramitochondrial reactions, particularly dehydrogenase regulation, although in some cell types mitochondria are considered to be involved in an $Ins1,4,5P_3$-insensitive Ca^{2+} pool (see later) (Pietrobon *et al.*, 1990). It is also relevant that Ca^{2+}-sensitive mutants of *S. cerevisiae* have no detectable mitochondrial defects (Ohya *et al.*, 1991a). However, in sporangiophores of *Phycomyces blakesleeanus*, Ca^{2+} was internally located in mitochondria, endoplasmic reticulum and vacuoles suggesting that all these organelles were involved in modulation of intracellular Ca^{2+} (Morales and Ruiz-Herrera, 1989).

Endoplasmic reticulum and other organelles

Schizosaccharomyces pombe contains a P-type ATPase gene, *cta3*, distinct from the *pma1* gene encoding the plasma membrane H^+-ATPase (Ghislain *et al.*, 1987, 1990). The deduced amino acid sequence of the cta3 protein is most closely related to that of the mammalian endo(sarco)plasmic reticulum Ca^{2+}-ATPases, and it is proposed that the Cta3 protein is involved in Ca^{2+} transport and homeostasis. The deduced amino acid sequence is also partly homologous to the products of the plasma membrane ATPase related genes, *PMR1* and *PMR2*, of *S. cerevisiae* which were proposed to be Ca^{2+}-ATPase pumps (Rudolph *et al.*, 1989). However, the novel fungal Ca^{2+}-ATPase encoded by *CTA3* appears different from *PMR1* and *PMR2* with the likely location being the endoplasmic reticulum (Goffeau *et al.*, 1990). Thus, in *S. pombe*, and possibly other fungi, Ca^{2+} is located in non-vacuolar organelles as well as the vacuole. The former may be analogous to the endo(sarco)plasmic reticulum of muscle cells or to calciosomes in non-muscle cells; the Cta3 protein may transport cytosolic Ca^{2+} into such organelles with high affinity (Ghislain *et al.*, 1990). The Ca^{2+}/nH^+ antiporter of the tonoplast has a relatively low affinity for Ca^{2+}, $K_m \sim 100$ μM in *S. cerevisiae* (Ohsumi and Anraku, 1983).

Halachmi *et al.* (1992) have demonstrated Ca^{2+} uptake in non-vacuolar Ca^{2+}-storing organelles in *S. pombe*. The *cta3* null mutation in the putative Ca^{2+}-ATPase gene (Ghislain *et al.*, 1990), reduced the level of ATP-dependent Ca^{2+} uptake into the non-vacuolar organelles and provides biochemical evidence for the existence of an intracellular Ca^{2+}-ATPase. Residual Ca^{2+}-uptake in the mutant indicated the presence of a second non-vacuolar Ca^{2+}-ATPase encoded by a different gene (Halachmi *et al.*, 1992). Sequestered Ca^{2+} in *N. crassa* was distributed along a gradient, maximal at the hyphal tip, with the morphology of CTC-fluorescent bodies possibly corresponding to endoplasmic reticulum rather than vacuoles (Schmid and Harold, 1988) (Chapter 5).

(c) Ca^{2+}-binding proteins

The Ca^{2+} homeostatic mechanisms discussed previously serve to (a) maintain a low $[Ca^{2+}]_c$ and (b) allow controlled changes in $[Ca^{2+}]_c$ when appropriate stimuli activate Ca^{2+}-dependent pathways. The ultimate targets of changes in $[Ca^{2+}]_c$ are cytosolic Ca^{2+} binding proteins (Pietrobon *et al.*, 1990). A variety of Ca^{2+}-binding proteins have been identified in eukaryotic microbes. Some have enzymatic functions where binding of Ca^{2+} serves to activate the enzyme, e.g. Ca^{2+}/phospholipid-dependent lipid and protein kinase (analogous to protein kinase C (PKC)) of *N. crassa* (Turian and Favre, 1990). Many cytoskeletal elements or subunits bind calcium which regulates their formation and/or organization (O'Day, 1990). However, of the various $[Ca^{2+}]_c$ regulators known, calmodulin is probably the most important and has received the most attention (Pitt and Kaile, 1990; Means *et al.*, 1991; Ohya and Anraku, 1992).

Calmodulin (CaM) is a ubiquitous eukaryotic Ca^{2+}-binding protein named after its Ca^{2+}-modulating function (Cheung, 1980; Anraku *et al.*, 1991). Calmodulins have been identified in many fungi and yeasts (Hubbard *et al.*, 1982; Trinci *et al.*, 1990) and there is increasing evidence for its involvement in several Ca^{2+}-dependent events (Pitt and Kaile, 1990; Turian and Favre, 1990; Paranjape and Datta, 1990). (Table 9.1). A

Ca^{2+}-induced conformational change in calmodulin enables CaM to interact with 'calmodulin target proteins' which include protein kinases, phosphatases, adenylate cyclases, phosphodiesterases and Ca^{2+}-ATPases. It is commonly found that Ca^{2+}-CaM has a much higher affinity (four orders of magnitude) for the target enzyme than Ca^{2+}-free CaM with the enzyme complex with Ca^{2+}-free CaM being inactive (Pietrobon *et al.*, 1990).

The most common trigger for activation of a Ca^{2+}-CaM-dependent reaction is an increase in $[Ca^{2+}]_c$ which results in the formation of more Ca^{2+}-CaM which activates target enzymes. An additional mechanism for activation of Ca^{2+}-CaM-dependent reactions may involve increasing the CaM concentration at a constant $[Ca^{2+}]_c$. This is a slower and energetically more expensive means of activating Ca^{2+}-CaM-dependent reactions but the increased CaM content may have longer lasting effects on cellular functions (Pietrobon *et al.*, 1990).

Calmodulins from invertebrates and plants show $> 90\%$ homology with bovine calmodulin, whereas those from fungi show only 60–80% homology. *S. cerevisiae* CaM (the most distantly related to mammalian CaM) shows 59% sequence identity with the protein containing three Ca^{2+}-binding sites in contrast with mammalian CaM which has four (Luan *et al.*, 1987). However, *S. cerevisiae* CaM has similar properties to other CaMs *in vitro* (Anraku *et al.*, 1991).

It is now known that CaM is required for several important cell functions in fungi and yeasts, including cell proliferation, cell cycle control and nuclear division (Anraku *et al.*, 1991). Calmodulin genes have been isolated from, e.g. *S. cerevisiae*, *S. pombe*, *Aspergillus nidulans* and *Candida albicans* (Davis *et al.*, 1986; Takeda and Yamamoto, 1987; Means *et al.*, 1991; Saporito and Sypherd, 1991). Disruption of the gene results in a loss of vegetative growth indicating the essentiality of CaM for cell proliferation (Davis *et al.*, 1986; Ohya and Anraku, 1989). CaM is also involved in the control and initiation of the cell division cycle. Yeast CaM increases parallel with budding and reaches twice the initial level before nuclear division takes place suggesting that nuclear division is the cell cycle step which requires most CaM (Uno *et al.*, 1989). Evidence for the positive involvement of Ca^{2+} in the progression of yeast mitosis includes growth arrest at G_2/M caused by Ca^{2+} deprivation, possibly because of adverse effects on a regulatory cascade of CaM caused by disruption of $[Ca^{2+}]_c$ homeostasis (Anraku *et al.*, 1991). In growing yeast, CaM appears to be involved in bud growth, cytokinesis, and chromosome segregation (Davis, 1992). CaM has an asymmetric cellular distribution, which changes through the cell cycle, and is concentrated at presumptive sites of bud formation. Such localization of CaM does not depend on high-affinity Ca^{2+} binding (Brockerhoff and Davis, 1992). Yeast mutants with a disrupted CaM function delocalize actin and contain a random distribution of CaM; CaM and actin distributions are therefore interdependent. It was also found that CaM localizes in the shmoo tips of α-factor-treated cells. CaM is therefore implicated in polarized cell growth in yeast (Brockerhoff and Davis, 1992; see also Chapter 13).

It should be noted that mutant strains of *S. cerevisiae*, where the Ca^{2+} binding loops of CaM were altered by site-directed mutagenesis and impaired in their ability to bind Ca^{2+}, were capable of good growth at normal rates. Thus, although CaM is required for growth, it can perform its essential function without binding Ca^{2+} (Geiser *et al.*, 1991). It is possible that certain essential enzymes require CaM as a subunit for full activity, independent of Ca^{2+}-binding, and/or that some of the essential Ca^{2+}-binding functions associated with CaM are carried out by different CaM-like proteins (Rose and Vallen, 1991). A novel gene fusion approach has been described by Stirling *et al.* (1992) to investigate the function of CaM in *S. cerevisiae*. A chimeric protein A (ProtA)-CaM polypeptide functions *in vivo* and exhibits Ca^{2+}-dependent binding to CaM target proteins. By using ProtA-CaM(fus2) to screen yeast expression libraries in the absence of Ca^{2+}, it should be possible to identify genes whose products mediate such essential functions (Stirling *et al.*, 1992).

(d) Ca²⁺-CaM-dependent protein phosphorylation

As described earlier, external stimuli may induce a rise in $[Ca^{2+}]_c$ which can trigger a variety of biochemical events connected with cell differentiation and proliferation. The mobilized Ca^{2+} may target key enzymes directly such as protein kinase C (Bazzi and Nelsestuen, 1988) though evidence indicates a more common role for Ca^{2+}-CaM as an enzyme and cytoskeletal modulator (Anraku *et al.*, 1991). Regulatory roles ascribed in eukaryotic cells

for Ca^{2+}-CaM include those in glycolysis and modulation of the activities of adenylate cyclase, cyclic nucleotide phosphodiesterase, several protein kinases and phosphatases. Thus, the action of the Ca^{2+}-CaM complex can directly influence cAMP-mediated pathways (Means *et al.*, 1991). All these enzymes are likely to be important in the regulation of cell proliferation since they directly or indirectly change the phosphorylation levels of their target proteins (Anraku *et al.*, 1991). However, although enzyme dependence on Ca^{2+}-CaM is widely demonstrated *in vitro*, unambiguous demonstration of Ca^{2+}-CaM-dependent physiological pathways in intact cells has been more difficult to demonstrate. The general use of anticalmodulin drugs is problematic because many inhibitors are relatively non-specific in their effects (Pietrobon *et al.*, 1990). Nevertheless, in many cases it appears that amplification of the Ca^{2+} signal is mediated by the Ca^{2+}-CaM complex *via* activation of protein phosphorylation (Cheung, 1980; Miyakawa *et al.*, 1989; Paranjape and Datta, 1990).

As well as CaM, other novel Ca^{2+}-binding proteins and a protein kinase C have been detected in *S. cerevisiae* (Ohya and Anraku, 1992). Analyses of Ca^{2+}-sensitive mutants revealed defects in bud emergence and reproductive organization (Ohya *et al.*, 1986a). Genetic analysis has shown that *CLS4* and *CDC24* are allelic (Ohya *et al.*, 1986b) with the *CLS4/CDC24* gene encoding a putative Ca^{2+}-binding protein (Miyamoto *et al.*, 1987). Molecular cloning of *CDC* genes has revealed another yeast CaBP encoded by the *CDC31* gene. A *cdc31* mutation exhibits defects in duplication of the spindle pole body, the organizing centre for yeast microtubules on the nuclear envelope (Baum *et al.*, 1986). Thus, Ca^{2+} may regulate bud formation and spindle pole body duplication by interplay between the *CLS4/CDC24* and *CDC31* gene products (Anraku *et al.*, 1991). Protein kinase C (PKC) has been purified from *S. cerevisiae* (Ogita *et al.*, 1990). The *PKC1* gene is essential; *PKC1*-depleted cells show cell cycle arrest (Levin *et al.*, 1990; see also section 13.3.1).

CaM-dependent protein kinases, including multifunctional protein kinase II, have been purified from *S. cerevisiae* (Londesborough, 1989; Ohya *et al.*, 1991b), *N. crassa* (Van Tuinen *et al.*, 1984) and *Fusarium oxysporum* (Hoshino *et al.*, 1991). Substrates for these, by analogy with mammalian cells, may include membrane-bound and cyto-skeletal proteins as well as enzymes (Colbran *et al.*, 1989a,b). Such mechanisms underlie the means by which transient Ca^{2+} signals can effect prolonged cellular responses. A Ca^{2+}-CaM-dependent protein kinase from *S. cerevisiae* had a broad substrate specificity *in vitro* which included two proteins which yielded phosphopolypeptides of 50 kDa and 200 kDa respectively (Londesborough, 1989). Ca^{2+} and CaM are required for protein synthesis and phosphorylation during germination of the insect pathogen *Metarhizium anisopliae* with evidence indicating involvement of Ca^{2+}-CaM-dependent protein kinase(s) as in other eukaryotic cell systems (St Leger *et al.*, 1989). CaM was located at conidial poles near to sites of germ-tube emergence and it seems that Ca^{2+} may be involved in establishment of apical dominance; cAMP may potentiate the effects of small changes in Ca^{2+} concentration (St Leger *et al.*, 1990). In the entomopathogen *Zoophthora radicans*, a Ca^{2+}-CaM system is again believed to be involved in appressorium formation with a requirement for Ca^{2+} influx (Magalhaes *et al.*, 1991). Germ-tube formation in *C. albicans* is also accompanied by an increase in protein phosphorylation, which could be inhibited by CaM antagonists and Ca^{2+} ionophores. Germ tubes/hyphae possessed more CaM activity than yeast cells with two germ tube-specific and three yeast-specific phosphoproteins being revealed by SDS-PAGE analysis (Paranjape *et al.*, 1990).

9.3   ADENOSINE 3',5'-CYCLIC MONOPHOSPHATE (CYCLIC AMP)

The cyclic nucleotides, adenosine 3',5'-cyclic monophosphate (cyclic AMP, cAMP) and guanosine 3',5'-cyclic monophosphate (cyclic GMP, cGMP), with Ca^{2+}, were the first recognized secondary messengers which modulate eukaryotic cell function (Laychock, 1989). In fungi, many functions have been ascribed to cAMP with a postulated involvement in nutrition, reproduction and morphogenesis, including dimorphism (Pall, 1981, 1984) (Chapter 19), although the evidence for direct and/or indirect roles in such processes is often contradictory and inconclusive.

9.3.1   Involvement of cAMP in fungal growth and differentiation

Evidence for the involvement of cAMP in fungal growth and differentiation frequently involves studies of morphological effects of exogenously supplied cAMP, precursors and analogues, and correlation with changes in cellular cAMP (which may also be achieved by cAMP enzyme inhibitors) as well as changes in cellular cAMP in response to nutritional and physicochemical morphogenetic and biochemical stimuli. Although cGMP has been detected in fungi (Eckstein, 1988), information on specific roles in fungal metabolism and growth is still very sparse.

The key enzymes of cAMP metabolism, adenylate cyclase (synthesis) and the phosphodiesterase (breakdown) have been detected in all fungi examined (Rosenberg and Pall, 1978; Suoranta and Londesborough, 1984; Cooper *et al.*, 1985; Brunton and Gadd, 1989; Pitt and Kaile, 1990). The latter enzyme is inhibited by low concentrations of Ca^{2+} as in *N. crassa* (Shaw and Harding, 1983). A rapid and transient increase in cellular cAMP concomitant with spore germination has been observed in several Zygomycetes, e.g. *Mucor rouxii*, *Phycomyces blakesleeanus*, *Pilobolus longipes* and in ascospores of *S. cerevisiae* (Cantore *et al.*, 1980; Moreno *et al.*, 1982; Dewerchin and Van Laere, 1984; Bourret and Smith, 1987; Beullens *et al.*, 1988) (Chapter 19). In germinating spores of *M. rouxii*, the rapid cAMP increase was accompanied by an increase in the level of cAMP phosphodiesterase. It was proposed that cAMP-dependent phosphorylation was the mechanism of shut-off of the glucose-induced cAMP signal (Tomes and Moreno, 1990). Several results have indicated that cAMP may be a trigger for spore germination. A variety of physical or chemical germination inducers result in transient elevation of cAMP and germination is induced by inhibition of cAMP phosphodiesterase (thereby elevating cellular cAMP) (Van Laere, 1986a). Cyclic AMP can regulate several regulatory enzymes by phosphorylation, e.g. trehalase and glycerol-3-phosphatase in *P. blakesleeanus* (Van Laere, 1986b). The effect of glucose in elevating cAMP has been proposed to be mediated by effects on the membrane potential, since depolarization can cause cAMP increase in *N. crassa* (Trevillyan and Pall, 1979), as well as changes in cytoplasmic pH (Caspani *et al.*, 1985). It is known that cAMP can act as a positive effector of the plasma membrane H^+-ATPase activity in *S. cerevisiae* (Ulaszewski *et al.*, 1989). However, membrane depolarization and intracellular acidification have now been discounted as triggers for the cAMP signal (section 9.4.2).

In *N. crassa*, substantial levels of cAMP are required for normal growth, although such a role may be indirect (Pall and Robertson, 1986). *N. crassa cr-1* lacks cAMP, because of a lack of adenylate cyclase, and does not exhibit size hierarchy of hyphae when grown in the absence of cAMP. During growth on high (millimolar) concentrations of cAMP or cAMP analogues, e.g. 8-bromocyclic AMP, hyphal morphology was partially restored after a long delay (Pall and Robertson, 1986). Supplementation of culture medium with cAMP largely reversed phenotypic anomalies of the *cr-1* mutant which suggested a relationship between low cellular levels of cAMP and the abnormal phenotype (Terenzi *et al.*, 1976, 1979; Rosenberg and Pall, 1983). However, it is known that the suppressor mutations may 'normalize' the *cr-1* phenotype without affecting adenylate cyclase activity (Terenzi *et al.*, 1979). It was therefore suggested that cAMP may primarily affect development and cell organization in *N. crassa* and that the abnormalities of *cr-1* were probably indirect consequences of low cellular cAMP. It is now known that invertase activity in wild-type and mutant strains of *N. crassa* is enhanced by exogenous cAMP, an effect dependent on cell wall integrity. It is therefore possible that exogenous cAMP may influence cell wall polysaccharide synthesis (Terenzi *et al.*, 1992). Addition of the lipophilic cAMP analogue N^6, O^2-dibutyryl adenosine 3',5'-cyclic monophosphate (dbcAMP) to several *Mucor* species also resulted in marked morphological changes including hyphal dilation, apical swelling, septation and multiple branching (Jones and Bu'Lock, 1977). In a wild-type strain of *Fusarium graminearum*, exogenous cAMP caused significant decreases in the hyphal extension rate (E) and the hyphal growth unit length (G) (Steele and Trinci, 1975) (Chapter 14) while a highly branched mutant strain was unaffected (Robson *et al.*, 1991c). It should be noted that the high concentrations of exogenous cAMP required to induce changes in fungal differentiation probably reflect the relative impermeability of fungal membranes to cAMP and related

compounds. More lipophilic cAMP analogues have often proved to be more effective morphogens than cAMP itself, e.g. dibutyryl cAMP (Jones and Bu'Lock, 1977; Cooper *et al.*, 1985; Brunton and Gadd, 1989). Both cAMP and cGMP may be released from growing fungi into external medium with possible roles including a signal for limiting mycelial density, and/or the regulation of intracellular levels of these cyclic nucleotides (Shaw and Harding, 1987).

Compounds which affect the activity of enzymes of cAMP metabolism may also induce changes in morphology, which can be correlated with changes in intracellular cAMP, although it is often unclear whether such effects are indirect since some of the compounds are potentially toxic. Inhibitors of adenylate cyclase, e.g. atropine and quinidine, caused a reduction in cAMP in *N. crassa*, whereas theophylline, a phosphodiesterase inhibitor, increased cAMP levels and increased branching (Scott and Solomon, 1975). Theophylline and caffeine (another phosphodiesterase inhibitor) also increased cAMP in *N. crassa* (Scott and Solomon, 1975), *Penicillium notatum* (Pitt and Kaile, 1989), *Aureobasidium pullulans* (Cooper *et al.*, 1985), *Ophiostoma ulmi* (Brunton and Gadd, 1989) and *C. albicans* (Sabie and Gadd, 1992).

The phenomenon of dimorphism exhibited by several fungi has often been related to changes in cellular cAMP (Chapter 19). In *C. albicans*, several reports have demonstrated a rise in intracellular cAMP which accompanies germ tube formation as well as promotion of germ tube formation by dbcAMP and theophylline (Niimi *et al.*, 1980; Chattaway *et al.*, 1981; Sabie and Gadd, 1992). However, Egidy *et al.* (1989) found that germ tube formation in *C. albicans* was preceded by a significant decrease in intracellular cAMP levels which increased as germ tube formation proceeded. Germ tube formation was inhibited by the addition of exogenous cAMP and compounds known to elevate intracellular cAMP (Egidy *et al.*, 1989). Such contradictory findings are frequently ascribed to differences in strains and methods of germ tube induction, as well as the use of cells from different phases of growth. Intracellular cAMP in *C. albicans* is dependent on the growth phase with higher cAMP being present in stationary phase cells than those at earlier stages of growth (Niimi, (1984). However, a yeast-to-mycelium (Y–M) transition in *Mucor racemosus* and

M. rouxii was also preceded by a decrease in cAMP levels (Larsen and Sypherd, 1974; Paveto *et al.*, 1975; Orlowski and Ross, 1981; Cantore *et al.*, 1983). Elevation of intracellular cAMP during a Y–M transition has been noted for other polymorphic organisms including *A. pullulans* (Cooper *et al.*, 1985) and *O. ulmi* (Brunton and Gadd, 1989). In *A. pullulans*, provision of exogenous adenosine and AMP also induced germ tube formation, these substances being metabolized to cAMP after entry into cells (Cooper *et al.*, 1985). Exogenous cAMP (10 mM) or cGMP induced bean rust (*Uromyces phaseoli*) uredospores to undergo mitotic division and septum formation, processes usually associated with appressorium formation. Similarly, inhibitors of cAMP phosphodiesterase and activators of adenylate cyclase induced both nuclear division and appressorium formation (Hoch and Staples, 1984). An 8-azido-[^{32}P]cAMP photoaffinity probe was used to identify three cyclic nucleotide binding peptides which bound either cAMP or cGMP. Phosphorylation of one peptide by [δ^{32}P]ATP was stimulated by cAMP or cGMP in the presence of Na_2MoO_4, a phosphatase inhibitor (Epstein *et al.*, 1989).

Exogenous cAMP can induce sclerotium formation in non-sclerotial isolates of *Rhizoctonia solani* (Hashiba, 1982). The existence of a cAMP-dependent protein kinase has been reported in *R. solani* (Uno *et al.*, 1985a,b) and the role of cAMP as a second messenger in sclerotium production has been suggested since hormone-induced sclerotium formation resulted in elevated endogenous cAMP levels, possibly by phosphodiesterase inhibition or adenylate cyclase activation (Sharada *et al.*, 1992). Maximum levels of cAMP coincided with the time of sclerotial initiation (Hashiba and Ishikawa, 1978). In *Trichoderma viride*, exposure to light can trigger conidiation in dark-grown mycelia. This was accompanied by an initial hyperpolarization of the membrane followed by a period of depolarization. The hyperpolarization resulted in a rise in cell ATP that was accompanied by an increase in cellular cAMP (Farkas *et al.*, 1985; Gresik *et al.*, 1988); it is known that ATP can activate adenylate cyclase (Rosenberg and Pall, 1983). It is possible that light stimulated the activity of the H$^+$-ATPase thus elevating transiently the membrane potential. An eventual decrease in cAMP may be a result of light stimulation of phosphodiesterase activity as shown in *N. crassa* (Sokolovsky and Kritsky, 1985).

9.3.2   cAMP-dependent protein kinases

Although cAMP is clearly involved in many aspects of fungal differentiation, it essentially has only one biochemical activity – the binding to and regulation of cAMP-dependent protein kinases (Pall, 1981; Gancedo *et al.*, 1985). Cyclic AMP-dependent protein kinase (PK-A) mediates virtually all the actions of hormones which exert their effects by elevation of intracellular cAMP. After activation, PK-A phosphorylates regulatory proteins which results in conformational changes and altered biochemical properties (Cohen, 1992; Cohen and Hardie, 1991). Cyclic AMP-dependent protein kinases have been identified in several fungi including *N. crassa* (Powers and Pall, 1980; Trevillyan and Pall, 1982). It should be noted that evidence from higher eukaryotic systems suggests that whereas catabolic pathways may be activated by protein kinase-mediated phosphorylation, inhibition of catabolic pathways is often more indirect due to the effect of PK-A resulting from inhibition of protein phosphatases which dephosphorylate sites phosphorylated by other protein kinases (Cohen and Hardie, 1991). Finally, it is significant that polyphosphoinositide synthesis in *S. cerevisiae* is stimulated by cAMP-dependent protein kinase by activation of phosphatidylinositol kinase and phosphatidylinositol 4-phosphate kinase which may lead to enhanced production of the other signalling molecules, Ins1,4,5 P$_3$ and diacylglycerol (DAG) (see later) (Kato *et al.*, 1989; Uno and Ishikawa, 1990).

9.4   G PROTEIN-LINKED RECEPTORS

Signal molecules may be detected by specific receptors on the plasma membrane; stimulation of such receptors results in eventual expression of an intracellular signal which can effect changes in, for example, growth, differentiation and metabolism. Some receptors operate by activating a membrane-bound GTP-binding protein (G-protein) and these can regulate different intracellular pathways that result in changes in the activity of specfic effectors, e.g. ion channels, adenylate cyclase, or enzymes of phosphoinositide metabolism (Neer and Clapham, 1988) (Figure 9.3). In recent years, G proteins have been found to be highly conserved components of signalling systems. G protein genes have been identified in many eukaryotic systems, including fungi, with the implication that all eukaryotes are

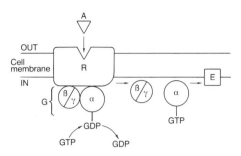

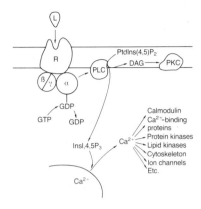

Figure 9.3 Simplified scheme for signal transduction through G protein-linked receptors. The agonist (A) induces a conformational change in the receptor (R) which is transmitted to the G protein (G), which dissociates into the α-subunit and the βγ-subunit, the α-subunit exchanging GDP for GTP in the activated state. The activated α-subunit and the βγ-subunits dissociate and one or both interacts with the effector(s) (E), for example, adenylate cyclase, K^+ channels, phospholipase C (Figure 9.4). Intrinsic GTPase activity of the α-subunit hydrolyses GTP to GDP releasing inorganic phosphate (P_i); α-GDP then recombines with βγ, thus ending the activation cycle (Neer and Clapham, 1988; Kurjan, 1992; Berridge, 1993).

Figure 9.4 Simplified diagram of the phosphoinositol signalling system leading to Ca^{2+} mobilization from intracellular compartment(s). L, ligand (or agonist); R, receptor; G, G protein composed of G_α and $G_{\beta\gamma}$ subunits. The α-subunits exchange bound GDP for GTP before interacting with phospholipase C (PLC). Evidence from other eukaryotic systems indicates that different PLC isozymes can be activated by both G_α and $G_{\beta\gamma}$ subunits. $PtdIns(4,5)P_2$, phosphatidylinositol 4,5-bisphosphate; DAG, diacylglycerol; PKC, protein kinase C; $Ins1,4,5P_3$, inositol 1,4,5-trisphosphate. The prime intracellular Ca^{2+} store may be the vacuole in fungi with some involvement of the endoplasmic reticulum (see text) (adapted from O'Day, 1990). It should also be noted that another pathway for intracellular release of $Ins1,4,5P_3$ and mobilization of intracellular Ca^{2+} in eukaryotic cells is mediated by tyrosine kinase-linked receptors (Berridge, 1993) although no detailed information is available for fungi.

likely to utilize G protein-mediated signalling systems (Kurjan, 1992). In fungi, the most understanding of G protein-mediated systems has been derived from studies on phosphoinositol signalling, the RAS adenylate cyclase signal transmission pathway and the pheromone response. Most of this work has used *S. cerevisiae*, although there are some other indirect examples of G protein-linked receptor systems in filamentous fungi (Trinci *et al.*, 1990).

9.4.1 Phosphoinositol signalling system: agonist-stimulated calcium mobilization

A stimulated catabolism of inositol lipids occurs in many eukaryotic cells in response to a wide range of external stimuli (Michell, 1975; Downes and Michell, 1982; Berridge, 1984, 1993; Berridge and Irvine, 1989). This phosphoinositide metabolism was subsequently proposed to be linked with the mode of action of those agonists that mobilize Ca^{2+} from internal stores, thus raising $[Ca^{2+}]_c$. There is now a large amount of information which demonstrates that inositol 1,4,5-trisphosphate (Ins $1,4,5P_3$) is the diffusible second messenger molecule that links the plasma membrane (where the relevant receptors to external stimuli are

located) and intracellular stores of Ca^{2+} (Michell *et al.*, 1981; Berridge, 1983, 1984; Berridge and Irvine, 1984, 1989). Phosphatidylinositol 4,5-bisphosphate ($PtdIns(4,5)P_2$; PIP_2), formed by a two-stage phosphorylation of phosphatidylinositol to phosphatidylinositol-4-phosphate ($PtdIns(4)P$) and then $PtdIns(4,5)P_2$, is the immediate precursor used by the receptor-mediated mechanisms in the plasma membrane. When agonists interact with the external receptors, a phosphodiesterase (phospholipase C, PLC) is activated *via* a G protein-mediated process and this cleaves $PtdIns(4,5)P_2$ into $Ins(1,4,5)P_3$ and diacylgylcerol (DAG) (Berridge, 1984, 1987, 1993) (Figure 9.4). Both these molecules have second messenger functions. The initial and primary effect of $Ins(1,4,5)P_3$ is to stimulate Ca^{2+} release from $Ins(1,4,5)P_3$-sensitive intracellular Ca^{2+} stores thus elevating $[Ca^{2+}]_c$ which can activate enzymes, modify CaBPs etc. as already discussed. An additional effect of Ins1,4,5P$_3$ may be to promote Ca^{2+} influx from the exterior, possibly after conversion to Ins1,3,4,5P$_4$ (Berridge, 1993); the importance of Ca^{2+} influx in signal transduction has been described (Iida *et al.*, 1990a,b; Gadd and Brunton, 1992). The other product of PtdIns(4,5)P$_2$ hydrolysis, DAG, has an important second messenger function in that it activates protein kinase C (PKC) which, in turn, phosphorylates certain proteins (Berridge, 1984, 1987; Nishizuka, 1984, 1992a,b). PKC is activated by phorbol esters which may mimic the stimulatory effect of DAG; the use of phorbol esters is one means of indirectly demonstrating PKC involvement in a cell response to external stimuli. Thus, hydrolysis of PtdIns(4,5)P$_2$ results in the formation of two second messengers, one pathway depending on DAG activating PKC to phosphorylate specific cellular proteins, whereas the other pathway depends on Ins(1,4,5)P$_3$ mediated elevation of [Ca^{2+}]$_c$, which may act *via* CaM, in activating and/or modulating target enzymes and protein phosphorylation. These two interacting pathways appear to function synergistically in stimulating a variety of eukaryotic cellular responses including growth and differentiation (Berridge, 1984, 1987, 1993; Berridge and Irvine, 1984, 1989). It may be relevant that the ability of inositol phosphates, including Ins1,4,5P$_3$, to bind Ca^{2+} may have a role in modulation of increases in [Ca^{2+}]$_c$ (Luttrell, 1993). Other mechanisms responsible for stimulating release of Ins1,4,5P$_3$ in eukaryotic cells, e.g. tyrosine kinase-linked receptors (Berridge, 1993), have received negligible attention in fungi as yet.

(a) Phosphoinositol signalling in fungi

There is accumulating evidence for the involvement of phosphoinositides in signal transduction in fungi. As described previously, several elements of eukaryotic signalling pathways have been identified in yeasts and fungi including CaM (Ortega Perez *et al.*, 1981; Muthukumar *et al.*, 1987; Laccetti *et al.*, 1987), CaM-dependent protein kinase (van Tuinen *et al.*, 1984; Bartelt *et al.*, 1988; Londesborough, 1989; Miyakawa *et al.*, 1989; St Leger *et al.*, 1989, 1990; Liu *et al.*, 1990), protein kinase C (PKC) (Favre and Turian, 1987; Thorner *et al.*, 1988; Fields and Thorner, 1989; Ogita *et al.*, 1990) and G proteins (St Leger *et al.*, 1990; Kozak and Ross, 1991). Phosphoinositide turnover in *S.*

cerevisiae is regulated by glucose (Kaibuchi *et al.*, 1986) whereas an essential role for PtdIns (4,5)P$_2$ in yeast cell proliferation has been reported (Uno *et al.*, 1988). The plasma membrane of *S. cerevisiae* is capable of synthesizing both CDP-diacylglycerol and phosphatidylinositol as well as phosphorylating phosphatidylinositol (Kinney and Carman, 1990; Flanagan and Thorner, 1992). More recently, *PKC* genes in *S. cerevisiae* have been characterized (Levin *et al.*, 1990; Yoshida *et al.*, 1992) and a gene which encodes a putative phosphoinositide-specific phospholipase C (PLC) has been isolated. This gene (*PLC1*) is important for cell growth and further provides evidence for signal transduction systems mediated by phospholipid hydrolysis being important in yeast cell growth (Yoko-o *et al.*, 1993). Kato *et al.* (1989) have shown that cAMP-dependent phosphorylation in *S. cerevisiae* activates phosphatidylinositol kinase and phosphatidylinositol-4-phosphate kinase which may lead to production of Ins(1,4,5)P$_3$ and DAG. Such observations also provide a link between inositol phospholipid-mediated and cAMP-mediated signalling systems (Figure 9.5) (see later).

Rather less work has been carried out with fungi other than *S. cerevisiae*. PtdIns(4,5)P$_2$ has been detected in *Fusarium graminearum* (Robson *et al.*, 1991d) and *N. crassa* (Hanson, 1991), cAMP- and inositol phospholipid-linked signalling pathways are involved in the mechanism of EtOH-induced filamentous growth in *Candida tropicalis* (Kamihara and Omi, 1989). Evidence has also been obtained for the involvement of an inositol lipid signalling system, including protein kinase C, in the yeast–mycelium transition of *Ophiostoma ulmi* (Dutch elm disease) (Brunton and Gadd, 1991). Cornelius *et al.* (1989) have demonstrated that Ins(1,4,5)P$_3$ stimulates Ca^{2+} release from vacuoles in *N. crassa*, organelles which are prime intracellular Ca^{2+} stores in fungi (Miller *et al.*, 1990). Hanson (1991) has demonstrated that all the components and enzymes essential for a PtdIns cycle are present in *N. crassa* and they respond to lithium in a similar manner as do animal cells, overall levels of phosphoinositides declining with concomitant disruption in cell activities. If such a fungal PtdIns cycle exists it may differ from the animal system in that it would include inositol-containing sphingolipids that are apparently unique to plants and fungi (Becker and Lester, 1980). In *S. cerevisiae*, the synthesis of sphingolipids occurs *via* the transfer of

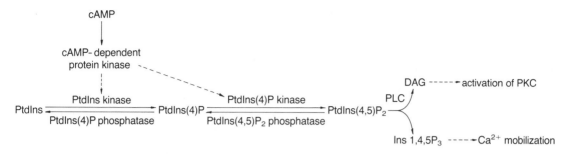

Figure 9.5 Inositol phospholipid cascade in *S. cerevisiae* showing interaction with cAMP (adapted from Uno and Ishikawa, 1990). DAG, diacylglycerol; Ins1,4,5P$_3$, inositol 1,4,5-trisphosphate; PLC, phospholipase C; PtdIns(4,5)P$_2$, phosphatidylinositol 4,5-bisphosphate; PtdIns, phosphatidylinositol; PtdIns(4)P, phosphatidylinositol 4-phosphate; PKC, protein kinase C.

inositol monophosphate from PtdIns to ceramides with the release of DAG and this may also occur in *N. crassa* (Hanson, 1991). It is noteworthy that sphingolipids are known to possess important enzyme regulatory properties in yeast (Hanson, 1984) whereas in animal cells, a sphingolipid cycle may act to inhibit PKC (Bell *et al.*, 1988). Prior *et al.* (1993) have shown that key components and intermediates of the phosphatidylinositol signal system exist in *F. graminearum*, *N. crassa* and *Phanerochaete chrysosporium* with a wide diversity in the number of isomers of inositol bis- and tris-phosphates. Inositol phosphate metabolism has been suggested as a component of the circadian oscillator and blue-light phototransduction in fungi. However, although protein kinases, Ca^{2+} and ion fluxes have been implicated in resetting of the oscillator of *N. crassa* (Lakin-Thomas *et al.*, 1990) and blue light can activate fungal G proteins (Kozak and Ross, 1991), there is little evidence as yet for the involvement of phosphoinositide signalling (Lakin-Thomas, 1993).

As described earlier, PKC plays an important role in eukaryote signal transduction pathways (Nishizuka, 1988). The existence of multiple sub-species of PKC has been demonstrated in *N. crassa* (Favre and Turian, 1987) and *S. cerevisiae* (Fields and Thorner, 1989; Ogita *et al.*, 1990). Yeast PKC is activated by the simultaneous addition of Ca^{2+}, DAG and phosphatidylserine but did not respond significantly to tumour-promoting phorbol esters. GTP did not serve as a phosphate donor and yeast PKC exhibited different substrate specificity to that of mammalian PKCs. H1 histone and protamine were poor substrates, and, using myelin basic protein as model substrates, yeast PKC preferentially phosphorylated threonyl residues (rat brain PKC preferentially phosphorylated seryl residues) (Ogita *et al.*, 1990). Induction of germ tube formation by the PKC-activating phorbol ester, phorbol 12-myristate 13-acetate (PMA) in *O. ulmi*, and the reversal of the inhibitory effect of the anti-CaM drug R24571 by PMA inferred the involvement of PKC, and therefore DAG, in the Y–M transition of this organism (Brunton and Gadd, 1991). Similarly, morphological effects of staurosporine, a potent protein kinase inhibitor, on *Pleurotus ostreatus* implied protein kinase involvement in the regulation of apical growth (Magae and Magae, 1993). Protein kinase C was subsequently purified from this organism, with activation resulting from the presence of PMA (Magae, 1993).

9.4.2 The RAS–adenylate cyclase signal transmission pathway of yeast

In eukaryotes, *RAS* genes encode low-molecular-weight GTP-binding, GTP-hydrolysing, (G) proteins, RAS proteins, that are postulated to play key roles in signal transduction (Broach, 1991). Most fungal work has been carried out with *S. cerevisiae* since it contains two genes that are structurally and functionally homologous to human *RAS* genes. RAS proteins probably function via a cycle of GTP/GDP exchange and GTP hydrolysis, the replacement of bound GDP with GTP resulting in activation and resultant stimulation of the activity of the target protein (Broach, 1991).

In *S. cerevisiae*, RAS proteins activate adenylate cyclase activity in an analogous way to the G_s proteins of mammalian adenylate cyclase (Gibbs and Marshall, 1989). The resulting endogenous cAMP activates cAMP-dependent protein kinase (PK-A) and this, as mentioned, via phosphorylation of target proteins, enhances several activities associated with growth. Cyclic AMP, and therefore the RAS proteins, appeared to be involved in control of progression over the start point in the G_1 phase of the yeast cell cycle and yeast strains with homologues of mammalian *RAS* oncogenes, elevated cAMP, or elevated cAMP dependent protein kinase activity could not arrest at the start point of the cell cycle under nutrient depletion (Toda *et al.*, 1985; Sass *et al.*, 1986). Furthermore, temperature-sensitive mutations in RAS or adenylate cyclase arrested at the start when shifted to the restrictive temperature (Matsumoto *et al.*, 1985; De Vendittis *et al.*, 1986). However, whether RAS control of the cell cycle is direct or is an indirect consequence of RAS effects on metabolism remains to be clarified (Broach, 1991). Several components of the RAS – adenylate cyclase activity have been identified. These include *CDC25*, required for activation of the RAS proteins, the RAS-GTPase activating proteins IRAI and IRA2, the *SRV2* gene product ('CAP'), a subunit of adenylate cyclase (Field *et al.*, 1990), and genes encoding for adenylate cyclase (*CYR1* = *CDC35*), cAMP phosphodiesterase (*PDE1*, *PDE2*) and the catalytic (*TPK1*, *TPK2*, *TPK3*) and regulatory (*BCY1*) subunits of cAMP-dependent protein kinase (Schmitt *et al.*, 1986; Broach and Deschenes, 1990; Thevelein, 1991, 1992). However, the signalling function of the RAS signal-transmission pathway is poorly understood although it is postulated that the presence of adequate nutrition is signalled to the cyclase by means of this pathway (Dumont *et al.*, 1989).

(a) Glucose-induced cAMP signalling and protein phosphorylation

Addition of readily fermentable sugars, e.g. glucose or fructose, to derepressed (respiring) *S. cerevisiae* results in a variety of rapid metabolic changes. These include inhibition of gluconeogenesis, inhibition of galactose and maltose transport, activation of glycolysis, and trehalose mobilization (Gadd, 1988; Thevelein, 1988; Gancedo and Serrano, 1989). Such phenomena were found to be effected by cAMP-dependent protein phosphorylation (Holzer, 1984; Thevelein, 1984, 1988). The relationship of this process to glucose repression is unclear although cAMP appears not to be involved as a second messenger (Matsumoto *et al.*, 1982). However, glucose repression of certain enzymes, e.g. alcohol dehydrogenase II and catalase T, appeared to involve the RAS–adenylate cyclase pathway (Thevelein, 1991).

Glucose addition to derepressed yeast causes a rapid rise in cellular cAMP which appears to trigger protein phosphorylation. This induction of a cAMP signal is mediated by the *CDC25-RAS*–adenylate cyclase pathway in yeast (Munder and Küntzel, 1989; Van Aelst *et al.*, 1990, 1991) which is activated by sugar concentrations appropriate for fermentation; low concentrations do not activate the pathway (Mbonyi *et al.*, 1988, 1990). The RAS pathway appears to function therefore as a sensor for detection of fermentable sugars external to the cells (Thevelein, 1991). As mentioned previously, glucose-induced cAMP signalling has been observed in spore germination in fungi other than yeast including various Zygomycetes (Van Mulders and Van Laere, 1984; Dewerchin and Van Laere, 1984) and *Pilobolus* (Bourret, 1986). Membrane depolarization, intracellular acidification and increased energy supply appear not to be triggers for the cAMP signal (Caspani *et al.*, 1985; Eraso *et al.*, 1987; Thevelein *et al.*, 1987a,b).

Although it is established that glucose can activate the plasma membrane H^+-ATPase, it is not known whether it is triggered by cAMP-dependent protein phosphorylation (dos Passos *et al.*, 1992). It now seems that the cAMP-protein kinase A signal pathway is not required for glucose activation of the H^+-ATPase although the presence of a kinase capable of phosphorylating the added sugar was required. It appears that glucose-induced activation pathways of cAMP synthesis and the H^+-ATPase have a common initiation point (dos Passos *et al.*, 1992).

(b) Acidification-induced cAMP signalling

Intracellular acidification also stimulates the RAS adenylate cyclase pathway which may constitute a response to conditions of stress (Thevelein, 1991). Acidification-induced cAMP synthesis has been described in fungi other than yeast including *N.*

crassa (Pall, 1977), *Mucor racemosus* (Trevillyan and Pall, 1979), *Coprinus macrorhizus* (Uno and Ishikawa, 1981) and *Phycomyces blakesleaanus* (Van Mulders and Van Laere, 1984).

(c) Glucose-induced signalling

Activation of adenylate cyclase by glucose and intracellular acidification implied nutritional stimulation of the RAS pathway in yeast (Gibbs and Marshall, 1989; Broach, 1991). However, the RAS activation pathway induced by fermentable sugar is glucose-repressible, and is not required for progression over the start point of the cell cycle (Thevelein, 1991; Hirimburegama *et al.*, 1992). Its function may be limited to metabolic control following the transition from gluconeogenic to fermentative growth (Thevelein, 1991). Work to date suggests the presence of a glucose-repressible protein in the pathway (Thevelein, 1991). It has been proposed that in glucose-repressed cells, a nitrogen source-induced signalling pathway (dependent on glucose) leading to activation of cAMP dependent protein kinase A triggers progression over the start point of the yeast cell cycle (Thevelein, 1992).

Another postulated glucose control system involving the RAS–adenylate cyclase pathway is the glucose-stimulation of phosphatidylinositol turnover in yeast (Kaibuchi *et al.*, 1986) although this has not been confirmed by others (Frascotti *et al.*, 1990). Nevertheless, the kinases responsible for phosphatidylinositol 4,5-bisphosphate (PtdIns $(4,5)P_2$) formation are stimulated by cAMP-dependent protein kinase (Kato *et al.*, 1989) (see earlier). Furthermore, Schomerus and Küntzel (1992) have shown that the CDC25 protein controls a nitrogen-specific signalling pathway in *S. cerevisiae* involving the effector phosphoinositide-specific phospholipase C (PI-PLC) as well as the glucose-induced activation of adenylyl cyclase (AC). The addition of $(NH_4)_2SO_4$ to starved yeast cells leads to a 3–4-fold increase in $Ins1,4,5P_3$ and DAG, products of PI-PLC (Schomerus and Küntzel, 1992).

(d) Nitrogen source-induced signalling

Nitrogen sources can cause rapid activation or inactivation (in the presence of glucose) of enzymes regulated by cAMP-dependent protein phosphorylation, e.g. trehalase (Thevelein, 1988), phosphofructokinase 2, glycogen synthase and phosphorylase (François *et al.*, 1988), although nitrogen source does not affect cAMP levels in vegetative or N-starved yeast cells (Thevelein and Beullens, 1985). Starvation of yeast cells for nitrogen causes arrest at the start point in G_1 of the cell cycle (Pringle and Hartwell, 1981). An explanation for the interaction between nitrogen source and cAMP for control of progression over the start point of the yeast cell cycle infers that an unidentified factor generated by the nitrogen source-induced pathway activates protein kinase A synergistically with cAMP (Thevelein, 1991). Alternatively, the nitrogen source-induced pathway may only activate free catalytic subunits (Figure 9.6). Such activation would also be indirectly dependent on the cAMP level. The main difference between these two models is that in the first model, maximum protein kinase activity is not modified by the nitrogen source, whereas in the second model, the maximum activity would be higher (Thevelein, 1991). Work with cell-cycle mutants, e.g. *cdc33* or *cdc60*, which arrest at the same point in the cell cycle as nutrient-starved or cAMP-deficient cells but are not themselves affected in cAMP metabolism (Verdier *et al.*, 1989) supports the existence of a separate nitrogen source-induced pathway leading to activation of protein kinase A and not the sugar-induced RAS–adenylate cyclase pathway (Thevelein, 1991). The function of the RAS–adenylate cyclase system in glucose-repressed cells may be to provide a base level of cAMP rather than the second messenger which triggers this process in response to an extracellular signal (Thevelein, 1991, 1992; Hirimburegama *et al.*, 1992).

Sulphate addition also has no effect on cAMP levels in sulphate-starved cells whereas a small increase in cAMP is caused by phosphate addition to phosphate-starved yeast cells. The latter phenomenon is difficult to interpret because of the phosphate requirement for ATP metabolism (Hirimburegama *et al.*, 1992). However, these nutrients, as well as the nitrogen source, can activate trehalase at low cAMP levels implying that cAMP is not involved as a second messenger, but rather that signalling depends on a basal cAMP level. Such results support the model whereby the nutrient sources activate the free catalytic subunits of cAMP-dependent protein

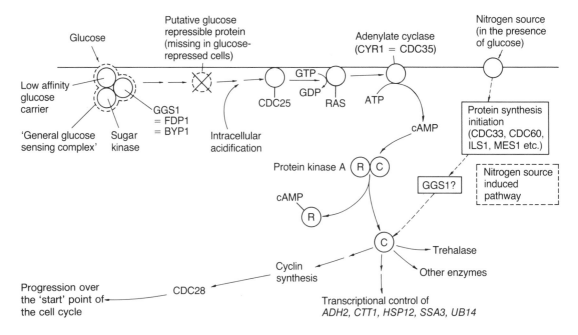

Figure 9.6 Outline model of the glucose- and acidification-induced signal-transmission pathway resulting in activation of the RAS–adenylate cyclase system in yeast and its relationship to the nitrogen source-induced, glucose-dependent signalling pathway (adapted from Thevelein, 1992). The general glucose sensing complex and the putative glucose-repressible protein are only required for activation by glucose. Intracellular acidification activates the pathway downstream of the putative glucose-repressible protein. Because of the presence of the glucose-repressible protein, glucose-induced activation of the RAS–adenylate cyclase system is no longer functional once the cells have become glucose-repressed. Hence, only during the transition period from derepression to repression is cAMP-dependent protein kinase being activated by the glucose-dependent RAS-adenylate cyclase pathway. Nitrogen signalling uses a separate pathway which appears to involve the initiation of protein synthesis and which is suggested to cause activation of free catalytic subunits of cAMP-dependent protein kinase. This nitrogen-induced pathway is dependent on the presence of glucose which might indicate that it requires activity of the general glucose sensing complex. (Not shown are the *IRA1* and *IRA2* gene products which stimulate the RAS-GTPase gene product which acts as a stimulator of the IRA proteins, the *PDA1*- and *PDA2*-encoded low- and high-affinity phosphodiesterases, feedback inhibition of protein kinase A on cAMP synthesis, the SRV2 or 'CAP' subunit of adenylate cyclase and the SCH9 and YAK1 protein kinases which appear to be involved downstream of cAMP-dependent protein kinase.)

kinase. Since the number of free catalytic subunits depends on the cAMP level, such signalling is only indirectly dependent on cAMP. The nature of the nitrogen-induced signalling pathway remains to be elucidated. It should be noted that the glucose-induced activation of trehalase is similar to that for activation of the RAS–adenylate cyclase pathway (Beullens *et al.*, 1988) indicating that the glucose-sensing system of the latter is also required for nitrogen-induced trehalase activation. It is possible that all glucose-induced regulatory effects in yeast are dependent on the same glucose-sensing system (Thevelein, 1992). The glucose requirement of the nitrogen-signalling pathway implies that the pathway could 'sense' both glucose, nitrogen, and possibly other nutrients, providing an explanation for why starvation of these nutrients causes arrest at the same point in the G_1 phase of the cell cycle (Hirimburegama *et al.*, 1992).

In conclusion, current opinion suggests that the RAS–adenylate cyclase pathway is not the mediator of nutrient-control of growth and cell cycle progression but is involved in 'sensing' the level of fermentable sugar during the transition from gluconeogenic (derepressed) to fermentative (repressed) growth. It is proposed that a 'general glucose sensor' (which may be involved in other glucose-induced regulatory phenomena in yeast) involving a low-affinity glucose carrier, a sugar kinase and the *GGS1* (=*FDP1*) gene product initially activates the pathway which, in repressed cells, provides a basal level of cAMP which liberates part of the catalytic subunits of cAMP-dependent protein kinase. These can be activated by a glucose-dependent nitrogen source signalling pathway implying that cAMP-PK is the integrator of nutrient availability in yeast rather than cAMP (Thevelein, 1992).

Pseudohyphal growth in *S. cerevisiae* is induced by nitrogen starvation and is controlled directly or indirectly by the RAS signal transduction pathway. Strains carrying the dominant *RAS2*[val19] mutation, which results in a constructively activated RAS-signal transduction pathway, elevated intracellular cAMP (Toda *et al.*, 1985) and sensitivity to nitrogen starvation (Toda *et al.*, 1987), show enhanced pseudohyphal formation (Gimeno *et al.*, 1992) (Chapter 19). The invasive nature of this type of proliferation enhances nutrient acquisition by the cells and may have implications for the pathogenicity of animal and plant dimorphic fungal pathogens.

9.4.3 Pheromone response in yeast

In *S. cerevisiae*, pheromone communication ensures cell diploidization in the mating process. Each haploid cell may exhibit mating type **a** or α, each excreting a signal peptide **a**-or α-factor, respectively, and possessing a surface receptor which can bind the pheromone produced by the 'opposite' mating type (Liao and Thorner, 1981) (Chapter 4). Receptors are coupled to the same G protein which is activated by conformational change. It seems that, in yeast, the $G_{\beta\gamma}$ subunit rather than the G_α subunit activates downstream components of the signal pathway (Gooday and Adams, 1993). The immediate target for the liberated G protein subunit is not known, with, so far, no evidence for involvement of, e.g. cAMP (Gooday and Adams, 1993). However, four different protein kinases (STE7, STE11, KSS1, FU33) as well as the *STE5* gene product are required for signal transduction, the kinases possibly constituting a phosphorylation cascade eventually activating the *STE12* transcription factor and allowing transcriptional induction of pheromone-responsive genes and expression of the diverse physiological changes required for conjugation, e.g. induction of surface agglutinins, cell division arrest in G1, and the formation of a conjugation tube (Kurjan, 1992; Konopka and Fields, 1992). Mating cells of the fission yeast, *S. pombe*, can also communicate by diffusible pheromones and certain components of this system resemble those found in budding yeast (Styrkarsdottir *et al.*, 1992). It should be noted that a rapid and transient increase in cellular Ca^{2+} occurs in yeast species in response to mating pheromone (Tachikawa *et al.*, 1987) (Chapter 4).

9.4.4 Osmosensing signal transduction

Glycerol is a major compatible solute in many fungi and is synthesized to high intracellular concentrations at high external osmolarities (Meikle *et al.*, 1988). In *S. cerevisiae*, genes have been isolated that are necessary for restoring the osmotic gradient across the cell membrane in response to elevated external osmolarity with two of these genes (*HOG1* and *PBS2*) encoding mitogen-activated protein kinase (MAP kinase) and MAP kinase gene families respectively (Brewster *et al.*, 1993). MAP kinases are activated by extracellular ligands, e.g. growth factors, via tyrosine kinase-linked receptors and function in protein phosphorylation cascades (Berridge, 1993). Brewster *et al.* (1993) found that rapid, *PBS2*-dependent tyrosine phosphorylation of HOG1 protein occurred in response to elevated external osmolarity, thus defining an osmosensitive signal transduction pathway. The mechanism of sensing osmotic stress is unclear. Although a decrease in turgor pressure, with a possible involvement of plasma membrane mechanosensitive ion channels (Gustin *et al.*, 1988; Zhou *et al.*, 1991) may function in the initial signal, other possibilities may include a ligand–receptor interaction where changes in turgor pressure alter the interaction between a wall-bound ligand and a plasma membrane receptor (Brewster *et al.*, 1993).

9.5   INTRACELLULAR pH

There is accumulating evidence that cytoplasmic pH (pH_i) may have an important role in eukaryotic cell signalling pathways, including those involved with cellular differentiation and metabolism (Busa and Nuccitelli, 1984). It therefore follows that cellular control mechanisms for pH_i which include activities of the plasma membrane- and vacuolar

H$^+$-ATPases, are also linked to cellular responses to external stimuli. In *C. albicans*, an organism where external pH can be used to control dimorphic transitions (Chapter 19), cytoplasmic alkalinization accompanies germ tube formation, pH$_i$ rising from pH 6.8 to pH 7.5 within 2 min and to > pH 8.0 after 20 min (Kaur *et al.*, 1988; Stewart *et al.*, 1988). Non-dimorphic variants grown under germ tube-inducing conditions, or cells grown in conditions that do not induce germ tube formation, do not exhibit such cytoplasmic alkalinization (Stewart *et al.*, 1988, 1989). Diethylstilboestrol (DES), an inhibitor of plasma membrane H$^+$-ATPase, inhibited dimorphism and cytoplasmic alkalinization thus implicating involvement of pH changes in morphogenesis (Stewart *et al.*, 1988). Kaur and Mishra (1991) have shown that *C. albicans* mycelium exhibits a higher H$^+$-ATPase activity than yeast cells although this does not result from *de novo* synthesis of enzyme protein (Gupta and Prasad, 1993). Regulation of H$^+$-pumping ATPase activity may arise from metabolic factors acting on gene expression and by signals generated from glucose metabolism/starvation which may modify the carboxyl-terminal domain of the enzyme and/or be specific to germ tube formation as well as being coupled with higher ratios of cell surface to effective cytosolic volume during germ tube formation (Monk *et al.*, 1991, 1993). Some pH$_i$ effects are clearly indirect because of the influence of transmembrane electrochemical H$^+$ gradients in fluxes of ions including Ca^{2+}, and the influence of pH$_i$ on levels and activities of, e.g. calmodulin and adenylate cyclase (Busa and Nuccitelli, 1984). For *C. albicans*, although cytoplasmic alkalinization may be necessary for the control and continued development of germ tubes and cell polarity, it seems unlikely to be the primary determinant in morphogenesis (Monk *et al.*, 1993).

9.6 PROTEIN PHOSPHATASES

The important role of phosphatases and the balance between phosphorylation and dephosphorylation in eukaryotic cell regulation is now appreciated (Cohen, 1992). Since protein phosphorylation by means of protein kinases is involved in signal transduction in fungi, it follows that protein phosphatases must also have a significant involvement in these processes and

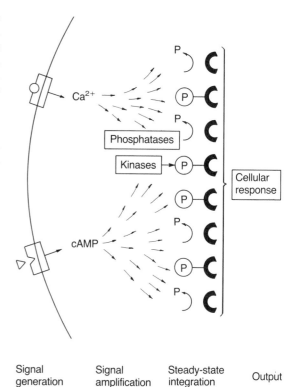

Signal generation   Signal amplification   Steady-state integration   Output

Figure 9.7 Simplified diagram of second messengers (Ca^{2+} and cAMP) and phosphoproteins in signal transduction in a eukaryotic cell. Signal transduction is mediated by cAMP-and Ca^{2+}-regulated phosphorylation/ dephosphorylation cascades (adapted from Kincaid, 1993).

indeed, the reversible phosphorylation of proteins is central for the integration of cellular events (Kincaid, 1993) (Figure 9.7). However, little is known of fungal protein phosphatases with most work being carried out with yeasts (Van Zyl *et al.*, 1992). Several genes which encode the catalytic subunits of protein serine–threonine phosphatases have been characterized in *S. cerevisiae* and *S. pombe* with a clear involvement in cell division cycle regulation and cell shape (Sneddon and Stark, 1991; Shimanuki *et al.*, 1993). Calcineurin (phosphoprotein phosphatase 2B, PP2B) is a phosphatase that is regulated by Ca^{2+} and CaM, with several of its substrates known to regulate other phosphatases and kinases *in vitro*, thus implying a role in the mediation of protein phosphorylation in response to fluctuations in [Ca^{2+}]$_c$. It is now known that *S. cerevisiae* possesses a calcineurin-like activity and contains a protein with properties similar to the B subunit of mammalian calcineurin (Cyert and Thorner, 1988; Hiraga *et al.*, 1993). Two genes, *CNA1* and *CNA2* (*CMP1*, *CMP2*), which encode CaM-binding proteins homologous to mammalian calcineurin A subunits have been identified which confirms that *S. cerevisiae* contains a Ca^{2+}-CaM-dependent phosphoprotein phosphatase related to mammalian calcineurin. Further, yeast *MATa* cells lacking these calcineurin A homologues are hypersensitive to pheromone action and impaired in their ability to re-enter the cell cycle after growth arrest in G1 in response to α-factor (Cyert *et al.*, 1991; Cyert and Thorner, 1992). The basic relationships of the deduced structure of the *N. crassa* and *A. niger* PP2B catalytic subunit are similar to those of *S. cerevisiae*, although structural elements of the *N. crassa* protein are more related to those found in mammals. It is possible that filamentous fungi and vertebrates share a common ancestry regarding the catalytic subunit of PP2B. Current data suggest that filamentous fungi have only one gene for this subunit and it is essential for growth. It has been postulated that PP2B is necessary for hyphal extension with involvement in processes such as regulation of Ca^{2+} fluxes and ion currents from the exterior (Kincaid, 1993).

9.7 CONCLUSIONS

Signal transduction is an integral component of virtually all cellular processes in fungi including growth, metabolism, differentiation, zoospore motility and pathogenesis (Hoch and Staples, 1991) yet is still poorly understood at cellular, physiological and molecular levels, especially in filamentous fungi. Although several components of the major eukaryotic signalling pathways have been identified and characterized in fungi, unified models of signal transduction and integration with cellular processes are still largely hypothetical. Much interpretation relies on the extensive information that exists for 'higher' (mammalian) eukaryotic cells although it is debatable whether fungi can be directly compared with such systems. Most fungal progress has depended on the increasing use of *S. cerevisiae* as a model eukaryote (Janssens, 1987) by non-mycologists though extrapolation of findings to filamentous or polymorphic fungi is again debatable. Some problems are clearly methodological. The filamentous growth habit can make culture and biomass separation difficult, compared with unicellular yeasts, and cell disruption methods can be inefficient and dependent on wall structure. Investigations on the role of Ca^{2+} as a second messenger are hampered by difficulties in measuring $[Ca^{2+}]_c$ meaning that much information has been obtained by indirect means, while inositol phosphate metabolism in fungi appears at least as complex as it is in mammalian systems. Several other inositol phosphates besides $Ins1,4,5P_3$ may have a second messenger function, including the modulation of Ca^{2+} fluxes (Berridge and Irvine, 1989). Much work needs to be carried out though future progress in all the areas described in this chapter will undoubtedly depend on the use of *S. cerevisiae*. The significance and relevance of such work to filamentous and polymorphic fungi must, however, be treated with the utmost caution.

REFERENCES

Alexandre, J. and Lassalles, J.-P. (1992) Intracellular Ca^{2+} release by InsP$_3$ in plants and effect of buffers on Ca^{2+} diffusion. *Philosophical Transactions of the Royal Society, London*, **338**, 53–61.

Anraku, Y., Umemoto, N., Hitara, R. and Wada, Y. (1989) Structure and function of the yeast vacuolar membrane proton ATPase. *Journal of Bioenergetics and Biomembranes*, **21**, 589–603.

Anraku, Y., Ohya, Y and Iida, H. (1991) Cell cycle control by calcium and calmodulin in *Saccharomyces cerevisiae*. *Biochimica et Biophysica Acta*, **1093**, 169–77.

Bartelt, D.C., Fidel, S., Farber, L.H. *et al.* (1988) Calmodulin-dependent multifunctional protein kinase in *Aspergillus nidulans*. *Proceedings of the National Academy of Sciences of the USA*, **85**, 3279–83.

Baum, P.R., Furlong, C. and Byers, B. (1986) Yeast gene required for spindle pole body duplication: homology of its product with Ca^{2+}-binding proteins. *Proceedings of the National Academy of Sciences of the USA*, **83**, 5512–16.

Bazzi, M.D. and Nelsestuen, G.L. (1988) Properties of membrane-inserted protein kinase-C. *Biochemistry*, **27**, 7589–93.

Becker, G. and Lester, R. (1980) Biosynthesis of phosphoinositol-containing sphingolipids from phosphatidylinositol by a membrane preparation from *Saccharomyces cerevisiae*. *Journal of Bacteriology*, **142**, 747–54.

Bell, R., Loomis, C. and Hannun, Y. (1988) Protein kinase C regulation by sphingosine/lysosphingolipids. *Cold Spring Harbor Symposia on Quantitative Biology*, **53**, 103–110.

Berridge, M.J. (1983) Rapid accumulation of inositol tris-phosphate reveals that agonists hydrolyse poly-phosphoinositides instead of phosphatidylinositol. *Biochemical Journal*, **212**, 849–58.

Berridge, M.J. (1984) Inositol trisphosphate and diacyl-glycerol as second messengers. *Biochemical Journal*, **220**, 345–60.

Berridge, M.J. (1987) Inositol trisphosphate and diacyl-glycerol: two interacting second messengers. *Annual Review of Biochemistry*, **56**, 159–93.

Berridge, M.J. (1993) Inositol trisphosphate and calcium signalling. *Nature*, **361**, 315–25.

Berridge, M.J. and Irvine, R.F. (1984) Inositol trisphos-phate, a novel second messenger in cellular signal transduction. *Nature*, **312**, 315–21.

Berridge, M.J. and Irvine, R.F. (1989) Inositol phos-phates and cell signalling. *Nature*, **341**, 197–205.

Bertyl, A. and Slayman, C.L. (1990) Cation-selective channels in the vacuolar membrane of *Saccharomyces*: dependence on calcium, redox state, and voltage. *Proceedings of the National Academy of Sciences of the USA*, **87**, 7824–8.

Beullens, M., Mbonyi, K., Geerts, L. *et al.* (1988) Studies on the mechanism of the glucose-induced cAMP-signal in glycolysis- and glucose repression mutants of the yeast *Saccharomyces cerevisiae*. *European Journal of Biochemistry*, **172**, 227–31.

Bourret, J.A. (1986). Evidence that a glucose-mediated rise in cyclic-AMP triggers germination of *Pilobolus longipes* spores. *Experimental Mycology*, **10**, 60–6.

Bourret, J.A. and Smith, C.M. (1987) Cyclic AMP regulation of glucose transport in germinating *Pilobolus longipes* spores. *Archives of Microbiology*, **148**, 29–33.

Boutry, M., Foury, F., Goffeau, A. (1977) Energy-dependent uptake of calcium by the yeast *Schizo-saccharomyces pombe*. *Biochimica et Biophysica Acta*, **464**, 602–12.

Brewster, J.L., De Valoir, T., Dwyer, N.D. *et al.* (1993) An osmosensing signal transduction pathway in yeast. *Science*, **259**, 1760–3.

Broach, J.R. (1991) RAS genes in *Saccharomyces cerevisiae*: signal transduction in search of a pathway. *Trends in Genetics*, **7**, 28–33.

Broach, J.R. and Deschenes, R.J. (1990) The function of *RAS* genes in *Saccharomyces cerevisiae*. *Advances in Cancer Research*, **54**, 79–139.

Brockerhoff, S.E. and Davis, T.N. (1992) Calmodulin concentrates at regions of cell growth in *Saccharomyces cerevisiae*. *Journal of Cell Biology*, **118**, 619–29.

Brunton, A.H. and Gadd, G.M. (1989) The effect of exoogenously-supplied nucleosides and nucleotides and the involvement of adenosine 3':5'-cyclic mono-phosphate (cyclic AMP) in the yeast-mycelium transi-tion of *Ceratocystis* (=*Ophiostoma*) *ulmi*. *FEMS Micro-biology Letters*, **60**, 49–54.

Brunton, A.H. and Gadd, G.M. (1991) Evidence for an inositol lipid signal pathway in the yeast-mycelium transition of *Ophiostoma* (*Ceratocystis*) *ulmi*, the Dutch elm disease fungus. *Mycological Research*, **95**, 484–91.

Busa, W.B. and Nuccitelli, R. (1984) Metabolic regulation via intracellular pH. *American Journal of Physiology*, **246**, R409–38.

Cantore, M.L., Galvagno, M.A. and Passeron, S. (1980) Variations in the levels of cyclic adenosine 3':5'-monophosphate and in the activities of adenylate cyclase and cyclic adenosine 3':5'-monophosphate phosphodiesterase during aerobic morphogenesis of *Mucor rouxii*. *Archives of Biochemistry and Biophysics*, **199**, 312–20.

Cantore, M.L., Galvagno, M.A. and Passeron, S. (1983) cAMP levels and *in situ* measurement of adenylate cyclase and cAMP phosphodiesterase activities during yeast-to-hyphae transition in the dimorphic fungus *Mucor rouxii*. *Cell Biology International Reports*, **7**, 947–54.

Carafoli, E. (1987) Intracellular calcium homeostasis. *Annual Review of Biochemistry*, **56**, 395–433.

Carafoli, E. and Penniston, J.T. (1985) The calcium signal. *Scientific American*, **253**, 70–116.

Carafoli, E., Balcavage, W.X., Lehninger, A.L. and Mattoon, J.R. (1970) Ca^{2+} metabolism in yeast cells and mitochondria. *Biochimica et Biophysica Acta*, **205**, 18–26.

Caspani, G., Tortora, P., Hanozet, G.M. and Guerritore, A. (1985) Glucose-stimulated cAMP increase may be mediated by intracellular acidification in *Saccharomyces cerevisiae*. *FEBS Letters*, **186**, 75–9.

Chattaway, F.W., Wheeler, P.R. and O'Reilly, J. (1981) Involvement of adenosine 3':5'-cyclic monophosphate in the germination of blastospores of *Candida albicans*. *Journal of General Microbiology*, **123**, 233–40.

Cheung, W.Y. (1980) Calmodulin plays a pivotal role in cellular regulation. *Science*, **207**, 19–27.

Cohen, P. (1992) Signal integration at the level of protein kinases, protein phosphatases and their substrates. *Trends in Biochemical Sciences*, **17**, 408–13.

Cohen, P. and Hardie, D.G. (1991) The actions of cyclic AMP on biosynthetic processes are mediated in-directly by cyclic AMP-dependent protein kinase. *Biochimica et Biophysica Acta*, **1094**, 292–9.

Colbran, R.J., Smith, R.K., Schworer, C.M. *et al.* (1989a). Regulatory domain of calcium calmodulin-dependent protein kinase-II-mechanism of inhibition and regula-tion by phosphorylation. *Journal of Biological Chemistry*, **264**, 4800–4.

Colbran, R.J., Schworer, C.M., Hashimoto, Y. *et al.* (1989b) Calcium calmodulin-dependent protein kinase II. *Biochemical Journal*, **258**, 313–25.

Cooper, L.A., Edwards, S.W. and Gadd, G.M. (1985) Involvement of adenosine 3':5'-cyclic monophosphate in the yeast–mycelium transition of *Aureobasidium pullulans*. *Journal of General Microbiology*, **131**, 1589–93.

Cornelius, G. and Nakashima, H. (1987) Vacuoles play a decisive role in calcium homeostasis in *Neurospora crassa*. *Journal of General Microbiology*, **133**, 2341–7.

Cornelius, G., Gebauer, G. and Techel, D. (1989) Inositol trisphosphate induces calcium release from *Neurospora crassa* vacuoles. *Biochemical and Biophysical Research Communications*, **162**, 852–6.

Corzo, A. and Sanders, D. (1992) Inhibition of Ca^{2+} uptake in *Neurospora crassa* by La^{3+}: mechanistic study. *Journal of General Microbiology*, **138**, 1791–5.

Crabbe, J.M., Iser, J.R. and Grant, B.R. (1988) Calcium control of differentiation in *Phytophthora palmivora*. *Archives of Microbiology*, **149**, 565–71.

Cyert, M.S. and Thorner, J. (1988) Identification of a calcineurin-like activity in the yeast *Saccharomyces cerevisiae*. *Journal of Cell Biology*, **107**, 841a.

Cyert, M.S. and Thorner, J. (1992) Regulating subunit (*CNB1* gene product) of yeast Ca^{2+} calmodulin-dependent phosphatases is required for adaptation to pheromone. *Molecular and Cellular Biology*, **12**, 3460–9.

Cyert, M.S., Kunisawa, R., Kaim, D. and Thorner, J. (1991) Yeast has homologs (*CNA1* and *CNA2* gene products) of mammalian calcineurin, a calmodulin-regulated phosphoprotein phosphatase. *Proceedings of the National Academy of Sciences of the USA*, **88**, 7376–80.

Davis, T.N. (1992) A temperature-sensitive calmodulin mutant loses viability during mitosis. *Journal of Cell Biology*, **118**, 607–17.

Davis, T.N., Urdea, M.S., Masiarz, F.R. and Thorner, J. (1986) Isolation of the yeast calmodulin gene: calmodulin is an essential protein. *Cell*, **47**, 423–31.

Deacon, J.W. and Donaldson, S.P. (1993) Molecular recognition in the homing responses of zoosporic fungi, with special reference to *Pythium* and *Phytophthora*. *Mycological Research*, **97**, 1153–71.

De Vendittis, E., Vitelli, A., Zahn, R. and Fasano, O. (1986) Suppression of defective RAS1 and RAS2 functions in yeast by an adenylate cyclase activated by a single amino acid change. *EMBO Journal*, **5**, 3657–63.

Dewerchin, M.A. and Van Laere, A.J. (1984) Trehalase activity and cyclic AMP content during early development of *Mucor rouxii* spores. *Journal of Bacteriology*, **158**, 575–9.

Dicker, W. and Turian, G. (1990) Calcium deficiencies and apical hyperbranching in wild-type and 'frost' and 'spray' morphological mutants of *Neurospora crassa*. *Journal of General Microbiology*, **136**, 1413–20.

Donaldson, S.P. and Deacon, J.W. (1993) Changes in motility of *Pythium* zoospores induced by calcium and calcium-modulating drugs. *Mycological Research*, **97**, 877–83.

Downes, P. and Michell, R.H. (1982) Phosphatidylinositol 4-phosphate and phosphatidylinositol 4,5-bisphosphate: lipids in search of a function. *Cell Calcium*, **3**, 467–502.

Dumont, J.E., Jauniaux, J.C. and Roger, P.P. (1989) The cyclic AMP-mediated stimulation of cell proliferation. *Trends in Biotechnology*, **14**, 67–71.

Eckstein, H. (1988) 3':5'-cyclic GMP in the yeast *Saccharomyces cerevisiae* at different metabolic conditions. *FEBS Letters*, **232**, 121–4.

Egidy, G.A., Paveto, M.C., Passeron, S. and Galvagno, M.A. (1989) Relationship between cyclic adenosine 3':5'-monophosphate and germination in *Candida albicans*. *Experimental Mycology*, **13**, 428–32.

Eilam, Y. (1982) The effect of monovalent cations on calcium efflux in yeast. *Biochimica et Biophysica Acta* **687**, 8–16.

Eilam, Y. and Chernichovsky, D. (1987), Uptake of Ca^{2+} driven by the membrane potential in energy-depleted yeast cells. *Journal of General Microbiology*, **133**, 1641–9.

Eilam, Y. and Othman, M. (1990) Activation of $Ca^{2+}+$ influx by metabolic substrates in *Saccharomyces cerevisiae*: role of membrane potential and cellular ATP levels. *Journal of General Microbiology*, **136**, 861–6.

Eilam, Y., Lavi. H. and Grossowicz, N. (1985) Cytoplasmic Ca^{2+} homeostasis maintained by a vacuolar Ca^{2+} transport system in the yeast *Saccharomyces cerevisiae*. *Journal of General Microbiology*, **31**, 623–9.

Eilam, Y., Othman, M. and Halachmi, D. (1990) Transient increase in Ca^{2+} influx in *Saccharomyces cerevisiae* in response to glucose: effects of intracellular acidification and cAMP levels. *Journal of General Microbiology*, **136**, 2537–43.

Epstein, L., Staples, R.C. and Hoch, H.C. (1989) Cyclic AMP, cyclic GMP, and bean rust uredospore germlings. *Experimental Mycology*, **13**, 100–4.

Eraso, P., Mazon, M.J. and Gancedo, J.M. (1987) Internal acidification and cAMP increase are not correlated in *Saccharomyces cerevisiae*. *European Journal of Biochemistry*, **165**, 671–74.

Farkaš, V., Sulová, Z. and Lehotský, J. (1985) Effect of light on the concentration of adenine nucleotides in *Trichoderma viride*. *Journal of General Microbiology*, **131**, 317–20.

Favre, B. and Turian, G. (1987) Identification of a calcium- and phospholipid-dependent protein kinase (protein kinase C) in *Neurospora crassa*. *Plant Science*, **49**, 15–21.

Field, J., Vojtek, A., Ballester, R. *et al.* (1990) Cloning and characterization of CAP, the *S. cerevisiae* gene encoding the 70 kd adenylyl cyclase-associated protein. *Cell*, **61**, 319–27.

Fields, F.O. and Thorner, J. (1989) Biochemical and genetic analysis of protein kinase C (PK-C) function in the yeast *Saccharomyces cerevisiae*. *Genetics Society of America, Yeast Genetics and Molecular Biology Meeting, Atlanta, Georgia*, 27 June–1 July 1989, p. 22.

Flanagan, C.A. and Thorner, J. (1992) Purification and characterization of a soluble phosphatidylinositol 4-kinase from the yeast *Saccharomyces cerevisiae*. *Journal of Biological Chemistry*, **267**, 24117–25.

François, J., Villaneuva, M.E. and Hers, H.G. (1988) The control of glycogen metabolism in yeast. I. Interconversion *in vivo* of glycogen synthase and glycogen phosphorylase induced by glucose, a nitrogen source or uncouplers. *European Journal of Biochemistry*, **174**, 551–9.

Frascotti, G., Baroni, D. and Martegani, E. (1990) The glucose-induced polyphosphoinositides turnover in *Saccharomyces cerevisiae* is not dependent on the CDC25-RAS mediated signal transduction pathway. *FEBS Letters*, **274**, 19–22.

Gadd, G.M. (1988) Carbon nutrition and metabolism, in *Physiology of Industrial Fungi*, (ed. D.R. Berry), Blackwell Scientific, Oxford, pp. 21–57.

Gadd, G.M. and Brunton, A.H. (1992) Calcium involvement in dimorphism of *Ophiostoma ulmi*, the Dutch elm disease fungus, and characterization of calcium uptake by yeast cells and germ tubes. *Journal of General Microbiology*, **138**, 1561–71.

Gancedo, C. and Serrano, R. (1989) Energy-yielding metabolism, in *The Yeasts*, Vol. 3, *Metabolism and Physiology of Yeasts*, (eds A.H. Rose and J.S. Harrison) Academic Press, London, pp. 205–59.

Gancedo, J.M., Mazon, M.J. and Eraso, P. (1985) Biological roles of cyclic AMP: similarities and differences between organisms. *Trends in Biochemical Sciences*, **10**, 210–12.

Garrill, A., Lew, R.R. and Brent Heath, I. (1992) Stretch-activated Ca^{2+} and Ca^{2+}-activated K^+ channels in the hyphal tip plasma membrane of the oomycete *Saprolegnia ferax*. *Journal of Cell Science*, **101**, 721–30.

Garrill, A., Jackson, S.L., Lew, R.R. and Brent Heath, I. (1993) Ion channel activity and tip growth: tip localised stretch-activated channels generate an essential Ca^{2+} gradient in the oomycete *Saprolegnia ferax*. *European Journal of Cell Biology*, **60**, 358–65.

Geiser, J.R., Van Tuinen, D., Brockerhoff, S.E. *et al.* (1991) Can calmodulin function without binding calcium? *Cell*, **65**, 949–59.

Ghislain, M., Schlesser, A. and Goffeau, A. (1987) Mutation of a conserved glycine residue modifies the vanadate sensitivity of the plasma-membrane H^+-ATPase from *Schizosaccharomyces pombe*. *Journal of Biological Chemistry*, **262**, 17549–55.

Ghislain, M., Goffean, A., Halachmi, D. and Eilam, Y. (1990) Calcium homeostasis and transport are affected by disruption of *eta3*, a novel gene encoding Ca^{2+}-ATPase in *Schizosaccharomyces pombe*. *Journal of Biological Chemistry*, **265**, 18400–7.

Gibbs, G.B. and Marshall, M.S. (1989) The *ras* oncogene – an important regulatory element in lower eucaryotic organisms. *Microbiological Reviews*, **53**, 171–85.

Gimeno, C.J., Ljungdahl, P.O., Styles, C.A. and Fink, G.R. (1992) Unipolar cell divisions in the yeast *S. cerevisiae* lead to filamentous growth: regulation by starvation and *RAS*. *Cell*, **68**, 1077–90.

Goffeau, A., Ghislain, M., Navarre, C. *et al.* (1990) Novel transport ATPases in yeast. *Biochimica et Biophysica Acta*, **1018**, 200–2.

Gooday, G.W. and Adams, D.J. (1993) Sex hormones and fungi. *Advances in Microbial Physiology*, **34**, 69–145.

Grešík, M., Nadežda, K. and Farkaš, V. (1988) Membrane potential, ATP and cyclic AMP changes induced by light in *Trichoderma viride*. *Experimental Mycology*, **12**, 295–301.

Griffith, J.M., Iser, J.R. and Grant, B.R. (1988) Calcium control of differentiation in *Phytophthora palmivora*. *Archives of Microbiology*, **149**, 565–71.

Gupta, P. and Prasad, R. (1993) Levels of plasma membrane H^+-ATPase do not change during growth and morphogenesis of *Candida albicans*. *FEMS Microbiology Letters*, **106**, 165–70.

Gustin, M.C., Martinac, B., Saimi, Y. *et al.* (1986) Ion channels in yeast. *Science*, **233**, 1195–7.

Gustin, M.C., Zhou, X.-L., Martinac, B. and Kung, C. (1988) A mechanosensitive ion channel in the yeast plasma membrane. *Science*, **242**, 762–5.

Halachmi, D. and Eilam, Y. (1989) Cytosolic and vacuolar Ca^{2+} concentrations in yeast-cells measured with the Ca^{2+}-sensitive fluorescence dye Indo-1. *FEBS Letters*, **256**, 55–61.

Halachmi, D. and Eilam, Y. (1993) Calcium homeostasis in yeast cells exposed to high concentrations of calcium. *FEBS Letters*, **316**, 73–8.

Halachmi, D., Ghislain, M. and Eilam, Y. (1992) An intracellular ATP-dependent calcium pump within the yeast *Schizosaccharomyces pombe*, encoded by the gene *cta3*. *European Journal of Biochemistry*, **207**, 1003–8.

Hanson, B.A. (1984) Role of inositol-containing sphingolipids in *Saccharomyces cerevisiae* during inositol starvation. *Journal of Bacteriology*, **159**, 837–42.

Hanson, B.A. (1991) The effects of lithium on the phosphoinositides and inositol phosphates of *Neurospora crassa*. *Experimental Mycology*, **15**, 76–90.

Hashiba, T. (1982) Sclerotial morphogenesis in the rice sheath blight fungus (*Rhizoctonia solani*). *Bulletin of the Hokuriku National Agricultural Experimental Station (Japan)*, **24**, 29–83.

Hashiba, T. and Ishikawa, T. (1978) Effect of adenosine 3',5'-cyclic monophosphate on induction of sclerotia in *Rhizoctonia solani*. *Phytopathology*, **65**, 159–62.

Hiraga, K., Suzuki, K., Tsuchiya, E. and Miyakawa, T. (1993) Identification and characterization of nuclear calmodulin-binding proteins of *Saccharomyces cerevisiae*. *Biochimica et Biophysica Acta*, **1177**, 25–30.

Hirimburegama, K., Durnez, P., Keleman, J. *et al.* Nutrient-induced activation of trehalase in nutrient-starved cells of the yeast *Saccharomyces cerevisiae*: cAMP is not involved as second messenger. *Journal of General Microbiology*, **138**, 2035–43.

Hoch, H.C. and Staples, R.C. (1984) Evidence that cyclic AMP initiates nuclear division and infection structure formation in the bean rust fungus *Uromyces phaseoli*. *Experimental Mycology*, **8**, 37–46.

Hoch, H.C. and Staples, R.C. (1991) Signaling for infection structure formation in fungi, in *The Fungal Spore and Disease Initiation in Plants and Animals* (eds G.T. Cole and H.C. Hoch), Plenum Press, New York, pp. 25–46.

Holmes, A.R., Cannon, R.D. and Shepherd, M.G. (1991) Effect of calcium ion uptake on *Candida albicans* morphology. *FEMS Microbiology Letters*, **77**, 187–94.

Holzer, H. (1984) Mechanism and function of reversible phosphorylation of fructose-1,6-bisphosphate in yeast, in *Molecular Aspects of Cellular Regulation*, vol. 3 (ed. P. Cohen), Elsevier, Amsterdam, pp. 143–54.

Hoshino, T., Mizutani, A., Shimizu, S. *et al* (1991) Calcium ion regulates the release of lipase of *Fusarium oxysporum*. *Journal of Biochemistry*, **110**, 457–61.

Hubbard, M., Bradley, N., Sullivan, P. *et al.* (1982) Evidence of the occurrence of calmodulin in the yeasts *Candida albicans* and *Saccharomyces cerevisiae*. *FEBS Letters*, **317**, 85–8.

Iida, H., Yagaura, Y. and Anraku, Y. (1990a) Essential role for induced Ca^{2+} influx followed by $[Ca^2]_i$ rise in maintaining viability of yeast cells late in the mating pheromone response pathway. *Journal of Biological Chemistry*, **265**, 13391–9.

Iida, H., Sakaguchi, S., Yagawa, Y. and Anraku, Y. (1990b) Cell cycle control by Ca^{2+} in *Saccharomyces cerevisiae*. *Journal of Biological Chemistry*, **265**, 21216–22.

Jackson, S.L. and Heath, I.B. (1989) Effects of exogenous calcium ions on tip growth, intracellular Ca^{2+} concentration, and actin arrays in hyphae of the fungus *Saprolegnia ferax*. *Experimental Mycology*, **13**, 1–12.

Jackson, S.L. and Heath, I.B. (1993) Roles of calcium ions in hyphal tip growth. *Microbiological Reviews*, **57**, 367–82.

Janssens, P.M.W. (1987) Did vertebrate signal transduction mechanisms originate in eukaryotic microbes? *Trends in Biochemical Sciences*, **12**, 456–9.

Johannes, E., Brosnan, J.M. and Sanders, D. (1991) Calcium channels and signal transduction in plant cells. *BioEssays*, **13**, 331–6.

Johannes, E., Brosnan, J.M. and Sanders, D. (1992) Calcium channels in the vacuolar membrane of plants: multiple pathways for intracellular calcium mobilization. *Philosophical Transactions of the Royal Society, London*, **338**, 105–12.

Jones, B.E. and Bu'Lock, J.D. (1977) The effect of N^6, O^2, -dibutyryl adenosine-3′,5′-cyclic monophosphate on morphogenesis in Mucorales. *Journal of General Microbiology*, **103**, 29–36.

Jones, R.P. and Gadd, G.M. (1990) Ionic nutrition of yeast – the physiological mechanisms involved and applications for biotechnology. *Enzyme and Microbial Technology*, **12**, 402–18.

Kaibuchi, K., Miyajima, A., Arai, K.-I. and Matsumoto, K. (1986) Possible involvement of *RAS*-encoded proteins in glucose-enduced inositol phospholipid turnover in *Saccharomyces cerevisiae*. *Proceedings of the National Academy of Sciences of the USA*, **83**, 8172–6.

Kamihari, T. and Omi, K. (1989) Increase in cyclic AMP content with enhanced phosphatidylinositol turnover in the cells of *Candida tropicalis* during mycelial growth caused by ethanol. *Yeast*, **5**, 437–40.

Kato, H., Uno, I., Ishikawa, T. and Takenawa, T. (1989) Activation of phosphatidylinositol kinase and phosphatidylinositol-4-phosphate kinase by cAMP in *Saccharomyces cerevisiae*. *Journal of Biological Chemistry*, **264**, 3116–21.

Kaur, S. and Mishra, P. (1991) Dimorphism-associated changes in plasma membrane H^+-ATPase activity of *Candida albicans*. *Archives of Microbiology*, **156**, 412–15.

Kaur, S., Mishra, P. and Prasad, R. (1988) Dimorphism-associated changes in intracellular pH of *Candida albicans*. *Biochimica et Biophysica Acta*, **972**, 277–82.

Kincaid, R. (1993) Calmodulin-dependent protein phosphatases from microorganisms to man. A study in structural conservatism and biological diversity. *Advances in Second Messenger and Phosphoprotein Research*, **27**, 1–23.

Kinney, A.J. and Carman, G.M. (1990) Enzymes of phosphoinositide synthesis in secretory vesicles destined for the plasma membrane in *Saccharomyces cerevisiae*. *Journal of Bacteriology*, **172**, 4115–17.

Kitamoto, K., Yoshizawa, K., Ohsumi, Y. and Anraku, Y. (1988) Mutants of *Saccharomyces cerevisiae* with defective vacuolar function. *Journal of Bacteriology*, **170**, 2687–91.

Klionsky, D.J., Herman, P.K. and Emr, S.D. (1990) The fungal vacuole: composition, function and biogenesis. *Microbiological Reviews*, **54**, 266–92.

Konopka, J.B. and Fields, S. (1992) The pheromone signal pathway in *Saccharomyces cerevisiae*. *Antonie van Leeuwenhoek*, **62**, 95–108.

Kovac, L. (1985) Calcium and *Saccharomyces cerevisiae*. *Biochimica et Biophysica Acta*, **840**, 317–23.

Kozak, K.R. and Ross, I.K. (1991) Signal transduction in *Coprinus congregatus* – evidence for the involvement of G-proteins in blue-light photomorphogenesis. *Biochemical and Biophysical Research Communications*, **179**, 1225–31.

Kurjan, J. (1992) Pheromone response in yeast. *Annual Review of Biochemistry*, **61**, 1097–129.

Laccetti, L., Staples, R.C. and Hoch, H.C. (1987) Purification of calmodulin from bean rust uredospores. *Experimental Mycology*, **11**, 231–5.

Lakin-Thomas, P.L. (1993) Evidence against a direct role for inositol phosphate metabolism in the circadian oscillator and the blue-light signal transduction pathway in *Neurospora crassa*. *Biochemical Journal*, **292**, 813–18.

Lakin-Thomas, P.L., Cote, G.G. and Brody, S. (1990) Circadian rhythyms in *Neurospora crassa* – biochemistry and genetics. *Critical Reviews in Microbiology*, **17**, 365–416.

Larsen, A.D. and Sypherd, P.S. (1974) Cyclic adenosine 3′,5′-monophosphate and morphogenesis in *Mucor racemosus*. *Journal of Bacteriology*, **117**, 432–8.

Laychock, S.G. (1989) Coordinate interactions of cyclic nucleotide and phospholipid metabolizing pathways in calcium-dependent cellular processes. *Current Topics in Cellular Regulation*, **30**, 203–42.

Levin, D.E., Fields, F.O., Kunisawa, R. *et al.* (1990) A candidate protein kinase C gene (*PKC1*) is required for the *S. cerevisiae* cell cycle. *Cell*, **62**, 213–24.

Liao, H.H. and Thorner, J. (1981) Adenosine 3′,5′-phosphate phosphodiesterase and pheromone response in the yeast *Saccharomyces cerevisiae*. *Journal of Bacteriology*, **148**, 919–25.

Lichko, L. and Okorokov, L. (1991) Purification and some properties of membrane-bound and soluble pyrophosphatases of yeast vacuoles. *Yeast*, **7**, 805–12.

Liu, Y., Ohki, Y.A., Azuma, Y. *et al.* (1990) Calmodulin and calmodulin-binding proteins of *Rhododsporidium toruloides*, a basidiomycetous yeast. *Journal of General Microbiology*, **136**, 131–6.

Londesborough, J. (1989) Purification of a Ca^{2+}/calmodulin-dependent protein kinase from baker's yeast. *Journal of General Microbiology*, **135**, 3373–83.

Luan, Y., Matsuura, I., Yazawa, M. *et al.* (1987) Yeast calmodulin: structural and functional differences compared with vertebrate calmodulin. *Journal of Biochemistry*, **102**, 1531–7.

Luttrell, B.M. (1993) The biological relevance of the binding of calcium ions by inositol phosphates. *Journal of Biological Chemistry*, **268**, 1521–4.

McGillivray, A.M. and Gow, N.A.R. (1987) The transhyphal electrical current of *Neurospora crassa* is carried principally by protons. *Journal of General Microbiology*, **133**, 2875–81.

Magae, Y. (1993) Purification and characterization of protein kinase C from *Pleurotus ostreatus*. *Journal of General Microbiology*, **139**, 165–9.

Magae, Y. and Magae, J. (1993) Effect of staurosporine on growth and hyphal morphology of *Pleurotus ostreatus*. *Journal of General Microbiology*, **139**, 161–4.

Magalhães, B.P., Wayne, R., Humber, R.A. *et al.* (1991) Calcium-regulation appressorium formation of the entomopathogenic fungus *Zoophthora radicans*. *Protoplasma*, **160**, 77–88.

Matsumoto, K., Uno, I., Toh-e, A. *et al.* (1982) Cyclic AMP may not be involved in catabolic repression in *Saccharomyces cerevisiae*: evidence from mutants capable of utilizing it as an adenine source. *Journal of Bacteriology*, **150**, 277–85.

Matsumoto, K., Uno, I. and Ishikawa, T. (1985) Genetic analysis of the role of cAMP in yeast. *Yeast*, **1**, 15–24.

Mbonyi, K., Beullens, M., Detremerie, K. *et al.* (1988) Requirement of one functional *RAS* gene and inability of an oncogenic *ras* variant to mediate the glucose-induced cyclic AMP signal in the yeast *Saccharomyces cerevisiae*. *Molecular and Cellular Biology*, **8**, 3051–7.

Mbonyi, K., Van Aelst, L., Arguelles, J.C. *et al.* (1990) Glucose-induced hyperaccumulation of cyclic AMP and defective glucose repression in yeast strains with reduced activity of cyclic AMP-dependent protein kinase. *Molecular and Cellular Biology*, **10**, 4518–23.

Means, A.R., Van Berkum, M.F.A., Bagchi, I. *et al.* (1991). Regulatory function of calmodulin. *Pharmacological Therapy*, **50**, 255–70.

Meikle, A.J., Reed, R.H. and Gadd, G.M. (1988) Osmotic adjustment and the accumulation of organic solutes in whole cells and protoplasts of *Saccharomyces cerevisiae*. *Journal of General Microbiology*, **134**, 3049–60.

Michell, R.H. (1975) Inositol phospholipids and cell surface receptor function. *Biochimica et Biophysica Acta*, **415**, 81–147.

Michell, R.H., Kirk, C.J., Jones, L.M. *et al.* (1981) The stimulation of inositol phospholipid metabolism that accompanies calcium mobilization in stimulated cells: defined characteristics and unanswered questions. *Philosophical Transactions of the Royal Society of London, Series B*, **296**, 123–37.

Miller, A.J., Vogg, G. and Sanders, D. (1990) Cytosolic calcium homeostasis in fungi: roles of plasma membrane transport and intracellular sequestration of calcium. *Proceedings of the National Academy of Sciences of the USA*, **87**, 9348–52.

Miyakawa, T., Oka, Y., Tsuchiya, E. and Fukui, S. (1989) *Saccharomyces cerevisiae* protein kinase dependent on Ca^{2+} and calmodulin. *Journal of Bacteriology*, **171**, 1417–22.

Miyamoto, S., Ohya, Y., Ohsumi, Y. and Anraku, Y. (1987) Nucleotide sequence of the *CLS4* (*CDC24*) gene of *Saccharomyces cerevisiae*. *Gene*, **54**, 125–32.

Monk, B.C., Kurtz, M.B., Marrinan, J.A. and Perlin, D.S. (1991) Cloning and characterization of the plasma membrane H$^+$-ATPase from *Candida albicans*. *Journal of Bacteriology*, **173**, 6826–36.

Monk, B.C., Niimi, M. and Shepherd, M.G. (1993) The *Candida albicans* plasma membrane and H$^+$-ATPase during yeast growth and germ tube formation. *Journal of Bacteriology*, **175**, 5566–74.

Morales, M. and Ruiz-Herrera, J. (1989) Subcellular localization of calcium in sporangiospores of *Phycomyces blakesleeanus*. *Archives of Microbiology*, **152**, 468–72.

Moreno, S., Galvagno, M.A. and Passeron, S. (1982) Control of *Mucor rouxii* adenosine 3':3'-monophosphate phosphodiesterase by phosphorylation-dephosphorylation and proteolysis. *Archives of Biochemistry and Biophysics*, **214**, 573–80.

Munder, T. and Kuntzel, H. (1989) Glucose-induced cAMP signaling in *Saccharomyces cerevisiae* is mediated by the CDC25 protein. *FEBS Letters*, **242**, 341–5.

Muthukumar, G. and Nickerson, K.W. (1984) Ca(II)-calmodulin regulation of fungal dimorphism in *Ceratocystis ulmi*. *FEMS Microbiology Letters*, **159**, 309–92.

Muthukumar, G. and Nickerson, K.W. (1985) Ca(II)-calmodulin regulation of morphological commitment in *Ceratocystis ulmi*. *FEMS Microbiology Letters*, **27**, 199–202.

Muthukumar, G., Luby, M.T. and Nickerson, K.W. (1986) Calmodulin activity in yeast and mycelial phases of *Ceratocystis ulmi*. *FEMS Microbiology Letters*, **37**, 313–16.

Muthukumar, G., Nickerson, A.W. and Nickerson, K.W. (1987) Calmodulin levels in yeasts and filamentous fungi. *FEMS Microbiology Letters*, **41**, 253–8.

Neer, E.J. and Clapham, D.E. (1988) Role of G protein subunits in transmembrane signalling. *Nature*, **333**, 129–34.

Niimi, M. (1984) The glucose effect in *Candida albicans*. *Fukuoka Acta Medica*, **75**, 356–65.

Niimi, N., Niimi, K., Tokunaga, J. and Nakayama, H. (1980) Changes in cyclic nucleotide levels and dimorphic transition in *Candida albicans*. *Journal of Bacteriology*, **142**, 1010–14.

Nishizuka, Y. (1984) Protein kinases in signal transduction. *Trends in Biochemical Sciences*, **4**, 163–6.

Nishizuka, Y. (1988) The molecular heterogeneity of protein kinase C and its implications for cellular regulation. *Nature*, **334**, 661–5.

Nishizuka, Y. (1992a) Signal transduction: crosstalk. *Trends in Biochemical Sciences*, **17**, 367.

Nishizuka, Y. (1992b) Intracellular signalling by hydrolysis of phospholipids and activation of protein kinase C. *Science*, **258**, 607–14.

O'Day, D.H. (1990) Calcium as an intracellular messenger in eucaryotic microbes, in *Calcium as an Intracellular Messenger in Eucaryotic Microbes*, (ed. D.H. O'Day), American Society for Microbiology, Washington, DC, pp. 3–13.

Ogita, K., Miyamoto, S., Koide, H. *et al.* (1990) Protein kinase C in *Saccharomyces cerevisiae*: comparison with the mammalian enzyme. *Proceedings of the National Academy of Sciences of the USA*, **87**, 5011–15.

Ohsumi, Y. and Anraku, Y. (1983) Calcium transport driven by a proton motive force in vacuolar membrane vesicles of *Saccharomyces cerevisiae*. *Journal of Biological Chemistry*, **41**, 17–22.

Ohsumi, Y., Kitamoto, K. and Anraku, Y. (1988) Changes induced in the permeability barrier of the yeast plasma membrane by cupric ion. *Journal of Bacteriology*, **170**, 2676–82.

Ohya, Y. and Anraku, Y. (1989) Functional expression of chicken calmodulin in yeast. *Biochemical and Biophysical Research Communications*, **158**, 541–7.

Ohya, Y. and Anraku, Y. (1992) Yeast calmodulin: structural and functional elements essential for the cell cycle. *Cell Calcium*, **13**, 445–55.

Ohya, Y., Ohsumi, Y. and Anraku, Y. (1986a) Isolation and characterization of Ca^{2+}-sensitive mutants of *Saccharomyces cerevisiae*. *Journal of General Microbiology*, **132**, 979–88.

Ohya, Y., Miyamoto, Y., Ohsumi, Y. and Anraku, Y. (1986b) Calcium-sensitive *cls4* mutant of *S. cerevisiae* with a defect in bud formation. *Journal of Bacteriology*, **165**, 28–33.

Ohya, Y., Umemoto, N., Tanida, I. *et al.* (1991a). Calcium-sensitive *c/s* mutants of *Saccharomyces cerevisiae* showing a Pet-phenotype are ascribable to defects of vacuolar membrane H^+-ATPase activity. *Journal of Biological Chemistry*, **266**, 13971–7.

Ohya, Y., Kawasaki, H., Suzuki, K. *et al.* (1991b). Two yeast genes encoding calmodulin-dependent protein kinases. *Journal of Biological Chemistry*, **266**, 12784–94.

Okorokov, L.A. (1985) Main mechanism of ion transport and regulation of ion concentrations in the yeast cytoplasm, in *Environmental Regulation of Microbial Metabolism*, (eds I.S Kulaev, E.A. Dawes and D.W. Tempest), Academic Press, London, pp. 339–49.

Orlowski, M. and Ross, J.R. (1981) Relationship of internal cyclic AMP levels, rates of protein synthesis and *Mucor* dimorphism. *Archives of Microbiology*, **129**, 353–6.

Ortega Perez, R., Van Tuinen, D., Marme, D. *et al.* (1981) Purification and identification of calmodulin from *Neurospora crassa*. *FEBS Letters*, **133**, 205–8.

Pall, M. (1977) Cyclic AMP and the plasma membrane potential in *Neurospora crassa*. *Journal of Biological Chemistry*, **252**, 7146–50.

Pall, M.L. (1981), Adenosine 3′,5′-phosphate in fungi. *Microbiological Reviews*, **45**, 462–480.

Pall, M.L. (1984), Is there a general paradigm of cyclic AMP action in eukaryotes? *Molecular and Cellular Biochemistry*, **58**, 187–91.

Pall, M.L. and Robertson, C.K. (1986) Cyclic AMP control of hierarchical growth pattern of hyphae in *Neurospora crassa*. *Experimental Mycology*, **10**, 161–5.

Paranjape, V. and Datta, A. (1990) Role of calcium and calmodulin in morphogenesis of *Candida albicans*, in *Calcium as an Intracellular Messenger in Eukaryotic Cells*, (ed. D.H. O'Day), American Society for Microbiology, Washington DC, pp. 362–74.

Paranjape, V., Roy, B.G. and Datta, A. (1990) Involvement of calcium, calmodulin and protein phosphorylation in morphogenesis of *Candida albicans*. *Journal of General Microbiology*, **136**, 2149–54.

dos Passos, J.B., Vanhalewyn, M., Brandao, R.L. *et al.* (1992) Glucose-induced activation of plasma membrane H^+-ATPase in mutants of the yeast *Saccharomyces cerevisiae* affected in cAMP metabolism, cAMP-dependent protein phosphorylation and the initiation of glycolysis. *Biochimica et Biophysica Acta*, **1136**, 57–67.

Paveto, C., Epstein, A. and Passeron, S. (1975) Studies on cyclic adenosine 3′,5′-monophosphate levels, adenylate cyclase and phosphodiesterase activities in the dimorphic fungus *Mucor rouxii*. *Archives of Biochemistry and Biophysics*, **169**, 449–57.

Pietrobon, D., Di Virgilio, F.D. and Pozzan, T. (1990) Structural and functional aspects of calcium homeostasis in eukaryotic cells. *European Journal of Biochemistry*, **193**, 599–62.

Pitt, D. and Ugalde, U.O. (1984) Calcium in fungi. *Plant, Cell and Environment*, **7**, 467–75.

Pitt, D. and Kaile, A. (1990) Transduction of the calcium signal with special reference to Ca^{2+} induced conidiation in *Penicillium notatum*, in *Biochemistry of Cell Walls and Membranes of Fungi*, (eds, P.J. Kuhn, A.P.J. Trinci, M. Jung *et al.*) Springer Verlag, Berlin, pp. 283–98.

Powers, P.A. and Pall, M.L. (1980) Cyclic AMP-dependent protein kinase of *Neurospora crassa*. *Biochemical and Biophysical Research Communications*, **95**, 701–6.

Pringle, J.R. and Hartwell, L.H. (1981) The *Saccharomyces cerevisiae* cell cycle, in, *The Molecular Biology of the Yeast Saccharomyce, Metabolism and Gene Expression* (eds J.N. Strathern, E.W. Jones and J.R. Broach), Cold Spring Harbour Laboratory, Cold Spring Harbor, pp. 97–142.

Prior, S.L., Cunnliffe, B.W., Robson, G.D. and Trinci, A.P.J. (1993) Multiple isomers of phosphatidyl inositol monophosphate and inositol bis- and trisphosphates from filamentous fungi. *FEMS Microbiology Letters*, **110**, 147–52.

Rasmussen, H. and Rasmussen, J.E. (1990) Calcium as intracellular messenger: from simplicity to complexity. *Current Topics in Cellular Regulation*, **31**, 1–109.

Rea, P.A., Kim, Y., Sarafian, V. *et al.* (1992) Vacuolar H$^+$-translocating pyrophosphatases – a new category of ion translocase. *Trends in Biochemical Sciences*, **17**, 348–53.

Reissig, J.L. and Kinney, S.G. (1983) Calcium as a branching signal in *Neurospora crassa*. *Journal of Bacteriology*, **154**, 1397–402.

Rivera-Rodriguez, N. and Rodriguez-del Valle, N. (1992) Effects of calcium ions on the germination of *Sporothrix schenckii* conidia. *Journal of Medical and Veterinary Mycology*, **30**, 185–95.

Robson, G.D., Wiebe, M.G. and Trinci, A.P.J. (1991a), Involvement of Ca^{2+} in the regulation of hyphal extension and branching in *Fusarium graminearum* A3/5. *Experimental Mycology*, **15**, 263–72.

Robson, G.D., Wiebe, M.G. and Trinci, A.P.J. (1991b) Low calcium concentrations induce increased branching in *Fusarium graminearum*. *Mycological Research*, **95**, 561–5.

Robson, G.D., Wiebe, M.G. and Trinci, A.P.J. (1991c) Exogenous cAMP and cGMP modulate branching in *Fusarium graminearum*. *Journal of General Microbiology*, **137**, 963–9.

Robson, G.D., Trinci, A.P.J., Wiebe, M.G. and Best, L.C. (1991d) Phosphatidylinositol 4,5-bisphosphate (PIP$_2$) is present in *Fusarium graminearum*. *Mycological Research*, **95**, 1082–4.

Rose, M.D. and Vallen, E.A. (1991) Acid loops fail the acid test. *Cell*, **65**, 919–20.

Rosenberg, G.B. and Pall, M.L. (1978) Cyclic AMP and cyclic GMP in germinating conidia of *Neurospora crassa*. *Archives of Microbiology*, **118**, 87–90.

Rosenberg, G.B. and Pall, M.L. (1979) Properties of two cyclic nucleotide deficient mutants of *Neurospora crassa*. *Journal of Bacteriology*, **137**, 1140–4.

Rosenberg, G.B. and Pall, M.L. (1983) Characterization of an ATP-Mg^{2+} dependent guanine nucleotide-stimulated adenylate cyclase from *Neurospora crassa*. *Archives of Biochemistry and Biophysics*, **221**, 243–53.

Rudolph, H.K., Antebi, A., Fink, G.R. *et al.* (1989) The yeast secretory pathway is perturbed by mutations in PMR1, a member of a Ca^{2+}-ATPase family. *Cell*, **58**, 133–45.

Sabie, F.T. and Gadd, G.M. (1989) Involvement of a Ca^{2+}-calmodulin interaction in the yeast–mycelial transition of *Candida albicans*. *Mycopathologia*, **108**, 47–54.

Sabie, F.T. and Gadd, G.M. (1992) Effect of nucleosides and nucleotides and the relationship between cellular adenosine 3':5'-cyclic monophosphate (cyclic AMP) and germ tube formation in *Candida albicans*. *Mycopathologia*, **119**, 147–56.

Sanders, D. (1990) Kinetic modelling of plant and fungal membrane transport systems. *Annual Review of Plant Physiology and Plant Molecular Biology*, **41**, 77–107.

Sanders, D., Miller, A.J., Blackford, S. *et al.* (1990) Cytosolic free calcium homeostasis in plants. *Current Topics in Plant Biochemistry and Physiology*, **9**, 20–37.

Saporito, S.M. and Sypherd, P.S. (1991) The isolation and characterization of a calmodulin-encoding gene (*CMD1*) from the dimorphic fungus *Candida albicans*. *Gene*, **106**, 43–9.

Sass, P., Field, J., Nikawa, J. *et al.* (1986) Cloning and characterization of the high-affinity cAMP phosphodiesterase of *S. cerevisiae*. *Proceedings of National Academy of Sciences of the USA*, **83**, 9303–7.

Schmid, J. and Harold, F.M. (1988) Dual roles for calcium ions in apical growth of *Neurospora crassa*. *Journal of General Microbiology*, **134**, 2623–31.

Schmitt, H.D., Wagner, P., Pfaff, E. and Gallwitz, D. (1986) The *ras*-related *YPT1* gene product in yeast: a GTP-binding protein that might be involved in microtubule organization. *Cell*, **47**, 401–12.

Schomerus, C. and Kuntzel, H. (1992) *CDC25*-dependent induction of inositol 1,4,5-trisphosphate and diacylglycerol in *Saccharomyces cerevisiae* by nitrogen. *FEBS Letters*, **307**, 249–52.

Scott, W.A. and Solomon, B. (1975) Adenosine 3',5'-cyclic monophosphate and morphology in *Neurospora crassa*: drug-induced alterations. *Journal of Bacteriology*, **122**, 454–63.

Sharada, K., Ikegami, H. and Hyakumachi, M. (1992) 2;4-D induced c-AMP mediated, sclerotial formation in *Rhizoctonia solani*. *Mycological Research*, **96**, 863–6.

Shaw, N.M. and Harding, R.W. (1983) Calcium inhibition of a heat-stable cyclic nucleotide phosphodiesterase from *Neurospora crassa*. *FEBS Letters*, **152**, 295–9.

Shaw, N.M. and Harding, R.W. (1987) Intracellular and extracellular cyclic nucleotides in wild-type and white collar mutant strains of *Neurospora crassa*. Temperature dependent efflux of cyclic AMP from mycelia. *Plant Physiology*, **83**, 377–83.

Shimanuki, M., Kinoshita, N., Ohruka, H. *et al.* (1993) Isolation and characterization of the fission yeast protein phosphatase gene *ppe1$^+$* involved in cell shape control and mitosis. *Molecular Biology of the Cell*, **4**, 303–13.

Slayman, C.L., Kaminski, P. and Stetson, D. (1990) Structure and function of fungal plasma-membrane ATPases, in *Biochemistry of Cell Walls and Membranes in Fungi*, (eds P.J. Kuhn, A.P.J. Trinci, M.J. Jung, *et al.*), Springer-Verlag, Berlin, pp. 299–316.

Sneddon, A.A. and Stark, M.J.R. (1991) Yeast-protein serine-threonine phosphatase genes and cell division cycle control. *Advances in Protein Phosphatases*, **6**, 307–30.

Sokolovsky, V.S. and Kritsky, M.S. (1985) Photoregulation of cyclic AMP phosphodiesterase activity of *Neurospora crassa*. *Biochemistry* (Russian), **282**, 1017–20.

St Leger, R.J., Roberts, D.W. and Staples, R.C. (1989) Calcium- and calmodulin-mediated protein synethesis and protein phosphorylation during germination, growth and protease production by *Metarhizium anisopliae*. *Journal of General Microbiology*, **135**, 2141–54.

St Leger, R.J., Butt, T.M., Staples, R.C. and Roberts, D.W. (1990) Second messenger involvement in differentiation of the entomopathogenic fungus *Metarhizium anisopliae*. *Journal of General Microbiology*, **136**, 1779–89.

Steele, G.C. and Trinci, A.P.J. (1975) Morphology and growth kinetics of differentiated and undifferentiated mycelia of *Neurospora crassa*. *Journal of General Microbiology*, **91**, 362–8.

Stewart, E., Gow, N.A.R. and Bowen, D.V. (1988) Cytoplasmic alkalinization during germ tube formation in *Candida albicans*. *Journal of General Microbiology*, **134**, 1079–87.

Stewart, E., Hawser, S. and Gow, N.A.R. (1989) Changes in internal and external pH accompanying growth of *Candida albicans*: studies of non-dimorphic variants. *Archives of Microbiology*, **151**, 149–53.

Stirling, D.A., Petrie, A., Pulford, D.J. *et al.* (1992) Protein A – calmodulin fusions: a novel approach for investigating calmodulin function in yeast. *Molecular Microbiology*, **6**, 703–13.

Styrkársdóttir, U., Egel, R. and Nielsen, O. (1992) Functional conservation between *Schizosaccharomyces pombe ste8* and *Saccharomyces cerevisiae STE11* protein kinases in yeast signal transduction. *Molecular and General Genetics*, **235**, 122–30.

Suoranta, K. and Londesborough, J. (1984) Purification of intact and nicked forms of a zinc-containing, Mg^{2+}-dependent, low K_m cyclic AMP phosphodiesterase from baker's yeast. *Journal of Biological Chemistry*, **259**, 6964–71.

Tachikawa, T., Miyakawa, T., Tsuchiya, E. and Fukiu, S. (1987) A rapid and transient increase of cellular Ca^{2+} in response to mating pheromone in *Saccharomyces cerevisiae*. *Agricultural and Biological Chemistry*, **51**, 1209–10.

Takeda, T. and Yamamoto, M. (1987) Analysis and *in vivo* disruption of the gene coding of calmodulin in *Schizosaccharomyces pombe*. *Proceedings of the National Academy of Sciences of the USA*, **84**, 3580–4.

Terenzi, H.F., Flawia, M.M., Tellez-Inon, M.T. and Torres, H.N. (1976) Control of *Neurospora crassa* morphology by cyclic adenosine 3',5'-monophosphate and dibutyryl cyclic adenosine 3',5'-monophosphate. *Journal of Bacteriology*, **126**, 91–9.

Terenzi, H.F., Jorge, J.A., Roselino, J.E. and Migliorini, R.H. (1979) Adenylyl cyclase deficient *cr-l* (crisp) mutant of *Neurospora crassa*: cyclic AMP-dependent nutritional deficiencies. *Archives of Microbiology*, **123**, 251–8.

Terenzi, H., Terenzi, H.F. and Jorge, J.A. (1992) Effect of cyclic AMP on invertase activity in *Neurospora crassa*. *Journal of General Microbiology*, **138**, 2433–9.

Thevelein, J.M. (1984) Regulation of trehalose mobilization in fungi. *Microbiological Reviews*, **48**, 42–59.

Thevelein, J.M. (1988) Regulation of trehalase activity by phosphorylation–dephosphorylation during developmental transitions in fungi. *Experimental Mycology*, **12**, 1–12.

Thevelein, J.M. (1991) Fermentable sugars and intracellular acidification as specific activators of the RAS adenylate cyclase signalling pathway in yeast – the relationship to nutrient induced cell cycle control. *Molecular Microbiology*, **5**, 1301–7.

Thevelein, J.M. (1992) The RAS-adenylate cyclase pathway and cell cycle control in *Saccharomyces cerevisiae*. *Antonie van Leeuwenhoek*, **62**, 109–30.

Thevelein, J.M. and Beullens, M. (1985) Cyclic AMP and the stimulation of trehalase activity in the yeast *Saccharomyces cerevisiae* by carbon sources, nitrogen sources and inhibitors of protein synthesis. *Journal of General Microbiology*, **131**, 3199–209.

Thevelein, J.M., Beullens, M., Honshoven, F. *et al.* (1987a) Regulation of the cAMP level in the yeast *Saccharomyces cerevisiae*: intracellular pH and the effect of membrane depolarizing compounds. *Journal of General Microbiology*, **133**, 2191–6.

Thevelein, J.M., Beullens, M., Honshoven, F. *et al.* (1987b) Regulation of the cAMP signal in the yeast *Saccharomyces cerevisiae*: the glucose-induced cAMP signal is not mediated by a transient drop in the intracellular pH. *Journal of General Microbiology*, **133**, 2197–205.

Thiel, R., Schreurs, W.J.A. and Harold, F.M. (1988) Transcellular ion currents during sporangium development in the water mould *Achlya bisexualis*. *Journal of General Microbiology*, **134**, 1089–97.

Thorner, J., Fields, F.O., Kunisawa, R. and Levin, D. (1988) Protein kinase C (PKC) is an essential enzyme in the yeast *Saccharomyces cerevisiae*. *Journal of Cell Biology*, **107**, 443a.

Toda, T., Uno, I., Ishikawa, T. *et al.* (1985) In yeast, Ras proteins are controlling elements of adenylate cyclase. *Cell*, **40**, 27–36.

Toda, T., Cameron, S., Sass, P. *et al.* (1987) Three different genes in *Saccharomyces cerevisiae* encode the catalytic subunits of the cAMP-dependent protein kinase. *Cell*, **50**, 277–87.

Tomes, C. and Moreno, S. (1990) Phosphodiesterase activity and cyclic AMP content during early germination of *Mucor rouxii* spores. *Experimental Mycology*, **14**, 78–83.

Trevillyan, J.M. and Pall, M.L. (1979) Control of cyclic adenosine 3',5' monophosphate levels by depolarizing agents in fungi. *Journal of Bacteriology*, **1**, 397–403.

Trevillyan, J.M. and Pall, M.L. (1982) Isolation and properties of a cyclic AMP-binding protein from *Neurospora* – evidence for its role as the regulatory subunit of cyclic AMP-dependent protein kinase. *Journal of Biological Chemistry*, **257**, 3978–86.

Trinci, A.P.J., Robson, G.D., Wiebe, M.G. *et al.* (1990) Growth and morphology of *Fusarium graminearum* and other fungi in batch and continuous culture, in *Microbial Growth Dynamics*, (eds R.K. Poole, M.J. Bazin and C.W. Keevil), IRL Press, Oxford, pp. 17–38.

Tsunoda, Y. (1991). Oscillatory Ca^{2+} signalling and its cellular functions. *New Biologist*, **3**, 3–17.

Turian, G. and Favre, B. (1990) Calcium- and phospholipid-phospholipid-dependent protein and lipid kinases in *Neurospora crassa*, in *Calcium as an Intracellular Messenger in Eucaryotic Microbes*, (ed D.H. O'Day), American Society for Microbiology, Washington, DC, pp. 165–91.

Ugalde, U.O., Virto, M.D. and Pitt, D. (1990) Calcium binding and induction of conidiation in protoplasts of *Penicillium cyclopium*. *Antonie van Leeuwenhoek*, **57**, 43–9.

Ulaszewski, S., Hilger, F. and Goffeau, A. (1989) Cyclic AMP controls the plasma membrane H^+-ATPase activity from *Saccharomyces cerevisiae*. *FEBS Letters*, **245**, 131–6.

Uno, I. and Ishikawa, T. (1981) Control of adenosine 3',5'-monophosphate level and protein phosphorylation by depolarizing agents in *Coprinus macrorhizus*. *Biochimica et Biophysica Acta*, **672**, 108–13.

Uno, I. and Ishikawa, T. (1990) Role of phosphatidyl-inositol metabolites in proliferation of yeast cells, in *Calcium as an Intracellular Messenger in Eukaryotic Cells*, (ed. D.H. O'Day), American Society for Microbiology, Washington, DC, pp. 52–64.

Uno, L., Ishikawa, T. and Hashiba, T. (1985a) Cyclic AMP-dependent phosphorylation of proteins in mycelia of *Rhizoctonia solani* forming sclerotia. *Annals of the Phytopathological Society of Japan*, **51**, 190–8.

Uno, L., Ishikawa, T. and Hashiba, T. (1985b) Detection of cAMP levels and cyclic AMP-receptor protein in mycelia of *Rhizoctonia solani* forming sclerotia. *Annals of the Phytopathological Society of Japan*, **51**, 183–9.

Uno, I., Fukami, K., Kato, H. *et al.* (1988) Essential role for phosphatidylinositol 4,5-bisphosphate in yeast cell proliferation. *Nature*, **33**, 188–90.

Uno, I., Ohya, Y., Anraku, Y. and Ishikawa, T. (1989) Cell cycle-dependent regulation of calmodulin levels in *Saccharomyces cerevisiae*. *Journal of General and Applied Microbiology*, **35**, 59–63.

Van Aelst, L., Boy-Marcotte, E., Camonis, J.H. *et al.* (1990) The C-terminal part of the *CDC25* gene product plays a key role in signal transduction in the glucose-induced modulation of cAMP level in *Saccharomyces cerevisiae*. *European Journal of Biochemistry*, **193**, 675–80.

Van Aelst, L., Jans, A.W.H. and Thevelein, J.M. (1991) Involvement of the *CDC25* gene product in the signal transmission pathway of the glucose-induced RAS-mediated cAMP signal in the yeast *Saccharomyces cerevisiae*. *Journal of General Microbiology* **137**, 341–9.

Van Laere, A.J. (1986a) Cyclic AMP, phosphodiesterase, and spore activation in *Phycomyces blakesleeanus*. *Experimental Mycology*, **10**, 52–9.

Van Laere, A.J. (1986b) Biochemistry of spore germination in *Phycomyces*. *FEMS Microbiology Reviews*, **32**, 189–98.

Van Mulders, A.R. and Van Laere, A.J. (1984) Cyclic AMP, trehalase and germination of *Phycomycees blakesleanus* spores. *Journal of General Microbiology*, **130**, 541–7.

Van Tuinen, D., Ortega Perez, R., Marme, D. and Turian, G. (1984) Calcium, calmodulin-dependent protein phosphorylation in *Neurospora crassa*. *FEBS Letters*, **176**, 317–20.

Van Zyl, W., Huang, W., Sneddon, A.A. *et al.* (1992) Inactivation of the protein phosphatase 2A regulating subunit A results in morphological and transcriptional defects in *Saccharomyces cerevisiae*. *Molecular and Cellular Biology*, **12**, 4946–59.

Verdier, J.M., Camonis, J.H. and Jacquet, M. (1989) Cloning of *CDC33*: a gene essential for growth and sporulation which does not interfere with cAMP production of *Saccharomyces cerevisiae*. *Yeast*, **5**, 79–90.

Wada, Y., Ohsumi, Y. and Anraku, Y. (1986) Mechanisms of acidification of the vacuolar membrane vesicles from *Saccharomyces cerevisiae*: effects of inorganic anions. *Cell Structure and Function*, **11**, 533.

Wada, Y., Ohsumi, Y., Tanifuji, M. *et al.* (1987) Vacuolar channel of the yeast *Saccharomyces cerevisiae*. *Journal of Biological Chemistry*, **262**, 17260–3.

White, C. and Gadd, G.M. (1987)/ The uptake and cellular distribution of zinc in *Saccharomyces cerevisiae*. *Journal of General Microbiology* **133**, 727–37.

Yoko-o, T., Matsui, Y., Yagisawa, H. (1993) The putative phosphoinositide-specific phospholipase C gene, *PLCl*, of the yeast *Saccharomyces cerevisiae* is important for cell growth. *Proceedings of the National Academy of Sciences of the USA*, **90**, 1804–8.

Yoshida, S., Ikeda, E., Uno, I. and Mitsuzawa, H. (1992) Characterization of a staurosporine- and temperature-sensitive mutant, *stt1*, of *Saccharomyces cerevisiae*: *STT1* is allelic to *PKC1*. *Molecular and General Genetics*, **231**, 337–44.

Youatt, J. (1993) Calcium and microorganisms. *Critical Reviews in Microbiology*, **19**, 83–97.

Zhou, X.-L., Stumpf, M.A., Hoch, H.C. and Kung, C. (1991) A mechanosensitive channel in whole cells and in membrane patches of the fungus *Uromyces*. *Science*, **253**, 1415–17.

INTERMEDIARY METABOLISM

10

A. Van Laere

Katholieke Universiteit Leuven, Laboratory for Developmental Biology, Botany Institute,
Kardinaal Mercierlaan 92, B 3001 Heverlee-Leuven, Belgium

10.1 INTRODUCTION

In attempting to review the metabolism of fungi in a concise manner one is confronted with both a plethora and a lack of information. Fungi are a very diverse kingdom of organisms, members of which have been evolving independently for over a billion years with often short generation times. Although their morphology is rather similar, one might expect a metabolic diversity that is at least as significant as the diversity in the animal kingdom. This diversity is reflected in the variety of substrates fungi can metabolize and use as carbon sources as well as in the chemicals they can produce as fermentation products or so-called secondary metabolites. These properties have been exploited in a large number of industrial applications. Growth requirements of others such as VAM-forming fungi or some biotrophic pathogens can be so intricate that it has proven impossible to grow them in culture. A discussion of fungal metabolism therefore has to be restricted to the most important or most general pathways.

This chapter will focus mainly on carbohydrate metabolism, the subsequent tricarboxylic acid cycle and associated anaplerotic reactions. Metabolism of lipids, nitrogenous compounds, phosphate and sulphur will be treated briefly. It is important to note that several aspects of metabolism have only been investigated in detail in a limited number of organisms, particularly in *Saccharomyces cerevisiae*. Because of the enormous evolutionary diversity of the fungi one should be cautious in extrapolating, explicitly or implicitly, from one or a few organisms to fungi in general.

10.2 CARBOHYDRATE METABOLISM

Carbohydrates occupy a central position in metabolism. Various monosaccharides, oligosaccharides and polysaccharides are excellent carbon sources for most fungi. Carbohydrates are also accumulated as reserves under certain conditions and they form major components of the cell wall. Moreover, intermediary metabolism is centred around phosphorylated sugars, both in glycolysis and the pentose-phosphate pathway.

10.2.1 Carbohydrate reserves

Trehalose and glycogen are the main carbohydrate reserves in fungi and have long been viewed as more or less interchangeable. More recently, however, glycogen and trehalose were shown to be, if not mutually exclusive, at least clear alternatives. In general, glycogen is synthesized during periods of active growth, especially when there is a high C/N ratio, and is broken down during the stationary phase or during the production of resting structures. In spores and stationary phase cells, it is apparently replaced by trehalose that is again broken down during resumption of growth or germination (Thevelein, 1984). Indeed, trehalose appears to be very important as a protectant against heat, dehydration and possibly other stresses (Van Laere, 1989; Wiemken, 1990). This protection is apparently not afforded by glycogen.

Glycogen has not been detected in Oomycetes where it is replaced by the $\beta(1-3)$, $\beta(1-6)$-linked glucan mycolaminaran (Coulter and Aronson, 1977). Trehalose has also not been reported in Oomycetes except in one case where two out of five *Pythium* species where reported to contain trehalose (Yoshida *et al.*, 1984). However, identification was only based on a single NMR peak and confirmation by independent methods is necessary. Trehalase has been described in the Oomycete *Lagenidium* sp. (McInnis and Domnas, 1973) but it

The Growing Fungus. Edited by Neil A.R. Gow and Geoffrey M. Gadd. Published in 1994 by Chapman & Hall, London. ISBN 0 412 46600 7

is probably involved in breakdown of trehalose from its insect host.

(a) Glycogen metabolism

Simultaneous synthesis and breakdown of glycogen would result in a futile cycle; glycogen metabolism therefore requires stringent regulation (Figure 10.1). The regulation of glycogen synthesis and breakdown has been investigated in detail only in *S. cerevisiae* (François and Hers, 1988; François *et al.*, 1988b). Both glycogen synthase and glycogen phosphorylase occur in an active (a) and an inactive (b) form interconvertible by phosphorylation and dephosphorylation. High glucose-6-phosphate concentrations stimulate phosphorylation (and concomitant activation) of glycogen synthase but dephosphorylation (and concomitant inactivation) of glycogen phosphorylase. There is yet no evidence for the involvement of cAMP but Ca^{2+} (and calmodulin) might be important for stimulation of kinase activities. Cyclic AMP may be involved in the regulation of glycogen synthesis and breakdown in *Coprinus cinereus* (Kuhad *et al.*, 1987). The glycogen phosphorylase from *Neurospora crassa* occurs as two interconvertible forms (Tellez-Inon and Torres, 1970).

In addition, $\alpha(1\text{--}4)$ (and $\alpha(1\text{--}6)$) glucosidases might be important for glycogen to glucose (and trehalose) conversion in yeasts (Colonna and Magee, 1978) and other fungi. Amylases, however, are probably more important as secretory enzymes for the mobilization of starch in saprophytic (Ueda, 1981) as well as phytopathogenic fungi (Krause *et al.*, 1991). In *Morchella vulgaris* a lyase, producing 1,5-D-anhydrofructose, uses $\alpha(1\text{--}4)$-glucan as substrate (Baute *et al.*, 1989).

(b) Trehalose metabolism

The synthesis and breakdown of trehalose could also lead to futile cycles (Figure 10.1) and therefore needs to be regulated tightly to avoid wasting ATP. Trehalose is synthesized from UDP-glucose (UDPG) and glucose-6-phosphate. The trehalose-6-phosphate generated by trehalose-6-phosphate synthase is hydrolysed to trehalose by trehalose-6-phosphate phosphatase. Both enzymatic activities appear to be properties of the same protein in yeast (Vandercammen *et al.*, 1989a; Londesborough and Vuorio, 1991).

It has been argued that trehalose-6-phosphate synthase is inactivated by cAMP-dependent phosphorylation in *S. cerevisiae* (Panek *et al.*, 1987) and *Candida utilis* (Vicente-Soler *et al.*, 1989). However, the results with yeast could not be duplicated by Vandercammen *et al.* (1989a) using apparently more reliable methods. Indeed, measuring trehalose-6-phosphate synthase in the presence of glycogen synthase is very difficult due to the stimulation of glycogen synthase by glucose 6-phosphate. Nevertheless, Panek *et al.* (1990) have maintained their original conclusions after a reinvestigation of their methods. The data of Londesborough and Vuorio (1991) suggest possible regulation of trehalose synthesis by a protein activator and/or limited proteolysis. According to François *et al.* (1991), cAMP might be involved in catabolite repression but not in inactivation of trehalose-6-phosphate synthase/trehalose-6-phosphate phosphatase. The recent isolation and sequencing of the gene for a small subunit of trehalose-phosphate synthase from *S. cerevisiae* (Bell *et al.*, 1992) will contribute to the elucidation of its regulation

Although other trehalose-degrading enzymes such as phosphotrehalase and trehalose phosphorylase have been described in plants and bacteria (e.g. Salminen and Streeter, 1986), only trehalase appears to be involved in trehalose breakdown in fungi. Two types of trehalase have been described (Thevelein, 1984). Trehalases with a neutral pH optimum occur in the cytosol and can be activated by cAMP-dependent phosphorylation. Such trehalases have been found, e.g. in the Zygomycetes *Phycomyces blakesleeanus* and *Mucor rouxii* (Van Laere *et al.*, 1987), *S. cerevisiae* (Uno *et al.*, 1983) and *C. utilis* (Arguelles *et al.*, 1986). Their activity increases, together with cAMP content, in germinating spores of *P. blakesleeanus* and *M. rouxii* (Van Laere, 1986; Van Laere *et al.*, 1987), or after addition of glucose to stationary phase cells of *S. cerevisiae* (Thevelein, 1984) and *C. utilis* (Arguelles *et al.*, 1986). Although trehalose content also decreases in germinating spores of *Pilobolus longipes* (Bourret *et al.*, 1989) and *N. crassa* (Hecker and Sussman, 1973) no increase in trehalase activity is found, even though cAMP content increases in *P. longipes* (Bourret *et al.*, 1991). However, it is known that the regulatory trehalase is rather unstable and that activity rapidly disappears in dilute solutions or during purification. Therefore, negative reports do not necessarily mean that trehalase does not occur in a particular fungus.

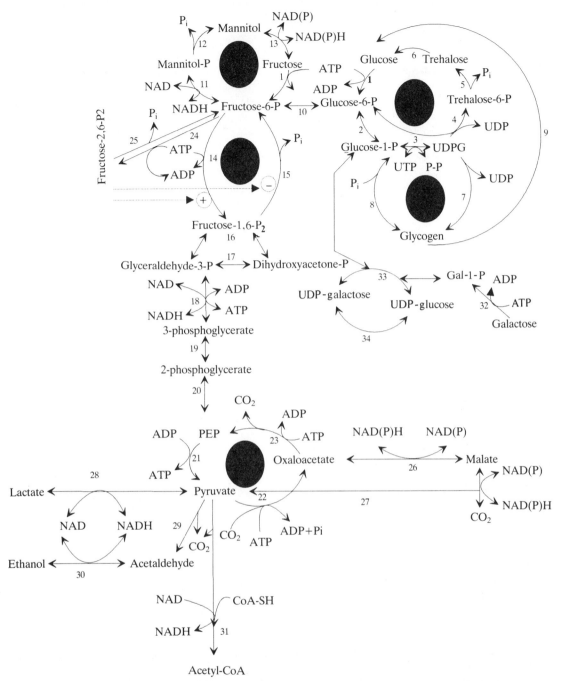

Figure 10.1 Scheme illustrating the synthesis and breakdown of carbohydrate reserves as well as glycolysis and gluconeogenesis. Circles emphasize some possible futile cycles. Enzymes involved are: 1, hexokinase; 2, phosphoglucomutase; 3, UDPG-pyrophosphorylase; 4, trehalose-phosphate synthase; 5, trehalose-phosphate phosphatase; 6, trehalase; 7, glycogen synthase; 8, glycogen phosphorylase; 9, α(1–4),(1–6) glucosidase; 10, phosphoglucose isomerase; 11, mannitol-phosphate dehydrogenase; 12, mannitol-phosphate phosphatase; 13, mannitol dehydrogenase; 14, phosphofructokinase 1; 15, fructose 1,6-bisphosphatase; 16, aldolase; 17, triose-phosphate isomerase; 18, glyceraldehyde-phosphate dehydrogenase; 19, phosphoglycerate mutase; 20, enolase; 21, pyruvate kinase; 22, pyruvate carboxylase; 23, phosphoenol-pyruvate carboxykinase; 24, fructose-6-phosphate 2-kinase (PFK2); 25, fructose 2,6-bisphosphatase (FBPase2); 26, malate dehydrogenase; 27, malic enzyme; 28, lactate dehydrogenase; 29, pyruvate decarboxylase; 30, alcohol dehydrogenase; 31, pyruvate dehydrogenase.

Some of the trehalases with acidic pH optima are lysosomal or vacuolar in nature. This is the case for, e.g. *S. cerevisiae* (Keller *et al.*, 1982) and *C. utilis* (Arguelles and Gacto, 1988). Their function is puzzling since trehalose apparently occurs in the cytosol, at least in *S. cerevisiae* (Keller *et al.*, 1982). Other acidic trehalases have been shown to be secreted, e.g. *Humicola grisea* (Zimmerman *et al.*, 1990), or to be attached to the cell wall or localized in the 'periplasmic space', e.g. *N. crassa* (Hecker and Sussman, 1973), *Schizosaccharomyces pombe* (Inoue and Shimoda, 1981) and *P. longipes* (Bourret *et al.*, 1989). The latter are probably involved in degradation of extracellular trehalose. It is possible that lysosomal trehalases are stored for secretion at a later time since, at least in *S. cerevisiae*, the secretory pathway is involved in vacuolar trehalase synthesis (Mittenbuhler and Holzer, 1991).

(c) Mannitol

Mannitol is a common polyol in Chytridiomycetes, Ascomycetes, Basidiomycetes and Deuteromycetes (Pfyffer and Rast, 1980; Pfyffer *et al.*, 1990). Mannitol does not appear to play an important role in osmotic adjustment of fungal cells (Beever and Laracy, 1986). Mannitol can be synthesized from fructose 6-phosphate by a NAD^+ dependent mannitol-phosphate dehydrogenase, followed by hydrolysis by a specific mannitol 1-phosphatase. Alternatively, in some fungi mannitol can be produced from fructose by a mannitol dehydrogenase.

The simultaneous presence of NAD^+-mannitol 1-phosphate dehydrogenase, mannitol 1-phosphatase and $NADP^+$-mannitol dehydrogenase in a number of fungi led Hult and Gatenbeck (1978) to suggest a mannitol cycle that would be responsible for ATP-dependent transhydrogenation of NADH + H^+ to NADPH + H^+ (Figure 10.1). A slightly different mannitol cycle including glucitol and glucose (Jennings and Burke, 1990) could transform two NADH into two NADPH per ATP. Whereas Hult *et al.* (1980) consider the mannitol cycle as a mechanism to generate additional reducing equivalents for biosynthetic purposes, Jennings and Burke (1990) prefer to see the wastage of ATP in the cycle as an overflow mechanism under stress conditions.

It is unlikely that the mannitol cycle occurs in a large number of fungi (Hult *et al.*, 1980). Moreover, no $NADP^+$- but an NAD^+-dependent

mannitol dehydrogenase was found in a number of non-mycorrhizal fungi (Ramstedt *et al.*, 1987). Even when all the enzymes for the functioning of the cycle are present, as in *Aspergillus nidulans*, the contribution of the cycle to the overall production of NADPH appears poor, even when demand is high, e.g. during nitrate reduction (Singh *et al.*, 1988). Nevertheless in the Ascomycete *Cenococcum graniforme* the cycle appears to be responsible for rapid label randomization in glucose *via* the symmetrical mannitol molecule (Martin *et al.*, 1985).

Mannitol kinases have not been detected in fungi despite several attempts (Ramstedt *et al.*, 1986). In *Agaricus bisporus*, which contains up to 35% mannitol, $NADP^+$-dependent mannitol dehydrogenase appears to be the only enzyme that can metabolize mannitol (Hammond, 1985).

10.2.2 Glycolysis

Glycolysis or the Embden–Meyerhof–Parnas (EMP) pathway occurs in the cytosol and is the ubiquitous means to convert glucose into pyruvate, providing the cell with energy, precursors and NADH. Most enzymes of the pathway catalyse reactions that are readily reversible *in vivo*. This is not the case for hexokinase, phosphofructokinase and pyruvate kinase which are believed to be the main control points of the pathway. The latter two genes are maximally induced during growth of *S. cerevisiae* on glucose (Moore *et al.*, 1991).

(a) Hexokinase

After breakdown of polymers to their constituent monosaccharides (Chapter 7) and transport (Chapter 8), further metabolism of carbohydrates is generally initiated by phosphorylation. Although UTP, CTP, GTP and ITP can function as phosphate donors, all sugar kinases described to date are about 2 to 50 times more active with ATP as substrate.

The higher fungi (Chytridiomycetes, Zygomycetes, Ascomycetes, Basidiomycetes, Deuteromycetes) all contain a hexokinase that phosphorylates glucose, fructose, mannose and glucosamine (as well as 2-deoxyglucose) (Table 10.1). Up to four different hexokinases have been described in *N. crassa* (Lagos and Ureta, 1980). Two hexokinases from *S. cerevisiae* have been cloned and sequenced

Table 10.1 Properties of some fungal hexose kinases

	Type	Glucose		Fructose		Mannose		2-DOG		Gluc NH2		ATP	GlucNAX	Xylose
		K_m	A_r^2	K_m	A_r	K_m	A_r	K_m	A_r	K_m	A_r	K_m	K_i	K_i
Oomycetes														
Pythium debaryanum	MFK	80												
Saprolegnia litoralis	MFK	0.15	5	1	100	0.06	30			0.25	10			
	GK		100		<1		<1		<1	0.5	75	0.6	0.3	30
Hypochytridiomycetes														
Hypochytrium catenoides	GK	0.15	100		<1		<1		<1	0.5	65	1.5	0.4	50
Chytridiomycetes														
Allomyces arbuscula	HK	0.06	100	15	100	0.1	45	0.2	100	0.8	30	0.5	0.1	80
Zygomycetes														
Mucor hiemalis	HK	0.08	100	0.9	230	0.04	90	3	100	0.5	35	0.6	2	15
Phycomyces blakesleeanus	HK	0.09	100	25	85	0.1	40	0.4	75	0.8	35	0.2	2	100
Ascomycetes														
Neurospora crassa	HK α	0.061	100	22.7	75	0.2	72	0.185	78			1.42		
	HK β	0.056	100	12.7	73	0.158	72	0.4	80			1.88		
	HK γ	0.077	100	2.9	138	0.27	100	7.9	67			0.75		
	HK δ	0.18	100	2.5	196	0.3	100	7.14	60			0.56		
Saccharomyces cerevisiae	HK 1	0.3	100	4.1	300		60					0.3		
	HK 2	0.6	100	2.5	130		28					0.29		
	GK	0.03	100	'31	0.4	0.12	20	1.46	45	0.74	9	0.05	1.7	
Basidiomycetes														
Schizophyllum commune	HK	0.3	100	2.5	175	0.2	80	1.5	95	0.9	50	0.1	2	65
	GMK	0.08	100	<1	<1	0.4	100	0.15	80	0.4	20	0.4	0.5	75
Deuteromycetes														
Candida tropicalis	HK 1	0.34	100	2.2	180									
	GK	0.29	100											
Rhodotorula glutinis	HK	0.1	100	2	310	0.1	80	0.3	75	4	7	0.5	8	300
	GMK	0.1	100	–	–	0.2	33	0.3	60	5	7	0.6	4	400
Aspergillus oryzae	HK	0.05	100	30	120	0.075	80	0.4	80	2	45		1.5	15
Aspergillus parasiticus	HK		100		28		64				55			

From Delvalle and Assensio (1978) and Lagos and Ureta (1980).
A_r^2, activity relative to best substrate (mostly glucose).

(Stachelek *et al.*, 1986). Some fungi also contain more specific kinases that are very similar to the hexokinases but are not (or only poorly) active on fructose; they have been called glucomanno-kinases or glucokinases. These hexose kinases are inhibited by *N*-acetylglucosamine and xylose but, unlike the enzymes from mammalian origin, are not inhibited by their product glucose-6-P.

In the Hyphochytridiomycetes a glucokinase was found that does not phosphorylate fructose or mannose whereas in the Oomycetes a manno-fructokinase (as well as a specific glucokinase in *Saprolegnia*) was detected (Delvalle and Assensio, 1978). It is not known whether these acivities will prove to be diagnostic for these taxonomic groupings.

Other kinases may well occur in fungi that can metabolize a variety of monosaccharides and their derivatives such as polyols, pentoses, aldonic acids, uronic acids and aminosugars. Whether the utilization of these compounds requires specific kinases, e.g. glycerol kinase, mannitol kinase (Jennings, 1984), poorly specific kinases, or an initial epimerization, reduction and/or oxidation has not always been investigated in detail. Some examples are given in Figure 10.3. In *Candida tropicalis*, metabolism of *N*-acetylglucosamine requires a specific kinase; *N*-acetylglucosamine-6-P is then deacetylated, and deaminated to fructose-6-P (Shepherd *et al.*, 1980). *Candida albicans* also metabolizes *N*-acetylmannosamine but no specific kinase seems to be involved since *N*-acetylmannosamine is first epimerized to *N*-acetylglucosamine (Biswas *et al.*, 1979). Galactose is first phosphorylated on C-1, transformed to UDP-galactose and epimerized to UDP-glucose in those organisms that can metabolize it (Gadd, 1988). Nevertheless, the hexokinase from *Aspergillus parasiticus* was reported to phosphorylate galactose and galactosamine to their respective 6-phosphate esters (Davidson, 1960).

(b) Phosphofructokinase

Since fungal hexokinases do not appear to be regulated tightly, the hexose-phosphate concentration can change quite dramatically under a variety of conditions. However, phosphofructokinase is the main regulatory enzyme of glycolysis in both prokaryotic and eukaryotic organisms (Uyeda, 1979). Phosphofructokinase forms also,

with its gluconeogenic counterpart, a possible futile cycle (Figure 10.1). It is therefore not surprising that most phosphofructokinases show a complex pattern of regulation by both their sub-strates as well as a number of positive and negative effectors.

Apart from *S. cerevisiae* (e.g., Kopperschläger *et al.*, 1977) little information is available on the kinetic properties and structure of purified fungal phosphofructokinases. Phosphofructokinases from *N. crassa*, *A. parasiticus* and other Ascomycetes, Basidiomycetes and Deuteromycetes (Table 10.2) are very unstable, probably because of their extreme sensitivity to proteolysis. This property is probably responsible for a number of older reports stating the absence of phosphofructokinase, e.g. in *Penicillium chrysogenum* and several yeasts (see Van Laere and Joosen, 1991).

Phosphofructokinase from the Oomycete *Pythium ultimum* has characteristics that differ greatly from other fungi and are more like the enzyme from *Dictyostelium discoïdeum* (Table 10.2). It has hyper-bolic kinetics towards ATP and fructose 6-phosphate and is not stimulated by AMP. In addition fructose 2, 6-bisphosphate has no effect on the enzyme from *P. ultimum* (Van Laere and Joosen, 1991), although it is a potent stimulator of animal and fungal (but not plant) phosphofructo-kinases (Van Schaftingen, 1986).

Table 10.2 shows that all phosphofructokinases from the Eumycota, as defined by Whittaker and Margulis (1978), have a number of characteristics in common. These include inhibition by high ATP concentrations, cooperativity towards fructose 6-phosphate and stimulation by fructose 2,6-bisphosphate. Moreover phosphofructokinases from almost all Amastigomycotina are inhibited by citrate and PEP and stimulated by phosphate and ammonium ions. Stimulation by AMP is a more variable property in the fungal kingdom. Whether aberrant characteristics are genuine or due to proteolytic breakdown or inappropriate assay conditions (pH, concentrations of substrates and effectors) cannot be ascertained.

(c) Pyruvate kinase

Pyruvate kinase might be another control point of glycolysis and here futile cycling with gluco-neogenic enzymes is also possible. The enzyme was shown to have hyperbolic kinetics towards

Table 10.2   Properties of fungal phosphofructokinases

	ATP	Fru-6-P	Fru 2,6-P₂	AMP	Citrate	PEP	NH₄⁺	Phosphate	Stability
Oomycete									
P. ultimum	Hᵃ	H	N	N	I	I	N	S	G
Chytridiomycete									
B. emersonii	M	S	A	A	N	–	N	N	G
Zygomycetes									
P. blakesleeanus	M	S	A	A	I	–	S		G
P. longipes			A				S		
M. rouxii	M	S	A	N	I	N	A		G
Ascomycetes									
N. crassa	M	S	–	N	–	–	–	–	U
Neurospora tetrasperma		–	A	–	–	–	–	–	–
S. cerevisiae	M	S	A	A	I	–	A	A	U
Basidiomycetes									
Ustilago nuda	–	–	A	–	–	–	–	–	U
A. bisporus	M	S	A	N	I	I	N	A	U
Rhodosporidium toruloides	H	S	A	N	I	–	S	–	U
Deuteromycetes									
P. notatum	M	S	–	A	I	I	–	–	U
Rhodotorula gracilis	–	S	A	–	–	–	–	–	–
C. albicans	M	S	–	A	–	–	A	I	U
C. parapsilosis	M	S	A						
A. niger	M	S	A	A	I	I	A	A	–

See Caubet *et al.* (1988), Vandercammen *et al*, (1989b), Van Laere and Joosen (1991) and Bourret *et al.* (1991) for references.
ᵃH, hyperbolic; M, inhibitory in high concentration; S, sigmoïdal dependence on substrate concentration; N, no effect; I, inhibitory; A, stimulatory; G, good stability; U, very unstable.

phosphoenolpyruvate and to be stimulated by small concentrations of fructose 1, 6-bisphosphate in the Ascomycetes *N. crassa* (Yeung and Kapoor, 1983) and *S. cerevisiae* (Morris *et al.*, 1984), the Deuteromycete *Aspergillus niger* (Meixner-Monori *et al.*, 1984) as well as the Zygomycetes *P. blakesleeanus* (De Arriaga *et al.*, 1989), *M. rouxii* (Passeron and Roselino, 1971) and *Mucor racemosus* (Hohn and Paznokas, 1987). The *P. blakesleeanus* enzyme is inhibited by alanine and by high ATP concentrations (De Arriaga *et al.*, 1989). In *Coprinus lagopus* pyruvate kinase is also highly regulated (Stewart and Moore, 1971). Different isoenzymes, one of them only poorly stimulated by fructose 1, 6-bisphosphate, have only been described in the *Mucor* species and their expression depends on culture conditions (Hohn and Paznokas, 1987). In *N. crassa* two different pyruvate kinase mRNAs were found but their origin is unknown (Devchand and Kapoor, 1987).

10.2.3   Gluconeogenesis

Gluconeogenesis, essentially the reversal of glycolysis, is essential for fungi growing on substrates such as ethanol, acetate, and lipids (Gadd, 1988). The main regulatory enzymes of gluconeogenesis are PEP carboxykinase and fructose 1,6-bisphosphatase and both form futile cycles with glycolytic counterparts (Figure 10.1). Isocitrate lyase and cytoplasmic malate dehydrogenase are also important for gluconeogenesis and are highly regulated in yeast (Holzer, 1989).

(a)   PEP carboxykinase

Phosphoenolpyruvate carboxykinase catalyses the ATP-dependent formation of PEP from oxaloacetate:

$$\text{oxaloacetate} + \text{ATP} \rightarrow \text{PEP} + \text{ADP} + CO_2$$

The enzymes from *Verticillium albo-atrum* (Hartman and Keen, 1974) and *P. blakesleeanus* (Sandmann and Hilgenberg, 1978) have been studied in some detail. The kinetic properties of the enzyme and its changes in activity during shift from glycolytic (e.g., glucose) to gluconeogenic (e.g., acetate) C-sources in for example, *S. cerevisiae* (Wilson and Bhattacharjee, 1986) are consistent with a gluconeogenic role. In yeast, glucose-induced inactivation of the enzyme appears to involve cAMP- and ATP-dependent modification, followed by cAMP-independent phosphorylation and subsequent limited proteolysis (Burlini *et al.*, 1989)

(b)   Fructose 1,6-bisphosphatase

Whereas phosphofructokinase is stimulated by micromolar concentrations of fructose 2,6-bisphosphate, fructose 1,6-bisphosphatase is very sensitive to inhibition by the same compound. This has been shown in, *Kluyveromyces fragilis* (Toyoda and Sy, 1984), *S. cerevisiae, Candida parapsilosis* (Caubet *et al.*, 1988), *A. niger* (Kubicek-Pranz *et al.*, 1990), *Blastocladiella emersonii* (Vandercammen *et al.*, 1989b) and *P. blakesleeanus* (Van Laere, unpublished). Moreover, fructose 2,6-bisphosphate has been shown to stimulate cyclic AMP-dependent phosphorylation and concomitant inactivation of fructose 1,6-bisphosphatase in *S. cerevisiae* (Gancedo *et al.*, 1983; Pohlig *et al.*, 1983). A similar mechanism might operate in *Candida maltosa* (Polnish and Hofmann, 1989) and *K. fragilis* (Toyoda and Sy, 1984) but has little effect on activity in the latter case.

(c)   Control of fructose 2,6-bisphosphate concentration

By its antagonistic effect on phosphofructokinase and fructose 1,6-bisphosphatase, fructose 2,6-bisphosphate is a key compound in the regulation of glycolysis and gluconeogenesis. A correlation between cyclic AMP content and fructose 2,6-phosphate concentration was found in germinating spores of *P. blakesleeanus* (Van Laere, 1986), and *P. longipes* (Bourret *et al.*, 1991), or after feeding starved cells of *S. cerevisiae* (François *et al.*, 1984) and *N. crassa* (Dumbrava and Pall, 1987). In *P. blakesleeanus, P. longipes* and *N. crassa*, this correlation proved to be fortuitous since it could be uncoupled under some conditions but in *S. cerevisiae* it was shown to be due to cAMP-dependent activation of phosphofructokinase 2 (fructose-6-phosphate 2-kinase) (François *et al.*, 1984, 1988a). Contrary to mammalian tissue, fructose-6-phosphate 2-kinase (PFK2) and fructose 2,6-bisphosphate 2-phosphatase (FBPase2) reside on different proteins and can be separated (François *et al.*, 1988a). Besides a specific FBPase2 some less

specific enzymes, some producing fructose-2-phosphate, have been described (Plankert *et al.*, 1988).

10.2.4   Non-glycolytic carbohydrate metabolism

(a)   Pentose-phosphate pathway

Whenever investigated, radiorespirometric as well as enzymatic evidence points toward the operation of the pentose-phosphate (PP) pathway in fungi although the quantitative importance can vary widely (Cochrane, 1976). The main role of the pentose phosphate or hexose monophosphate pathway is the production of NADPH for biosynthetic purposes such as synthesis of lipids, and assimilation of nitrate and ammonia. A number of intermediates of this pathway such as ribose-5-phosphate (nucleic acids) and erythrose-4-phosphate (aromatic compounds) are diverted as building blocks of cellular components. These compounds can, however, also be formed from fructose-6-phosphate and glyceraldehyde-3-phosphate without the intervention of the oxidative part of the pathway (Figure 10.2). In addition, numerous polyols can be produced from these intermediates by the action of more or less specific isomerases, dehydrogenases and phosphatases of varying specificity. The quantitative importance of the PP-pathway appears to be controlled by the demand for NADPH and media increasing the demand for NADPH (such as nitrate as nitrogen source) increase the percentage glucose metabolized along this pathway (Berry, 1975).

(b)   The Entner–Doudoroff (ED) pathway

The ED pathway is an alternative to glycolysis and produces pyruvate and glyceraldehyde-3-phosphate from glucose-6-phosphate (Figure 10.2). Two enzymes, 6-phosphogluconate dehydratase and 2-keto-3-deoxygluconate-6-phosphate aldolase are unique to this pathway. There are some early radiorespirometric data from *Tilletia caries* and *Caldariomyces fumago* which suggest but do not prove operation of the ED-pathway in these fungi (Cochrane, 1976). Since then, despite a number of attempts (Cochrane, 1976; Held and Goldman, 1986), corroborative evidence for this pathway has only been obtained for *Penicillium notatum* again by radiorespirometric methods (Pitt and Mosley,

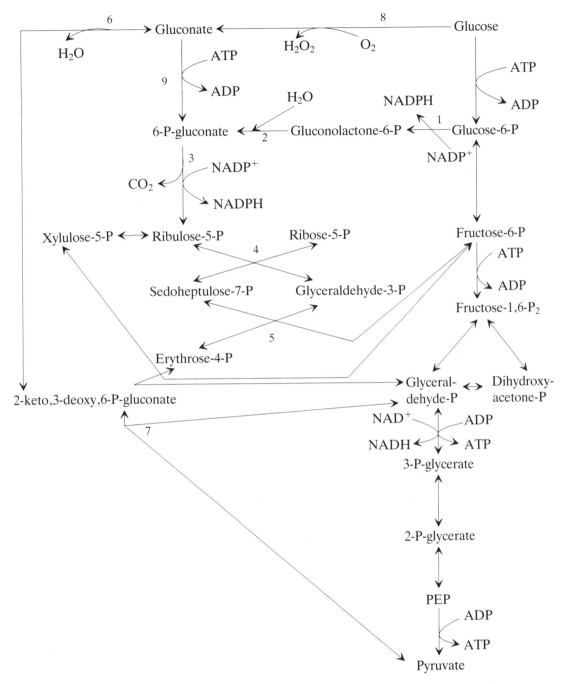

Figure 10.2 Alternative pathways of glucose metabolism. Enzymes of the pentose-phosphate pathway are: 1, glucose-6-phosphate dehydrogenase; 2, lactonase; 3, 6-phosphogluconate dehydrogenase; 4, transketolase; 5, transaldolase. Special enzymes of the Entner–Doudoroff pathway are: 6, 6-phosphogluconate dehydratase and 7, 2-keto, 3-deoxy, 6-phosphogluconate aldolase. Glucose and gluconate can also enter metabolism via: 8, glucose oxidase and 9, gluconokinase.

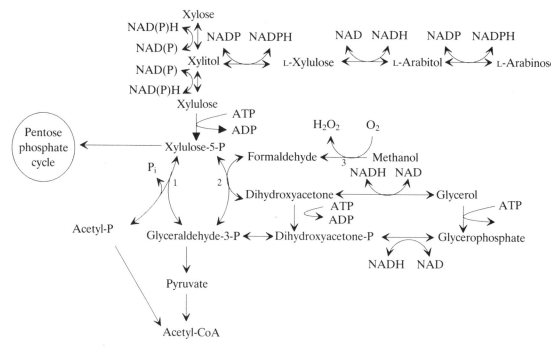

Figure 10.3 Metabolism of xylose, L-arabinose, glycerol and methanol. Special enzymes are: 1, xylulose-5-phosphate phosphoketolase; 2, dihydroxyacetone synthase; 3, alcohol oxidase.

1985a) but also by *in vitro* enzyme assays (Pitt and Mosley, 1985b). The enzymatic activities of the two key enzymes were, however, very low and variable and only crude extracts have been assayed where a number of artefacts could occur. *In vitro* evidence suggesting a non-phosphorylative alternative of the ED pathway was reported for *P. notatum* (Pitt and Mosley, 1985b) and *A. niger* (Elzainy *et al.*, 1973). The available evidence supports the conclusion that ED metabolism does not occur in the majority of fungi and the few special cases where it might occur await final confirmation.

(c) Other pathways

An alternative to the last part of the EMP pathway has been proposed. Dihydroxyacetone-phosphate could be transformed to methylglyoxal (pyruvaldehyde) and phosphate by methylglyoxal synthetase. Glyoxylase could transform methylglyoxal to D-lactate which can then be oxidized to pyruvate. Some of the enzymes involved occur or can be induced in a number of yeasts (Babel and

Hofmann, 1981). The biological significance of this phenomenon remains obscure.

Metabolism of D-xylose (and L-arabinose) involves an initial reduction to xylitol, followed by an oxidation to xylulose and a subsequent phosphorylation to xylulose 5-phosphate (McCraken and Gong, 1983) (Figure 10.3). Only when the initial reduction can be carried out with NADH as reductor is anaerobic metabolism of xylose possible; a large overproduction of NADH would otherwise occur (Van Dijken and Scheffers, 1986). An alternative pathway might be induced in some yeasts grown on xylose. Under these conditions a phosphoketolase is induced which converts xylulose-5-phosphate into triose phosphate and acetylphosphate (Ratledge and Holdsworth, 1985).

Glycerol can be metabolized, after phosphorylation to glycerophosphate, by glycerophosphate dehydrogenase or a mitochondrial glycerophosphate oxidase in, for example *C. utilis*, *S. cerevisiae*, *N. crassa*, *A. niger* and *Fusarium oxysporum* (Castro and Loureiro-Dias, 1991). Alternatively, glycerol is first oxidized to dihydroxyacetone by a

NAD⁺dependent glycerol dehydrogenase and subsequently phosphorylated to dihydroxyacetone-phosphate in, for example *S. pombe* or *Candida valida* (Castro and Loureiro-Dias, 1991) (Figure 10.3).

Methylotrophic yeasts such as *Candida boidinii* and *Hansenula polymorpha* are able to grow on methanol as sole carbon source. Under these conditions well-developed peroxisomes, with large amounts of alcohol oxidase (producing formaldehyde) and dihydroxyacetone synthase, for assimilation are induced (Goodman, 1985; Veenhuis *et al.*, 1992) (Figure 10.3). Dissimilation of the formaldehyde (after binding to glutathione) yields formic acid and CO_2 as well as NADH + H^+ (Gadd, 1988). Microbodies (glyoxysomes and peroxisomes) are also important for the metabolism of other xenobiotics such as D-amino acids and n-alkanes (Veenhuis and Harder, 1989). Several *Rhodotorula* species are even able to oxidize phenol to catechol (phenolhydroxylase) and break the aromatic ring (catechol 1,2-dioxygenase) (Katayama-Hirayama *et al.*, 1991). Several fungi, e.g. *A. nidulans*, can break down aromatic acids resulting from lignin degradation (Kuswandi and Roberts, 1992)

(d)   Estimation of relative contributions of pathways

The relative importance of the different pathways can be estimated by radiochemical methods. Several improvements on the original method of Bloom and Stetten (1953) have been published (e.g. Katz and Rognstad, 1967; Wang, 1972; Van Laere and Carlier, 1974) which all take into account a number of difficulties encountered with, for example the contribution of other pathways and/or the randomization of label by the PP-pathway, gluconeogenesis, mannitol or glycerol cycling. Contributions of different pathways in fungi vary between species and can alter with the growth medium (Cochrane, 1976). Although changes in the relative contribution of the different pathways have been noted, e.g. during germination or sporulation (Pitt and Mosley, 1985a; Van Laere *et al.*, 1987), no general pattern has emerged from these studies (Cochrane, 1976). It is likely that these changes are the consequence rather than a cause of the differentiation processes.

10.3   LIPIDS

Lipid biosynthesis and degradation has been only sparingly investigated in fungi but is reviewed by Weete (1980) and Chopra and Khuller (1984). Most of our knowledge comes from extrapolation of metabolic data from other organisms supported by investigations mainly with *Saccharomyces* or *Candida* sp.

Biosynthesis of fatty acids (Figure 10.4) takes place in the cytosol and starts with carboxylation of acetyl-CoA to malonyl-CoA. From this malonyl-CoA consecutive C_2-units are added to acetyl-CoA or the growing fatty-CoA ester chain by a complex fatty acid synthase complex harbouring seven different enzymatic activities. Introduction of double bonds is a microsomal process requiring, besides NADH and oxygen as cosubstrates, cytochrome b_5 and other electron transport components. CoA esters as well as polar lipids such as phosphatidylcholine are substrates for desaturation. Some uncommon fatty acids carrying hydroxy-, epoxy- or methyl groups, or acetylenic bonds also occur in some fungi (Chopra and Khuller, 1984). Biosynthesis of phospholipids is dealt with in Chapter 4. Phosphatidic acid is also the primary intermediate in glycerolipid synthesis. After hydrolysis to diacylglycerol, a third fatty acid is transferred from fatty-acyl-CoA by acyltransferase. Synthesis of triacylglycerols was shown to occur in lipid particles of yeast (Christiansen, 1978).

Lipids and phospholipids can be degraded by lipases and a variety of phospholipases to constituent fatty acids, glycerol, phosphate and bases (Chopra and Khuller, 1984). Fatty acid metabolism can occur using three different pathways.

1. Oxidation in the α-position requires a peroxidase and an aldehyde dehydrogenase. It is probably of limited importance but was shown to occur in *C. utilis* (Fulco, 1967).
2. As far as is investigated, the classical β-oxidation is most common but in fungi the process does not occur in the mitochondria but in the peroxisomes (Kawamoto *et al.*, 1978; Park *et al.*, 1991). The $FADH_2$ generated in the process is not oxidized by the respiratory chain but via an oxidase whereas NADH + H^+ is brought to the mitochondria by the glycerophosphate/dihydroxyacetone-phosphate shuttle (Fukui and Tanaka, 1981; Boulton and Ratledge, 1984).

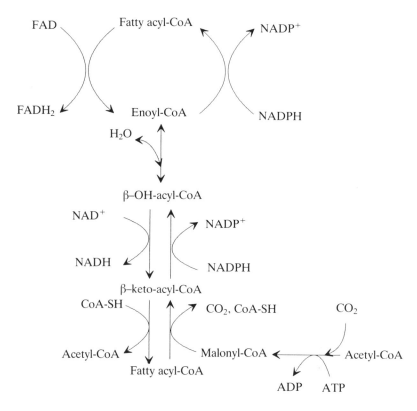

Figure 10.4 Synthesis and β-oxidation of fatty acids.

Acetylcarnitine might represent a shuttling vector of acetyl-units between the peroxisome and mitochondrion (Ueda *et al.*, 1982) and carnitine acetyltransferase is essential for growth on acetate or fatty acids in *A. nidulans* (Midgley, 1993). Propionate units originating from odd-numbered fatty acids may be metabolized by a methyl-citric acid cycle in fungi (Uchiyama *et al.*, 1982).

3. Oxidation in the ω-position is probably also of limited importance for fatty acid metabolism. However, a number of fungi such as *Cladosporium resinae* are able to use n-alkanes or even more recalcitrant hydrocarbons as a carbon and energy source (Walker and Cooney, 1973; Griffin and Cooney, 1979). Metabolism of these products (as well as other xenobiotics) can be initiated by NADP$^+$-dependent ω-oxidation using cytochrome P_{450} (Sariaslani, 1991). The resulting alcohols can be oxidized by alcohol

and aldehyde dehydrogenases or oxidases (Fukui and Tanaka, 1981; Boulton and Ratledge, 1984). The resulting fatty acids are broken down by classical β-oxidation.

10.4 METABOLISM OF PYRUVATE AND ACETATE

10.4.1 The TCA or Krebs cycle

The pyruvate generated in glycolysis can be converted into acetyl-CoA by pyruvate dehydrogenase. Under aerobic conditions this acetyl-CoA (as well as that generated from fatty acid breakdown) can be oxidized to CO_2 in the mitochondria via the Krebs- or tricarboxylic acid (TCA) cycle (Figure 10.5). This cycle yields some ATP but mainly NADH + H$^+$ and FADH$_2$ which are reoxidized by O$_2$ via the respiratory chain, yielding large amounts of ATP. Intermediates of the

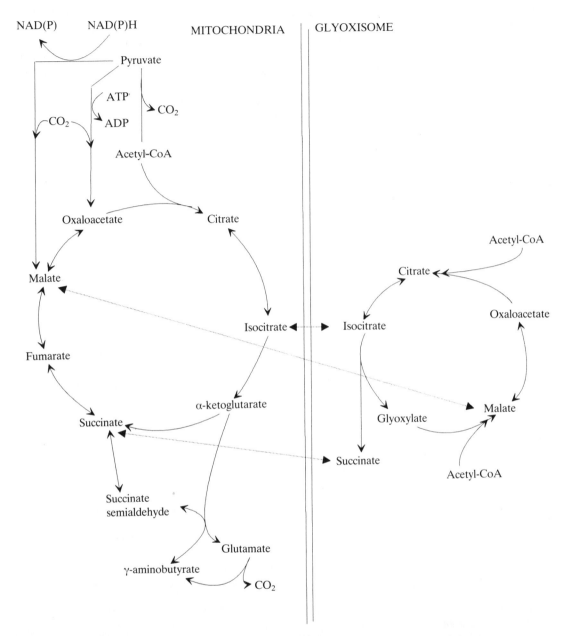

Figure 10.5 Metabolism of acetate by the TCA cycle (left) and glyoxylate cycle (right). The figure also illustrates possible shunting of the TCA cycle over γ-aminobutyrate or the glyoxylate cycle and anaplerotic reactions.

cycle can also be withdrawn for biosynthetic purposes, especially α-ketoglutarate for glutamate and oxaloacetate for aspartate synthesis.

Due to the instability of a number of TCA-cycle enzymes, especially α-ketoglutarate dehydrogenase (Meixner-Monori *et al.*, 1985), the operation of a complete TCA cycle has long been in doubt for a number of fungi. The TCA-cycle and its regulation in filamentous fungi have been the subject of a recent review (Kubicek, 1988). Interpretation of results is often complicated by the fact that the coexistence of mitochondrial and cytosolic (and even glyoxisomal or peroxisomal) isoenzymes has not always been taken into account. Kubicek (1988) concluded that present evidence supports operation of the TCA cycle in fungi. Available evidence from a limited number of fungi further suggests that the activity of the cycle is regulated by the NADH/NAD$^+$ ratio (and hence by the activity of the respiratory chain and the AMP/ATP ratio) at the level of isocitrate dehydrogenase and α-ketoglutarate dehydrogenase. A major difficulty is the lack of understanding of the relative contributions by mitochondrial NAD$^+$- and the cytoplasmic NADP$^+$-dependent isocitrate dehydrogenases.

The glyoxylate cycle can of course be seen as an alternative to part of the TCA cycle circumventing α-ketoglutarate dehydrogenase which is difficult to detect in fungi. It supposes massive shuttling of C$_6$ and C$_4$ compounds between compartments. Another bypass of the same part of the TCA cycle is provided along glutamate, γ-aminobutyrate, succinate semialdehyde and succinate (Guignard and Brody, 1983) (Figure 10.5).

10.4.2 Anaplerotic reactions

During growth on glycolytic substrates, TCA cycle intermediates are withdrawn from the cycle, e.g. for the synthesis of aspartate and glutamate. In order to avoid halting of the cycle, C$_4$ compounds have to be provided by alternative means (Figure 10.5).

(a) CO$_2$-fixation

Pyruvate carboxylase (catalysing pyruvate + CO$_2$ + ATP → oxaloacetate + ADP + P$_i$) appears to be the main anaplerotic reaction in fungi. The enzyme is inhibited by aspartate and activated by acetyl-CoA in all fungi examined (Hilgenberg *et al.*, 1987). CO$_2$- fixation is very common in fungi (Hilgenberg *et al.*, 1987). Its contribution to organic carbon is very small, even under oligotrophic conditions, and label distribution from ^{14}CO$_2$ can be explained by anaplerotic reactions (Parkinson *et al.*, 1991).

Another possible anaplerotic reaction is malic enzyme: pyruvate + CO$_2$ + NAD(P)H + H$^+$ →

malate + NAD(P)$^+$. Although the enzyme occurs in a number of fungi, its properties do not suggest that its function is anaplerotic *in vivo* (Casselton, 1976). In *A. nidulans* however, it is present in high concentrations in acetate-grown cells (Dijkema and Visser, 1987). The glyoxylate cycle can also function anaplerotically.

(b) The glyoxylate cycle

During metabolism of fat or C$_2$-compounds, the glyoxylate cycle can also generate C$_4$ compounds to replenish the TCA-cycle (Figure 10.5). Depending on the nature of the reserve (fat or carbohydrate) and external substrate (glycolytic or gluconeogenic), the glyoxylate cycle can fulfil different roles:

1. gluconeogenic: producing C$_4$ units destined for PEP formation by PEP carboxykinase;
2. anaplerotic: preventing exhaustion of TCA cycle intermediates;
3. catabolic: generating C$_4$ units for production of reducing power and ATP for biosynthesis.

Galons *et al.* (1990) estimated that in wild type *S. cerevisiae*, 58% of acetate flux passed through the glyoxylate shunt versus 42% through oxidative reactions.

The two enzymes unique to the glyoxylate cycle, i.e. isocitrate lyase and malate synthase, occur in all classes of fungi and are probably located in glyoxisome or peroxisome-like microbodies (Maxwell *et al.*, 1975). Some fungi, e.g. *S. pombe* (Tsai *et al.*, 1989) lack glyoxylate cycle enzymes and cannot grow on acetate as sole carbon source, although they can assimilate acetate. In some organisms such as *N. crassa* (Schwitzguebel *et al.*, 1981) and *C. tropicalis* (Fukui and Tanaka, 1979) these organelles appear to be unable to synthesize citrate and cooperation between mitochondria and glyoxisomes is necessary with a shuttling of C$_6$ and C$_4$ compounds between the two organelles (Casselton, 1976). In *S. cerevisiae*, however, the citrate synthase isoenzyme coded by *CIT2* is peroxisomal (Lewin *et al.*, 1990) and yeast can probably complete the glyoxylate cycle within the peroxisome. The activity of glyoxylate cycle enzymes has been shown to increase under gluconeogenic conditions, e.g. with acetate as substrate, and to decrease in the presence of glucose, e.g. in *M. racemosus* (O'Connell and Paznokas, 1980), *S.*

cerevisiae (Lopez Boado *et al.*, 1988) *C. maltosa* (Polnish and Hofmann, 1989) and *P. blakesleeanus* (Rua *et al.*, 1990a).

The fungal isocitrate lyases investigated to date are inhibited by their products, succinate and glyoxylate, or gluconeogenic derivatives (PEP and fructose 1,6-bisphosphate) (Rua *et al.*, 1990b). Initial (partial) inactivation of isocitrate lyase after addition of glucose depends on (cAMP-dependent ?) phosphorylation of the enzyme in yeast (Lopez-Boado *et al.*, 1988) and results with *P. blakesleeanus* (Rua *et al.*, 1990a) and *C. maltosa* (Polnish and Hofmann, 1989) suggest a similar mechanism. In bacteria such as *Escherichia coli*, however, flux through the glyoxylate shunt is mainly regulated by phosphorylation (and inactivation) of isocitrate dehydrogenase (Holms, 1987).

10.4.3 Respiration

The large amounts of NADH + H$^+$ (and FADH$_2$) should be reoxidized to allow continued operation of glycolysis and the TCA-cycle. Under aerobic conditions this is performed with oxygen as electron acceptor via the respiratory electron transport chain localized in the inner mitochondrial membrane. Besides the ubiquitous cytochrome-dependent and ATP-generating path for the oxidation of NADH + H$^+$, an alternative pathway branches from ubiquinone and transfers electrons via an unknown oxidase towards oxygen in many fungi. Only heat, and not ATP, is generated in this last step. This cyanide- and azide-insensitive pathway can be inhibited by salicylhydroxamic acid and is rather common in plants and fungi. It has been described in all major classes of fungi (Henry and Nyns, 1975). The quantitative importance of this path varies greatly depending on culture conditions and developmental stage, e.g. in *Aureobasidium pullulans* (Gadd and Edwards, 1981), *C. albicans* (Aoki and Ito-Kuwa, 1984) and *M. rouxii* (Cano-Conchola *et al.*, 1988). It can be induced by factors inhibiting functioning of the normal cytochrome-dependent pathway or inhibitors of mitochondrial RNA and protein synthesis. Spores and stationary phase cells may also exhibit cyanide-insensitive respiration. The biological function of this pathway is not clear. In some plants it might be related to heat generation, e.g. in the spadix of *Arum*, or cold resistance, e.g. in *Cornus*, but in other plants (Laties, 1982) and in

fungi (Akimenko *et al.*, 1983) it looks more like an overflow mechanism.

10.4.4 Fermentation

In the absence of oxygen or when, as in *S. cerevisiae*, respiration is repressed by an excess of glucose, electrons from NADH + H$^+$ produced in glycolysis are transferred to organic electron acceptors yielding fermentation products. Although fungi can produce a variety of substances in rather large quantities, the main fermentation products are lactate and more commonly ethanol. Besides the work of Lejohn (1971) on lactate dehydrogenase in the 'Phycomycetes' and a number of studies on the enzyme from *P. blakesleeanus* (Busto *et al.*, 1984) few data have been presented on lactate dehydrogenase. The enzyme is inhibited by GTP in Oomycetes and by ATP in Chytridiomycetes and Zygomycetes (Lejohn, 1971).

A number of anaerobic chytridiomycete-like fungi, isolated from the digestive tract of herbivores produce formate, lactate, acetate, ethanol and hydrogen as fermentation products (Teunissen *et al.*, 1991; Kostyukovsky *et al.*, 1991). Although the biochemistry of these processes has not been elucidated, it is very similar to fermentation by *E. coli* and probably involves splitting of pyruvate in acetate and formate moieties; formate can then be split into hydrogen and CO$_2$.

Glycerol is a rather special fermentation product since glycerol fermentation does not yield energy. Nevertheless, some organisms produce great quantities of glycerol even under aerobic conditions. Glycerol production is often correlated with osmotic stress and many fungi appear to rely at least partially on glycerol for turgor restoration (Beever and Laracy, 1986; Van Laere and Hulsmans, 1987; André *et al.*, 1988; Bellinger and Larher, 1988; Yagi *et al.*, 1992). Glycerol production in *Phycomyces* spores is controlled by the activity of a specific glycerol-3-phosphatase. The activity of this extremely unstable enzyme is controlled by cAMP-dependent phosphorylation. Phosphorylation makes the enzyme less sensitive to inhibition by physiological concentrations of phosphate (Van Schaftingen and Van Laere, 1985). Glycerol-3-phosphatase also hydrolyses dihydroxyacetone phosphate, so that NADPH can contribute to glycerol production via the abundantly present dihydroxyacetone reductase (Van Laere, 1986)

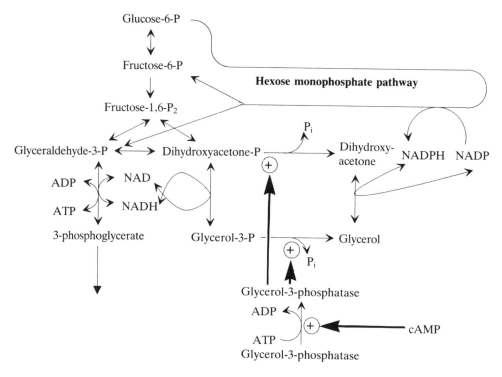

Figure 10.6 Regulation of glycerol production in germinating spores of *Phycomyces blakesleeanus.*

(Figure 10.6). Little is known about the regulation of glycerol production in other organisms. We also do not know whether the synthesis of other polyols is regulated by the action of specific isomerases, dehydrogenases or phosphatases. Further discussion of polyol metabolism and function is available (Jennings, 1984; Rast and Pfyffer, 1989; Jennings and Burke, 1990; Lewis, 1991).

10.5 NITROGEN METABOLISM

10.5.1 Nitrogen sources

Since Millbank (1969) concluded that fungi cannot fix atmospheric nitrogen, two later reports concerning *Pleurotus ostratus* (Gunterova and Gallon, 1979) and *P. sajor-caju* (Thayumanavan, 1980) claimed that these fungi fixed nitrogen and had nitrogenase activity. However, the evidence is not completely irrefutable and, according to Wainwright (1988), nitrogen fixation in fungi remains unproven. Some fungi might contribute to (non-energy yielding) nitrification or even to denitrifi-

cation but their contribution is likely to be small (Dighton and Boddy, 1989). However, extensive production of N_2O and N_2 by *F. oxysporum* and many other species has been reported (Shoun *et al.*, 1992).

Most fungi can grow on nitrate as a nitrogen source. Nitrate assimilation involves reduction by NADPH-dependent nitrate reductase (2-electron transfer) and subsequent reduction of nitrite by NAD(P)H-dependent nitrite reductase (6-electron transfer) (Dunn-Coleman *et al.*, 1984; Tomsett, 1989). Some nitrate reductase genes have been cloned and sequenced (Okamoto *et al.*, 1991)

A number of fungi, e.g. *S. cerevisiae*, lack the ability to metabolize nitrate and require reduced nitrogen for growth. Ammonium salts generally fulfil this requirement but they can be replaced by a number of organic nitrogen compounds ranging from urea and amino acids to proteins which are good nitrogen and carbon sources for a number of fungi (Kalisz *et al.*, 1986). Some fungi, such as *P. blakesleeanus*, prefer asparagine (Hilgenberg *et al.*, 1987) or other amino acids as N-source whereas

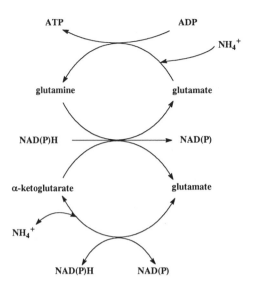

Figure 10.7 Possible pathways of ammonia incorporation: 1, glutamate dehydrogenase; 2, glutamine synthetase; 3, glutamine: α-ketoglutarate-aminotransferase (GOGAT).

others, such as the yeast form of *M. racemosus* (Peters and Sypherd, 1979), require glutamate for growth. For a more complete review of regulatory aspects of nitrogen metabolism see Pateman and Kinghorn (1975) and Marzluf (1981).

10.5.2 Ammonia assimilation

Whatever the nature of the nitrogen in the medium, ammonia is ultimately generated, be it by the action of nitrate and nitrite reductase, urease, proteases, oxidases or dehydrogenases. Even when glutamate is used as carbon and nitrogen source, some of the nitrogen has to be converted into ammonia in order to generate glutamine. In principle, there are four ways to fix ammonia in organic molecules (Figure 10.7).

1. NAD$^+$-dependent glutamate dehydrogenase is a cytosolic enzyme in *A. nidulans* (Stevens *et al.*, 1989) but is not believed to play an important role in ammonia fixation in higher fungi. The kinetic properties of the enzyme, the equilibrium constant and the evolution of activities in different conditions suggest that the enzyme is only involved in glutamate breakdown *in vivo* (Pateman and Kinghorn, 1975; Marzluf, 1981; Vierula and Kapoor, 1987; Miller and Magasanik,

1990). Nevertheless, it might be the main entry of ammonia in lower fungi since, according to Lejohn (1971), NADP$^+$-dependent glutamate dehydrogenases are not present in these fungi. However, a NADP$^+$-specific glutamate dehydrogenase has since been described in *M. racemosus* (Peters and Sypherd, 1979) and different regulatory properties have been ascribed to the purified *P. blakesleeanus* enzyme (Van Laere, 1988). The NAD$^+$-dependent glutamate dehydrogenase genes from *N. crassa* (Vierula and Kapoor, 1989) and *S. cerevisiae* (Frederick and Kinsey, 1990; Miller and Magasanik, 1990) have been cloned and sequenced. The possible relationship of this enzyme with dimorphism in *Mucor* spp. is discussed by Orlowski (1991).

2. NADP$^+$-dependent glutamate dehydrogenase has been detected in all Ascomycetes, Basidiomycetes and Deuteromycetes investigated (e.g. Lejohn, 1971). This enzyme is more likely to be responsible for ammonia fixation taking into account its kinetic properties and the more favourable NADPH/NADP$^+$ ratio. Moreover, the evolution of activity in a number of fungi is consistent with a biosynthetic role which is not the case for the NAD$^+$-dependent enzyme (Pateman and Kinghorn, 1975; Marzluf, 1981). Several NADP$^+$-dependent glutamate dehydrogenase genes have been cloned and sequenced (Hawkins *et al.*, 1989).

3. Glutamine synthetase uses the energy from ATP-hydrolysis to produce the amide-nitrogen in glutamine. In plants this is the main, if not the only entry port of ammonia. From this more 'energetic' amide group, N is incorporated in a plethora of other compounds such as glutamate, glucosamine, purines, pyrimidines, asparagine, histidine or tryptophan. The genes from *S. cerevisiae* (Benjamin *et al.*, 1989) and *S. pombe* (Barel *et al.*, 1988) have been cloned and sequenced. In *N. crassa* two different glutamine synthetase genes and polypeptides have been identified (Calderon *et al.*, 1990).

4. Although glutamate synthase does not fix ammonia it is intimately associated with glutamine synthetase in the production of glutamate and together these enzymes aminate α-ketoglutarate to glutamate. However, since ATP is used in this two-step process the equilibrium is far more into the biosynthetic direction. Moreover, the affinity of glutamate

synthase for ammonia is much higher and the enzyme may become very important at low ammonium concentrations (Hummelt and Mora, 1980). NADH-dependent glutamate synthase has been described in many fungi including *S. pombe, S. cerevisiae, N. crassa, A. nidulans* and several *Candida* species (Kusnan *et al.*, 1989; Holmes *et al.*, 1989).

NADPH-dependent glutamate synthase has been described in *K. fragilis* (Nisbet and Slaughter, 1980) and *A. niger* (Savov *et al.*, 1986). However, care must be taken in measuring such an enzyme since small amounts of ammonia in the glutamine substrate might result in problems with abundantly present NADPH-dependent glutamate dehydrogenase. It has been suggested on the basis of genetic data that there are two glutamate synthase activities in yeast (Folch *et al.*, 1989). Whether the enzyme will prove to be ubiquitous in fungi, especially in the lower fungi lacking NADPH-dependent glutamate dehydrogenase, remains to be investigated.

The relative importance of the different pathways of ammonia fixation has been estimated in *N. crassa* and *A. nidulans* and might vary between fungi and in different developmental stages or culture conditions (Hummelt and Mora, 1980; Kusnan *et al.*, 1989; Wilcock *et al.*, 1992).

10.5.3 Amino acid metabolism

From glutamate and glutamine, nitrogen can be incorporated into all other nitrogen compounds by transaminases and other enzymes. Glutamine can be hydrolysed to glutamate and ammonia by the action of glutaminase. In *S. cerevisiae* (Soberon *et al.*, 1989) and *N. crassa* (Calderon *et al.*, 1985) two glutamine aminotransferases (pyruvate- and glyoxylate-dependent) have been described. The α-ketoglutaramate resulting from this reaction is converted by ω-amidase to α-ketoglutarate and ammonium. Glutamate can also be metabolized along the γ-aminobutyric acid shunt by the action of glutamate decarboxylase, γ-aminobutyrate transaminase and succinate semialdehyde dehydrogenase (Christensen and Schmit, 1980).

It is beyond the scope of this chapter to survey the synthesis and breakdown of all amino acids, nucleotides and other nitrogen-containing compounds. It is, however, worthwhile mentioning

that lysine can be synthesized along two different pathways, the α-aminoadipic acid (AAA) pathway or the diaminopimelic acid (DAP) pathway. Oomycetes and Hyphochytridiomycetes, which in a number of other characteristics are rather similar to plants, use the DAP-pathway whereas the 'true fungi' synthesize lysine along the AAA-pathway (Vogel, 1965). Enzymes of the tryptophan synthase complex are also aggregated in a different way in the Oomycetes (Hütter and De Moss, 1967).

The urea cycle plays a pivotal role in N-metabolism since it is at the crossroads between vacuolar storage, formation of urea and metabolism to polyamines (Figure 10.8). Urea can be broken down to CO_2 and NH_3 by urease; in *S.cerevisiae*, however, urea is first carboxylated to alophanate and then hydrolysed to CO_2 and NH_3 (Slaughter, 1988). Arginine metabolism in *N. crassa* and *S. cerevisiae* has been reviewed by Davis (1986). A role of urea as a signal molecule in *Mucor* hyphal development was proposed by Inderlied *et al.* (1985). The evidence is, however, largely indirect and it is not clear how urea could fulfil its role as a morphogenetic signal (Orlowski, 1991). The synthesis of aromatic amino acids in yeast has been reviewed by Braus (1991).

10.5.4 Polyamines

Stevens and Winter (1979) and Tabor and Tabor (1985) have covered the earlier work on biosynthesis and effects of polyamines on fungi. It is now clear that fungi, in contrast to plants and bacteria, use only ornithine decarboxylase (ODC) and not arginine decarboxylase for the production of putrescine. Consequently difluoromethylornithine (DFMO) and diaminobutanone but not difluoromethyl-arginine inhibit polyamine biosynthesis. It remains to be investigated whether the Oomycetes, which have a number of other plant-like characteristics, also only use the ODC pathway. ODC activity has already been demonstrated in the Oomycete *Achlya ambisexualis* (Wright *et al.*, 1982).

There are a large number of fungi where it has been demonstrated that spore germination or growth is accompanied by increases in polyamine content and/or that inhibitors of ODC can interfere with these processes. These include a number of Mucorales (Ruiz-Herrera and Calvo-Mendez, 1987;

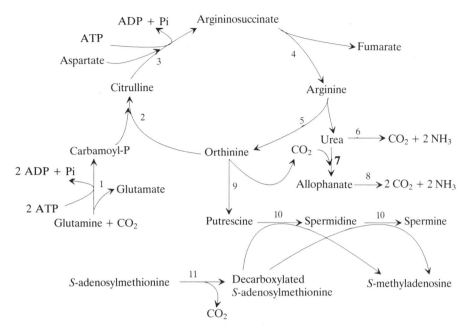

Figure 10.8 The urea cycle and the production of polyamines in fungi. Enzymes involved are: 1, CO_2: glutamine amidoligase; 2, ornithine carbamoyltransferase; 3, citrulline: aspartate ligase; 4, argininosuccinate: arginine lyase; 5, arginase; 6, urease; 7 and 8, urea: carbon dioxide ligase; 9, ornithine decarboxylase; 10, aminopropyltransferase; 11, *S*-adenosylmethionine decarboxylase

Calvo-Mendez *et al.*, 1987) and several other pathogenic and saprophytic species (Smith *et al.*, 1990; Garcia *et al.*, 1991; Singhania *et al.*, 1991). In a number of cases these increases in polyamine content were shown to correspond with increases in ODC activity (Stevens and Stevens, 1981; Shapira *et al.*, 1989).

The increase in spermidine content in germinating *M. rouxii* spores coincided with an increase in *S*-adenosylmethionine decarboxylase activity (Calvo-Mendez and Ruiz-Herrera, 1991). Yeast–mycelium (Y–M) transitions in a number of *Mucor* species are also accompanied by increases in polyamine content and ODC activity (Calvo-Mendez *et al.*, 1987). Y–M transitions and also sporulation can be inhibited by diaminobutanone, a competitive inhibitor of ODC, in a number of Mucorales (Martinez-Pacheco *et al.*, 1989). The role of polyamines in *Mucor* dimorphism is reviewed by Orlowski (1991) as discussed in section 19.4.4.

One major problem in assessing the role of polyamines is that their cellular distribution is not known except in *N. crassa* where large vacuolar concentrations appear to be associated with polyphosphate anions (Davis, 1986). Another problem is that large quantities can be bound to other molecules and it is not known whether the free or bound polyamines are most important. Although the role of polyamines in cell differentiation seems well established, their mode of action remains largely speculative. Recently published data suggest that polyamines might be involved in regulation of gene expression through decreasing methylation of DNA (Cano *et al.*, 1988) or metabolic regulation by stimulating cAMP-dependent protein kinase (Guthmann *et al.*, 1990).

10.6 PHOSPHATE METABOLISM

Polyphosphates covering a wide range of molecular sizes can be found in a variety of fungi (Harold, 1966). However, polyphosphates could not be detected in any of the Oomycetes investigated (Chilvers *et al.*, 1985). Whether the greater sensitivity of Oomycetes to phosphonates (Coffey and Ouimette, 1989) may be due to their inability of

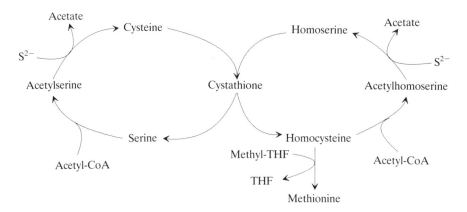

Figure 10.9 Incorporation of sulphide into amino acids. THF, tetrahydrofolate.

detoxifying them (perhaps into polyphosphonates) is not clear.

Synthesis of polyphosphates is catalysed by a ATP-polyphosphate-phosphotransferase (polyphosphate kinase) and polyphosphate is broken down by an (exo-) polyphosphatase in fungi. In some fungi a 1,3-phosphoglycerate polyphosphate phosphotransferase (see Wood and Clark, 1988) and a polyphosphate depolymerase (endopolyphosphatase) were described and the operation of a pyrophosphate polyphosphate phosphotransferase was suggested (Kulaev, 1975). Polyphosphate glucokinase or AMP:polyphosphate phosphotransferses have not been detected in fungi (Wood and Clark, 1988), nor pyrophosphate dependent fructose-6-phosphate kinase. Mutations in three phosphatase genes had no effect on polyphosphate metabolism in *S. pombe* and *S. cerevisiae* (Müller *et al.*, 1992).

Suggestions for the roles of polyphosphate include neutralization of cell wall cations (Sypherd *et al.*, 1978), neutralization of vacuolar arginine (Davis, 1986), a phosphate reserve, reserve for ATP synthesis, metal chelation, and glucose uptake (Wood and Clark, 1988). Jakubowski (1986) found an accumulation of adenosine 5'-tetra and pentaphosphates during sporulation of yeast cells and suggested that they might originate from polyphosphates. However, a possible signalling role has to be confirmed. Lejohn (1983) has discussed the occurrence of polyphosphorylated compounds in relation to sporulation of *Achlya* sp.

10.7 SULPHUR METABOLISM

Although SO_2 and the related chemicals sulphite, bisulphite and metabisulphite are toxic to fungi (Rose, 1989) a number of fungi can oxidize elemental sulphur to thiosulphate, tetrathionate or sulphate (Wainwright, 1989).

$$S \rightarrow S_2O_3^{2-} \rightarrow S_4O_6^{2-} \rightarrow SO_4^{2-}$$

In some fungi, thiosulphate oxidation appears to contribute to the energy balance since it causes an increase in biomass production under carbon-limiting conditions (Jones *et al.*, 1991). Normal sulphur metabolism, however, involves the reduction of sulphate to sulphide. After binding of sulphate into adenosine 5'-phosphosulphate (APS) and probably phosphorylation to PAPS (3'-phosphoadenosine-5'-phosphosulphate) sulphate is reduced to sulphite and carrier bound (or free) sulphide (Slaughter, 1989). Reduced sulphur is further incorporated in organic compounds by reaction of the free or carrier-bound sulphide with acetylserine or acetylhomoserine to yield cysteine and homocysteine. Cysteine and homocysteine can also be interconverted *via* cystathione. Methylation of homocysteine yields methionine (Figure 10.9). Glutathione has been proposed as an endogenous reserve of reduced sulphur in *S. cerevisiae* (Elskens *et al.*, 1991).

S-containing amino acids are of course alternative sources of sulphur. Exceptionally sulphate does not support growth but sulphide, thiourea or S-containing amino acids are necessary (Ingraham and Emerson, 1954). Some fungi produce volatile sulphur compounds such as hydrogen sulphide, methyl mercaptan or dimethylsulphide (Slaughter, 1989). Although most of these products probably originate by 'accidental losses', some of them may act as attractants for animals and assist in the distribution of spores (Bellina-Agostinone *et al.*, 1987).

10.8 CONCLUSIONS

Much of our knowledge of fungal metabolism results from extrapolation of a number of model systems which are generally of industrial importance, e.g. *S. cerevisiae* and *A. niger*. Fungi promise to become increasingly important for agriculture (e.g. mycorrhiza, mycoherbicides, entomophagous and nematophagous fungi) or biotechnology and a more profound investigation in their potential is certainly warranted. The advent of molecular biological techniques and the use of specific probes derived, e.g. from yeast, should soon boost investigations into specific enzymes, pathways and their regulation at the genomic level in different classes of fungi. This will ultimately result in a better appreciation of the similarities and differences in the fungal kingdom.

REFERENCES

Akimenko, V.K., Trutko, S.M., Medentsev, A.G. and Korobov, V.P. (1983) Distribution of cyanide -resistant respiration among yeasts and bacteria and its relation to oversynthesis of metabolites. *Archives of Microbiology*, **136**, 234–41.

André, L., Nilson, A. and Adler, L. (1988) The role of glycerol in the osmotolerance of the yeast *Debaryomyces hansenii*. *Journal of General Microbiology*, **134**, 669–77.

Aoki, S. and Ito-Kuwa, S. (1984) The appearance and characterization of cyanide-resistant respiration in the fungus *Candida albicans*. *Microbiology and Immunology*, **28**, 393–406.

Arguelles, J.C. and Gacto, M. (1988) Differential location of regulatory and nonregulatory trehalases in *Candida utilis*. *Canadian Journal of Microbiology*, **31**, 529–37.

Arguelles, J.C., Vicente-Soler, J. and Gacto, M. (1986) Protein phosphorylation and trehalase activation in *Candida utilis*. *FEMS Microbiology Letters* **34**, 361–5.

Babel, and Hofmann, K.H. (1981) The conversion of triosephosphate via methylglyoxal, a bypass to the glycolytic sequence in methylotrophic yeasts? *FEMS Microbiology Letters*, **10**, 133–6.

Barel, I., Bignell, G., Simpson, A. and MacDonald, D. (1988) Isolation of a DNA fragment which complements glutamine synthetase deficient strains of *S. pombe*. *Current Genetics*, **13**, 487-94.

Baute, M.A., Baute, R. and Deffieux, G. (1989) Fungal enzymic activity degrading 1,4-α-D-glucans to 1,5-D-anhydrofructose. *Phytochemistry*, **27**, 3401–3.

Beever, R.E. and Laracy, E.P. (1986) Osmotic adjustment in the filamentous fungus *Aspergillus nidulans*. *Journal of Bacteriology*, **168**, 139–42.

Bell, W., Klaassen, P., Ohnacker, M. *et al.* (1992) Characterization of the 56-kDa subunit of yeast trehalose-6-phosphate synthase and cloning of its gene reveal its identity with the product of CIF1, a regulator of carbon catabolite inactivation. *European Journal of Biochemistry*, **209**, 951–9.

Bellina-Agostinone, C., D'Antonio, M. and Pacioni, G. (1987) Odour composition of the summer truffle, *Tuber aestivum*. *Transactions of the British Mycological Society*, **88**, 568–9.

Bellinger, Y. and Larher, F. (1988) A ^{13}C comparative nuclear magnetic resonance study of organic solute production and excretion by the yeasts *Hansenula anomala* and *Saccharomyces cerevisiae* in saline media. *Canadian Journal of Microbiology*, **34**, 605–12.

Benjamin, P.M., Wu, J.L., Mitchell, A.P. and Magasanik, B. (1989) Three regulatory systems control expression of glutamine synthetase in *Saccharomyces cerevisiae* at the level of transcription. *Molecular and General Genetics*, **217**, 370–7.

Berry, D.R. (1975) The environmental control of the physiology of filamentous fungi, in *The Filamentous Fungi*, Vol. 1, *Industrial Mycology*, (eds. J.E. Smith and D.R. Berry), Edward Arnold, London, pp. 16–32.

Biswas, M., Singh, B. and Datta, A. (1979) Induction of *N*-acetylmannosamine catabolic pathway in yeast. *Biochimica et Biophysica Acta*, **585**, 535–42.

Bloom, B. and Stetten, D. (1953) Pathways of glucose catabolism. *Journal of the American Chemical Society*, **75**, 5446.

Boulton, C.A. and Ratledge, C. (1984) The physiology of hydrocarbon utilizing microorganisms, in *Topics in Enzyme and Fermentation Biotechnology*, Vol. 9, (ed. H. Wiseman), Ellis Horwood, Chichester, pp. 11–77.

Bourret, J.A., Flora, L.L. and Carnell, L. (1991) Cyclic AMP regulation of fructose metabolism in germinating *Pilobolus longipes* spores. *Experimental Mycology*, **15**, 44–54.

Bourret, J.A., Flora, L.L. and Ferrer, L.M. (1989) Trehalose mobilization during early germination of *Pilobolus longipes* sporangiospores. *Experimental Mycology*, **13**, 140–48.

Braus, G.H. (1991) Aromatic amino acid biosynthesis in the yeast *Saccharomyces cerevisiae*; a model system for the regulation of a eukaryotic biosynthetic pathway. *Microbiological Reviews*, **55**, 349–70.

Burlini, N., Morandi, S., Pellegrini, R. *et al.* (1989) Studies on the degradative mechanism of phosphoenolpyruvate carboxykinase from the yeast *Saccharomyces cerevisiae*. *Biochimica et Biophysica Acta*, **1014**, 153–61.

Busto, F., de Arriaga, D. and Soler, J. (1984) The kinetic mechanism of pyruvate reduction of lactate dehydrogenase from *Phycomyces blakesleeanus*. *International Journal of Biochemistry*, **16**, 171–6.

Calderon, J., Martinez, L.M. and Mora, J. (1990) Isolation and characterisation of a *Neurospora crassa* mutant altered in the alfa polypeptide of glutamine synthetase. *Journal of Bacteriology*, **172**, 4996–5000.

Calderon, J., Morett, E. and Mora, J. (1985) ω-amidase pathway in the degradation of glutamine in *Neurospora crassa*. *Journal of Bacteriology*, **161**, 807–9.

Calvo-Mendez, C., Martinez-Pacheco, M. and Ruiz-Herrera, J. (1987) Regulation of ornithine decarboxylase activity in *Mucor bacilliformis* and *Mucor rouxii*. *Experimental Mycology*, **11**, 270–7.

Calvo-Mendez, C. and Ruiz-Herrera, J. (1991) Regulation of S-adenosylmethionine decarboxylase during the germination of sporangiospores of *Mucor rouxii*. *Journal of General Microbiology*, **137**, 307–14.

Cano, C., Herrera-Estrella, L. and Ruiz-Herrera, J. (1988) DNA methylation and polyamines in regulation of development of the fungus *Mucor rouxii*. *Journal of Bacteriology*, **170**, 5946–8.

Cano-Canchola, C., Escamilla, E. and Ruiz-Herrera, J. (1988) Environmental control of the respiratory system in the dimorphic fungus *Mucor rouxii*. *Journal of General Microbiology*, **134**, 2993–3000.

Casselton, P.J. (1976) Anaplerotic pathways, in *The Filamentous Fungi*, Vol. 2, *Biosynthesis and metabolism*, (eds. J.E. Smith and D.R. Berry), Edward Arnold, London, pp. 121–36.

Castro, I.M. and Loureiro-Dias, M.C. (1991) Glycerol utilization in *Fusarium oxysporum* var. *lini*: regulation of transport and metabolism. *Journal of General Microbiology* **137**, 1497–502.

Caubet, R., Guerin, B., and Guerin, M. (1988) Comparative studies on the glycolytic and hexose monophosphate pathways in *Candida parapsilosis* and *Saccharomyces cerevisiae*. *Archives of Microbiology*, **149**, 324–9.

Chilvers, G.A., Lapeyrie, F.F. and Douglass, P.A. (1985) A contrast between oomycetes and other taxa of mycelial fungi in regard to metachromatic granule formation. *New Phytologist*, **99**, 203–10.

Chopra, A. and Khuller, G.K. (1984) Lipid metabolism in fungi. *Critical Reviews in Microbiology*, **11**, 209–50.

Christensen, R.L. and Schmit, J.C. (1980) Regulation of glutamic acid decarboxylase during *Neurospora crassa* conidial germination. *Journal of Bacteriology*, **144**, 983–90.

Christiansen, K. (1978) Triacylglycerol synthesis in lipid particles from baker's yeast (*Saccharomyces cerevisiae*). *Biochimica et Biophysica Acta*, **530**, 78–90.

Cochrane, V.W. (1976) Glycolysis, in *The Filamentous Fungi*, Vol. 2, *Biosynthesis and metabolism*, (eds. J.E. Smith and D.R. Berry), Edward Arnold, London, pp. 65–91.

Coffey, M.D. and Ouimette, D.G. (1989) Phosphonates: antifungal compopunds against Oomycetes, in *Nitrogen, Phosphorus and Sulphur Utilization by Fungi*, (eds.

L. Boddy, R. Marchant and D.J. Read), Cambridge University Press, Cambridge, pp. 107–30

Colonna, W.J. and Magee, P.T. (1978) Glycogenolytic enzymes in sporulating yeasts. *Journal of Bacteriology*, **134**, 844–53.

Coulter, P.B. and Aronson, J.M. (1977) Glycogen and other soluble glucans from Chytridiomycete and Oomycete species. *Archives of Microbiology*, **115**, 317–22.

Davidson, E.A. (1960) Hexokinase from *Aspergillus parasiticus*. *Journal of Biological Chemistry*, **235**, 23–5.

Davis, R.H. (1986) Compartmental and regulatory mechanisms in the arginine pathways of *Neurospora crassa* and *Saccharomyces cerevisiae*. *Microbiological Reviews*, **50**, 280–313.

De Arriaga, D., Busto, F., Del Valle, P. and Soler, J. (1989) A kinetic study of the pH effect on the allosteric properties of pyruvate kinase from *Phycomyces blakesleeanus*. *Biochimica et Biophysica Acta*, **998**, 221–30.

Delvalle, J.A. and Assensio, C. (1978) Distribution of adenosine 5'-triphosphate (ATP)-dependent hexose kinases in microorganisms. *BioSystems*, **10**, 265–82.

Devchand, M.R. and Kapoor, M. (1987) Some aspects of the regulation of pyruvate kinase levels in *Neurospora crassa*. *Canadian Journal of Microbiology*, **33**, 322–6.

Dighton, J. and Boddy, L. (1989) Role of fungi in nitrogen, phosphorus and sulphur cycling in temperate forest ecosystems, in *Nitrogen, Phosphorus and Sulphur Utilization by Fungi*. (eds L. Boddy, L. Marchant and D.J. Read), Cambridge University Press, Cambridge, pp. 269–98.

Dijkema, C. and Visser, J. (1987) ^{13}C-NMR analysis of *Aspergillus* mutants disturbed in pyruvate metabolism. *Biochimica et Biophysica Acta*, **931**, 311–19.

Dumbrava, V.A. and Pall, M.L. (1987) Regulation of fructose 2,6-bisphosphate levels in *Neurospora crassa*. *Biochimica et Biophysica Acta*, **925**, 210–17.

Dunn-Coleman, N.S., Smarelli, J.Jr. and Garrett, R.H. (1984) Nitrate assimilation in eukaryotic cells. *International Review of Cytology*, **92**, 1–50.

Elskens, M.T., Jaspers, C.J. and Penninckx, M.J. (1991) Glutathione as an endogenous sulphur source in the yeast *Saccharomyces cerevisiae*. *Journal of General Microbiology*, **137**, 637–44.

Elzainy, T.A., Hassan, M.M. and Allam, A.M. (1973) New pathway of non-phosphorylated degradation of gluconate by *Aspergillus niger*. *Journal of Bacteriology*, **114**, 457–9.

Folch, J.L., Antaramian, A., Rodriguez, L. *et al.* (1989) Isolation and characterisation of a *Saccharomyces cerevisiae* mutant with impaired glutamate synthase activity. *Journal of Bacteriology*, **171**, 6776–81.

François, J. and Hers, H. (1988) The control of glycogen metabolism in yeast II. A kinetic study of the two forms of glycogen synthase and glycogen phosphorylase and an investigation of their interconversion in a cell-free extract. *European Journal of Biochemistry*, **174**, 561–7.

François, J., Neves, M.J. and Hers, H.G. (1991) The control of trehalose biosynthesis in *Saccharomyces cerevisiae*: evidence for a catabolite inactivation and repression of trehalose-6-phosphate synthase and trehalose-6-phosphate phosphatase. *Yeast*, **7**, 575–87.

François, J., Van Schaftingen, E. and Hers, H. (1984) The mechanism by which glucose increases fructose 2, 6-bisphosphate concentration in *Saccharomyces cerevisiae*. A cyclic AMP-dependent activation of phospho-fructokinase-2. *European Journal of Biochemistry*, **145**, 187–93.

François, J., Van Schaftingen, E. and Hers, H. (1988a) Characterization of phosphofructokinase 2 and of the enzymes involved in the degradation of fructose 2, 6-bisphosphate in yeast. *European Journal of Biochemistry*, **171**, 599–608.

François, J., Villanueva, M.E. and Hers, H. (1988b) The control of glycogen metabolism in yeast. I. Interconversion *in vivo* of glycogen synthase and glycogen phosphorylase induced by glucose, a nitrogen source or uncouplers. *European Journal of Biochemistry*, **174**, 551–9.

Frederick, G.D. and Kinsey, J.A. (1990) Nucleotide sequence and nuclear protein binding of the two regulatory sequences upstream of the am (GDH) gene in *Neurospora*. *Molecular and General Genetics*, **221**, 148–54.

Fukui, S. and Tanaka, A. (1979) Yeast peroxisomes. *Trends in Biochemical Sciences*, **4**, 246–9.

Fukui, S. and Tanaka, A. (1981) Metabolism of alkanes by yeast. *Advances in Biochemical Engineering*, **19**, 217–37.

Fulco, A.J. (1967) Chain-elongation, 2-hydroxylation and decarboxylation of long chain fatty acids by yeast. *Journal of Biological Chemistry*, **242**, 3608–14.

Gadd, G.M. (1988) Carbon nutrition and metabolism, in *Physiology of Industrial Fungi*, (ed D.R. Berry), Blackwell, Oxford, pp. 21–57.

Gadd, G.M and Edwards, S.W. (1981) Changes in the cytochrome levels during growth of the yeast-like fungus *Aureobasidium pullulans*. *Current Microbiology*, **6**, 161–6.

Galons, J.P., Tanida, I., Ohya, Y. *et al.* (1990) A multinuclear magnetic resonance study of a cls11 mutant showing the Pet-phenotype of *Saccharomyces cerevisiae*. *European Journal of Biochemistry*, **193**, 111–19.

Gancedo, J.M., Mazon, M.J. and Gancedo, C. (1983) Fructose 2, 6-bisphosphate activates the cAMP-dependent phosphorylation of yeast fructose 1, 6-bisphosphatase *in vitro*. *Journal of Biological Chemistry*, **258**, 5998–9.

Garcia, I.J., Nicolas, G. and Valle, T. (1991) Effect of difluoromethylornithine on growth, cell size and germination of *Ceratocystis ulmi* spores. *Plant Science*, **77**, 131–6.

Goodman, J.M. (1985) Dihydroxyacetone synthase is an abundant constituent of the methanol-induced peroxisome of *Candida boidinii*. *Journal of Biological Chemistry*, **260**, 7108–13.

Griffin, W.M. and Cooney, J.J. (1979) Degradation of model recalcitrant hydrocarbons by microorganisms from freshwater ecosystems. *Development in Industrial Microbiology*, **20**, 479–88.

Guignard, R. and Brody, S. (1983) Conidium formation and germination in *Neurospora crassa*: glutamic acid metabolism. *Experimental Mycology*, **7**, 133–40.

Gunterova, A. and Gallon, J. (1979) *Pleurotus ostratus*: a nitrogen fixing fungus? *Biochemical Society Transactions*, **7**, 1293–5.

Guthmann, M., Pastori, R. and Moreno, S. (1990) Polyamines and basic proteins stimulate activation by cAMP and catalytic activity of *Mucor rouxii* cAMP-dependent protein kinase. *Cellular Signalling*, **2**, 395–402.

Hammond, J.B.W. (1985) The biochemistry of *Agaricus* fructification, in *Developmental Biology of the Higher Fungi*, (eds. D. Moore, L.A. Casselton, D.A. Wood and J.C. Frankland), Cambridge University Press, Cambridge, pp. 389–401

Harold, F.M. (1966) Inorganic polyphosphates in biology: structure, metabolism, and function. *Bacteriological Reviews*, **30**, 772–94.

Hartman, R.E. and Keen, N.T. (1974) The phopsphoenol-pyruvate carboxykinase of *Verticillium albo-atrum*. *Journal of General Microbiology*, **81**, 21–6.

Hawkins, A.R., Gurr, S.J., Montague, P. and Kinghorn, J.R. (1989) Nucleotide sequence and regulation of expression of the *Aspergillus nidulans* gdhA gene encoding NADP dependent glutamate dehydrogenase. *Molecular and General Genetics*, **218**, 105–11.

Hecker, L.I. and Sussman, A.S. (1973) Localization of trehalase in the ascospores of *Neurospora*. *Journal of Bacteriology*, **115**, 592–9

Held, G. and Goldman, M. (1986) Pathways of glucose catabolism in the smut fungus *Ustilago violacea*. *Canadian Journal of Microbiology*, **32**, 56–61.

Henry, M.F. and Nyns, E.J. (1975) Cyanide insensitive respiration. An alternative mitochondrial pathway. *Sub-Cellular Biochemistry*, **4**, 1–65.

Hilgenberg, W., Burke, P.V., and Sandmann, G. (1987) Metabolic pathways, in *Phycomyces*, (eds. E. Cerda-Olmedo and E.D. Lipson), Cold Spring Harbor Laboratory, Cold Spring Harbor, pp. 155–98.

Hohn, T.M. and Paznokas, J.L. (1987) Purification and properties of two isozymes of pyruvate kinase from *Mucor racemosus*. *Journal of Bacteriology*, **169**, 3525–30.

Holmes, A.R., Collings, A., Farnden, K.J.F. and Shepherd, M.G. (1989) Ammonium assimilation by *Candida albicans* and other yeasts: evidence for activity of glutamate synthase. *Journal of General Microbiology*, **135**, 1423–30.

Holms, W.H. (1987) Control of flux through the citric acid cycle and the glyoxylate bypass in *Escherichia coli*. *Biochemical Society Symposia*, **5**, 17–31.

Holzer, H.J. (1989) Proteolytic catabolite inactivation in *Saccharomyces cerevisiae*. *Revisiones Sobre Biologia Cellular*, **21**, 305–19.

Hult, K. and Gatenbeck, S. (1978) Production of NADPH in the mannitol cycle and its relation to polyketide formation in *Alternaria alternata. European Journal of Biochemistry*, **88**, 607–12.

Hult, K., Veide, A. and Gatenbeck, S. (1980) The distribution of the NADPH regenerating mannitol cycle among fungal species. *Archives of Microbiology*, **128**, 253–5.

Hummelt, G. and Mora, J. (1980) Regulation and function of glutamate synthase in *Neurospora crassa*. *Biochemical and Biophysical Research Communications*, **96**, 1688–94.

Hütter, R. and De Moss, J.A. (1967) Organization of the tryptophan pathway: a phylogenetic study of the fungi. *Journal of Bacteriology*, **94**, 1896–907.

Inderlied, C.B., Peters, J. and Cihlar, R.L. (1985) *Mucor racemosus*, in *Fungal dimorphism* (ed. P.S. Szanislo), Plenum Press, New York, pp. 337–59

Ingraham, J.L. and Emerson, R. (1954) Studies on the nutrition and metabolism of the aquatic phycomycete *Allomyces. American Journal of Botany*, **14**, 146–52.

Inoue, H. and Shimoda, C. (1981) Changes in trehalose content and trehalase activity during spore germination in fission yeast, *Schizosaccharomyces pombe. Archives of Microbiology*, **129**, 19–22.

Jakubowski, H. (1986) Sporulation of the yeast *Saccharomyces cerevisiae* is accompanied by synthesis of adenosine 5′ tetraphosphate and adenosine 5′ pentaphosphate. *Proceedings of the National Academy of Sciences USA*, **83**, 2378–82.

Jennings, D.H. (1984) Polyol metabolism in fungi. *Advances in Microbial Physiology*, **25**, 149–93.

Jennings, D.H. and Burke, R.M. (1990) Compatible solutes – the mycological dimension and their role as physiological buffering agents. *New Phytologist*, **114**, 277–83.

Jones, R., Parkinson, S.M., Wainwright, M. and Kilham, K. (1991) Oxidation of thiosulphate by *Fusarium oxysporum* grown under oligotrophic conditions. *Mycological Research*, **95**, 1169–74.

Kalisz, H.M., Moore, D. and Wood, D.A. (1986) Protein utilization by basidiomycete fungi. *Transactions of the British Mycological Society*, **86**, 519–25.

Katayama-Hirayama, K., Tobita, S. and Hirayama, K. (1991) Degradation of phenol by yeast *Rhodotorula. Journal of General and Applied Microbiology*, **37**, 147–56.

Katz, J. and Rognstad, R. (1967) The labeling of pentose phosphate from glucose-^{14}C and estimation of the rates of transaldolase, transketolase, the contribution of the pentose cycle, and ribose phosphate synthesis. *Biochemistry*, **6**, 2227–47.

Kawamoto, S., Nozaki, C., Tanaka, A. and Fukui, S. (1978) Fatty acid β-oxidation system in microbodies of n-alkane grown *Candida tropicalis. European Journal of Biochemistry*, **83**, 609–15.

Keller, F., Schellenberg, M. and Wiemken, A. (1982) Localization of trehalase in vacuoles and of trehalose in the cytosol of yeast (*Saccharomyces cerevisiae*). *Archives of Microbiology*, **131**, 298–301.

Kopperschläger, G., Bär, J., Nissler, K. and Hofmann, E. (1977) Physicochemical parameters and subunit composition of yeast phosphofructokinase. *European Journal of Biochemistry*, **81**, 317–25.

Kostyukovsky, V.A., Okunev, O.N. and Tarakanov, B.V. (1991) Description of two anaerobic fungal strains from the bovine rumen and influence of diet on the fungal population *in vivo. Journal of General Microbiology*, **137**, 1759–64.

Krause, D.R., Wood, C.J. and MacLean, D.J. (1991) Glucoamylases (exo-1, 4-α-D-glucan glucanohydrolase, EC 3.2.1.3) is the major starch degrading enzyme secreted by the phytopathogenic fungus *Colletotrichum gloeosporides. Journal of General Microbiology*, **137**, 2463–8.

Kubicek, C.P. (1988) Regulatory aspects of the tricarboxylic acid cycle in filamentous fungi. A review. *Transactions of the British Mycological Society*, **90**, 339–49.

Kubicek-Pranz, E.M., Mozelt, M., Röhr, M. *et al.* (1990) Changes in the concentration of fructose 2,6-bisphosphate in *Aspergillus niger* during stimulation of acidogenesis by elevated sucrose concentrations. *Biochimica et Biophysica Acta*, **1033**, 250–5.

Kuhad, R.C., Rosin, I.V. and Moore, D. (1987) A possible relation between cyclic AMP levels and glycogen mobilization in *Coprinus cinereus. Transactions of the British Mycological Society*, **88**, 229–36.

Kulaev, I.S. (1975) Biochemistry of inorganic polyphosphates. *Reviews of Physiological Biochemistry and Pharmacology*, **73**, 131–58.

Kusnan, M.B., Klug, K. and Fock, H.P. (1989) Ammonia assimilation by *Aspergillus nidulans*: (^{15}N) ammonia study. *Journal of General Microbiology*, **135**, 729–38.

Kuswandi, C. and Roberts, C.F. (1992) Genetic control of the protocatechuic acid pathway in *Aspergillus nidulans. Journal of General Microbiology*, **138**, 817–23.

Lagos, R. and Ureta, T. (1980) The hexokinases from the wild type and morphological mutant strains of *Neurospora crassa. European Journal of Biochemistry*, **104**, 357–65

Laties, G.G. (1982) The cyanide resistant alternative path in higher plant respiration. *Annual Review of Plant Physiology*, **33**, 519–55.

Lejohn, H.B. (1971) Enzyme regulation, lysine pathways and cell wall structures as indicators of major lines of evolution in fungi. *Nature*, **231**, 164–8.

Lejohn, H.B. (1983) L-Glutamine alteration of gene expression, not of polyphosphate and calcium metabolism, is a key event in arresting fungal sporulation. *Canadian Journal of Biochemistry and Cell Biology*, **61**, 262–23.

Lewin, A.S., Hines, V. and Small, G.M. (1990) Citrate synthase encoded by the CIT2 gene of *Saccharomyces cerevisiae* is peroxisomal. *Molecular and Cellular Biology*, **10**, 1399–405.

Lewis, D.H. (1991) Fungi and sugars-a suite of interactions. *Mycological Research*, **95**, 897–904.

Londesborough, J. and Vuorio, O. (1991) Trehalose-6-phosphate synthase/phosphatase complex from bakers' yeast: purification of a proteolytically activated form. *Journal of General Microbiology*, **137**, 323–30.

Lopez-Boado, Y.S., Herrero, P., Fernandez, T., Fernandez, R. and Moreno, F. (1988) Glucose stimulated phosphorylation of yeast isocitrate lyase *in vivo*. *Journal of General Microbiology*, **134**, 2499–505.

Martin, F., Canet, D. and Marchal, J.P. (1985) [13]C Nuclear magnetic resonance study of mannitol cycle and trehalose synthesis during glucose utilization by the ectomycorrhizal Ascomycete *Cenococcum graniforme*. *Plant Physiology*, **77**, 499–502.

Martinez-Pacheco, M., Rodriguez, G., Reyna, G., Calvo-Mendez, C. and Ruiz-Herrera, J. (1989) Inhibition of the yeast-mycelial transition and the morphogenesis of mucorales by diamino-butanone. *Archives of Microbiology*, **151**, 10–14.

Marzluf, G.A. (1981) Regulation of nitrogen metabolism and gene expression in fungi. *Microbiological Reviews*, **45**, 437–61.

Maxwell, D.P., Maxwell, M.D., Hänssler, G. *et al.* (1975) Microbodies and glyoxylate-cycle enzyme activities in filamentous fungi. *Planta*, **124**, 109–23.

McCraken, L.D. and Gong, C. (1983) D-Xylose metabolism by mutant strains of *Candida* sp. *Advances in Biochemical Engineering/Biotechnology*, **27**, 33–56.

McInnis, T. and Domnas, A. (1973) The properties of trehalase from the mosquito parasitizing water mold *Lagenidium* species. *Journal of Invertebrate Pathology*, **22**, 313–20.

Meixner-Monori, B., Kubicek, C.P. and Röhr, M. (1984) Pyruvate kinase from *Aspergillus niger*: a regulatory enzyme in glycolysis? *Canadian Journal of Microbiology*, **30**, 16–22.

Meixner-Monori, B., Kubicek, C.P., Habison, A. *et al.* (1985) Presence and regulation of the α-ketoglutarate dehydrogenase multienzyme complex in the filamentous fungus *Aspergillus niger*. *Journal of Bacteriology*, **161**, 265–71.

Midgley, M. (1993) Carnitine acetyltransferase is absent from *acuJ* mutants of *Aspergillus nidulans*. *FEMS Microbiology Letters*, **108**, 7–10.

Millbank, J.W. (1969) Nitrogen fixation in moulds and yeasts – a reappraisal. *Archiv für Mikrobiologie*, **68**, 32–9.

Miller, S.M. and Magasanik, B. (1990) Role of NAD-linked glutamate dehydrogenase in nitrogen metabolism in *Saccharomyces cerevisiae*. *Journal of Bacteriology*, **172**, 4927–35.

Mittenbühler, K. and Holzer, H. (1991) Characterization of different forms of trehalase in the secretory pathway. *Archives of Microbiology*, **155**, 217–20.

Moore, P.A., Sagliocco, F.A., Wood, R.M.C. and Brown, A.W.P. (1991) Yeast glycolytic mRNA's are differentially regulated. *Molecular and Cellular Biology*, **11**, 5330–7.

Morris, N., Ainsworth, S. and Kinderlerer, J. (1984) The regulatory properties of yeast pyruvate kinase. *Biochemical Journal*, **217**, 641–7.

Müller, J., Westenberg, B., Boller, T. and Wiemken, A. (1992). Synthesis and degradation of polyphosphate in the fission yeast *Schizosaccharomyces pombe*: mutations in phosphatase genes do not affect polyphosphate metabolism. *FEMS Microbiology Letters*, **92**, 151–6.

Nisbet, B.A. and Slaughter, J.C. (1980) Glutamate dehydrogenase and glutamate synthase from the yeast *K. fragilis*: variability in occurrence and properties. *FEMS Microbiology Letters*, **7**, 319–21.

O'Connell, B.T. and Paznokas, J.L. (1980) Glyoxylate cycle in *Mucor racemosus*. *Journal of Bacteriology*, **143**, 416–21.

Okamoto, P.M., Fu, Y.H. and Marzluf, A. (1991) *Nit-3*, the structural gene of nitrate reductase in *Neurospora crassa*: nucleotide sequence and regulation of mRNA synthesis and turnover. *Molecular and General Genetics*, **227**, 213–23.

Orlowski, M. (1991) *Mucor* dimorphism. *Microbiological Reviews*, **55**, 234–58.

Panek, A.C., Araujo, P.S., Moura-Neto, V. and Panek, A.D. (1987) Regulation of the trehalose-6-phosphate synthase complex in *Saccharomyces cerevisiae*. *Current Genetics*, **11**, 459–65.

Panek, A.C., Araujo, P.S., Poppe, S.C. and Panek, A.D. (1990) On the determination of trehalose-6-phosphate synthase in *Saccharomyces*. *Biochemistry International*, **21**, 695–704.

Park, W.S., Murphy, P.A. and Glatz, B.A. (1991). Evidence of peroxisomes and peroxisomal enzymes activities in the oleaginous yeast *Apiotrichum curvatum*. *Canadian Journal of Microbiology*, **37**, 361–7.

Parkinson, S.A., Jones, K., Mehang, A.A. *et al.* (1991) The quantity and fate of carbon assimilated from [14]CO$_2$ by *Fusarium oxysporum* grown under oligotrophic and near oligotrophic conditions. *Mycological Research*, **95**, 1345–9.

Passeron, S. and Roselino, E. (1971) A new form of pyruvate kinase in mycelium of *Mucor rouxii*. *FEBS Letters*, **18**, 9–12.

Pateman, J.A. and Kinghorn, J.R. (1975) Nitrogen metabolism, in *The Filamentous Fungi*, Vol. 2 *Biosynthesis and metabolism*, (eds. J.E. Smith and D.R. Berry), Edward Arnold, London, pp. 159–238.

Peters, J. and Sypherd, P.S. (1979) Morphology associated expression of nicotinamide adenine dinucleotide dependent glutamate dehydrogenase in *Mucor racemosus*. *Journal of Bacteriology*, **137**, 1134–9.

Pfyffer, G.E. and Rast, D.M. (1980) The polyol pattern of some fungi not hitherto investigated for sugar alcohols. *Experimental Mycology*, **4**, 160–70.

Pfyffer, G.E., Boraschi-Gaia, C., Weber, B. *et al.* (1990) A further report on the occurrence of acyclic sugar alcohols in fungi. *Mycological Research*, **94**, 219–22.

Pitt, D. and Mosley, M.J. (1985a) Pathways of glucose catabolism and the origin and metabolism of pyruvate during calcium induced conidiation of *Penicillium notatum*. *Antonie Van Leeuwenhoek*, **51**, 365–84.

Pitt, D. and Mosley, M.J. (1985b) Enzymes of gluconate metabolism and glycolysis in *Penicillium notatum*. *Antonie Van Leeuwenhoek*, **51**, 353–64.

Plankert, U., Purwin, C. and Holzer, H. (1988) Characterization of yeast fructose-2,6-bisphosphate 6-phosphatase. *FEBS Letters*, **239**, 69–72.

Pohlig, G., Wingender-Drissen, R., Noda, T. and Holzer, H. (1983) Cyclic AMP and fructose 2,6-bisphosphate stimulated *in vitro* phosphorylation of yeast fructose-1,6-bisphosphatase. *Biochemical and Biophysical Research Communications*, **115**, 317–24.

Polnish, E. and Hofmann, K. (1989) Cyclic AMP, fructose 2,6-bisphosphate and catabolite inactivation of enzymes in the hydrocarbon assimilating yeast *Candida maltosa. Archives of Microbiology*, **152**, 269–72.

Ramstedt, M., Niehaus, W.G. and Söderhall, K. (1986) Mannitol metabolism in the mycorrhizal fungus *Piloderma croceum. Experimental Mycology*, **10**, 9–18.

Ramstedt, M., Jirjis, R. and Söderhall, K. (1987) Metabolism of mannitol in mycorrhizal and non-mycorrhizal fungi. *New Phytologist*, **105**, 281–7.

Rast, D.M. and Pfyffer, G.E. (1989) Acyclic polyols and higher taxa of fungi. *Botanical Journal of the Linnean Society*, **99**, 39–57.

Ratledge, C. and Holdsworth, J.E. (1985) Properties of a pentulose-5-phosphate phosphoketolase from yeast grown on xylose. *Applied Microbiology and Biotechnology*, **22**, 217–21.

Rose, A.H. (1989) Transport and metabolism of sulphur dioxide in yeasts and filamentous fungi, in *Nitrogen, Phosphorus and Sulphur Utilization by Fungi*, (eds. L. Boddy, R. Marchant and D.J. Read), Cambridge University Press, Cambridge, pp. 59–72.

Rua, J., De Arriaga, D., Busto, F. and Soler, J. (1990a) Effect of glucose on isocitrate lyase in *Phycomyces blakesleeanus. Journal of Bacteriology*, **171**, 6391–3.

Rua, J., De Arriaga, D., Busto, F. and Soler, J. (1990b) Isocitrate lyase from *Phycomyces blakesleeanus*. The role of Mg^{2+} ions, kinetics and evidence for two classes of modifiable thiol groups. *Biochemical Journal*, **272**, 359–67.

Ruiz-Herrera, J. and Calvo-Mendez, C. (1987) Effect of ornithine decarboxylase inhibitors on the germination of sporangiospores of Mucorales. *Experimental Mycology*, **11**, 287–96.

Salminen, S.O. and Streeter, J.G. (1986) Enzymes of α, α-trehalose metabolism in soybean nodules. *Plant Physiology*, **81**, 538–41.

Sandmann, G. and Hilgenberg, W. (1978) Phosphoenolpyruvatcarboxykinase aus *Phycomyces blakesleeanus* Bgff. *Zeitschrift für Naturforschung*, **33c**, 667–70.

Sariaslani, F.S. (1991) Microbial cytochromes P-450 and xenobiotic metabolism. *Advances in Applied Microbiology*, **36**, 133–78.

Savov, V.A., Kuyumdzieva, A., Atev, A.P. and Panaiotov, H.A. (1986) Influence of nitrogen limitation on the activity of certain nitrogen metabolic enzymes in the *Aspergillus niger* A3 strain. *Doklady Bolgarskoi Akademii Nauk*, **39**, 101–3.

Schwitzguebel, J.P., Möller, I.M. and Palmer, J.M. (1981) Changes in density of mitochondria and glyoxysomes from *Neurospora crassa*: a reevaluation utilizing silica sol gradient centrifugation. *Journal of General Microbiology*, **126**, 289–95.

Shapira, R., Altman, A., Henis, Y. and Chet, I. (1989) Polyamines and ornithine decarboxylase activity during growth and differentiation in *Sclerotium rolfsii. Journal of General Microbiology*, **135**, 1361–7.

Shepherd, M.G., Ghazali, H.M. and Sullivan, P.A. (1980) N-acetyl D-glucosamine kinase and germ tube formation in *Candida albicans. Experimental Mycology*, **4**, 147–59.

Shoun, H., Kim, D., Uchiyama, H. and Sugiyama, J. (1992) Denitrification by fungi. *FEMS Microbiology Letters*, **94**, 277–82.

Singh, M., Scrutton, N.S. and Scrutton, M.C. (1988) NADPH-generation in *Aspergillus nidulans*: is the mannitol cycle involved? *Journal of General Microbiology*, **134**, 643–54.

Singhania, S., Satyanarayana, T. and Rajam, M.V. (1991) Polyamines of thermophilic moulds: distribution and effect of polyamine biosynthesis inhibitors on growth. *Mycological Research*, **95**, 915–17.

Slaughter, J.C. (1988) Nitrogen metabolism, in *Physiology of Industrial Fungi*, (ed. D.R. Berry), Blackwell Scientific Publications, Oxford, pp. 58–76.

Slaughter, J.C. (1989) Sulphur compunds in fungi, in *Nitrogen, Phosphorus and Sulphur Utilization by Fungi*, (eds. L. Boddy, R. Marchant and D.J. Read), Cambridge University Press, Cambridge, pp. 91–106

Smith, T.A., Barker, J.H.A. and Jung, M. (1990) Growth inhibition of *Botrytis cinerea* by compounds interfering with polyamine metabolism. *Journal of General Microbiology*, **136**, 985–92.

Soberon, M., Olamendi, J., Rodriguez, L. and Gonzalez, A. (1989) Role of glutamine aminotransferase in glutamine catabolism by *Saccharomyces cerevisiae* under microaerophylic conditions. *Journal of General Microbiology*, **135**, 2693–7.

Stachelek, C., Stachelek, J., Swan, J. *et al.* (1986) Identification, cloning and sequence determination of the genes specifying hexokinase A and B from yeast. *Nucleic Acids Research*, **14**, 945–63.

Stevens, L. and Stevens, E. (1981) Regulation of ornithine decarboxylase activity during the germination of conidia of *Aspergillus nidulans. FEMS Microbiology Letters*, **11**, 229–32.

Stevens, L. and Winter, M.D. (1979) Spermine, spermidine and putrescine in fungal development. *Advances in Microbial Physiology*, **19**, 63–148.

Stevens, L., Duncan, D. and Robertson, P. (1989) Purification and characterisation of NAD-glutamate dehydrogenase from *Aspergillus nidulans. FEMS Microbiology Letters*, **48**, 173–7.

Stewart, G.R. and Moore, D. (1971) Factors affecting the level and activity of pyruvate kinase from *Coprinus lagopus* sensu Buller. *Journal of General Microbiology*, **66**, 361–70.

Sypherd, P.S., Borgia, P.T. and Paznokas, J.L. (1978) Biochemistry of dimorphism in the fungus *Mucor. Advances in Microbial Physiology*, **18**, 67–104.

Tabor, C.W. and Tabor, H. (1985) Polyamines in microorganisms. *Microbiological Reviews*, **49**, 81–99.

Tellez-Inon, M.T. and Torres, H.N. (1970) Interconvertible forms of glycogen phosphorylase in *Neurospora crassa*. *Proceedings of the National Academy of Sciences USA*, **66**, 459–63.

Teunissen, M.J., Op Den Camp, H.J.M., Orpin, C.G. *et al.* (1991) Comparison of growth characteristics of anaerobic fungi isolated from ruminant and non-ruminant herbivores during cultivation in a defined medium. *Journal of General Microbiology*, **137**, 1401–8.

Thayumanavan, B. (1980) Nitrogen fixation by the fungus *Pleurotus sajor-caju* (Fr) Singer. *Indian Journal of Biochemistry and Biophysics*, **17**, 75–7.

Thevelein, J.M. (1984) Regulation of trehalose mobilization in fungi. *Microbiological Reviews*, **48**, 42–59.

Tomsett, A.B. (1989) The genetics and biochemistry of nitrate assimilation in ascomycete fungi, in *Nitrogen, Phosphorus and Sulphur Utilization by Fungi*, (eds. L. Boddy, R. Marchant and D.J. Read), Cambridge University Press, Cambridge, pp. 33–37.

Toyoda, Y. and Sy, J. (1984) Purification and phosphorylation of fructose 1,6-bisphosphatase from *Kluyveromyces fragilis*. *Journal of Biological Chemistry*, **259**, 8718–23.

Tsai, C.S., Mitton, K.P. and Johnson, B.F. (1989) Acetate assimilation by the fission yeast, *Schizosaccharomyces pombe*. *Biochemistry and Cell Biology*, **67**, 464–7.

Uchiyama, H., Ando, M., Toyonaka, Y. and Tabuchi, T. (1982) Subcellular localization of the methylcitric-acid-cycle enzymes in propionate metabolism of *Yarrowia lipolytica*. *European Journal of Biochemistry*, **125**, 523–7.

Ueda, S. (1981) Fungal glucoamylases and raw starch digestion. *Trends in Biochemical Sciences*, **6**, 89–90.

Ueda, M., Tanaka, A. and Fukui, S. (1982) Peroxisomal and mitochondrial carnitine acetyltransferase in alkane-grown yeast *Candida tropicalis*. *European Journal of Biochemistry*, **124**, 205–10.

Uno, I. Matsumoto, K., Adachi, K. and Ishikawa, T. (1983) Genetic and biochemical evidence that trehalase is a substrate for cAMP-dependent protein kinase in yeast. *Journal of Biological Chemistry*, **258**, 10867–72.

Uyeda, K. (1979) Phosphofructokinase. *Advances in Enzymology and Related Areas of Molecular Biology*, **48**, 193–244.

Vandercammen, A., François, J. and Hers, H. (1989a) Characterisation of trehalose 6-phosphate synthase and trehalose-6-phosphate phosphatase in *Saccharomyces cerevisiae*. *European Journal of Biochemistry*, **132**, 613–20.

Vandercammen, A., François, J.M., Torres, B.B. *et al.*, (1989b) Fructose 2,6-bisphosphate and carbohydrate metabolism during the life cycle of the aquatic fungus *Blastocladiella emersonii*. *Journal of General Microbiology*, **136**, 137–46.

Van Dijken, J.P. and Scheffers, W.A. (1986) Redox balances in the metabolism of sugars by yeasts. *FEMS Microbiology Reviews*, **32**, 199–224.

Van Laere, A.J. (1986) Biochemistry of spore germination in *Phycomyces*. *FEMS Microbiology Reviews*, **32**, 189–98.

Van Laere, A.J. (1988) Purification and properties of NAD-dependent glutamate dehydrogenase from *Phycomyces* spores. *Journal of General Microbiology*, **134**, 1597–601.

Van Laere, A.J. (1989) Trehalose, reserve and/or stress metabolite? *FEMS Microbiology Reviews*, **63**, 201–10.

Van Laere, A.J. and Carlier, A.R. (1974) Compartmentation and respiration pathways in *Avena* coleoptile segments. *Zeitschrift für Pflanzenphysiology*, **71**, 163–74.

Van Laere, A.J. and Hulsmans, E. (1987) Water potential, glycerol synthesis, and water content of germinating *Phycomyces* spores. *Archives of Microbiology*, **147**, 257–62.

Van Laere, A.J. and Joosen, H. (1991) Properties of phosphofructokinase 1 from *Pythium ultimum* and *Agaricus bisporus* and comparison with other fungal phosphofructokinases 1. *Experimental Mycology*, **15**, 223–31.

Van Laere, A.J., Van Assche, J.A. and Furch, B. (1987) The sporangiospore: dormancy and germination, in *Phycomyces*, (eds. E. Cerda-Olmedo and E. Lipson), Cold Spring Harbor Laboratory, Cold Spring Harbor, pp. 247–79.

Van Schaftingen, E. (1986) Fructose 2,6-bisphosphate. *Advances in Enzymology and Related Areas of Molecular Biology*, **59**, 315–95.

Van Schaftingen, E. and Van Laere, A.J. (1985) Glycerol formation and the breaking of dormancy of *Phycomyces blakesleeanus* spores. Role of an interconvertible glycerol-3-phosphatase. *European Journal of Biochemistry*, **148**, 399–404.

Veenhuis, M. and Harder, W. (1989) Occurrence, proliferation and metabolic function of yeast microbodies. *Yeast*, **5**, S517–24.

Veenhuis, M., van der Klei, I.J., Titorenko, V. and Harder, W. (1992) *Hansenula polymorpha*: an attractive organism for molecular studies of peroxisome biogenesis and function. *FEMS Microbiology Letters*, **100**, 393–404.

Vicente-Soler, J. Arguelles, J.C. and Gacto, M. (1989) Presence of two trehalose-6-phosphate synthase enzymes in *Candida utilis*. *FEMS Microbiology Letters* **61**, 273–8.

Vierula, P.J. and Kapoor, M. (1987) Antibiotic induced derepresssion of the NAD-specific glutamate dehydrogenase of *Neurospora crassa*. *Journal of Bacteriology*, **169**, 5022–7.

Vierula, P.J. and Kapoor, M. (1989) NAD-specific glutamate dehydrogenase of *Neurospora crassa*. cDNA cloning and gene expression during derepression. *Journal of Biological Chemistry* **264**, 1108–14.

Vogel, H.J. (1965) Lysine biosynthesis and evolution, in *Evolving Genes and Proteins*, (eds. V. Brysen and H.J. Vogel), Academic Press, New York, pp. 25–40.

Wainwright, M. (1988) Metabolic diversity of fungi in relation to growth and mineral cycling in soil – a review. *Transactions of the British Mycological Society* **90**, 159–70.

Wainwright, M. (1989) Inorganic sulphur oxidation by fungi, in *Nitrogen, Phosphorus and Sulphur Utilization by Fungi*, (eds. L. Boddy, R. Marchant and D.J. Read), Cambridge University Press, Cambridge, pp. 73–90.

Walker, J.D. and Cooney, J.J. (1973) Pathway of n-alkane oxidation in *Cladosporium resinae*. *Journal of Bacteriology*, **115**, 635–9.

Wang, C.H. (1972) Radiorespirometric methods, in *Methods in Microbiology*, Vol 6B, (eds. J.R. Norris and D.W. Ribbons), Academic Press, New York, pp. 185–230.

Weete, J.D. (1980) *Lipid Biochemistry of Fungi and Other Organisms*. Plenum, New York.

Whittaker, R.H. and Margulis, L. (1978) Protist classification and the kingdoms of organisms. *BioSystems* **10**, 3–18.

Wiemken, A. (1990) Trehalose in yeast, stress protectant rather than reserve carbohydrate. *Antonie van Leeuwenhoek*, **58**, 209–17.

Wilcock, M.J., McDougall, B.M. and Seviour, R.J. (1992) Enzymes involved in ammonia assimilation in the fungus *Acremonium persicinum*. *FEMS Microbiology Letters*, **97**, 67–72.

Wilson, A.J. and Bhattacharjee, J.K. (1986) Regulation of phosphoenolpyruvate carboxykinase and pyruvate kinase in *Saccharomyces cerevisiae* grown in the presence of glycolytic and gluconeogenic carbon sources and the role of mitochondrial function on gluconeogenesis. *Canadian Journal of Microbiology*, **32**, 969–72.

Wood, H.G. and Clark, J.E. (1988) Biological aspects of inorganic polyphosphates. *Annual Review of Biochemistry*, **57**, 235–60.

Wright, J.M., Gulliver, W.P., Michalski, C.J. and Boyle, S.M. (1982) Ornithine decarboxylase activity and polyamine content during zoospore germination and hormone-induced sexual differentiation of *Achlya ambisexualis*. *Journal of General Microbiology*, **128**, 1509–15.

Yagi, T., Nogami, A. and Nishi, T. (1992) Salt tolerance and glycerol accumulation of a respiration-deficient mutant isolated from the petite-negative, salt-tolerant yeast *Zygosaccharomyces rouxii*. *FEMS Microbiology Letters*, **92**, 289–93.

Yeung, M.C. and Kapoor, M. (1983) A study of the properties of pyruvate kinase isolated from a mutant of *Neurospora crassa*: a comparison with the parental enzyme. *International Journal of Biochemistry*, **15**, 523–9.

Yoshida, M., Murai, T. and Moriya, S. (1984) ^{13}C-NMR spectra of plant pathogenic fungi. *Agricultural and Biological Chemistry*, **48**, 909–14.

Zimmerman, A.L.S., Terenzi, H.F. and Jorge, J.A. (1990) Purification and properties of an extracellular conidial trehalase from *Humicola grisea* var. *thermoidea*. *Biochimica et Biophysica Acta*, **1036**, 41–6.

GENETICS OF FUNGI

11

A.J. Clutterbuck
Genetics Building, Institute of Biomedical and Life Sciences, University of Glasgow, Glasgow, UK

11.1 INTRODUCTION

The genetics of fungi impinges on all mycologists, if for no other reason than that they need to be aware of variation in the species they are studying. For many fungi, variation also includes genetic instability of the individual isolate. However, genetics can do much more than issue caveats; its techniques, especially when combined with those of molecular biology, are invaluable tools for investigating problems ranging from cell function on the one hand to population dynamics on the other.

This chapter is in two parts, the first discussing natural variation and gene exchange in the wild; and the second dealing with fungal genetics as a laboratory tool. Inevitably genetics and molecular biology are intertwined, and some genetical topics overflow into Chapter 12.

11.2 NATURAL VARIATION

Although morphology is an important and readily applied taxonomic criterion for the higher fungi, it is less applicable to more microbial species. Indeed, there are isolates classified as *mycelia sterilis*, some of which may be of economic importance, that are morphologically featureless since they lack both sexual and asexual reproductive structures. In characterizing these, and providing precise information for population and taxonomic studies on any species, molecular characterization of the genetic material or its immediate products provides the ultimate test of identity and relatedness. It is even possible to classify unseen organisms without ever isolating them by using the polymerase chain reaction (PCR, section 11.2.1g) to amplify and characterize their nucleic acids (e.g. Schmidt *et al.*, 1991).

11.2.1 Methods of molecular characterization

A variety of methods for characterizing DNA and proteins are valuable in fungal taxonomy at all levels: see reviews by Kurtzman (1985), Bruns *et al.* (1991) and Kohn (1992). Kohn lists as taxonomic criteria nine types of biological feature, three ways in which proteins can be characterized, and seven methods of DNA analysis.

(a) Proteins

Characterization of wild isolates in terms of their proteins is a well established practice and need not be discussed in detail here. In summary, proteins may be analysed *en masse*, separating them by SDS-PAGE (sodium dodecyl sulphate polyacrylamide gel electrophoresis). Multiple protein samples may be treated simultaneously and the resulting complex patterns provide a plentiful basis for distinguishing strains (e.g. Brasier, 1991), although they are difficult to analyse quantitatively. Isozyme analysis, in which specific enzymes are characterized, again usually in terms of electrophoretic mobility, gives much more limited, and therefore more precise information, and immunological detection methods similarly provide relatively simple plus-or-minus scoring of response to a particular antibody.

(b) Chromosomes: electrophoretic karyotyping

Standard gel electrophoresis can only separate DNA molecules of sizes less than about 20 kilobases (kb), but it is possible to separate whole chromosomes up to about 10 megabases (Mb) (Figure 11.5a) using a variety of current switching systems which make the DNA snake through a gel (reviewed by Mills and McClusky, 1990; Skinner *et al.*, 1991). Each system has a different acronym (CHEF, TAFE etc.) within the general heading of

The Growing Fungus. Edited by Neil A.R. Gow and Geoffrey M. Gadd. Published in 1994 by Chapman & Hall, London. ISBN 0 412 46600 7

PFGE (Pulse Field Gel Electrophoresis). These methods show that fungal karyotypes, even in comparisons of closely related strains, are surprisingly variable (e.g. Miao *et al.* 1991). This may come as no surprise to fungal geneticists who have had to cope with strains in which translocations have been induced during mutagenesis, and it correlates with the apparent ease with which gene order is rearranged, for instance in gene clusters, as discussed in Chapter 12. For example, in *Cochliobolus heterostrophus* the majority of chromosomes differed in size between the two isolates compared, and one of the 16 chromosomes was deemed to be a dispensable 'B' chromosome (see Figure 11.5) since neither it, nor its DNA, could be found in one of the two strains (Tzeng *et al.*, 1992).

Provided the chromosomes to be resolved are small enough, and for some species such as *Phytophthora infestans* they are not, electrophoretic karyotyping is useful for identifying individual strains. In some cases it can show up gross differences between isolates assumed to belong to the same species e.g. races of *Leptosphaeria maculans* previously believed to differ only in degree of virulence (Taylor *et al.*, 1991; Chapter 12, Table 12.1).

(c)   DNA: restriction fragment length polymorphism (RFLP)

DNA is cut with a particular restriction enzyme and the fragments separated, on the basis of size, by gel electrophoresis. Particular fragments are then revealed by hybridizing with an appropriate radiolabelled probe, i.e. a short length of DNA from the region to be tested. Strain-specific variation in the positions of restriction sites, or insertion or deletion of DNA between the sites, will show up as fragment size differences (Figure 11.1). By using a range of probes, closely related strains can be distinguished or, if it is possible to carry out appropriate crosses, fragment size differences can be used as markers on a linkage map (see below) of a chromosomal region, or indeed, of the complete genome (Figure 11.5b). The mitochondrial DNA, because of its limited size and ease of isolation, and ribosomal RNA genes, because of their universality, are favourite substrates for this purpose, but any cloned sequence, whether its function is known or not, is a potential probe.

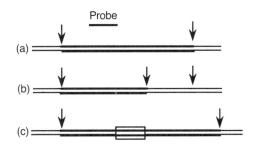

Figure 11.1 Restriction fragment length polymorphism (RFLP). (a), (b) and (c) represent three versions of a stretch of DNA; the arrows indicate sites at which a particular restriction enzyme cuts. (b) differs from (a) in having an additional restriction site, whereas (c) has an insertion of an extra piece of DNA represented by the box. The DNAs are cut with the restriction enzyme, separated on an electrophoretic gel and transferred to a membrane where they are hybridized to a radiolabelled copy of the short segment of DNA shown. The probe will hybridize with a different length fragment in each case, as shown by the heavy lines.

(d)   DNA fingerprinting

This name can apply to any method which reveals enough differences between organisms to allow them to be classified as individuals, or for fungi, clones. On this basis, RFLP analysis using a combination of separate probes has been described as fingerprinting (McDonald and Martinez, 1991). However, the procedure first devised for human DNA is more specific and is distinguished by the use of probes to 'microsatellite' DNA. Microsatellites consist of clusters of simple, repetitive sequences and most higher eukaryote genomes include a variety of such simple sequences, each represented in more than one cluster. Mispaired recombination between homologous clusters readily generates microsatellites of different sizes (Figure 11.2), with the result that if a set of microsatellites is probed, every sexually generated individual of a complex species can be shown to have inherited a unique combination.

Most fungi contain little repetitive DNA and there is no direct evidence that they have microsatellites. Nevertheless a human microsatellite DNA was successfully used as a probe for fingerprinting pathotypes of *Ophiostoma ulmi*, which were also characterized by mitochondrial DNA restriction mapping (Hinz *et al.*, 1991). Meyer *et al.* (1991) also used simple sequences such as

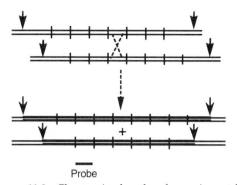

Probe

Figure 11.2 Change in length of a microsatellite sequence by recombination. The short stretches of DNA separated by vertical lines are repeats of the same simple sequence. Recombination between two identical clusters can generate versions which will be seen to be of different lengths when cut at the restriction sites indicated by arrows, separated by electrophoresis and probed with a copy of the sequence.

(GACA)$_4$ to fingerprint a number of fungi. The repetitive 'MGR' element has proved to be a valuable fingerprinting probe for the cereal pathogen *Magnaporthe grisea* (Hamer *et al.*, 1989). However, genetic analysis by Romao and Hamer (1992) showed that copies of the element are distributed around the genome rather than clustered, and as a result they could also be used as markers for an RFLP map comprising eight linkage groups. Both the structure of the MGR element, and its scattered distribution make it likely that it is a transposon (Chapter 12) rather than a simple sequence repeat (Valent and Chumley, 1991).

(e) DNA hybridization

The double strandedness of DNA makes it possible to use the ability of the single strands of denatured DNAs of two strains to hybridize as a direct test of genetic identity. Furthermore, the stability of the hybrid, as determined by its melting temperature, can also be used as a quantitative measure of sequence similarity. This is a sensitive method and shows up a high degree of divergence both within and between fungal species. For instance Vilgalys and Johnson (1987) concluded that some sexually compatible strains of the basidiomycete *Collybia dryophila* group show only 75% DNA similarity. Furthermore, sexually incompatible, although morphologically similar, strains showed as little as 37% similarity, a degree

of divergence enough to define different orders of birds or superfamilies of primates.

(f) DNA sequence comparisons

Non-coding DNA can change its base composition rapidly on an evolutionary time scale. Furthermore the composition of coding DNA is not as rigidly fixed as might be supposed, since a change in the third base of many codons does not alter the amino acid specified, and even if it does, some changes can be regarded as conservative substitutions, entailing no significant change in protein function. For this reason DNA sequence comparisons of protein coding genes are appropriate for monitoring relatively short-term changes, but comparisons of the DNA sequences of ribosomal RNA genes, or of derived amino acid sequences of protein coding genes are more appropriate for broader taxonomic studies.

The ultimate level of genetic distance can be estimated as the number of genetic substitutions which have occurred between one evolutionary line and another. This can be estimated for fully sequenced DNA, provided that the degree of divergence is small enough for double changes at the same site to be negligible. Generally such studies are made on genes sequenced in the first place for other reasons, when the most interesting comparisons are often those between widely differing organisms, comparing the positions of introns and other major features, e.g. for the triosephosphate isomerase gene of *Aspergillus nidulans* (McKnight *et al.*, 1986).

Two important results from such comparisons are first that Oomycetes are related to plants whereas other fungi are closer to animals (Förster *et al.*, 1990a). This was deduced from studies of the DNA sequence of the gene for the small subunit of ribosomal RNA. Second, comparison of the deduced amino acid sequences for glyceraldehyde-3-phosphate dehydrogenase (GAPD) came to the conclusion that some yeasts are evolutionarily quite distant from filamentous fungi, and more controversially, that they are closer to bacteria than they are to other Ascomycetes (Smith, 1989). Methods of analysis of sequence comparisons are complex, and have to take into account varying rates of change in different evolutionary branches. The evidence suggests that change is faster in the fungi than in some other biological kingdoms, and since GAPD genes show signs of even more rapid change, these conclusions remain to be confirmed (Bruns *et al.*, 1991). That comparisons using only one gene can sometimes give bizarre results is shown by Harmsen *et al.* (1992) who confirmed that the yeast GAPD gene differs significantly from that of filamentous fungi, but also found that the *Agaricus bisporus* gene was even less closely related, despite the fact that other Basidiomycete GAPDs were quite similar to those of filamentous Ascomycetes. If we take into account the additional possibility of rare horizontal gene transfer between unrelated species (Smith *et al.*, 1992; see also discussion of the β-lactam biosynthesis gene cluster in Chapter 12), it is clearly unwise to construct a taxonomic tree on data from only one gene.

Mitochondria have their own ribosomes encoded by the mitochondrial DNA (Chapter 12). Comparisons of nuclear and mitochondrial ribosomal RNA genes suggest that in the true fungi, as in animals, the mitochondrial versions evolve considerably faster than the nuclear ones (Bruns and Szaro, 1992) whereas the reverse is true for Oomycetes (Förster *et al.*, 1990b), which adds further evidence to the view that they should be included in the same category as plants.

(g) Polymerase chain reaction (PCR) sequencing

It is unusual for the DNA sequence of a specific gene to be available for more than a very limited number of species. However, if one sequence is known, it is now possible to use PCR to provide information on the homologous sequence from a series of related organisms. The basis of the PCR reaction is shown diagrammatically in Fig. 11.3. It uses two short synthetic DNA primers, based on the ends of a known sequence whose homologues are to be amplified, and it results in microgram quantities of a segment of DNA. This is a rapid procedure that can provide a great deal of information very quickly, provided that highly conserved sequences can be identified for use as primers. Because of the near freedom of substitution at the third codon base, coding sequences can evolve rapidly, even if the corresponding amino acid sequence is highly conserved. For this reason essential regions of a gene for functional RNA, such as ribosomal RNA, may be preferable to a protein-coding gene. Since the method is so powerful, safeguards are also necessary to avoid

Figure 11.3 Polymerase chain reaction (PCR) amplification of DNA. Two short segments of DNA (heavy lines A and B), corresponding to known or predicted sequences in the target DNA to be amplified (solid line), are annealed to denatured DNA to serve as primers for DNA polymerase. During successive steps of denaturation, reannealing of primers and polymerization, the primers will anneal to the newly made strands as well as to the original DNA, leading to exponential amplification of the target.

Figure 11.4 Random amplified polymorphic DNA (RAPD). A single primer, short and non-specific enough to correspond to many sites in the genome, is used in the PCR reaction (Figure 11.3). Most of these sites (e.g. a and b) will be unproductive, but if two sites (e.g. c and d) happen to be within a few kilobases of each other and on opposite strands, amplification of the intervening DNA will proceed. The size of the segments amplified will depend, as for RFLPs (Figure 11.1), on insertions or deletions in the interval between two sites, whereas mutations to either priming sequence may lead to loss of the band.

contamination. A specific protocol for PCR amplification of fungal ribosomal DNA is given by White *et al.* (1990), and Erlich and Arnheim (1992) have reviewed the uses of PCR amplification in genetics.

(h) Random amplified polymorphic DNA (RAPD) markers

If instead of using two primers corresponding to a known DNA sequence to prime the PCR reaction, a single, shorter, random sequence is used, it will anneal to a variety of sites in the target DNA (Figure 11.4), and in some cases two such sites will be in opposite orientation and close enough together to allow amplification of the intervening stretch (Nicholson *et al.*, 1991). In this way a variety of unrelated segments of DNA will be amplified, and the PCR reaction can be prolonged to give enough DNA of each fragment to be visualized directly on a gel without the need for blotting or use of a radioactive probe. The resulting bands may then be treated much as RFLP markers, e.g. for distinguishing subspecies (Strongman and MacKay, 1993). Williams *et al.* (1991) also describe methods for genetic analysis of *Neurospora crassa* using RAPD markers. However, it should be remembered that since the method depends on juxtaposition of two priming sites, a change in either priming site will result in loss of a band. Furthermore, reproducible results require very stringent standardization and interlaboratory differences are common. The RAPD method also has the disadvantage that whereas well-chosen PCR primers should be able to discriminate between the fungal target and contaminating DNA, RAPD primers are intentionally non-specific. This may be important for instance, where DNA from a pathogenic fungus could be confused with that of its host.

11.2.2 Variation in natural populations

There are few reports of extensive fungal population surveys, but this situation is likely to be rectified with the availability of molecular methods and their application to populations of phytopathogenic fungi. For wild populations of the classical laboratory fungi, Perkins and Turner (1988) reviewed quite extensive studies of *Neurospora* heterothallic species, and more limited studies of *Aspergillus nidulans* populations were reviewed by Croft and Jinks (1977). Examples of population studies of pathogenic fungi are reported by Valent and Chumley (1991) for *M. grisea* and Brasier (1991, 1992), Brasier and Hansen (1992) and Fry *et al.* (1992) for *Phytophthora* species.

The general conclusion from studies of fungal populations, irrespective of the methods used, is that they are extremely variable. This implies very rapid (or very prolonged) molecular evolution, accompanied by much more limited morphological change. There may be many reasons for this, some of them related to the relative simplicity of fungal biology. Simple organisms should be able to evolve rapidly since they have fewer interacting pathways which would require adjustment when one is modified. On the other hand simple genomes contain little redundant DNA (Chapter 12) that is under minimal evolutionary constraint

and in more complex organisms can evolve faster than essential coding sequences.

It is also inevitable that the inclination of a taxonomist to subdivide a biological group will depend on the number of differences detectable between individuals. Organisms with simple morphology may falsely be regarded as more uniform than they really are.

11.3 GENETIC RECOMBINATION IN NATURE

Fungi provide fine examples of the biological paradox that whereas some species seem to go to great lengths to encourage outbreeding, closely related forms apparently get by without sexual reproduction of any kind. Many important fungi are designated as members of the Deuteromycetes meaning that no sexual stage has been found. In some cases this may only reflect failure to find conditions for sexual reproduction, but there can be little doubt that many long-studied strains, e.g. of commercial fungi such as *Aspergillus niger*, and human pathogens such as *Candida albicans* are constitutionally asexual.

The selective forces for and against sexual reproduction have been discussed in detail (e.g. Maynard Smith, 1978). Although the adaptability conferred by outbreeding may be important in the long term, it has the obvious disadvantage as a method of reproduction that it requires the cooperation of two compatible individuals. Its abandonment, on the grounds of economy, would therefore be tempting in the short term, particularly for an opportunistic, 'r-selected' weed-like species.

The classical explanation for the existence of sexual reproduction is to allow recombination, which for genes on the same chromosome necessitates the remarkably complex process of crossing over. It is probable that the events of breakage and rejoining of chromosomes which this entails, arose in the first place as a mechanism for repairing damaged DNA. Crossing-over itself can also be seen as a repair system; in vegetative cells of a diploid organism an intact homologous chromosome can be used as a template for repair of its damaged relative, or during sexual reproduction of any organism, two inadequate genomes can give rise to one perfect one. Here the role of natural selection is the avoidance of deterioration of the species by 'Müller's ratchet' i.e. the accumulation of deleterious mutations if they are not regularly weeded out.

The opposite side of this coin is that novel gene combinations which may be required for survival in a changing environment can also be constructed. In some cases the environmental change may be physical, but an important component of the environment of most species is other organisms, and evolution of more competitive forms by predatory, rival, or, in the case of parasitic fungi, host species, requires the organism in question to evolve merely to stand still: this is the 'Red Queen' hypothesis of evolution (Van Valen, 1973).

One form of predator to which all organisms appear to be at risk is 'parasitic', or to be slightly less rude to it, 'selfish' DNA (Doolittle and Sapienza, 1980; Orgel and Crick, 1980). The most striking instances of selfish DNA are transposons, which can propagate themselves in the genome with no benefit to the host (Chapter 12). It is evident that in some cases sexual reproduction could play a role in their elimination, but unfortunately it is equally clear that sexual interaction can encourage their spread!

Other sorts of intracellular parasite may be equally pernicious. Viruses and rogue mitochondria are also discussed in Chapter 12. For these cytoplasmic elements the value to the host species of sexual reproduction is perhaps better supported, since an advantageous nucleus can leave its renegade cytoplasm behind if it takes the male role.

Perkins and Turner (1988) noted that all homothallic *Neurospora* species were aconidial, which led them to suggest that an important function of sexual reproduction in this genus is as a route to the production of refractory spores. Heterothallic *Neurospora* species, on the other hand retain the ability to make asexual conidia which, although they are not so tough or long-lived as ascospores, do provide a means of distribution without the cooperation of two mating types. However, most heterothallic isolates studied were genetically unique, suggesting that they had indeed arisen from ascospores rather than by vegetative reproduction of nearby clones.

Much of our knowledge of fungal populations comes, not surprisingly, from fungi of economic importance, particularly plant pathogens. It is striking that widespread and successful populations of plant pathogens may be deprived of sexual reproduction either because they consist of only one mating type (e.g. for *Phytophthora infestans*, Fry *et al.*, 1992) or because of the absence of an obligate host for the sexual stage (*Puccinia graminis*, Burdon and Roelfs, 1985). Cereal rust *P. graminis* populations have been surveyed in America, where in some states the *Berberis* host for the sexual stage has been eradicated, and in Australia, where *Berberis* was never present. These revealed considerable isozyme variability in areas where sexual reproduction was possible, but in areas where only asexual reproduction was possible a few widespread clones predominated. However, this, did not appear to limit *Puccinia* species as pathogens in Australia, where pathotypes for new wheat varieties had apparently arisen by mutation (Burdon *et al.*, 1983). A point to remember, however, is that the plant crops on which they prey are themselves unnaturally monomorphic, so that a full range of variability in the predator may be less important.

11.3.1 Heterokaryosis

At one time it was supposed that fungi were able to increase their versatility by making use of a feature unique to coenocytic filamentous organisms, namely heterokaryosis (Pontecorvo, 1946). On the contrary, it now appears that fungal individuality is the rule (Rayner 1991, and Chapter 2) and most confronting pairs of Ascomycete or Basidiomycete isolates are likely to differ in heterokaryon incompatibility factors (e.g. Caten and Jinks, 1966; Croft and Jinks, 1977; Perkins and Turner, 1988; Correll, 1991). Exceptions to this rule are fungi which may have been spread as vegetative clones along with the host plants for which they are pathogenic (e.g. Joaquim and Rowe, 1991). Incompatibility may manifest itself as failure of hyphae to fuse (prefusion incompatibility) or, more commonly, as inviability of the fusion product (postfusion incompatibility), symptoms of failure being vacuolation, dissolution of nuclei and generation of pigment (described as a 'barrage' reaction). The net result is that heterokaryosis is not often likely to be the ideal system of cooperation it was once thought to be (Carlile, 1987).

Rather, we must suppose that the fungi are as wary of the perils of casual intercourse for the transfer of viruses or rogue mitochondria as are human beings (Caten, 1972). It should be noted that heterokaryon incompatibilities do not appear to inhibit sexual interactions (Butcher 1968) but would prevent parasexual recombination.

Why then is hyphal fusion so common? It may be that self-fusion has a role in hyphal growth. The vigour of cytoplasmic flow in the hyphae of *Neurospora crassa* suggests that in addition to flowing towards the growing tips, it also circulates. In this species mycelium as much as 1 cm inside a colony can contribute to hyphal extension at the colony edge (Ryan *et al.*, 1943) so cytoplasmic circulation may be part of a distribution system giving support from the centre of the colony to hyphal extension at the periphery (Chapter 14). This may be even more important in wood-rotting basidiomycetes which invade by a collectivist strategy, forming fungal ropes that can transport nutrients over long distances (Rayner, 1991; and Chapter 2). At the other extreme, in heterokaryons of fungi such as *Verticillium dahliae* nuclear mixing may be confined to a few cells (Puhalla and Mayfield, 1974) and a physiological role for hyphal fusion is less obvious.

11.3.2 Mating types

The biological cunning exemplified by mating systems is only to be matched by our exasperation at the fact that some species do not bother to mate at all. Simple bipolar mating systems are readily understandable to the human mind, since we operate one ourselves. Multiple mating types, generated by multiple alleles at two or even more loci, have the clear advantage, especially if populations are sparse, that an individual will be compatible with a large proportion of the other individuals that it meets (Chapter 18). Still more cunning are the fungi, that exploit mating-type switching systems, or other mechanisms of secondary homothallism, to allow self-fertilization in the absence of a mate. However, it is undoubtedly strange that a mechanism should have evolved to encourage outbreeding, and that a further complex mechanism should then evolve to negate the effects of the first. We must suppose that this reflects the fluctuating demands of selective influences in the wild.

11.3.3 Parasexuality

Parasexuality is the name given to the series of steps: heterokaryon formation, diploidization, somatic crossing-over and haploidization, which can together be seen as an alternative to the sexual cycle. Each of these occurs at such low frequencies that we are inclined to think of their occurrence as rare accidents. Although its ability to generate recombinants may be limited, even rare parasexual interactions could be an important source of phenotypic variability. It is possible that imperfect fungi make up for their asexuality by use of the parasexual cycle, but in order to do so with any frequency, such fungi must not be subject to heterokaryon incompatibility. Parasexual exchange can also be assumed to be ineffective in 'cytoplasmic cleansing' which was put forward as one of the advantages of sexual reproduction in the previous section.

Evidence for the occurrence of parasexual interactions in nature is limited. Vegetative diploids of *Aspergillus nidulans* are found in nature (Upshall, 1981). It has also been concluded that a virulent Australian strain of *Puccinia graminis* could only have arisen as a somatic hybrid in the absence of sexual reproduction (Burdon *et al.*, 1981).

11.3.4 Causes of variation and instability

Kistler and Miao (1992), reviewing causes of fungal variation, listed chromosomal polymorphism, changes in repetitive DNA, transposons, virus-like elements and mitochondrial plasmids as the prevalent causes: these are discussed in this and the following chapter. To this list may be added gene conversion, inversion and other standard switch mechanisms and spontaneous chromosomal loss; this may be of inessential B-chromosomes (section 11.2.1b), or of duplicate chromosomes present in aneuploids (Käfer, 1961). Non-Mexican populations of *P. infestans* include interfertile diploid and tetraploid strains whose cross progeny are of variable ploidy (Whittaker *et al.*, 1991). It must be expected that some of these will have genetically unbalanced chromosome constitutions which will be open to selection for improved variants. Duplications for chromosome segments can also result from crosses between strains differing with respect to chromosome translocations (Bainbridge and Roper, 1966; Perkins and Barry, 1977). These are also subject to spontaneous deletion, usually resulting in an improved growth rate.

In most instances, the cause of instability of a fungal isolate remains unknown, if for no other reason that the natural tendency of laboratory workers is to accept a defined stock only if it is uniform and stable. However, some unstable systems may represent an important feature of the biology of the fungus, for instance sexually fertile isolates of *M. grisea* frequently show marked morphological instability, whereas their stable subcultures have lost the capacity for sexual reproduction (Valent and Chumley, 1991). *A. nidulans* cultures show a similar but less extreme phenomenon, exhibiting continuous vegetative variation in frequency of sexual fruiting bodies over the surface of a colony (Croft, 1966).

Switching between an impressive list of different colony phenotypes is one of the complicating factors in *Candida* genetics (Scherer and Magee, 1990; Soll *et al.*, 1993). The yeast–hyphal dimorphism switch appears to be of the nature of an induction system, responding fairly predictably to specific environmental changes. Switches between a variety of yeast morphologies, on the other hand, although occurring with different frequency in different environments, are less predictable, and are due to an unknown switch mechanism. To complicate matters, the propensity to undergo specific switches is also under genetic control, differing markedly between strains, and it is postulated that variants in propensity to switch may themselves be uncovered by genetic events such as mitotic recombination or chromosome loss (Chapter 19), as has been demonstrated for recessive auxotrophic mutations (see below).

11.4   FUNGAL GENETICS IN THE LABORATORY

11.4.1   Mutant isolation

The creation of mutants in haploid fungi is easy; purification of mutants and definition of a screen to select only those mutants which will be useful or informative may take a little extra trouble.

Mutants are required for two main reasons; first, as markers for genetic analysis, and second, as potent tools for investigating a range of biological processes from simple biochemical pathways to development. They also have an important role in strain identification and the avoidance of contamination.

The choice of mutagen is largely a matter of convenience; ultraviolet light is simple to use, but has the drawback that it may induce chromosome aberrations as well as point-mutations, whereas chemical mutagens, unless used with care, can be so powerful that they induce multiple mutations. For work with a virgin organism, there is a lot to be said for minimizing mutagen doses (Bos, 1987), or even relying on spontaneous mutation (Teow and Upshall, 1983).

If simple genetic markers are required, RFLPs or RAPD variants now give an immediate supply of alleles for a species with any natural variability. However, certain classes of induced mutant may make more convenient starting points. Spore colour in particular, or other immediately visible markers, are invaluable in any sort of genetic analysis since they provide an immediate indication of segregation, whether this is expected or not. Drug-resistant mutants are second choice, mainly because of their ease of isolation, and auxotrophs the third. Certain auxotrophs can also be selected by virtue of their resistance to anti-metabolites; a proportion of mutants resistant to chlorate, fluoroacetate, 5-fluoro-orotic acid, and selenate are respectively deficient in nitrate reductase (Cove, 1976), acetate utilization (Apirion, 1962), uracil biosynthesis (Boeke *et al.*, 1984) and sulphate uptake (Arst, 1968). For isolation of auxotrophs from conidial fungi, velvet pad replication is a very useful procedure, and its power can be magnified by the use of compounds which induce microcolonial growth form, e.g. detergents (Mackintosh and Pritchard, 1963; Maleszka and Pieniazek, 1981) or the non-metabolizable sugar sorbose (Tatum *et al.*, 1949). A variety of mutant enrichment techniques have also been devised to remove non-mutant mycelium which can grow without supplements; a recent example is the use of lytic enzymes to select non-growing mutants or segregants (Debets *et al.*, 1989).

A novel method of generating point mutations in *N. crassa* uses the RIP phenomenon. If a cloned gene is reintroduced into the genome by transformation, the result is usually a duplication (Chapter 12). If the resulting strain is then passed through meiosis, one or both copies will be mutated by the RIP mechanism. It is likely to be most useful for cloned genes of complex or non-selectable function; recent examples are the generation of mating type A point mutants (Glass and Lee, 1992) and disruption of the chitin synthase *chs-1* gene (Yarden and Yanofsky, 1991).

Transposon mutagenesis has not often been used in fungi, perhaps because transposons are not ubiquitous in fungi, or because so many other methods are available. However, a comparable procedure has been used to mutate, and subsequently clone, the *wA* gene from *A. nidulans* by insertion of a transforming plasmid (Tilburn *et al.*, 1990). Specific mutants may also be generated by gene disruption following cloning (Chapter 12).

Mutant isolation is a greater problem in diploid fungi, but it is one that has been solved for a great variety of higher organisms and is also possible for diploid fungi, e.g. *Phytophthora cactorum* (Elliott and MacIntyre, 1973), provided that they have a manageable sexual cycle. The genetics of *Candida albicans* (reviewed by Scherer and Magee, 1990) is a problem in this respect because it is imperfect, and the isolation of auxotrophic mutants has been shown to depend on the fact that many of the wild strains are already heterozygous for certain mutations.

An important step in isolating a mutant is its purification. This is a simple step for fungi with uninucleate vegetative spores, but may require more care for other fungi. In *N. crassa*, whose conidia are multinucleate, passage through the sexual cycle is a standard procedure for new mutants, and for other species it may be necessary to recover mutants from mutated protoplasts or fragmented hyphae.

Not all mutants can be isolated in a pure state, for example engineered null mutants of the *A. nidulans bimE* gene, which are unable to complete the cell cycle, are lethal but could be recovered as components of a heterokaryon (Osmani *et al.*, 1988). The state of the cell cycle resulting from the mutation could then be examined in uninucleate germinating conidia.

When mutants are used as a method of analysis rather than merely as genetic markers, the critical step is the definition of mutants of interest. Intensive investigations of single mutants have sometimes proved disappointing because the mutant studied was in a gene with only a peripheral effect on the pathway under investigation. The method is therefore most fruitful if a large-scale, if not exhaustive, survey of mutant phenotypes is

possible, in order to give the experimenter a good idea of the range of phenotypes to look out for. A survey of conidiation-deficient mutants of *A. nidulans* showed a wide spectrum of phenotypes (Martinelli and Clutterbuck, 1971), from which only those critical for conidiation were selected for further study. These subsequently proved to belong to genes essential for specific developmental steps (Clutterbuck, 1969; Clutterbuck and Timberlake, 1992). Similarly the large collection of conditional lethal mutants described by Morris (1975) has given a lot of information about both the mechanics and control of the of cell cycle, as each gene has been cloned and sequenced (summarized by Morris and Enos, 1992).

An essential procedure in sorting a collection of mutants is to determine by complementation tests which of them are repeat mutations at the same loci. Where loci have complex functions, this may also reveal unexpected relationships. Heterokaryons are usually ideal for this purpose, but in some cases, for instance where complementation has to take place in uninucleate phialides (Clutterbuck, 1969), the ability to construct heterozygous diploids is essential. Traditionally, the next step is genetic mapping.

11.4.2 Genetic mapping

(a) Conventional genetic mapping

Genetic mapping might be thought redundant in a molecular age, but in practice there is a certainty about a mapped gene for which there is no substitute. First if a gene is mapped, its identity with, or distinction from, other genes of similar function is established. Second, in the course of the manipulations involved in analysis of crosses, the investigator is likely to discover any important gene interactions contributing to the mutant phenotype, and finally the products of crosses and the map information gained can greatly simplify strain construction and analysis of manipulated constructs.

The scope for genetic mapping clearly depends on the sexual status of the species concerned. Heterothallic fungi have the advantage that the two parents involved in a cross can be specified precisely, but it turns out that homothallic species, such as *A. nidulans*, also outcross with little difficulty, and many auxotrophic strains tend to

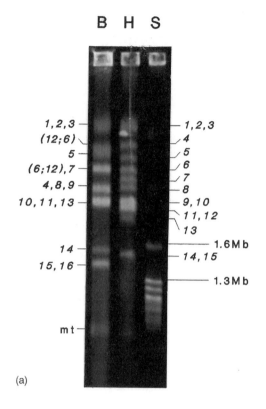

Figure 11.5 Electrophoretic karyotype and RFLP linkage map of *Cochliobolus heterostrophus*. (a) Electrophoretic karyotype: chromosomes were separated by transverse alternating field electrophoresis (TAFE). Lanes B and H contain DNA from two strains of the fungus which show an overall similarity, but also some differences, as indicated by the chromosome numbers to the left and right of the figure respectively. Lane S contains *Saccharomyces cerevisiae* DNA, included as size markers for the smaller chromosomes. Chromosomes that coincide after separation by this method could be distinguished using a different current switching regime. Note that chromosome 16 is totally absent from strain H, and is therefore designated as a supernumerary or B chromosome. mt = mitochondrial DNA. (b) RFLP linkage map: the majority of markers are RFLP sites; they are joined by lines of different density depending on the statistical significance of linkage between them. Map distances are given to the left of each chromosome. The thin line to the left indicates chromosomes of strain B and the dashed line to the right that of strain H; they diverge only for chromosomes 6 and 12 where the two strains differed by a reciprocal translocation. DNA probes hybridizing to chromosome 16 in strain B did not hybridize to anything in strain H. Reproduced, with permission, from Tzeng *et al.* (1992).

form more hybrid than selfed fruiting bodies. Homothallism also has the advantage that, barring other complications, any strain can be crossed with any other.

Where many mutant alleles of one gene have been generated, intragenic mapping is a preliminary to sequencing the sites of mutation. This has proved a valuable step in a number of genes; for example, the *prnA* (proline catabolism) gene of *A. nidulans* was mapped by a combination of conventional mapping of point mutants and deletion mapping (Sharma and Arst, 1985). The resulting map was important in choosing segments to amplify by PCR for sequencing. Similarly Clutterbuck *et al.* (1992) were able to make a conventional intragenic map of the *A. nidulans brlA* gene, which suggested that most of the mutations

previously isolated were concentrated towards the 3' end of the gene, in a region known to encode a DNA binding domain (Chapter 17). PCR amplification and sequencing directed towards this region (G.W. Griffith and A.J. Clutterbuck, unpublished) has confirmed this conclusion.

Intragenic mapping in *A. nidulans* is conducted much like standard meiotic mapping, monitoring crossovers between mutant sites in relation to markers on either side of the gene. However, in some fungi (e.g. *Ascobolus immersus*: Paquette and Rossignol, 1978) crossing over within genes is rare, but mutant sites can still be mapped on a gradient of gene conversion, whose frequency is highest at one end of the gene, probably at a site at which the recombination process is initiated.

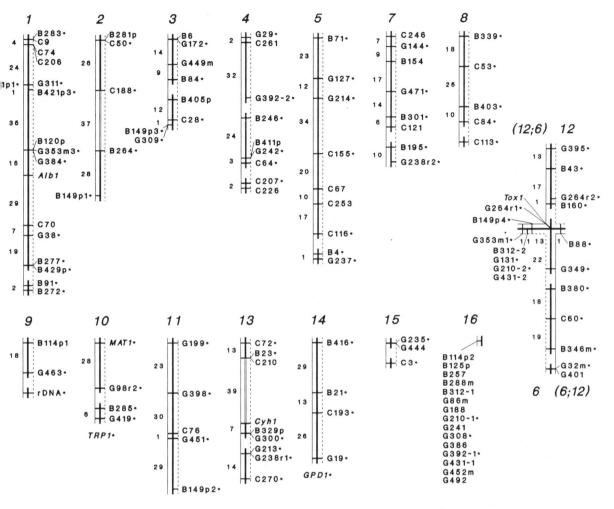

(b) RFLP mapping

For a number of economically important fungi genetic maps based mainly on restriction fragment length polymorphisms (RFLPs) have recently been published. In many cases, moreover, the genetic maps have been correlated with chromosomes using specific probes to electrophoretic karyotypes. In *Phanerochaete chrysosporium* RFLP mapping made it possible to locate genes responsible for lignin degradation in this genetically unexplored fungus (Raeder *et al.*, 1989). In the maize pathogen *Cochliobolus heterostrophus*, 126 markers, mostly RFLPs, were mapped to linkage groups (Figure 11.5b) which could be then correlated with chromosomes separable by PFGE (Figure 11.5a) (Tzeng *et al.*, 1992).

Despite the economic prominence of the button mushroom *Agaricus bisporus*, its genetics has been almost non-existent until recently. The problems with this fungus are that it exhibits secondary homothallism and has no uninucleate spore stage for isolation of pure strains. Moreover, the spontaneously dikaryotized progeny of crosses were rarely homoallelic for any RFLP markers examined. The mycelium was also, until recently, resistant to protoplasting, but solution of this problem provided the answer to the others, since homoallelic, presumably monokaryotic haploid strains could then be isolated. Comparison of a detailed RFLP map with the electrophoretic karyotype (which incidentally also required protoplasts as a source of intact chromosomal DNA: Royer *et al.*,

1992) led to the conclusion that recombination rates, at least for centromeric regions of the chromosomes, were low. This, in turn, if combined with a postulated tendency for non-sister spores to fuse in the formation of new dikaryons, explained the unexpectedly high level of heteroallelism of daughter mycelia (Kerrigan *et al.*, 1993).

(c) DNA and map distance

Tzeng *et al.* (1992) compared the total number of map units for *C. heterostrophus* estimated from genetic mapping with the DNA content as determined by electrophoretic karyotyping. They concluded that one map unit (centiMorgan) represented about 23 kb of DNA. For many fungi the range is 10–60 kb/map unit. They suggest that this ratio may be influenced by the degree of similarity of the parent strains. Where RFLP mapping is used, the parent strains are likely to be selected to be as dissimilar as possible, with the result that the total map length obtained is likely to be a minimum. For example, Raeder *et al.* (1989) have calculated that the ratio is 59 kb/map unit for *Phanerochaete chrysosporium*, based on comparison of the RFLP genetic map and DNA quantities calculated from the electrophoretic karyotype. In contrast, for *A. nidulans* laboratory strains derived from a single isolate the ratio of DNA to linkage distance has been calculated to be about 5 kb/map unit when account is taken of incompleteness of the current linkage map. This adds up to over 4000 map units (Clutterbuck, 1992, 1993) but comparison of individual chromosomes with their counterparts in the electrophoretic karyotype suggests that only 70% of the genome is mapped. The calculated ratio agrees reasonably well with some more precise estimates for cloned segments, although these are quite heterogeneous. Such heterogeneity is not surprising, and indeed in *Saccharomyces cerevisiae* the ratios for various segments of chromosome III differ by factors of over 100 (Oliver *et al.*, 1993). *N. crassa* experimental stocks are more widely based than those of *A. nidulans* and here the total linkage map is about 1000 map units (Perkins, 1993), although the genome size is similar to that for *A. nidulans*.

(d) Parasexual mapping

Whether or not the parasexual cycle is important in nature, it can be useful as a mapping tool for both sexual and asexual fungi. Parasexual recombination in phytopathogenic fungi has been reviewed by Tinline and McNeil (1969) and in *Aspergillus* species by Clutterbuck (1992). In *A. nidulans* it is an efficient method for locating new mutations to chromosomes before using meiotic crosses for more precise mapping. It is standard practice to make a heterokaryon between a strain carrying a new mutant and a master strain with markers on all eight chromosomes (McCully and Forbes, 1965). From this heterokaryon, a diploid is selected, and this is haploidized using benlate or other mitotic poisons. As a rule, mitotic recombination is a rare event, and independent of haploidization, therefore each chromosome segregates in one piece, and any new markers should be seen to be linked to one of the eight chromosome-specific markers. Details of this linkage are then explored using meiotic mapping.

In meiotic mapping visible markers such as vegetative spore colour mutants are advantageous in providing instantly monitored evidence of segregation. This is even more true of parasexual analysis, and it is doubtful if the parasexual cycle would have been discovered without the conspicuous yellow, white and green conidial mutants of *A. nidulans*.

Mitotic crossing over can also be used to locate markers within chromosomes (Käfer, 1961), but requires a series of markers on the same chromosome and in coupling with the mutant to be mapped. Since construction of such a chromosome would require a prior series of crosses, the method is not often used. In *A. niger*, mitotic recombination appears to be more frequent, and mitotic linkage maps are generated directly, due to occasional mitotic recombination events before haploidization (Debets *et al.*, 1989). The result is a remarkably detailed genetic map (Bos *et al.*, 1993).

11.5 CONCLUDING REMARKS

Much of the fungal genetics literature has concentrated on model species, but with the advent of molecular methods for characterizing DNA, both analytical and population genetics are ready to move beyond this stage to apply the lessons from the models to a wider range of species. The limitations on this expansion are often biological; problems with growth on defined medium or spore germination are examples, but if these are overcome, even for a species without a sexual cycle, parasexual genetics may be fruitful, and surveys of population diversity and genetic identity are possible even in the absence of knowledge of the life cycle. In the meantime studies of the laboratory-tamed models can be predicted to continue to throw up both problems and solutions for application in the wider mycological world.

REFERENCES

Apirion, D. (1962) A general system for the automatic selection of auxotrophs from prototrophs and vice versa in micro-organisms. *Nature*, **195**, 959–61.

Arst, H.N. Jr. (1968) Genetic analysis of the first steps of sulphate metabolism on *Aspergillus nidulans*. *Nature*, **219**, 268–70.

Bainbridge, B.W. and Roper, J.A. (1966) Observations on the effects of a chromosome duplication in *Aspergillus nidulans*. *Journal of General Microbiology*, **42**, 417–24.

Boeke, J.D., LaCroute, F. and Fink, J.R. (1984) A positive selection for mutants lacking orotidine-5'-phosphate decarboxylase in yeast: 5-fluoro-orotic acid resistance. *Molecular and General Genetics*, **197**, 345–6.

Bos, C.J. (1987) Induction and isolation of mutants in fungi at low mutagen doses. *Current Genetics*, **12**, 471–4.

Bos, C.J., Debets, F. and Swart, K. (1993) *Aspergillus niger*, in *Genetic Maps*, 6th edn, Vol. 3 *Lower Eukaryotes*, (ed. S.J. O'Brien), Cold Spring Harbor Laboratory, Cold Spring Harbor, pp. 3.87–91.

Brasier, C.M. (1991) Current questions in *Phytophthora* systematics: the role of the population approach, in *Phytophthora*, (eds J.A. Lucas, R.C. Shattock, D.S. Shaw and L.R. Cooke), Cambridge University Press, Cambridge, pp. 104–28.

Brasier, C.M. (1992) Evolutionary biology of *Phytophthora*: I Genetic systems, sexuality and the generation of variation. *Annual Review of Phytopathology*, **30**, 153–71.

Brasier, C.M. and Hansen, E.M. (1992) Evolutionary biology of *Phytophthora*: II Phylogeny, speciation and population structure. *Annual Review of Phytopathology*, **30**, 173–200.

Bruns, T.D. and Szaro, T.M. (1992) Rate and mode differences between nuclear and mitochondrial small-subunit rRNA genes in mushrooms. *Molecular Biology and Evolution*, **9**, 837–55.

Bruns, T.D., White, T.J. and Taylor J.W. (1991) Fungal molecular systematics. *Annual Review of Ecology and Systematics*, **22**, 525–64.

Burdon, J.J. and Roelfs, A.P. (1985) The effect of sexual and asexual reproduction on the isozyme structure of populations of *Puccinia graminis*. *Phytopathology*, **75**, 1068–73.

Burdon, J.J., Marshall, D.R. and Luig, N.H. (1981) Isozyme analysis indicates that a virulent cereal rust pathogen is a somatic hybrid. *Nature*, **293**, 565–6.

Burdon, J.J., Luig, N.H. and Marshall, D.R. (1983) Isozyme uniformity and virulence variation in *Puccinia graminis* f. sp. *tritici* and *P. recondita* f.sp. *tritici* in Australia. *Australian Journal of Biological Sciences*, **36**, 403–10.

Butcher, A.C. (1968) The relationship between sexual outcrossing and heterokaryon incompatibility in *Aspergillus nidulans*. *Heredity*, **23**, 443–52.

Carlile, M.J. (1987) Genetic exchange and gene flow: their promotion and prevention, in *British Mycological Society Symposium*, Vol. 12 (eds A.D.M. Rayner, C.M. Brasier and D. Moore), Cambridge University Press, Cambridge, pp. 203–214.

Caten, C.E. (1972) Heterokaryon incompatibility and cytoplasmic infection in fungi. *Journal of General Microbiology*, **72**, 221–9

Caten, C.E. and Jinks, J.L. (1966) Heterokaryosis: its significance in wild homothallic ascomycetes and fungi imperfecti. *Transactions of the British Mycological Society*, **49**, 81–93.

Clutterbuck, A.J. (1969) A mutational analysis of conidial development in *Aspergillus nidulans*. *Genetics*, **63**, 317–27.

Clutterbuck, A.J. (1992) Sexual and parasexual genetics of *Aspergillus* species, in *Aspergillus: Biology and Industrial Applications*, (eds J.W. Bennett and M.A. Klich), Butterworth-Heinemann, Boston, pp. 3–18.

Clutterbuck, A.J. (1993) *Aspergillus nidulans*, nuclear genes, in *Genetic Maps*, 6th edn, Vol. 3, *Lower Eukaryotes*, (ed. S.J. O'Brien), Cold Spring Harbor Laboratory, Cold Spring Harbor, pp. 371–84

Clutterbuck, A.J., Stark, M.S. and Gupta, G. (1992) An intragenic map of the *brlA* locus of *Aspergillus nidulans*. *Molecular and General Genetics*, **231**, 212–16.

Clutterbuck, A.J. and Timberlake, W.E. 1992 Genetic regulation of sporulation in the fungus *Aspergillus nidulans*, in *Development, the Molecular Approach*, (eds V.E.A. Russo, S. Brody, D. Cove and S. Ottolenghi), Springer-Verlag, Berlin, pp 103–20.

Correll, J.C. (1991) The relationship between formae speciales, races and vegetative compatibility groups in *Fusarium oxsporium*. *Phytopathology*, **81**, 1061–4.

Cove, D.J. (1976) Chlorate toxicity in *Aspergillus nidulans*. *Molecular and General Genetics*, **146**, 147–59.

Croft, J.H. (1966) A reciprocal phenotypic instability affecting development in *Aspergillus nidulans*. *Heredity*, **21**, 565–79.

Croft, J.H. and Jinks, J.L. (1977) Aspects of the population genetics of *Aspergillus nidulans*, in *Genetics and Physiology of Aspergillus*, (eds J.A. Pateman and J.E. Smith), Academic Press, London, pp. 339–60.

Debets, A.J.M., Swart, K. and Bos, C.J. (1989) Mitotic mapping in linkage group V of *Aspergillus niger* based on selection of auxotrophic recombinants by Novozym enrichment. *Canadian Journal of Microbiology*, **35**, 982–8.

Doolittle, W.F. and Sapienza, C. (1980) Selfish genes, the phenotype paradigm and genome evolution. *Nature*, **284**, 618–19.

Elliott, C.G. and MacIntyre, D. (1973) Genetical evidence on the life-history of *Phytophthora*. *Transactions of the British Mycological Society*, **60**, 311–16.

Erlich, H.A. and Arnheim, N. (1992) Genetic analysis using the polymerase chain reaction. *Annual Review of Genetics*, **26**, 479–506.

Förster, H., Coffey, M.D., Elwood, H. and Sogin, M.L. (1990a) Sequence analysis of the small subunit ribosomal RNAs of three zoosporic fungi and implications for fungal evolution. *Mycologia*, **82**, 306–12.

Förster, H, Oudemans, P. and Coffey, M.D. (1990b) Mitochondrial diversity within six species of *Phytophthora*. *Experimental Mycology*, **14**, 18–31

Fry, W.E., Goodwin, S.B., Matuszak, J.M. *et al.* (1992) Population genetics and intercontinental migrations of *Phytophthora infestans*. *Annual Review of Phytopathology*, **30**, 107–29.

Glass, N.L. and Lee, L. (1992) Isolation of *Neurospora crassa* mating type mutants by repeat induced point (RIP) mutation. *Genetics*, **132**, 125–33.

Hamer, J.E., Farrell, L. Orbach, M.J. *et al.* (1989) Host species-specific conservation of a family of repeated DNA sequences in the genome of a fungal plant pathogen. *Proceedings of the National Academy of Sciences of the United States of America*, **86**, 9981–5.

Harmsen, M.C., Schuren, F.H.J., Moukha, S.M. *et al.* (1992) Sequence analysis of the glyceraldehyde-3-phosphate dehydrogenase genes from the basidiomycetes *Schizophyllum commune*, *Phanerochaete chrysosporium* and *Agaricus bisporus*. *Current Genetics*, **22**, 447–54

Hinz, W.E., Jeng, R.S., Hubbes, M.M. and Horgen, P.A. (1991) Identification of three populations of *Ophiostoma ulmi* (aggressive subgroup) by mitochondrial DNA restriction-site mapping and nuclear DNA fingerprinting. *Experimental Mycology*, **15**, 316–25.

Joaquim, T. and Rowe, R.C. (1991) Vegetative compatibility and virulence of strains of *Verticillium dahliae* from soil and potato plants. *Phytopathology*, **81**, 552–8.

Käfer, E. (1961) The processes of spontaneous and induced somatic segregation in vegetative nuclei of *Aspergillus nidulans*. *Genetics*, **46**, 1581–609.

Kerrigan, R.W., Royer, T.C., Baller, L.M. *et al.* (1993) Meiotic behavior and linkage relationships in the secondarily homothallic fungus *Agaricus bisporus*. *Genetics*, **133**, 225–36.

Kistler, H.C. and Miao, V.P.W. (1992) New modes of genetic change in filamentous fungi. *Annual Review of Phytopathology*, **30**, 131–52.

Kohn, L.M. (1992) Developing new characters for fungal systematics; an experimental approach for determining the rank of resolution. *Mycologia*, **84**, 139–63

Kurtzman, C.P. (1985) Molecular taxonomy of the fungi, in *Gene Manipulations in Fungi*, (eds J.W. Bennett and L.L. Lasure), Academic Press, Orlando, pp. 35–63

Mackintosh, M.E. and Pritchard, R.H. (1963) The production and replica plating of micro-colonies of *Aspergillus nidulans*. *Genetical Research*, **4**, 320–2.

Maleszka, R. and Pieniazek, N.H. (1981) Modified replica plating technique of microcolonies of *Aspergillus nidulans* using Triton X-100. *Aspergillus News Letter*, **15**, 36–7.

Martinelli, S.D., and Clutterbuck, A.J. (1971) A quantitative survey of conidiation mutants in *Aspergillus nidulans*. *Journal of General Microbiology*, **69**, 261–8.

Maynard Smith, J. (1978) *The Evolution of Sex*, Cambridge University Press, Cambridge.

McCully, K.S. and Forbes, E. (1965) The use of p-fluorophenylalanine with 'master strains' of *Aspergillus nidulans* for assigning genes to linkage groups. *Genetical Research*, **6**, 352–9.

McDonald, B.A. and Martinez, J.P. (1991) DNA fingerprinting of the plant pathogenic fungus *Mycosphaerella graminicola* (anamorph *Septoria tritici*). *Experimental Mycology*, **15**, 146–58.

McKnight, G.L., O'Hara, P.J. and Parker, M.L. (1986) Nucleotide sequence of the triosephosphate isomerase gene from *Aspergillus nidulans*: implications for a differential loss of introns. *Cell*, **46**, 143–7.

Meyer, W., Koch, A., Niemann, C. *et al.* (1991) Differentiation of species and strains among filamentous fungi. by DNA fingerprinting. *Current Genetics*, **19**, 239–42.

Miao, V.P.W., Matthews, D.E. and VanEtten, H.D. (1991) Identification and chromosomal locations of a family of cytochrome P-450 genes for pisatin detoxification in the fungus *Nectria haematococca*. *Molecular and General Genetics*, **226**, 214–23.

Mills, D. and McClusky, K. (1990) Electrophoretic karyotyping of fungi: the new cytology. *Molecular Plant-Microbial Interactions*, **3**, 351–7.

Morris, N.R. (1975) Mitotic mutants of *Aspergillus nidulans*. *Genetical Research*, **26**, 237–54.

Morris, N.R. and Enos, A.P. (1992) Mitotic gold in a mold. *Trends in Genetics*, **8**, 32–7.

Nicholson, P., Brown, J. and Atkinson, M. (1991) Fingerprinting fungi, in *Genes in Ecology*, (Eds R.J. Berry, T.J. Crawford and G.M. Hewitt). Blackwell Scientific Publications, London, pp. 477–80.

Oliver, S.G., James, C.M., Gent, M.E. and Indge, K.J. (1993) Yeast genome organization and evolution, in *The Eukaryote Genome, Organisation and Regulation; 50th Symposium of the Society for General Microbiology*, (eds P. Broda, S.G. Oliver and P.F.G. Sims), Cambridge University Press, Cambridge, pp 1–17.

Orgel, L.E. and Crick, F.H.C. (1980) Selfish DNA: the ultimate parasite. *Nature*, **284**, 604–7

Osmani, S.A., Engle, D.B., Doonan, J.H. and Morris, N.R. (1988) Spindle formation and chromatin condensation in cells blocked at interphase by mutation of a negative cell cycle control gene. *Cell*, **52**, 241–51.

Paquette, N. and Rossignol, J.L. (1978) Gene conversion spectrum of 15 mutants giving post-meiotic segregation in the b_2 locus of *Ascobolus immersus*. *Molecular and General Genetics*, **163**, 313–26.

Perkins, D.D. (1993) *Neurospora crassa* nuclear genes, in *Genetic Maps*, 6th edn (eds S.J. O'Brien) Cold Spring Harbor Laboratory, Cold Spring Harbor, New York, pp. 3.11–20.

Perkins, D.D. and Barry, E.G. (1977) The cytogenetics of *Neurospora*. *Advances in Genetics*, **19**, 133–285.

Perkins, D.D. and Turner, B. (1988) *Neurospora* from natural populations: towards the population biology of a haploid eukaryote. *Experimental Mycology*, **12**, 91–131.

Pontecorvo, G. (1946) Genetic systems based on heterokaryosis. *Cold Spring Harbor Symposium on Quantitative Biology*, **11**, 193–201.

Puhalla, J.E. and Mayfield, J.E. (1974) The mechanism of heterokaryotic growth in *Verticillium dahliae*. *Genetics*, **76**, 411–22.

Raeder, U., Thomson, W. and Broda, P. (1989) RFLP-based genetic maps of *Phanerochaete chrysosporium* ME446: lignin peroxidase genes occur in clusters. *Molecular Microbiology*, **3**, 911–8.

Rayner, A.D.M. (1991) The challenge of the individualistic mycelium. *Mycologia*, **83**, 48–71.

Romao, J. and Hamer, J.E. (1992) Genetic organization of a repeated DNA sequence family in the rice blast fungus. *Proceedings of the National Academy of Sciences of the United States of America*, **89**, 5316–20.

Royer, J.C., Hintz, W.E., Kerrigan, R.W. and Horgen, P.A. (1992) Electrophoretic karyotype analysis of the button mushroom, *Agaricus bisporus*. *Genome*, **35**, 694–8.

Ryan, J.F., Beadle, G.W. and Tatum, E.L. (1943) The tube method of growth measurement of *Neurospora*. *American Journal of Botany*, **30**, 784–99.

Scherer, S. and Magee, P.T. (1990) Genetics of *Candida albicans*. *Microbiological Reviews*, **54**, 226–41.

Schmidt, T.M., DeLong, E.F. and Pace, N.R. (1991) Analysis of a marine picoplankton community by 16S rRNA gene cloning and sequencing. *Journal of Bacteriology*, **173**, 4371–8.

Sharma, K.K. and Arst, H.N. Jr (1985) The product of the regulatory gene of the proline catabolism cluster of *Aspergillus nidulans* is a positive-acting protein. *Current Genetics*, **9**, 299–304.

Skinner, D.Z., Budde, A.D. and Leong, S.A. (1991) Molecular karyotype analysis of fungi, in *More Gene Manipulations in Fungi*, (eds J.W. Bennett and L. Lasure), Academic Press, Orlando, pp. 86–103.

Smith, M.W., Feng, D.-F. and Doolittle, R.F. (1992) Evolution by acquisition: the case for horizontal gene transfers. *Trends in Biochemical Sciences*, **17**, 489–93.

Smith, T.L. (1989) Disparate evolution of yeasts and filamentous fungi indicated by phylogenetic analysis of glyceraldehyde-3-phosphate dehydrogenase genes. *Proceedings of the National Academy of Sciences of the United States of America*, **86**, 7063–6.

Soll, D.R., Morrow, B. and Srikantha, T. (1993) Switching and gene regulation in *Candida albicans*, in *The Eukaryote Genome, Organisation and Regulation; 50th Symposium of the Society for General Microbiology*, (eds P. Broda, S.G. Oliver and P.F.G. Sims), Cambridge University Press, Cambridge, pp. 211–40.

Strongman, D.B. and MacKay, R.M. (1993) Discrimination between *Hirsutella longicolla* var. *longicolla* and *Hirsutella longicolla* var. *cornuta* using random amplified polymorphic DNA fingerprinting. *Mycologia*, **85**, 65–70.

Tatum, E.L., Barratt, R.W. and Cutter, V.M., Jr (1949) Chemical induction of colonial paramorphs of *Neurospora* and *Syncephalastrum*. *Science*, **109**, 509–11.

Taylor, J.L., Borgmann, I. and Séguin-Swartz, G. (1991) Electrophoretic karyotyping of *Leptosphaeria maculans* differentiates highly virulent and weakly virulent isolates. *Current Genetics*, **19**, 273–7.

Teow, S.C. and Upshall, A. (1983) A spontaneous mutation approach to genetic study of filamentous fungi. *Transactions of the British Mycological Society*, **81**, 513–21.

Tilburn, J., Roussel, F. and Scazzocchio, C. (1990) Insertional inactivation and cloning of the *wA* gene of *Aspergillus nidulans*. *Genetics*, **126**, 81–90.

Tinline, R.D. and MacNeil, B.H. (1969) Parasexuality in plant pathogenic fungi. *Annual Review of Phytopathology*, **7**, 147–70.

Tzeng, T.-H., Lyngholm, I.K., Ford, C.F. and Bronson, C.R. (1992) A restriction fragment length polymorphism map and electrophoretic karyotype of the fungal maize pathogen *Cochliobolus heterostrophus*. *Genetics*, **130**, 81–96.

Upshall, A. (1981) Naturally occurring diploid isolates of *Aspergillus nidulans*. *Journal of General Microbiology*, **122**, 7–11.

Valent, B. and Chumley, F.G. (1991) Molecular genetic analysis of the rice blast fungus, *Magnaporthe grisea*. *Annual Review of Phytopathology*, **29**, 443–67.

Van Valen, L. (1973) A new evolutionary law. *Evolutionary Theory*, **1**, 1–30.

Vilgalys, R. and Johnson, J.L. (1987) Extensive genetic divergence associated with speciation in filamentous fungi. *Proceedings of the National Academy of Sciences of the United States of America*, **84**, 2355–8.

White, T.J., Bruns, T., Lee, S. and Taylor, J. (1990) Amplification and direct sequencing of fungal ribosomal DNA for phylogenetics, in *PCR Protocols, a Guide to the Methods and Applications*, (eds M.A. Innes, D.H. Gelfand, J.J. Sniansky and T.J. White), Academic Press, San Diego, pp. 315–22.

Whittaker, S.L., Shattock, R.C. and Shaw, D.S. (1991) Inheritance of DNA content in sexual progenies of *Phytophthora infestans*. *Mycological Research*, **95**, 1094–100.

Williams, J.G.K., Kubelik, A.R., Rafalski, J.A. and Tingey, S.V. (1991) Genetic analysis with RAPD markers, in *More Gene Manipulations in Fungi*, (eds J.W. Bennett and L.L. Lasure), Academic Press, San Diego pp. 432–9.

Yarden, O. and Yanofsky, C. (1991) Chitin synthase 1 plays a major role in cell wall biosynthesis in *Neurospora crassa*. *Genes and Development*, **5**, 2420–30.

MOLECULAR BIOLOGY

A.J. Clutterbuck

Genetics Building, Institute of Biomedical and Life Sciences, University of Glasgow, Glasgow, UK

12.1 INTRODUCTION

Molecular biology provides some powerful tools for the analysis of any organism, and fungi have many advantages as subjects of molecular investigation. Molecular biology has made sense of many features of biochemistry, cell biology and genetics, from which the discipline evolved, but its methods are also applicable to much wider fields. Among these are investigations of individuality and stability of isolates, their interrelationships in populations, and their identity in taxonomic and evolutionary terms. This chapter deals first with a description of the molecular features peculiar to fungi and then with appropriate molecular methods. Methods for identifying molecular variation in fungi have been described in Chapter 11.

12.2 THE MOLECULAR FEATURES OF FUNGI

12.2.1 DNA and chromosomes

Most filamentous fungi have small genomes, in the region of 30 Mb (Table 12.1), that is about twice as much DNA as the yeast *Saccharomyces cerevisiae* and about 10 times that of *Escherichia coli*, but one hundredth of the mammalian genome. Oomycetes have larger genomes, in addition to which, those investigated so far are diploid or polyploid. The estimates quoted in Table 12.1 have been obtained by a variety of methods which, as can be seen for *Neurospora crassa*, do not necessarily agree with each other. The more modern methods of DNA renaturation kinetics and electrophoretic karyotyping are probably more accurate, but all are subject to provisos. Chromosome numbers are generally in the range 7 to 20 (Skinner *et al.*, 1991; Turner 1993) but there are some notable exceptions: the yeast *Schizosaccharomyces*

pombe has only 3, whereas the chromosomes of *Basidiobolus ranarum* (see below) are too numerous to count.

There are reasons why small nuclei may be an advantage for fungi which are developmentally simple and can probably manage with small genomes. One advantage is the physical one that, at least in some species nuclei are required to pass through septal pores, and also through narrow bridges between anastomosed mycelia. Second, small genomes can be replicated quickly and cheaply, so are convenient for nutritional opportunists competing for a new food source. The third is the evolutionary argument, applicable to any organism, that small genomes minimize mutation. This is not quite as simple as it may seem; much of the excess DNA in larger genomes appears to be non-functional, and such DNA can, by definition, change its composition without affecting the phenotype, so it carries no mutational burden. However, if this unnecessary DNA includes degenerated repeats of coding genes, these may impose a 'recombinational load' by their propensity to destroy functional gene copies by recombining with them. Repeats of non-coding DNA may also lead to chromosomal aberrations as a result of recombination between dispersed copies. Furthermore, any unnecessary DNA is a potential site for chromosome breakage.

(a) Haploids, diploids, dikaryons and polyploids

Why are most fungi haploid? Diploids have the obvious advantage that recessive mutations are harmless, but this is only so until the number of mutations has increased to a point where homozygotes become a problem. Kondrashov and Crow (1991) calculate that diploids are favoured by high mutation rates, recessiveness of mutations, and, less obviously, by epistatic interactions. The argument

The Growing Fungus. Edited by Neil A.R. Gow and Geoffrey M. Gadd. Published in 1994 by Chapman & Hall, London. ISBN 0 412 46600 7

Table 12.1 DNA amounts per genome in fungi

Species	n^a	DNA (Mb)	Method[b]	References
Ascomycetes				
Saccharomyces cerevisiae	16	14	EK	Oliver *et al.* (1993)
Schizosaccharomyces pombe	3	14	EK (*not*1)	Fan *et al.* (1989)
Neurospora crassa	7	45	Cp	Horowitz and Macleod (1960)
Neurospora crassa	7	34	DRK	Dutta and Ohja (1972)
Neurospora crassa	7	27	DRK	Krumlauf and Marzluf (1979)
Neurospora crassa	7	47	EK	Orbach *et al.* (1988)
Podospora anserina	7	33	EK	Osiewacz *et al.* (1990)
Septoria tritici	18	31	EK	McDonald and Martinez (1991)
Leptosphaeria maculans				
highly virulent strains	6–8	9	EK	Taylor *et al.* (1991)
weakly virulent strains	12–14	16		
Deuteromycetes				
Aspergillus flavus group	c.8	36	EK	Keller *et al.* (1992)
Candida albicans	7–9	14–18	DRK	Riggsby *et al.* (1982)
Basidiomycetes				
Agaricus bisporus	13	34	EK	Royer *et al.* (1992)
Coprinus cinereus	12	39	DRK	Dutta and Ohja (1972)
Schizophyllum commune	11	35	EK	Horton and Raper (1991)
Zygomycetes				
Absidia glauca	10	42	EK	Kayser and Wöstermeyer (1991)
Mucor azygospora	?	31	KRK	Dutta and Ohja (1972)
Oomycetes				
Achlya bisexualis	?	42	DRK	Hudspeth *et al.* (1977)
Bremia lactucae	?	50	DRK	Francis *et al.* (1990)
Phytophthora megasperma	12–30	62	KRK	Mao and Tyler (1991)
Phytophthora megasperma	12–30	46	EK	Tooley and Carras (1992)
Phytophthora infestans	8–12	260	Cp	Tooley and Therrien (1987)

[a] Haploid chromosome number.
[b] Abbreviations: Cp: cytophotometry, DRK: DNA renaturation kinetics, EK: electrophoretic karyotype (*not*1: of *not*1 digest).

is complex, but in summary such gene interaction implies 'truncated selection' in which an individual carrying only a few deleterious mutations is relatively fit, but in an individual carrying more than that limited number, the mutations interact to curtail survival or reproduction. They suggest that such epistasis may be typical of 'K-selected' populations, where individuals compete over a prolonged period for limited resources, whereas in exponentially growing 'r-selected' populations each deleterious gene has its own slight and independent effect on reproduction.

This analysis seems appropriate for many fungi (see discussion by Rayner, 1991, and Chapter 2), which are r-selected opportunists and limit their mutation rates by having small genomes and few or no transposons (see below). On the other hand, some Basidiomycetes have adopted a K-selected

sedentary life style and appropriately enough are diploids, e.g. for an extreme version, the 1500-year-old, 10 000 kg thallus of *Armillaria bulbosa* reported by Smith *et al.* (1992).

Dikaryons have only some of the properties of diploids, and may be considered as an intermediate form. Duplicate genomes in separate nuclei allow the accumulation of recessive mutations, but not repair of DNA by homologues. It also seems that the protein synthesis potential of a dikaryotic cell is less than would be expected if both nuclei were fully active; a *Schizophyllum commune* dikaryon cell has only 1.3 times the volume, protein and RNA content of a haploid cell (de Vries and Reddingius, 1984). By comparison, diploids of *Aspergillus nidulans* have vegetative cells with double the volume per nucleus of haploid ones (Clutterbuck, 1969a).

According to Caten and Day (1977) sporadic vegetative diploids occur in all fungal groups except the Phycomycetes, whereas in the Oomycetes and a few other fungi, the vegetative phase is naturally diploid or polyploid. It could be argued that diploidy might relate to the often parasitic life style of Oomycetes, but an equally good argument is that diploidy is an accident of ancestry in that the Oomycetes are more closely related to algae and Myxomycetes than to other fungi (Förster *et al.*, 1990). It is also curious that the yeasts, including parasitic species such as *Candida albicans*, which appear to be extremely economical in their genome size (Table 12.1) are also found in nature as diploids. Among the more orthodox fungi, the Ascomycete *Ascocybe grovesii* also has a diploid vegetative phase (Dixon, 1959) and a small proportion of wild isolates of *A. nidulans* (all belonged to one heterokaryon compatibility group) are diploid (Upshall, 1981).

In a few cases, diploidy has been postulated on the basis of cytogenetics, e.g. for *Phytophthora infestans* (Sansome, 1977), but in many instances the first evidence in favour of diploidy is the difficulty of obtaining mutants. Those mutants that are obtained can be explained by the hypothesis that the starting strain was already heterozygous for a defective allele (e.g. for *C. albicans*: Whelan *et al.*, 1980). Homozygosity for these mutants can result from a second mutation, or more frequently, from mitotic recombination. Diploidy for *C. albicans* was confirmed by comparison of DNA amount and complexity, as determined by renaturation kinetics (Riggsby *et al.*, 1982). In the sexual species *Phytophthora cactorum*, Elliott and MacIntyre (1973) were able to show that a mutagen-treated isolate segregated one quarter methionine-requiring progeny on selfing, as predicted for a recessive mutant in a diploid. The diploid status of the Oomycete *Bremia lactucae* was confirmed by the observation (Hulbert and Michelmore, 1988) of heterozygosity for RFLP markers (Chapter 11).

Natural polyploids have been postulated mainly in fungi which are commonly found as diploids; these include *S. cerevisiae* (Hinchcliffe, 1991), non-Mexican isolates of *P. infestans* (Tooley and Therrien, 1987), and *C. albicans* (Suzuki *et al.*, 1986). The king of fungal nuclei, however, at least in terms of size, is found in the parasitic Zygomycete *Basidiobolus ranarum*. The interphase nucleus of this species is about 300 μm^3 in volume (Robinow, 1963) and sits in state in a $3 \times 10^4 \times \mu m^3$ cell, giving a cell/nuclear volume ratio of 100. Compare this with *A. nidulans* in which such a cell would contain about 350 nuclei, each approximately 1 μm^3 in volume (Clutterbuck, 1969a). The implication is that *B. ranarum* nuclei are highly polyploid and/or contain a great deal of non-coding DNA.

(b)   Repetitive DNA

Detailed analysis of DNA repetition by hybridization has shown that in both *A. nidulans* (Timberlake, 1978) and *N. crassa* (Krumlauf and Marzluf, 1980) repetitive DNA is minimal and can largely be accounted for by ribosomal RNA genes of which there are an estimated 185 copies in *N. crassa*. Hybridization of cosmids to electrophoretically separated chromosomes of *A. nidulans* (Brody *et al.*, 1991) showed that 64% of cosmids hybridized to only one chromosome. However, 1% of cosmids hybridized strongly to all chromosomes suggesting that they contain common elements which might be telomeric or centromeric. No repetition was detected by hybridization to genomic blots of clones totalling 100 kb from *Coprinus cinereus* (Wu *et al.*, 1983), which has a similar sized genome to *A. nidulans* and *N. crassa* (about 30 Mb).

In contrast, the size of the larger genomes of Oomycetes can probably be accounted for by their repetitive DNA content, e.g. *Achlya bisexualis*: 42 Mb, 18% repetitive (Hudspeth *et al.*, 1977), *Bremia lactucae*: 50 Mb, 60% repetitive (Francis *et al.*, 1990) and *Phytophthora megasperma*: 62 Mb, 53% repetitive (Mao and Tyler, 1991).

A variety of repetitive elements have also been described in the yeast *Candida albicans* (Lasker *et al.*, 1992), some of which can be used for genetic fingerprinting as discussed in Chapter 11.

(c)   Transposable elements

The best understood repetitive DNA, other than that representing gene families, belongs to transposable elements (transposons), either functional or degenerate. These are segments of DNA, of no evident value to their hosts, which have the ability to reproduce themselves by insertion of replicate copies at random sites in the genome. The best analysed fungal example is the yeast Ty element (Kingsman *et al.*, 1988). Despite supposed selective pressures keeping fungal genome size to a minimum, transposons have also been detected in some filamentous fungi. In *N. crassa* the 7 kb Tad element, originally found in the Adiopodoumé isolate from the Ivory Coast, was shown to transpose in progeny of crosses with laboratory strains (Kinsey and Helber, 1989). A second but smaller (1.5 kb) *N. crassa* repetitive element called Pogo, which has a transposon-like structure, has been found, in varying numbers in telomere-proximal DNA (Schechtman, 1990). It has not been shown to transpose, however, and some copies look as if they may have been disabled by RIPping (see below). Since similar sequences have been found in other fungi, Schechtman suggests that this element may be a recent introduction from another species. The MGR repetitive element of *Magnaporthe grisea*, which has proved to be an invaluable feature for identification of pathogenic races of this fungus (Chapter 11), also has many of the characteristics of a transposable element (Hamer *et al.*, 1989; Valent and Chumley, 1991).

Other repetitive DNAs found in a variety of fungi are simple sequences which may also be telomeric (Rodriguez and Yoder, 1991).

(d) Centromeres

Although centromeres figure on the genetic maps of both *N. crassa* and *A. nidulans*, located in the one fungus by tetrad analysis and in the other by mitotic crossing-over, in neither case has it yet proved possible to clone a centromere. This suggests that centromeres of filamentous fungi are probably more complex than those of yeast, and may contain repeats which would make them difficult to propagate in bacteria. A yeast centromere did not disturb chromosomal segregation when inserted into an *A. nidulans* chromosome (Boylan *et al.*, 1986) implying that it was not recognized by this fungus. The autonomously replicating *Aspergillus* plasmid ARp1 (Gems *et al.*, 1991) might now overcome the problem of defining centromeric clones by providing an assay for centromeric activity.

(e) Telomeres

Telomeres in eukaryotic microbes can be more complex than might be supposed for entities whose functions are merely to be chromosome ends (Greider, 1993). Two telomeres have been cloned from *N. crassa* (Schechtman, 1990). Both carry multiple repeats of the simple sequence (TTAGGG), which is the same as that found at the ends of human chromosomes. In *N. crassa* and *A. nidulans* (Conelly and Arst, 1991) a probe complementary to this sequence hybridized to approximately the expected number of telomeres (14 and 16 respectively). In both organisms these sequences were sensitive to BAL-31 exonuclease digestion of the native DNA implying that they occupied terminal chromosomal locations. As is common with the telomeric DNA of other eukaryotes, one of the *N. crassa* clones also included multiple copies of a transposon-like sequence (see above). The approach to the *A. nidulans* telomere was unusual, involving a translocation strain which had lost a considerable length of the chromosome. The missing terminal segment proved to be at least 240 kb in length and must be assumed to be dispensable since the phenotypic effect of the deletion was minimal. It may therefore be relevant that the ends of yeast chromosomes include a number of repeats of dispensable genes (Charron *et al.*, 1989).

In some instances at least, telomeres may include origins of replication: see the section on autonomously replicating plasmids below (section 12.3.1b).

(f) Chromatin

The small size of fungal chromosomes suggests that they may not conform to the standard eukaryote composition. For some time, histones could not be isolated from fungi, but the apparent deficiency proved to be technical and now histones, and the genes which encode them, have been characterized from both *N. crassa* (Woudt *et al.*, 1983) and *A. nidulans* (Ehinger *et al.*, 1990). Histone genes of most eukaryotes are present in multiple copies and are unusual in that they lack introns. However in both fungal species most of the histone genes are present as single copies (in *A. nidulans* one is duplicated and in *N. crassa*, two are), and they all contain introns.

(g) DNA methylation

Antequera *et al.* (1984) found that cytosine methylation levels in most fungi were below their detection limit of 0.1%. *Sporotrichum dimorphosporum* and *Phycomyces blakesleeanus* were exceptions, the latter having 2.9% of its cytosines methylated. Russell *et al.* (1987) also found 0.24–0.4 mol% 5-methylcytosine in various developmental stages of *N. crassa*. Although these values are low in comparison with many higher organisms, it seems probable that methylation has a function, particularly in repetitive DNA (see the RIP and MIP phenomena below). In *Schizophyllum commune*, a species in which Antequera *et al.* were unable to detect methylation, Buckner *et al.* (1988) found that ribosomal DNA repeats were heavily methylated, especially in the dikaryon, suggesting that methylation is confined to specialized sequences and to stages of the life cycle when these sequences are less actively transcribed (see above).

(h) Recombination

Although fungi have contributed more than any other group of eukaryotes to investigations of the mechanism of recombination (Orr-Weaver and Szostak, 1985; Klein, 1993), this is too complex a subject to review here. In any case, the assumption made in these studies is that fungi are not distinctive in their recombination procedures, but models for all eukaryotes.

(i) RIPping

Neurospora crassa has a special mechanism for dealing with repetitive DNA (reviewed by Selker, 1990). When transformation became a standard tool in *Neurospora* genetics, it was discovered that if a transformant containing duplicated genes was passed through the sexual cycle, one or both copies of the gene became mutated. The acronym RIP (Repeat Induced Point mutation) was adopted for this phenomenon, which is stimulated by repeated sequences in the dikaryotic cells of the sexual fruiting body. The RIP process makes GC to AT base changes consistent with the passage of a processive DNA-cytidine deaminase along the duplicated segment. Tandem duplications are more prone to RIP than scattered ones, but none are immune.

This process seems suited for inactivating unwanted repetitive elements such as transposons. Of course, the fact that a few transposons have been described in *Neurospora* and other genera (see

above) testifies on the one hand, to the necessity for such a mechanism, and on the other to the incomplete protection afforded by the RIP mechanism as a guard dog against these particular (white) elephants.

Few fungi, however, seem to employ the RIP system; it has not been reported for the other genetical workhorses, yeast and *Aspergillus* and it has specifically failed to show up in tests in *Sordaria* (Le Chevanton *et al.*, 1989) and *Podospora* (Coppin-Raynal *et al.*, 1989). *Ascobolus immersus* does have a comparable, but different system; duplicate genes are inactivated, but this inactivation is always reversible at low frequency. It has been shown to be due to base methylation ('MIP'), not accompanied by mutation (Rhounin *et al.*, 1992). Cytosine methylation is also a feature of RIP-mutated DNA in *Neurospora*, but surprisingly, it seems to be a consequence rather than a cause of RIP-mutation, and does not persist through the vegetative cycle (Selker 1990).

12.2.2 Fungal genes

Gene structure has been reviewed in detail by Gurr *et al.* (1987), Rambosek and Leach (1987) and Unkles (1992). As in other features of genome structure, the filamentous fungi are economical in the size of their genes. Whereas introns have been found in more than half the genes sequenced (in contrast to yeast and yeast-like dimorphic fungi such as *Candida albicans*), these introns are predominantly short, most being around 50 bp in length. Some introns may be regarded as exceptions, e.g. the two introns of 497 and 392 bp in the *A. nidulans stuA* and *brlA* genes respectively, which may be part of complex transcription control systems associated with differentiation (Chapter 17). Approximately a third of genes analysed in *A. nidulans* and *N. crassa* have no introns, but others have many, e.g. the *A. nidulans niaD* and *niiA* genes contain six and seven introns respectively, the longest of which is 91 bp, the remainder all being between 46 and 66 bp (Johnstone *et al.*, 1990). As in yeast though less markedly so, there is a tendency for intron sites to cluster near the 5' half of the gene. Intron splice sites follow the standard pattern for spliceosomal introns, and although an internal splice site can usually be recognized, it is more variable than the precise sequence demanded by *Saccharomyces* and *Candida*.

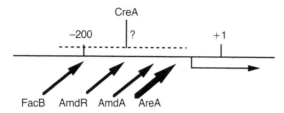

Figure 12.1 Regulation of transcription of the *Aspergillus nidulans amdS* gene. The *A. nidulans amdS* gene encodes a catabolic acetamidase. It is regulated by the products of one repressor and four positively acting regulators. The product of the *creA* gene represses transcription in the presence of a preferred carbon source; its site action is not yet known. The product of the *areA* gene is required for transcription, and is only active in the absence of ammonium ions which provide a simpler nitrogen source. Transcription is also dependent on the activity of at least one of the remaining regulators: the *amdR* gene product mediates induction by ω-amino acids, and the *amdA* and *facB* gene products mediate induction by acetate. Modified from Hynes *et al.* (1988). The horizontal arrow represents the start of the transcript, the +1 coordinate being the translation start point.

Since there are 61 sense codons, but only 20 amino acids, alternative codons are available for most amino acids. Again, non-randomness in codon preference is less extreme in filamentous fungi than it is in yeast, and the *A. nidulans* 'dialect' is as distinct from that of yeast as the latter is from *E. coli* (Lloyd and Sharp, 1991). *N. crassa* shows slightly greater bias in codon usage than *A. nidulans*, but both follow the standard pattern in which more highly expressed genes show the greatest bias (Gurr *et al.*, 1987).

(a) Gene transcription and regulation

Many, but by no means all fungal genes analysed carry recognizable 5' transcription start signals known as TATA and CAAT boxes (Gurr *et al.*, 1987). In some cases upstream sequences have also been identified as recognition sites for known regulatory gene products. Probably the most detailed picture of upstream regulatory sites is for the *A. nidulans amdS* gene (Hynes *et al.*, 1988), which codes for acetamidase, an enzyme which can be involved in both carbon and nitrogen supply and is regulated by at least five regulatory proteins (Figure 12.1). The sites of action of the different regulatory proteins have been identified by specific mutations and manipulated deletions,

but also by titration of the different activator proteins using transformants containing multiple copies of putative activator binding sites. These, by absorbing the activator protein, diminish the inducibility of the *amdS* gene (Kelly and Hynes, 1987).

Fungi are particularly attractive for studies of gene regulation since they are eukaryotes displaying considerable metabolic, especially catabolic, versatility. They also have the microbial advantages for handling of large numbers of individuals, ease of mutant isolation if they are haploid, and the availability of diploids or heterokaryons for dominance and allelism tests, not to mention facilities for genetic analysis and gene cloning. Many reviews are available (e.g. Marzluf, 1981; Davis and Hynes, 1991; Scazzocchio, 1992; Caddick and Turner, 1993). The last of these lists 25 known regulator genes for catabolic pathways in *A. nidulans* and *N. crassa*. The majority of these act positively, that is the gene product enhances transcription of its target gene while responding to either inductive or repressive signals from the environment, but a proportion act negatively, blocking transcription, and some may do either according to the circumstances.

Developmental control of gene expression can be regarded as a special case of regulation, and has provided some detailed examples of cascade regulation (Chapter 17). Developmental regulation is expected to differ from its metabolic cousin in that many of the signals are internal, i.e. one developmental step will depend on completion of previous steps as much as on environmental conditions, and secondly, development must allow for maintenance of a differentiated state, thus conidiation in *A. nidulans* involves positive feedback loops for the three main regulators identified, *brlA*, *abaA* and *wetA* (Chapter 17, Figure 17.4).

The value of detailed genetic mapping for elucidation of the functional elements of a regulator gene is well illustrated by Kudla *et al.* (1990) who used PCR sequencing (see below) on appropriate sections of the *A. nidulans areA* gene to pinpoint sites of mutants with particular phenotypes. This gene is particularly revealing in this respect since it regulates a spectrum of structural genes responding to nitrogen availability, and particular *areA* mutants may fail to activate one responder, while up-regulating another.

(b) Gene clusters

As the chromosomal location of most eukaryote genes seems quite random, it is of some curiosity when clusters of genes of related function are encountered. The prokaryote explanation for gene clusters, that they are transcribed as a poly-cistronic unit, has never been found to apply in fungi. Moreover the necessity for clustering does not seem to be constant, e.g. in *A. nidulans* nitrate and nitrate reductase genes are clustered, along with an uptake gene (Johnstone *et al.*, 1990), but in *N. crassa* these genes are scattered, although in other respects their regulation seems to be very similar, and indeed components are interchangeable (Hawker *et al.*, 1991).

It is difficult to avoid the conclusion however, that the advantage of clustering lies in the ability to respond to shared enhancers acting on the whole cluster. Comparison of the seven-gene quinate utilization clusters (*qut/qa*) of *A. nidulans* and *N. crassa* (Figure 12.2a) suggests that the selective forces in favour of clustering must have been strong enough to maintain this organization despite a minimum of three DNA rearrangements that have shuffled the gene order during evolutionary separation of the two species (Geever *et al.*, 1989; Lamb *et al.*, 1990).

Another interesting example is the penicillin biosynthesis cluster (Figure 12.2b), which is similar in *Penicillium chrysogenum* and *A. nidulans*. It consists of three genes, two of which (*acvA/pcbAB* and *ipnA/pcbC*), because of their distinctive codon usage, lack of introns and the absence of homologues in related species, are postulated to have been imported by an unknown horizontal gene transfer mechanism from a Streptomycete (Aharonowitz *et al.*, 1992). One explanation offered for their presence as a cluster is that they were imported as an operon, although they would have required to be split into two transcription units with the addition of appropriate eukaryotic expression signals. Furthermore, the tendency for genes to congregate, as opposed to a historical reason for their juxtaposition, is demonstrated by the fact that these two genes have been joined in the cluster by a third (*acyA/pcbDE*) which contains introns, and is found only in fungi.

The significance of clustering for gene expression has been tested for one of the *A. nidulans* Spo clusters (Figure 12.2c). Orr and Timberlake (1982) estimated from DNA hybridization experiments that of genes whose transcripts are confined to conidia, about 80% existed as clusters of two or more. Insertion of the *argB*, arginine biosynthesis gene, into one of these clusters (at sites indicated by arrows in Figure 12.2c) brought this gene under partial developmental control, while transfer of one cluster component, SpoC1C, to unrelated sites partially relieved it of such control (Miller *et al.*, 1987). It must be concluded that transcription of these genes responds to multiple influences, some local to the gene, others cluster-wide. No function has been determined for the SpoC1 cluster (Aramayo *et al.*, 1989), but it seems logical that in a situation such as a dormant spore, where little transcription occurs, the few active genes should be assigned to a limited number of chromosomal domains. Perhaps the same may be true of other genes whose transcription is crucial in adverse circumstances.

A different form of clustering is exhibited by single genes which encode multiple enzyme activities. The classic example of this is the aromatic biosynthesis (*arom*) complex of *N. crassa* (Coggins *et al.*, 1987) and *A. nidulans* (Hawkins, 1987). In addition to complete transcriptional coordination, it has been suggested that this arrangement might allow channelling of intermediates to enzyme domains performing successive steps in the pathway, thus avoiding loss of intermediates to enzymes of the *qut/qa* pathway which degrade the same products. The first experimental tests of channelling showed that, contrary to the hypothesis, the Arom multienzyme was thoroughly leaky to intermediates, but more recent work (Lamb *et al.*, 1992b) suggests that, once again, adverse conditions should be considered, since channelling may be more important when metabolite flux is at a minimum.

12.2.3 Extrachromosomal inheritance

(a) Nuclear plasmids

It has been suggested on the basis of electrophoretic karyotyping of a variety of fungi (e.g. Miao *et al.*, 1991) that some small chromosomes are inessential and can be regarded as equivalent to the supernumerary 'B' chromosomes found in some plants and animals. They might equally well

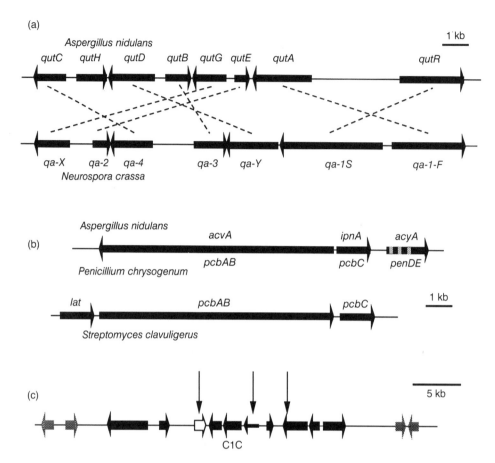

Figure 12.2 Fungal gene clusters. (a) The quinate utilization gene clusters of *Aspergillus nidulans* (Lamb *et al.*, 1990) and *Neurospora crassa* (Geever *et al.*, 1989). Large arrows indicate transcripts. Dashed lines join the six genes of comparable sequence in the two fungi; only the *Aspergillus qutH* gene has no known homologue in *N. crassa* (Lamb *et al.* 1992a). (b) The β-lactam biosynthetic clusters in *A. nidulans*, *Penicillium chrysogenum* and the bacterium *Streptomyces clavuligerus*. The *acvA/pcbAB* and *ipnA/pcbC* genes are shared by all three organisms, but are transcribed in the same direction in *Streptomyces* while they are divergent in the fungi. In *Streptomyces*, a third β-lactam gene, *lat*, is adjacent to the other two, and four other functionally related genes (not shown) are linked nearby. It is not clear whether the three *Streptomyces* genes shown are transcribed on a single mRNA. In the two fungi, the *acyA/penDE* gene, which has no homologue in bacteria joins the cluster. This gene is also distinctive in including introns (indicated as stippled blocks). Redrawn from Aharonowitz *et al.* (1992) (c) The *A. nidulans* SpoC1 cluster. Dark horizontal arrows indicate spore-specific transcripts, the open arrow a conidiophore-specific transcript, and grey arrows transcripts which show only partial developmental regulation. The thinner arrow indicates a low level transcript. Vertical arrows mark positions at which insertion of an *argB* gene (normally unregulated) brought it under developmental control. The SpoC1C gene which partially lost its developmental specificity when inserted elsewhere in the genome, is also labelled. Redrawn from Miller *et al.* (1987).

be described as linear plasmids. Circular plasmids are unusual in eukaryotes, and it has long been known that circular chromosomes are unstable (McClintock, 1932). However, evidence for a natural nuclear plasmid in the Basidiomycete *Phanerochaete chrysosporium* has been obtained from transformation experiments (see below), which gave rise to products interpreted as recombinants between the added vector and a resident replicating plasmid (Randall *et al.*, 1991).

(b)   Mitochondrial genomes

The fungal mitochondrial genomes are perhaps less bizarre than those of protozoa, but they are, nevertheless remarkably diverse, indeed perverse, in their organization and behaviour (Williamson, 1993). Some mitochondrial genomes have regularly been seen as DNA circles (Wolf and Del Giudice, 1988), whereas others are genome-length linear mitochondrial DNA molecules (Weselowski and Fukuhara, 1981; Kovac *et al.*, 1982). Yet further species have a mixture of circular and linear molecules of diverse lengths. Still more confusingly, related species were reported to differ in whether they had linear or circular DNAs. This paradox has now been explained for both yeasts and filamentous fungi, including Oomycetes, by the hypothesis that most copies of the mitochondrial genome are indeed linear molecules of varying length thrown off by a rolling circle replicative form (Malezska *et al.*, 1991; Maleszka and Clark-Walker, 1992).

In addition to their regular genomes and some aberrant derivatives described below, mitochondria often contain further circular or linear molecules with no homology to standard mitochondrial genes. These are usually regarded as plasmids. For example, a variety of circular plasmids are found in *Neurospora* species; Labelle and Fiji plasmids (Stohl *et al.*, 1982) appear to be harmless, but modifications of Mauriceville and Varkud plasmids are able to invade and disrupt the mitochondrial genome (Akins *et al.*, 1986)

Linear mitochondrial plasmids appear to be a separate class. In some cases they have been shown to encode their own polymerases and they are possibly derived from viruses (Sakaguchi, 1990). The best investigated is the *N. intermedia kalilo* plasmid (Chan *et al.*, 1991) which again makes its presence felt when it disrupts the mitochondrial genome proper. In other fungi such as *Fusarium solani* linear plasmids may be important if they are associated with host specificity in plant pathogenesis: Samac and Leong (1988) were able to demonstrate that curing a strain of a linear mitochondrial plasmid, by growing the fungus on ethidium bromide, reduced its pathogenicity.

Ascomycete mitochondrial genomes also vary considerably in size (Wolf and Del Giudice, 1988). A few yeasts have no more than the extremely economical 19 kb found in mammals, but for most species the mitochondrial genome is larger, ranging up to the 115 kb for *Cochliobolus heterostrophus* (Garber and Yoder, 1984).

Despite these size variations, differences in genetic content appear to be slight. Most mitochondrial DNAs include genes for two ribosomal subunits, most, but not necessarily all, tRNAs, some subunits of cytochrome oxidase, ATPase and a number of open reading frames, including some of unidentified function, adding up to 11–19 coding genes in all (Wolf and Del Giudice, 1988). Whereas most mitochondrial genomes encode NADH dehydrogenase subunits, those of *Saccharomyces cerevisiae* and *Schizosaccharomyces pombe* do not. The ATPase subunit 9 gene is nuclear in mammals, mitochondrial in yeast, and has copies in both locations in *A. nidulans* and *N. crassa*, although in the latter, it has been shown that only the nuclear copy is active. In general it is no clearer for fungi than for other organisms why particular genes should be in one place or the other.

Much greater differences in mitochondrial genome size are due to non-coding DNA. *N. crassa* has a series of GC-rich palindromes (Yin *et al.*, 1981), whereas other species differ in the number of introns present. Mitochondrial introns are of interest as being of the self-splicing type, characterized by their large size and complex potential secondary structure, on which basis they can be subdivided into types I and II, and the fact that they often contain internal open reading frames. Some of these introns are demonstrably mobile (Lambowitz, 1989), e.g. two mitochondrial DNA insertions of *N. crassa* appeared to be infectious. In another instance, Earl *et al.* (1981) crossed two *A. nidulans* subspecies whose mitochondrial DNA differed in size by 20%, due entirely to six extra introns in one strain. The cross yielded mitochondria carrying a resistance marker from the smaller DNA parent, but three introns from the larger.

Fungal mitochondria, unlike those of higher eukaryotes, also recombine readily; Rowlands and Turner (1975) found a high proportion of recombinants among vegetative segregants from an *A. nidulans* heterokaryon between parents differing at three mitochondrial loci, implying frequent exchange between these markers. It is postulated that the presence of self-splicing introns and the occurrence of recombination are connected: mammalian mitochondria contain no introns and do not recombine, whereas certain intron-encoded products of yeast have been shown, not only to cut specific sites in intron-free genes leading to intron insertion, but also to cause recombination in adjacent regions (Katylak *et al.*, 1985).

Mitochondrial introns can also cause disturbances to the mitochondrial genome which are ultimately lethal. The phenomenon of senescence in *Podospora anserina* results from free replication of mitochondrial DNA fragments containing an excised 'mobile intron' I from the *coxI* gene (Kück, 1989). The excised intron alone is capable of replication, but the rogue elements often include lengthy neighbouring sequences, and they do their damage by outreplicating the normal mitochondrial DNA.

Another rogue mitochondrial genome derivative causes the *N. crassa* 'stopper' phenomenon (Bertrand *et al.*, 1980). Here the mitochondrial DNA recombines with itself to form two circles, one of which replicates faster than the other, resulting in erratic growth of the fungus. Other fungal degenerative disorders with a mitochondrial basis are 'poky' in *Neurospora* (Mitchell and Mitchell, 1952), 'vegetative death' in *A. nidulans* (Arlett, 1957) and 'ragged' in *A. amstelodami* (Lazarus *et al.*, 1980); these have been reviewed by Griffiths (1992).

(c) Viruses

Fungal viruses have been extensively reviewed in Lemke (1979) and Buck (1986) and there is a recent review of yeast viruses by Wickner (1992). Most fungal viruses have single- or double-stranded RNAs as their genetic material, and it is convenient to include among them virus-like particles (VLPs) which are not given the full status of viruses for two rather different reasons: either, as is general for fungal viruses, extracellular transmission has not been proved, or they are unencapsidated. The unencapsidated versions, at least in some cases, have a mitochondrial location (Barroso and Labarère, 1993). Interest in fungal viruses naturally centres on their debilitating effect on the host, which in the case of pathogenic fungi such as *Ceratocystis ulmi* (Rogers *et al.*, 1987) and *Cryphonectria parasitica* (Nuss and Koltin, 1990; Nuss, 1992) might afford means of biological control. On the other hand, for commercial fungi, viruses are obviously a pest (Romaine and Schlagnhaufer, 1990), and, rather astonishingly,

fungi are also blamed for acting as vectors for plant viruses to which they are not themselves susceptible (Teakle, 1983). In the laboratory, viruses and VLPs are two of the many possible causes of irregular colony growth.

12.3 MOLECULAR METHODS FOR FUNGI

Experimental molecular methods for fungi have been summarized in a number of reviews (e.g. Timberlake, 1991; May, 1992).

12.3.1 Transformation

Transformation of many fungi is now possible, although efficiency varies; for reviews see Fincham (1989), van den Hondel and Punt (1991). The critical steps in transformation are usually the formation of healthy protoplasts, and the choice of an effective selective marker. The most commonly used method of DNA transformation employs polyethylene glycol in combination with calcium ions. Alternative methods include electroporation of protoplasts (Ward *et al.*, 1989; Chakraborty and Kapoor, 1990) and 'biolistic' (Armaleo *et al.*, 1990) or lithium chloride-mediated transformation of whole cells (Allison *et al.*, 1992).

For the first transformation experiments with a particular species, it is often the case that the only genes available for selection of transformants are those cloned from other species. However, plasmids containing heterologous selective markers are less effective than those containing genes from the species concerned; for example for *Trichoderma viride* (Gruber *et al.*, 1990) the homologous *pyrG* gene was 15 times better than that from *Neurospora*. Nevertheless, a plasmid such as pAN7, which contains a bacterial hygromycin gene coupled to an *A.nidulans* promoter, has been used to develop transformation systems for a wide variety of fungi (Punt *et al.*, 1987). The *N.crassa* benomyl resistance gene is another example (McClung *et al.*, 1989). The advantage of resistance genes as selective markers is that the recipient can be genetically wild-type, but their effectiveness depends on the sensitivity of the host to the particular fungicide, a factor which may vary widely. A different sort of universal selective marker is the *A. nidulans amdS* gene, a homologue of which does not exist in many fungi, but when introduced enables the recipient to grow on acetamide as carbon and/or nitrogen source (e.g. Turgeon *et al.*,1986). Other widely applicable systems are those for which mutants can easily be selected (Chapter 11). The *A. nidulans niaD* (nitrate reductase) and *sC* (sulphate uptake) genes are available for transformation of such mutants (Daboussi *et al.*,1989; Buxton *et al.*,1989)

(a) Integrating plasmids

In the filamentous fungi for which transformation systems were first devised (*A. nidulans and N. crassa*), early experience showed that plasmids were only maintained in the fungus if they recombined with a chromosome. Comparisons with replicating plasmids (below) suggest that this is often the limiting step in obtaining high-frequency transformation. Three types of recombination event were originally described for yeast (Hinnen *et al.*, 1978), and also seen in filamentous fungi. If a mutant allele of the selected marker exists in the recipient, this may be changed to wild-type by double crossing-over or gene conversion, without integration of the plasmid responsible. Alternatively, a single homologous (Figure 12.3) or non-homologous crossover will result in integration of the plasmid. In *N. crassa*, non-homologous integration predominates, as in transformation of vertebrate cells, but the proportion of homologous integration events can be increased if the region of homology is large (Asch and Kinsey, 1990). In *A. nidulans* homologous and non-homologous integrations are of similar frequency and the site may depend on the selective marker employed. For instance, whereas *argB* most frequently integrates homologously, the *amdS* gene commonly integrates non-homologously, and in multiple copies. It is possible that this gene, which is not highly expressed in its home environment, may be more effective if integrated ectopically. The opposite experience was obtained in attempting to transform *S. commune* to hygromycin resistance, when it was concluded that the *hygB* gene was often poorly expressed, probably because it integrated into regions of methylated DNA which silenced its expression (Mooibroek *et al.*, 1990).

Plasmid integration may in some cases be accompanied by rearrangements to the neighbouring chromosomal regions (Durrens *et al.*, 1986), perhaps reflecting unexpected complexity in the process of plasmid-chromosomal recombination,

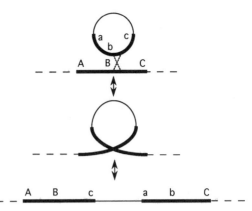

Figure 12.3 Homologous plasmid integration and two-step gene replacement. Recombination between a plasmid and a homologous chromosomal region will give rise to tandem repeats of the region of homology (**A/a, B/b, C/c**), separated by the remainder of the plasmid. If integration is non-homologous, no tandem repetition will result (however, there may be a homologue elsewhere in the genome). The homologous integration process is reversible, and if the second crossover is at the opposite end of the region of homology from the first, the overall result may be two-step replacement of a gene or part thereof. For example, if the first crossover is as diagrammed to the right of the **B/b** gene, and the second is to its left, the result will be replacement in the chromosome of the **B** allele with **b**.

or equally likely, implying that plasmids preferentially recombine with chromosomal regions which are already broken. Chromosome damage in such cases might well be the consequence of DNA repair systems induced by the presence of fragmented exogenous DNA.

(b) Autonomously replicating plasmids

Standard *E. coli* shuttle vectors used to transform *Mucor* and other Zygomycetes give unstable transformants in which the plasmid replicates autonomously (van Heeswijk, 1986). Although this is a useful system for complementation experiments, stably integrated plasmids would be more useful for expression of heterologous genes, and are necessary if gene replacement is required. For this reason, in the reverse of the type of experiment designed to construct autonomously replicating plasmids for Ascomycetes (see below), DNA fragments which promote chromosomal integration have been inserted into Zygomycete vectors (Arnau *et al.*, 1991).

Ustilago maydis plasmids preferentially integrate homologously if linearized and ectopically if not (Fotheringham and Holloman, 1990). Sequences conferring autonomous plasmid replication have also been isolated from *Ustilago* (Tsukuda *et al.*, 1988) and also from *Nectria haematococca* (Samac and Leong, 1989). In the latter case, the effective sequences consisted of telomeric DNA from a mitochondrial plasmid. This material enhanced transformation and promoted autonomous replication in *Ustilago*, which was used as the test organism, but perversely enough, it did so only for closed circular plasmids.

Whereas yeast autonomously replicating sequences (ARS) are well characterized and DNA fragments with such activity can be isolated from a variety of sources, yeast ARSs are not usually active in the parent species, an exception being the Ascomycete *Ashbya gossypii* (Wright and Philippsen, 1991). A fragment from *A. nidulans* which behaved as an ARS in yeast was named *ans1* (Ballance and Turner, 1985). It enhanced transformation in *A. nidulans* but, unexpectedly, the resulting transformants had all resulted from plasmid integration.

We have isolated an unrelated *A. nidulans* sequence designated *AMA1* which does promote autonomous plasmid replication in that fungus (Gems *et al.*, 1991). The plasmid containing it is designated ARpl, and under the best circumstances can transform close to 100% of nucleated protoplasts. Arpl is maintained at a copy number of 5–10 per genome, but is rapidly lost in the absence of selection and even under selection, more than 50% of conidia have lost it. Plasmids can be isolated from the fungus as monomers, dimers and higher multimers, and we suspect that they may replicate infrequently, producing large numbers of copies which are then gradually diluted out. *AMA1* consists of inverted repeats of a 3 kb sequence and while some subclones of ARp1 are able to replicate autonomously, all transform less efficiently than the original and we have not been able to identify a simple origin of replication.

The meiotic instability of *Neurospora* transformants was initially attributed to autonomously replicating plasmids (Grant *et al.*, 1984), but has now been explained as a result of destruction of transformation-generated repeats by the RIP process (see above).

Transformation of the Basidiomycete *Phanerochaete chrysosporium* resulted in an unstable transformant containing a plasmid that, on re-isolation, proved to be substantially rearranged (Randall *et al.*, 1991). It was concluded that the transforming vector had recombined with a resident 8.5 kb plasmid designated pME, acquiring an extra 2 kb which conferred the ability to replicate. Replication or partition of the new plasmid was inefficient, however, since it was maintained at only one copy per 5–10 nuclei.

A more remarkable instance of *in vivo* conversion of a vector into a replicating form occurred in *Fusarium oxysporium* (Powell and Kistler, 1990). In this case the introduced plasmid acquired chromosomal material acting as telomeres, with the result that the plasmid was able to replicate as an autonomous linear molecule. This parallels the deliberate creation of a semi-stably replicating linear plasmid for *Podospora anserina* by addition of telomeres (Perrot *et al.*, 1987).

(c) Cotransformation

If a fungus is transformed simultaneously with two DNAs, one selected, the other not, a considerable proportion of transformants score positively for markers on both DNAs. There are two possible explanations for this phenomenon, one is that one exogenous DNA, after integration into the genome, becomes a target for further integration events. Support for a limited (twofold) increase in transformation rates and targeting of a previously integrated plasmid has been reported for *Leptosphaeria maculans* (Farman and Oliver, 1992).

The alternative theory is that exogenous DNAs tend to recombine with each other more readily than with the chromosome, with the result that what eventually becomes integrated is likely to be a product of previous recombination of two or more plasmids (Figure 12.4). These may be copies of the same plasmid, leading to tandem plasmid repeats, as frequently observed, or if the plasmids are different the result will be cotransformation, markers from both DNA species becoming integrated at the same chromosomal site. Support for plasmid–plasmid recombination comes from experiments with the *Aspergillus* replicating plasmid ARp1 (Gems and Clutterbuck, 1993); if ARp1 is added along with a selectable plasmid, markers from the two plasmids will often exhibit linked

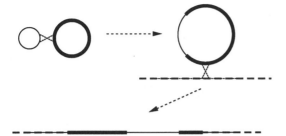

Figure 12.5 One-step gene replacement by double crossing-over between a linear DNA fragment and homologous flanking regions in a chromosome. The central region need not be homologous and the box in the diagram will usually contain a selectable marker in addition to the manipulated version of the gene to be replaced.

Figure 12.4 Integrative cotransformation resulting from recombination between two plasmids followed by recombination of the resulting complex with a chromosome. If the two plasmids recombining are identical, the result will be the integration of tandem repeats.

instability, implying that the plasmids have recombined to form a replicating dimer. Preferential recombination with a resident plasmid is also invoked as the explanation for the autonomous replication of transforming plasmids in *Phanerochaete chrysosporium* (Randall *et al.*, 1991). *A. nidulans* can recombine two circular plasmids, or a circular plasmid plus a linear DNA molecule, and ligation of two linear molecules seems to be equally efficient. The formation of replicating concatemers from linearized plasmids also occurs in *U. maydis* (Fotheringham and Holloman, 1990), and ligation of linear DNAs is even more efficient in *P. infestans* (Judelson, 1993) although in this species recombination between circular plasmids is rare.

(d) Transformant stability

Most transformation experiments yield, in addition to recoverable transformants, slow-growing colonies characterized as 'abortives' which cannot be subcultured on the selective medium. These are believed to contain some plasmids which are able to express the selective marker but do not replicate. Next in stability are transformants with plasmids which replicate or partition with varying efficiency, giving vegetatively unstable transformants. Finally, integrative transformants are generally fairly stable, at least in vegetative growth. Chromosomal integration is mediated by vegetative recombination, normally a rare event, and excision using the same mechanism (Figure 12.3) occurs at a similar frequency during vegetative growth (e.g. Dunne and Oakley, 1988).

Many transformants are in fact stable enough to be used in industrial-scale fermenters (Dunn-Coleman *et al.*, 1992). At meiosis, recombination is greatly increased, and a variety of events are possible (Petes and Hill, 1988); tandem repeats produced by homologous integration can recombine with each other, leading to plasmid excision (Figure 12.3), or they may recombine with those on a homologous chromosome. They can also undergo gene conversion, i.e. where the tandem repeats originally differed from each other, they may end up identical. As a result, transformants that are vegetatively reasonably stable may be quite unstable at meiosis (Upshall, 1986).

In *N. crassa* the meiotic instability of transformants due to recombination is masked by the RIP process (see above).

(e) Gene replacement

Gene replacement is an important tool in molecular analysis of gene function. For its success, it requires homologous chromosomal integration of manipulated versions of the gene. This can be achieved in one step in which the introduced version replaces the resident copy by double crossover or gene conversion (Figure 12.5). Alternatively the initial step may be additive integration, which is then followed by mitotic or meiotic recombination to eliminate one of the copies (Figure 12.3). With luck the previously resident copy will be the one lost, leaving the new version. Homologous plasmid integration occurs at varying frequencies in different fungi, probably depending not only on the species, but on the selective marker and the size and possibly the recombination characteristics of the DNA involved. In *A. nidulans* gene replacement is a reliable technique

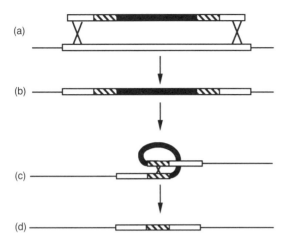

Figure 12.6 Repeatable gene disruption. The open box represents the gene to be disrupted, the filled box a bidirectionally selectable marker and the hatched segment a direct repeat of any convenient DNA sequence. (a) double crossing-over between the target gene and a linear disrupting DNA; (b) the disrupted product containing a selectable marker; (c) mitotic recombination to eliminate the now counter-selected marker; (d) the gene remains disrupted but has lost the selectable marker, so that in a diploid fungus the same construct can be used to disrupt the second copy of the gene.

(Miller *et al.*, 1985); in *N. crassa* it is less easily achieved because plasmid integration is predominantly non-homologous, but is nevertheless an important tool (Case *et al.*, 1992). Gene replacement is also possible in *Podospora anserina* using cosmid vectors (Coppin-Raynal *et al.*, 1989).

In diploid fungi such as *Candida*, two copies of a gene need to be replaced. For this purpose Alani *et al.* (1987) devised an ingenious technique for yeast which uses a bidirectionally selectable marker for gene disruption. The marker is flanked by direct repeats (Figure 12.6) so that when reverse selection is applied, mitotic recombination between these repeats can lead to elimination of the selected marker. The same DNA construct can then be used again to disrupt the second copy of the gene. This technique has been adapted for use with the *C. albicans GAL1* marker (Gorman *et al.*, 1991) and has also been used with the *C. albicans URA3* marker by Gow *et al.* (1993) to disrupt a hypha-specific chitin synthase gene.

12.3.2   Gene cloning

Fungal genes can be cloned using any of the methods used in other organisms; however, the fungi stand out as having special advantages for application of certain methods. First, because fungi can be treated as microbes, genes can be cloned by complementation of mutants, and second, because of the compactness of fungal genomes, chromosome walking is much easier than in more complex organisms. Cloning methods for fungi have been reviewed by Timberlake (1991) and Turner (1991).

Examples of the effective use of mutants for cloning are to be found elsewhere in this book. For instance, analysis of the conidiation pathway of *A. nidulans* (Chapter 17) started with the isolation and genetic analysis of a large set of conidiation mutants (Clutterbuck, 1969b). A more complex example is the study of tubulin genes in the same fungus (Chapter 6). The first mutants in this study were isolated on the basis of their resistance to the antimicrotubule fungicide benomyl, which led to the discovery that the *benA* gene coded for β-tubulin (Sheir-Neiss *et al.*, 1978). A second step depended on one *benA* mutant (*benA33*) which was also temperature sensitive for growth. Suppressors of this defect included mutants at the *tubB* and *mipA* genes which code for proteins interacting with β-tubulin, namely α-tubulin and the previously unknown γ-tubulin, respectively (Oakley *et al.*, 1987, 1990).

The first step in cloning by complementation is construction of a gene bank consisting of fragmented chromosomal DNA inserted into a suitable vector. Plasmids, which will hold around 10 kb of DNA, and cosmids taking up to 40 kb, each have their own advantages for this purpose. A mutant for the gene to be cloned is transformed with members of the gene bank and transformants in which the mutant phenotype under study has been complemented to wild-type are selected. The second step is recovery of the complementing DNA fragment. If the vector is present in the transformant as a free molecule by virtue of autonomous replication, or as a result of reversible integration into a chromosome (Figure 12.3), it may be possible to rescue it simply by transforming *E. coli* with transformant DNA. Alternatively, the plasmid may be recoverable from restricted and re-ligated transformant DNA. If this is not successful, the alternative is to start the process with batteries of ordered gene-bank members, rather than with mass transformation. In this case the complementing member can be recovered by sib-selection, retesting subsections of the array until the active member is pin-pointed.

Where the phenotype of interest is too complex to be determined for a large number of transformants (mating-type genes are a good example: Chapter 18), a relevant gene is more likely to be cloned by first determining its genetic map position (Chapter 11) and then starting the cloning process from a linked marker which can be selected directly. Chromosome walking, that is the detection by hybridization of a series of overlapping clones along the chromosome, will then lead to the required gene. For *A. nidulans* cloning by this method is simplified if the gene is located to a chromosome, either genetically or by electrophoretic karyotyping, after which the search can be confined to a subset of chromosome-specific cosmids (Brody *et al.*, 1991).

12.4 CONCLUSIONS

The rate of advance in molecular analysis is such that our picture of molecular organization is becoming both more complete and more complex. In some cases the new information provides the solution to longstanding mysteries, whereas in others it reveals unexpected problems. In either case it is something that no biologist can afford to ignore, and the power of molecular tools is such that, while they are developing with sometimes bewildering rapidity, there are few areas of biology that will not benefit from their use.

REFERENCES

Aharonowitz, Y., Cohen, G. and Martin, J.F. (1992) Penicillin and cephalosporin biosynthetic genes: structure, organization, regulation and evolution. *Annual Review of Microbiology*, **46**, 461–95.

Akins, R.A., Kelly, R.L. and Lambowitz, A.M. (1986) Mitochondrial plasmids of *Neurospora*: integration into mitochondrial DNA and evidence for reverse transcription in mitochondria. *Cell*, **47**, 505–16.

Alani, E., Cao, L. and Kleckner, N. (1987) A method for gene disruption that allows repeated use of *URA3* selection in the construction of multiply disrupted yeast strains. *Genetics*, **116**, 541–5.

Allison, D.S., Rey, M.W., Berka, R.M. *et al.* (1992) Transformation of the thermophilic fungus *Humicola*

grisea var. *thermoidea* and overproduction of *Humicola* glucoamylase. *Current Genetics*, **21**, 225–9.

Antequera, F., Tamame, M., Villanueva, J.R. and Santos, T. (1984) DNA methylation in the fungi, *Journal of Biological Chemistry*, **259**, 8033–6.

Aramayo, R., Adams, T.H. and Timberlake, W.E. (1989) A large cluster of highly expressed genes is dispensable for growth and development in *Aspergillus nidulans*. *Genetics*, **122**, 65–71.

Arlett, C.F. (1957) Induction of cytoplasmic mutations in *Aspergillus nidulans*. *Nature*, **179**, 1250–1.

Armaleo, D., Yee, G.-N., Klein, T.M. *et al.* (1990) Biolistic nuclear transformation of *Saccharomyces cerevisiae* and other fungi. *Current Genetics*. **17**, 97–103.

Arnau, J., Jepsen, L.P. and Trøman, P. (1991) Integrative transformation by homologous recombination in the zygomycete *Mucor circinelloides*. *Molecular and General Genetics*, **225**, 193–8.

Asch, D.K. and Kinsey, J.A. (1990) Relationship of vector size to homologous integration during transformation of *Neurospora crassa* with the cloned *am* (GDH) gene. *Molecular and General Genetics*, **221**, 37–43

Ballance, D.J. and Turner, G. (1985) Development of a high-frequency transforming vector for *Aspergillus nidulans*. *Gene*, **36**, 321–31.

Barroso, G. and Labarère, J. (1993) Transcription of naked double-stranded RNA molecules in a fraction containing large vesicles plus mitochondria from the basidiomycete *Agrocybe aegerita*. *Journal of General Microbiology*, **139**, 287–93.

Bertrand, H., Collins, R.A., Stohl, L.L. *et al.* (1980) Deletion mutants of *Neurospora crassa* mitochondrial DNA and their relationship to the 'stop-start' growth phenotype. *Proceedings of the National Academy of Sciences of the United States of America*, **77**, 6032–6.

Boylan, M.T., Holland, M.J. and Timberlake, W.E. (1986) *Saccharomyces cerevisiae* centromere *CEN11* does not induce chromosomal instability when integrated into the *Aspergillus nidulans* genome. *Molecular and Cellular Biology*, **6**, 3621–5.

Brody, H., Griffith, J., Cuticchia, A.J. *et al.* (1991) Chromosome-specific recombinant DNA libraries from the fungus *Aspergillus nidulans*. *Nucleic Acids Research*, **19**, 3105–9.

Buck, K.W. (ed.) (1986) *Fungal Virology*, CRC Press, Boca Raton, FL.

Buckner, B., Novotny, C.P. and Ullrich, R.C. (1988) Developmental regulation of the methylation of the ribosomal DNA in the basidiomycete fungus *Schizophyllum commune*. *Current Genetics*, **14**, 105–11.

Buxton, F.P., Gwynne, D.I. and Davies, R.W. (1989) Cloning of a new bidirectionally selective marker for *Aspergillus* strains. *Gene*, **84**, 329–34.

Caddick, M.X. and Turner, A.S. (1993) The control of gene expression in filamentous fungi, in *The Eukaryote Genome, Organisation and Regulation*, 50th Symposium of the Society for General Microbiology (eds P. Broda, S.G. Oliver and P.F.G. Sims), Cambridge University Press, Cambridge, pp. 241–73.

Case, M.E., Geever, R.F. and Asch, D.K. (1992) Use of gene replacement transformation to elucidate gene function in the *qa* gene cluster of *Neurospora crassa*. *Genetics*, **130**, 729–36.

Caten, C.E. and Day, P.R. (1977) Diploidy in plant pathogenic fungi. *Annual Review of Phytopathology*, **15**, 295–318.

Chakraborty, B.N. and Kapoor, M. (1990) Transformation of filamentous fungi by electroporation. *Nucleic Acids Research*, **18**, 6737.

Chan, B.S.-S., Court, D.A., Vierula, P.J., and Bertrand, H. (1991) The *Kalilo* linear senescence-inducing plasmid of *Neurospora* is an invertron and encodes DNA and RNA polymerases. *Current Genetics*, **20**, 225–37.

Charron, M.J., Read, E., Haut, S.R. and Michels, C.A. (1989) Molecular evolution of the telomere-associated *MAL* loci of *Saccharomyces*. *Genetics*, **122**, 307–16

Clutterbuck, A.J. (1969a) Cell volume per nucleus in haploid and diploid strains of *Aspergillus nidulans*. *Journal of General Microbiology*, **55**, 291–9.

Clutterbuck, A.J. (1969b) A mutational analysis of conidial development in *Aspergillus nidulans*. *Genetics*, **63**, 317–27.

Coggins, J.R., Duncan, K., Anton, I.A. *et al.* (1987) The anatomy of a multifunctional enzyme. *Biochemical Society Transactions*, **15**, 754–9.

Conelly, J.C. and Arst, H.N., Jr (1991) Identification of a telomeric fragment from the right arm of chromosome III of *Aspergillus nidulans*. *FEMS Microbiology Letters*, **80**, 295–7

Coppin-Raynal, E., Picard, M. and Arnaise, S. (1989) Transformation by integration in *Podospora anserina* III. Replacement of a chromosome segment by a two-step procedure. *Molecular and General Genetics*, **219**, 270–6.

Daboussi, M.J., Djebelli, A., Gerlinger, C. *et al.* (1989) Transformation of seven species of filamentous fungi using the nitrate reductase gene of *Aspergillus nidulans*. *Current Genetics*, **15**, 453–6.

Davis, M.A. and Hynes, M.J. (1991) Regulatory circuits in *Aspergillus nidulans*, in *More Gene Manipulations in Fungi*, (eds J.W. Bennett and L.L. Lasure), Academic Press, San Diego, pp. 151–89.

de Vries, O.M.H. and Reddingius, J. (1984) Synthesis of macromolecules and compartment size in monokaryotic and dikaryotic hyphae of *Schizophyllum commune*, *Experimental Mycology*, **8**, 378–81.

Dixon, P.A. (1959) Life history and cytology of *Ascocybe grovesii* Wells. *Annals of Botany*, NS **23**, 509–20.

Dunn-Coleman, N.S., Brodie, E.A., Carter, G.L. and Armstrong, G.L. (1992) Stability of recombinant strains under fermentation conditions, in *Applied Molecular Genetics of Filamentous Fungi*, (eds J.R. Kinghorn and G. Turner), Chapman and Hall, London, pp. 152–74.

Dunne, P.W. and Oakley, B.R. (1988) Mitotic gene conversion, reciprocal recombination and gene replacement at the *benA*, beta-tubulin, locus of *Aspergillus nidulans*. *Molecular and General Genetics*, **213**, 339–45.

Durrens, P., Green, P.M., Arst, H.N. Jr and Scazzocchio, C. (1986) Heterologous insertion of transforming DNA and generation of new deletions associated with transformation in *Aspergillus nidulans*. *Molecular and General Genetics*, **203**, 544–9.

Dutta, S.K. and Ojha, M. (1972) Relatedness between major taxonomic groups of fungi based on the measurement of DNA nucleotide sequence homology. *Molecular and General Genetics*, **114**, 232–40.

Earl, A.J., Turner, G., Croft, J.H. *et al.* (1981) High frequency transfer of species specific mitochondrial DNA sequences between members of the Aspergillaceae. *Current Genetics*, **3**, 221–8.

Ehinger, A., Denison, S.H. and May, G.S. (1990) Sequence organization and expression of the core histone genes of *Aspergillus nidulans*. *Molecular and General Genetics* **222**, 416–24.

Elliott, C.G. and MacIntyre, D. (1973) Genetical evidence on the life-history of *Phytophthora*. *Transactions of the British Mycological Society*, **60**, 311–16.

Fan, J.-B., Chikashige, Y., Smith, C.L. *et al.*, (1989) Construction of a *Not*I1 restriction map of the fission yeast *Schizosaccharomyces pombe* genome. *Nucleic Acids Research*, **17**, 2801–18.

Farman, M.L. and Oliver, R.P. (1992) Transformation frequencies are enhanced and vector DNA is targeted during retransformation of *Leptosphaeria maculans*, a fungal plant pathogen. *Molecular and General Genetics*, **231**, 243–7.

Fincham, J.R.S. (1989) Transformation in fungi. *Microbiological Reviews*, **53**, 148–70.

Förster, H., Coffey, D., Elwood, H. and Sogin, M.L. (1990) Sequence analysis of the small subunit ribosomal RNAs of three zoosporic fungi and implications for fungal evolution. *Mycologia*, **82**, 306–12.

Fotheringham, S. and Holloman, W.K. (1990) Pathways of transformation in *Ustilago maydis* determined by DNA conformation. *Genetics*, **124**, 833–43.

Francis, D.M., Hulbert, S.A. and Michelmore, R.W. (1990) Genome size and complexity of the obligate fungal pathogen *Bremia lactucae*. *Experimental Mycology*, **14**, 299–309.

Garber, R.C. and Yoder, O.C. (1984) Characterization of the mitochondrial chromosome and population genetics of a restriction length polymorphism. *Current Genetics*, **8**, 621–8.

Geever, R.F., Huiet, L., Baum, J.A. *et al.*, (1989) DNA sequence, organization and regulation of the *qa* gene cluster of *Neurospora crassa*. *Journal of Molecular Biology*, **207**, 15–34.

Gems, D.H. and Clutterbuck, A.J. (1993) Co-transformation with autonomously-replicating helper plasmids facilitates gene cloning from an *Aspergillus nidulans* gene library. *Current Genetics*, **24**, 520–4

Gems, D.H., Johnstone, I.L., and Clutterbuck, A.J. (1991) An autonomously replicating plasmid transforms *Aspergillus nidulans* at high frequency. *Gene*, **98**, 61–7.

Gorman, J.A., Chan, W. and Gorman, J.W. (1991) Repeated use of *GAL1* for gene disruption in *Candida albicans*. *Genetics*, **129**, 19–24.

Gow, N.A.R., Swoboda, R., Bertram, G. *et al.*, (1993) Key genes in the regulation of dimorphism in *Candida albicans*, in *Dimorphic Fungi in Biology and Medicine*, (eds H. Vanden Bossche *et al.*), Plenum Press, New York, pp. 61–71.

Grant, D.M., Lambowitz, A.M., Rambosek, J.A. and Kinsey, J.A. (1984) Transformation of *Neurospora* with recombinant plasmids containing the cloned glutamate dehydrogenase (*am*) gene. Evidence for autonomous replication. *Molecular and Cellular Biology*, **4**, 2041–51.

Greider, C.W. (1993) Telomeres and telomerase in small eukaryotes, in *The Eukaryote Genome, Organisation and Regulation; 50th Symposium of the Society for General Microbiology*, (eds P. Broda, S.G. Oliver and P.F.G. Sims), Cambridge University Press, Cambridge, pp. 31–42.

Griffiths, A.J.F. (1992) Fungal senescence. *Annual Review of Genetics*, **26**, 351–72.

Gruber, F., Visser, J., Kubicek, C.P. and de Graaf, L.H. (1990) Cloning the *Trichoderma reesei pyrG* gene and use as a homologous marker for a high frequency transformation system. *Current Genetics*, **18**, 447–57.

Gurr, S.J., Unkles, S.E. and Kinghorn, J.R. (1987) The structure and organization of nuclear genes of filamentous fungi, in *Gene Structure in Eukaryote Microbes* (ed. J.R. Kinghorn), IRL Press, Oxford, pp. 93–139.

Hamer, J.E., Farrell, L. Orbach, M.J. *et al.*, (1989) Host species-specific conservation of a family of repeated DNA sequences in the genome of a fungal plant pathogen. *Proceedings of the National Academy of Sciences of the United States of America*, **86**, 9981–5.

Hawker, K.L., Montague, P., Marzluf, G.A. and Kinghorn, J.R. (1991) Heterologous expression and regulation of the *Neurospora crassa nit-4* pathway-specific regulatory gene for nitrate assimilation in *Aspergillus nidulans. Gene*, **100**, 237–40.

Hawkins, A.R. (1987) The complex arom locus of *Aspergillus nidulans*: evidence for multiple gene fusions and convergent evolution. *Current Genetics*, **11**, 491–8.

Hinchcliffe, E. (1991) Strain improvement of brewing yeast, in *Applied Molecular Genetics*, (eds J.F. Peberdy, C.E. Caten, J.E. Ogden and J.W. Bennett), Cambridge University Press, Cambridge, pp. 129–45.

Hinnen, A., Hicks, J.B. and Fink, G.R. (1978) Transformation of yeast. *Proceedings of the National Academy of Sciences of the United States of America*, **75**, 1929–33.

Horowitz, N.H. and Macleod, H. (1960) The DNA content of *Neurospora crassa. Microbial Genetics Bulletin*, **17**, 6–7.

Horton, J.S. and Raper, C.A. (1991) Pulse-field gel electrophoretic analysis of *Schizophyllum commune* chromosomal DNA. *Current Genetics*, **19**, 77–80.

Hudspeth, M.E.S., Timberlake, W.E. and Goldberg, R.B. (1977) DNA organization in the water mold *Achlya. Proceedings of the National Academy of Sciences of the United States of America*, **74**, 4332–6

Hulbert, S.H. and Michelmore, R.W. (1988) DNA restriction fragment length polymorphism and osmotic variation in the lettuce downy mildew fungus, *Bremia lactucae. Molecular Plant Microbe Interactions*, **1**, 17–24.

Hynes, M.J., Corrick, C.M., Kelly, J.M. and Littlejohn, T.G. (1988) Identification of the sites of action for regulatory genes controlling the *amdS* gene of *Aspergillus nidulans. Molecular and Cellular Biology*, **8**, 2589–96.

Johnstone, I.L., McCabe, P.C., Greaves, P. *et al.* (1990) Isolation and characterisation of the *crnA-niiA-niaD* gene cluster for nitrate assimilation in *Aspergillus nidulans. Gene*, **90**, 181–92.

Judelson, H.S. (1993) Intermolecular ligation mediates efficient cotransformation in *Phytophthora infestans. Molecular and General Genetics*, **239**, 241–50.

Katylak, Z., Lazarowska, J. and Slonimski, P.P. (1985) Intron encoded proteins of mitochondria: key elements of gene expression and genomic evolution, in *Achievements and Perspectives of Mitochondrial Research*, vol. 2, (eds E. Quagliariello, E.C. Slater, F.M. Palmieri *et al.*) Elsevier, Amsterdam, pp. 1–21.

Kayser, T. and Wöstermeyer, J. (1991) Electrophoretic karyotype of the zygomycete *Absidia glauca*: evidence for differences between mating types. *Current Genetics*, **19**, 279–84.

Keller, N.P., Cleveland, T.E. and Bhatnagar, D. (1992) Variable electrophoretic karyotype of members of *Aspergillus* section *flavi. Current Genetics*, **21**, 371–5.

Kelly, J.M. and Hynes, M.J. (1987) Multiple copies of the *amdS* gene of *Aspergillus nidulans* cause titration of trans-acting regulatory proteins. *Current Genetics*, **12**, 21–31.

Kingsman, A.J., Adams, S.E., Fulton, S.M. *et al.* (1988) The yeast retrotransposon Ty and related elements, in *Transposition*, 43rd Symposium of the Society for General Microbiology, (eds A.J. Kingsman, K.F. Chater and S.M. Kingsman), Cambridge University Press, Cambridge, pp. 223–46.

Kinsey, J.A. and Helber, J. (1989) Isolation of a transposable element from *Neurospora crassa. Proceedings of the National Academy of Sciences of the United States of America*, **86**, 1929–33.

Klein, H.L. (1993) Current issues in homologous recombination, in *The Eukaryote Genome, Organisation and Regulation*, 50th Symposium of the Society for General Microbiology, (eds P. Broda, S.G. Oliver and P.F.G. Sims), Cambridge University Press, Cambridge, pp 161–84.

Kondrashov, A.S. and Crow, J.F. (1991) Haploidy or diploidy: which is better? *Nature*, **251**, 314–5.

Kovac, L., Lazowska, J. and Slonimski, P.P. (1982) A yeast with linear molecules of mitochondrial DNA. *Molecular and General Genetics*, **197**, 420–4.

Krumlauf, R. and Marzluf, G.A. (1979) Characterization of the genome complexity and organization of the *Neurospora crassa* genome. *Biochemistry*, **18**, 3705–13.

Krumlauf, R. and Marzluf, G.A. (1980) Genome organization and characterization of the repetitive and inverted DNA sequences in *Neurospora crassa. Journal of Biological Chemistry*, **255**, 1138–45.

Kück, U. (1989) Mitochondrial DNA rearrangements in *Podospora anserina*. *Experimental Mycology*, **13**, 111–20.

Kudla, B., Caddick, M.X., Langdon, T. *et al.* (1990) The regulatory gene *areA* mediating nitrogen metabolic repression in *Aspergillus nidulans*. Mutations affecting specificity of gene activation alter a loop residue of a putative zinc finger. *EMBO Journal*, **9**, 1355–64.

Lamb, H.K., Hawkins, A.R., Smith, M. *et al.* (1990) Spatial and biological characterisation of the complete quinic acid utilisation gene cluster in *Aspergillus nidulans*. *Molecular and General Genetics*, **223**, 17–23.

Lamb, H.K., Roberts, C.F. and Hawkins, A.R. (1992a) A second gene (*qutH*) within the *Aspergillus nidulans*-quinic-acid utilisation gene cluster encodes a protein with a putative zinc-cluster motif. *Gene*, **112**, 219–24.

Lamb, H.K., van den Hornbergh, P.T.W., Newton, G.H. *et al.* (1992b) Differential flux through quinate and shikimate pathways: implications for the channelling hypothesis. *Biochemical Journal*, **284**, 181–7.

Lambowitz, A.M. (1989) Infectious introns. *Cell*, **56**, 323–6.

Lasker, B.A., Page, L.J. and Kobayashi, G.S. (1992) Isolation, characterization, and sequencing of *Candida albicans* repetitive element 2. *Gene*, **116**, 51–7.

Lazarus, C.M., Earl, A.J., Turner, G. and Kuntzel, H. (1980) Amplification of a mitochondrial DNA sequence in the cytoplasmically inherited 'ragged' mutant of *Aspergillus amstelodami*. *European Journal of Biochemistry*, **106**, 633–41.

Le Chevanton, C., Leblon, G. and Lebilcot, S. (1989) Duplications created by transformation in *Sordaria macrospora* are not inactivated during meiosis. *Molecular and General Genetics*, **218**, 390–6.

Lemke, P. (ed.) (1979) *Viruses and Plasmids in Fungi*, Marcel Dekker, New York.

Lloyd, A.T. and Sharp, P.M. (1991) Codon usage in *Aspergillus nidulans*. *Molecular and General Genetics*, **230**, 288–94.

Maleszka, R., Skelly, P.J. and Clark-Walker, G.D. (1991) Rolling circle replication of DNA in yeast mitochondria. *EMBO Journal*, **10**, 3923–9.

Maleszka, R, and Clark-Walker, G.D. (1992) *In vivo* configuration of mitochondrial DNA in fungi and zoosporic moulds. *Current Genetics*, **22**, 341–4.

Mao, Y. and Tyler, B.M. (1991) Genome organization of *Phytophthora megasperma* f.sp. *glycinea*. *Experimental Mycology*, **15**, 283–91.

Marzluf, G. (1981) regulation of nitrogen metabolism and gene expression in fungi. *Microbiological Reviews*, **45**, 437–61.

May, G. (1992) Fungal technology, in *Applied Molecular Genetics of Filamentous Fungi* (eds J.R. Kinghorn and G. Turner), Chapman and Hall, London, pp. 1–21.

McClintock, B. (1932) A correlation of ring-shaped chromosomes with variegation in *Zea mays*. *Proceedings of the National Academy of Sciences of the United States of America*, **18**, 677–81.

McClung, C.R., Phillips, J.D., Orbach, M.J. and Dunlap, J.C. (1989) New cloning vectors using benomyl resistance as a dominant marker for selection in *Neurospora crassa* and in other filamentous fungi. *Experimental Mycology*, **13**, 299–302

McDonald, B.A. and Martinez, J.P. (1991) Chromosome length polymorphisms in a *Septoria tritici* population. *Current Genetics*, **19**, 265–71.

Miao, V.P.W., Matthews, D.E. and VanEtten, H.D. (1991) Identification and chromosomal locations of a family of cytochrome P-450 genes for pisatin detoxification in the fungus *Nectria haematococca*. *Molecular and General Genetics*, **226**, 214–23.

Miller, B.M., Miller, K.Y. and Timberlake, W.E. (1985) Direct and indirect gene replacements in *Aspergillus nidulans*. *Molecular and Cellular Biology*, **5**, 1714–21.

Miller, B.M., Miller, K.Y., Roberti, K.A. and Timberlake, W.E. (1987) Position-dependent and -independent mechanisms regulate cell-specific expression of the SpoC1 cluster of *Aspergillus nidulans*. *Molecular and Cellular Biology*, **7**, 427–34.

Mitchell, M.B. and Mitchell, H.K. (1952) A case of 'maternal' inheritance in *Neurospora crassa*. *Proceedings of the National Academy of Sciences of the United States of America*, **38**, 442–9.

Mooibroek, H., Kuipers, A.G.J., Sietsma, J.H. *et al.* (1990) Introduction of hygromycin B resistance into *Schizophyllum commune*; preferential methylation of donor DNA. *Molecular and General Genetics*, **222**, 41–8.

Nuss, D.L. (1992) Biological control of chestnut blight: an example of virus-mediated attenuation of fungal pathogenesis. *Microbiological Reviews*, **56**, 561–76

Nuss, D.L. and Koltin, Y. (1990) Significance of dsRNA genetic elements in plant pathogenic fungi. *Annual Review of Phytopathology*, **28**, 37–58.

Oakley, B.R., Oakley, C.E. and Rinehart, J.E. (1987) Conditionally lethal *tubA* α-tubulin mutation in *Aspergillus nidulans*. *Molecular and General Genetics*, **208**, 135–44.

Oakley, B.R., Oakley, C.E., Yoon, Y. and Jung, M.K. (1990) γ-tubulin is a component of the spindle pole body that is essential for microtubule function in *Aspergillus nidulans*. *Cell*, **61**, 1289–301.

Oliver, S.G., James, C.M., Gent, M.E. and Indge, K.J. (1993) Yeast genome organization and evolution, in *The Eukaryote Genome, Organisation and Regulation*; 50th Symposium of the Society for General Microbiology, (eds P. Broda, S.G. Oliver and P.F.G. Sims), Cambridge University Press, Cambridge, pp. 1–7.

Orbach, M.J., Vollrath, D., Davis, R.W. and Yanofsky, C. (1988) An electrophoretic karyotype of *Neurospora crassa*. *Molecular and Cellular Biology*, **8**, 1469–73.

Orr, W.C. and Timberlake, W.E. (1982) Clustering of spore-specific genes in *Aspergillus nidulans*.. *Proceedings of the National Academy of Sciences of the United States of America*, **79**, 5976–80.

Orr-Weaver, T.L. and Szostak, J.W. (1985) Fungal recombination. *Microbiological Reviews*, **49**, 33–58.

Osiewacz, H.D., Clairmont, A. and Huth, M. (1990) Electrophoretic karyotype of the ascomycete *Podospora anserina*. *Current Genetics*, **18**, 481–3.

Perrot, M., Barreau, C. and Bégueret, J. (1987) Non-integrative transformation in the filamentous fungus *Podospora anserina*: stabilization of a linear vector by the chromosomal ends of *Tetrahymena thermophila*. *Molecular and Cellular Biology*, **7**, 1725–30.

Petes, T.D. and Hill, C.W. (1988) Recombination between repeated genes in microorganisms. *Annual Review of Genetics*, **22**, 147–68.

Powell, W.A. and Kistler, H.C. (1990) In vivo rearrangement of foreign DNA by *Fusarium oxysporium* produces linear self-replicating plasmids. *Journal of Bacteriology*, **172**, 3163–71.

Punt, P.J., Oliver, R.P., Dingemanse, M.A. *et al.* (1987) Transformation of *Aspergillus* based on the hygromycin resistance marker from *Escherichia coli*. *Gene*, **56**, 117–24.

Rambosek, J. and Leach, J. (1987) Recombinant DNA in filamentous fungi: progress and prospects. *Critical Reviews in Biotechnology*, **6**, 357–93.

Randall, J., Reddy, C.A. and Boominathan, K. (1991) A novel extrachromosomally maintained transformation vector for the lignin-degrading basidiomycete *Phanerochaete chrysosporium*. *Journal of Bacteriology*, **173**, 776–82.

Rayner, A.D.M. (1991) The phytopathological significance of mycelial individualism. *Annual Review of Phytopathology*, **29**, 305–23.

Rhounin, L., Rossignol, J.-L. and Faugeron, G. (1992) Epimutation of repeated genes in *Ascobolus immersus*. *EMBO Journal*, **11**, 4451–7.

Riggsby, W.S., Torres-Bauza, L.J., Wills, J.W. and Townes, T.M. (1982) DNA content, kinetic complexity, and the ploidy question in *Candida albicans*. *Molecular and Cellular Biology*, **2**, 853–62.

Robinow, C.F. (1963) Observations on cell growth, mitosis, and division in the fungus *Basidiobolus ranarum*. *Journal of Cell Biology*, **17**, 123–52.

Rodriguez, R.J. and Yoder, O.C. (1991) A family of conserved repetitive elements from the fungal plant pathogen *Glomerella cingulata* (*Colletotrichum lindemuthianum*). *Experimental Mycology*, **15**, 232–42.

Rogers, H.J., Buck, K.W. and Brasier, C.M. (1987) A mitochondrial target for double-stranded RNA in diseased isolates of the fungus that causes Dutch elm disease. *Nature*, **329**, 558–60.

Romaine, C.P. and Schlagnhauffer, B. (1990) Prevalence of double stranded RNAs in healthy and La France disease-infected basidiocarps of *Agaricus bisporus*. *Mycologia*, **81**, 822–5.

Rowlands, R.T. and Turner, G. (1975) Three-marker extranuclear mitochondrial crosses in *Aspergillus nidulans*. *Molecular and General Genetics*, **141**, 69–79.

Royer, J.C., Hintz, W.E., Kerrigan, R.W. and Horgen, P.A. (1992) Electrophoretic karyotype analysis of the button mushroom, *Agaricus bisporus*. *Genome*, **35**, 694–8.

Russell, P.J., Rodland, K.D., Rachlin, E.M. and McCloskey, J.A. (1987) Differential DNA methylation during the vegetative life cycle of *Neurospora crassa*. *Journal of Bacteriology*, **169**, 2902–5.

Sakaguchi, K. (1990) Invertrons, a class of structurally and functionally related genetic elements that includes linear DNA plasmids, transposable elements, and genomes of adeno-type viruses. *Microbiological Reviews*, **54**, 66–74.

Samac, D.A. and Leong, S.A. (1988) Two linear plasmids in mitochondria of *Fusarium solani* f. sp. *cucurbitae*. *Plasmid*, **19**, 57–67.

Samac, D.A. and Leong, S.A. (1989) Characterization of the termini of linear plasmids from *Nectria haematococca* and their use in construction of an autonomously replicating transformation vector. *Current Genetics*, **16**, 187–94.

Sansome, E. (1977) Polyploidy and induced gametangial formation in British isolates of *Phytophthora infestans*. *Journal of General Microbiology*, **99**, 311–16.

Scazzocchio, C. (1992) Control of gene expression in the catabolic pathways of *Aspergillus nidulans*: a personal and biased account, in *Aspergillus: Biology and Industrial Applications*, (eds J.W. Bennett and M.A. Klich) Butterworth-Heinemann, Boston, pp 43–68.

Schechtman, M.G. (1990) Characterization of telomere DNA from *Neurospora crassa*. *Gene*, **88**, 159–65.

Selker, E.U. (1990) Premeiotic instability of repeated sequences in *Neurospora crassa*. *Annual Review of Genetics*, **24**, 579–613.

Sheir-Neiss, G., Lai, M.H. and Morris, N.R. (1978) Identification of a gene for β-tubulin in *Aspergillus nidulans*. *Cell*, **15**, 639–47.

Skinner, D.Z., Budde, A.D. and Leong, S.A. (1991) Molecular karyotype analysis of fungi, in *More Gene Manipulations in Fungi*, (eds J.W.Bennett and L. Lasure), Academic Press, Orlando, pp. 86–103.

Smith, M.L., Bruhn, J.N. and Anderson, J.B. (1992) The fungus *Armillaria bulbosa* is among the largest and oldest living organisms. *Nature*, **356**, 428–31.

Stohl, L.H., Collins, J.C., Cole, M.D. and Lambowitz, A.M. (1982) Characterization of two new plasmid DNAs found in wild-type *Neurospora intermedia* strains. *Nucleic Acids Research*, **10**, 1439–58.

Suzuki, T., Kanbe, T., Kuroiwa, T. and Tanaka, K. (1986) Occurrence of a ploidy shift in a strain of the imperfect yeast *Candida albicans*. *Journal of General Microbiology*, **132**, 443–53.

Taylor, J.L., Borgmann, I. and Séguin-Swartz, G. (1991) Electrophoretic karyotyping of *Leptosphaeria maculans* differentiates highly virulent and weakly virulent isolates. *Current Genetics*, **19**, 273–7.

Teakle, D.S. (1983) Zoosporic fungi and viruses: double trouble, in *Zoosporic Plant Pathogens: a Modern Perspective*, (ed. S.T. Buczacki), Academic Press, London, pp. 233–48.

Timberlake, W.E. (1978) Low repetitive DNA content in *Aspergillus nidulans*. *Science*, **202**, 973–4.

Timberlake, W.E. (1991) Cloning and analysis of fungal genes, in *More Gene Manipulations in Fungi*, (eds J.W. Bennett and L. Lasure), Academic Press, Orlando, pp. 51–85.

Tooley, P.W. and Carras, M.M. (1992) Separation of chromosomes of *Phytophthora* species using CHEF gel electrophoresis. *Experimental Mycology*, **16**, 188–96.

Tooley, P.W. and Therrien, C.D. (1987) Cytophotometric determination of the nuclear DNA content of 23 Mexican and 18 non-Mexican isolates of *Phytophthora infestans*. *Experimental Mycology*, **11**, 19–26.

Tsukuda, T., Carleton, S., Fotheringham, S. and Holloman, W.K. (1988) Isolation and characterization of an autonomously self-replicating sequence form *Ustilago maydis*. *Molecular and Cellular Biology*, **8**, 3703–9.

Turgeon, B.G., Garber, R.C. and Yoder, O.C. (1986) Transformation of the fungal maize pathogen *Cochliobolus heterostrophus* using the *Aspergillus nidulans amdS* gene. *Molecular and General Genetics*, **210**, 450–3.

Turner, G. (1991) Strategies for cloning genes from filamentous fungi, in *Applied Molecular Genetics*, (eds J.F. Peberdy, C.E. Caten, J.E. Ogden and J.W. Bennett), Cambridge University Press, Cambridge, pp. 29–43.

Turner, G. (1993) Gene organisation in filamentous fungi, in *The Eukaryote Genome, Organisation and Regulation, 50th Symposium of the Society for General Microbiology*, (eds P. Broda, S.G. Oliver, and P.F.G. Sims), Cambridge University Press, Cambridge, pp. 107–25.

Unkles, S.E. (1992) Gene organization in industrial filamentous fungi, in *Applied Genetics of Filamentous Fungi*, (eds J.R. Kinghorn and G. Turner), Blackie Academic and Professional, London, pp. 28–53.

Upshall, A. (1981) Naturally occurring diploid isolates of *Aspergillus nidulans*. *Journal of General Microbiology*, **122**, 7–11.

Upshall, A. (1986) Genetic and molecular characterization of *argB*+ transformants of *Aspergillus nidulans*. *Current Genetics*, **10**, 593–9.

Valent, B. and Chumley, F.G. (1991) Molecular genetic analysis of the rice blast fungus, *Magnaporthe grisea*. *Annual Review of Phytopathology*, **29**, 443–67.

Van den Hondel, C.A.M.J.J. and Punt, P.J. (1991) Gene transfer systems and vector development for filamentous fungi, in *Applied Molecular Genetics*, (eds J.F. Peberdy, C.E. Caten, J.E. Ogden and J.W. Bennett), Cambridge University Press, Cambridge, pp. 1–28.

van Heeswijk, R. (1986) Autonomous replication of plasmids in *Mucor* transformants. *Carlsberg Research Communications*, **51**, 433–43.

Ward, M., Kodama, K.H. and Wilson, L.J. (1989) Transformation of *Aspergillus awamori* and *A. niger* by electroporation. *Experimental Mycology*, **13**, 289–93.

Weselowski, M. and Fukuhara, H. (1981) Linear mitochondrial deoxyribonucleic acid from the yeast *Hansenula mrakii*. *Molecular and Cellular Biology*, **1**, 387–93.

Whelan, W.L., Partridge, R.M. and Magee, P.T. (1980) Heterozygosity and segregation in *Candida albicans*. *Molecular and General Genetics*, **180**, 107–13.

Wickner, R.B. (1992) Double-stranded and single-stranded RNA viruses of *Saccharomyces cerevisiae*. *Annual Review of Microbiology*, **46**, 347–75.

Williamson, D.H. (1993) Microbial mitochondrial genomes – windows on other worlds, in *The Eukaryote Genome, Organisation and Regulation, 50th Symposium of the Society for General Microbiology*, (eds P. Broda, S.G. Oliver and P.F.G. Sims), Cambridge University Press, Cambridge, pp. 73–106.

Wolf, K. and Del Giudice, L. (1988) The variable mitochondrial genome of ascomycetes: organization, mutational alterations, and expression. *Advances in Genetics*, **25**, 186–308.

Woudt, L.P., Pastink, A., Kempers-Vaenstra, A.E. *et al.* (1983) The genes coding for H3 and H4 in *Neurospora crassa* are unique and contain intervening sequences. *Nucleic Acids Research*, **11**, 5347–60.

Wright, M.C. and Philippsen, P. (1991) Replicative transformation of the filamentous fungus *Ashbya gossypii* with plasmids containing *Saccharomyces cerevisiae* ARS elements. *Gene*, **109**, 99–105.

Wu, M.M.J., Cassidy, J.R. and Pukkila, P.J. (1983) Polymorphism in DNA of *Coprinus cinereus*. *Current Genetics*, **7**, 385–92.

Yin, S., Heckman, J. and RajBhandary, U.L., (1981) Highly conserved G.C. rich palindromic DNA sequences which flank tRNA genes in *Neurospora crassa* mitochondria. *Cell*, **26**, 326–32.

PART FOUR

COORDINATION OF GROWTH AND DIVISION

N.A.R. Gow
Department of Molecular and Cell Biology, Marischal College, University of Aberdeen, Aberdeen, UK

13.1 INTRODUCTION

Cells are said to exhibit polarized cell development or tip growth when the expansion of the cell surface is focused at one location. The term polarity has also been used in the context of mechanisms that establish new sites for growth, either by budding, germ tube formation or by branching. This chapter considers how tip growth and cell polarity are established, maintained and regulated in fungal cells.

The problem of polarized cell development has perplexed developmental biologists for many years. Analysis of the regulation of cell polarity is difficult because it depends on the integrated workings of the intact cell. New insights have been gained through the analysis of polarity mutants in yeast (discussed later) which help focus attention on genes and hence gene products that are important for the establishment and maintenance of polarized growth. Other advances have been made through the analysis of filamentous fungi where the cells are large enough to be able to physically manipulate them or make use of patch-clamp technologies or various microelectrodes. Thus fungi prove to be convenient models for both genetic and physiological analyses of the phenomenon of polarity.

13.1.1 Patterns of cell growth of fungi

Hyphal extension is an extreme example of polarized cell growth since cell extension is restricted to a narrow zone defined by the tapering hyphal apex. Indeed, the rate of wall synthesis in the apical 1 μm of a hypha may be 50 times that 50 μm behind it (Gooday, 1971; Gooday and Trinci, 1980; Figure 15.2). Although this apical growth pattern has been known for more than a century (Reinhardt, 1892) the underlying mechanisms that account for polarized hyphal growth are not yet understood. The growth of other fungal cell types including yeast forms can also be shown to be polarized to some degree. For example, during the early stages in the formation of a bud of a yeast cell growth is initially polar. As the bud continues to expand growth becomes isotropic. This has been demonstrated for budding of the yeast form of *Candida albicans* by labelling the cell surface with polylysine-coated microspheres and observing their relative displacement during cell expansion (Staebell and Soll, 1985; Merson-Davies and Odds, 1992) (Chapter 19). In *Mucor rouxii* the pattern of cell wall growth has been followed during sporangiospore germination by autoradiographic imaging of sites of N-acetylglucosamine incorporation into chitin (Bartnicki-Garcia and Lippman, 1969). This showed that cell wall biosynthesis occurred randomly over the surface of spores undergoing swelling then became highly polarized at the time when the germ tube was formed. Autoradiographic studies of budding in *Saccharomyces cerevisiae* again suggest that cell wall expansion is predominantly by tip growth (Johnson and Gibson, 1966) and that cell expansion of the fission yeast *Schizosaccharomyces pombe* occurs only at one cell pole for most of the cell cycle (Johnson, 1965; Mitchison and Nurse, 1985). Autoradiography has also been used to demonstrate that chitin synthesis was highly polarized during hyphal extension of a wide range of fungal hyphae (Gooday, 1971). Indeed the phenomenon of intercalary growth, where extension is by the equal expansion of all regions of the cell surface, is very rare, although it occurs apparently during the elongation of stipes of certain mushrooms (Gooday, 1979).

Cell wall expansion is the ultimate manifestation of the cellular machinery that brings about polarized growth. In order for the wall in a hyphal tip or a bud to grow the cell wall biosynthetic enzymes must be deposited at just that location. In fungi

The Growing Fungus. Edited by Neil A.R. Gow and Geoffrey M. Gadd. Published in 1994 by Chapman & Hall, London. ISBN 0 412 46600 7

these enzymes are membrane-associated and are transported to the cell surface in microvesicles (Bracker *et al.*, 1976; Bartnicki-Garcia *et al.*, 1978; Bartnicki-Garcia and Bracker, 1984). Fusion of the microvesicles with the surface provides new membrane for expansion and enzymes for wall synthesis (Chapter 3). Thus polarized growth can be reduced to a problem of localized vesicle exocytosis. The vesicles are themselves transported to the surface by the filamentous components of the cytoskeleton (Chapter 6) and so vectorial cytoplasmic transport of vesicles is also polarized and carefully controlled. Thus in order to address how polarized cell wall synthesis arises we need also to look at the arrangement and mechanics of the cytoskeleton.

13.2   CELL STRUCTURE AND POLARITY

There are many structural features of fungal cells that are important for polarized growth. The process of growth is multifactorial but the principal participants are the vesicles that supply the biosynthetic enzymes for cell wall synthesis, the cytoskeleton that transports them and the cell wall where the shape of the cell is set.

13.2.1   Vesicles and the Spitzenkörper

The ultrastructure of apical microvesicles in fungal hyphae has been reviewed in Chapter 5. These vesicles supply enzymes and membrane for cell expansion. In *Neurospora crassa* over 38 000 vesicles fuse with the hyphal tip every minute (Collinge and Trinci, 1974) and these vesicles must be prevented from fusing with the lateral cell wall prior to arrival at the extension zone at the tip. Using phase contrast microscopy the cytoplasm of the fungal hyphal apex is sometimes seen to contain a refractile or phase dark region called the Spitzenkörper (= apical body) which is due to the high concentration of microvesicles in the cytoplasm (Girbardt, 1969; Grove and Bracker, 1970; Grove *et al.*, 1970). Electron microscopical examination of thin sections of fixed fungal material (Girbardt, 1969; Grove and Bracker, 1970; Grove *et al.*, 1970), and more recently of freeze-substituted material (Howard and Aist, 1980; Howard, 1981; Heath *et al.*, 1985; Howard and O'Donnell, 1987; Roberson and Fuller, 1988) showed that the organization of the vesicles had a distinct architecture in different fungal groups. The microvesicles

come in two general sizes (Chapter 5). The smallest vesicles (less than 100 nm) have been called chitosomes that may be specialized vehicles for the transport of chitin synthase to the surface (Bracker *et al.*, 1976; Bartnicki-Garcia and Bracker, 1984) (Chapter 5).

A recent mathematical model (reviewed in Chapter 15) suggests that this concentration of vesicles acts as the terminus for the forward moving cytoplasmic membrane vesicles, and the source for the vesicles that will go on to fuse with the surface (Bartnicki-Garcia *et al.*, 1989). Thus, it is suggested that vesicles are first collected in a local depository in the tip, called the vesicle supply centre, before being distributed to the cell surface (Chapter 15). For this model to be credible the Spitzenkörper must be actively involved in the process of localizing vesicle exocytosis. This view of the Spitzenkörper as a fungal-specific organelle is supported by observations of their phylogenetic-specific morphology in the electron microscope. Even more compelling evidence that the Spitzenkörper functions as a distinct organelle comes from recent experiments in which video-enhanced microscopy has been used to study dynamic changes in the Spitzenkörper of growing hyphae (Lopez-Franco and Bracker, unpublished) (Figures 13.1 and 13.2). Using this technique and a chamber that enabled the undisturbed hyphae to be observed at high magnification, eight different Spitzenkörper arrangements were described based on the observed vesicle cluster and core morphology (Figure 13.1). A total of 34 fungi from all the major fungal groups were examined, excluding Oomycetes (for which apical bodies are less evident). Remarkable correlations were found between the precise location and architecture of the Spitzenkörper and the shape and growth of individual hyphae. For example, when a hyphal tip changed direction the Spitzenkörper was seen to be displaced towards the side of the tip where growth was reoriented. One or more satellite Spitzenkörpers were observed in some specimens. These could move away from the main apical body then return and fuse with it (Figure 13.2). The presence of these satellites also had observable effects on hyphal extension causing bulging of the hyphal apex immediately above. Finally, careful examination of the extension rates of the fungi showed that hyphal extension rate was not constant but occurred in pulses of definable amplitude and

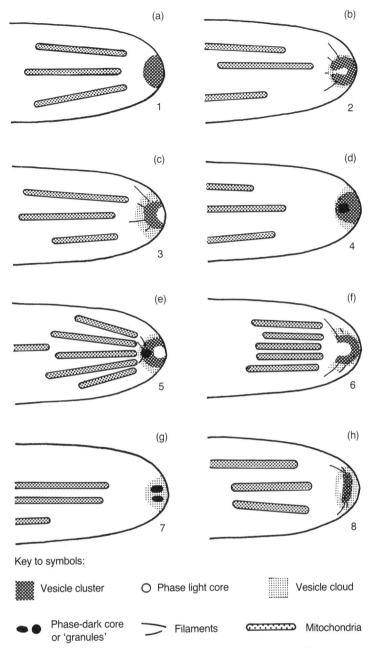

Figure 13.1 Eight distinct morphologies in the hyphal apices of filamentous fungi as seen with phase contrast microscopy. Examples of fungi exhibiting each Spitzenkörper pattern are given in parentheses. (a) Pattern 1, dark vesicle cluster without a visible core (*Agaricus brunnescens, Gaeumannomyces graminis*). (b) Pattern 2, dark vesicle cluster open towards the back with a narrow light core to the rear of the cluster (*Sclerotinium rolfsii*). (c) Pattern 3, dark vesicle cluster with an eccentric light core oriented towards the apical pole (*Fusarium culmorum, Bipolaris maydis, Magnaporthe grisea, Trichoderma viride*). (d) Pattern 4, grey vesicle cluster with an eccentric dark core located near the back of the cluster (*Aspergillus niger, Sclerotinia sclerotiorum*). (e) Pattern 5, grey vesicle cluster containing a light core oriented towards the apical pole and a dark core towards the back (*Galactomyces citri-aurantii*). (f) Pattern 6, a cup-shaped dark vesicle cluster surrounding a large phase-light core (*Rhizoctonia solani, Gelasinospora* sp.). (g) Pattern 7, a grey vesicle cloud containing two or more dark vesicle 'granules' (*Colletotrichum graminïcola, Leptosphaerulina briosana*). (h) Pattern 8, highly pleomorphic Spitzenkörper, often appearing as a thick dark band within a vesicle cloud and with a small light core behind the vesicle cluster (*Neurospora crassa*). Key to symbols (see insert). From Rosamaria Franco and Charles Bracker (unpublished) with permission.

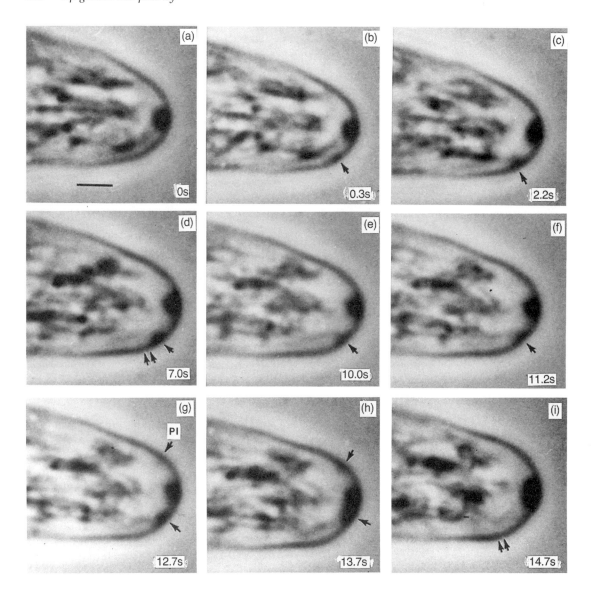

Figure 13.2   Dynamic behaviour of a Spitzenkörper and a Spitzenkörper satellite (arrow) in a growing hyphal apex of *Fusarium culmorum*. The position of the satellite Spitzenkörper affects the shape of the cell apex. In (a–d) the satellite stopped during migration, for more than 5 s, and then later resumed migration towards the main Spitzenkörper. The cell wall outgrew slightly next to the stationary satellite (double arrows). While the satellite merges, a new satellite begins to develop. Scale bar 2 μm. From Rosamaria Franco and Charles Bracker (unpublished) with permission.

duration. These pulses were shown to be reproducible features of the growth of the hyphae of each species and were not experimental artefacts due to the method of imaging (Lopez-Franco and Bracker, unpublished).

These data support the view that the Spitzenkörper is a distinct fungal-specific organelle that plays a critical role in hyphal morphogenesis and polarity. Now the nature of this dynamic structure has been revealed it is incumbent on future research investigations to attempt to establish the molecular composition and mechanics of the Spitzenkörper and hence its function.

13.2.2 Cytoskeleton

A comprehensive analysis of the fungal cytoskeleton is given in Chapter 6. The cytoskeleton has a major role in the process of tip growth and discussion here is limited to a summary of how the cytoskeleton moves and shapes the hyphal apex or buds of yeasts.

The principal components of the cytoskeleton are microfilaments of F-actin composed of G-actin monomeric protein (molecular weight 43 kD) and microtubules which are helical assemblies of a family of homologous proteins, α, β and γ tubulin (molecular weight 50–55 kD). Microfilaments and cytoplasmic microtubules composed of β-tubulin are collectively responsible for the movement of organelles to the apex and the localization of vesicle exocytosis (McKerracher and Heath, 1987; Gow, 1989a; Heath, 1990; Harold, 1990; Madden *et al.*, 1992).

(a) *Microtubules and tip growth*

The general organization of cytoplasmic microtubules in fungi is discussed elsewhere (Chapter 6). In terms of the polarity of growth it is relevant to consider where these cytoskeletal elements are nucleated. Studies in *Uromyces phaseoli* showed that depolymerized microtubules repolymerized first at the apex then proceeded backwards, suggesting that the organizing centres may be at the hyphal tip (Hoch and Staples, 1985). An apical organizing centre has been described for the ascus of *Sordaria macrospora* (Thompson-Coffe and Zickler, 1992) and a spindle pole antigen in *Saccharomyces cerevisiae* (SPA2) co-localized with the tips of buds and shmoo projections and predicted the site of bud emergence (Snyder, 1989; Snyder *et al.*, 1991). However, no support was found for the hypothesis that microtubules are nucleated at the hyphal apex of germ tubes of *C. albicans* (Barton and Gull, 1988).

The role of microtubules in organelle transport is an area where reports are contradictory. Convincing physiological and genetic evidence has been obtained that suggests that some fungal organelles such as the nucleus are moved forward by attachment to molecular motors attached to the cytoplasmic microtubules (Oakley and Morris, 1980; Oakley and Rinehart, 1985). A kinesin-like gene has been identified in *Aspergillus nidulans*, but this does not apparently function as a motor since mutants in this gene (*bimC*) still transport nuclei (Enos and Morris, 1990). Mutations in β-tubulin and treatments with microtubule inhibitors, such as benomyl, eliminated nuclear migration in *A. nidulans* without affecting mitochondrial cytoplasmic transport suggesting that microtubules were important for nuclear migration but not cytoplasmic transport of mitochondria (Oakley and Morris, 1980; Oakley and Rinehart, 1985). Benomyl also inhibited nuclear migration in *N. crassa* (That *et al.*, 1988). This contrasts with the findings of McKerracher and Heath (1986a) who used UV microbeams to study the role of microtubules in *Basidiobolus magnus*. Irradiation of small portions of the hypha was shown to cause local destruction of microtubules and led to the contraction of the cytoplasm and displacement of the nucleus (McKerracher and Heath, 1985, 1986a). However, the displaced nucleus was still able to migrate forward suggesting the positioning of the nucleus, but not nuclear migration was not due to interactions with microtubules (McKerracher and Heath, 1986a,b).

There are, however, contradictions in the literature, in particular with respect to the role of cytoplasmic microtubules in vesicle translocation and cell growth. In *Saprolegnia ferax* no preferential association was found between cytoplasmic microvesicles and microtubules (Heath *et al.*, 1985). In contrast, in *Gilbertella persicaria* close physical associations between vesicles and microtubules were seen using freeze-substitution fixation of hyphae (Howard, personal communication). Treatment of fungal hyphae with a wide variety of microtubule inhibitors (benomyl, nocodazole, colchicine and others) showed a variety of effects and sensitivities in different species. Microtubule inhibitors led to the dispersal of the Spitzenkörper and reduction in linear growth rate of hyphae of *Fusarium acuminatum* (Howard and Aist, 1977, 1980) and caused inhibition of germ tube growth and the formation of multiple germ tubes in *N. crassa* (That *et al.*, 1988). However, microtubule inhibitors had little or no measurable effect on the growth of hyphae of *Candida albicans* (Yokoyama *et al.*, 1990), *Uromyces phaseoli* (Herr and Heath, 1982) and *Phytophthora infestans* (Temperli *et al.*, 1991). Growth of *A. nidulans* was affected at high benomyl concentrations but was possible at lower concentrations and in strains harbouring mutations in the β-tubulin structural gene *benA* (Oakley and Rinehart, 1985). In *S. cerevisiae*, experiments using antimicrotubule drugs and β-tubulin mutants again suggested that cytoplasmic microtubules were not necessary for bud-site selection and expansion or for vesicle transport (Huffaker *et al.*, 1988; Jacobs *et al.*, 1988; Matsuzaki *et al.*, 1988). There is also some evidence for microtubule-based growth of shmoo cell projections during mating of *S. cerevisiae* (Gehrung and Snyder, 1990; Flescher *et al.*, 1993). Also, a multicopy suppresser of an actin binding protein (Myo2, see below) has been isolated that encoded a kinesin-like mechanoenzyme that mediates microtubule-based vesicle transport (Lillie and Brown, 1992). Thus microtubule-based systems may play some role in vesicle transport in yeast and seem likely to be involved in the transport of some but not all organelles in fungi.

(b) Actin and tip growth

There is good ultrastructural, physiological and genetic evidence that actin plays a vital role in the process of tip growth. Cells stained with rhodamine phalloidin reveal actin rich zones at the sites of growth – the apex of germ tubes, hyphae and buds of yeast cells (e.g. Adams and Pringle, 1984; Kilmartin and Adams, 1984; Marks and Hyams, 1985; Anderson and Soll, 1986; Heath, 1987; Alfa and Hyams, 1990; Jackson and Heath, 1990a; Yokoyama *et al.*, 1990; Kwon *et al.*, 1991; Roberson, 1992, Jackson and Heath, 1993a). In Oomycetes such as *S. ferax* the actin at the hyphal apex takes the form of an actin cap of true filaments whereas the distal actin is mainly in the form of actin cables and punctate actin plaques (Heath, 1987). The actin cap is thought to have a structural role in stabilizing the plastic hyphal tip as well as being involved in localizing vesicle exocytosis (Jackson and Heath, 1990b; Heath, 1990). Rhodamine stained yeast cells and hyphae of true fungi reveal a background of diffuse staining with punctate actin plaques and a few filamentous actin cables (Chapter 6). The formation of an actin rich zone predicts where growth will occur and the dispersal of actin occurs when tip growth ceases (e.g. Marks and Hyams, 1985;

Anderson and Soll, 1986; Alfa and Hyams, 1990; Heath 1990). When actin is disrupted with cytochalasins, growth, vesicle arrangement and tip shape are invariably affected (Betina *et al.*, 1972; Allen *et al.*, 1980; Yokoyama *et al.*, 1990). In addition, actin mutants and null mutations in or overexpression of actin binding protein (Apb1) led to delocalized cell wall growth (Novick and Botstein, 1985; Drubin *et al.*, 1988; Adams *et al.*, 1991). Actin mutants also have defects in the secretion of enzymes such as invertase and these cells accumulate microvesicles, again suggesting that the mutant phenotype was due to disruption of the mechanism that localized exocytosis (Novick and Botstein, 1985). Thus actin is involved directly or indirectly in the translocation of vesicles through the fungal cell and the localization of the site of exocytosis and hence cell growth.

The existence of an actin–myosin based system for vesicle translocation is suggested from the above experiments and from the observation that disruption of yeast tropomyosin genes leads to the disappearance of actin cables in yeast (Liu and Bretscher, 1989). Also the *ABP1* gene of yeast has a C-terminus with a SH3 domain that is closely homologous to the actin-binding SH3 domain at the C-terminus of myosin-I (Drubin *et al.*, 1990). Biochemical evidence has also been obtained for the presence of myosin heavy chain in yeast (Drubin *et al.*, 1988) and a novel myosin gene *MYO2* has been identified that is necessary for bud growth and vesicle transport (Johnston *et al.*, 1991). Some ultrastructural studies appear to show close physical association of microtubules with actin suggesting that microtubules may further interact with this actin–myosin based system (Hoch and Staples, 1985; McKerracher and Heath, 1986b; Heath and Kaminskyj, 1989). Thus some of the growth effects that are apparently due to effects on microtubules could be explained on the basis of secondary effects on the actin–myosin based cytoskeleton. It is also possible that some components of the cytoplasm could be transported apically simply on the basis that they are swept along in the cytoplasmic flow and not because they are tethered to molecular motors that travel along the filamentous elements of the fungal cytoskeleton.

In summary, both microtubules and actin-based components have important roles in tip growth and polarity of fungi. The dominant role for the cytoplasmic microtubules would seem to be in the movement of the nucleus. An actin–myosin–based system would seem to be important in the establishment and maintenance of polarity through the localization of sites of vesicle exocytosis and the transport of microvesicles to the cell surface. Further aspects of the role of the cytoskeleton in growth are discussed in Chapter 6 and in section 13.3.1.

13.2.3 Cell wall

The biochemistry of cell wall structure and its synthesis is discussed in Chapter 3. Some key points are worth reiterating here in the context of tip growth. The main structural elements in the cell wall are chitin and glucans (for other reviews see Gow, 1989a; Gooday, 1993; Wessels, 1993). At the apex the wall is plastic because the nascent chitin chains have not yet become thickened into distinct microfibrils by hydrogen bonding (Burnett, 1979; Vermeulen and Wessels, 1984, 1986) and because the chitin chains have not become covalently cross-linked via amino acid linkages with cell wall glucans (Sietsma and Wessels, 1979, 1981; Sonnenberg *et al.*, 1982, 1986; Wessels *et al.*, 1983; Sietsma *et al.*, 1985; Wessels, 1986) (Figure 3.5). These processes occur as the cell wall matures and is lateralized by expansion of the tip. A controversial idea is the plasticity of the apex which also reflects the presence of cell wall lytic enzymes such as chitinase and cellulase in Oomycetes. Evidence has been found for the parallel regulation of chitinase and chitin synthase (Rast *et al.*, 1991) and for a membrane-bound zymogenic chitinase that could modulate chitin assembly and tip growth (Humphreys and Gooday, 1984). However, it has also been argued that the presence of lytic enzymes is not necessary to explain the plasticity of the apical wall (Wessels, 1984) and that these enzymes might function as exoenzymes rather than growth-associated autolysins. However, chitinase, β-glucanase or cellulase in Oomycetes (Thomas and Mullins, 1967) are likely to be important in the formation of branches where a rigid cell wall must first be softened.

13.3 REGULATION OF CELL POLARITY

The problem of how the components of the cell become polarized has been investigated in cells from bacteria to mammals (for a review see Nelson, 1992). The problem appears to the investigator as a Russian doll, with each level of organization being regulated at a finer level, each regulatory element having itself to be regulated. Regulatory elements must involve controlling the orientation of the cytoskeleton so that protein traffic becomes focused to certain sites of the cell surface.

It should be noted that many mutations and experimental treatments influence the formation and selection of sites for polarized growth as well as growth itself. However, site selection and subsequent polar growth processes can be dissociated genetically and physiologically. For example some yeast mutants (discussed below) cannot regulate the sites at which buds are formed, but the buds that emerge randomly grow normally. For filamentous fungi choline addition can affect branch initiation without influencing hyphal extension (Weibe *et al.*, 1992; Markham *et al.*, 1993) again suggesting that the establishment and maintenance of growth can be regulated independently.

13.3.1 Genetic regulation of polarized growth

In fungi, the genetic analysis of polarity has been carried out almost exclusively with *S. cerevisiae* (reviewed by Drubin, 1991; Madden *et al.*, 1992; Nelson 1992). It is likely, though not proven, that aspects of the site selection for buds of *S. cerevisiae* will also be relevant to the regulation of germ tube formation and branching of filamentous fungi. This, however, is a question for future research.

The control of polarity in yeast can be broken down into consideration of the regulation of the pattern of budding, the mechanism of assembly of the bud site, the effect of this on the orientation of the cytoskeleton and the effect of the cytoskeleton on polar growth (Figure 13.3).

In first considering the regulation of budding of *S. cerevisiae* it must be appreciated that this varies with strain mating type. For haploid cells (**a** or α cells) or diploid **a**/**a** or α/α cells, bud formation occurs close to the site of the previous bud (axial budding). In diploid (**a**/α) cells the pattern is said to be bipolar so that mother cells bud either from a site near the previous bud site or near the opposite pole and daughter cells usually bud from the opposite pole from the site where they were born (also see Chapter 19). Direct examination of microcolonies of mutagenized cells identified five

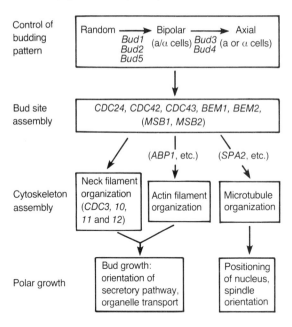

Figure 13.3 Genetic regulation of the morphogenetic pathway for bud assembly and growth as discussed in the text. From Drubin (1991) with permission.

genes (*BUD1 – BUD5*) which affected the normal budding pattern (Chant and Herskowitz, 1991; Chant *et al.*, 1991; Powers *et al.*, 1991; Chenevert *et al.*, 1992). Mutations in *BUD1*, *BUD2* or *BUD5* caused random budding patterns, irrespective of mating type. Mutations in *BUD3* and *BUD4* led to bipolar budding in **a** or α cells which normally had an axial budding pattern but had no effect on **a**/α strains that were normally bipolar (Figure 13.3). It was therefore suggested that *BUD1*, *BUD2* and *BUD5* were active in all yeast strains but that *BUD3* and *BUD4* were inactive for strains exhibiting bipolar growth (Chant and Herskowitz, 1991). These five genes were only concerned with the regulation of polarity since none of the respective mutations had any effect on bud assembly and growth. This is in contrast with the bud site assembly genes (*CDC24*, *CDC42*, *CDC43*, *BEM1* and *BEM2*) which have a common mutant phenotype where no buds are formed and the cell increases in size by non-localized growth (Bender and Pringle, 1989, 1991; Adams *et al.*, 1990).

Three of the *BUD* genes and some bud assembly genes encode proteins that are components of a GTPase cycle (Figure 13.4). Bud1 is a small

GTP-binding protein of the ras family (Chant *et al.*, 1991), Bud2 is a GTP activating protein (Nelson, 1992) and Bud5 appears to be an activator of RAS proteins that catalyses GDP–GTP exchange (Bender and Pringle, 1991; Chant *et al.*, 1991; Powers *et al.*, 1991). Although the function of Bud3 and Bud4 are not yet known they may be involved in defining the membrane target for vesicle exocytosis during budding. *CDC42* encodes a small GTP-binding protein and *CDC43* encodes a geranyl-geranyltransferase that is homologous to *DPR1/RAM1* that prenylates the C-terminus of small GTP-binding proteins (Finegold *et al.*, 1991; Ohya *et al.*, 1991). In addition, Sec4 and Ypt1 are yeast GTPases that are required for membrane traffic and polarized vesicle exocytosis (Novick and Brennwald, 1993). Sec4 is localized to the apex of buds (Brennwald and Novick, 1993). Thus there is considerable evidence that GTP-cycling is an important regulatory process is the formation of the buds of yeast.

Genetic studies suggest that there is an interaction between the *BUD* genes and the genes concerned with the assembly of the bud site. For example overexpression of *BUD1* can suppress temperature-sensitive mutations in *CDC24* (Bender and Pringle, 1989) which can lead to random budding or failure to form buds. Also, strains that are constructed with mutations in both *BUD5* and *BEM1* are non-viable although single mutations in either gene are non-lethal. The implication of this negative synergism leading to 'synthetic lethality' of the double mutant is that there is a functional interaction between *BUD5* and *BEM1* gene products (Chant *et al.*, 1991). Further genetic interactions suggest that the bud site assembly genes function together and are linked to the functions of the *BUD* genes (Drubin, 1991).

In contrast to the studies above that demonstrate an interaction between bud selection and bud assembly genes, there are few indications of how bud assembly genes interact with the cytoskeleton. However, *BEM1* has two SH3 domains similar to the SH3 domain of Apbl that binds actin and is important for bud growth (Chenevert *et al.*, 1992) (section 13.2.2). *BEM1* mutations also lead to large non-budded cells with abnormal actin distributions (Chenevert *et al.*, 1992). Thus *ABP1* and *BEM1* may interact with the actin component of the cytoskeleton and thereby influence vesicle

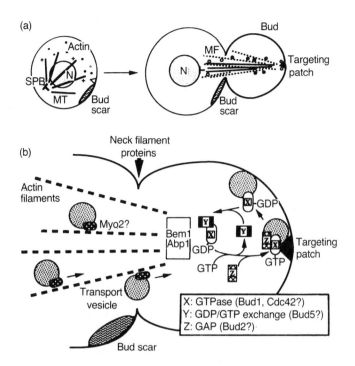

Figure 13.4 Regulation of bud formation in *S. cerevisiae* by the cytoskeleton and components of the GTPase cycle. (a) An unbudded cell with microtubules (MT) nucleated from a spindle pole body (SPB) and random plaques of actin (spots). Polarization of growth is proposed to be initiated by activation of a membrane targeting patch (triangle), near the bud scar. During polarization the microtubules and actin become oriented towards the targeting patch. In the growing bud (b) vesicles are delivered towards the patch by an actin–myosin-based transport along microfilaments (MF) perhaps via Myo2 motors. Microtubules may play a secondary role in vesicle transport but are not required obligatorily. Beml and Abpl may link the cytoskeleton to the target site on the cell membrane as discussed in the text. Finally *BUD* gene products and other components of a putative GTPase cycle may mediate the docking of the vesicles to the surface and hence determine the site of bud formation. From Nelson (1992), with permission.

exocytosis and hence cell wall growth and polarity (Figures 13.3 and 13.4). *SPA2*, which encodes a spindle pole antigen located at the poles of buds and shmoo projections is not essential for budding but seems to play an important role in shmoo formation during mating and sites of cytokinesis (Gehrung and Snyder, 1990; Costigan *et al.*, 1992; Flescher *et al.*, 1993). Genetic evidence also been produced that links *SPA2* function with GTP-binding proteins (Flescher *et al.*, 1993). Thus *SPA2* may be important in linking the microtubule cytoskeleton with genetic determinants of polarity.

Finally it should be noted that the gene products of *CDC24* (*CLS4*) and *CDC43* may be modulated by calcium ions and that tip-high gradients of calcium have been found at the apex of a variety of polarized cell types including fungal hyphae, as discussed below.

13.3.2 Calcium gradients

Many eukaryotic cells that grow by apical extension have tip-high cytoplasmic calcium gradients (Reiss and Herth, 1979; Brownlee and Wood, 1986; Kropf and Quatrano, 1987; Yuan and Heath, 1991; Miller *et al.*, 1992; Read *et al.*, 1992; Berger and Brownlee, 1993; Jackson and Heath, 1993b; Knight *et al.*, 1993). The function of the calcium gradients has not been established unequivocally (discussed below) but there is mounting evidence that they are important for polarized tip growth. Calcium gradients have been demonstrated using fluorescent dyes such as chlorotetracycline (CTC) (e.g. Reiss and Herth, 1979; Kropf and Quatrano, 1987; Yuan and Heath, 1991) which visualizes membrane-associated calcium; Indo-1, Fura-2, Fluo-3, Calcium Green-1 (e.g. Read *et al.*, 1992; Miller *et al.*, 1992; Berger and Brownlee, 1993; Knight *et al.*, 1993; Jackson and Heath, 1993b) which can be used to quantify free cytoplasmic calcium ion concentrations by ratiometric methods (Read *et al.*, 1992 for a review) and calcium-selective micro-electrodes (e.g. Brownlee and Wood, 1986). Fungi have proven to be rather difficult specimens in which to measure calcium ion distributions. The hyphal tip is fragile and therefore difficult to pierce with a microelectrode without causing it to burst or stop growing. Microinjection has, however, been used successfully to introduce calcium-dyes into hyphae (Read *et al.*, 1992; Knight *et al.*, 1993). Electroporation can also be used to permeabilize fungal cells to these dyes but the process stops growth and a period of recovery is necessary before observations can be made (Jackson and Heath, 1990b; Read *et al.*, 1992). A further complication is that living fungal hyphae have a tendency to sequester dyes rapidly into vesicular or reticular organelles (Read *et al.*, 1992; Knight *et al.*, 1993). This occurs after microinjection or electroporation and is not prevented by conjugation of the dyes to high-molecular-weight dextrans which limits their diffusion across membranes. However, tip-high cytoplasmic gradients of calcium ions have been reported in the hyphae of the oomycete *Saprolegnia ferax* using Fluo-3 or Indo-1 (Yuan and Heath, 1991; Jackson and Heath, 1993b; Garrill *et al.*, 1993) and CTC-gradients have been reported for *S. ferax* (Jackson and Heath, 1989; Yuan and Heath, 1991), *N. crassa* (Schmid and Harold, 1988), *Achlya* species (Reiss and Herth, 1979) and *Basidiobolus magnus* (McKerracher and Heath, 1986b). In *S. ferax* where the free cytoplasmic calcium ion concentration and the membrane-associated calcium have both been measured, the free Ca^{2+} gradient is very steep and largely confined to the extreme apex while the membrane-associated calcium is maximal subapically (Figure 13.5). It therefore seems likely that polarised hyphal extension of fungal hyphae is associated with locally high calcium concentrations.

Recent evidence using patch-clamp techniques suggests that the tip-high gradient reflects the spatial organization of calcium channels in the cell membrane. In *S. ferax* the arrangement and concentration of calcium channels in the apical and subapical membrane were examined by collecting protoplasts as they were liberated from the tips of hyphae that were gently treated with protoplasting enzymes. The protoplasts were liberated as a 'string of pearls' so that the origin of each protoplast relative to the original hyphal tip could be determined (Garrill *et al.*, 1992). Using patch-clamp electrophysiology two types of channels were identified: (a) a Ca^{2+}-activated K^+ channel that was thought to be involved in turgor regulation, but was not obligatory for growth and (b) a stretch-activated Ca^{2+} channel that was activated by K^+ ions and which may be essential for apical extension (Garrill *et al.*, 1992, 1993). The stretch-activated channels were concentrated at the hyphal apex and were blocked by Gd^{3+} ions which also inhibited hyphal extension and dissipated the tip-high Ca^{2+} gradient revealed by Indo-1 (Garrill *et al.*, 1993). In contrast to the stretch-activated channels the Ca^{2+}-activated K^+ channels were uniformly distributed along the hyphal cell membrane. These could be inhibited by tetraethylammonium, which only caused a transient effect on growth. Stretch-activated Ca^{2+} channels have also been identified in the germ tube apices of the plant pathogen *Uromyces appendiculatus* where they are thought to play an important role in orientation and differentiation responses of the fungus in relation to topographical features of the plant leaf surface (thigmotropism) (Hoch *et al.*, 1987; Zhou *et al.*, 1991). These data suggest that the tip-high calcium gradient is important for polarized hyphal extension and is generated by a locally high concentration of stretch-activated calcium channels in the hyphal apex. It is presumed that the channels are delivered to the surface in microvesicles. They may be maintained there by anchoring them to the cytoskeleton or by membrane recycling as they are lateralized by tip expansion.

Other evidence suggests strongly that calcium ions play important roles in the regulation of the growth of hyphal apices and the formation of branches. When mycelia of *S. ferax* (Jackson and Heath, 1989), *N. crassa* (Schmid and Harold, 1988), *Fusarium graminearum* (Robson *et al.*, 1991a) and *Aspergillus fumigatus* (Gow *et al.*, 1992) were grown in media containing low concentrations of Ca^{2+} the rate of tip extension was reduced markedly.

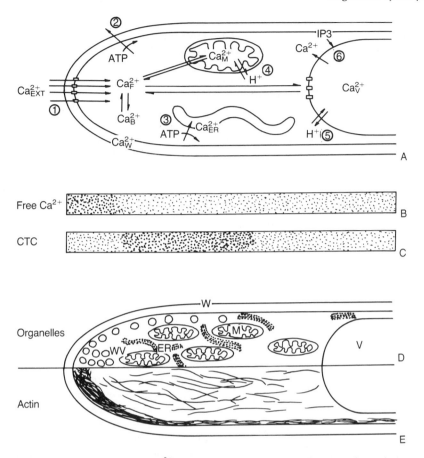

Figure 13.5 Gradients of free cytoplasmic Ca^{2+} (revealed by imaging with Fluo-3) and membrane associated Ca^{2+} (revealed by CTC staining) in a hypha of *Saprolegnia ferax* and the possible origin of these two gradients related to calcium conducting transport proteins, location of organelles and organization of actin. The vacuole (V), mitochondrion (M), endoplasmic reticulum (ER), cell wall (W) and cell wall microvesicles (WV) are shown. Calcium transport proteins are (1) stretch-activated calcium channels, (2) plasma membrane Ca^{2+}-ATPase, (3) ER-sequestering Ca^{2+}-ATPase, (4) H^+/Ca^{2+} exchanger in the ER, (5) H^+/Ca^{2+} antiporter of the vacuole and (6) IP_3-releasable Ca^{2+} transport system of the vacuole. From Jackson and Heath (1993b), with permission.

However branch formation of *F. graminearum* (Robson *et al.*, 1991a) and *A. fumigatus* (Gow *et al.*, 1992) was stimulated by low calcium concentrations and the specific growth rate of the mycelium as a whole was not affected except at nanomolar calcium concentrations. Thus at low calcium concentrations hyperbranched colonies with slowly extending tips were formed. In *N. crassa* the hyperbranching mutants 'frost' and 'spray' could be restored to a wild type branching pattern by the addition of high concentrations (50–500 mM) of Ca^{2+} and wild type hyphae could be induced to a hyperbranching morphology

using the calcium-channel blocker verapamil (Dicker and Turian, 1990). Calcium channel blockers also stimulated branching in *F. graminearum* (Robson *et al.*, 1991b).

Further evidence for a role for calcium ions in branch formation comes from experiments with the ionophore A23187 which conducts Ca^{2+} and other ions (see later). In *N. crassa* 0.1 mM ionophore stimulated branching (Reissig and Kinney, 1983) whereas in *F. graminearum* the effect of the ionophore was concentration dependent; concentrations above 0.1 mM stimulated branching, but low concentrations (30–80 nM) decreased branching frequency (Robson *et al.*, 1991a). Branch formation in *F graminearum* has also been shown to be enhanced by exogenous supply of calmodulin inhibitors and by cAMP and cGMP (Robson *et al.*, 1991b,c). Thus calcium mediated effects on growth and branching may be mediated in turn via calmodulin and/or cAMP, cGMP or inositol phosphate second messenger systems (Chapter 9).

The calcium gradient within a hypha must be regulated carefully if it has a role in controlling tip extension. The regulation of cytoplasmic calcium concentration via transport across the cell membrane and sequestration in organelles such as the vacuole, endoplasmic reticulum and mitochondria (Yuan and Heath, 1991; Jackson and Heath, 1993b) (Figure 13.5) is considered in Chapters 8 and 9. Given the presence of a standing gradient in the cytoplasm several calcium-dependent processes could be affected. First, it is known that the polymerization of monomeric G-actin to filamentous F-actin is a calcium-dependent process. This may in turn modulate tip growth by affecting the site at which microvesicles are brought to the cell surface (Jackson and Heath, 1989; Garrill *et al.*, 1993; Jackson and Heath, 1993b) (Chapter 6). Polarized cytoplasmic flow towards the hyphal apex is another cytoskeletal-based process and this has been shown in *B. magnus* and *S. ferax* to be regulated in part by calcium ions (McKerracher and Heath, 1986b; Jackson and Heath, 1992). The process of vesicle fusion with membranes is also calcium-dependent for a wide range of systems and the locally high calcium concentration at the extreme hyphal apex may therefore provide a permissive environment for vesicle exocytosis. The calcium ionophore A23187 was shown to alter the apical pattern of microvesicle accumulation and hence hyphal elongation in *Phycomyces blakesleeanus* (Ruiz-Herrera *et al.*, 1989). In addition, native chitin synthase activity from mycelial extracts of this fungus was activated by Ca^{2+} and calmodulin and inhibited by EGTA and EDTA (which chelated calcium ions) and by the calmodulin inhibitor TFP (Martinez-Cadena and Ruiz-Herrera, 1987).

A note of caution has been raised about the methodology employed in many investigations of calcium-based fungal physiology when using EGTA, A23187 and other 'calcium reagents' (Youatt, 1993). EGTA is not a specific calcium chelator and has a high affinity for other divalent cations in particular Mn^{2+}, Fe^{2+} and Zn^{2+}. Therefore, suitable controls must be included to exclude possible effects due to chelations of other ions. The ionophore A23187 is similarly non-specific and is capable of conducting several divalent cations. Many other calcium reagents are not wholly specific and so experiments must be interpreted with caution (Youatt, 1993). Several examples exist where addition of Mn^{2+} can compensate for Ca^{2+} depletion of the medium suggesting that Mn^{2+} may be able to compensate for some Ca^{2+}-dependent functions (Szaniszlo *et al.*, 1993; Youatt and McKinnon, 1993; Miller and Gow, unpublished). Some fungi including *S. ferax* and *Achlya bisexualis* can grow temporally in the absence of added calcium ions (Kropf *et al.*, 1984; Jackson and Heath, 1989) when they presumably use calcium from intracellular stores. For *N. crassa* growth halts immediately when calcium ions are removed from the growth medium and EGTA is added (McGillivray and Gow, 1987; Takeuchi *et al.*, 1988). In the case of *Allomyces macrogynus* it is difficult to show calcium-dependency of growth since the fungus can complete its life cycle through several life cycles in media containing only nanomolar levels of exogenous calcium ions (Youatt, 1993; Youatt and McKinnon, 1993). Clearly fungi are rather different in their demand for the supply of calcium and this must be taken into account when considering general roles for calcium ions in tip growth.

In addition to the physiological data described in this section it should also be borne in mind that in *Saccharomyces cerevisiae* there is genetic evidence that the gene product of *CDC24* is a calcium binding protein which has a central role in the regulation of polarity (Ohya *et al.*, 1986, 1991; Madden *et al.*, 1992) (see above). Therefore, the cumulative evidence for calcium being important in polarized development of fungi is substantial. Calcium gradients could influence hyphal growth in a number of different ways which may not be the same for all fungal species. They may be involved in directing cytoplasmic flow and microvesicle guidance and fusion with the hyphal apex or yeast bud site and by regulating the activity of important wall biosynthetic enzymes at this location. Further clues that calcium ions are important for hyphal polarity come from experiments employing applied electrical fields which are described below.

13.3.3 Ion currents

The ionic traffic across the plasma membrane of eukaryotic cells is normally organized so that ion channels and ion pumps are spatially segregated (Gow, 1989b; Nuccitelli, 1990). As a result ion currents circulate between the sites of the cell membrane where there is maximal clustering of the sources and the sinks for ion transport. The net flow of electrical current carried by the circulating ions can be detected with an ultra-sensitive voltmeter called the vibrating micro-electrode (Jaffe and Nuccitelli, 1974). This type of microelectrode is sensitive enough to detect the minute electrical fields around individual cells, such as fungal hyphae, that drive electrical currents. Vibrating microelectrodes have been used to study electrical currents in many different cell and tissue types including plant, animal and microbial examples. A major finding in these experiments was that the three-dimensional pattern of these currents reflects the polarity and axes of growth of the cells (Gow, 1989b) thus it was of interest to try to establish whether ionic currents were important for the regulation of tip growth and polarity in fungi.

The first experiments on ionic currents of fungi focused on two water moulds, *Blastocladiella emersonii*, a chytrid with a sporangium and rhizoids, and *Achlya bisexualis*, a filamentous oomycete. The current pattern for *B. emersonii* was such that positive electrical current, carried mainly by protons, entered the rhizoids during vegetative growth (Stump *et al.*, 1980) (Figure 13.6). In *A. bisexualis* and filamentous fungi in general, positive proton-carried current normally entered the growing apex (Kropf *et al.*, 1984; Gow *et al.*, 1984; Gow, 1984; Horwitz *et al.*, 1984). In *A. bisexualis* the mechanism of current generation was further characterized. Inward current was shown to be due to amino acid–proton co-transport (symport) localized at the tip and the outward current was due to electrogenic proton efflux via a plasma membrane ATPase (Kropf *et al.*, 1984; Gow *et al.*, 1984; Schreurs and Harold, 1988). The possibility that the proton current was important for cell polarity was supported initially by the finding that there was a temporal and spatial correlation between the current and growth. For example, in *A. bisexualis* the removal of amino acids, from the growth medium or an increase in external pH,

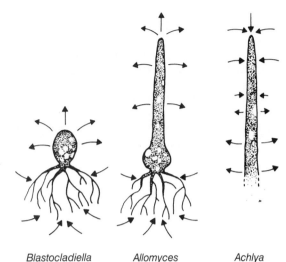

Blastocladiella   *Allomyces*   *Achlya*

Figure 13.6 Circulating proton currents in three water moulds. Inward currents are associated with the rhizoids of *Blastocladiella emersonii* and *Allomyces macrogynus* and the hyphal tip of the oomycete *Achlya bisexualis*. The inward currents are associated with chemotropic regions involved in proton-driven nutrient uptake and are not directly related to the process of tip growth. From Youatt *et al.* (1988) with permission.

reduced or abolished the current and halted extension simultaneously (Kropf *et al.*, 1984; Gow *et al.*, 1984). The proton current also established an extracellular pH gradient around the hypha, with the medium adjacent to the tip relatively alkaline (Gow *et al.*, 1984). This profile responded to changes in medium composition and pH in the same way as did the current suggesting the pH gradient was a direct manifestation of the proton current. Also new growing points such as vegetative or sexual antheridial branches were predicted by the location of new inwardly directed currents (Kropf *et al.*, 1983; Gow and Gooday, 1987). These early experiments therefore supported the hypothesis that growing hyphal tips were sites of proton influx and that this may be important for polarized hyphal growth. Fluorescence-ratio measurements of cytoplasmic pH in *Penicillium cyclopium* revealed an acidic apical cytoplasm which is again consistent with this statement (Roncal *et al.*, 1993).

However, an increasing body of evidence has shown that the proton current is in fact incidental

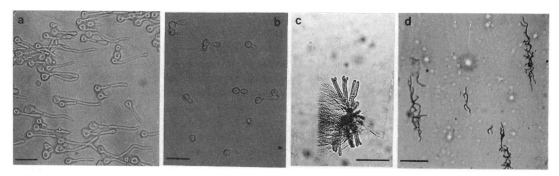

Figure 13.7   Fungal galvanotropism. Directional growth of (a) germ tubes and (b) buds of *C. albicans*, rhizoids of (c) *Allomyces macrogynus* and (d) hyphae and branches of colonies of *Neurospora crassa*. The anode is on the left and the cathode on the right in each photograph. The scale bars are 20 μm (a,b) 100 μm (c) and 300 μm (d). From (a,b) Crombie *et al.* (1990), (b) Youatt *et al.* (1988) and (d) McGillivray and Gow (1987), with permission.

to the process of tip growth and regulation of polarity. It was shown that the current at a hyphal tip could decline or even reverse direction when it formed a lateral branch (Kropf *et al.*, 1983). In *Neurospora crassa* non-growing hyphae were found with normal currents and examples of hyphae with normal growth rates but highly attenuated or reversed currents were also described (McGillivray and Gow, 1987; Takeuchi *et al.*, 1988; Schreurs and Harold, 1988; Cho *et al.*, 1991). Therefore, there are many examples where there is no correlation between the direction or magnitude of the current and the process of tip growth. In particular in the chytrid *Allomyces macrogynus* it was shown that hyphal growth was associated with a constitutive outwardly directed proton current (Youatt *et al.*, 1988; De Silva *et al.*, 1992). This pattern was preserved even when the hyphae underwent 'reverse development' when they widened their hypha by an active growth process that started at the tip then proceeded backwards down the hypha. De Silva *et al.* (1992) pointed out that a correlation could be drawn between sites of nutrient driven transport and sites of inwards currents in Chytridiomycetes and Oomycetes (Figure 13.6). The rhizoids of *B. emersonii*, and *A. macrogynus* and the hyphal apex of *A. bisexualis* are all chemotropic and are therefore likely to be sites of nutrient uptake. The hypha of *A. macrogynus* is not chemotropic, however, and nutrients for hyphal growth are apparently provided by the rhizoids then transferred to the hypha (De Silva *et al.*, 1992). Thus it was suggested that the inward currents reveal sites where proton-driven nutrient

transport is localized in these fungi. Therefore the current was concerned with local nutrition and not the regulation of tip growth.

It should be noted that the electrical current of fungi as measured with a vibrating microelectrode represents the net direction of flow of all ions that pass across the cell membrane. For fungi, calcium ions are normally a small fraction of the total ionic traffic and do not contribute much to the electrical current (Kropf *et al.*, 1984; McGillivray and Gow, 1987; Takeuchi *et al.*, 1988; De Silva *et al.*, 1992). However, new generation vibrating, ion-selective microelectrodes, including calcium-specific vibrating electrodes, have now been developed that may facilitate examination of the role of calcium ion entry in tip growth (discussed above).

13.3.4   Applied electrical fields

Fungi grow towards the anode or cathode of an externally applied electrical field or they may align perpendicularly to it. This galvanotropic behaviour has been demonstrated for reverting protoplasts (De Vries and Wessels, 1982), germ tubes (McGillivray and Gow, 1986; Van Laere, 1988; Wittekindt *et al.*, 1989; Cho *et al.*, 1991; Lever *et al.*, 1993), the buds of yeast cells (Crombie *et al.*, 1990), rhizoids (Youatt *et al.*, 1988; De Silva *et al.*, 1992), hyphae and their branches (McGillivray and Gow, 1986; Wittekindt *et al.*, 1989; Lever *et al.*, 1993) (Figure 13.7). Galvanotropism is not directly related to the natural electrical currents generated by fungi since there is no obvious correlation between the natural electrical polarity of a fungus and its response to an applied electrical field (Gow, 1987, 1989b). However, because electrical fields can impose polarity on cells they can be used to investigate the mechanisms that result in directed cell growth.

The extent of electrical field-induced cell alignment is voltage-dependent (McGillivray and Gow 1986; Van Laere, 1988) and pH dependent (Van Laere, 1988; Lever *et al.*, 1993). Voltage-dependency suggests that the electric field may cause the redistribution of proteins in the cell membrane by electrophoresis resulting in asymmetries that lead to the polarization of growth (Jaffe, 1977). Direct evidence for electric-field induced asymmetries in several membrane proteins has been obtained for animal cells (Orida and Poo, 1978; Poo, 1981; Stollberg and Fraser, 1988). Cytoplasmic proteins or organelles may also be affected, but the high capacitance of cell membranes would be expected to shield the cytoplasm from most of the external field. *C. albicans* yeast cells that were exposed to an electrical field retained a memory of that experience even after the field was switched off since germ tubes that emerged from field-treated yeast cells were still polarized cathodally (Crombie *et al.*, 1990). This is again persuasive that field-induced asymmetries in the cellular organization were maintained, to some extent, even when the field was switched off.

The pH-dependence of fungal galvanotropism may therefore be due to pH-induced alterations in the charged residues of such proteins. Thus, in growth media of increasing alkalinity the net charge of a protein would be expected to become more negative and it may be expected that anodotropic hyphae may become more strongly oriented towards the anode (Lever *et al.*, 1993). This is in accord with findings for the galvanotropic behaviour of *N. crassa* which becomes increasingly polarized towards the anode at high pH, but it does not explain the pH-dependency of fungi such as *Aspergillus nidulans*, *Coprinus cinereus* and *Phycomyces blakesleeanus* which become increasingly oriented towards the cathode under these conditions (Van Laere, 1988; Lever *et al.*, 1993).

Imposed electrical fields would also be expected to affect tip growth and polarity via their effects on the membrane potential of the cell. Fungal cells have a negative resting membrane potential of around -150 mV (Slayman and Slayman, 1962; Kropf, 1986). The internal negativity is due for the most part to the electrogenic efflux of protons by

the membrane ATPase which generates a proton gradient that is used to drive nutrient uptake by secondary active transport (Chapter 8). In an electrical field the external potential at the cathodal end of the cell will be more negative and the membrane potential will be reduced (depolarized). At the anodal end of the cell the membrane potential will be hyperpolarized. Depolarization of the membrane potential at the cathodal end of a cell can result in the opening of voltage-gated calcium ion channels (Robinson, 1985; Tsien and Tsien, 1990). Electric-field-induced calcium uptake at the cathodal end of animal neurites cells has been demonstrated directly using fura-2 imaging (Bedlack *et al.*, 1992; Davenport and Kater, 1992). This is significant in the light of findings that the galvanotropic behaviour of *Aspergillus fumigatus* (Lever *et al.*, 1993), *Mycotypha africana* (Wittekindt *et al.*, 1989) and *Candida albicans* (Buchan, Gooday and Gow, unpublished) is Ca^{2+}-dependent. For *C. albicans*, calcium-channel blockers and calmodulin inhibitors also reduced galvanotropism (Buchan, Gooday and Gow, unpublished). These fungi ignore electrical fields in media containing very low calcium-ion concentrations. Therefore, the calcium-dependent galvanotropic behaviour of fungi would seem to highlight other findings (discussed above) that suggest an important role for calcium ions in the growth and polarity of fungi.

Two other observations suggest that electrical fields influence the membrane potential of fungi. Hyphae of *N. crassa* that grew for prolonged periods in electrical fields reoriented from an anodic to a perpendicular alignment (McGillivray and Gow, 1986) (Figure 13.7). This was interpreted as a response that would relieve the increasing magnitude of the perturbation of membrane potential at the cell poles of cells that were parallel to the field. Rhizoids of *A. macrogynus* grew anodally to fixed maximum length, suggesting that this length was set by the maximum hyperpolarization that could be tolerated by the extending rhizoid tips (De Silva *et al.*, 1992) (Figure 13.7).

Membrane potential effects may also provide a possible explanation for the pH dependency of cathodotropic fungi. It has been suggested that galvanotropism of anodotropic fungi may be due mainly to electrophoretic displacement of membrane proteins whereas cathodotropic hyphae may redirect growth in response to cathodal uptake of calcium ions. If anodotropic fungi employed voltage-insensitive calcium channels their anodal redistribution could also be explained on the basis of electric-field-induced Ca^{2+} uptake, this time at the anodic pole, since the hyperpolarization of the anode-facing cell membrane would increase calcium transport through non-voltage gated channels (Lever *et al.*, 1993). Electric-field regulation of calcium transport could in turn affect growth by influencing the organization of calcium-dependent processes such as actin polymerization (Onuma and Hui, 1988), the function of Ca^{2+}-binding proteins such as Cdc24 (Ohya *et al.*, 1986) or the activity of enzymes such as chitin synthase which have been shown to be influenced by calcium and calmodulin (Martinez-Cadena and Ruiz-Herrera, 1987). It is noteworthy that bacteria have also been shown to be galvanotropic suggesting that electrical field effects do not always modulate growth via changes in actin (Rajnicek *et al.*, 1994).

Electric field experiments are therefore serving as useful tools to explore the mechanistic basis of cell polarity in fungi.

13.3.5 Turgor pressure

It has long been assumed that the force that expands the apex of a hypha or the bud of a yeast is a positive internal turgor pressure. Because hyphae contain relatively high concentrations of solutes, water tends to enter the cell by osmosis and a positive turgor pressure is generated against the inner membrane face which is opposed by the non-elastic cell wall. As mentioned above, however, the apical wall is less rigid, perhaps due to the immature nature of the nascent chitin microfibrils which are not fully thickened (Burnett, 1979), because glucan and chitin are not covalently cross-linked, as they are in the subapical hypha and because there may be lytic enzymes softening the hyphal tip (section 13.2.3). Therefore, turgor pressure provides a global force that may be used to expand the cell where the wall is locally distensible. Apart from a role in expanding the apex, positive turgor pressure can also be used by fungi in long distance nutrient translocation (Eamus *et al.*, 1985; Jennings, 1987) and in the processes of sporangial emptying, spore discharge, the expansion of nematode capturing structures and penetration of plant cell surfaces by appressoria.

Despite the evident importance of turgor pressure to tip growth and assumed positive correlations between the magnitude of the turgor pressure and hyphal extension rate (Saunders and Trinci, 1979; Koch, 1982) (Chapter 15) few attempts have been made to measure it directly in fungi. Turgor pressure can be measured in at least three ways (Money, 1990). First, osmometry can be used to determine the difference in osmotic potential between the cytoplasm and growth medium and hence the turgor pressure. Internal osmotic potential is difficult to determine accurately since the mycelial cytoplasm must be extracted by freezing and thawing (Luard and Griffin, 1981; Luard 1982a; Woods and Duniway, 1986). Second, by exposing the cell to a range of solute concentrations the point of incipient plasmolysis can be determined – at this point the internal and external osmotic potentials can be presumed to be the same. Third, pressure probe micropipettes can be inserted into a hypha and the pressure that must be applied to prevent a low-viscosity oil droplet from being forced back up the pipette can be measured (Money, 1990; Money and Harold, 1992). The last of these techniques is the most direct method and was found to give consistently higher estimates for turgor pressures than those obtained by osmometry (Money, 1990).

Osmometry has been used to demonstrate correlations between hyphal extension rates and turgor pressure in *Serpula lacrymans* (Eamus and Jennings, 1986), *Chrysosporium fastidium, Eurotium amsteldami, Fusarium equiseti, Phellinus noxius, Phytophthora cinnamomi, Penicillium crysosporium* and *Xeromyces bisporus* (Luard and Griffin, 1981) and for *Phytophthora cryptogea* and *Fusarium moniliforme* (Woods and Duniway, 1986). Further experiments have shown that filamentous fungi respond to increases in external osmotic pressure by accumulating compatible solutes including K^+, glycerol, mannitol, erythritol and arabitol (Luard, 1982a,b,c; Pfyffer and Rast, 1988; Pfyffer and Rast, 1989).

The most detailed analyses of the relationship between hyphal extension and turgor pressure have been carried out on hyphae of *Achlya bisexualis* and *Saprolegnia ferax* (Money and Harold, 1992; Kaminskyj *et al.*, 1992). Interestingly significantly different conclusions were derived from these two closely related oomycetes. Money and Harold (1992) showed that the extension rate of the hyphae of *A. bisexualis* was not affected until turgor was reduced to less than a third of the normal level of 6–8 bars (0.6–0.8 MPa) and that a slow growth rate was still possible in the apparent absence of turgor. In the short term, superfusion of medium with an external osmoticum often reduced extension rate transiently but the hyphae then resumed growth at the same extension rate as before. Surprisingly, resumption of growth during superfusion treatment was probably due to softening of the apex, rather than regulation of turgor pressure. The evidence for this was that the pressure required to burst hyphal tips via a pressure probe, decreased in proportion to the external osmotic pressure, but the measured internal turgor was not affected by changes to the external osmoticum (at least in the short term). Thus *A. bisexualis* appears not to respond to changes in external osmotic pressure by controlling the concentration of internal compatible solutes (regulation of turgor), instead the plasticity of the wall is modulated to balance the force applied against it.

S. *ferax* hyphae required turgor for hyphal growth but the magnitude of the turgor pressure was not always found to correlate with the hyphal extension rate (Kaminskyj *et al.*, 1992). Proportionality of turgor and extension rate was only apparent at fast extension rates and high turgor. This correlation did not occur in media of high osmotic pressures where a more or less constant turgor pressure was found. Addition of the K^+ channel agonist tetrapentyl ammonium chloride to growing hyphae reduced markedly hyphal growth without affecting the turgor pressure measured by osmometry. At face value these studies suggest that despite the phylogenetic similarity of *S. ferax* and *A. bisexualis* and the fact that both organisms grow in aquatic environments of low external osmotic pressure, *S. ferax* appears to regulate turgor pressure whereas *A. bisexualis* does not.

13.4 CONCLUSIONS

Growth polarity is a complex multifactorial property of cells and is fundamental to the way in which almost all cells grow and develop. Fungi are excellent systems for exploring basic mechanisms of polarity since they exhibit well-defined patterns of polarized growth, because they are readily manipulated *in vitro* and because polarity mutants

are available which can illuminate underlying molecular mechanisms. For fungi it seems likely that actin–myosin based cytoskeleton plays a dominant role in bringing growth-supporting microvesicles to the cell surface and that the fusion of these vesicles seems to involve GTP-cycling and locally high calcium concentrations. These vesicles carry the enzymes that form the load-bearing polysaccharides that are acted against by turgor pressure and which ultimately form the scaffolding that determines the shape of the fungal cell.

REFERENCES

Adams, A.E.M. and Pringle, J.R. (1984) Relationship of actin and tubulin distribution to bud growth in wild-type and morphogenetic mutant *Saccharomyces cerevisiae*. *Journal of Cell Biology*, **98**, 934–45.

Adams, A.E.M., Johnson, D.I., Longnecker, R.M. *et al.* (1990) *CDC24* and *CDC43*, two additional genes involved in budding and the establishment of cell polarity in the yeast *Saccharomyces cerevisiae*. *Journal of Cell Biology*, **111**, 131–42.

Adams, A.E.M., Botstein, D. and Drubin, D.G. (1991) Requirement of yeast fimbrin for actin organization and morphogenesis in vivo. *Nature*, **354**, 404–8.

Alfa, C.E. and Hyams, J.S. (1990) Distribution of tubulin and actin through the cell division cycle of the fission yeast *Schizosaccharomyces japonicus* var. *versitalis*: comparison with *Schizosaccharomyces pombe*. *Journal of Cell Science*, **96**, 71–7.

Allen, E.D., Aiuto, R. and Sussman, A.S. (1980) Effects of cytochalasins on *Neurospora crassa* I. growth and ultrastructure. *Protoplasma*, **102**, 63–75.

Anderson, J. and Soll, D.R. (1986) Differences in actin localization during bud and hypha formation in the yeast *Candida albicans*. *Journal of General Microbiology*, **132**, 2035–47.

Bartnicki-Garcia, S. and Bracker, C.E. (1984) Unique properties of chitosomes, in *Microbial Cell Wall Synthesis and Autolysis*, (ed. C. Nombela) Elsevier Science Publishers, Amsterdam, pp. 101–12.

Bartnicki-Garcia, S. and Lippman, E. (1969) Fungal morphogenesis: cell wall construction in *Mucor rouxii*. *Science*, **165**, 302–4.

Bartnicki-Garcia, S., Bracker, C.E., Reyes, E. and Ruiz-Herrera, J. (1978) Isolation of chitosomes from taxonomically diverse fungi and synthesis of chitin microfibrils *in vitro*. *Experimental Mycology*, **2**, 173–92.

Bartnicki-Garcia, S., Hergert, F. and Gierz, G. (1989) Computer simulation of fungal morphogenesis and the mathematical basis for hyphal (tip) growth. *Protoplasma*, **153**, 46–57.

Barton, R. and Gull, K. (1988) Variation in the cytoplasmic microtubule organization and spindle length between the two forms of the dimorphic fungus *Candida albicans*. *Journal of Cell Science*, **91**, 211–20.

Bedlack, Jr, R.S., Wei, M-D. and Loew, L.M. (1992) Localized membrane potential depolarizations and localized calcium influx during electric field-guided neurite growth. *Neuron*, **9**, 393–403.

Bender, A. and Pringle, J.R. (1989) Multicopy suppression of the *cdc24* budding defect in yeast by *CDC42* and three newly identified genes including the ras-related gene *RSR1 Proceedings of the National Academy of Sciences of the USA*, **86**, 9976–80.

Bender, A. and Pringle, J.R. (1991) Use of a screen for synthetic lethal and multicopy suppresser mutants to identify two new genes involved in morphogenesis in *Saccharomyces cerevisiae*. *Molecular and Cellular Biology*, **11**, 1295–305.

Berger, F. and Brownlee, C. (1993) Ratio confocal imaging of free cytoplasmic calcium gradients in polarising and polarised *Fucus* zygotes. *Zygote*, **1**, 9–15.

Betina, V., Micekova, D. and Nemec, P. (1972) Antimicrobial properties of cytochalasins and their alteration of fungal morphology. *Journal of General Microbiology*, **71**, 343–9.

Bracker, C.E., Ruiz-Herrera, J. and Bartnicki-Garcia, S. (1976) Structure and transformation of chitin synthetase particles (chitosomes) during microfibril synthesis *in vitro*. *Proceedings of the National Academy of Sciences of the USA*, **73**, 4570–4.

Brennwald, P. and Novick, P. (1993) Interactions of three domains distinguishing the ras-related GTP-binding proteins Ypt1 and Sec4. *Nature*, **362**, 560–3.

Brownlee, C. and Wood, J.W. (1986) A gradient of cytoplasmic free calcium in growing rhizoid cells of *Fucus serratus*. *Nature*, **320**, 624–6.

Burnett, J.H. (1979) Aspects of the structure and growth of hyphal walls, in *Fungal Walls and Hyphal Growth*, (eds J.H. Burnett and A.P.J. Trinci) Cambridge University Press, Cambridge, pp. 1–15.

Chant, J. and Herskowitz, I. (1991) Genetic control of bud site selection in yeast by a set of gene products that constitute a morphogenetic pathway. *Cell*, **65**, 1203–12.

Chant, J., Corrado, K., Pringle, J.R. and Herskowitz, I. (1991) Yeast *BUD5*, encoding putative GDP–GTP exchange factor, is necessary for bud site selection and interacts with bud formation gene *BEMI*. *Cell*, **65**, 1213–24.

Chenevert, J., Corrado, K., Bender, A. *et al.* (1992) A yeast gene (*BEM1*) necessary for cell polarization whose gene product contains two SH3 domains. *Nature*, **356**, 77–9.

Cho, C., Harold, F.M. and Scheurs, W.J.A. (1991) Electric and ionic dimensions of apical growth in *Achlya* hyphae. *Experimental Mycology*, **15**, 34–43.

Collinge, A.J., and Trinci, A.P.J. (1974) Hyphal tips of wild type and spreading colonial mutants of *Neurospora crassa*. *Archives of Microbiology*, **99**, 353–68.

Costigan, C., Gehrung, S. and Synder, M. (1992) A synthetic lethal screen identifies *SLK1*, a novel protein kinase homolog implicated in yeast cell morphogenesis and cell growth. *Molecular and Cellular Biology*, **12**, 1162–78.

Crombie, T., Gow, N.A.R. and Gooday, G.W. (1990) Influence of applied electrical fields on yeast and hyphal growth of *Candida albicans*. *Journal of General Microbiology*, **136**, 311–17.

Davenport, R.W. and Kater, S.B. (1992) Local increases in intracellular calcium elicit local filopodial responses in helisoma neuronal growth cones. *Neuron*, **9**, 405–416.

De Silva, L.R., Youatt, J., Gooday, G.W. and Gow, N.A.R. (1992) Inwardly directed ionic currents of *Allomyces macrogynus* and other water moulds indicate sites of proton-driven nutrient transport but are incidental to tip growth. *Mycological Research*, **96**, 925–31.

De Vries, S.C. and Wessels, J.G.H. (1982) Polarized outgrowth of hyphae by constant electrical fields during reversion of *Schizophyllum commune* protoplasts. *Experimental Mycology*, **6**, 95–8.

Dicker, J.W. and Turian, G. (1990) Calcium deficiencies and apical hyperbranching in wild-type and the 'frost' and 'spray' morphological mutants of *Neurospora crassa*. *Journal of General Microbiology*, **136**, 1413–20.

Drubin, D.G. (1991) Development of cell polarity in budding yeast. *Cell*, **65**, 1093–6.

Drubin, D.G., Miller, K.G. and Botstein, D. (1988) Yeast actin-binding proteins: evidence for a role in morphogenesis. *Journal of Cell Biology*, **107**, 2551–61.

Drubin, D.G., Mullholland, J., Zhu, Z. and Botstein, D. (1990) Homology of a yeast actin-binding protein to signal trasnduction proteins and myosin-I. *Nature*, **343**, 288–90.

Eamus, D. and Jennings, D.H. (1986) Turgor and fungal growth: studies on water relations of mycelia of *Serpula lacrimans* and *Phallus impudicus*. *Transactions of the British Mycological Society*, **86**, 527–35.

Eamus, D., Thompson, W., Cairney, J.W.G. and Jennings, D.H. (1985) Internal structure and hydraulic conductivity of basidiomycete translocating organs. *Journal of Experimental Botany*, **36**, 1110–16.

Enos, A.P. and Morris, N.R. (1990) Mutation of a gene that encodes a kinesin-like protein blocks nuclear division in *A. nidulans*. *Cell*, **60**, 1019–27.

Finegold, A.A., Johnson, D.I., Fransworth, C.C. *et al.* (1991) Protein geranylgeranylytransferase of *Saccharomyces cerevisiae* is specific for Cys-Xaa-Xaa-Leu motif proteins and requires the *CDC43* gene product but not the *DPR1* gene product. *Proceedings of the National Academy of Sciences of the USA*, **88**, 4448–52.

Flescher, E.G., Madden, K. and Snyder, M. (1993) Components required for cytokinesis are important for bud site selection. *Journal of Cell Biology*, **122**, 373–86.

Garrill, A., Lew, R.R. and Heath, I.B. (1992) Stretch-activated Ca^{2+} and Ca^{2+}-activated K^+ channels in the hyphal tip plasma membrane of the oomycete *Saprolegnia ferax*. *Journal of Cell Science*, **101**, 721–30.

Garrill, A, Jackson, S.L., Lew, R.R. and Heath, I.B. (1993) Ion channel activity and tip growth: tip localised stretch-activated channels generate an essential Ca^{2+} gradient in the oomycete *Saprolegnia ferax. European Journal of Cell Biology*, **60**, 358–65.

Gehrung, S. and Snyder, M. (1990) The *SPA2* gene of *Saccharomyces cerevisiae* is important for pheromone-induced morphogenesis and efficient mating. *Journal of Cell Biology*, **111**, 1451–64.

Girbardt, M. (1969) Die ultrastruktur der apikalregion von pilzhyphen. *Protoplasma*, **67**, 413–41.

Gooday, G.W. (1971) An autoradiographic study of hyphal growth of some fungi. *Journal of General Microbiology*, **67**, 125–33.

Gooday, G.W. (1979) Chitin synthesis and differentiation in *Coprinus cinereus*, in *Fungal Walls and Hyphal Growth* (eds J.H. Burnett and A.P.J. Trinci), Cambridge University Press, Cambridge, pp. 203–23.

Gooday, G.W. (1993) Cell envelope diversity and dynamics in yeasts and filamentous fungi. *Journal of Applied Bacteriology Symposium Supplement*, **74**, 12S–20S.

Gooday, G.W. and Trinci, A.P.J. (1980) Wall structure and biosynthesis in fungi, in *The Eukaryotic Microbial Cell, Society for General Microbiology Symposium, vol.30* (eds G.W. Gooday, D. Lloyd and A.P.J. Trinci), Cambridge University Press, Cambridge, pp. 207–51.

Gow, N.A.R. (1984) Transhyphal electrical currents in fungi. *Journal of General Microbiology*, **130**, 3313–18.

Gow, N.A.R. (1987) Polarity and branching in fungi induced by electrical fields, in *Spatial Organization in Eukaryotic Microbes* (Volume 23 of Special publications of the Society for General Microbiology) (eds R.K. Poole and A.P.J. Trinci). IRL Press, Oxford, pp. 25–41.

Gow, N.A.R. (1989a) Control of the extension of the hyphal apex. *Current Topics in Medical Mycology*, **3**, 109–52.

Gow, N.A.R. (1989b) The circulating ionic currents of microorganisms. *Advances in Microbial Physiology*, **30**, 89–123.

Gow, N.A.R. and Gooday, G.W. (1987) Effects of antheridiol on growth, branching and electrical currents of *Achlya ambisexualis. Journal of General Microbiology*, **133**, 3531–35.

Gow, N.A.R., Kropf, D.L. and Harold, F.M. (1984) Growing hyphae of *Achlya bisexualis* generate a longitudinal pH gradient in the surrounding medium. *Journal of General Microbiology*, **130**, 2967–74.

Gow, N.A.R., Miller, P.F.P. and Gooday, G.W. (1992) Life at the apex: growth of the hyphal tip. *Journal of Chemical Technology and Biotechnology*, **56**, 217–19.

Grove, S.M. and Bracker, C.E. (1970) Protoplasmic organization of hyphal tips among fungi: vesicles and Spitzenkorper. *Journal of Bacteriology*, **104**, 989–1009.

Grove, S.M., Bracker, C.E. and Morré, D.J. (1970) An ultrastructural basis for hyphal tip growth in *Pythium ultimum. American Journal of Botany*, **57**, 245–66.

Harold, F.M. (1990) To shape a cell: an inquiry into the causes of morphogenesis of microorgansims. *Microbiological Reviews*, **54**, 381–431.

Heath, I.B. (1987) Preservation of a labile cortical array of actin microfilaments in growing hyphal tips of the fungus *Saprolegnia ferax. European Journal of Cell Biology*, **44**, 10–16.

Heath, I.B. (1990) The roles of actin in tip growth in fungi. *International Review of Cytology*, **123**, 95–127.

Heath, I.B. and Kaminskyj, S.G.W. (1989) The organisation of tip-growth related organelles and microtubules revealed by quantitative analysis of freeze-substituted oomycete hyphae. *Journal of Cell Science*, **93**, 41–52.

Heath, I.B., Rethoret, K. Arsenault, A.L. and Ottensmeyer, F.P. (1985) Improved preservation of the form and contents of wall vesicles and the Golgi apparatus in freeze substituted hyphae of *Saprolegnia. Protoplasma*, **128**, 81–93.

Herr, F.B. and Heath, M.C. (1982) The effects of antimicrotubule agents on organelle positioning in the cowpea rust fungus, *Uromyces phaseoli* var. *vignae. Experimental Mycology*, **6**, 15–24.

Hoch, H.C. and Staples, R.C. (1985) The microtubule cytoskeleton in hyphae of *Uromyces phaseoli* germlings: its relationship to the region of nucleation and to the F-actin cytoskeleton. *Protoplasma*, **124**, 112–22.

Hoch, H.C., Staples, R.C., Whitehead, B. *et al.*, (1987) Signaling for growth orientation and cell differentiation by surface topography in *Uromyces. Science*, **235**, 1659–62.

Horwitz, B.A., Weisenseel, M.H., Dorm, A. and Gressel, J. (1984) Electric currents around growing *Trichoderma* hyphae: before and after photoinduction of conidiation. *Plant Physiology*, **74**, 912–16.

Howard, R.J. (1981) Ultrastructural analysis of hyphal tip cell growth in fungi: Spitzenkörper, cytoskeleton and endomembranes after freeze-substitution. *Journal of Cell Science*, **48**, 89–103.

Howard, R.J and Aist, J.R. (1977) Effects of MBC on hyphal tip organisation, growth and mitosis of *Fusarium accuminatum*, and their antagonism by D_2O. *Protoplasma*, **92**, 195–210.

Howard, R.J and Aist, J.R. (1980) Cytoplasmic microtubules and fungal morphogenesis: ultrastructural effects of methyl benzimidazole-2-ylcarbamate determined by freeze-substitution of hyphal tip cells. *Journal of Cell Biology*, **87**, 55–64.

Howard, R.J. and O'Donnell, K. (1987) Freeze substitution of fungi for cytological analysis. *Experimental Mycology*, **11**, 250–69.

Huffaker, T.C., Thomas, J.H. and Botstein, D. (1988) Diverse effects of β-tubulin mutations on microtubule formation and function. *Journal of Cell Biology*, **106**, 1997–2010.

Humphreys, A.M. and Gooday, G.W. (1984). Properties of chitinase activities from *Mucor mucedo*: evidence for a membrane-bound zymogenic form. *Journal of General Microbiology*, **130**, 1359–66.

Jackson, S.L. and Heath, I.B. (1989) Effects of exogenous calcium ions on tip growth, intracellular Ca^{2+} concentration, and actin arrays in hyphae of the fungus *Saprolegnia ferax. Experimental Mycology*, **13**, 1–12.

Jackson, S.L. and Heath, I.B. (1990a) Visualization of actin arrays in growing hyphae of the fungus *Saprolegnia ferax*. *Protoplasma*, **154**, 66–70.

Jackson, S.L. and Heath, I.B. (1990b) Evidence that actin reinforces the extensible hyphal apex of the oomycete *Saprolegnia ferax*. *Protoplasma*, **157**, 144–53.

Jackson, S.L. and Heath, I.B. (1992) UV microirradiations elicit Ca^{2+} dependent apex-directed cytoplasmic contractions in hyphae. *Protoplasma*, **170**, 46–52.

Jackson, S.L. and Heath, I.B. (1993a) The dynamic behavior of cytoplsamic F-actin in growing hyphae. *Protoplasma*, **173**, 23–34.

Jackson, S.L. and Heath, I.B. (1993b) Roles of calcium ions in hyphal tip growth. *Microbiological Reviews*, **57**, 367–82.

Jacobs, C.W., Adams, A.E.M., Szaniszlo, P.J. and Pringle, J.R. (1988) Functions of microtubules in the *Saccharomyces cerevisiae* cell cycle. *Journal of Cell Biology*, **107**, 1409–26.

Jaffe, L.F. and Nuccitelli, R. (1974) An ultrasensitive vibrating probe for measuring extracellular electrical currents. *Journal of Cell Biology*, **63**, 614–28.

Jaffe, L.F. (1977) Electrophoresis along cell membranes. *Nature*, **265**, 600–2.

Jennings, D.H. (1987) Translocation of solutes in fungi. *Biological Reviews*, **62**, 215–43.

Johnson, B.F. (1965) Autoradiographic analysis of regional wall growth of yeast, *Schizosaccharomyces pombe*. *Experimental Cell Research* **39**, 613–24.

Johnson, B.F. and Gibson, E.J. (1966) Autroradiographic analysis of regional wall growth of yeasts III, *Saccharomyces cerevisiae*. *Experimental Cell Research*, **41**, 580–91.

Johnston, G.C., Prendergast, J.A. and Singer, R.A. (1991) The *Saccharomyces cerevisiae MYO2* gene encodes an essential myosin for vectorial transport of vesicles. *Journal of Cell Biology*, **113**, 539–51.

Kaminskyj, S.G.W., Garrill, A. and Heath, I.B. (1992) The relation between turgor and tip growth in *Saprolegnia ferax*: turgor is necessary, but not sufficient to explain apical extension rates. *Experimental Mycology*, **16**, 64–75.

Kilmartin, J.V. and Adams, A.E.M. (1984) Structural rearrangments of tubulin and actin during the cell cycle of the yeast *Saccharomyces*. *Journal of Cell Biology*, **98**, 922–33.

Koch. A.L. (1982) The shape of hyphal tips of fungi. *Journal of General Microbiology*, **128**, 947–51.

Knight, H., Trewavas, A.J. and Read, N.D. (1993) Confocal microscopy of living fungal hyphae microinjected with Ca^{2+}-sensitive fluorescent dyes. *Mycological Research*, **97**, 1505–15.

Kropf, D.L. (1986) Electrophysiological studies of *Achlya* hyphae: ionic currents studies by intracellular recording potentials. *Journal of Cell Biology*, **102**, 1209–16.

Kropf, D.L. and Quatrano, R.S. (1987) Localization of membrane-associated calcium development of fucoid algae using chlorotetracycline. *Planta*, **171**, 158–70.

Kropf, D.L., Lupa, M.D., Caldwell, J.C. and Harold, F.M. (1983) Cell polarity: endogenous ion currents precede and predict branching in the water mold *Achlya*. *Science*, **220**, 1385–87.

Kropf, D.L., Caldwell, J.C., Gow, N.A.R. and Harold, F.M. (1984) Transcellular ion currents in the water mould *Achlya*. Amino acid symport as a mechanism of current entry. *Journal of Cell Biology*, **99**, 486–96.

Kwon, Y.H., Hoch, H.C. and Staples, R.C. (1991) Cytoskeletal organization in *Uromyces* urediospore germling apices during appressorium formation. *Protoplasma*, **165**, 37–50.

Lever, M.,C., Robertson, B.E.M., Buchan, A.D.B. *et al* (1993) pH and Ca^{2+} dependent galvanotropism of filamentous fungi: implications and mechanisms. *Mycological Research*, **98**, 301–6.

Lillie, S.H. and Brown, S.S. (1992) Suppression of a myosin defect by a kinesin-related gene. *Nature*, **256**, 358–61.

Liu, H. and Bretscher, A. (1989) Disruption of the single tropomyosin gene in yeast results in the disappearance of actin cables from the cytoskeleton. *Cell*, **57**, 233–42.

Luard, E. (1982a) Accumulation of intracellular solutes by two filamentous fungi in response to growth at low steady state osmotic potential. *Journal of General Microbiology*, **128**, 2563–74.

Luard, E. (1982b) Effect of osmotic shock on some intracellular solutes in two filamentous fungi. *Journal of General Microbiology*, **128**, 2575–81.

Luard, E. (1982c) Growth and accumulation of solutes by *Phytophthora cinnamomi* and other lower fungi in response to changes in external osmotic potential. *Journal of General Microbiology*, **128**, 2583–90.

Luard, E. and Griffin, D.M. (1981) Effect of water potential on fungal growth and turgor. *Transactions of the British Mycological Society*, **76**, 33–40.

Madden, K., Costigan, C. and Snyder, M. (1992) Cell polarity and morphogenesis in *Saccharomyces cerevisiae*. *Trends in Cell Biology*, **2**, 22–9.

Markham, P., Robson, G.D., Bainbridge, B.W. and Trinci, A.P.J. (1993) Choline: its role in the growth of filamentous fungi and the regulation of mycelial morphology. *FEMS Microbiology Reviews*, **104**, 287–300.

Marks, J. and Hyams, J.S. (1985) Localization of F-actin through the cell division cycle of *Schizosaccharomyces pombe*. *European Journal of Cell Biology*, **39**, 27–32.

Martinez-Cadena, G. and Ruiz-Herrera, J. (1987) Activation of chitin synthetase from *Phycomyces blakesleeanus* by calcium and calmodulin. *Archives of Microbiology*, **148**, 280–5.

Matsuzaki, F., Matsumoto, S. and Yahara, I. (1988) Truncation of the carboxy-terminal domain of yeast β-tubulin causes temperature-sensitive growth and hypersensitivity to antimitotic drugs. *Journal of Cell Biology*, **107**, 14427–35.

McGillivray, A.M. and Gow, N.A.R. (1986) Applied electrical fields polarize the growth of mycelial fungi. *Journal of General Microbiology*, **132**, 2515–25.

McGillivray, A.M. and Gow, N.A.R. (1987) The transhyphal electrical current of *Neurospora crassa* is carried principally by protons. *Journal of General Microbiology*, **133**, 1875–81.

McKerracher, L.J. and Heath, I.B. (1985) Microtubules around migrating nuclei in conventionally-fixed and freeze-substituted cells. *Protoplasma*, **125**, 162–72.

McKerracher, L.J. and Heath, I.B. (1986a) Fungal nuclear behaviour analysed by ultraviolet microbeam irradiation. *Cell Motility and the Cytoskeleton*, **6**, 35–47.

McKerracher, L.J. and Heath, I.B. (1986b) Polarized cytoplasmic movement and inhibition of saltations induced by calcium-mediated effects of microbeams in fungal hyphae. *Cell Motility and the Cytoskeleton*, **6**, 136–45.

McKerracher, L.J. and Heath, I.B. (1987) Cytoplasmic migration and intracellular movements of organelles during tip growth of fungal hyphae. *Experimental Mycology*, **11**, 79–100.

Merson-Davies, L. and Odds, F.C. (1992) Expansion of the *Candida albicans* cell envelope in different morphological forms of the fungus. *Journal of General Microbiology*, **138**, 461–6.

Miller, D.R., Callahan, D.A., Gross, D.J. and Hepler, P. (1992) Free Ca^{2+} gradient in growing pollen tubes of *Lilium*. *Journal of Cell Science*, **101**, 7–12.

Mitchison, J.M. and Nurse, P. (1985) Growth in cell length in the fission yeast *Schizosaccharomyces pombe*. *Journal of Cell Science*, **75**, 357–76.

Money, N.P. (1990) Measurement of turgor pressure. *Experimental Mycology*, **14**, 416–25.

Money, N.P. and Harold, F.M. (1992) Extension of the water mold *Achlya*: interplay of turgor and wall strength. *Proceedings of the National Academy of Sciences of the USA*, **89**, 4245–9.

Nelson, W.J. (1992) Regulation of cell surface polarity from bacteria to mammals. *Science*, **258**, 948–55.

Novick, P. and Botstein, D. (1985) Phenotypic analysis of temperature-sensitive yeast actin mutants. *Cell*, **40**, 405–16.

Novick, P. and Brennwald, P. (1993) Friends and family: the role of the rab GTPases in vesicular traffic. *Cell*, **75**, 597–601.

Nuccitelli, R. (1990) Vibrating probe technique for studies of ion transport, in *Noninvasive Techniques in Cell Biology*, (eds J.K. Foskett and S. Grinstein), Wiley-Liss, New York, pp. 273–310.

Oakley, B.R. and Morris, N.R. (1980) Nuclear movement is β-tubulin-dependent in *Aspergillus nidulans*. *Cell*, **19**, 255–62.

Oakley, B.R. and Rinehart, J.E. (1985) Mitochondria and nuclei move by different mechanisms in *Aspergillus nidulans*. *Journal of Cell Biology*, **101**, 2392–7.

Ohya, Y., Miyamoto, S., Ohsumi, Y. and Anraku, Y. (1986) Calcium-sensitive *cls4* mutants of *Saccharomyces cerevisiae* with a defect in bud formation. *Journal of Bacteriology*, **165**, 28–33.

Ohya, Y., Goebel, M., Goodman, L.E. *et al.* (1991) Yeast *CAL1* is a structural and functional homologue to the *DPR1* (RAM) gene involved in ras processing. *Journal of Biological Chemistry*, **266**, 12356–60.

Onuma, E.K. and Hui, S.-W. (1988) Electric-field directed cell shape changes, displacement, and cytoskeletal reorganization are calcium dependent. *Journal of Cell Biology*, **106**, 2067–75.

Orida, N. and Poo, M.-M. (1978) Electrophoretic movement and localization of acetylcholine receptors in the embryonic muscle cell membrane. *Nature*, **275**, 31–5.

Pfyffer, G.E. and Rast, D.M. (1988) The polyol pattern of fungi as influenced by the carbohydrate nutrient source. *New Phytologist*, **109**, 321–6.

Pfyffer, G.E. and Rast, D.M. (1989) Accumulation of acyclic polyols and trehalose as related to growth form and carbohydrate source in the dimorphic fungi *Mucor rouxii* and *Candida albicans*. *Mycopathologia*, **105**, 25–33.

Poo, M-M. (1981) *In situ* electrophoresis of membrane components. *Annual Review of Biophysics and Bioengineering*, **10**, 245–76.

Powers, S., Gonzales, E., Christensn, T. *et al.* (1991) Functional cloning of *BUD5*, a *CDC25*-related gene from *S. cerevisiae* that can suppress a dominant-negative *RAS2* mutant. *Cell*, **65**, 1225–31.

Rajnicek, A.M., McCaig, C.D. and Gow, N.A.R. (1994) Electric fields induce curved growth of *Enterobacter clocacae*, *Escherichia coli* and *Bacillus subtilis* cells: impliations for mechanisms of galvanotropism and bacterial growth. *Journal of Bacteriology*, **176**, 702–13.

Rast, D.R., Horsch, M., Furter, R. and Gooday, G.W. (1991) A complex chitinolytic system in exponentially growing mycelium of *Mucor rouxii*: properties and function. *Journal of General Microbiology*, **137**, 2797–810.

Read, N.D., Allan, W.T.G., Knight, H. *et al.* (1992) Imaging and measurement of cytosolic free calcium in plant and fungal cells. *Journal of Microscopy*, **166**, 57–86.

Reinhardt, M.O. (1892) Das Wachstum der Pilzhphen. Ein Beitrag zur Kenntnis des Flächenwachstums vegetalischer Zellmembranen. *Jahrbucher für Wissenschaftliche Botanik*, **23**, 479–566.

Reiss, H-D. and Herth, W. (1979) Calcium gradients in tip growing plants visualized by chlorotetracycline-fluorescence. *Planta*, **146**, 615–21.

Reissig, J.L. and Kinney, S.G. (1983) Calcium as a branching signal in *Neurospora crassa*. *Journal of Bacteriology*, **154**, 1397–402.

Roberson, R.W. (1992) The actin cytoskeleton in hyphal cells of *Sclerotium rolfsii*. *Mycologia*, **84**, 41–51.

Roberson, R.W. and Fuller, M.S. (1988) Ultrastructural aspects of the tip of *Sclerotium rolfsii*. preserved by freeze substitution. *Protoplasma*, **146**, 143–9.

Robinson, K.R. (1985) The responses of cells to electrical fields: a review. *Journal of Cell Biology*, **101**, 2023–7.

Robson, G.D., Wiebe, M.G. and Trinci, A.P.J. (1991a) Involvement of Ca^{2+} in the regulation of hyphal extension and branching in *Fusarium graminearum* A3/5. *Experimental Mycology*, **15**, 263–72.

Robson, G.D., Wiebe, M.G. and Trinci, A.P.J. (1991b) Low calcium concentrations induce increased branching in *Fusarium graminearum*. *Mycological Research*, **95**, 561–5.

Robson, G.D., Wiebe, M.G. and Trinci, A.P.J. (1991c) Exogenous cAMP and cGMP modulate branching in *Fusarium graminearum*. *Journal of General Microbiology*, **137**, 963–9.

Roncal, T., Ugalde, U.O. and Irastorza, A. (1993) Calcium-induced conidiation in *Penicillium cyclopium*: calcium triggers cytosolic alkalinization at the hyphal tip. *Journal of Bacteriology*, **175**, 879–86.

Ruiz-Herrera, J., Valenzuela, C., Martinez-Cadena, G. and Obregon, A. (1989) Alterations in the vesicular pattern of and wall growth of *Phycomyces* induced by the calcium ionophore A23197. *Protoplasma*, **148**, 15–25.

Saunders, P.T. and Trinci, A.P.J. (1979) Determination of tip shape in fungal hyphae. *Journal of General Microbiology*, **110**, 469–73.

Schmid, J. and Harold, F.M. (1988) Dual role for calcium ions in apical growth of *Neurospora crassa*. *Journal of General Microbiology*, **134**, 2623–31.

Schreurs, W.J.A. and Harold, F.M. (1988) Transcellular proton current in *Achlya bisexualis* hyphae: relationship to polarized growth. *Proceedings of the National Academy of Sciences of the USA*, **85**, 1534–8.

Sietsma, J.H. and Wessels, J.G.H. (1979) Evidence for covalent linkages between chitin and β-glucan in a fungal wall. *Journal of General Microbiology*, **114**, 99–108.

Sietsma, J.H. and Wessels, J.G.H. (1981) Solubility of (1→3)-β-D-(1→6)-β-D-glucan in fungal walls: importance of presumed linkage between glucan and chitin. *Journal of General Microbiology*, **125**, 209–21.

Sietsma, J.H., Sonnenberg, A.S.M. and Wessels, J.H. (1985) Localization by autoradiography of (1→3)-β-D-(1→6)-β-linkages in a wall glucan during hyphal growth of *Schizophyllum commune*. *Journal of General Microbiology*, **131**, 1331–7.

Slayman, C.L. and Slayman, C.W. (1962) Measurements of membrane potential in *Neurospora*. *Science*, **136**, 876–7.

Sonnenberg, A.S.M., Sietsma, J.H. and Wessels, J.G.H. (1982) Biosynthesis of alkali-insoluble cell-wall glucan in *Schizophyllum commune* protoplasts. *Journal of General Microbiology*, **128**, 2667–74.

Sonnenberg, A.S.M., Sietsma, J.H. and Wessels, J.G.H. (1985). Spatial and temporal differences in the synthesis of (1→3)-β-D-(1→6)-β linkages in a wall glucan of *Schizophyllum commune*. *Experimental Mycology*, **9**, 141–8.

Snyder, M. (1989) The *SPA2* protein of yeast localizes to sites of cell growth. *Journal of Cell Biology*, **108**, 1419–29.

Snyder, M., Gehrung, S. and Page, B.D. (1991) Studies concerning the temporal and genetic control of cell polarity in *Saccharomyces cerevisiae*. *Journal of Cell Biology*, **114**, 515–32.

Staebell, M and Soll, D.R. (1985) Temporal and spatial differences in cell wall expansion during bud and mycelium formation in *Candida albicans*. *Journal of General Microbiology*, **131**, 1467–80.

Stollberg, J. and Fraser, S.E. (1988) Acetylcholine receptors and concanavalin A-binding sites on cultured *Xenopus* muscle cells: electrophoresis, diffusion and aggregation. *Journal of Cell Biology*, **107**, 1397–408.

Stump, R.F., Robinson, K.R., Harold, R.L. and Harold, F.M. (1980) Endogenous electrical currents in the water mould *Blastocladiella emersonii* during growth and sporulation. *Proceedings of the National Academy of Sciences of the USA*, **77**, 6673–7.

Szaniszlo, P.J., Mendoza, L. and Karuppayil, S.M. (1993) Clues about chromoblastomycotic and other dermatiaceous fungal pathogens based on *Wangiella* as a model, in *Dimorphic Fungi in Biology and Medicine*, (eds H. Vanden Bossche, F.C. Odds and D. Kerridge) Plenum Press, New York, pp. 241–55.

Takeuchi, Y., Schmid, J., Caldwell, J.H. and Harold, F.M. (1988) Transcellular ion currents and extension of *Neurospora crassa* hyphae. *Journal of Membrane Biology*, **101**, 33–41.

Temperli, E., Roos, U-P. and Hohl, H.R. (1991) Germ tube growth and the microtubule cytoskeleton in *Phytophthora infestans*: effects of antagonists of hyphal growth, microtubule inhibitors, and ionophores. *Mycological Research*, **95**, 611–17.

That, T.C.C.-T, Rossier, C., Barja, F. *et al.* (1988) Induction of multiple germ tubes in *Neurospora crassa* by antimicrotubulin agents. *European Journal of Cell Biology*, **46**, 68–79.

Thomas, des S. and Mullins, J.T. (1967) Role of enzymatic wall softening in plant morphogenesis. *Science*, **156**, 84–5.

Thompson-Coffe, C. and Zickler, D. (1992) Three microtubule-organising centers are required for ascus growth and sporulation in the fungus *Sordaria macrospora*. *Cell Motility and the Cytoskeleton*, **22**, 257–73.

Tsien, R.W. and Tsien, R.Y. (1990) Calcium channels, stores, and oscillations. *Annual Review of Cell Biology*, **6**, 715–60.

Van Laere, A.J. (1988) Effect of electrical fields on polar growth of *Phycomyces blakesleeanus*. *FEMS Microbiological Letters*, **49**, 111–16.

Vermeulen, C.A. and Wessels, J.G.H. (1984) Ultrastructural differences between wall apices of growing and non-growing hyphae of *Schizophyllum commune*. *Protoplasma*, **120**, 123–31.

Vermeulen, C.A. and Wessels, J.G.H. (1986) Chitin biosynthesis by a fungal membrane preparation. Evidence for a transient non-crystalline state of chitin. *European Journal of Biochemistry*, **158**, 411–15.

Wessels, J.G.H. (1984) Apical hyphal wall extension: do lytic enzymes play a role?, in *Microbial Cell Wall Synthesis and Autolysis* (ed. C. Nombela) Elsevier, Amsterdam, pp. 31–42.

Wessels, J.G.H. (1986) Cell wall synthesis in apical hyphal growth. *International Review of Cytology*, **104**, 37–79.

Wessels, J.G.H. (1993) Wall growth, protein excretion and morphogenesis in fungi. *New Phytologist*, **123**, 387–413.

Wessels, J.G.H., Sietsma, J.H. and Sonnenberg, A.S.M. (1983) Wall synthesis and assembly during hyphal morphogenesis in *Schizophyllum commune*. *Journal of General Microbiology*, **129**, 1607–16.

Wiebe, M.G., Robson, G.D., and Trinci, A.P.J. (1992) Evidence for the independent regulation of hyphal extension and branch initiation in *Fusarium graminearum* A3/5. *FEMS Microbiology Letters*, **90**, 179–84.

Wittekindt, E., Lamprecht, I. and Kraepelin, G. (1989) DC electrical fields induce polarization effects in the dimorphic fungus *Mycotypha africana*. *Endocytobiosis and Cell Research*, **6**, 41–56.

Woods and Duniway, J.M. (1986) Some effects of water potential on turgor, and respiration of *Phytophthora cryptogea* and *Fusarium moniliforme*. *Phytopathology*, **76**, 1248–54.

Yaun, S. and Heath, I.B. (1991) Chlortetracycline staining patterns of growing hyphal tips of the oomycete *Saprolegnia ferax*. *Experimental Mycology*, **15**, 91–102.

Yokoyama, K., Kaji, H., Nishimura, K and Miyaji, M. (1990) The role of microfilaments and microtubules in apical growth and dimorphism of *Candida albicans*. *Journal of General Microbiology*, **136**, 1067–75.

Youatt, J. (1993) Calcium and microorganisms. *Critical Reviews in Microbiology*, **19**, 83–97.

Youatt, J. and McKinnon, I. (1993) Manganese (Mn^{2+}) reverses the inhibition by EGTA of the growth of fungi. *Microbios*, **74**, 181–6.

Youatt, J., Gow, N.A.R. and Gooday, G.W. (1988) Bioelectric and biosynthetic aspects of cell polarity in *Allomyces macrogynus*. *Protoplasma*, **146**, 118–26.

Zhou, X-L., Stumpf, A., Hoch, H.C. and Kung, C. (1991) A mechanosensitive channel in whole cells and in membrane patches of the fungus *Uromyces*. *Science*, **253**, 1415–17.

KINETICS OF FILAMENTOUS GROWTH AND BRANCHING

14

J.I. Prosser
Department of Molecular and Cell Biology, Marischal College, University of Aberdeen, Aberdeen, UK

14.1 INTRODUCTION

Both Chapters 14 and 15 are concerned with quantification of fungal growth and morphology. This is frequently necessary for purely practical or technical reasons. For example, many physiological studies require information on the specific rate of growth or biomass yield. Similarly, large-scale industrial fermentations involving filamentous fungi can only be operated and controlled efficiently on the basis of quantitative information on, for example, changes in substrate, biomass and product concentrations. The ability to predict such changes accurately is a reflection of our understanding of the physiological mechanisms controlling growth. An elementary knowledge of microbial growth would enable the prediction that, during growth, biomass concentration will increase with time and substrate concentration will decrease. Accurate description and prediction of the rates of change in these concentrations, however, requires detailed knowledge of the mechanisms involved in transport of substrate into the cell, its conversion into biomass and the ways in which other factors, such as temperature, pH, oxygen concentration and accumulation of inhibitors, control these processes.

14.1.1 Definitions

Growth may be defined as the orderly increase in cell components leading to an increase in biomass. Chapters 14 and 15 distinguish between kinetics and mathematical modelling of growth. This distinction is artificial and is made for convenience. A mathematical model is a description of a process or system in a mathematical form. Mathematical equations describing growth kinetics are concerned with temporal changes in biomass, or other features of growth and are therefore, themselves, mathematical models.

Three distinctions may be drawn between kinetics and other forms of mathematical models of fungal growth. The first, and most important, concerns the function of the model. Kinetic equations are usually used to describe changes in properties of a culture in the simplest and most convenient form. Consequently they are frequently empirical in nature and may be based solely on experimental data. They need not be based on any knowledge of underlying growth mechanisms, although such a knowledge will increase their reliability. The models to be considered in Chapter 15 generally function as quantitative hypotheses, based on sets of assumptions regarding a particular aspect of fungal growth. The predictions of the model then represent a basis for experimental work. Agreement between experimental and predicted data will provide support for the assumptions underlying the model, whereas discrepancies may lead to modification or rejection of these assumptions. Experimental work therefore follows this form of mechanistic modelling, and it is the starting point for more descriptive, empirical modelling. This distinction leads to differences in complexity. Kinetic equations are generally relatively simple, consisting frequently of a single differential or algebraic equation. Mechanistic models are more complex to enable quantitative consideration of underlying assumptions. They usually consist of sets of differential equations and are frequently insoluble analytically and require computer simulations for generation of predictions. Finally, the range of models described in Chapter 15 is much

The Growing Fungus. Edited by Neil A.R. Gow and Geoffrey M. Gadd. Published in 1994 by Chapman & Hall, London. ISBN 0 412 46600 7

greater than that of growth kinetics, considering, for example, morphology and differentiation.

14.1.2 Measurement of growth

Many of the concepts and kinetic equations regarding microbial growth are derived from studies of unicellular microorganisms. As such they concentrate on population growth, which can be measured most easily, and rarely consider individuals. The growth form of filamentous fungi is more complex. For example, extension of individual hyphae is localized at the tip, whereas biomass synthesis supporting that growth may take place throughout the mycelium. Population studies must consider this heterogeneity, which will depend to some extent on the growth medium. In liquid culture heterogeneity will increase if pellet formation or differentiation occurs. On solid media, heterogeneity will increase towards the centre of colonies. Fungal growth is also considered at different levels, ranging from that of individual hyphae, through mycelia to populations of mycelia and each level requires different techniques for assessment.

(a) Growth of individual hyphae

Growth of individual hyphae is measured microscopically on solid medium as an increase in length. Hyphae are usually grown on the surface of a Cellophane membrane, overlaying the agar, to maintain hyphae in a single plane of focus. Traditionally measurements were carried out *in situ* using eyepiece micrometers or projected images, or by time-lapse photography. Analysis is now greatly facilitated by the increased availability of image analysis systems which enable automated measurement of hyphal lengths and subsequent kinetic analysis of data (Adams and Thomas, 1988; Wiebe and Trinci, 1990; Gray and Morris, 1992).

(b) Growth of mycelia

Image analysis techniques are also now used to study the growth of fungal mycelia both in liquid culture and on solid media. In liquid culture this involves measurement of hyphal growth unit length (section 14.3) and interbranch distances for populations of young mycelia, whereas on solid media, changes in total hyphal length are also measured. If biomass density and hyphal diameter are assumed to be constant, measurements of total hyphal length provide an estimate of biomass.

(c) Population growth

Population growth is measured most directly by measurement of biomass dry weight. This requires separation of biomass from the growth medium which is relatively straightforward in liquid culture but is slow, tedious and requires relatively large amounts of biomass for accuracy. Turbidimetric techniques are less reliable due to the heterogeneous nature of liquid cultures of filamentous fungi. They have, however, been used to measure growth of populations of fungal pellets, although settling of pellets can introduce inaccuracies. More detailed analysis of pellet populations includes determination of pellet number and size (Edelstein and Hadar, 1983) and analysis of their structure. Turbidometry can be used if mycelia are macerated before analysis (Pegg and Jonlaekha, 1981; Granade *et al.*, 1985), although this increases the time required for analysis and physical disruption and damage to hyphae may affect estimation. *In situ* measurement of absorbance by microspectroscopy following growth in the wells of microtitre plates (Granade *et al.*, 1985) has also been used and gives similar estimates to those obtained from absorbance measurements of macerated material.

If a cell component constitutes a constant proportion of the biomass during growth, chemical estimation of its concentration can be used to estimate biomass concentration. The disadvantage of this group of techniques is variation in cell composition with growth conditions, specific growth rate, medium composition and other factors but Whipps *et al.* (1982) suggest potentially suitable compounds which are relatively specific and easy to assay. These include cell wall components, storage compounds, lipids and enzymes, of which cell wall components such as chitin, mannan and glucan have been used most widely. More general but indirect techniques may also be used, for example measurement of rates of decrease in substrate concentration or increase in product concentration (frequently using radiolabelled substrates).

The most common technique for estimation of

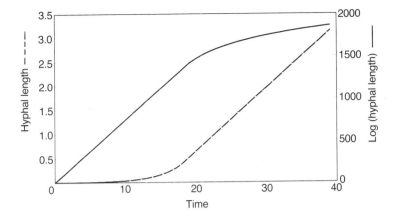

Figure 14.1 Idealized growth kinetics of germ tube hyphae, showing switch from exponential to linear kinetics.

growth of filamentous fungi on solid media is measurement of the increase in radii of circular colonies, and subsequent calculation of colony radial growth rate. The applicability and validity of this technique for estimation of fungal growth is discussed in section 14.4.2. Measurement of total population biomass on solid medium is more difficult due to difficulties in separating mycelia from the medium. The most common method is to grow colonies on the surface of a Cellophane membrane overlaying the agar. Mycelium can then be scraped off for analysis, although significant growth is required for accurate measurement of biomass.

14.2 GROWTH KINETICS OF INDIVIDUAL HYPHAE

14.2.1 Hyphal extension

During growth on solid medium, vegetative hyphae extend by incorporation of new wall material and membrane in the apical, tapered region of the apex, termed the extension zone (Steele and Trinci, 1975a), the precise mechanisms being discussed in Chapter 13. This material is synthesized in distal hyphal regions and packaged in membrane bound vesicles with subsequent transport to the tip. The rate of tip extension is therefore determined by the flux of material to the tip and the rate at which it can be incorporated. Detailed models relating the shape of the extension zone to mechanisms of tip growth are discussed in Chapter 15 but tip growth of individual hyphae has consequences for their

growth kinetics. Thus, spore germination results in formation of a germ tube, whose early growth is supported by mobilization and utilization of storage compounds in the spore. As the germ tube develops, it contributes to biosynthesis and extension by uptake and metabolism of nutrients from the medium. Extension rate accelerates as germ-tube length increases and growth becomes auto-catalytic. Consequently, hyphal length increases exponentially at a constant specific rate (Figure 14.1) which may be significantly greater than the maximum specific growth rate in the equivalent liquid medium because of contribution from endogenous spore reserves (but see section 15.3.3).

Exponential growth cannot proceed indefinitely and extension rate eventually reaches a constant value, i.e. extension is linear (Figure 14.1). This occurs when the tip can no longer incorporate the increasing amount of material being supplied or, more likely, when transport of material from regions distant from the tip is limited. The latter may result from a breakdown in apical polarity, such that transport occurs at a rate less than the hyphal extension rate. A more common explanation is the formation of septa which prevent transport of material to the tip. Extension rate will then be dependent on biosynthesis within the apical compartment only. The growth kinetics of individual hyphae can therefore be represented by two simple equations, the first (equation 14.1) describing exponential increase in length and the second (equation 14.2) a constant extension rate:

$$L = L_0 e^{k_1 t} \tag{14.1}$$

$$L = L_0 + k_2t \qquad (14.2)$$

where L and L_0 represent hyphal length at time t and at the beginning of the appropriate growth phase and k_1 and k_2 are constants.

Both hyphal extension rate and the length of hypha supporting tip growth vary considerably within fungi. For example, linear growth occurs in *Rhizopus stolonifer* when the germ tube is only 40 μm in length, whereas exponential growth of sporangiophores of *Phycomyces blakesleeanus* continues until they are 4 mm in length (Trinci, 1969). In *Candida albicans* vacuolation behind the tip results in linear growth immediately following spore germination, with no discernible exponential phase (Gow and Gooday, 1982).

Hyphal extension rate will depend on the amount of material supplied to the tip and on the surface area of the extension zone, which will increase with hyphal diameter and with increased tapering of the tip. Little work has been done on the relationship between tip shape and extension rate but extension rate is generally found to increase with hyphal diameter within a single organism. The precise relationship is not always well defined. For example, Zhu and Gooday (1992) found a direct relationship between hyphal extension rate and the square of the diameter for hyphae of *Botrytis cinerea*, suggesting extension rate to be dependent on the rate of supply of material to the tip. In *Mucor rouxii*, however, although extension rate increased with diameter, no precise quantitative relationship could be identified (Zhu and Gooday, 1992).

Although hyphal extension rate reaches a constant value, this is not necessarily the maximum possible rate for the organism and growth conditions studied. There is evidence (section 14.3.4) for a gradual developmental process in the hyphae of some fungi leading to increased extension rate. Thus, although short-term measurements indicate constant extension rate, long-term changes occur such that the properties of individual hyphae in mature colonies differ from those in young mycelia. The mechanisms associated with this developmental process are unclear but presumably involve improvement in transport mechanisms and/or changes in cellular mechanisms leading to incorporation of material in the extension zone.

14.2.2 The direction of hyphal growth

The kinetics of hyphal growth on Cellophane membranes covering agar medium are well-characterized and relatively simple but it must be remembered that in their natural environment hyphae will penetrate solid media. This occurs during normal growth on agar, and colonies must be considered as lens shaped rather than disc shaped. The kinetics of growth into solid substrates is largely unexplored but is of great significance both for ecological studies, e.g. on penetration of degrading woody tissue or growth through soil, and in biotechnological processes associated with solid-state fermentations.

Even on the surface of cellophane membranes, hyphae rarely grow in straight lines, in part due to unevenness in the growth surface. Tropic responses will be determined by extension zone behaviour, and are discussed in section 14.3.3. with respect to spatial organization within colonies. Significant deviations occur during spiral growth, which is relatively common and involves bending or coiling of hyphae. Madelin *et al.* (1978) found that of 157 fungi examined, 21 showed pronounced spiralling and 39 weak spiralling in either clockwise or counter clockwise directions, depending on the strain. Spiralling is not observed in aerial hyphae or hyphae growing within agar or in liquid medium and is most pronounced in young colonies or sparsely branched colonies on poor nutritional media. It is believed to result from rotation of the extension zone wall which can be observed most readily in large aerial hyphae, e.g. in sporangiophores of *Phycomyces* and *Mucor mucedo* and conidiophores of *Aspergillus giganteus* (Gooday and Trinci, 1980). Rotation will cause the tip to roll over the agar surface as it extends and the proportion of hyphae spiralling increases with the degree of hardness of the agar where frictional forces will be greatest, and slippage will be less.

14.3   GROWTH OF MYCELIA

14.3.1   Exponential growth

The above discussion implies that, after the phase of linear extension, biomass is synthesized in hyphae at a rate greater than the rate of incorporation at the tip. In addition, microbial growth is normally associated with exponential increases in biomass when conditions are favourable for growth

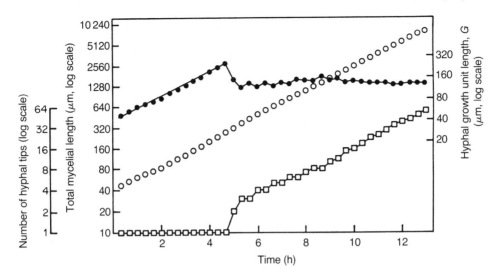

Figure 14.2   Changes in total mycelial length (O), number of hyphal tips (□) and hyphal growth unit length (●) for *Geotrichum candidum* growing on solid medium. (From Trinci, 1974, with permission.)

and when nutrients are in excess. Although individual hyphae extend at a constant, linear rate, exponential growth of filamentous fungi is possible through formation of lateral or dichotomous branch hyphae. The first branch is usually formed from the germ tube towards the end of the period of exponential extension. This rarely affects the extension rate of the parent hypha, even though early growth of the branch is supported by material provided from the parent hypha. Branch length increases at an accelerating rate before reaching a constant value which, in young mycelia, is equal to that of the parent hypha. In terms of growth kinetics, branch formation may be considered equivalent to cell division in unicellular microorganisms.

Exponential growth therefore occurs through an exponential increase in the number of branches, each of which extends at the same constant rate. This was first demonstrated experimentally for *Geotrichum candidum*, *Neurospora crassa*, *Penicillium chrysogenum* and *Aspergillus nidulans* by Trinci (1974), and data for *G. candidum* are illustrated in Figure 14.2. This work also demonstrated that the specific rates of increase in total mycelial length and in the total number of branches were equal to the specific growth rate of these organisms growing in equivalent media but in liquid culture, with biomass concentrations determined by dry weight measurements. If hyphal diameter and the density

of the biomass are assumed to be constant, total mycelial length will be directly related to biomass concentration and changes in total mycelial length are described by equation 14.3.

$$H = H_o e^{\mu t} \tag{14.3}$$

where H and H_o represent total mycelial lengths at times t and 0 and μ is the specific growth rate.

14.3.2   Hyphal growth unit

A further growth characteristic plotted in Figure 14.2 is the length of the hyphal growth unit, G. This is the ratio of total mycelial length to the total number of branches and is therefore the average length of hypha associated with a growing tip. Values of G increase during mycelial development and then oscillate until branches are formed continuously, when G reaches a constant value. Thus, whereas the specific growth rate provides an indication of the kinetics of branch formation, the hyphal growth unit provides information on the branching density, increasing as branching becomes more sparse. The hyphal growth unit is a property of the mycelium and its mathematical properties and its relationship to other hyphal and colony growth parameters have been extensively analysed by Kotov and Reshetnikov (1990). The relative constancy of the hyphal growth unit length indicates a regulatory mechanism within the colony, in that a branch is formed somewhere in the mycelium when the value of G, characteristic for the organism and growth conditions, is exceeded. In addition, Katz *et al.* (1972) demonstrated a relationship between specific growth rate, hyphal growth unit length and mean extension rate, E, which is important in highlighting quantitatively the way in which mycelial growth may be regulated to optimize utilization and colonization of solid substrates.

Mean extension rate is calculated as:

$$E = \frac{2(H_t + H_o)}{(B_o + B_t)} \tag{14.4}$$

where H_o and H_t represent hyphal lengths at time 0 and 1 h later and B_o and B_t are the respective numbers of hyphal tips. This is related to specific growth rate and hyphal growth unit length by equation 14.5.

$$E = \mu G \tag{14.5}$$

G is defined here as a hyphal length. This assumes implicitly that hyphal diameter is constant but variations can occur with different growth and environmental conditions, even within the peripheral growth zone (section 14.4.1). G should, therefore, be redefined as a volume of hypha (Robinson and Smith, 1976) and equation 14.5 can be rewritten as:

$$E = \frac{\mu V_g}{r^2} \tag{14.6}$$

where V_g represents hyphal growth unit volume and r equals hyphal radius.

During colonization of solid substrates, hyphal extension provides a means of exploring new regions for fresh nutrients, whereas branch formation enables full utilization of medium already colonized. If nutrient becomes exhausted or hyphae encounter nutrient-poor regions, it is important that maximum hyphal extension rates are maintained to increase the probability of finding regions of fresh substrate, despite reduction in specific growth rate through nutrient limitation. This is achieved by regulating the distribution of biomass such that it is diverted from branch formation to tip growth, i.e. if μ is decreased, E can only be maintained if G is increased. Thus, under nutrient-poor conditions, branching is sparser, whereas nutrient-rich conditions lead to extensive branch formation to utilize fully available substrate. Equation

14.5 describes how this is achieved in quantitative terms. Similar effects can be mimicked by treatment of mycelia with paramorphogens, e.g. L-sorbose, which increase branching but do not affect specific growth rate. In one such experiment Trinci and Collinge (1973) found that addition of 20 g sorbose l^{-1} led to a reduction in G from 323 to 40 μm while extension rate was reduced from 1001 to 100 μm h^{-1}.

14.3.3 Assessment of morphology using fractal geometry

Fractal analysis is a means of determining the degree to which a pattern or structure is 'self-similar', i.e. similar at different scales of measurement. The fractal dimension obtained from fractal analysis quantifies self-similarity and also provides a measure of the space-filling capacity of the structure. This approach has been used to analyse branching patterns in young mycelia, colonies growing on solid media and mycelial fragments in liquid culture. All three applications will be discussed here, although the last two concern aspects of morphology discussed in detail elsewhere.

Fractal analysis of branching patterns may be illustrated by the work of Obert *et al.* (1990) on mycelia of *Ashlya gossypii*. Following image analysis, the colonies were covered by a grid of boxes, side length ε, and the number of boxes intersected by the mycelium (N_{box}) was determined. If the colony is fractal, the following power law should be followed:

$$N_{box}(\varepsilon) = C\varepsilon^{-D} \tag{14.7}$$

where C is a constant and D is the fractal dimension. N_{box} is therefore determined for a range of box sizes delimited by hyphal diameter and the total size of the colony and C and D are determined by regression analysis. In fact, two analyses were carried out, one for the whole colony, mass fractals, and the second for the colony margin, surface fractals. These gave fractal dimensions of 1.94 and 1.45, respectively. A single unbranched hypha will have a fractal dimension of 1 and D therefore increases as a colony develops. A value of 2 reflects homogeneous distribution within the colony and the mass fractal dimension of 1.94 indicates complete filling of the centre of a colony by branching hyphae. Similar analysis has been carried out by Ritz and Crawford (1990) and Jones *et al.* (1993a) for colonies of *Trichoderma viride* and *Pycnoporus cinnabarinus* respectively. Both found fractal behaviour with fractal dimensions in the range 1.4–2.0.

Matsuura and Miyazima (1993) used fractal geometry to characterize the margins of colonies of *Aspergillus oryzae* growing at different incubation temperatures and on media of different nutrient composition and different agar concentrations. Growth was monitored following germination of spores inoculated in a straight line across solidified medium. For medium containing high concentrations of nutrients and 1.5% (w/v) agar, and at an incubation temperature of 24°C, growth of colonies continued as a smooth front, until half of the medium was colonized. At lower incubation temperatures, lower nutrient concentrations or lower agar concentrations, the front became rough and mycelia split into frond-like formations, with different degrees of thickness and branching. Roughness was quantified by determining the mean square of deviation, $\sigma(l, h)$, of distance of the colony margin from the point of inoculation (h) as a function of horizontal distance, l, along the margin. Roughness is then described in the parameter α, calculated from the equation:

$$\sigma(l,h) \sim l^\alpha \tag{14.8}$$

Growth on high nutrient concentrations at 24°C gave a value of $\alpha = 0.62$ and values increased to 0.70 and 0.71 at 18°C and lower nutrient concentration respectively, as margin roughness increased. The method therefore provides a description of colony margin morphology which may then be used to quantify changes resulting from different growth conditions. The different colony morphologies presumably result from effects of growth conditions on the relationships between hyphal extension rate, branch formation, diffusion of and competition for nutrients and accumulation of waste metabolites. It is not yet possible, however, to link these effects quantitatively to the value of α.

The production of mycelial inocula for fermentation processes involves application of shear forces of different magnitude and duration. Jones *et al.* (1993b) used an image analysis system to determine the size distribution of mycelial fragments of the ligninolytic fungus *P. cinnabarinus* and quantified the effect on fractal fragmentation

dimension. Log-log plots of frequency size distributions were linear, with increased coefficient of variation as shear forces increased, indicating a more uniform size distribution. The fractal dimension, D, is calculated from the slope of the log-log plot and decreases with increasing shear rate or increasing time of treatment, indicating decreased distribution size probability. Shear forces of 13.9×10^3, 27.8×10^3 and 41.7×10^3 were used, each for either 30 s or 60 s. However, a threshold value of D was reached at the intermediate shear force, imposed for 30 s, further increases giving no detectable increase in fragmentation.

Fractal measures of hyphal, mycelial or colony morphology therefore have a quantitative descriptive function and their use has been encouraged by the availability of image analysis technology which greatly facilitates their measurement (Jones *et al.*, 1993a). The biological meaning of the calculated parameters is, however, generally not clear and currently provides no advantage in comparison with traditional measures, whose definitions and derivations are more apparent.

14.3.4 Spatial organization

The microscopic appearance of mycelia on solid medium indicates mechanisms for the development of a well-ordered mycelium in which hyphae avoid each other. Hyphae tend not to cross and a hypha approaching a second hypha will often turn or grow alongside. This requires consideration of interactions between hyphae, in addition to the rate at which they grow and the rates of branch formation. There is evidence of avoidance reactions from a number of studies. For example, Trinci *et al.* (1979) observed avoidance reactions between two hyphae approaching in *N. crassa*, *A. nidulans* and *Mucor hiemalis* operating over distances of 30, 27 and 24 μm respectively. Similar behaviour was found over a range of 10–20 μm when hyphae of *Mucor mucedo* were brought physically into close proximity (Hutchinson *et al.*, 1980), but most hyphae in the colony were spaced further apart than this.

Avoidance reactions may result from movement away from regions in which nutrients are depleted, or where inhibitors have accumulated, or movement towards regions of higher nutrient concentration. Tropisms towards organic compounds have, however, only been reported in the Oomycetes and chytrids. The most likely reason for negative autotropism is the establishment of oxygen gradients (Robinson, 1973a, b) around growing hyphae. This could reduce the oxygen concentration around a hypha to concentrations low enough to reduce the extension rate of an approaching hypha, which would then either stop growing or turn away. The cellular mechanism for such tropic responses must reside in the tip, as wall material behind the tip is rigid, whereas the extension zone wall is extensible and capable of directional growth. Robinson (1973b) suggested a mechanism for oxygen regulated tropism which involved establishment of a gradient in respiratory activity across the extension zone. This could lead to increased production and concentration of vesicles at the side subject to a higher oxygen concentration. The greater wall synthesizing activity in this region would lead to increased extensibility, and subsequent bulging and extension towards the higher oxygen concentration. Oxygen may also be involved in spatial organization of hyphae within mycelia through its effects on branching as Robinson (1973b) found branch production by hyphae of *G. candidum* to be more frequent from sides of hyphae facing higher oxygen concentrations.

14.3.5 Mycelial differentiation

The above discussion refers to hyphae in young developing mycelia. As a colony forms, the kinetics of hyphal growth and branching at the centre of a colony will be influenced by reduction in concentrations of nutrients and oxygen, accumulation of inhibitory end products, production of secondary metabolites and changes in factors such as pH. The relative importance of these factors on biomass formation in developing mycelia is unknown, but they will obviously lead to a decrease in growth rate, with associated effects on branch formation.

The most obvious changes resulting from growth limitation are the development of differentiated structures, in particular spore-bearing aerial hyphae and fruiting bodies. Changes also occur in vegetative hyphae towards the colony centre. For example, hyphae exhibit signs of ageing, becoming swollen and vacuolated and increasingly compartmentalized through plugging of septal pores. In addition, hyphal fusions are formed,

potentially enabling protoplasmic continuity throughout the mycelium.

Mycelia at the colony margin are entering regions of nutrient excess and it might be expected that their growth would be unaffected by events at the colony centre. Nevertheless, in some fungi, notably *N. crassa*, it is possible to distinguish differences in growth kinetics of 'undifferentiated' mycelia found in young colonies and 'differentiated' mycelia found at the edge of mature colonies (Steele and Trinci, 1975b). The former have the properties of mycelia described above. All hyphae have the same extension rate and diameter, extension zones have the same shape and size and branches are subtended at an angle of 90° from the parent. In differentiated mycelia a hierarchy exists such that parental hyphae extend faster, have larger extension zones and are wider than branches which they subtend. In addition, branches are subtended at angles less than 90°, such that hyphae become oriented towards the colony margin. Although this distinction is not observed in some organisms, e.g. *G. candidum*, it is found in a number of fungi, in addition to *N. crassa*. For example, in *Coprinus disseminatus* a hierarchy exists such that extension rates of leading hyphae, primary branches and secondary branches vary in the proportion 100:66:18 respectively (Butler, 1961).

The development of differentiated mycelia has been followed during growth of *N. crassa* on solid medium (McLean and Prosser, 1987) and begins, 22 h after germ tube formation with a rapid reduction in branch angle from 90° to approximately 63° (Figure 14.3a). This is followed by an increase in extension rates and diameters of all hyphae and establishment of the hierarchy described above until extension rates and diameters reach new constant values (Figure 14.3b, c). At this point the ratio between diameters of leading hyphae, primary branches and secondary branches is 100:66:42, and the ratio for extension rates is 100:62:26. These values are similar to those found in *Coprinus disseminatus* where the differentiation process takes 6–7 days (Butler, 1984).

Two processes appear to be taking place in mycelial differentiation. Germ-tube hyphae extend more slowly than all hyphae in differentiated mycelia, even secondary branches. This is surprising when it is considered that germ-tube hyphae are growing under apparently ideal conditions,

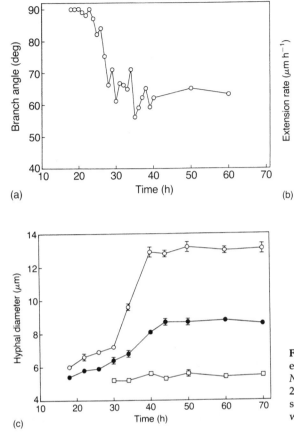

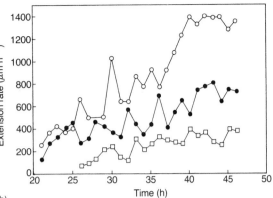

(a)

(b)

(c)

Figure 14.3 Changes in (a) branch angle, (b) hyphal extension rates and (c) hyphal diameter in colonies of *Neurospora crassa* following growth on solid medium for 20 h. (O, leading hyphae; ●, primary branches; □ secondary branches). (From McLean and Prosser, 1987, with permission.)

with no substrate limitation, no accumulation of inhibitors and negligible interactions. This phenomenon has also been observed in *Candida albicans* (Gow and Gooday, 1982) and must involve improvements in transport and incorporation of wall material (referred to in section 14.2.1).

The more dramatic and sudden changes which affect branch angle, hyphal diameter and extension rate, may result from production of paramorphogenetic compounds at the colony centre. These compounds may be secondary metabolites or staling products, produced when growth conditions become unfavourable, with subsequent diffusion to the colony margin. At the margin they result in orientation of branches and growth away from the colony centre. They also induce establishment of apical dominance, possibly by reducing the rate of branch formation or by direct effects on the extension zone. For example, they may reduce the rate of rigidification of apical wall material,

thereby increasing the diameter and size of the extension zone and increasing extension rate. Alternatively, internal control of hyphal diameter and extension rate may be exerted by cAMP as proposed by Pall and Robertson (1986).

An indication of the conditions experienced by mycelia at the margin of mature colonies was provided by Robson *et al.* (1987) who measured changes in glucose concentration and in agar beneath developing colonies of *Rhizoctonia solani*. Although there was no evidence of glucose limitation at the advancing colony margin, glucose concentration in the agar fell sharply behind the margin and was fully utilized beneath the colony centre after growth for 6 days (Figure 14.4). More significantly for mycelial differentiation processes, penicillin was present at significant concentrations in agar ahead of colonies of *Penicillium chrysogenum*. Glucose will diffuse rapidly through agar, and these observations therefore demonstrate the

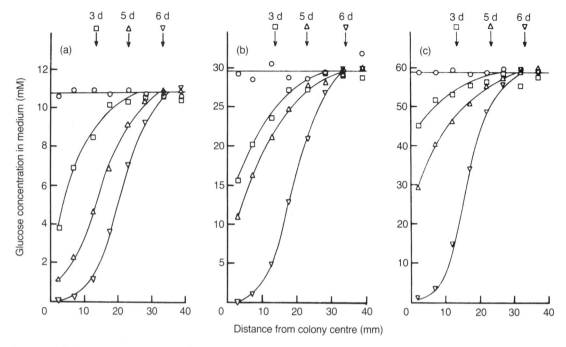

Figure 14.4 Glucose concentration of medium in uninoculated plates (O) and beneath colonies of *Rhizoctonia cerealis* growing on minimal medium containing (a) 10 mM, (b) 25 mM and (c) 50 mM glucose for 3 ($\square$), 5 ($\triangle$) and 6 ($\triangledown$) days. Arrows indicate the position of the colony margin 3, 5 and 6 days after inoculation. (From Robson *et al.*, 1987, with permission).

extent of substrate utilization by colonies growing on agar and the nutritional limitations to growth at the colony centre.

14.4   COLONY GROWTH KINETICS

The kinetics of growth of fungal colonies are effectively those of individual hyphae. As a mycelium develops it soon forms a circular colony. The radius of the colony initially increases exponentially but becomes constant when individual hyphae extend at a linear rate. Colony radial growth rate, K_r, may not be directly equivalent to maximum hyphal extension rate, which shows some variability and colony radial growth rate will reflect the average properties of individual hyphae. In addition, variation in the direction of hyphal growth will affect its contribution to colony radial expansion.

As indicated in section 14.1.2(c) it is difficult to determine accurately biomass concentrations of colonies growing on solid medium and the precise growth kinetics in terms of biomass production

have not been characterized. Nevertheless, exponential growth will cease when conditions at the colony centre become unfavourable but radial growth of the colony can continue indefinitely because of the ability to regulate the distribution of biomass between tip growth and branch production.

14.4.1   The peripheral growth zone

Pirt (1967) suggested that following initial exponential growth, colony expansion will result from growth of a ring of hyphae at the colony margin. This marginal ring of hyphae was termed the peripheral growth zone by Trinci (1971). Mycelium within this zone grows exponentially at a specific rate equivalent to the specific growth rate in liquid medium of equivalent composition. The width (w) of the zone remains constant and represents the maximum length of hypha supporting tip growth. Thus growth could still occur inside this ring, towards the colony centre, but would not contribute to radial expansion. K_r, w and specific growth rate (μ) are related by equation 14.9.

$$K_r = \mu w \qquad (14.9)$$

This is similar in form to equation 14.5, the difference being that equation 14.9 is concerned with maximum values of extension rate and hyphal length associated with tip growth, whereas equation 14.5 considers mean values. Morrison and Righelato (1974) combined equations 14.5 and 14.9 to give:

$$K_r = \mu G k \qquad (14.10)$$

where k is a constant.

The width of the peripheral growth zone was originally determined as the minimum length of colony margin whose growth was unaffected by severing hyphae with a razor blade (Trinci, 1971). It is now recognized that the peripheral growth zone width, at the colony level, is equivalent to the length of hypha contributing to tip growth. Its length may therefore be determined experimentally as the length of the germ tube at which growth switches from exponential to linear kinetics. Values of w determined using these techniques compare well with those calculated from independent measurements of K_r and w for a wide range of species and environmental conditions, providing good support for the concept. As discussed above, more general application of equation 14.10 requires consideration of w as hyphal volume, as hyphal diameter will vary with growth rate and environmental conditions.

14.4.2 Consequences for the use of K_r to measure fungal growth

The ease with which colony radial growth rate can be determined experimentally has led to its widespread use in assessing effects of environmental and other factors on fungal growth and in comparing growth of different fungi. Consideration of equations 14.5, 14.9 and 14.10 highlights the dangers in this approach. Specific growth rate is only directly related to colony radial growth rate if w is constant. This significantly limits its utility and applicability of K_r as a measure of growth rate as w varies considerably between species and strains, with many environmental and nutritional conditions and following treatment with inhibitors and paramorphogens. This does not always apply,

e.g. temperature (Trinci, 1971) and water activity (Inch and Trinci, 1987) affect K_r and μ to the same extent, implying no effect on w. Changes in specific growth rate may, as indicated in section 14.3.2, alter branching density and therefore w, with no associated change in K_r. Indeed this ability is an important advantage of the filamentous growth form. K_r is therefore only an indirect measure of μ, and its use requires careful consideration of associated effects on branching and on hyphal diameter.

14.5 GROWTH IN LIQUID CULTURE

Although growth of microorganisms on solid media has been studied most extensively using filamentous organisms, growth kinetics in liquid culture are based mainly on studies with unicellular organisms. This is due to the relative ease with which biomass concentrations of unicells can be measured, using turbidimetric techniques. Other practical difficulties hinder study of filamentous organisms in liquid culture. For example, heterogeneity within the biomass can be introduced by attachment and growth of organisms on the walls, agitators and probes in fermentor vessels. This occurs with unicells also, but is more widespread and extensive in fungal cultures. Heterogeneity is also introduced by the formation of pellets, discussed separately in section 14.5.4. Both of these factors lead to areas of growing and non-growing biomass, their relative proportions influencing overall growth kinetics. In addition, intrinsic heterogeneity arises from the mechanism for hyphal growth, with extension, but little *de novo* biosynthesis, at the tip, active biosynthesis behind the tip and reduced activity in more distal regions as hyphae age and vacuolate. Despite this, the growth kinetics of fungal cultures can be described to a large extent by kinetic equations derived from bacterial studies, particularly when growth occurs as dispersed mycelia.

14.5.1 Batch culture

Growth in batch cultures occurs on a limited amount of nutrient until growth is restricted by exhaustion of nutrients or development of unfavourable conditions. Batch growth is typically divided into a number of phases. The first, the lag phase, is a period of preparation for growth, which may require spore germination, physiological adaptations, e.g. synthesis of enzyme systems required for substrate utilization, or removal of inhibitory compounds carried over with the inoculum.

(a) Exponential growth

Growth of fungi as dispersed mycelia is assumed to be equivalent to that of unicellular microorganisms in that biomass, substrate and product concentrations are assumed to be uniform throughout the culture, and biomass is considered to be homogeneous with no areas of differentiation. Under these conditions, and assumptions, growth kinetics are similar to those of unicells and the lag phase is followed by an acceleration phase and a phase of exponential or logarithmic growth. As on solid media, exponential growth results from autocatalysis through exponential production of branches, each of which extends at a linear rate. The increase in biomass concentration (x) can therefore be represented by the equation:

$$\frac{dx}{dt} = \mu x \tag{14.11}$$

where μ is the specific growth rate. In fact, the specific growth rate during the exponential phase of growth is considered to be the maximum possible for the particular growth medium and cultural conditions and μ is therefore equivalent to the maximum specific growth rate, μ_m. Integration of equation 14.11 yields:

$$\ln x = \ln x_0 + \mu t \tag{14.12}$$

where x_0 is the biomass concentration at the start of exponential growth. This equation is equivalent to equation 14.3 and enables calculation of μ_m as the slope of a semi-logarithmic plot of x vs t. Although eukaryotes tend to grow more slowly than bacteria, filamentous fungi such as *Aspergillus* or *Penicillium* have maximum specific growth rates in the order 0.1–0.3 h^{-1}, equivalent to doubling times of 2–7 h.

(b) Deceleration phase

As conditions become unfavourable for growth, specific growth rate decreases from the maximum and the culture enters a deceleration phase. This is usually due to utilization of alternative components of the medium, accumulation of inhibitory compounds, oxygen limitation or reduction in the pH of the medium. If reduction in specific growth rate results solely from exhaustion of nutrients, the length of the deceleration phase will be negligible, as the time taken for high concentrations of biomass to utilize concentrations of substrate likely to lead to limitation would be small. The many and varied factors governing entry into, and growth during the deceleration phase have been given little attention. Growth kinetics during this period are therefore largely uncharacterized, despite the importance of the deceleration phase for biotechnological processes, as the period when secondary metabolite production begins. Differentiation structures, e.g. spores, may also appear during this phase, particularly at air–water interfaces, but differentiation and morphologies are less complex than those seen in colonies on solid media.

Deceleration in growth rate due to oxygen limitation is of particular importance for cultures of filamentous fungi because of the influence of their morphology on rheological properties. In cultures of unicellular microorganisms shear stress is directly proportional to shear rate, and cultures act as Newtonian fluids. Dispersed mycelial cultures, however, exhibit non-Newtonian behaviour in which apparent viscosity increases with agitation rate, reducing mass transport of nutrients, oxygen and heat (Banks, 1977; van Suijdam and Metz, 1981). The increased energy required for efficient mixing and oxygen transfer significantly increases costs of large-scale fungal fermentations and can accelerate entry into the deceleration phase, in comparison with unicellular cultures of equivalent biomass and activity.

Another important factor, which has received little attention, is the effect of physical factors, such as agitation, on the breakage and fragmentation of mycelia. Little is known of the physical properties of fungal hyphae, and of damage caused by breakage and consequent effects on growth kinetics.

(c) Stationary phase

As growth rate decreases the culture eventually enters a stationary phase in which biomass concentration is relatively constant. This constancy may represent a balance between growth and death of mycelia and activity may be maintained within the biomass by utilization of storage compounds. Activity during the stationary phase is of particular importance when associated with the production of storage compounds and secondary metabolites. When autolytic processes begin to dominate, the death phase begins and biomass concentration falls, although there may be renewed growth from the products of autolysis.

The amount of biomass formed at the stationary phase is described by the yield coefficient, Y, defined as the amount of biomass formed per unit substrate converted. If Y is assumed to be constant, the rate of substrate conversion during batch growth is described by the equation:

$$\frac{ds}{dt} = -\frac{1}{Y}\frac{dx}{dt} = -qx \qquad (14.13)$$

where q is the specific rate of substrate utilization. The values of both q and Y depend on the nature of the substrates and the cultural conditions and may be related to specific growth rate by combination with equation 14.11 to give:

$$\mu = qY \qquad (14.14)$$

14.5.2 Continuous culture

In batch culture, growth is only constant during the exponential phase. This prevents study of the physiology and properties of organisms at submaximal specific growth rates and of the effects of limitation by different substrates. This may be achieved using continuous culture, in which fresh medium is supplied to the culture at a constant rate, while culture fluid and biomass are removed at the same rate such that culture volume remains constant. Continuous culture studies have proved invaluable for physiological studies but the technique is not widely used in large-scale industrial processes, even though it can provide greater productivity than batch culture. The major problems are the need to sterilize large volumes of medium continuously, maintenance of axenic cultures and regression of producer strains.

(a) Substrate limitation

Continuous culture is usually preceded by growth of the organism in batch culture to stationary phase. When supply of fresh medium is initiated growth proceeds, and material from the vessel is washed out, until the concentration of a single

component of the medium is reduced to a level at which it limits specific growth rate. In physiological studies the medium is designed to determine which substrate limits growth by providing all other components in excess. The relationship between specific growth rate and the concentration of a limiting substrate, s, may be described by the Monod (1942) equation:

$$\mu = \frac{\mu_m s}{K_s + s} \qquad (14.15)$$

where K_s is the saturation constant for growth, equivalent to the substrate concentration at which $\mu = \mu_m/2$. Although the Monod equation is used most frequently to describe growth during substrate limitation, alternatives have been proposed, usually to describe experimental data which do not obey the Monod equation. Often the quality of experimental data is not sufficient to enable critical testing or distinction between different types of kinetics. It must also be remembered that equation 14.15 is essentially an empirical equation. Although it might be suggested that growth limitation may result from limitation to, for example, uptake of a substrate, and that growth kinetics will reflect the kinetics of action of a single enzyme reaction controlling uptake, the real situation is likely to be more complex as biomass production results from a complex set of enzymatic reactions. Deviations from the Monod equation may therefore provide useful information regarding the physiology of organisms growing under substrate limitation. Alternative descriptions of growth limitation are usually more complex than equation 14.15 and are therefore of less descriptive and practical use, although some (e.g. Dabes *et al.*, 1973) may have a more sound mechanistic basis. Care must be taken, however, in the application and use of purely empirical relationships. Deviations from normal kinetics may reflect deficiencies in the experimental system and, although alternative kinetics may provide a better description of growth under the particular set of experimental conditions, they may mask such deficiencies and, in themselves, provide no additional information on the physiology of fungal growth.

(b) Establishment of a steady state

Increases in biomass lead to a decrease in substrate concentration and specific growth rate until balance is reached between biomass production and removal of material through washout. The resultant equilibrium, or steady state, is characterized by constant concentrations of biomass, substrate and product which, in theory, remain constant indefinitely.

The changes in biomass and substrate concentrations during growth in continuous culture can be described by equations 14.16 and 14.17:

$$\frac{dx}{dt} = \mu x - Dx \qquad (14.16)$$

$$\frac{ds}{dt} = Ds_r - \frac{\mu x}{Y} - Ds \qquad (14.17)$$

where μ is described by equation 14.15, s_r is the concentration of the limiting substrate in the inflowing medium and D is the dilution rate, defined as flow rate divided by the volume of culture fluid. If D is less than μ_m a steady state is established. Under these conditions, equations 14.16 and 14.17 may be set equal to zero, providing expressions for steady-state biomass and substrate concentrations as functions of dilution rate. In addition, it can be seen from equation 14.16 that in the steady state, the specific growth rate will equal the dilution rate. Thus, specific growth rate may be controlled by altering the dilution rate, most conveniently by adjusting the rate of supply of fresh medium. This enables investigation of growth at submaximal growth rates. In addition, the composition of the inflowing medium may be manipulated to obtain limitation of growth by different nutrients, allowing investigation of their role in growth and product formation.

The problems of growth in liquid batch culture discussed above are exacerbated in continuous culture. For example, wall growth increases continuously leading to increased heterogeneity. Equations 14.16 and 14.17 then no longer describe growth accurately, as they assume biomass to be uniformly active and equally likely to be washed out. Maintenance of steady-state cultures also involves breakage of mycelia to provide new centres for growth. As discussed above, the mechanisms and effects of such processes are little understood.

Despite these problems, continuous culture studies have been used to determine values for growth parameters, e.g. Y, K_s, for filamentous fungi. The ability to control specific growth rate

has provided additional information on the kinetics of branch formation and more rigorous tests of equations 14.5 and 14.7. For example, when *Geotrichum candidum* is grown under conditions of glucose limitation, an increase in specific growth rate results in increased hyphal diameter and decreased hyphal growth unit length but has no significant effect on hyphal growth unit volume (Robinson and Smith, 1979). In addition, lateral branches are formed at specific growth rates less than 0.4 h^{-1} whereas apical branches and extension rate increase at higher dilution rates. Smith and Robinson (1980) also demonstrated similarities between cell dimensions and hyphal extension rates in glucose-limited continuous cultures and in cultures grown on solid media with low initial glucose concentrations.

(c)   Maintenance energy

At low dilution rates, steady-state biomass concentration decreases, whereas the Monod equation predicts effectively constant biomass concentrations as D approaches 0. This is due to consumption of substrate for cell maintenance, e.g. processes such as maintenance of ion, pH and osmotic gradients across the cell membrane and protein and nucleic acid turnover. This requires modification of equation 14.17 to give (Pirt, 1965):

$$\frac{ds}{dt} = Ds_r - \frac{\mu x}{Y_g} - Ds - mx \qquad (14.18)$$

where Y_g is the true growth yield and m is the maintenance coefficient. Both Y_g and m are assumed in equation 14.18 to be constant but this is not always the case for bacterial or fungal cultures, as m may vary with specific growth rate. Nevertheless equation 14.18 provides the simplest and most widely applicable description of changes in substrate concentration in continuous culture and enables calculation of the proportion of substrate contributing to maintenance. For example, during growth of *Penicillium chrysogenum* in glucose-limited chemostats, 10% of substrate is utilized for maintenance at the maximum specific growth rate (0.8 h^{-1}) and 70% at a specific growth rate of 0.05 h^{-1} (Righelato, 1979).

14.5.3   Fed-batch culture

The majority of large-scale industrial fermentations involving filamentous fungi involve fed-batch culture in which biomass is grown initially in batch culture until the substrate is fully utilized. Fresh nutrient is then added, and material may be removed, but in a different manner to that described in the previous section. For example, a concentrated form of a component, or components, of the original medium may be supplied, rather than complete medium. Similarly, a product precursor may be added. The aim is to promote product formation but not necessarily to increase biomass concentration. Substrate is converted immediately on entry into the fermentor by high levels of active biomass. Metabolism is directed towards product formation, rather than growth, and substrate inhibition and catabolite repression are minimized. In an alternative form, repeated fed-batch culture, substrate is added continuously and material removed at regular intervals. In both systems, the rate of addition differs from the rate of removal and volume is, therefore, never constant, unlike the situation in a chemostat. Maximum control of fed-batch cultures is achieved by addition of substrate at times determined by specific properties of the culture which are continuously monitored.

One major difference between fed-batch and continuous culture is that specific growth rate is not constant in the former. The physiology of growth of microorganisms under transient conditions is poorly understood and reliable quantitative growth kinetics have not been characterized. Nevertheless, Pirt (1975) provided an analysis of repeated fed-batch culture which predicted establishment of a pseudo-steady state in which the specific growth rate equals dilution rate. However, unlike the situation in a chemostat, volume is continually increasing in fed-batch culture and specific growth rate therefore decreases.

14.5.4 Pellet formation

The disadvantages of dispersed mycelial growth have been discussed, and include increased wall growth and reduction in efficiency of mixing and oxygen supply. These problems may be solved to some extent by growth in the form of pellets, which also improve harvesting through improved filtration characteristics of the broth.

Pellets are spherical or ellipsoidal masses of hyphae with variable internal structure, ranging from loosely packed hyphae, forming 'fluffy

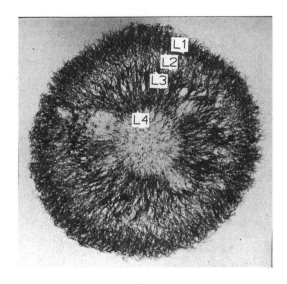

Figure 14.5 A cross section of a pellet of *Penicillium chrysogenum* stained with cresyl violet and showing the layers L1–L4 described in the text. (From Wittler *et al.*, 1986, with permission.)

pellets, to tightly packed, compact, dense pellets (Yanagita and Kogane, 1963). Wittler *et al.* (1986) proposed the existence of four regions (Figure 14.5). The outer region consists of viable hyphae and surrounds a layer of hyphae showing signs of autolysis. In hollow pellets, a third layer is found containing hyphae with irregular wall structure, while the centre of the pellet contains no recognizable mycelia. The density of hyphae within pellets is of significance for diffusion of nutrients and oxygen to the mycelial biomass, with consequent effects on growth, particularly at the centre of compact pellets. Theoretical models for the effect of pellet structure on oxygen uptake are discussed in Chapter 15.

Pellet formation results from aggregation of spores before germination, entrapment of spores by germ tubes or, less commonly, aggregation of young mycelia. Formation therefore depends to a large extent on physicochemical properties of spores and hyphae, and only indirectly on physiological properties. As a result, a wide range of factors have been implicated in pellet formation and structure (see Prosser and Tough, 1991 for recent review). For example, high rates of agitation and shear forces can reduce aggregation of spores, reducing pellet formation, and can also disrupt pellet structure. Similarly, anionic polymers such as carbopol and Junlon (polyacrylic acid) can prevent pellet formation by reducing spore aggregation in a number of fungi (Jones *et al.*, 1989). The composition of the growth medium can affect pellet formation through effects on branching patterns (Pirt and Callow, 1959) whereas production of extracellular polysaccharides can also affect pellet formation, through effects on the viscosity of the medium. The concentration of spores in the inoculum also affects the degree of spore aggregation, and hence pellet formation and generally pellet number decreases with increased spore concentration (Sharma and Padwal-Desai, 1985).

Pelleted cultures are traditionally assumed to follow cube-root kinetics, following the early observations of Emerson (1950), on growth of *N. crassa*, and Marshall and Alexander (1960) who investigated a number of fungi and actinomycetes. These kinetics are described by the equation:

$$M^{1/3} = M_0^{1/3} + kt \qquad (14.19)$$

where M represents biomass concentration, M_0 initial biomass concentration and k is a constant. These kinetics may be explained (Pirt, 1966) by considering the heterogeneity within pellets in a similar manner to that within colonies growing on solid medium. Thus, a pellet is assumed to increase in radius at a constant rate through exponential growth of mycelia within an outer shell of active hyphae. The width, w, of this shell is determined by the ability of material to diffuse through the mycelium. It will therefore depend on pellet structure and is not directly equivalent to the peripheral growth zone width. Thus steady state concentrations of biomass and substrate, and the critical dilution rate, depend not only on D but also on the stirrer speed, pellet size and fragmentation. Any growth occurring inside this region would not lead to pellet expansion and would be limited by diffusion of nutrients, in particular oxygen, through the outer growing shell. Changes in colony radius may therefore be described by the equation:

$$r = r_o + w\mu t \qquad (14.20)$$

where r and r_0 represent pellet radii at times t and 0. The outer growing shell has a width w and specific growth rate μ. If the pellet is assumed to be spherical and to have a constant biomass density, ρ, equation 14.20 can be rewritten in terms of biomass:

$$M = M_0 + \left(\frac{4}{3}\pi\rho n\right)^{1/3} w\mu t \qquad (14.21)$$

where M and M_0 are the biomass of pellets at times t and 0 respectively and n is the number of pellets. This is equivalent to equation 14.19 with:

$$k = \left(\frac{4}{3}\pi\rho n\right)^{1/3} \mu t \qquad (14.22)$$

The model therefore predicts exponential growth in batch culture until restrictions to diffusion of nutrients, or oxygen, through the outer pellet layers reduce growth rates in the centre of the pellet. Subsequent growth will then follow cube root kinetics. As with much work on fungal growth kinetics, it is frequently difficult to obtain experimental data of sufficient accuracy to distinguish different kinetic models and, although cube root kinetics are predicted for pellet growth, it is difficult to distinguish them from exponential growth kinetics. Koch (1975), however, provides a unified and general kinetic model for growth in liquid culture and on solid media, which predicts the majority of growth kinetics described in this chapter. The model, and its underlying assumptions, will be discussed in Chapter 15.

An important feature of pellet growth in liquid culture is its effect on kinetics in continuous culture. The assumptions on which equations 14.20 and 14.21 are based include homogeneity within the biomass. In pellets a proportion of the biomass is inactive and specific growth rate of the total population is less than the maximum specific growth rate. Consequently, the critical dilution rate, at which biomass washes out, is less than the maximum specific growth rate of the organism and will depend on factors controlling pellet size and structure.

14.6 CONCLUSIONS

It is possible to use simple kinetic equations to describe the growth of filamentous fungi at the hyphal, mycelial and colony levels when growth is unrestricted and when there is no heterogeneity. Under these conditions, there is conformity between growth kinetics at these three levels and similar kinetics are observed experimentally in liquid culture and on solid media. Where appropriate, kinetics of growth of populations, i.e. populations of hyphae, correspond to the kinetics of growth of populations of unicells. In addition, in many cases, the kinetics may be understood in terms of the cellular mechanisms controlling growth and branching. Problems arise when conditions become unfavourable for growth and heterogeneity in the growth medium or in the biomass becomes significant. Although growth limitation by substrate concentration is well described in continuous culture, other forms of limitation have been characterized poorly, both theoretically and experimentally. This reflects our lack of understanding of the complexity of the effects of such factors on the physiology of growth and branching. Simple, generalized descriptions of growth under such conditions may not exist but the more complex mechanistic models discussed in Chapter 15 may provide a basis for such studies.

REFERENCES

Adams, H.L. and Thomas, C.R. (1988) The use of image analysis for morphological measurements on filamentous microorganisms. *Biotechnology and Bioengineering*, **32**, 707–12.

Banks, G.T. (1977) Aeration of mould and streptomycete culture fluids, in *Topics in Enzyme and Fermentation Biotechnology*, Vol. 1, (ed. A. Wiseman) Ellis Horwood, Chichester pp. 72–110.

Butler, G.M. (1961) Growth of hyphal branching systems in *Coprinus disseminatus*. *Annals of Botany*, **25**, 341–52.

Butler, G.M. (1984) Colony ontogeny in basidiomycetes, in *The Ecology and Physiology of the Fungal Mycelium*, (eds. D.H. Jennings and A.D.M. Rayner), Cambridge University Press, Cambridge, pp. 53–71.

Dabes, J.N., Finn, R.K. and Wilke, C.R. (1973) Equations of substrate-limited growth: the case for Blackman kinetics. *Biotechnology and Bioengineering*, **15**, 1159–77.

Edelstein, L. and Hadar, Y. (1983) A model for pellet size distributions in submerged cultures. *Journal of Theoretical Biology*, **105**, 427–52.

Emerson, S. (1950) The growth phase in *Neurospora* corresponding to the logarithmic phase in unicellular organisms. *Journal of Bacteriology*, **60**, 221–3.

Gooday, G.W. and Trinci, A.P.J. (1980) Wall structure and biosynthesis in fungi. *Symposia of the Society for General Microbiology*, **30**, 207–51.

Gow, N.A.R. and Gooday, G.W. (1982) Growth kinetics and morphology of colonies of the filamentous form of *Candida albicans*. *Journal of General Microbiology*, **128**, 2187–94.

Granade, T.C., Hehmann, M.F. and Artis, W.M. (1985) Monitoring of filamentous fungal growth by *in situ* microspectrophotometry, fragmented mycelium absorbance density, and ¹⁴C incorporation: alternatives to mycelial dry weight. *Applied and Environmental Microbiology*, **49**, 101–8.

Gray, D.I. and Morris, B.M. (1992) A low cost video analysis system for the BBC Master computer. *Binary*, **4**, 58–61.

Hutchinson, S.A., Sharma, P., Clarke, K.R. and Macdonald, I. (1980) Control of hyphal orientation in colonies of *Mucor hiemalis*. *Transactions of the British Mycological Society*, **75**, 177–91.

Inch, M.M. and Trinci, A.P.J. (1987) Effects of water activity on growth and sporulation of *Paecilomyces farinosus* in liquid and solid media. *Journal of General Microbiology*, **133**, 247–52.

Jones, C.L., Lonergan, G.T. and Mainwaring, D.E. (1993a) A rapid method for the fractal analysis of fungal colony growth using image processing. *Binary*, **5**, 171–80.

Jones, C.L., Lonergan, G.T. and Mainwaring, D.E. (1993b) Mycelial fragment size distribution: an analysis based on fractal geometry. *Applied Microbiology and Biotechnology*, **39**, 242–9.

Jones, P., Shahab, B.A., Trinci, A.P.J. and Moore, D. (1989) Effect of polymeric additives, especially Junlon and Hostacerin, on the growth of some basidiomycetes in submerged culture. *Transactions of the British Mycological Society*, **90**, 577–83.

Katz, D., Goldstein, D. and Rosenberger, R.F. (1972) Model for branch initiation in *Aspergillus nidulans* based on measurement of growth parameters. *Journal of Bacteriology*, **109**, 1097–100.

Koch, A.L. (1975) The kinetics of mycelial growth. *Journal of General Microbiology*, **89**, 209–16.

Kotov, V. and Reshetnikov, S.V. (1990) A stochastic model for early mycelial growth. *Mycological Research*, **94**, 577–86.

Madelin, M.F., Toomer, D.J. and Ryan, J. (1978) Spiral growth of fungus colonies. *Journal of General Microbiology*, **106**, 73–80.

Marshall, K.C. and Alexander, M. (1960) Growth characteristics of fungi and actinomycetes. *Journal of Bacteriology*, **80**, 412–16.

Matsuura, S. and Miyazima, S. (1993) Colony of the fungus *Aspergillus oryzae* and self-affine fractal geometry of growth fronts. *Fractals*, **1**, 11–19.

McLean, K.M. and Prosser, J.I. (1987) Development of vegetative mycelium during colony growth of *Neurospora crassa*. *Transactions of the British Mycological Society*, **88**, 489–95.

Monod, J. (1942) *Recherches sur la Croissance des Cultures Bacteriennes*, 2nd edn. Hermann, Paris.

Morrison, K.B. and Righelato, R.C. (1974) The relationship between the hyphal branching, specific growth rate and colony radial growth rate in *Penicillium chrysogenum*. *Journal of General Microbiology*, **81**, 517–20.

Obert, M., Pfeifer, P. and Sernetz, M. (1990) Microbial growth patterns described by fractal geometry. *Journal of Bacteriology*, **172**, 1180–5.

Pall, M.L. and Robertson, C.K. (1986) Cyclic AMP control of hierarchical growth pattern of hyphae in *Neurospora crassa*. *Experimental Mycology*, **10**, 161–5.

Pegg, G.F. and Jonglaekha, N. (1981) Assessment of colonisation in chrysanthemum grown under different photoperiods and infected with *Verticillium dahliae*. *Transactions of the British Mycological Society*, **76**, 353–60.

Pirt, S.J. (1965) The maintenance energy of bacteria in growing cultures. *Proceedings of the Royal Society Series B*, **163**, 224–31.

Pirt, S.J. (1966) A theory of the mode of growth of fungi in the form of pellets in submerged culture. *Proceedings of the Royal Society Series B*, **166**, 369–73.

Pirt, S.J. (1967) A kinetic study of the mode of growth of surface colonies of bacteria and fungi. *Journal of General Microbiology*, **47**, 181–97.

Pirt, S.J. (1975) *Principles of Microbe and Cell Cultivation*, Blackwell Scientific Publications, Oxford.

Pirt, S.J. and Callow, D.S. (1959) Continuous flow culture of the filamentous mould *Penicillium chrysogenum* and the control of its morphology. *Nature*, **184**, 307–10.

Prosser, J.I. and Tough, A.J. (1991) Growth mechanisms and growth kinetics of filamentous microorganisms. *CRC Critical Reviews in Biotechnology*, **10**, 253–74.

Righelato, R.C. (1979) The kinetics of mycelial growth, in *Fungal Walls and Hyphal Growth*, (eds J.H. Burnett and A.P.J. Trinci), Cambridge University Press, Cambridge, pp. 385–401.

Ritz, K. and Crawford, J. (1990) Quantification of the fractal nature of colonies of *Trichoderma viride*. *Mycological Research*, **94**, 1138–52.

Robinson, P.M. (1973a) Autotropism in fungal spores and hyphae. *Botanical Review*, **39**, 367–84.

Robinson, P.M. (1973b) Chemotropism in fungi. *Transactions of the British Mycological Society*, **61**, 303–13.

Robinson, P.M. and Smith, J.M. (1976) Morphogenesis and growth kinetics of *Geotrichum candidum* in continuous culture. *Transactions of the British Mycological Society*, **66**, 413–20.

Robinson, P.M. and Smith, J.M. (1979) Development of cells and hyphae of *Geotrichum candidum* in chemostat and batch culture. *Transactions of the British Mycological Society*, **72**, 39–47.

Robinson, P.M. and Smith, J.M. (1980) Apical branch formation and cyclic development in *Geotrichum candidum*. *Transactions of the British Mycological Society*, **75**, 233–8.

Robson, G.D., Bell, S.D., Kuhn, P.J. and Trinci, A.P.J. (1987) Glucose and penicillin concentrations in agar medium below fungal colonies. *Journal of General Microbiology*, **133**, 361–7.

Sharma, A. and Padwal-Desai, S.R. (1985) On the relationship between pellet size and aflatoxin yield in *Aspergillus parasiticus*. *Biotechnology and Bioengineering*, **27**, 1577–80.

Smith, J.M. and Robinson, P.M. (1980) Development of somatic hyphae of *Geotrichum candidum*. *Transactions of the British Mycological Society*, **74**, 159–165.

Steele, G.C. and Trinci, A.P.J. (1975a) The extension zone of mycelial hyphae. *New Phytologist*, **75**, 583–7.

Steele, G.C. and Trinci, A.P.J. (1975b) Morphology and growth kinetics of hyphae of differentiated and undifferentiated mycelia of *Neurospora crassa*. *Journal of General Microbiology*, **91**, 362–8.

Trinci, A.P.J. (1969) A kinetic study of the growth of *Aspergillus nidulans* and other fungi. *Journal of General Microbiology*, **57**, 11–24.

Trinci, A.P.J. (1971) Influence of the peripheral growth zone on the radial growth rate of fungal colonies. *Journal of General Microbiology*, **67**, 325–44.

Trinci, A.P.J. (1974) A study of the kinetics of hyphal extension and branch initiation of fungal mycelia. *Journal of General Microbiology*, **81**, 225–36.

Trinci, A.P.J. and Collinge, A.J. (1973) Influence of L-sorbose on the growth and morphology of *Neurospora crassa*. *Journal of General Microbiology*, **78**, 179–92.

Trinci, A.P.J., Saunders, P.T., Gosrani, R. and Campbell, K.A.S. (1979) Spiral growth of mycelial and reproductive hyphae. *Transactions of the British Mycological Society*, **73**, 283–92.

van Suijdam, J.C. and Metz, B. (1981) Influence of engineering variables on the morphology of filamentous molds. *Biotechnology and Bioengineering*, **23**, 111–48.

Whipps, J.M., Haselwandter, K., McGee, E.E.M. and Lewis, D.H. (1982) Use of biochemical markers to determine growth, development and biomass of fungi in infected tissues, with particular reference to antagonistic and mutualistic biotrophs. *Transactions of the British Mycological Society*, **79**, 385–400.

Wiebe, M.G. and Trinci, A.P.J. (1990) Dilution rate as a determinant of mycelial morphology in continuous culture. *Biotechnology and Bioengineering*, **38**, 75–81.

Wittler, R., Baumgartl, H., Lubbers, D.W. and Schugerl, K. (1986) Investigations of oxygen transfer into *Penicillium chrysogenum* pellets by microprobe measurement. *Biotechnology and Bioengineering*, **28**, 1024–6.

Yanagita, T. and Kogane, F. (1963) Cytochemical and physiological differentiation of mould pellets. *Journal of General and Applied Microbiology*, **9**, 171–87.

Zhu, W.-Z. and Gooday, G.W. (1992) Effects of nikkomycin and echinocandin on differentiated and undifferentiated mycelia of *Botrytis cinerea* and *Mucor rouxii*. *Mycological Research*, **96**, 371–7.

MATHEMATICAL MODELLING OF FUNGAL GROWTH

15

J.I. Prosser

Department of Molecular and Cell Biology, Marischal College, University of Aberdeen, Aberdeen, UK

15.1 INTRODUCTION

A mathematical model was defined in Chapter 14 as a description of a process or system in a mathematical form. The kinetic equations described in Chapter 14 therefore represent a type of mathematical model and Chapters 14 and 15 broadly distinguish two modelling approaches, that of the former chapter being largely descriptive and empirical. Kinetic equations were relatively simple, based on experimental observations and designed to be of practical use, e.g. in comparing quantitatively growth of different fungi, or growth under different environmental or nutritional conditions. In this chapter, more complex models will be discussed. Most are based on biological assumptions regarding mechanisms controlling particular aspects of fungal growth and generate predictions which may be tested experimentally.

Both modelling approaches involve close association between theoretical and experimental work but neither approach is universally applicable or correct. Indeed many of the mechanistic models discussed below contain empirical components and may be used to explain simpler kinetics of growth. In any modelling study, however, the role of the model and its relationship to experimental work must be clarified to prevent confusion. For example, a good fit between experimental data on hyphal growth and predictions of a mathematical model based solely on those data in itself does not further our knowledge of the mechanisms underlying hyphal growth, and those data which do not fit should be made explicit. It provides an accurate description of those data, which may not be applicable to other organisms, or to the same organism under different growth conditions. On the other hand, critical experimental testing of

mechanistic models, and good fit between experimental and predicted data, increases confidence in the assumptions underlying the model. In this respect mechanistic models are of greater value in quantifying growth mechanisms which may be universal.

Agreement between experimental and predicted data will provide support for the assumptions underlying the model, whereas discrepancies may lead to modification or rejection of these assumptions. Experimental work, therefore, follows this form of mechanistic modelling, whereas it is the starting point for more descriptive, empirical modelling. This distinction leads to differences in complexity. Kinetic equations are generally relatively simple, consisting frequently of a single differential or algebraic equation. Mechanistic models are more complex to enable quantitative consideration of underlying assumptions. They usually consist of sets of differential equations and are frequently insoluble analytically and require computer simulations for generation of predictions. Finally, the range of models described in Chapter 15 is much greater than that for growth kinetics, considering, for example, tip morphology and mycelial differentiation.

15.2 GROWTH AND SHAPE OF HYPHAL TIPS

Hyphal extension occurs by expansion of the hyphal tip or extension zone resulting from incorporation, in that region, of material synthesized in and transported from distal hyphal regions. The extension zone is also the region of hypha capable of responding to external stimuli and determining the direction of hyphal extension. An understanding of events occurring in the

The Growing Fungus. Edited by Neil A.R. Gow and Geoffrey M. Gadd. Published in 1994 by Chapman & Hall, London. ISBN 0 412 46600 7

tip region is therefore essential for quantification of hyphal and mycelial growth kinetics.

15.2.1 Early models

Early models for tip growth focused on the relationship between tip shape and applied models used to describe growth of plant tips. For example, Green and King (1966) and Green (1974) applied a model for growth of tips of the alga *Nitella*. The tip was assumed to be hemispherical in shape, with the same radius (*R*) as the cylindrical hypha, and tip expansion was described by considering the change in shape of a small circular region of wall material as it moved backwards from the tip and outwards from the central longitudinal axis of the hypha. If growth is isotropic, i.e. occurring equally in both planes, the circular shape of such a region will be maintained, whereas anisotropic growth, in which longitudinal and meridional growth differ, will result in elliptical shapes. The degree of anisotropy is quantified by the allometric coefficient, which is the ratio of rates of growth in the meridional and longitudinal axes. The movement of a region of wall material over the surface of the tip, as the tip apex extends, is related to the allometric coefficient (*K*) by the equation:

$$\frac{\mathrm{d}m}{\mathrm{d}t} = A\frac{r^K}{R} \qquad (15.1)$$

where *m* is the distance from the tip, *A* is a constant and *r* is the radius of the extension zone at the appropriate point on the tip wall. *r* is 0 at the hyphal apex and equal to *R* at the base of the extension zone.

The effects of anisotropy may be investigated by varying *K* and Green and King (1966) compared predictions of equation 15.1 with results from a physical model of the extension zone consisting of an expanded rubber membrane constrained at the rim. Anisotropy could be modelled by reinforcing the membrane at different points and the model was tested for values of *K* equal to 0.5, 2, and 1 (isotropic growth). This work therefore introduced the concept of a relationship between tip shape and growth mechanisms mediated through expansion of the tip acting against wall material with different degrees of extensibility. Green (1974) proposed a more direct application of equation 15.1, relating the different degrees of extensibility to the mechanism for wall growth proposed by

Bartnicki-Garcia (1973) in which new 'packets' of wall material are incorporated from membrane-bound vesicles, after softening of existing wall by the action of wall-lytic enzymes.

The assumption of hemispherical shape leads to the prediction from equation 15.1 that the specific rate of wall expansion at any point on the extension zone wall is proportional to the cosine of the angle subtended by a line drawn from that point to the base of the extension zone and the longitudinal axis. The data quoted by Green (1974) do not show such a relationship. This represents a deficiency in the assumption of hemispherical shape, rather than a deficiency in the approach adopted by the model, as Trinci and Saunders (1977) demonstrated fungal hyphal tips to be non-hemispherical. They assumed fungal hyphal tips to be elliptical in longitudinal cross-section and quantified the degree of eccentricity of tips as the ratio between the distance from the base of extension zone to the hyphal apex and the radius of the extension zone at its base. For a hemispherical tip this ratio would have a value of 1, but Trinci and Saunders (1977) observed values ranging from 2.48 to 20.45 for fungal tips. They therefore extended and generalized Green and King's (1966) model for tips with semi-elliptical cross section with varying degrees of eccentricity. This model predicted the cosine relationship for hemispherical tips, as in the original model, but as tip eccentricity increased a cotangent relationship was predicted between the angle subtended by a point on the extension zone wall and the specific rate of wall expansion.

The cotangent relationship fits better with data quoted in Green (1974) and also with experimental data on wall expansion in a number of fungi. Three types of data were used to test the model. First, Castle (1958) measured longitudinal and meridional wall expansion in tips of sporangiophores of *Phycomyces blakesleeanus* by microscopic observation of starch grains placed on the wall surface. Second, Gooday (1971) used microautoradiography to determine the rate of incorporation of radiolabelled *N*-acetylglucosamine, a precursor of chitin, into the extension zone wall of several fungi. Third, if membrane-bound vesicles carry precursors to the extension zone, then variation in their concentration will provide a measure of variation in the rate of wall synthesis. Data of this sort were obtained by Collinge and Trinci (1974) from electron micrographs of transverse cross-sections of hyphal tips of *Neurospora crassa*. All three sets of data indicate the cotangent relationship predicted by the model of Trinci and Saunders (1977), based on the assumption of semi-elliptical tips, although direct quantitative fit between predictions and experimental data was not attempted.

Similar predictions were made by the model of da Riva Ricci and Kendrick (1972). This was essentially a computer model, which used an iterative process to simulate expansion of the extension zone, and of the hypha. The model was based on the growth of a region of wall material following incorporation and movement away from the extending tip. The increase in area of the region was proportional to its existing area and to an expansion function. When a constant expansion function was incorporated, i.e. a constant specific rate of area increase with uniform expansion over the whole tip, hyphae were generated whose diameter decreased with continued growth. Hyphae with constant diameter were generated when the specific rate of area expansion decreased with increasing distance from the hyphal apex. A decrease in expansion rate could be due to reduced supply of material to the wall, i.e. reduced vesicle incorporation, or a reduction in the ability of existing wall to incorporate such material. The latter could be due to increased rigidity or 'setting' of the wall, as suggested by da Riva Ricci and Kendrick (1972), again providing a link between tip shape, tip extension and wall expansion to be exploited in subsequent models.

15.2.2 Surface Stress Theory

The approach of Trinci and Saunders (1977), described above, involves assumptions regarding the mode of wall expansion and generation of shapes which are then compared qualitatively with those observed. An alternative approach was adopted by Saunders and Trinci (1979) and Koch (1982) and has been developed and applied widely by Koch, to describe growth of a range of unicellular microorganisms, and is termed the Surface Stress Theory (Koch, 1983). Its application to fungal growth is based on the concept of apical plasticity, proposed by Robertson (1965). The major feature of the theory is that the shape adopted by fungal hyphae results from the action of internal turgor pressure on the wall of the extension zone. Whereas turgor is considered to be evenly distributed, the composition of the tip wall will vary as a result of mechanisms by which new wall material is incorporated. Variation in wall elasticity and surface tension forces will then lead to observed hyphal tip shape.

Saunders and Trinci (1979) compared two mechanisms for tip growth, which generated different tip shapes. In the first model, wall precursors were supplied to the tip by longitudinal flow along the hypha. The density of flow was a function of distance from the longitudinal axis and growth of the wall was assumed to be proportional to the amount of material supplied, as in the models of Green (1974). They demonstrated mathematically that variation in the density of flow determined the value of the allometric coefficient but had no influence on tip shape. The second approach considered incorporation of new wall material in an elastic form, which then gradually rigidifies as the tip continues to extend, becoming completely rigid at the base of the extension zone. The extension zone was considered to be in a steady state, with internal turgor pressure balanced by surface tension forces in the wall. The pressure difference, P, across the wall was given by:

$$P = \alpha\left(\frac{1}{R_1} + \frac{1}{R_2}\right) \qquad (15.2)$$

where α represents the resistance to stretching and R_1 and R_2 are the principal radii of curvature of the surface of the extension zone. Hyphal diameter will be determined by the time taken for unset material to become rigid and extension rate will be a function of rigidification and the rate of addition of new material. The theory therefore predicts that tip shape will reflect the variation in surface tension, and will be inversely related to the rate of incorporation of new material. On the assumption that hyphal tips were semi-ellipsoidal, predictions were made of the variation in specific rates of wall expansion as a function of distance from the tip for semi-ellipsoids with different degrees of eccentricity (Figure 15.1). For a hemispherical tip, in which tip radius and length are equal, their model predicted that wall elasticity will be constant throughout the extension zone. For tips with elliptical cross-section of increasing eccentricity, a cotangent relationship was predicted, as is observed experimentally. The model therefore provides evidence that hyphal extension

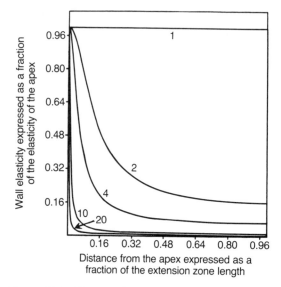

Wall elasticity expressed as a fraction of the elasticity of the apex

Distance from the apex expressed as a fraction of the extension zone length

Figure 15.1 The specific rate of wall expansion (expressed as a proportion of the maximum and equivalent to wall elasticity) as a function of distance from the apex of the extension zone for extension zones with different ratios of length to radius. (From Saunders and Trinci, 1979, with permission.)

is driven by turgor pressure acting on a cell wall within which surface tension varies due to a gradient in the rate of incorporation of new material in an elastic or 'unset' form. Rigidification of wall material was suggested to result from an increase in the thickness, length or density of microfibril components within the wall or formation of cross-linkages. There is now strong experimental evidence (Wessels, 1986) for addition of new material in a plastic form with subsequent rigidification through covalent cross-linking of wall components.

Koch (1982) made no assumptions regarding idealized tip shape but measured hyphal tip shape from electron micrographs. The surface tension (T) within the wall was then predicted using the equation:

$$S = Pr/2T \qquad (15.3)$$

where S, is the slope of the tip surface, P is the internal hydrostatic pressure and r is tip radius. Predicted values of $1/T$, proportional to S/r were then compared with experimental data on incorporation of wall material in tips from which tip-shape data had been obtained (Figure 15.2). The

high quality of fit provided good evidence for the theory although microfilaments and microtubules within the extension zone may also influence tip shape in addition to variation in surface tension.

15.2.3   The VSC model

Bartnicki-Garcia *et al.* (1989,1990) proposed an alternative to the Surface Stress Theory with an attractive model based on a vesicle supply centre (VSC) which coordinates the distribution of vesicles to the wall of the extension zone. The VSC is thought to be equivalent to the Spitzenkörper, which is seen in some fungi as an accumulation of vesicles within the extension zone (Chapter 13). Unfortunately, in other fungi the Spitzenkörper is characterized by an absence of vesicles. Vesicles are assumed to travel from distal regions to the VSC. This acts as a point source of vesicles which are projected at random and travel in straight lines and constant velocity until they hit the wall of the extension zone, where they are incorporated and increase surface area. The model assumes vesicles to be of one type and to contain all requirements for wall growth. The VSC can be stationary or can move with the tip. In the former case, the model predicts, in two dimensions, formation of a circle. In three dimensions, this is equivalent to spherical growth as seen, for example, prior to spore germination. Movement of the VSC from the centre of the spore will lead to localized bulging, as occurs in germ tube formation. If the VSC then moves in the same direction, and at a constant rate, continued hyphal extension will occur, at the same constant rate (Figure 15.3). Polarity will also be maintained as a greater proportion of vesicles will reach the wall of the extension zone than other regions of the hypha. Other assumptions regarding VSC movement lead to formation of other structures, e.g. sporangia. The above assumptions lead to the following equation for the curve (termed hyphoid) describing tip shape:

$$y = x \cot \frac{Vx}{N} \qquad (15.4)$$

where y and x are distances along longitudinal and radial axes respectively, V is the velocity of the VSC and N is the rate of vesicle production. The ratio V/N is the distance, d, between the VSC and the tip wall. Tip shape therefore depends on a single parameter which defines the shape of the entire hypha, with no discontinuity between the

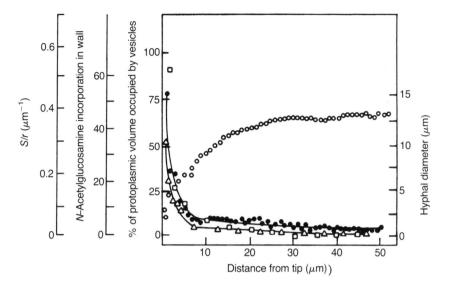

Figure 15.2 Changes in hyphal diameter (O), vesicle concentration (●) (the percentage of protoplasmic volume occupied by vesicles) and specific rate of incorporation of the wall precursor *N*-acetylglucosamine (△) (number of silver grains per 10 μm² incorporated in 1 min) with distance from the tip of *Neurospora crassa*. Open squares are values of *S/r*. (From Koch, 1982, with permission.)

extension zone and the cylindrical non-extending hypha.

Although derived in a different manner, the cotangent relationship predicted by the VSC model is similar to that of Saunders and Trinci (1979) and Bartnicki-Garcia *et al.* (1989) found a good fit between the predicted hyphoid shape and those observed experimentally, although the fit was less good for Oomycetes (e.g. *Saprolegnia*) and for tips fixed by freeze substitution rather than by conventional techniques. A further test of their model was that the VSC should be situated a distance *d* from the hyphal tip. This in fact appears to be the case for a number of fungi. Koch (personal communication) however, in a critical examination of electron micrographs of a number of fungi did not find this prediction to be borne out. In a critique of the VSC model he also identified further limitations, in its treatment of two-dimensional growth and in calculation of the increase in the region occupied by the material incorporated rather than the increase in wall area. Modification of the model to account for these deficiencies produces a more complex description of tip shape which depends on several parameters and increases discrepancies between the predicted and observed position of the Spitzenkörper. A

more fundamental problem is the assumption that wall material, once incorporated, remains at a fixed angle from the origin, rather than moving outwards as the tip extends forward.

The VSC model is attractive in its simplicity and its ability to predict a variety of observed morphologies from a limited number of parameters. Problems arise when its assumptions are examined in more detail, when critical experimental testing is performed and when detailed mechanisms for its operation are considered. Regarding the last of these, the generation of continued hyphal growth at a constant rate depends on a constant rate of movement of the VSC. There is, however, no indication how this might occur or how, as is necessary, movement is linked to mycelial growth rate, a basic feature of the Surface Stress Theory. As pointed out by Koch (personal communication), the two models are not entirely incompatible but the situation is considerably more complex than that indicated by the original VSC model.

15.3 GROWTH OF HYPHAE AND MYCELIA

Mathematical models of fungal growth operate at different levels, each level providing the higher

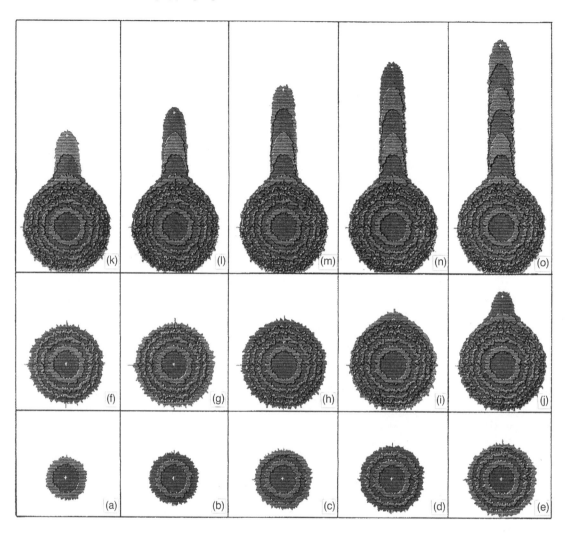

Figure 15.3   A computer simulation of spherical growth (a–h), germ tube formation (i–k) and germ tube growth (l–o) generated using the model of Bartnicki-Garcia *et al.* (1989, 1990). The position of the VSC is indicated by the cross. (From Bartnicki-Garcia *et al.*, 1989, with permission.)

level with basic information, and enabling omission of that which would merely complicate description. Thus, models of hyphal and mycelial growth may utilize information on tip growth arising from the models described above but need not concern themselves with parameters such as wall extensibility, except in situations which these are known to have influence. Other models attempt to use information at one level to describe growth at a higher level. If successful, this approach increases confidence in the mechanisms proposed at the lower level.

15.3.1   Vesicular model

Prosser and Trinci (1979) constructed a model for early growth of fungal mycelia on solid media which attempted to link proposed cellular mechanisms for hyphal growth and branching with experimentally observed growth kinetics and branching patterns. The model treats a hypha as a cylinder divided into a number of compartments. Vesicles are produced at a constant rate in distal hyphal regions and travel to the tip, at a constant rate, by movement between compartments. They

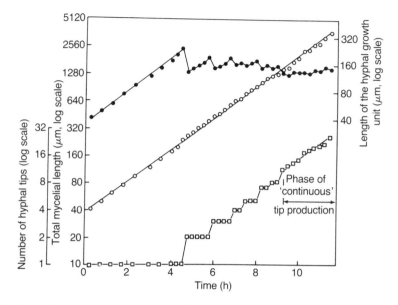

Figure 15.4   Predicted changes in total mycelial length (O), number of hyphal tips (□) and hyphal growth unit length (●) for *Geotrichum candidum* growing on solid medium. Compare with Figure 14.2. (From Prosser and Trinci, 1979, with permission.)

accumulate in the tip compartment, equivalent to the extension zone, where they fuse with existing wall and membrane. Incorporation of vesicles within the extension zone is assumed to follow saturation kinetics and leads to an increase in wall surface area and hyphal extension, dependent on the surface area of vesicle membrane and hyphal diameter. The duplication cycle concept (Trinci, 1979) is introduced by consideration of nuclear division and segregation and septation within apical compartments. Branch formation is assumed to occur following accumulation of vesicles behind septa, producing lateral branches which extend following the same growth laws as the parent hypha. Evidence for this comes from electron micrographs showing accumulation of vesicles behind septa (Trinci and Collinge, 1974) and from septation mutants of *Aspergillus nidulans* (Trinci and Morris, 1979). The latter form septa at 25°C but not at 37°C. Growth at the latter temperature is characterized by formation of both apical and lateral branches, but at 25°C only lateral branches are formed. Apical branching at 37°C could therefore have resulted from transport to the tip of vesicles whose movement would otherwise have been prevented by septa.

The model predicts changes in the length of individual hyphae, the number of branches, total mycelial length and interbranch distances. Predicted mycelial growth kinetics were similar to those observed experimentally for *Aspergillus nidulans* and *Geotrichum candidum* (Figure 15.4), indicating the validity of the assumptions on which the model was based. The model also has the flexibility for investigation of the effects of environmental parameters on cellular processes, e.g. vesicle production, septation, and determination of the ways in which such factors might affect mycelial development and differentiation.

15.3.2   Hyphal population model

Prosser and Trinci (1979) modelled mycelial growth by describing properties of individual hyphae, each growing and branching in the same manner. Edelstein (1982) and Edelstein and Segal (1983) adopted a different approach in considering the average properties of all hyphae within a mycelium. They also emphasized the role of the hyphal tip in controlling and regulating mycelial growth and were the first to consider the effects of hyphal death, hyphal fusion and different forms of branching on mycelial growth and development. Their model considered growth in two dimensions and was based on two properties of the mycelium, hyphal density (ρ) defined as hyphal length per unit area and tip density (*n*) defined as the number of tips per unit area. Changes in these properties are functions of both time and distance, and the model was formulated as a set of two partial differential equations:

$$\frac{\delta\rho}{\delta t} = nv - d(\rho) \tag{15.5}$$

$$\frac{\delta n}{\delta t} = \frac{\delta nv}{\delta x} + \sigma(\rho, n) \tag{15.6}$$

where *v* is tip extension rate, *x* is distance and $d(\rho)$ and $\sigma(\rho,n)$ are functions defining, respectively, hyphal death through autolysis (rate constant, γ_1) and production of new tips through branch formation. Thus, within a certain area, hyphal density will increase due to tip growth at a rate equal to the product of tip density and tip extension rate (*nv*, equivalent to tip flux) and will decrease through hyphal death (*d*). Changes in tip density will be determined by flux of tips within a specific area and will increase due to branching and/or decrease due to tip death or anastomosis, all of which are represented by the function σ. An important feature of the model is the use of different forms of this function to describe dichotomous and lateral branching, tip to hypha and tip to tip anastomosis and tip death due to atrophy or overcrowding. For the case of dichotomous branching, σ is equal to the product of the rate of branch formation and the number of daughters produced per tip. For lateral branching σ is the number of branches produced per unit length of hypha per unit time. Edelstein's (1982) basic model was extended by Edelstein and Segal (1983) to consider changes in substrate concentration during colony growth. Substrate concentration within the medium was described by simple diffusion and uptake by mycelium was described by Michaelis–Menten kinetics. Changes within the mycelium were also considered, being dependent on uptake, diffusive and convective flux and consumption for growth and maintenance.

Predictions of the original model were represented using phase plane analysis, in which two dependent variables were plotted against each other for a number of time-dependent solutions. This allowed investigation of the ability of colonies adopting different branching strategies to continue development. The model predicted that mycelia in which hyphae branched dichotomously or formed tip to tip anastomoses would produce colonies which were self-limiting and did not spread, with growth occurring throughout the colony and hyphal density increasing significantly at the centre. Spreading colonies were predicted to result from mycelia forming tip to hypha anastomoses, with uniform hyphal density and tip density maintained at a maximum immediately behind the colony margin. The latter is typical of mycelial growth on agar. The former may reflect growth on small, concentrated nutrient sources and provides localized increases in hyphal and tip density which act as precursors for fruiting bodies and other secondary structures. Simulations were carried out with constant values for rate constants. Within developing mycelia rate constants will change and may lead to changes in hyphal and tip density associated with differentiation. The model allows simulation of the effects of such changes but the increased complexity necessitates simulation using numerical approximation methods, and does not allow for phase plane analysis.

One frequently observed form of colony differentiation is rhythmic growth, in which alternate bands of densely and sparsely branched mycelia are formed. This is predicted by the model by considering the effect of four levels of intracellular metabolite concentration on extension rate and branching frequency. Hyphal extension does not occur below the first concentration, and reaches a maximum rate at the second. Similarly, branching frequency switches from a basal level, below a third concentration, to a higher maximum level above a fourth concentration. Predictions of the model are sensitive to the choice of the four regulating metabolite concentrations but combinations were found which predicted the regular occurrence of regions of high and low hyphal and tip density.

Predictions of the model were tested specifically with experimental data on colony growth of *Sclerotium rolfsii* (Edelstein *et al.*, 1983), incorporating lateral branching, tip-hypha anastomoses and hyphal autolysis, all of which are characteristic of mycelia of this organism. Hyphal and tip densities were predicted as functions of distance from the colony margin. These were compared with experimental data from mycelia growing on solid medium containing 0.5% and 2% glucose by weighing portions of mycelium removed from growing colonies. At both glucose concentrations hyphal density near the colony centre was low, but density near the margin increased during growth. At a glucose concentration of 0.5%, but not 2%, hyphal density decreased, in particular at the colony centre. Qualitative agreement with experimental data was good and the model also predicted that the hyphal growth unit was constant within the peripheral growth zone. Predictions of the model incorporating substrate uptake and utilization also appear qualitatively similar to experimental data of Robson *et al.* (1987) on changes in glucose concentration beneath colonies of *Rhizoctonia cerealis* (Figure 14.4).

This model represents a significant advance on previous descriptions, through its treatment of substrate uptake and utilization. Consideration of mechanisms controlling mycelial development increases the complexity of the model, necessitating numerical approximation methods for simulation, but provides much greater flexibility and wider application of the model. It also provides a basis for studying colony development and differentiation, with the capacity to predict conditions required for formation of mycelial aggregates.

As indicated above, the complexity of fungal colonies and mycelia makes difficult the formulation of mechanistic models as sets of differential equations capable of analytical solution. In addition, the requirement for consideration of spatial and temporal changes leads to use of partial differential equations, which often must be solved using numerical approximation methods. For these reasons, the finite element approach of Prosser and Trinci (1979), and other computer-based approaches, are attractive for modelling complex systems and for increasing flexibility in simulating growth under changing conditions.

15.3.3 A symmetric branching model

A link between mycelial population models and macroscopic growth was recently provided by the model of Viniegra-Gonzalez *et al.* (1993), which is based on the mathematics of symmetric trees. This considers a mycelium as a population of interbranch segments of average length L_{av}, defined as:

$$L_{av} = \frac{L_t}{N_s} \tag{15.7}$$

where L_t is the total hyphal length and N_s is the

number of segments, and equals $2(N_t - 1)$, where N_t is the total number of tips. The hyphal growth unit length is therefore approximately twice L_{av}. The model also considers the branching level, k, defined as the number of interbranch segments from the spore, minus one. If μ is the hyphal extension rate an expression can be derived for the specific growth rate of the mycelium, μ_x:

$$\mu_x = \frac{(2 \ln 2)\bar{\mu}_r}{\bar{\gamma} L_{av}} \tag{15.8}$$

where

$$\bar{\gamma} = \frac{2^{k+1} - 1)}{2^k} \tag{15.9}$$

(A bar above a character indicates an average property throughout the mycelium.) Incorporation of the definition of the hyphal growth unit with equation 15.8 leads to two relationships:

$$G = \gamma L_{av} \tag{15.10}$$

$$\mu_x = \frac{(1.386)\bar{\mu}_r}{G} \tag{15.11}$$

Equation 15.10 predicts that G will be greater at higher levels of branching. This had been suggested by Caldwell and Trinci (1973) but this expression provides a quantitative relationship which shows good agreement with their experimental data on growth of young mycelia of *G. candidum*. Equation 15.11 also allows consideration of the extent of mycelial branching and its effect on the specific growth rate.

This model is developed further to allow prediction of specific growth rate, determined in terms of biomass, incorporating a frequency distribution for the proportion of biomass which is inactive. This 'macroscopic' model predicts the observed difference in specific growth rate between germ tubes and exponentially growing mycelia. It is also used to predict growth during batch culture of *Aspergillus niger* and provides a more accurate description than the logistic equation. The model therefore provides a new approach to descriptions of mycelial growth and is valuable in linking morphological properties to kinetics. Testing of the model will be facilitated with the image analysis techniques now available for quantification of mycelial morphology.

15.3.4 Colony development

Georgiou and Shuler (1986) presented a model for colony growth and differentiation to form conidia. The model was similar to that of Edelstein (1982) in considering fungal biomass as a whole, rather than as individual hyphae, and distinguished four biomass components: vegetative biomass, competent biomass (capable of differentiation), conidiophore biomass and conidial biomass. Changes in nutrient concentration within the agar were described by simple diffusion equations. These were incorporated into the basic model of seven differential equations, describing rates of change of these four biomass components, the increase in colony radius and changes in concentrations of glucose and nitrate. Predictions were generated for the effect of glucose concentration on extension rate, which agreed favourably with available published data and predicted changes in nutrient concentrations within the agar which are qualitatively similar to those observed experimentally by Robson *et al.* (1987). In both cases, nutrient concentrations were depleted well in advance of the extending colony. It was not possible to test other predictions, particularly those regarding conidiation, through lack of experimental data.

15.3.5 Stochastic models for mycelial growth

As discussed in Chapter 14, the early development of a mycelial colony is a well ordered and regulated process in which hyphae appear to avoid each other and become oriented radially towards the colony margin, where they are evenly spaced. This leads, eventually, to the formation of a circular colony. Circular morphology might result from chemotropic responses of individual hyphae, but Hutchinson *et al.* (1980) used a stochastic model to determine whether the observed statistical variability in the properties of individual hyphae could explain circular morphology, without the need for consideration of avoidance reactions.

Stochastic models consider the probabilities of events occurring, as opposed to deterministic models in which the behaviour of a system is determined entirely by the equations describing the system and the values of model parameters and initial conditions. Stochastic models can therefore take account of natural variability within a system, and the consequences of such variability.

Hutchinson *et al.* (1980) determined variability in colonies of *Mucor hiemalis* growing on solid media and used this as a basis for a stochastic model for colony growth. Variability was measured for three characteristics of mycelial growth: hyphal extension rate, interbranch distance (internodal length) and branch angle. From experimental data for each characteristic they determined the type of distribution which best described the data and the mean and distribution around the mean for a population of hyphae in a growing mycelium.

Variability in hyphal extension rate was found to fit a half-normal distribution, with probability density function:

$$p(\chi|\delta) = (2|\pi\delta)^{0.5}\exp(-\chi^2|2\delta^2) \quad (0<\chi<\infty) \quad (15.12)$$

The mean was found to be 85.3 μm h^{-1} and δ = 107. Interbranch distance followed a gamma distribution:

$$p(\chi|\alpha) = \theta^\alpha\chi^{\alpha-1}\exp[-\theta_\chi|\Gamma(\alpha)] \quad (0<\chi<\infty) \quad (15.13)$$

The shape parameter, δ, was 1.5 and the scale parameter, θ, was 0.011. Branch angle followed a normal distribution:

$$p(\chi|\mu,\delta) = (2|\pi\delta)^{-0.5}\exp[-(\chi-\mu)^2|2\delta^2]$$
$$(-\infty<\chi<\infty) \quad (15.14)$$

The mean, μ, was 50° and the standard deviation, μ was 170.

These three distributions were then used to generate a colony from a short length of hypha, equivalent to a germ tube. At each of a series of time intervals, random numbers were generated and used to predict, from the first distribution, the increase in hyphal length during that time interval. Similar examination of the second distribution, for distribution of interbranch distances, was used to determine whether a branch was formed during that time interval. If this was the case the third distribution was used to determine the angle which the branch formed with the parent hypha. Subsequent growth of the branch was determined as for the parent hypha. Iteration of this process generated mycelia (Figure 15.5) which were similar in appearance to observed mycelia, except that branches were assumed to grow in straight lines, as the model did not consider changes in the direction of hyphal growth through avoidance reactions or for other reasons. In addition, simulated colonies expanded faster than those observed experimentally. The most interesting finding, of hyphal extension is considered, rather than assuming growth in straight lines.

15.4 GROWTH OF FUNGAL PELLETS

Growth of filamentous fungi is frequently characterized by formation of pellets. The effects of pellet formation on basic growth kinetics have been discussed in Chapter 14 and consideration of a pellet as a spherical mass of non-growing mycelium surrounded by an outer shell of active hyphae leads to prediction cube-root kinetics.

15.4.1 Application of the logistic equation

Cube-root growth kinetics have been observed experimentally (Trinci, 1970) but practical difficulties in accurately measuring biomass concentrations in liquid make distinction of different types of growth kinetics difficult. For example, Koch (1975) fitted Trinci's data to the logistic equation, modified to account for mycelial growth. The logistic equation was originally constructed to describe growth of individuals within a population and is based on the assumption that specific growth rate decreases as a negative linear function of population size, leading to the equation:

$$\frac{dN}{dt} = rN - \frac{rN^2}{k} \tag{15.15}$$

where N is the number of individuals, r is maximum specific growth rate or intrinsic rate of increase and k is the yield or carrying capacity of the environment. Solution of this equation gives:

$$N = \frac{kN_0 \exp^{rt}}{k - N_0 + N_0 \exp^{rt}} \tag{15.16}$$

where N_0 represents initial cell number. The equation predicts a sigmoid growth curve. To describe mycelial growth, Koch replaced the number of individuals by the mycelial biomass, W, occupying unit volume of space, dV. N_0 and k were replaced by the initial (S) and maximum (K) mycelial biomass per unit volume and t was replaced by $T - t$. T represents the time since growth of the colony began and t is the time at which growth first occurred within the volume element being considered. Equation 15.15 can then be rewritten for $T - t > 0$:

$$\frac{dW}{dt} = \frac{SK \exp^{r(T-t)}}{K - S - S \exp^{r(T-t)}} \, dV \tag{15.17}$$

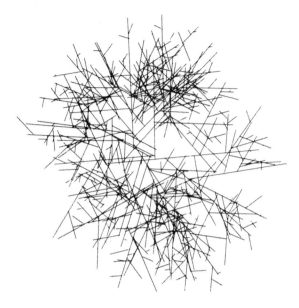

Figure 15.5 A mycelium simulated using the model of Hutchinson *et al.* (1980) after growth for 10 h. (From Hutchinson *et al.*, 1980, with permission.)

however, was that simulated colonies were circular. Thus, circular morphology may result solely from natural and observed variation in the three chosen properties of individual hyphae and does not require hyphal avoidance reactions or other mechanisms associated with regulation of hyphal extension and branch formation.

Yang *et al.* (1992) proposed a model for mycelial growth with combines the model of Prosser and Trinci (1979) with the stochastic approach adopted by Hutchinson *et al.* (1980) to describe the direction of tip growth and branching, and the site of branch formation. Frequency distributions for these mycelial properties were obtained experimentally for *Geotrichum candidum* but all were fitted to normal distributions only. The deterministic component of the model, which described hyphal extension and septation, was similar to that of Prosser and Trinci (1979), the major difference being the assumption that material moved towards the tip by diffusion, rather than by active transport. The model predicts exponential increase in total mycelial length and intercalary compartments of similar size to those observed experimentally. The model also predicts more realistic mycelial morphology as variation in the direction

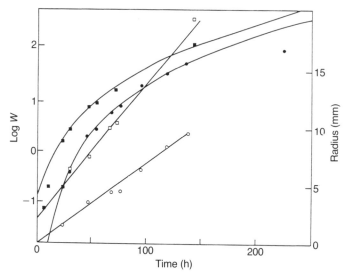

Figure 15.6 Changes in biomass (W) (solid symbols) and pellet radius (open symbols) during growth of *Aspergillus nidulans* in liquid culture at temperatures of 37°C (squares) and 25°C (circles). Symbols are experimental data of Trinci (1970) and solid lines are predictions of the model of Koch (1975). (From Koch, 1975, with permission.)

Restriction of growth to two dimensions, to form a colony of uniform height, h, extending at a constant radial growth rate, a, and with negligible lag and exponential growth periods leads to the expression:

$$W = \int_0^T \frac{SK\exp^{r(T-t)}}{K - S - S\exp^{r(T-t)}} \, 2\pi ha^2 t \, dt \qquad (15.18)$$

An equivalent expression was derived for three-dimensional growth, describing growth in liquid medium in the form of pellets and the model was solved using numerical approximation techniques. The kinetics of growth were found to depend on the ratio of final, or maximum, to initial mycelial biomass density, K/S. For values of K/S in the range 10^2–10^4, growth is predicted to be exponential and then curvilinear. Comparison of experimental data on fungal growth on solid media, where growth was restricted to two dimensions by covering agar with membranes, gave the best fit with $K/S > 100$. Biomass increased exponentially, with linear radial expansion, and the relatively high ratio reflects the need for surface colonization to enable growth above the surface and to supply aerial hyphae with nutrients from beneath the mycelium. Experimental data for pellets of *Aspergillus nidulans* were characterized best by a value of $K/S \approx 1$ and predicted exponential growth

followed by cube-root kinetics for total biomass and linear radial expansion. The fits between experimental and predicted data are illustrated in Figure 15.6. Pellet growth is characterized by rapid and simultaneous colonization of medium surrounding the pellet by densely packed hyphae covering the pellet surface. This effectively leaves no unoccupied space for subsequent growth, giving the low observed ratio of K/S.

This empirical model is important in providing a unified description of mycelial growth in liquid culture and on solid medium based on a limited number of parameters. It demonstrates the conditions required for the range of kinetics observed experimentally, linear colony expansion, exponential, square-root and cube-root kinetics for biomass and explains these in terms of the capacity of mycelia to colonize unoccupied regions of substrate.

15.4.2 Effects of oxygen limitation

Although the above approach provides information on changes in pellet biomass concentration, it provides little indication of the heterogeneity which is characteristic of pellets. The major cause of heterogeneity is diffusional limitation of both nutrients and oxygen which arises from dense hyphal packing. The extent of this limitation depends on the density of packing but in compact pellets biomass production will cease close to the surface and eventually autolysis will occur. In less dense pellets the actively growing shell is wider, substrates diffuse freely throughout the pellet and transfer also occurs via turbulent diffusion and convective flow (Schugerl et al., 1983; Wittler et al., 1986). Pirt's (1966) model of pellet growth (described in Chapter 14) was based on growth of an active peripheral shell surrounding a spherical pellet, and was used to derive an expression for the pellet radius, R_c, at which diffusion of nutrients to the pellet core is limited by hyphal packing:

$$R_c = m \left(\frac{6D'Ys_m}{\rho\mu} \right)^{1/2}$$

(15.19)

where D' represents the diffusion coefficient of the substrate, Ys_m is the biomass yield on that substrate, ρ is biomass density and μ is specific growth rate. The nutrient which is likely to limit growth first is oxygen and use of the diffusion coefficient for oxygen and growth parameters for *Penicillium chrysogenum* predicted a critical radius which compares well with those determined experimentally.

The importance of oxygen limitation in pellet growth, and in creating physiological heterogeneity within the biomass, has led to more detailed models of oxygen diffusion and utilization. For example, Aiba and Kobayashi (1971) described the specific rate of respiration within a pellet as a hyperbolic function of dissolved oxygen concentration:

$$q'_{O_2} = q_{O_2} \left(\frac{K_m + \bar{C}}{C} \right) \left(\frac{C}{K_m + C} \right)$$

(15.20)

C and $\bar{C}$ represent concentrations of oxygen within the pellet and in the surrounding medium respectively, q'_{O_2} and q_{O_2} are the respective respiration rates and K_m is a saturation constant. Calculation of the oxygen balance within the pellet requires consideration of both respiration and inward diffusion. The latter may be modelled by standard diffusion equations, leading to:

$$D \left(\frac{d^2C}{dr^2} + \frac{2dC}{rdr} \right) = 2\rho q_{O_2} \left(\frac{C}{K_m + C} \right)$$

(15.21)

where r is pellet radius and ρ represents biomass density within the pellet. This equation was used to determine the relative rates of respiration

within the pellet and in the surrounding medium (q'_{O_2}/q_{O_2}) and predictions agreed well with experimental data.

Kobayashi et al. (1973) used a similar model, assuming that the oxygen concentration at the outer pellet surface was held constant at the concentration within the medium and the concentration at the centre was zero. An effectiveness factor was introduced to describe the reduction in respiration with increasing distance from the pellet surface. They investigated several situations, with respiration constant within the pellet, varying with age or dependent on adaptation of the mycelium to oxygen limitation. Experimental data on pellet growth of *Aspergillus niger* indicated the last of these situations to be the most likely.

The predictions of models of oxygen transfer and utilization within pellets were tested using microelectrodes by Michel et al. (1992). Oxygen concentrations in pellets of *Phanerochaete chrysosporium*, 1.85 mm in diameter, decreased to undetectable levels within 0.4 mm of the surface. A model, similar to those described above, was used to determine V_{max} and K_m for oxygen within the pellet and these values were then used to predict oxygen limitation as a function of pellet size and oxygen concentration in the medium. Predictions of CO_2 evolution from populations of pellets during batch growth were also obtained and agreed well with experimental data.

15.4.3 Fragmentation of pellets

An important aspect of pellet growth is fragmentation due to shear forces. This may cause hyphal damage but fragments may also act as centres for new growth, enabling reseeding of the pellet population. Unfortunately, despite their importance, there have been few studies on the physical properties of pellets and dispersed mycelia, in part because of experimental and technical difficulties. Although, there is some information on the effects of mycelial broths on rheological characteristics little is known about hyphal tensile strength, susceptibility to breakage or fragmentation and the influence of shear stresses on hyphal activity and viability. Two theoretical approaches have been adopted, the first being purely empirical. For example, Taguchi (1971) used the following equations to describe the effect of impeller speed (N) and diameter (D_i) on pellet diameter (D_p) and the number of non-disrupted pellets (n), considering both removal of material from the pellet surface and complete rupture of pellets:

$$\frac{dD_P}{dt} = k_c(ND_i)^{5.5}D_P^{5.7} \tag{15.22}$$

$$\frac{dn}{dt} = -\alpha D_P^{3.2}N^{6.65}D_i^{8.75} \tag{15.23}$$

The parameters k_c and α are specific to the organisms and fermentor systems and are derived solely from the experimental data. Models of this type may be of practical use under these specific conditions, but do not provide insight into the mechanisms controlling pellet fragmentation and damage and are unlikely to be applicable in other situations. Van Suijdam *et al.* (1982) also used an effectiveness factor, defined as the ratio of the actual oxygen consumption rate at a particular point within the pellet to the maximum oxygen consumption rate achieved in the presence of excess oxygen. Assuming oxygen to be the limiting substrate, the effectiveness factor was used to correct specific growth rate and hence biomass production rate within pellets for the effects of diffusional limitation.

The second approach to modelling pellet growth illustrates the difference between empirical and mechanistic models. The empirical models above are relatively simple in form and are easily simulated. They are, however, based solely on experimental data, are descriptive, do not, in themselves, increase understanding of the physiology of pellet growth and are not widely applicable. The mechanistic approach, described below, is more fundamental in nature and attempts to describe growth with greater reference to the pellet physiology, but is necessarily more complex and more difficult to simulate.

A major assumption of the models described above is that cultures are homogeneous with respect to pellet size and density and that dispersed mycelial growth is negligible. This assumption is not valid and variation in pellet size will result from differences in the number of spores aggregating to give original pellets and from fragmentation processes, leading to production of smaller pellet fragments and to dispersed mycelia. The distribution of pellet size, and pellet number, is critical as it determines the proportion of actively growing mycelium, in contact with substrate, and the proportion of biomass within the pellet which may be active in secondary metabolite production, but not growth. In addition, dispersed mycelia, produced through fragmentation, will have different growth kinetics and substrate utilization characteristics to pellets.

Edelstein and Hadar (1983) were the first to consider pellet size distributions in modelling growth and biomass production by liquid cultures of *Sclerotium rolfsii*. They were also the first to consider reseeding of pellet populations by mycelial fragments produced through fragmentation due to shear damage. The model describes pellet growth as passage of spherical particles through a series of size classes, with the increase in pellet size modelled by partial differential equations describing changes in pellet radius (r). Radial growth rate is described by a logistic type equation:

$$\frac{dr}{dt} = K - \beta r \tag{15.24}$$

where K is the rate of increase in r due to growth and β is the rate of decrease in r due to shear. Pellets are lost from each size class through 'washout' or shear effects, the proportion being defined by a death term $d(r)$:

$$d(r) = C(K - \beta r)^n \tag{15.25}$$

where C and n are constants and $n \geqslant 1$. A dispersion coefficient, ε, is also introduced to account for the distribution in growth rates among pellets.

The complete model is represented by:

$$\frac{\delta p}{\delta t} = -\frac{[\delta p(K - \beta r)]}{\delta r} - d(r)p + \frac{\varepsilon \delta^2 p}{\delta r^2} \tag{15.26}$$

where p is equivalent to $p(r, t)$, the number of pellets of radius r at time t. At any time reseeding is determined by the proportion of viable fragments, ϕ, produced as a consequence of pellet break-up.

The complexity of this model prevents analytical solution but simulation was possible using numerical approximation methods. An initial size distribution was defined, based on experimental data, and size distributions were then derived for a series of time intervals during growth in batch culture. The value of ε affected total biomass and the rate of fragment accumulation. High values of φ, e.g. 50% reseeding, resulted in the development of a broad range of pellet sizes and the size distribution was characterized by a shoulder. At lower values of φ (10–30% reseeding) the size distribution showed a distinct peak and the rate of biomass accumulation was reduced.

In its basic form the model is relatively inflexible in that, for example, it is not possible to model the effects of substrate concentration on growth and the biological meaning of some of the model parameters is unclear. A more fundamental problem with the mathematical structure of the model is its instability (Tough, 1989) which prevents simulation outside a narrow range of model parameters. To solve these problems, Tough *et al.* (submitted) constructed a finite element model in which the pellet population is placed in compartments of particular size range. Pellet growth occurs by movement through compartments of increasing size and the process is reversed when fragmentation or pellet break-up occurs. Within each size class the pellet population is considered to be homogeneous and the increased flexibility of the model enables definition of growth, fragmentation and substrate utilization kinetics with greater biological meaning. Simulation of the model defines changes in biomass concentration, pellet size and substrate concentrations during batch and continuous culture growth and characterization of the effects of shear forces resulting from agitation. The model has yet to be tested for cultures of fungal pellets but predictions agree well with experimental data on pelleted growth of the filamentous actinomycete *Streptomyces coelicolor* (Tough, 1989).

15.5 CONCLUSIONS

The models described in this chapter illustrate the range of modelling approaches that have been adopted in describing fungal growth. For most, the aim has been to generate predictions, capable of experimental verification, which are based on assumptions regarding fundamental mechanisms of growth. The link between theory and experimental is essential. If predictions are not obeyed, assumptions must be modified or alternative mechanisms adopted. Successful prediction of observed behaviour, however, increases confidence in the mechanisms proposed by a model. This verification process is made more exacting by the quantitative approach, which provides more precise assumptions and more critical comparison of experimental and theoretical results.

This review is not exhaustive and has concentrated on vegetative growth and morphology of filamentous fungi. Although some aspects of fungal fermentations have been considered, these have concentrated on behaviour particular to the filamentous growth form. Models of product formation exist for both filamentous and unicellular fungi, but were considered to be outside the scope of this chapter. Shared regulatory mechanisms for unicellular and multicellular growth may also provide the link between models for these different growth forms. Models of a more ecological nature, e.g. those describing the epidemiology of fungal diseases, have also been omitted. The models discussed provide the basis for such studies and are seen to operate at different levels. These levels are linked by extraction of important and relevant information from models at adjacent levels. This prevents unmanageable complexity although mechanistic models generally consist of sets of equations that cannot be solved analytically. This complexity increases dependence on computer-based simulations, involving either numerical approximation of differential equations or initial construction of a computer model. The latter approach has advantages in flexibility and in transparency, which is particularly important when modelling is carried out by those not well-qualified in mathematics.

The formulation of hypotheses as mathematical models frequently exposes deficiencies in our understanding of biological processes. In this respect, it is interesting to note that there is still considerable debate regarding quantitative aspects of processes as fundamental as tip growth and branch formation. In many cases the challenge lies with the experimentalist to obtain relevant data with which to test models critically. In this respect, the modeller may be seen as directing experimental work by indicating key questions to be tackled. This challenge will become particularly important when the information on vegetative growth and branching, which forms the basis of this chapter, is extended to quantify the more complex processes involved in formation of secondary structures and in mycelial differentiation.

REFERENCES

Aiba, S. and Kobayashi, K. (1971) Comments on oxygen transfer within a mold pellet. *Biotechnology and Bioengineering*, **12**, 583–8.

Bartnicki-Garcia, S. (1973) Fundamental aspects of hyphal morphogenesis. *Symposia of the Society of General Microbiology*, **23**, 245–67.

Bartnicki-Garcia, S., Hergert, F. and Gierz, G. (1989) Computer simulation of fungal morphogenesis and the mathematical basis of hyphal (tip) growth. *Protoplasma*, **153**, 46–57.

Bartnicki-Garcia, S., Hergert, F. and Gierz, G. (1990) A novel computer model for generation of cell shape: application to fungal morphogenesis, in *Biochemistry of Cell Walls and Membranes in Fungi*, (eds P.J. Kuhn, A.P.J. Trinci, M.J. Jung and L.G. Copping), Springer-Verlag, Berlin, pp. 43–60.

Caldwell, I.Y. and Trinci, A.P.J. (1973) The growth unit of the mould *Geotrichum candidum*. *Archiv fur Mikrobiologie*, **88**, 1–10.

Castle, E.S. (1958) The topography of tip growth in a plant cell. *Journal of General Physiology*, **41**, 913–26.

Collinge, A.J., and Trinci, A.P.J. (1974) Hyphal tips of wild-type and spreading colonial mutants of *Neurospora crassa*. *Archiv fur Mikrobiologie*, **99**, 353–68.

da Riva Ricci, D. and Kendrick, B. (1972) Computer modeling of hyphal tip growth in fungi. *Canadian Journal of Botany*, **50**, 2455–62.

Edelstein, L. (1982) The propagation of fungal colonies: a model for tissue growth. *Journal of Theoretical Biology*, **98**, 697–701.

Edelstein, L. and Hadar, Y. (1983) A model for pellet size distributions in submerged cultures. *Journal of Theoretical Biology*, **105**, 427–52.

Edelstein, L. and Segal, L.A. (1983). Growth and metabolism in mycelial fungi. *Journal of Theoretical Biology*, **104**, 187–210.

Edelstein, L. Hadar, Y., Chet, I. *et al.* (1983) A model for fungal colony growth applied to *Sclerotium rolfsii*. *Journal of General Microbiology*, **129**, 1873–81.

Georgiou, G. and Shuler, M.L. (1986) A computer model for the growth and differentiation of a fungal colony on solid substrate. *Biotechnology and Bioengineering*, **28**, 405–16.

Gooday, G.W. (1971) An autoradiographic study of hyphal growth of some fungi. *Journal of General Microbiology*, **67**, 125–33.

Green, P.B. (1974) Morphogenesis of the cell and organ axis – biophysical models. *Brookhaven Symposium on Biology*, **25**, 166–90.

Green, P.B. and King, A. (1966) A mechanism for the origin of specifically oriented texture in development with special reference to *Nitella* wall texture. *Australian Journal of Biological Science*, **19**, 421–37.

Hutchinson, S.A., Sharma, P., Clarke, K.R. and Macdonald, I. (1980) Control of hyphal orientation in colonies of *Mucor hiemalis*. *Transactions of the British Mycological Society*, **75**, 177–91.

Kobayashi, T., van Dedem, G. and Moo-Young, M. (1973) Oxygen transfer into mycelial pellets. *Biotechnology and Bioengineering*, **15**, 27–45.

Koch, A.L. (1975) The kinetics of mycelial growth. *Journal of General Microbiology*, **89**, 209–16.

Koch, A.L. (1982) The shape of hyphal tips of fungi. *Journal of General Microbiology*, **128**, 947–51.

Koch, A.L. (1983) The surface stress theory of microbial morphogenesis. *Advances in Microbial Physiology*, **29**, 301–66.

Michel, F.C., Grulke, E.A. and Reddy, C.A. (1992) Determination of the respiration kinetics for mycelial pellets of *Phanerochaete chrysosporium*. *Biotechnology and Bioengineering*, **58**, 1740–5.

Pirt, S.J. (1966) A theory of the mode of growth of fungi in the form of pellets in submerged culture. *Proceedings of the Royal Society Series B*, **166**, 369–73.

Prosser, J.I. and Trinci, A.P.J. (1979) A model for hyphal growth and branching. *Journal of General Microbiology*, **111**, 153–64.

Robertson, N.F. (1965) The fungal hypha. *Transactions of the British Mycological Society*, **48**, 1–8.

Robson, G.D., Bell, S.D., Kuhn, P.J. and Trinci, A.P.J. (1987) Glucose and penicillin concentrations in agar medium below fungal colonies. *Journal of General Microbiology*, **133**, 361–367.

Saunders, P.T. and Trinci, A.P.J. (1979) Determination of tip shape in fungal hyphae. *Journal of General Microbiology*, **110**, 469–73.

Schugerl, K., Wittler, R. and Lorentz, T. (1983) The use of molds in pellet form. *Trends in Biotechnology*, **1**, 120–2.

Taguchi, H. (1971) The nature of fermentation fluids. *Advances in Biochemical Engineering*, **1**, 1–30.

Tough, A. (1989) A theoretical model of streptomycete growth in submerged culture. PhD Thesis, University of Aberdeen, Aberdeen, Scotland.

Trinci, A.P.J. (1970) Kinetics of the growth of mycelial pellets of *Aspergillus nidulans*, *Archiv fur Mikrobiologie*, **73**, 353–67.

Trinci, A.P.J. (1979) Duplication cycle and branching in fungi, in *Fungal Walls and Hyphal Growth* (eds J.H. Burnett and A.P.J. Trinci), Cambridge University Press, Cambridge, pp. 319–58.

Trinci, A.P.J. and Collinge, A. (1974) Occlusion of the septal pores of damaged hyphae of *Neurospora crassa* by hexagonal crystals. *Protoplasma*, **80**, 56–67.

Trinci, A.P.J. and Morris, N.R. (1979) Morphology and growth of a temperature-sensitive mutant of *Aspergillus nidulans* which forms aspetate mycelia at nonpermissive temperatures. *Journal of General Microbiology*, **114**, 53–9.

Trinci, A.P.J. and Saunders, P.T. (1977) Tip growth of fungal hyphae. *Journal of General Microbiology*, **113**, 243–8.

van Suijdam, J.C., Hols, H. and Kossen, N.W.F. (1982) Unstructured model for growth of mycelial pellets in submerged cultures. *Biotechnology and Bioengineering*, **24**, 177–91.

Viniegra-Gonzalez, G., Saucedo-Castaneda, G., Lopez-Isunza, F. and Favela-Torres, E. (1993) Symmetric branching model for the kinetics of mycelial growth. *Biotechnology and Bioengineering*, **42**, 1–10.

Wessels, J.G.H. (1986) Cell wall synthesis in apical hyphal growth. *International Review of Cytology*, **104**, 37–79.

Wittler, R., Baumgartl, H., Lubbers, D.W. and Schugerl, K. (1986) Investigations of oxygen transfer into *Penicillium chrysogenum* pellets by microprobe measurement. *Biotechnology and Bioengineering*, **28**, 1024–36.

Yang, H., King, R., Reichl, U. and Gilles, E.D. (1992) Mathematical model for apical growth, septation, and branching of mycelial microorganisms. *Biotechnology and Bioengineering*, **39**, 49–58.

PART FIVE

DIFFERENTIATION

SPORULATION OF LOWER FUNGI

G.W. Beakes

Department of Biological and Nutritional Sciences, The University, Newcastle upon Tyne, UK

16.1 INTRODUCTION

16.1.1 The lower fungi

Unlike the 'higher fungi' described in later chapters the 'lower fungi' are a very diverse and polyphyletic assemblage of organisms. Ultrastructural (Moss 1986; Beakes, 1987; Barr, 1992; Powell, 1994) and molecular studies (Bowman *et al.*, 1992) reveal that many of the zoosporic groups within the 'lower fungi' (Oomycetes, Plasmodiophorales, Thraustochytriales) show closer evolutionary affinities with the protists and chromophyte algae than they do with most other fungi. The uniflagellate Chytridiomycetes appear to be on the same evolutionary line as the terrestrial fungi, including the non-motile Zygomycetes. Because of shared morphological similarities (e.g. non-septate hyphae and the production of sporangiospores), the latter have traditionally been grouped in the lower fungi. Spore formation in the lower fungi involves the localized delimitation of cytoplasmically enriched regions of the thallus (Figures 16.1, 16.2, 16.12, 16.14, 16.19a, 16.21, 16.42, 16.44a and 16.45), which is either enveloped by a thick protective wall (Figures 16.44, 16.47) or cleaved into spores (Figures 16.1, 16.3, 16.15, 16.19b, 16.20b, 16.22a,b). Many of the group are aquatic (water moulds) and produce motile zoospores which may be uniflagellate (Chytridiomycetes, Hyphochytriomycetes), biflagellate (Figure 16.9; Oomycetes, Thraustochytriales) or even polyflagellate (as in the rumen chytrid *Neocallimastix*). In contrast the terrestrial mucoralean 'pin moulds' form non-motile sporangiospores (Figures 16.11 and 16.12), which are dispersed by wind and or water droplets. They typically have spores coated in a hydrophobic surface layer with a characteristic rodlet structure (Figure 16.46) which their aquatic counterparts lack.

Spore development in the lower fungi has been widely documented since sporulation in these organisms can usually be induced with a high degree of predictability and synchrony by the relatively simple manipulation of environmental conditions and most were amenable to fixation protocols for examination at the ultrastructural level (Gay and Greenwood, 1966; Bracker, 1968; Barstow and Lovett, 1975). Unfortunately, compared with the 'higher fungi' many of these fungi have proved less amenable for genetic studies and relatively little use has been made of developmental mutants. For instance oomycete fungi are either diploid or polyploid which makes the generation of mutants more difficult and their genetic analysis has been further hindered by the low levels of germination often obtained from the resulting sexual progeny. These organisms have also proved relatively difficult to manipulate using the tools of modern molecular biology, although successful transformation protocols have recently been reported in Peronosporales. This account, although concentrating upon the aquatic zoosporic fungi, will where appropriate, compare and contrast development with the terrestrial zygomycete fungi with which they have been traditionally grouped.

16.1.2 Different spore types and their ecological roles

Throughout the lower fungi three major types of spore are produced:

1. asexual dispersive spores, e.g. zoospores (Figure 16.9), aplanospores (Figure 16.4), sporangiospores (Figures 16.11 and 16.12),
2. sexual perennating resting spores, e.g. resting bodies/sporangia (Figure 16.39), oospores (Figure 16.38), zygospores (Figure 16.37) and,

The Growing Fungus. Edited by Neil A.R. Gow and Geoffrey M. Gadd. Published in 1994 by Chapman & Hall, London. ISBN 0 412 46600 7

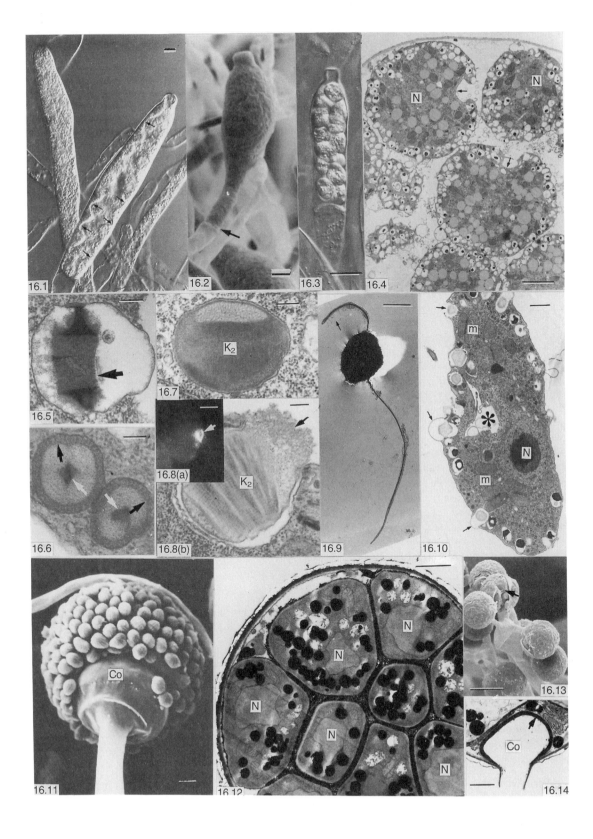

Figures 16.1–16.4   Stages of zoosporogenesis in the water moulds *Saprolegnia diclina* (Figures 16.1–16.3) and *Achlya flagellata* (Figure 16.4).

Figure 16.1   Zoosporangium initial and prespore stages. Note infurrowing of central vacuole to form cleavage furrows (arrows) delimiting spore initials (interference contrast). Bar = 10 μm.

Figure 16.2   SEM of differentiated sporangium, showing apical papillum and basal septum (arrowed). Bar = 5 μm.

Figure 16.3   Mature sporangium containing fully differentiated zoospores (interference contrast). Bar = 10 μm.

Figure 16.4   TEM of mature sporangium of *Achlya* containing differentiated primary 'aplanospores'. Note characteristic notch on one side of spore (arrowed). Bar = 3 μm.

Figures 16.5–16.8   Cytoplasmic organelles associated with sporogenesis in water moulds (Saprolegniaceae).

Figure 16.5   Fingerprint/dense body vesicle from a differentiating sporangium of *Achlya flagellata* (as in Figure 16.4). Note characteristic fingerprint-like lammellations of electron dense inclusion body (arrowed). Bar = 100 nm.

Figure 16.6   Encystment vesicles from the saprolegnian water mould *Calyptralegnia*. Note structured peripheral cortex (dark arrows) and electron-dense cross-sectional profile of short spine-like cyst coat ornamentation (white arrows). Bar = 100 nm.

Figure 16.7   Secondary kinetosome-associated K_2 body of *Achlya flagellata*. Note dense particulate matrix and upper tubule containing cavity. Bar = 200 nm.

Figure 16.8(a)   FITC-wheat-germ agglutinin (WGA) stained secondary zoospore of *S. diclina* showing ventral pad associated with discharging K_2-bodies in secondary zoospores. Bar = 5 μm. (b) Discharging K_2-body in a secondary zoospore of *S. diclina* showing the dispersing of the pad of adhesive material which appears to be derived from the tubule cotaining cavity. Bar = 200 nm.

Figure 16.9   Whole mount preparation of a secondary-type zoospore of the downy mildew pathogen *Pseudoperonospora humuli*. Note the fine hairs on the anterior 'flimmer' flagellum (arrowed). Bar = 5 μm.

Figure 16.10   Thin section LS profile of a secondary zoospore of *Achlya flagellata*. The spore contains a single nucleus (N) and many mitochondria (m). Note the cortical large peripheral vesicles which distend or even rupture the spore membrane (small arrows). The water expulsion apparatus is asterisked (*). Bar = 1.0 μm.

Figures 16.11–16.14   Sporogenesis in mucoralean Zygomycetes.

Figure 16.11   Mature sporangium of *Rhizopus* sp. following breakdown of outer sporangial wall. Note differentiated columella (Co) and distinctive ridges on surface of spores. Unpublished micrograph courtesy of T.A. Booth, BioMedical EM Laboratory, Newcastle University. Bar = 5 μm.

Figure 16.12   Section through a miniature sporangium (sporangiole) of the muroralean fungus *Thamnidium elegans*. Unlike sporangiospores of *Mucor*, each sporangiole spore is uninucleate (N). Note abundant electron-dense lipid globules in this spore. Bar = 2.0 μm.

Figure 16.13   SEM micrograph of branched sporangioles of *Thamnidium elegans*. A ruptured sporangiole wall revealing sporangiospores is arrowed. Bar = 10 μm.

Figure 16.14   Profile through base of *Thamnidium* sporangiole showing thickened wall (arrowed) to the columella (Co). Bar = 1.0 μm.

Figures 16.12–16.14   from Beakes and Campos-Takaki (1984), with permission.

3. an asexual perennating phase, e.g. gemmae (Figure 16.42), chlamydospores (Figure 16.44), thallo/arthrospores (Figures 16.43, 16.45 and 16.47).

Sporulation is typically induced by nutrient depletion (Cantino and Turian, 1958; Gay and Greenwood, 1966, Griffin and Breuker, 1969). The first phase of sporulation involves the production of large numbers of small, 3–10 μm diameter, uninucleate spores (Gay and Greenwood, 1966; Armbruster, 1982; Beakes and Campos-Takaki, 1984; Hyde *et al.*, 1991a; Figures 16.4, 16.10 and 16.12) although mucoralean sporangiospores are often multinucleate (Bracker, 1968; Hammill, 1981). These spores ensure the localized dispersion of the fungus over relatively short distances so that the organism can fully exploit a utilizable food base. Such asexual spores are usually relatively thin walled (Figure 16.12) or, in the case of motile zoospores naked (Figures 16.4 and 16.10), although even these often possess a glycocalyx

associated with their plasma membrane (Beakes *et al.*, 1993c; Powell, 1994). They contain endogenous storage reserves such as lipids and glycogen (Figures 16.4, 16.10 and 16.12) but have no endogenous dormancy mechanisms and germinate immediately once they locate a suitable substrate. Resting spores usually arise as a result of sexual reproduction following the fusion of male and female gametes (Figure 16.27c), whole thalli (Figures 16.27a, 16.34 and 16.36) or delimited gametangia (Figures 16.31, 16.32 and 16.28b). Following cytoplasmic fusion (plasmogamy) nuclear fusion (karyogamy) may take place immediately or may be delayed until later in differentiation. Typically spores which arise as a result of sexual interactions, such as oospores (Figure 16.38) resting sporangia (Figure 16.39) and zygospores (Figure 16.37) are large, 15–30 μm diameter, thick walled and packed with storage reserves such as lipids and glucans (Figures 16.35, 16.36, 16.38 and 16.39). Unlike their asexual spores, resting spores are not usually actively

---

Figure 16.15 Diagram illustrating main cytoplasmic cleavage mechanisms in the lower fungi. (a) Series of diagrams illustrating sporangial cleavage in *Phytophthora* based on rapid freeze substitution fixation. Specialized cleavage vesicles arise from Golgi dictyosomes associated with nucleus apices (A). The nuclei in the sporangial periphery are orientated towards the wall. Paired sheets of membrane reflex back from the nuclear apices, whilst a continuous sheet of membrane runs parallel to the sporangium wall and differentiates a thin peripheral periplasmic layer (B). These membranous sheets completely delimit the uninucleate spore initials and fuse with the cortical cisternum. The thin cortical shell of periplasmic cytoplasm disappears, presumably having disintegrated and become assimilated in the peripheral layer of matrical material (C). In the final stages of differentiation the fully formed zoospores assume their more rounded shape with a flagellar groove forming opposite each nuclear pole (D). Modified and adapted from Hyde *et al.* (1991) with permission. (b) Schematic summary diagrams illustrating variations in cleavage patterns reported in aquatic, zoosporic, fungi. The arrows indicate the suggested direction of growth of the cleavage furrows. (A), Centrifugal (outwardly progressing cleavage furrows) cleavage, exemplified by zoospore and oospore differentiation in *Saprolegnia*. Spores are delimited either by a redistribution of a pre-existing central vacuole or by the addition of smaller peripheral cleavage vesicles. (B), Delimitation of spore initials by the alignment of small cleavage vesicles along presumptive lines of cleavage. These vesicles grow outwards and coalesce with adjacent cleavage vesicles and perhaps also with the infurrowing plasma membrane. This pattern appears to be common in the chytridiales, although it is based largely on observations using conventially fixed material. (C), The centripetal differentiation of spores by an infurrowing cleavage system in which spore initials are defined by furrows derived from ingrowth of plasma membrane. In some *Pythium* sp. this cleavage membrane system is derived from the localized fusion of Golgi vesicles in the region of the flagellar groove. The furrows also coalesce with a central vacuole to delimit the spore initials (c upper segment). In the marine fungus *Thraustochytrium*, there is no central vacuole and the infurrowing membranes may be added to by the coaslescence of cytoplasmic vesicles (c lower segment). (D), A variant of the centripetal ingrowth system occurs in *Zygorhizidium* where the cytoplasm is divided into progressively smaller blocks until individual initials are differentiated. (c) Diagrammatic summary of cleavage vesicle disposition during sporangiospore formation in mucoralean fungi. In these fungi the cleavage vesicles can be identified because they are lined with sculptured material which eventually forms the outermost sculptured coat to the sporangiospores. In the early 'precleavage' phase (A) small spherical cleavage vesicles appear and increase in size. During the early stages of spore cleavage, the cleavage vesicles coalesce to give rise to cisternae which delimit both the sporangiospores and columella (co)(B). During the final stages of sporogenesis (C) the delimited spores acquire a thickened wall and their characteristic shape and the columella wall (cw) thickens. In the final stage of differentiation the outer sporangium wall (sw) ruptures. Based on diagrams in Bracker (1968) and Hammill (1981).

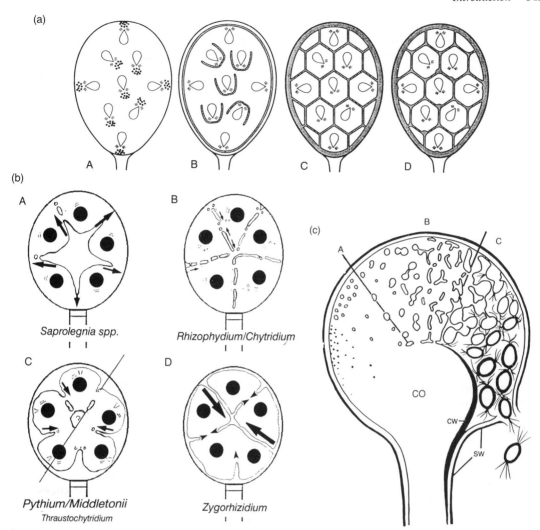

Figure 16.15

dispersed from their site of formation, but often remain *in situ* and are released into the soil or sediments after the rotting away of the host/ substrate tissues. Often these spores possess a constitutive (endogenous) dormancy mechanism and require a period of months or even years before germination can take place (Beakes and Gay, 1978a; Kerwin *et al.*, 1986; Beakes and Bartnicki-Garcia, 1989). The final group of asexual resting spore types (Figures 16.42–16.45 and 16.47) are somewhat intermediate between the two spores just outlined. They are asexual spores, but are often quite large, 10 < 50 μm, thick walled and multinucleate (Figures 16.42, 16.44 and 16.47). As

with resting spores they contain abundant reserves of glucans and lipids (Figures 16.44 and 16.47). These spores either remain *in situ* or are liberated from the parent thallus by disarticulation at their septa (Powell *et al.*, 1981; Barrera, 1983; Beakes *et al.*, 1984a) or by the lysis of their sub-tending hyphal walls (Powell and Blackwell, 1991). Such spores readily germinate when nutrients are replenished (Hemmes and Wong, 1975; Beakes *et al.*, 1984b). Formation of these spores enables the fungus to preserve as much of its viable mycelial cytoplasm as possible by packaging and protecting it in a thick cell wall. This strategy ensures short-term survival once thallus senescence has set in and environmental conditions have become less favourable for asexual sporulation.

Selected aspects of the differentiation of these three types of spore will now be considered in turn.

16.2　ASEXUAL SPORULATION

16.2.1　Sporulation specific organelles

Once the switch from vegetative growth to asexual sporulation has been triggered, hyphal extension ceases and cytoplasm accumulates in the tips of hyphae which swell to form typically cylindrical (Figure 16.1), ovoidal (Figure 16.21) or spheroidal (Figure 16.13) sporangia. As well as the accumulation and concentration of pre-existing cytoplasmic organelles such as nuclei, mitochondria and lipid globules, new sporulation specific organelles appear which are not present in the cytoplasm of vegetative hyphae. The possible origins of, inter-relationships between and fate of these various sporulation vesicle populations in an idealized 'sporulating oomycete' has been summarized diagrammatically in Figure 16.16. In oomycetes one of the first sporulation organelles to appear are the so-called 'densebody/fingerprint' vesicles (Gay and Greenwood, 1966) which contain electron-dense globular inclusions which develop a characteristic multilamellate sub-structure (Figures 16.5 and 16.16). These vesicles accumulate in developing sporangia and oogonia (Figure 16.30b) and their resulting zoospores (Figures 16.4 and 16.10; Beakes, 1983; Hemmes, 1983) and oospheres (Figure 16.33) (Beakes and Gay, 1978a; Beakes, 1980a; Hemmes, 1983). They contain phosphorylated β(1–3) glucans and are thought to be a significant carbohydrate storage reserve in sporangia, zoospores and young oogonia (Wang and Bartnicki-Garcia, 1980). In germinating cysts they are rapidly assimilated into the expanding somatic vesicle system and the electron-dense granules soon disappear (Figure 16.16).

In *Blastocladiella emersonii*, membrane bound vesicles containing electron-dense cup-shaped inclusion bodies, known as gamma-bodies, also accumulate during zoosporogenesis (Barstow and Lovett, 1975). It was thought that they contained chitin synthetase enzymes and were primarily involved in chitin biosynthesis and cell-wall biogenesis upon encystment (Mills and Cantino, 1981). However, subsequently it has been shown that these vesicles contain two major sporulation-specific proteins (41 000 and 43 000 mol wt.) but no chitin synthetase activity was detected (Hohn et al., 1984). This suggests that, as with oomycete dense body vesicles, this fraction is primarily a storage vesicle.

In saprolegnian oomycetes a series of membrane-bound organelles appear in the sporangial cytoplasm before or around the time of sporangium delimitation (Gay and Greenwood, 1966; Armbruster 1982; Beakes, 1983). These include (a) packets of proteinaceous flimmer tubules which eventually decorate the anterior flagellum of the zoospore (Figure 16.9), (b) bar-bodies/encystment vesicles (Figure 16.6) which secrete a preformed electron-dense outer layer to the cyst wall and associated spines (Beakes, 1983), (c) conspicuous kinetosome-associated vesicles (K-bodies; Figure 16.7) which give rise to a ventral pad of adhesive material (Figures 16.8a, b) (Lehnen and Powell, 1989) and (d) a vesicle fraction known in *Phytophthora* as the large-peripheral vesicles (LPV) and in *Saprolegnia* as the fibrillar vesicles (FV). The latter accumulate in the zoospore cortex and were originally thought to have an adhesive function (Beakes, 1983) but it now seems more likely that they contain glycoprotein storage proteins (Hardham et al., 1991; Burr and Beakes, 1994). Monoclonal antibody (MAb) markers to most of these vesicle fractions have now been obtained in *Phytophthora cinnamomi* (Hardham et al., 1991) (Figure 16.16) and have enabled the precise spatial and temporal functions of these various sporulation organelles to be elucidated (e.g. Hyde et al., 1991a). For instance, these antibody probes have also been used to explore the time at which the various vesicle fractions appear following mineral salts induction of sporulation. Both the LPV vesicles and the ventral vesicles (K-body equivalents) are synthesized early in differentiation and reach maximal levels within 5 h of induction, whereas the dorsal vesicles (encystment vesicle equivalents) appear somewhat later at around 9 h. Developing sporangia cannot be observed until around 7 h after induction which means that the synthesis of many specialized sporulation organelles in *Phytophthora* is taking place in the hyphae before there is any

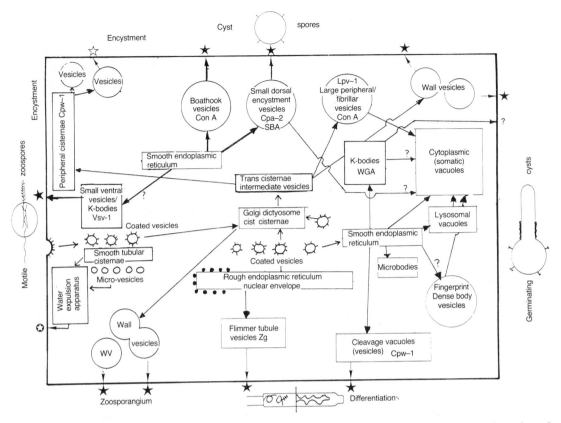

Figure 16.16 Summary diagram illustrating the complex interrelationships between the main membrane-bound organelles formed from the 'endomembrane system' during zoosporogenesis in oomycete fungi (largely based upon observations on *Saprolegnia, Achlya* and *Phytophthora*). The hub of the system is represented by the vesicle-generating organelles, the Golgi dictyosome and cisternae of smooth and rough endoplasmic reticulum. The lower segment of the diagram represents the developing sporangium, the left hand side the swimming zoospore, the upper segment the encysting zoospore and cyst and the right hand side the germinating cyst (and return to vegetative differentiation). The stars indicate the fusion of the various endomembrane components at various stages of differentiation. The names of various vesicle fractions are indicated as are lectin labelling properties and monoclonal antibodies which have been raised to these vesicles in *Phytophthora cinnamomi* (Hardham *et al.*, 1991). Unused sporogenesis organelles are mostly re-assimilated in the somatic vacuole system during cyst germination and the re-establishment of the vegetative growth phase. Based on personal interpretations and papers of Beakes (1983), Hardham *et al.* (1991), Lehnen and Powell (1989).

microscopic evidence of sporangium formation (Hardham, personal communication).

16.2.2 Biochemical changes associated with zoosporogenesis

Zoosporogenesis in aquatic fungi is associated with dramatic changes in the levels of gene expression. In *Blastocladiella* it has been estimated that 80% of the genome is transcribed during zoosporogenesis (Johnson and Lovett, 1984)! The highest level of complexity regarding gene expression was observed during the period of zoospore differentiation. In *Achlya ambisexualis* also, it has been shown that DNA-dependent RNA synthesis is essential for sporangium differentiation (Griffin and Breuker, 1969; Horgen and O'Day 1975; Timberlake *et al.*, 1973). A number of unusual phosphorylated compounds are involved in the transcriptional regulation of asexual differentiation in this species (Lejohn *et al.*, 1977; Horgen, 1981). If protein synthesis is blocked by the addition of inhibitors, such as cycloheximide, sporangium differentiation is rapidly inhibited (Timberlake *et al.*, 1973), indicating that development is largely dependent on newly transcribed and synthesized proteins. Sporangium differentiation in *A. ambisexualis* is associated with an increase in a variety of lysosomal enzymes (e.g. ribonucleases, α-mannosidases, acid phosphatases) whose production reaches a peak immediately preceding spore cleavage (Horgen and O'Day, 1975; Sutherland *et al.*, 1976). It is proposed that differentiation therefore relies on a rapid turnover of pre-existing cytoplasmic components, such as amino acids and nucleic acids, in the absence of endogenous nutrients and net synthesis (Timberlake *et al.*, 1973). In spite of much effort, it is disappointing that to date none of the genes controlling the synthesis of sporogenesis specific organelles has yet been identified. However, the gene encoding β-tubulin has recently be detected in cDNA libraries made from sporulating *Phytophthora cinnamomi* (Lehnen personal communication). In *A. ambisexualis*, lipid is accumulated during zoosporogenesis to a level about 40% above that in the undifferentiated mycelium (Law and Burton, 1976) and about 10% of the dry weight of the asexual spores is composed of lipid.

In contrast to the high rates of *de novo* protein synthesis in the differentiating sporangia, differentiated zoospores do not apparently synthesize proteins (Jaworski and Stumhofer, 1984), although they are rich in polyadenylated messenger RNA. Interestingly, most of this poly(A) mRNA was found to be already complexed with the ribosomes and it was suggested that protein synthesis in these motile spores is inhibited by an endogenous inhibitor. Zoospores of *Blastocladiella emersonii* encyst within minutes and this involves the rapid synthesis of a new chitinous wall (Selitrennikoff *et al.*, 1980). This rapid synthesis is actually achieved by the instantaneous alleviation of endproduct inhibition of the hexosamine biosynthetic pathway rather than by invoking *de novo* enzyme biosynthesis (Selitrennikoff *et al.*, 1980).

16.2.3 Spore formation and cytoplasmic cleavage

In zoosporic fungi, following the delimitation of the sporangium by a crosswall, the cytoplasmic mass differentiates into uninucleate, flagellate zoospores (Figures 16.9–16.11), although some genera such as *Achlya* produce non-motile aplano or sporangiospores (Figure 16.4) (Armbruster, 1982). In zygomycete fungi the differentiation of the cytoplasmic mass within the globular sporangium simultaneously delimits both the spores and septum/columella (Figures 16.14 and 16.15c, Bracker, 1968; Hammill, 1981). In the chytrid *Karlingia* (Figure 16.26) and the oomycete *Phytopththora* (Figures 16.21 and 16.22) the initiation of spore differentiation within the fully formed sporangium may require an additional stimulus such as a light or temperature shock. One of the fundamental questions relating to asexual sporogenesis is the mechanism by which the mass of cytoplasm becomes precisely differentiated into spores of uniform size and organelle complement. In oomycete fungi, the timing of various events varies sequentially and spatially from one genus to another (Figure 16.17). In some species flagellum differentiation precedes the cleavage of cytoplasm (e.g. *Lagenidium giganteum*, *Blastocladiella emersonii* and *Coelomomyces dodgei*) whereas in *Saprolegnia* spp. (Figure 16.17) and many chytridialian species (Figures 16.19 and 16.20) it occurs during the late stages of cleavage or even after spore differentiation (Gay and Greenwood, 1966; Beakes *et al.*, 1993a). In members of the Peronosporales such as *Phytophthora cinnamomi* (Figure 16.17) cleavage and flagellum differentiation are concurrent events (Hyde *et al.*, 1991a,b). A further source of variation in oomycete fungi is that some genera (e.g. *Lagenidium* and *Pythium*) discharge their sporangial contents into a thin-walled evanescent vesicle prior to cytokinesis (Figure 16.17).

The precise way in which the sporangium cytoplasm is cleaved into spores and the mechanism by which such a high degree of precision is attained has long fascinated mycologists and cell biologists. The mechanism of cytokinesis in *Phytophthora cinnamomi* has recently had to be re-evaluated in the light of significantly different patterns of development observed in rapid-freezing and freeze-substituted (RF-FS) fixed sporangia compared with conventionally fixed ones (Figures 16.15a A–D and 16.22) (Hyde *et al.*, 1991a,b). It had been widely reported that cleavage was brought about by the alignment of elongate or spherical electron-transparent vesicles derived from the Golgi along presumptive lines of cleavage (Figure 16.22a Hyde *et al.*, 1991a) followed by their

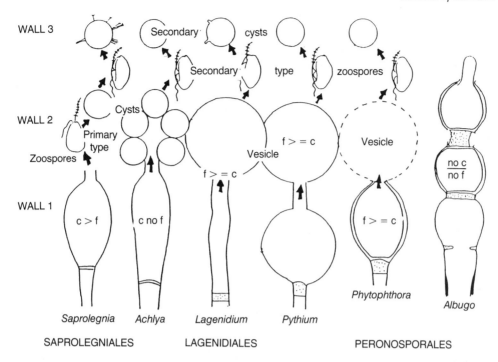

Figure 16.17 Main patterns of zoosporogenesis in oomycete fungi, indicating the temporal position of cytoplasmic cleavage (c) and flagellum (f) formation and spacial location of these events within or without original sporangium. In the plant pathogen *Albugo*, zoosporogenesis has been completely suppressed and the sporangia have become transformed into 'conidia'. From Beakes (1987).

bilateral fusion to delimit the zoospores (as in Figure 16.15b B). However, RF-FS reveals that zoospore initials are defined by a network of narrow cleavage cisternae (Figures 16.15a A–D and 16.22b) which are initiated in the region of the nuclear apex, from Golgi derived vesicles and associated transitional elements (Hyde *et al.*, 1991b). In addition this study revealed formation of a cortical cleavage cisternum which completely envelopes the differentiating spore mass (Figure 16.15a B). This procedure also revealed that far from being electron transparent, the developing cleavage furrows contain a dense granular matrix (Figure 16.22b). Recently chytrid sporangia fixed by high pressure freezing followed by freeze substitution have similarly revealed the presence of a dense interstitial material within the cleavage furrows (Powell, 1994). The *Phytophthora* pattern of cytokinesis may be widespread in sporangial fungi and cleavage patterns involving vesicle coalescence have been reported in conventionally fixed chytrids such as *Rhizophydium planktoninicum*

(Beakes *et al.*, 1993c) and *Chytridium confervae* (Powell, 1994) may have to be re-evaluated using RF-FS fixation. Other patterns of spore cytokinesis do exist in these fungi. The cleavage of zoospores (Figure 16.1) and oospores (Figure 16.32) in *Saprolegnia ferax* (Gay and Greenwood, 1966) is effected by the formation of a large central vacuole (as in Figure 16.30b) which becomes centrifugally enlarged and/or redistributed to delimit the uninucleate spore initials (Figures 16.1, and 16.15b, A). The spores are eventually formed when the tonoplast fuses with the plasma membrane as shown in Figure 16.32a, b. A number of other zoosporic fungi show a centripetal, infurrowing pattern of cytokinesis (Figure 16.15b C, D). This pattern is best shown in marine Thraustochytrid genera such as *Thraustochytrium* and *Shizochytrium* (Kazama, 1975; Moss, 1986). In these organisms cleavage furrows are initiated at the spore periphery and infurrow, perhaps by fusing cytoplasmic vesicles (Figure 16.15b C). A somewhat similar pattern was described in *Pythium middletonii* (Bracker and Heintz, unpublished; see Beakes, 1987) except that the infurrowing cleavage system eventually fuses with a central vacuole system to delimit the spores (Figure 16.15b C). A variation on this infurrowing pattern of cytokinesis has recently been illustrated in the chytrids *Zygorhizidium affluens* and *Z. planktonicum* (Beakes *et al.*, 1993a). In these species the cleavage system divides the cytoplasmic mass into blocks which become progressively divided into smaller units by the infurrowing of cleavage cisternae (Figures 16.15b D, 16.19b and 16.20b).

What defines these planes of cleavage? It has long been thought that elements of cytoskeleton probably play a crucial role in this. The addition of microtubule inhibitors, such as colchicine and oryzalin, result in aberrant patterns of zoospore cleavage, frequently resulting in the formation of abnormally large polyflagellate zoospores or inhibiting spore formation entirely (Kazama, 1975; Olson and Lange, 1983; Hyde and Hardham, 1993). The only direct evidence that microtubules may be providing the motive force for cleavage development comes from the marine fungus *Thraustochytrium*, where microtubule-cleavage furrow cross-bridges have been observed (Kazama, 1975). *Phytophthora* sporangia were induced to form sporangia in the presence of oryzalin and the disposition of mitochondria, nuclei, LPV, ventral and dorsal vesicles was determined using fluorescent dyes or antibody probes. Microtubules were found to be essential for the correct spatial distribution of the cortical organelles (LPVs, dorsal and ventral vesicles) although their peripheral accumulation was not prevented (Hyde and Hardham, 1993).

The role of actin microfilaments in this process has been explored using rhodamine-phalloidin labelling of differentiating sporangia. In *Saprolegnia ferax*, *Achlya ambisexualis* (Heath and Harold, 1992; Harold and Harold, 1992), *Phytophthora cinnamomi* (Hyde and Hardham, 1993) and rumen chytrids (Li and Heath, personal communication) the disposition of arrays of actin in the sporangial cytoplasm closely mirrored the presumptive cleavage furrows. Treatment of *Phytophthora* sporangia with the anti-microfibrillar drug, cytochalasin D resulted in abnormal cleavage patterns but had no effect on the precise spatial segregation of the main sporulation organelles (Hyde and Hardham, 1993).

Other inhibitors also can prevent or perturb normal cleavage of spores. Monensin, an ionophore which disrupts endomembrane activities in plant and animal cells, had its greatest effect on differentiating zoosporangia and gametangia of *Allomyces* at the time of cleavage resulting in the release of undifferentiated multinucleate cells, often with multiple flagella (Sewell *et al.*, 1986). Calcium ions also appear to be an essential requirement for normal cleavage. The microinjection of *Phytophthora* sporangia with a chelating agent specific for calcium inhibits both spore cleavage and zoospore release (Jackson personal communication). In *Saprolegnia*, streptomycin which is believed to be largely acting as a calcium antagonist, inhibits normal cleavage in sporangia and oogonia (Beakes and Gay, 1980). Calcium may therefore be regulating vesicle fusion and the generation of the cleavage apparatus.

16.2.4 Zoospore discharge

The final stage of zoosporogenesis involves the liberation of zoospores from the sporangium. In *Saprolegnia* the apex of the sporangium differentiates into a flattened nipple-like papillar projection (Figures 16.2, 16.3 and 16.23a). In thin section this papillum region is contiguous with the sporangium wall although the flattened end wall is significantly thinner than the walls on either side (Figure 16.23b). Immediately before discharge a zoospore moves into this papillum and shortly afterwards the wall is breached allowing a stream of spores to escape (Figure 16.23c).

In chytridiomycete fungi the mechanism of zoospore release has traditionally been an important taxonomic character and shows a range of patterns (Figure 16.18a–e). At it simplest part or whole of the sporangium wall disintegrates releasing the zoospores (Figure 16.18a). A variation on this mechanism has recently been described in the diatom parasite, *Zygorhizidium affluens* (Beakes *et al.*, 1993b), which has evolved an elaborate spring-loaded lid (operculum) which springs off, allowing the zoospore mass to escape, and leaving the discharged sporangium with a distinctive wide 'rimmed' aperture (Figure 16.25). In another algal parasite *Z. planktonicum*, the operculum lid appears to erode around its margin and is forced to one side as the zoospores escape, still encapsulated in a matrix of fibrillar material (Beakes *et al.*,

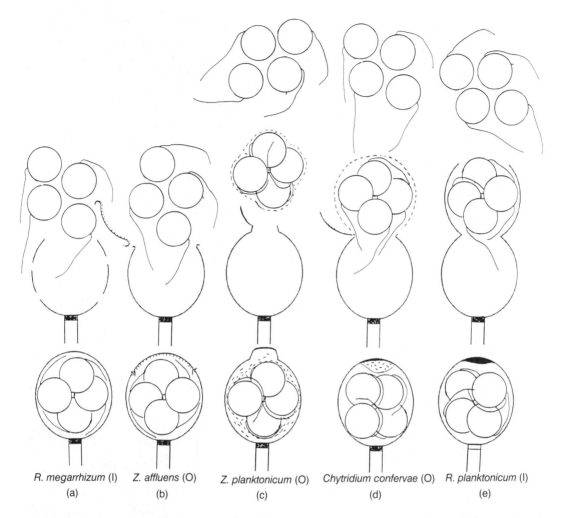

R. megarrhizum (I)   Z. affluens (O)   Z. planktonicum (O)   Chytridium confervae (O)   R. planktonicum (I)

(a)   (b)   (c)   (d)   (e)

Figure 16.18 Schematic diagrams summarizing the different patterns of zoospore release observed in the chytrids. In the simplest pattern (a, *Rhizophydium megarrhizum*), the sporangial wall simply ruptures or fragments releasing the zoospores. In some species (b, *Zygorhizidium affluens*) there is a spring-loaded lid (operculum) which flips off releasing the zoospores. A fibrillar matrix material coats the inside surface of the operculum. In *Zygorhizidium planktonicum* (c) a fibrillar matrix surrounds the zoospores and upon release, by rupture of the apical sporangium wall, the spores remain constrained by the matrix before this disperses releasing the zoospores. In *Chytridium confervae* (d) a fibrillar plug underlying the apical operculum, expands to form a constraining vesicle into which the zoospores flow, prior to its rupture and release. In *Rhizophydium planktonicum* (e) a thickened region of the apical wall expands and contains the escaping zoospores until it ruptures releasing the zoospores and forming an ephemeral collar around the neck of the sporangium. (I) = inoperculate, (O) = operculate.

1993b). This binds zoospores together in a mass for a brief time before the matrix material disperses and the zoospores are liberated to swim away (Figure 16.18c). In *Chytridium confervae* (Taylor and Fuller, 1981) a lens of fibrillar material forms part of the apical papillum similar to the more extensive papillar plugs of *Karlingia* (Figure 16.24) and *Phytophthora* (Figure 21; Hemmes, 1983). In *Chytridium*, the outer wall of the sporangium ruptures and is displaced to one side and the matrix material balloons out restraining the mass of escaping zoospores in a thin evanescent vesicle

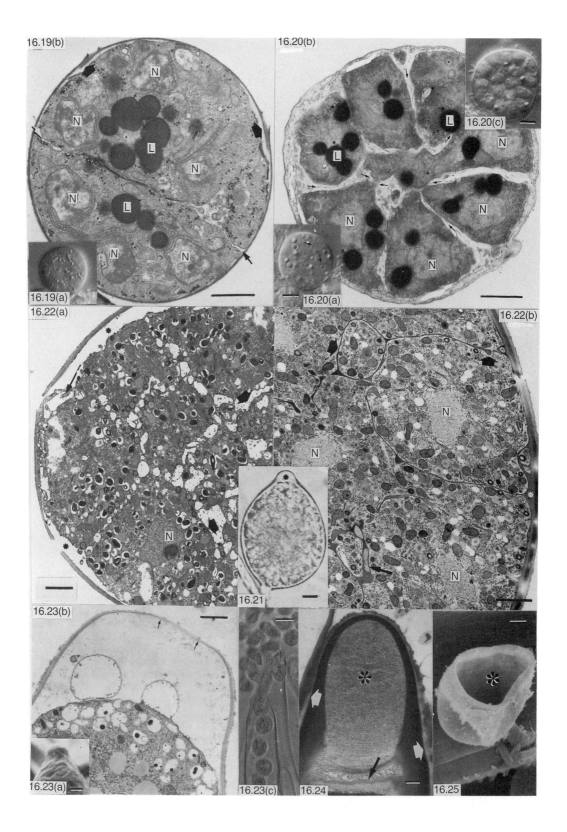

16.19(b)   16.20(b)   16.20(c)

16.19(a)   16.20(a)

16.22(a)   16.22(b)

16.21

16.23(b)

16.23(a)   16.23(c)   16.24   16.25

(Figure 16.18d), as occurs in many *Phyptophthora* spp. (Hyde *et al.*, 1991a). A variation on this theme, occurs in *Rhizophydium planktonicum*, where the papillum appears to be no more than a localized thickening of the sporangial apex which cannot be differentiated from the rest of the sporangium wall (Beakes *et al.* 1993c). Upon discharge this thickened wall stretches, restraining the discharging zoospores until it ruptures, releasing the zoospores, and forming a delicate collar around the neck of the sporangium (Figure 16.18e).

The forces or mechanisms which drive the discharge of zoospores from sporangia of aquatic fungi have been the subject of much speculation over the years (Gay and Greenwood, 1966). In a range of oomycete fungi (*Aphanomyces* Hoch and Mitchell, 1973; *Phytophthora* Gisi *et al.*, 1979; *Achlya*: Money and Webster, 1988; Money *et al.*, 1988a,b) it has been demonstrated that an osmotically generated pressure gradient drives spore discharge (Fig. 16.23c). In *Achlya* it is proposed that the sporangial walls are semipermeable with a pore diameter of around 2 nm which retain high-molecular-weight material in their intersporal space derived from the vacuole following cleavage (Money and Webster, 1988b). The presence of a clearly structured matrical material, derived from the cleavage furrow system

---

Figure 16.19 (a) Precleavage sporangium of the chytrid *Rhizophydium* showing uniform appearance of cytoplasm (interference contrast). Bar = 5 μm. (b) Section of a sporangium of *Zygorhizidium affluens* in the early stages of cleavage. The cytoplasmic mass has been cleaved into two large cytoplasmic blocks prior to further division (by infurrowing). This sporangium has been stained with a silver stain to reveal carbohydrate localization. The margins of the spring-loaded operculum are indicated by broad arrows. The nuclei (N) are peripherally distributed and the lipid (L) has become centrally concentrated at this stage. Bar = 1.0 μm.

Figure 16.20 (a) Cleaving sporangium of a *Rhizophydium* sp. showing prominent lipid globules and polygonal cleavage furrows (arrowed) in cytoplasm (interference contrast). Bar = 5 μm. (b) Thin section profile of a sporangium of *Zygorhizidium planktonicum*, late in cleavage showing the way in which the infurrowing cleavage system has delimited the uninucleate (N) spore initials. In this species at this stage the lipid (L) has still not coalesced into a single globule. Bar = 1.0 μm. (c) Mature sporangium of the chytrid *Rhizophydium* showing fully differentiated zoospore mass. Bar = 2 μm.

Figure 16.21 Mature sporangium of *P. cactorum* prior to induction of spore differentiation. Note prominent apical papillum (*). Bar = 3 μm

Figure 16.22 Details of differentiating zoospores of *P. cinnamomi* showing appearance of cleavage furrows in conventional glutaraldehyde/osmium fixed and rapid freezing-freeze substituted sporangia. Adapted from Hyde *et al.* (1991a,b), with permission.

Figures 16.23–16.26 Discharge apparatus in zoosporangia.

Figure 16.23 (a) SEM showing diffentiated apical papillum of *Saprolegnia diclina*. (b) TEM through apical papillum of *Saprolegnia diclina*. Note the absence of the inner wall layer (between arrows) in the middle of the thin flattened papillum wall. Bar = 1.0 μm. (c) Discharging primary zoospores of *S. diclina* showing liberation through the papillar ruptured pore (the wall of which has been completely degraded). Bar = 10 μm.

Figure 16.24 Complex multilayered operculum of the chytrid *Karlingia rosea*. Note electron-dense layer below outer sporangium wall (white arrows), thick fibrillar papillar plug (asterisked) and the additional underlying wall layers form what has been traditionally referred to as the endo-operculum (dark arrow). Bar = 1.0 μm.

Figure 16.25 SEM of a discharged sporangium of *Z. affluens*. Note the complete removal of the operculum (see in Figure 16.19) leaving a wide aperture, the wall of which has rolled back from the lip to form a rigid rim to discharged sporangium. Bar = 1.0 μm.

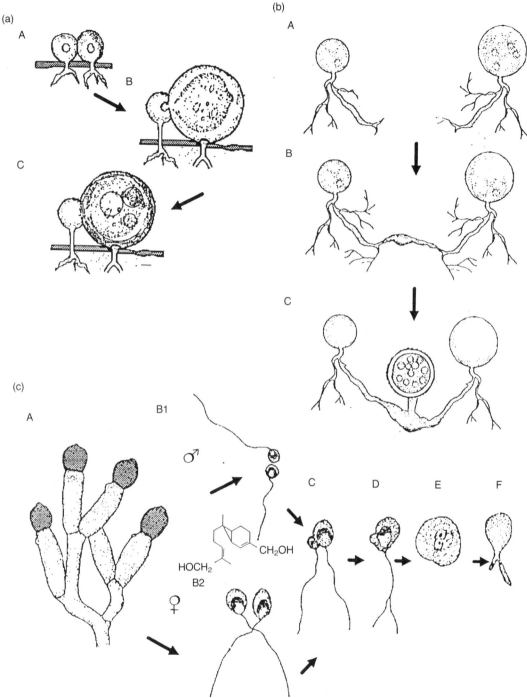

Figure 16.26 Semi-schematic diagrams illustrating some of the different types of sexual reproduction reported in chytridriomycete fungi. (a) A–C. Thallus fusion in *Rhizophydium* and *Zygorhizidium affluens*. (b) A–C Rhizoid fusion in *Chytriomyces halinus* (based on Miller and Dylewski, 1981). (c) A–F Motile anisogamous reproduction in *Allomyces macrogynus*. The male gametangium forms above the female in this species. The structure of the attractant pheromone sirenin, produced by the female gametes is also illustrated. (b and c adapted from Alexopoulos, 1962.)

surrounding differentiated spores in *Phytophthora* sporangia (Hyde *et al.*, 1991b), further supports the uniform applicability of this mechanism in oomycetes.

16.3 SEXUAL REPRODUCTION

16.3.1 Mating interactions and resting spore development

Sexual reproduction in the lower fungi also shows considerable diversity (Figures 16.27–16.39). A range of typical gametangial interactions in chytridiomycete fungi are illustrated in Figure 16.27 and range from the fusion of whole thalli, as exemplified by *Rhizophydium couchii* and *Zygorhizidium planktonicum* (Figures 16.27a, 16.34 and 16.36), through rhizoidal anastamosis as in *Chytriomyces hyalinus* and *Nowakowskiella elegans* (Figure 16.27) to anisogamous fusion observed in *Allomyces macrogynus* (Figure 16.27c). In oomycete fungi sexual reproduction is oogamous with the formation of female oogonia and male antheridia (Figures 16.28a, 16.29, 16.30a and 16.31). Unlike the other lower fungi the vegetative thallus of oomycete fungi is diploid and meiosis takes place in the delimited antheridium (Figure 16.31) and oogonium (Figure 16.30a, b) prior to cleavage (Beakes and Gay, 1977). The egg cells in *Saprolegnia furcata* are differentiated by a centrifugal cleavage (Figure 16.32a, e) process similar to that outlined in their zoosporangia (Figure 16.16.bA). Immediately following cleavage the egg cells are naked but in saprolegnian oomycetes following a brief (ca. 30 min) amoeboid phase (Figure 16.32b, c), round up and acquire a primary oosphere wall (Figure 16.32d) (Beakes and Gay, 1978*b*). Fertilization tube development is initiated shortly after cleavage and by the time the oospheres have become walled the fertilization tube has almost made contact with the first oosphere (Figure 16.32d). The fertilization tube then produces a peg which penetrates the oosphere and releases a single male gametic nucleus into the egg cell (Figure 16.33; Beakes and Gay, 1977). In Peronosporalean oomycetes the oosphere is differentiated by an internal cleavage furrow system which segregates the oogonium cytoplasm into a peripheral periplasm and central oosphere (Beakes, 1981; Hemmes, 1983). In many *Phytophthora* species the antheridium is actually penetrated by the oogonium initial cell, and forms a collar around the base of the expanding oogonium, in the so-called amphigynous pattern of differentiation (Hemmes, 1983; Beakes *et al.*, 1986). The egg cells then undergo a major reorganization of their cytoplasmic constituents eventually giving rise to a dormant oospore which is rich in lipid and other storage granules but apart from the single peripheral fusion nucleus lacks most other recognizable organelles (Figure 16.38a, b) (Beakes and Gay, 1978a,b; Beakes, 1980a; Hemmes, 1983; Beakes *et al.*, 1986). In *Achlya americana* lipids constitute 7% of the total dry weight of the mature oospores and show an increase in the proportion of unsaturated fatty acids compared with the oospheres (Fox *et al.*, 1983). Over 90% of this lipid fraction was triacyl glycerols, with phospholipids (membrane constituents) constituting less than 1%. In contrast to the zoospore the mature oospore wall constitutes over 30% of the total spore volume and is complex and multilayered (Beakes and Gay, 1978b). This wall layer appears to be the main repository of storage carbohydrate in these spores and its rapid erosion and assimilation (Figure 16.41) is one of the first recognizable events of germination (Beakes, 1980b; Beakes and Bartnicki-Garcia, 1989).

The biochemical switches which induced thick-walled resistant sporangium formation (RS) over the 'ordinary' thin-walled asexual sporangia (OS) have been investigated extensively in *Blastocladiella emersonii* (Cantino and Turian, 1958). The presence of bicarbonate or high levels of CO_2 induced the formation of RS sporangia. During RS differentiation there were significant differences in the activities of certain tricarboxylic acid cycle enzymes resulting in an accumulation of glyoxylate which, it was speculated, was required for amino acid and RNA synthesis. However, more recently it has been questioned whether changes in enzyme pathways regulate development or are a consequence of the switch to a different pattern of morphogenesis.

Syngamy and resting spore development has been described in *Chytriomyces hyalinus* (Miller and Dylewski, 1981), which is a species which reproduces by means of rhizoidal anastomosis followed by the development of a separate resting body (Figure 16.27). In the diatom parasite *Z. planktonicum* two adjacent thalli conjugate with the male thallus producing a fairly thick fertilization tube and making contact with the neighbouring

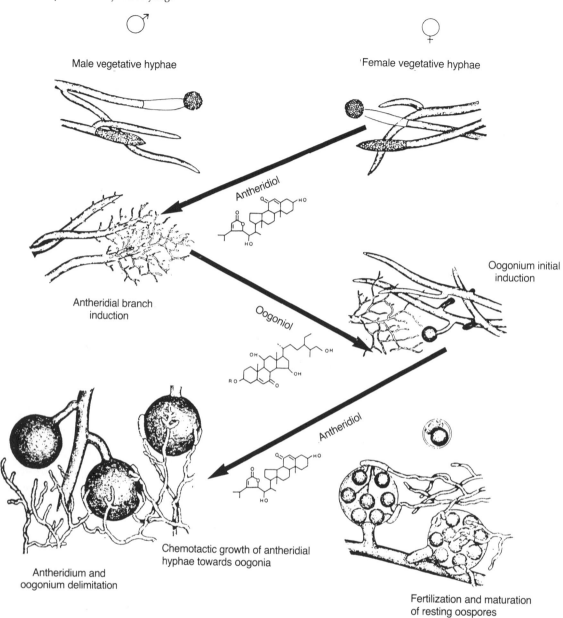

Figure 16.27 Oogenesis in *Achlya* ambisexualis showing co-ordinating function of the female hormone, antheridiol and the male hormone oogoniol. Modified from Raper (1940) and Horgen (1977, 1981).

'female' (Figure 16.34). Following plasmogamy the female thallus enlarges and the cytoplasm acquires large globular deposits of lipid and other storage material (Figure 16.35). At the most advanced stages the male thallus appears completely devoid of contents and the mature resting sporangium is thick walled and packed with dense lipid filled cytoplasm (Figure 16.36). It is not known whether specific mating types have to be adjacent to each other to induce this or whether spatial or temporal factors may govern which thallus acts as male and which as female. Superficially the resting sporangia

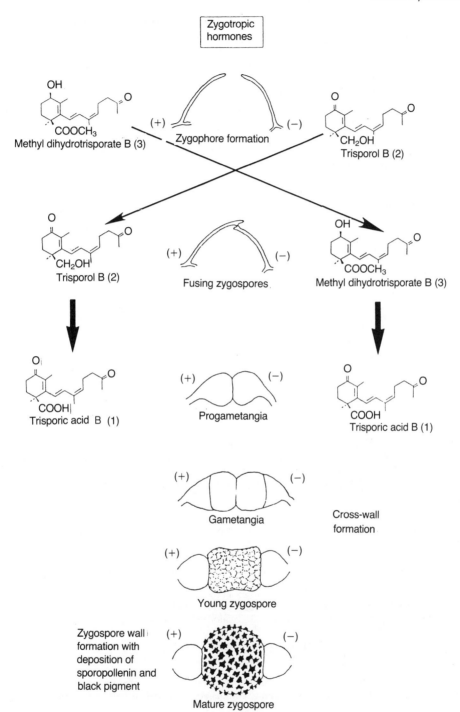

Figure 16.28 Schematic diagram summarizing zygosporogenesis in *Mucor mucedo*. The 'progamone' precursors trisporal B formed in (−) and metabolized by (+) and methyl dihydrotrisporate B formed in (+) and metabolized by (−) and the mating hormone trisporic acid B to which they are converted are illustrated. Based on Gooday (1978).

(body) of many chytrids, such as a *Rhizophydium* parasite of *Planktothrix* (Figure 16.39a, b) appear very similar to oomycete oospores. The lipid becomes largely peripherally distributed and the centre of the spore is rich in glycogen and electron-dense granules (Figure 16.39b). There is, however, no direct equivalent of the membrane bound ooplast vacuole present in oospores (Figure 16.38b). Even in oomycete fungi the biochemistry of the ooplast is still uncertain. Although it is derived from the coalescence of dense-body/fingerprint vesicles (Beakes and Gay, 1978a; Beakes 1980a, 1981) which in sporangia contain phosphorylated glucans, these cannot be extracted from mature oospores (Wang and Bartnicki-Garcia, 1980).

In heterothallic zygomycete fungi, genetically compatible strains are designated (+) and (−).

They produce special hyphal branches known as zygophores which are attracted to each other until they make contact (Figure 16.28b). There then follows a complex series of cytoplasmic events which ultimately leads to the formation of the thick-walled resting zygospore (Figures 16.28b and 16.37). Fine-structural details of zygosporogenesis have been described for a number of species (Hawker and Beckett, 1971; O'Donnell *et al.*, 1976, 1977a b,c). The (+) and (−) gametangial walls make contact and appear to lose their separate identity and form what is termed a fusion septum. The delimiting gametangial septa grow centripetally eventually delimiting the two progametangial cells (Figure 16.28b). Following progametangial formation the fusion septum dissolves centrifugally. Very soon after the dissolution of this separating wall a new thicker inner zygospore wall

---

Figure 16.29 An early stage of gametic copulation in *Saprolegnia diclina*. The antheridial hypha has been attracted to the enlarging oogonium (Og) and made contact with it (arrow), although neither organ has yet become delimited from their subtending hyphae. Bar = 1.0 μm.

Figure 16.30 (a,b) A delimited oogonium of *S. furcata* at time of meiosis, showing large transparent prophase I meiotic nuclei (A). Bar = 15 μm. Corresponding thin section profile of part of oogonium cytoplasm showing large prophase nuclei (N), and large central vacuole system (V). Bar = 1.0 μm.

Figure 16.31 A fully differentiated oogonium (Og) and antheridium (A). Note the swollen antheridium which is firmly bound to the oogonium wall. Bar = 10 μm.

Figure 16.32 Time lapse series of light micrographs of an oogonium (Og) of *S. furcata* showing differentiation of oospheres (egg initials) immediately after cleavage. (a) Immediate precleavage stage showing hemispherical oosphere initials. (b) Eggs at time of cleavage, is effected by fusion of vacuole and plasma membrane. (c) At 5 min after cleavage: note the rather large irregular oospheres showing masses of extruded protoplasm (*). (d) Fully differentiated oospheres, 40 min after cleavage showing the approaching fertilization tube (arrowed) with its gametic nucleus. From Beakes and Gay (1978a).

Figure 16.33 TEM profile of fertilization tube (FT) of *S. furcata* showing penetration of the thin-walled oosphere. Bar = 2 μm. From Beakes and Gay (1978a). A = anthendium.

Figures 16.34 and 16.35 Gametangial copulation in *Zygrohizidium planktonicum* (Chytridiales).

Figure 16.34 Two pairs of copulating thalli on the surface of the diatom *Synedra*. The larger thallus at this stage is the male which is producing the rather thick fertilization tube, and the smaller thallus is the female. Note how the wall of the fertilization tube appears continuous with the outer wall of the female thallus. Bar = 1.0 μm.

Figure 16.35 Section through a developing zygote, showing nucleus (N), large coalesced droplets of lipid bodies (L) and other 'storage' organelles. Bar = 2.0 μm.

Figure 16.36 Profile of a mature zygote developed from the female thallus, showing dense cytoplasm containing large lipid globules (L) and thick cell wall. Note empty conjugation tube and smaller empty male thallus (♂). Bar = 1.0 μm.

Figure 16.37 Mature resting zygospore of a *Rhizopus* sp. Note prominent stellate ornamentations (*) on the zygote wall. Bar = 1.0 μm.

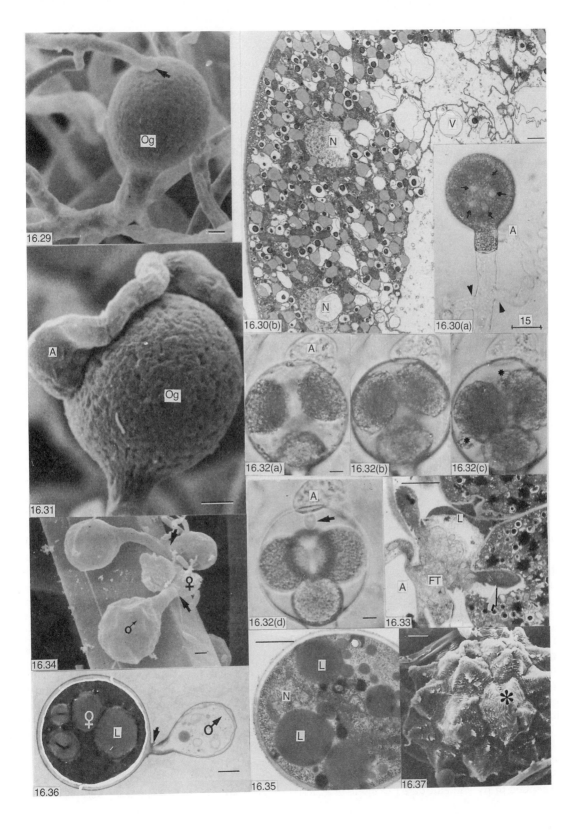

16.29

16.30(b)   16.30(a)

16.31

16.32(a)   16.32(b)   16.32(c)

16.32(d)   16.33

16.34

16.36   16.35   16.37

begins to be accreted equatorially. Prominent warty ornamentations are formed in the peripheral cytoplasm and are inserted into the wall early in the developmental process (Hawker and Beckett, 1971; O'Donnell 1976). As the developing zygospore expands the original outer primary wall of the fused progametangia ruptures and is pulled apart as the zygospore expands. Zygospore walls have interlocking warty spines, become heavily melanized (Figure 16.37) and have a highly complex sculptured surface which varies exquisitely from species to species (O'Donnell *et al.*, 1977a,b,c).

16.3.2   Hormonal and biochemical regulation of gametogenesis

In many species sexual differentiation appears to be regulated and co-ordinated by diffusible hormones. This topic has been reviewed in some detail previously (Barksdale, 1969; Horgen, 1977, 1981; Pommerville, 1981; Gooday, 1978, 1992) and will only be outlined in the broadest terms here. Perhaps the most extensively investigated system has been the co-ordination of antheridium and oogonium formation in the heterothallic oomycete *Achlya* (Figure 16.28a). The existence of these diffusible regulators of differentiation was first demonstrated by John Raper in a series of classic studies carried out in the late 1930s (Raper, 1939, 1940). Two hormones, antheridiol and oogoniol (Figure 16.28a) were later purified and characterized by Barksdale and her co-workers (Barksdale, 1969; McMorris, 1978). The female thallus produces the hormone antheridiol which acts on the male thallus and triggers the formation of antheridial branches and in turn induces the male plant to produce a second hormone, oogoniol which acts on the female thallus, inducing the production of oogonial initials (Figure 16.28a). Antheridiol also seems to be responsible for the directed growth of the antheridial branches towards the developing oogonial initials. These hormones are steroids and are very similar to mammalian sex hormones (Horgen, 1977, 1981). It has been demonstrated that antheridiol binds to a cytosolic protein in the male thallus which was similar to the steroid receptors described in other eukaryotes (Riehl and Toft, 1984; Riehl *et al.*, 1984). Throughout the 1970s considerable effort was expended to try and elucidate the mode of action

of these hormones at the molecular level but ultimately these studies have largely proved inconclusive. The first detectable effect when antheridiol is applied to a female thallus is an increase in rRNA synthesis and in ribosome number (Horgen *et al.*, 1975). Around three hours after antheridiol addition there is a dramatic increase in transcription-regulated messenger poly(A)RNA synthesis and 30 min or so later there is a large increase in protein synthesis and the appearance of inducible peptides (Kane *et al.*, 1972; Horgen *et al.*, 1975, 1983; Horton and Horgen, 1985; Timberlake, 1976). These events more or less coincide with the time of appearance of the fine antheridial branches (Horgen, 1977, 1981). The formation of the latter correlates with the synthesis of cellulase which softens the hyphal walls allowing side branch development. About 2 h after induction, prior to the dramatic increase in synthesis of mRNA, there is an acetylation of specific histones in the *Achlya* chromatin which is thought to bring about the activation/derepression of new genes (Horgen and Ball, 1974; Brunt and Silver, 1987). However, of 36 cDNA clones prepared from the male mycelium only three were regulated by the addition of antheridiol, and all of these were down-regulated rather than switched on (Horton and Horgen, 1989). Furthermore, when the mRNAs that were transcribed and accumulated in the male thallus in response to antheridiol were sequenced they showed no significant differences from the population present before induction (Rozek and Timberlake, 1980). Therefore in spite of all this effort no unique hormone-induced genes or gene product have been identified. As Rozek and Timberlake (1980) have pointed out, the qualitative changes in gene activity induced by antheridiol are probably quite subtle.

Other factors which appear to induce or be involved in the regulation of sexual differentiation in oomycetes, include a requirement for calcium (Fletcher, 1979; Cooper-Palomar and Powell, 1988), reduced levels of cyclic AMP (Kerwin and Washino, 1984; Herman *et al.*, 1990) and, in Peronosporalean oomycetes, the requirement of exogenous sterols and unsaturated fatty acids (Kerwin *et al.*, 1986). The suppression of the normal oxidation of arachidonic acid to lipoxygenases appears to be specifically associated with the induction of sexual reproduction in oomycetes (Herman and Hamberg, 1985). Antheridiol suppresses normal lipoxygenase activity in the male thalli of *A. ambisexualis* (Herman *et al.*, 1988) and has also been shown to be reduced once gametogenesis is initiated in a number of other oomycetes including *Saprolegnia* (Herman and Hamberg, 1987) and *Lagenidium giganteum* (Simmons *et al.*, 1987). This reduction in lipoxygenase activity has been shown to be concomitant with the starvation induced switch to sexual differentiation in *Saprolegnia ferax* but does not occur in similarly induced *S. parasitica* isolates which have lost their ability to reproduce sexually (Herman and Luchini, 1989). The lipoxygenase enzyme has been shown to be down-regulated in response to cues which induce sexual reproduction, including decreasing cAMP levels by addition of phosphodiesterase inhibitors such as theophylline (Herman *et al.*, 1990). However, it is still unclear exactly what the functional implications of the reduced levels of lipoxygenase activity are and how such changes are involved in the regulation of sexual differentiation.

The hormonal regulation of sexual differentiation in *Allomyces* is more straightforward since they act as pheromones attracting the motile male and female gametes to each other (Figure 16.26c) (Gooday, 1992). The female gametes secrete a chemotactic attractant sirenin, which is a sesquiterpene (Figure 16.26c) which only attracts male gametes (Pommerville, 1981, 1982). More recently a sesquiterpene male-produced pheromone, parisin has also been identified (Pommerville and Olson, 1987), which only attracts female gametes. This mating process is also regulated by divalent cations with increasing concentrations of Ca, Mg or Sr reducing the time of mating from just over a minute to 30 s or less (Pommerville, 1982). Once the gametes have made contact they become conjoined at a localized region of the zoospore plasma membrane towards the basal end of the spore (Figure 16.26c, C, D), towards the region of flagellum insertion (Pommerville and Fuller, 1976). Initially many small cytoplasmic bridges are formed at the fusion interface and a row of electron-dense vesicles become aligned adjacent to this zone. Once fusion has been completed the cells round up and nuclear fusion (karyogamy) then follows. This fusion is inhibited if the spores are trypsinated (suggesting the involvement of a glycoprotein receptor) or treated with agents such as diphenylhydramine and chloroquinine which reduce membrane fluidity (Pommerville, 1982). The actin inhibitor cytochalasin B also inhibits fusion if applied to the larger female gametes but has no effect when applied only to male gametes.

The final hormone regulating mating system which will be considered is the zygophore interactions during the early stages of zygosporogenesis in mucoralean fungi (Figure 16.28) (Gooday, 1978, 1992). In species such as *Mucor mucedo* zygophores of opposite (+/−) mating types are attracted to each other until contact is made and fusion takes place (Figure 16.28). The sex hormone which has been implicated in this mating interaction is a C_{18} terpenoid, trisporic acid. It elicits gametangial formation and increases the rate of accumulation of terpenoids such as β-carotene which are ultimately involved in the melanization of the resting spore wall (Gooday, 1978). This interaction provides an elegant example of biochemical collaboration between different mating types with the (+) and (−) thalli possessing complementary enzyme systems for processing different precursors in the collaborative synthesis of trisporic acid. The (+) thallus synthesizes the precursor methyl dihydrotrisporate B which has to diffuse into the (−) thallus before it can be processed to trisporic acid, whereas the (−) thallus synthesizes the precursor trisporal B which is taken up by the (+) thallus and processed into trisporic acid. It has been suggested that these precursors of trisporate synthesis probably act as the postulated volatile gamones which are responsible for the initiation and attraction of zygophores towards each other (Gooday, 1992). The zygophores of both (+) and (−) strains bind the lectin wheat-germ agglutinin (WGA) much more strongly than the vegetative hyphae (Gooday, 1992). Polyclonal antibodies were also raised to (+) and (−) zygophores which stained both (+) and (−) zygophores but not vegetative hyphae. The lectin concanavalin A (Con A) also showed much stronger binding to developing gametangia than to other parts of the mycelium. It was concluded that there are probably zygophore specific surface glycoproteins which modulate the successful binding fusion of hyphae during successful mating but the specific sexual agglutinins have yet to be identified and characterized (Gooday, 1992).

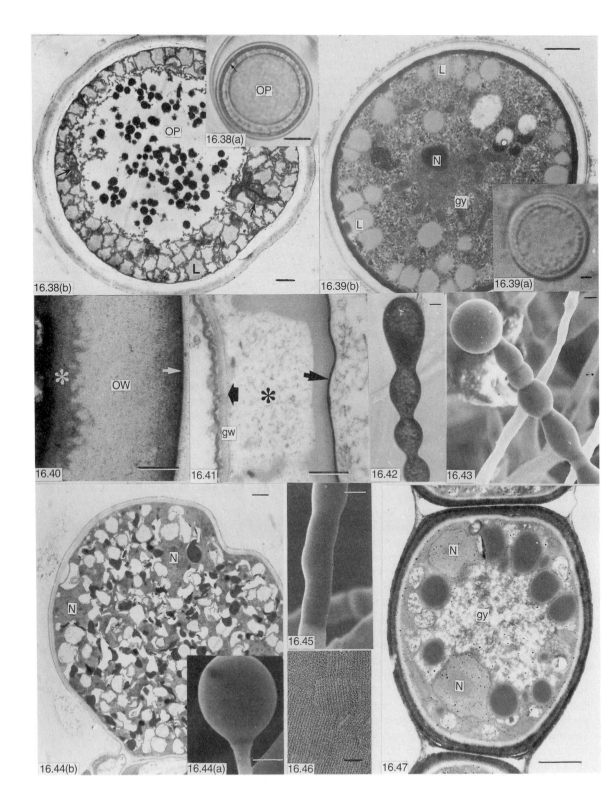

16.38(a)

16.38(b)

OP

OP

L

16.39(b)

L

L

N

gy

16.39(a)

16.40

OW

16.41

gw

16.42

16.43

16.44(b)

N

N

16.45

16.44(a)

16.46

16.47

N

gy

N

16.4 ASEXUAL RESTING SPORE FORMATION

16.4.1 Gemmae/chlamydospores in aquatic fungi

Compared with zoospore and sexual spore formation, the regulation and development of asexual resting structures in aquatic fungi has received far less attention. In saprolegnian oomycetes an initial burst of zoosporangium formation is often followed by the formation of large numbers of asexual gemmae, particularly in the older, inner regions of the colony. The appearance of these so-called gemmae (in Saprolegniales) or chlamydospores (in Peronosporales) is also quite variable. In the Saprolegniales typically chains of spherical, clavate or ovoid structures are formed (Figure 16.42) whereas in *Phytophthora*, single spherical structures are more typical (cf. Figure 16.44a, b). In *Phytophthora* these structures have been shown to contain all the typical sporulation organelles, including LPV and dorsal and ventral vesicles and confirmed by vesicle specific labelling with monoclonal antibodies (Hardham, personal communication). However, the most abundant cytoplasmic component in many Peronosporalean gemmae appears to be fingerprint type vesicles, in which the electron-dense component is frequently attached to the vesicular membrane (Figure 16.44b) and which may fuse to give rise to a large central vacuole (Hemmes, 1983; Hemmes and Wong, 1975). In *Saprolegnia* species which normally produce abundant sporangia, the gemmae contain fibrillar vesicles, lipid globules and dense body/finger print vesicles although encystment vesicles and K-bodies are not present. However, species, such as *S. furcata* which normally reproduce primarily sexually (forming mostly oogonia rather than zoosporangia) produce more spherical

Figure 16.38 Light (a) and electron micrographs (b) of mature oospores of *Saprolegnia*. Note large central ooplast vacuole (OP) containing peripheral lipid globules (L). Note the optically dense inner layer to the oospore wall (arrowed). Bars =(a) 5 μm (b) 1 μm. From Beakes and Gay (1978b).

Figure 16.39 Light (a) and electron micrographs (b) of mature zygotes (resting sporangia) of the chytrid *Rhizophydium* sp. Note peripheral lipid globules (L), central nucleus (N) and glycogen (gy) rich cytoplasm. Bars = (a) 5 μm and (b) 1 μm.

Figure 16.40 Mature oospore wall (OW) of *Saprolegnia furcata* showing thick multilayered appearance. There is an irregular electron dense inner layer (*) and a thin electron dense outer layer (white arrow). Bar = 0.5 μm. From Beakes and Gay (1978b).

Figure 16.41 Corresponding oospore wall in a germinating oospore of *Phytophthora megasperma*. Note the significant erosion of the inner wall (*) and the deposition of a new germination wall (gw) layer. The enzymic digestion of the wall is limited by the electron dense layer (arrowed). Bar = 0.5 μm. From Beakes and Bartnicki-Garcia (1989).

Figures 16.42–16.47 Asexual resting spores (gemma, chlamydospores, thallo or arthrospores)

Figure 16.42 Chain of terminal gemmae in *Saprolegnia ferax*. Bar = 5 μm.

Figure 16.43 Chain of terminal thallo/arthro spores in *Thamnidium elegans*. Bar = 10 μm.

Figure 16.44 (a,b) Corresponding SEM and TEM micrographs of a chlamydospore of *Pythium mammillatum*. Note uniform rounded shape, peripheral nuclei (N) and electron dense inclusion bodies. Scale bar =(a) 10 μm and (b) 1.0 μm.

Figure 16.45 SEM of a thallospore of *Thamnidium* from a septum delimited hyphal compartment. Bar = 5 μm.

Figure 16.46 Freeze fracture surface profile of an asexual sporangiospore wall of *Thamnidium* showing rodlet layer(−)typical of many hydrophobic terrestrial spores. Bar = 50 nm. From Beakes and Campos-Takaki (1984).

Figure 16.47 LS profile of a chain of thallospores of *Thamnidium*. Note thick electron dense wall, nuclei (N) and abundant glycogen (gy) and lipid. Bar = 2.0 μm. From Beakes *et al.* (1983a).

gemmae, which contain abundant lipid and dense body vesicles but no zoosporic organelles. It was found that the proportion of gemmae was increased in the presence of inhibitors of sporulation, such as streptomycin (Beakes and Gay, 1980). The cell walls of these spores are often greatly thickened compared with corresponding zoosporangia or oogonium walls although the extent of thickening is quite variable (Hemmes and Wong, 1975; Hemmes,1983). Although chlamydospore walls can be seen to be composed of outer and inner layers they lack the specialized outer electron-dense layer which characterizes mature oospore walls (Figures 16.40 and 16.41). In Oomycetes, therefore,asexual gemmae/chlamydospores may be considered to be partially differentiated sporangia or oogonia.

Some of the so-called 'resting bodies/sporangia' of chytridiomycetous fungi are also probably asexual perennating sporangia. One of the few ultrastructural accounts of these has been of the spores produced by the mycoparasite *Septoderma rhizophydii* (Powell and Blackwell, 1991). The cytoplasm contained large lipid globules typical of many resting spores. The wall was unusual for a chytrid in that it possessed knob-like tubercles, reminiscent of ornamentation seen in many terrestrial conidial fungi. These spores are released by the lysis and disintegration of the basal cell which subtends the developing spore.

16.4.2 Arthrospore/chlamydospore formation in the Mucorales

Many mucoralean fungi produce large numbers of thallic asexual spores (Figures 16.43 and 16.46) and in some species such as *Ellisomyces anomalus* they are the main propagative spores (Beakes *et al.*,1984a, b). In *Mucor*, arthrosporulation has been shown to be stimulated over sporangiosporogenesis by low pH and high glucose concentrations (Barrera,1983). In *Mucor rouxii* the maximum production of arthrospores occurred during growth in the absence of a fermentable carbon source in a medium containing peptone and yeast extract, supplemented with potassium acetate (Barrera, 1983). In *Gilbertella persicaria* large numbers of chlamydospores are formed in the hyphae when grown in submerged broth culture (Powell *et al.*, 1981).

The chlamydospores of *Gilbertella* were single intercalary structures and in their pattern of

formation were quite similar to those described for oomycete fungi above (Powell *et al.*, 1981).The hyphal compartment became swollen and accumulated lipid and glycogen and the hyphal wall became thickened. This spore was then delimited by the centrifugal insertion of crosswalls, which contained plasmodesmatal connections with the subtending hyphal compartments. During maturation the wall increased in thickness and three discrete layers were recognizable, the outer of which was the original hyphal wall.

In *Ellisomyces* (Beakes *et al.*, 1984 a,b) and *Mucor rouxii* (Barrera, 1983) these asexual spores are, in general, formed in chains usually of elliptical or spheroidal cells (Figure 16.43). In *Ellisomyces* the extent to which the cytoplasm of these spores accumulated lipid and glycogen (Figure 16.47) depended on the glucose concentration of the sporulation medium. On 4% glucose, lipid accounted for over 30% of the spore volume fraction whereas on 0.06% glucose, lipid was only about 5%. Interestingly, glycogen levels remained more or less constant at around 20%. The difference in spore volume was made up by increased vacuolation (50% volume fraction compared with 20%) of spores grown on the low glucose medium. In both *Ellisomyces* and *Mucor*, the arthrospores are released from the parent thallus by the disarticulation of the spore chains in the region of the septal crosswalls, although there were minor structural differences between the two systems (Barrera, 1983; Beakes *et al.*, 1984a).

16.5 CONCLUSIONS

It is apparent from this description of sporogenesis that there is a tremendous diversity of spore types to be found in the lower fungi.

Over the past three decades the developmental patterns and structure of these spores have been documented in considerable detail and precision. Because of the ease with which differentiation can be manipulated in the laboratory, organisms such as *Blastocladiella emersoni*, *Achlya ambisexualis* and *Mucor rouxii* have became favoured tools for fungal developmental biologists and biochemists. Although we now have a fairly comprehensive understanding of many of the morphogenetic, biochemical and regulative processes involved in sporulation these studies have not so far been complemented with a comparable understanding of the molecular control of differentiation. Although considerable effort has been expended in exploring molecular aspects of differentiation in *Blastocladiella* and *Achlya*, to date no one has succeeded in identifying the genes associated with the diverse array of sporulation specific organelles. The recent development of comprehensive panels of MAbs for organelles associated with zoosporogenesis in *Phytophthora cinnamomi* (Hardham *et al.*, 1991) provides an important tool for exploring organelle biogenesis and differentiation and is already yielding important and exciting discoveries (Hyde *et al.*, 1991a,b). With the recent development of a workable transformation system in *Phytophthora*, new tools for exploring regulation of spore differentiation in these fungi at the gene level may soon be available.

REFERENCES

Alexopoulos, C.J. (1962) *Introductory Mycology*. Wiley, New York.

Armbruster, B.L. (1982) Sporangiosporogenesis in three genera of the Saprolegniaceae. I Presporangium hyphae to early primary spore initial stage. *Mycologia*, **74**, 433–59.

Barr, D.J.S. (1992), Evolution and kingdoms of organisms from the perspective of a mycologist. *Mycologia*, **84**, 1–11.

Barrera, C.R. (1983) Formation and ultrastructure of *Mucor rouxii* arthrospores. *Journal of Bacteriology*, **155**, 886–95.

Barksdale, A.W. (1969) Sexual hormones of *Achlya* and other fungi. Science, **166**, 235–44.

Barstow, W.E. and Lovett, J.S. (1975) Formation of gamma particles during zoosporogenesis in *Blastocladiella emersonii*. *Mycologia*, **67**, 518–29.

Beakes, G.W. (1980a) Electron microscopic study of oospore maturation and germination in an emasculate isolate of *Saprolegnia ferax*. 1. Gross changes. *Canadian Journal of Botany*, **58**, 182–94.

Beakes, G.W. (1980b) Electron microscopic study of oospore maturation and germination in an emasculate isolate of *Saprolegnia ferax*. 2. Wall differentiation. *Canadian Journal of Botany*, **58**, 195–208.

Beakes, G.W. (1981) Ultrastructural aspects of oospore differentiation, in *The Fungal Spore: Morphogenetic Controls*, (eds H. Hohl and G. Turian), Academic Press, London and New York, pp. 71–94.

Beakes, G.W. (1983) A comparative account of cyst coat ontogeny in saprophytic and fish-lesion isolates (pathogenic) of the *Saprolegnia diclina-parasitica* complex. *Canadian Journal of Botany*, **61**, 603–25.

Beakes, G.W. (1987) Oomycete phylogeny: ultrastructural perspectives, in *Fungal Phylogeny and Evolution* (eds C.M. Brasier, A.D.M. Rayner and D. Moore), Cambridge University Press, Cambridge, pp. 405–421.

Beakes, G.W. and Bartnicki-Garcia, S. (1989) Ultrastructure of mature oogonium–oospore wall complexes in *Phytophthora megasperma*: a comparison of *in vivo* and *in vitro* dissolution of the oospore wall. *Mycological Research*, **93**, 321–34.

Beakes, G.W. and Campos-Takaki, G.M. (1984) Sporangiole ultrastructure in *Ellisomyces anomalus* (Thamnidiaceae). *Transactions of the British Mycological Society*, **83**, 607–13.

Beakes, G.W. and Gay, J.L. (1977) Gametangial nuclear division and fertilization in *Saprolegnia furcata* as observed by light and electron microscopy. *Transactions of the British Mycological Society*, **69**, 459–71.

Beakes, G.W. and Gay, J.L. (1978a) A light and electron microscopic study of oospore maturation in *Saprolegnia furcata*. 1. Cytoplasmic changes. *Transactions of the British Mycological Society*, **71**, 11–24.

Beakes, G.W. and Gay, J.L. (1978b) A light and electron microscopic study of oospore maturation in *Saprolegnia furcata*. 2. Wall changes. *Transactions of the British Mycological Society*, **71**, 25–35.

Beakes, G.W. and Gay, J.L. (1980) The effects of streptomycin on growth and sporulation of *Saprolegnia* spp. *Journal of General Microbiology*, **119**, 361–71.

Beakes, G.W., Campos-Takaki, G.M. and Takaki, M. (1984a) The effect of glucose concentration and light on thallospore (chlamydospore/arthrospore) differentiation in *Ellisomyces anomalus* (Thamnidiaceae, Mucorales). *Canadian Journal of Botany*, **62**, 2677–87.

Beakes, G.W., Campos-Takaki, G.M., Takaki, M. and Dietrich, S.M.C. (1984b) Cultural, physiological and structural aspects of thallospore formation in *Ellisomyces anomalus* (Thamnidiaceae). *Transactions of the British Mycological Society*, **83**, 593–605.

Beakes, G.W., El-Hamalawi, Z.A. and Erwin, D.C. (1986) Ultra-structure of mature oospores of *Phytophthora megasperma* f. sp. *medicaginis*: preparation protocols and the effects of MTT vital staining and permanganate pre-treatment. *Transactions of the British Mycological Society*, **86**, 195–206.

Beakes, G.W., Canter, H.M. and Jaworski, G.H.M. (1993a) Comparative ultrastructural ontogeny of zoosporangia of *Zygorhizidium affluens* and *Z. planktonicum* chytrid parasites of the diatom *Asterionella formosa*. *Mycological Research*, **96**, 1047–59.

Beakes, G.W., Canter, H.M. and Jaworski, G.H.M. (1993b) Ultrastructural study of operculation (discharge apparatus) and zoospore discharge in zoosporangia of *Zygorhizidium affluens* and *Z. planktonicum* chytrid parasites of the diatom *Asterionella formosa*. *Mycological Research*, **96**, 1060–67.

Beakes, G.W., Canter, H.M. and Jaworski, G.H.M. (1994) Sporangium differentiation and zoospore fine-structure of the chytrid *Rhizophydium planktonicum* Canter emmed., a fungal parasite of *Asterionella formosa*. *Mycological Research*, **97**, 1059–74.

Bowman, B.H., Taylor, J.W., Brownlee, A.G. *et al.* (1992) Molecular evolution of the fungi: relationship of

Basidiomycetes, Ascomycetes, and Chytridiomycetes. *Molecular and Biological Evolution*, **9**, 285–96.

Bracker, C.E., (1968) Ultrastructure and development of sporangia in *Gilbertella persicaria*. *Mycologia*, **60**, 1016–67.

Brunt, S.A. and Silver, J.C. (1987) Steroid hormone-regulated basic proteins in *Achlya ambisexualis*. *Experimental Mycology*, **11**, 65–9.

Burr, A.W. and Beakes, G.W. (1994) Characterization of zoospore and cyst surface structure in saprophyte and fish pathogenic *Saprolegnia* species (Oomycete fungal protists). *Protoplasma* (in press).

Cantino, E.C. and Turian, G.F. (1958) Physiology and development of lower fungi (Phycomycetes). *Annual Review of Microbiology*, **13**, 97–124.

Cooper-Palomar, J.L. and Powell, M.J. (1988) Sites of calcium concentration during sexual reproduction of *Achlya ambisexualis*. *Mycologia*, **80**, 783–789.

Fletcher, J. (1979) An ultrastructural investigation into the role of calcium in oosphere initial development in *Saprolegnia diclina*. *Journal of General Microbiology*, **113**, 316–26.

Fox, N.C., Coniglio, J.G. and Wolf, F.T. (1983) Lipid composition and metabolism in oospores and oospheres of *Atchlya americana*. *Experimental Mycology*, **7**, 216–26.

Gay, J.L. and Greenwood, A.D. (1966) Structural aspects of zoospore production in *Saprolegnia ferax* with particular reference to the cell and vacuolar membranes. *Colston Symposium Series*, **18**, 95–110.

Gisi, U., Schwinn, F.J. and Oertli, J.J. (1979) Dynamics of indirect germination in *Phytophthora cactorum* sporangia. *Transactions of the British Mycological Society*, **72**, 437–46.

Gooday, G.W. (1978) Functions of trisporic acid. *Philosophical Transactions of the Royal Society London. Series B*, **284**, 509–20.

Gooday, G.W. (1992) The fungal surface and its role in sexual interactions, in *Perspectives in Plant Cell Recognition*. (eds J.A. Callow and J.R. Green), Cambridge University Press, Cambridge, pp. 33–58.

Griffin, D.H. and Breuker, C. (1969) Ribonucleic acid synthesis during the differentiation of sporangia in the water mold *Achlya*. *Journal of Bacteriology*, **98**, 689–96.

Hammill, T. (1981) Mucoralean sporangiosporogenesis, in *The Fungal spore: Morphogenetic Controls*, (eds H. Hohl and G. Turian), Academic Press, London and New York, pp. 173–94.

Hardham, A.R., Gubler, F.J. and Duniec, J. (1991) Ultrastructural and immunological studies of zoospores of *Phytophthora*, in *Phytophthora* (eds J.A. Lucas, R.C. Shattock, D.S. Shaw and L.R. Cooke), Cambridge University Press, Cambridge, pp. 50–69.

Harold, R.L. and Harold, F.M. (1992) Configuration of actin microfilaments during sporangium development in *Achlya bisexualis* comparison of two staining protocols. *Protoplasma*, **171**, 110–16.

Hawker, L.E. and Beckett, A. (1971) Fine structure and development of the zygospore of *Rhizopus sexualis* (Smith) Callen. *Philosophical Transactions of the Royal Society London. Series B*, **263**, 71–100.

Heath, I.B. and Harold, R.L. (1992) Actin has multiple roles in the formation and architecture of zoospores of the oomycetes, *Saprolegnia ferax* and *Achlya bisexualis*. *Journal of Cell Science*, **102**, 611–27.

Hemmes, D.E. (1983) Cytology of *Phytophthora*, in *Phytophthora. Its Biology, Ecology, and Pathology*, (eds D.C. Erwin, S. Bartnicki-Garcia and P.H. Tsoa), APS Press, St Paul, MN, pp. 9–40.

Hemmes, D.E. and Wong, D.L.S. (1975) Ultrastructure of chlamydospores of *Phytophthora cinnamomi* during development and germination. *Canadian Journal of Botany*, **53**, 2945–57.

Herman, R.P. and Hamberg, M.M. (1985) Prostaglandins and prostaglandin-like substances are indicated in normal growth and development in oomycetes. *Prostaglandins*, **29**, 819–30.

Herman, R.P. and Hamberg, M.M. (1987) Properties of the soluble arachidonic acid 15-lipoxygenase and hydroperoxide isomerase from the oomycete *Saprolegnia parasitica*. *Prostaglandins*, **34**, 129–39.

Herman, R.P. and Luchini, M.M. (1989) Lipoxygenase activity in the oomycete *Saprolegnia* is dependent upon environmental cues and reproductive competence. *Experimental Mycology*, **13**, 372–9.

Herman, R.P., Luchini, M.M. and Herman, C.A. (1988) Hormone-dependent lipoxygenase activity in *Achlya ambisexualis*. *Experimental Mycology*, **13**, 95–9.

Herman, R.P., Luchini, M.M., Martinez, Y.M. et al. (1990) Cyclic nucleotides modulate lipoxygenase activity and reproduction in oomycetes. *Experimental Mycology*, **14**, 322–30.

Hoch, H.C. and Mitchell, J.E. (1973) The effects of osmotic water potentials on *Aphanomyces* euteiches during zoosporogenesis. *Canadian Journal of Botany*, **51**, 413–20.

Hohn, T.M., Lovett, J.S. and Bracker, C.E. (1984) Characterization of the major proteins in gamma particles, cytoplasmic organelles in *Blastocladiella emersonii* zoospores. *Journal of Bacteriology*, **58**, 253–63.

Horgen, P.A. (1977) Steroid induction of differentiation: *Achlya* as a model system, in *Eukaryote Microbes as Model Developmental Systems*, (eds D.H. O'Day and P.A. Horgen) Marcel Dekker, New York, pp. 272–94.

Horgen, P.A. (1981) The role of the steroid sex pheromone antheridiol in controlling the development of male sex organs in the water mold, *Achlya*. in *Sexual Interactions in Eukaryote Microbes*, (eds D.H. O'Day and P.A. Horgen), Academic Press, New York, pp. 155–78.

Horgen, P.A. and Ball, S.F. (1974) Nuclear protein acetylation during hormone-induced sexual differentiation in *Achlya ambisexualis*. *Cytobios*, **10**, 181–5.

Horgen, P.A. and O'Day, D.H. (1975) The developmental patterns of lysosomal enzyme activities during Ca^{2+}-induced sporangium formation in *Achlya bisexualis*. III α mannosidase. *Archives of Microbiology*, **102**, 9–12.

Horgen, P.A. Smith, R., Silver, J.C. and Craig, G. (1975) Hormonal stimulation of ribsomal RNA synthesis in *Achlya ambisexualis*. *Canadian Journal of Biochemistry*, **53**, 1341–5.

Horgen, P.A. Iwanochko, M. and Bettiol, M.F. (1983) Antheridiol, RNA polymerase II and sexual development in the aquatic fungus, *Achlya*. *Archives of Microbiology*, **134**, 314–19.

Horton, J.S. and Horgen P.A. (1985) Synthesis of an antheridiol-inducible peptide during sexual morphogenesis of *Achlya ambisexualis* E87. *Journal of Biochemistry and Cell Biology*, **63**, 355–65.

Horton, J.S. and Horgen, P.A. (1989) Molecular cloning of cDNAs regulated during the steroid-induced sexual differentiation in the aquatic fungus *Achlya*. *Experimental Mycology*, **13**, 263–73.

Hyde, G.J., Gubler, F. and Hardham, A.R. (1991a) Ultrastructure of zoosporogenesis in *Phytophthora cinnamomi*. *Mycological Research*, **95**, 577–91.

Hyde, G.J., Lancelle, S., Hepler, P.K. and Hardham, A.R. (1991b) Freeze substitution reveals a new model for sporangial cleavage in *Phytophthora*, a result with implications for cytokinesis in other eukaryotes. *Journal of Cell Science*, **100**, 735–46.

Hyde, G.J. and Hardham, A.R. (1993) Microtubules regulate the generation of polarity in zoospores of *Phytophthora cinnamomi*. *European Journal of Cell Biology*, **62**, 75–85.

Jaworski, A.J. and Stumhofer, P. (1984) Dormant ribosomes in *Blastocladiella emersonii* zoospores are arrested in elongation. *Experimental Mycology*, **8**, 13–24.

Johnson, S.A. and Lovett, J.S. (1984) Gene expression during development of *Blastocladiella emersonii*. *Experimental Mycology*, **8**, 132–8.

Kane, B.E. Reiskind, J.B. and Mullins, J.B. (1972) Hormonal control of sexual morphogenesis in *Achlya*: dependence on protein and ribonucleic acid syntheses. *Science*, **180**, 1192–3.

Kazama, F.Y. (1975) Cytoplasmic cleavage during zoosporogenesis in *Thraustochytrium* sp.: ultrastructure and the effects of colchicine and D_2O. *Journal of Cell Science* **17**, 155–70.

Kerwin, J.L. and Washino, R.L. (1984) Cyclic nucleotide of oosporogenesis by *Lagenidium giganteum*. *Experimental Mycology*, **8**, 215–24.

Kerwin, J.L., Simmons, C.A. and Washino, R.K. (1986) Oosporogenesis by *Lagenidium giganteum* in liquid culture. *Journal of Invertebrate Pathology*, **47**, 258–70.

Law, S.T. and Burton, D.N. (1976) Lipid metabolism in *Achlya*: studies of lipid turnover during development. *Canadian Journal of Microbiology*, **22**, 1710–15.

LeJohn, H.B., Klassen, G.R., McNaughton, D.R. *et al.* (1977) Unusual phosphorylated compounds and transcriptional control in *Achlya* and other aquatic molds, in *Eukaryotic Microbes as Model Developmental Systems* (eds D.H. O'Day and P.A. Horgen) Marcel Dekker, New York, pp. 69–96.

Lehnen, L.J. and Powell, M.P. (1989) The role of kinetosome-associated organelles in the attachment of encysting secondary zoospores of *Saprolgnia ferax* to substrates. *Protoplasma*, **149**, 163–74.

McMorris, T.C. (1978) Sex hormones of the aquatic fungus *Achlya*. *Lipids*, **13**, 716–22.

Miller, C.E. and Dylewski, D.P. (1981) Syngamy and resting body development in *Chytriomyces hyalinus* (Chytridiales). *American Journal of Botany*, **68**, 342–9.

Mills, G.L. and Cantino, E.C. (1981) Chitosome-like vesicles from gamma particles of *Blastocladiella emersonii* synthesise chitin. *Archives of Microbiology*, **130**, 72–7.

Money, N.P., Webster, J. and Ennos, R. (1988a) Dynamics of sporangial emptying in *Achlya intricata*. *Experimental Mycology*, **12**, 13–27.

Money, N.P., Webster, J. and Ennos, R. (1988b) Cell wall permeability and its relationship to spore release in *Achlya intricata*. *Experimental Mycology*, **12**, 169–79.

Moss, S.T. (1986) Biology and phylogeny of the Labyrinthulales and Thraustochytriales, in *Zoosporic Marine Fungi*, (ed. S.T. Moss), Academic Press, London, pp. 105–29.

O'Donnell, K.L., Hooper, G.R. and Fields, W.G. (1976) Zygosporogenesis in *Phycomyces blakesleeanus*. *Canadian Journal of Botany*, **54**, 2573–86.

O'Donnell, K.L., Ellis, J.J., Hesseltine, C.W. and Hooper, G.R. (1977a) Morphogenesis of azygospores induced in *Gilbertella persicaria* (+) by imperfect hybridization with *Rhizopus stolonifera* (−). *Canadian Journal of Botany*, **58**, 2721–7.

O'Donnell, K.L., Ellis, J.J., Hesseltine, C.W. and Hooper, G.R. (1977b) Zygosporogenesis in *Gilbertella persicaria*. *Canadian Journal of Botany*, **58**, 2712–20.

O'Donnell, K.L., Flegler, S.L., Ellis, J.J. and Hesseltine, C.W. (1977c) The *Zygorhynchus* zygosporangium and zygospore. *Canadian Journal of Botany*, **58**, 2721–7.

Olson, L.W. and Lange, L. (1983) Abnormal spore cleavage: abnormal spores of *Allomyces macrogynus*. *Nordic Journal of Botany*, **3**, 657–64.

Pommerville, J. (1981) The role of sexual pheromones in *Allomyces*, in *Sexual Interactions in Eukaryotic Microbes*, (eds D.H. O'Day and P.A. Horgen), Academic Press, New York, pp. 53–77.

Pommerville, J. (1982) Morphology and physiology in gamete mating and gamete fusion in the fungus *Allomyces*. *Journal of Cell Science*, **53**, 193–209.

Pommerville, J. and Fuller, M.S. (1976) The cytology of the gametes and fertilization of *Macrogynus macrogynus*. *Archives of Microbiology*, **109**, 21–30.

Pommerville, J. and Olson, L.W. (1987) Evidence for a male-produced pheromone in *Allomyces macrogynus*. *Experimental Mycology*, **11**, 245–8.

Powell, M.J. (1994) Production and modifications of extracellular structures during development of Chytridiomycetes. *Protoplasma* (in press).

Powell, M.J. and Blackwell, W.H. (1991) A proposed dispersal mechanism for *Septoderma rhizophydii*. *Mycologia*, **83**, 673–80.

Powell, M.J., Bracker, C.E. and Sternshein, D.J. (1981) Formation of chlamydospores in *Gilbertella persicaria*. *Canadian Journal of Botany*, **59**, 908–28.

Raper, J.R. (1939) Sexual hormones in *Achlya*. I. Indicative evidence for a hormone co-ordinating mechanism. *American Journal of Botany*, **26**, 639–50.

Raper, J.R. (1940) Sexual hormones in *Achlya*. II. Distance reactions, conclusive evidence for a hormonal co-ordinating mechanism. *American Journal of Botany*, **27**, 162–73.

Riehl, R.M. and Toft, D.O. (1984) Analysis of the steroid receptor in *Achlya ambisexualis*. *Journal of Biological Chemistry*, **259**, 15324–30.

Riehl, R.M., Toft, D.O., Meyer, M.D. *et al.* (1984) Detection of a pheromone-binding protein in the aquatic fungus *Achlya ambisexualis*. *Experimental Cell Research*, **53**, 544–9.

Rozek, C.E. and Timberlake, W.E. (1980) Absence of evidence for changes in messenger RNA populations during steroid induced cell differentiation in *Achlya*. *Experimental Mycology*, **4**, 33–47.

Selitrennikoff, C.P., Dalley, N.E. and Sonneborn, D.R. (1980) Regulation of the hexosamine biosynthetic pathway in the water mold *Blastocladiella emersonii*: sensitivity to endproduct inhibition is dependent upon the life cycle phase. *Proceedings of the National Academy of Sciences of the USA*, **77**, 5999–6002.

Sewell, T., Olson, L., Lange, L. and Pommerville, J. (1986) The effect of monensin on gametogenesis and zoosporogenesis in the aquatic fungus, *Allomyces macrogynus*. *Protoplasma*, **113**, 129–39.

Simmons, C.A., Kerwin, J.L. and Washino, R.K. (1987) Preliminary characterization of lipoxygenase from the entomopathogenic fungus *Lagenidium giganteum*, in *The Metabolism, Structure and Function of Plant Lipids*, (eds P.K. Stumpf, J.B. Mudd and W.D. Nes), Plenum, New York, pp. 421–23.

Sutherland, R.B., Schuerch, B.M., Ball, S.F. and Horgen, P.A. (1976) The developmental patterns of lysosomal enzyme activities during Ca^{2+}–induced sporangium formation in *Achlya bisexualis*. III Ribonucleases. *Archives of Microbiology*, **109**, 289–94.

Taylor, J.W. and Fuller, M.S. (1981) The Golgi apparatus, zoosporogenesis, and development of the zoospore discharge apparatus in *Chytridium confervae*. *Experimental Mycology*, **5**, 35–59.

Timberlake, W.E. (1976) Alterations in RNA and protein synthesis associated with steroid hormone-induced sexual morphogenesis in the water mold *Achlya*. *Developmental Biology*, **51**, 202–14.

Timberlake, W.E., McDowell, L., Cheney, J. and Griffin, D.H. (1973) Protein synthesis during the differentiation of sporangia in the water mold *Achlya*. *Journal of Bacteriology*, **116**, 67–73.

Wang, M.C. and Bartnicki-Garcia, S. (1980) Distribution of mycolaminarins in cell wall beta-glucans in the life cycle of *Phytophthora*. *Experimental Mycology*, **4**, 269–80.

ASEXUAL SPORULATION IN HIGHER FUNGI

T.H. Adams

Department of Biology, Texas A&M University, College Station, Texas, USA

17.1 INTRODUCTION

The characteristic asexual spore in higher fungi is the conidium. Conidia are classically defined as specialized, non-motile, asexual propagules that are usually formed from the side or tip of a sporogenous cell and do not develop by progressive cytoplasmic cleavage (Alexopoulos and Mims, 1979). For members of the subdivision Deuteromycotina, or Fungi Imperfecti, conidiation represents the primary means of reproduction, and for taxonomists, the main characteristic for classification. The Deuteromycetes are generally considered to be anamorphs of the other conidial fungi (Ascomycetes and Basidiomycetes) that have permanently lost their sexual stage or teleomorph (Cole, 1986).

There is a huge variety of morphologically distinct conidial types produced by the higher fungi ranging from simple spheres to spirally curved and star-shaped structures. The many mechanisms through which this morphological diversity in conidial types arises have been reviewed in detail by Cole (1986) and will not be examined here. It is important to note, however, that there are two basic modes to conidial morphogenesis termed blastic and thallic development (Cole, 1986; Figure 17.1). Blastic conidia differentiate from hyphae by growth of an enlarged, recognizable proconidium that is then separated from the parent hypha by a basal septum. Many cytological aspects of blastic conidiation resemble yeast budding. Thallic conidia differentiate from preformed fertile hyphae by either converting an entire hyphal segment into a single conidium (holothallic development) or by converting a single hyphal segment into several conidia by making additional septa (arthric development). In

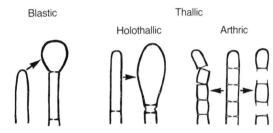

Figure 17.1 Diagrammatic summary of different modes of conidial and conidiogenous cell development. Reproduced with permission from Cole (1986).

thallic conidiation, enlargement of the recognizable proconidium occurs only after septum formation delimits the spore. The wide variety of conidial form is achieved through variations in these general developmental themes.

Conidial development is the result of an intricate series of tightly regulated biochemical events. This chapter will describe the molecular genetic approaches that have been taken towards understanding the mechanisms controlling conidial morphogenesis. Such detailed analyses of conidial development have been primarily limited to two ascomycetous fungal species, *Aspergillus nidulans* and *Neurospora crassa* (Matsuyama *et al.*, 1974; Clutterbuck, 1977; Timberlake and Marshall, 1988, 1989; Springer and Yanofsky, 1989; Timberlake, 1990). These two fungi are extremely well suited for studies of fungal sporulation because of the ability to carry out sophisticated classical and molecular genetic analyses including the use of DNA-mediated transformation (Pontecorvo *et al.*, 1953; Perkins *et al.*, 1982; Timberlake and Marshall, 1989). It is presumed that the biochemical mechanisms regulating conidial morphogenesis in these fungi will apply in many cases to other fungi as well.

The Growing Fungus. Edited by Neil A.R. Gow and Geoffrey M. Gadd. Published in 1994 by Chapman & Hall, London. ISBN 0 412 46600 7

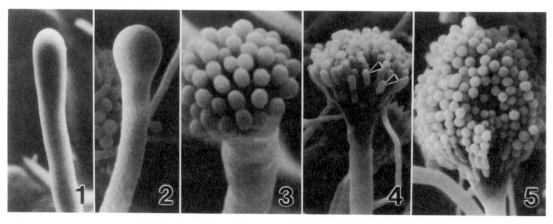

Figure 17.2 Scanning electron micrographs showing stages in conidiophore development (Reproduced with permission from Mims *et al*. 1988). (1) Young conidiophore stalk just prior to vesicle differentiation. (2) Developing vesicle. (3) Developing metulae. (4) Developing phialides (arrowheads). (5) Tip of a mature conidiophore bearing numerous conidial chains.

17.2   CONIDIATION IN *ASPERGILLUS NIDULANS*

17.2.1   Growth versus development

The asexual lifecycle of *A.nidulans* can be divided into two distinct phases, growth and reproduction. The first phase includes all those events leading up to and including the initiation of development as an asexually derived spore, or conidium, germinates to produce hyphae that grow and branch outwards to form an undifferentiated network of interconnected cells called the mycelium. When the appropriate conditions are met, some of the hyphal cells within the mycelium stop growing and initiate the second phase of the life cycle. This phase involves the execution of the developmental programme that directs formation of specialized reproductive structures known as conidiophores that then produce thousands of conidiospores (Figure 17.2).

(a)   Initiation of development: environmental requirements and acquisition of developmental competence

Under normal conditions at least three criteria must be met before conidiophore development will occur. First, colonies must be exposed to an air interface (Morton,1961; Axelrod *et al*., 1973; Timberlake,1980). Except under special circumstances, cells grow vegetatively when in submerged cul-

ture but rapidly undergo development when exposed to air. The precise nature of this inductive signal is poorly understood. Available evidence suggests that the conidiation signal is not associated with changes in O_2 or CO_2 concentration (Morton, 1961). One suggestion is that conidiation occurs in response to cell surface changes induced by the abrupt formation of an air/water interface at the hyphal surface, but the question remains unresolved.

The second requirement for conidiation to take place is light (Mooney *et al*.,1990; Mooney and Yager, 1990). Wild type *A.nidulans* strains produce few conidiophores in the dark even if hyphae are exposed to air. The developmental programme for conidiation is activated specifically by exposure to red light but activation is at least partially suppressed if cells are rapidly shifted from red to far-red light. The red/far-red photoreversibility of conidiation control is reminiscent of phytochrome-mediated responses in higher plants leading to the suggestion that a phytochrome-like molecule may regulate asexual reproduction (Pratt, 1982; Mooney and Yager, 1990). The requirement for light in conidiation is dependent upon the *veA* gene (Mooney and Yager, 1990). Whereas development in *veA*+ wild-type strains is light dependent, *veA1* mutant strains conidiate equally well in the light and the dark.

The final requirement for conidiation to occur is that cells first undergo a defined period of vegetative growth (Axelrod *et al.*, 1973; Champe *et al.*, 1981). This need can be observed when cultures are grown in submerged conditions for various times and then induced to conidiate by exposure to air. As the submerged growth period is increased, the length of time a culture must be exposed to air before conidiophores are observed decreases. If development is induced after at least 18 h of submerged growth, the time required for development remains constant at about 4 h. These results have been interpreted to mean that vegetative cells are incapable of responding to the inductive signal until they have grown for at least 18 h. Cells that have acquired the ability to initiate development immediately when presented with the appropriate environment are said to be developmentally competent.

The mechanisms that control the acquisition of developmental competence and the switch from relatively undifferentiated vegetative growth to conidiophore development are largely unknown but are apparently genetically determined rather than environmentally dependent (Pastushok and Axelrod, 1976; Champe *et al.*, 1981). *Aspergillus* conidiophore development is thus proposed to occur as a programmed event in the life cycle rather than as a response to unfavourable environmental conditions. Neither the concentration of a limiting nutrient such as glucose, nor continuous transfer to fresh medium alters the timing of conidiation. On the other hand, precocious conidiation mutants with a decreased time requirement for acquisition of competence have been described (Axelrod *et al.*, 1973). The mutations in these strains could define aspects of the programme that determine the time conidiation begins. Unfortunately, precocious mutants have not been characterized further.

(b) Morphological changes in Aspergillus conidiophore development

Detailed descriptions of the development of conidiophores and conidia can be found in an ultrastructural investigation of *A. nidulans* conidiation by Mims *et al.* (1988; Figure 17.2). Conidiophore development in *A. nidulans* begins with differentiation of hyphal elements into thick-walled foot cells that eventually support mature conidiophores. Foot cells produce aerial conidiophore stalk initials that, like hyphae, grow by apical extension, but unlike hyphae, to a genetically predetermined height of about 100 μm. At this point, there is an important change in growth pattern as apical extension ceases and the conidiophore stalk tip swells to produce the multinucleate conidiophore vesicle. A yeast-like budding mode follows resulting in production of a layer of uninucleate cells called primary sterigmata or metulae. Cytoplasmic continuity between the metulae and the conidiophore vesicle is maintained by means of septal pores, like those in hyphal septa, that are guarded by spherical structures called Woronin bodies. Each of the metulae, in turn, buds to produce a second tier of uninucleate sterigmata termed phialides. Long chains of conidia are produced through repeated asymmetric divisions of the sporogenous phialides. This is a classic example of blastic conidiation.

17.2.2 Genetics of Aspergillus conidiation

Asexual sporulation is a dispensable function in the *Aspergillus* lifecycle (for review see Clutterbuck, 1974, 1977; Timberlake and Marshall, 1988; Timberlake, 1990). Developmental mutants can easily be isolated following mutagenesis by using simple visual screens to identify strains that fail to produce normal developmental structures. Mutant strains that lack asexual spores can usually reproduce through the sexual cycle and are easily maintained and manipulated as ascospore (meiotic spore) stocks. Mutants lacking both sexual and asexual spores can be maintained indefinitely as vegetative cultures. The ability to isolate and maintain developmental mutants with relative ease has allowed successful identification and characterization of several regulatory and non-regulatory loci that function specifically in conidiation.

(a) Number and functions of genes involved in development

Differentiation of the multiple cell types that make up the *A. nidulans* conidiophore has been demonstrated to involve the sequential activation of several hundred genes (Timberlake, 1980). However, not all of these genes are required for discrete developmental events (Aramayo *et al.*, 1989). Whereas Timberlake (1980) demonstrated that approximately 1200 diverse mRNAs accumulate to varying concentrations only during conidiation,

Martinelli and Clutterbuck's (1971) quantitative survey of conidiation mutants led to an estimate of 45–100 loci that are uniquely involved in asexual sporulation. This large discrepancy in predicted gene number may be explained by the possibility that genes detected based only on expression patterns may encode redundant or incremental functions that would not be detectable by simple visual examination of mutants. For example, a deletion of a 38 kbp region from *A. nidulans* containing numerous spore-specific genes did not result in detectable changes in development (Aramayo *et al.*, 1989). Furthermore, subtle defects in sporulation like the spore wall defect resulting from deletion of the *rodA* gene might be difficult to detect in broad screens based on tedious visual examination of colonies (Stringer *et al.*, 1991). Finally, Martinelli and Clutterbuck (1971) limited their analysis to mutations that did not alter vegetative growth rates. mRNAs present at high levels during sporulation may also be present at low levels in vegetative cells and may be required for some aspect of normal growth and metabolism. These mutants would have been excluded by Martinelli and Clutterbuck. Regardless of the actual number of loci involved it is certain that numerous genes are specifically required for sporulation and that at least in many cases their transcription is developmentally controlled.

Clutterbuck proposed three functional definitions for genes identified as specifically affecting development (Clutterbuck, 1977). Strategic loci determine the switch from vegetative growth to conidiophore development. This group includes genes required for acquisition of developmental competence or for responding to induction stimuli. Tactical loci regulate the orderly progression of conidiophore assembly without affecting the ability to initiate the developmental pathway. Finally, auxiliary loci determine secondary aspects of the structural or physiological characters of the conidiophore or spores such as pigmentation. These genetic loci are discussed below based on whether developmental failure occurs before or after conidiation commences.

Mutations that affect acquisition of developmental competence or induction

The largest class of mutants identified by Martinelli and Clutterbuck (1971) were unable to initiate the developmental pathway and were therefore characterized as strategic loci. Mutants in this category are broadly categorized as 'flat' or 'fluffy' depending on whether they produce aerial structures or not. Flat mutants fail to produce aerial structures and therefore never conidiate. By contrast, fluffy mutants produce large masses of vegetative aerial hyphae giving the colony a cotton-like appearance. Some fluffy mutants produce small numbers of conidiophores from aerial hyphae and are therefore classified as oligosporogenous whereas others are totally aconidial. In most cases it has not been possible to demonstrate whether these mutants fail to become competent or lack other functions necessary for developmental initiation.

Champe's group used conditional mutations to show that three different *aco* (aconidial) genes, *acoA*, *acoB* and *acoC*, are blocked in conidiation before acquisition of developmental competence (Champe *et al.*, 1981; Yager *et al.*, 1982; Butnick *et al.*, 1984a,b). These precompetence mutants were isolated based on the idea that genes required for competence may no longer be essential once a strain has already become competent. It was reasoned that temperature-sensitive precompetence mutants could be distinguished from aconidial mutants affected in postcompetence events by growing at the permissive temperature in submerged culture past the time wild-type requires for competence and then simultaneously shifting to the restrictive temperature and inducing conditions. Precompetence mutants conidiate because the functions they are missing at the restrictive temperature are no longer needed. Each of the three preinduction mutants identified by this screen are non-allelic and all are blocked in both asexual and sexual sporulation.

acoA, *acoB* and *acoC* mutants grown at the restrictive temperature all overproduce abundant quantities of a complex set of metabolites that absorb ultraviolet light. At least some of these compounds are diphenyl ethers that probably arise from acetate through the polyketide pathway (Butnick *et al.*, 1984a; Champe and Simon, 1992). Although *acoA*, *acoB* and *acoC* mutants are sexually sterile at the restrictive temperature, they also produce large quantities of a hormone-like fatty acid metabolite called psi factor (precocious sexual inducer) that stimulates sexual sporulation while inhibiting asexual sporulation in wild-type strains (Champe *et al.*, 1987; Champe and El-Zayat, 1989;

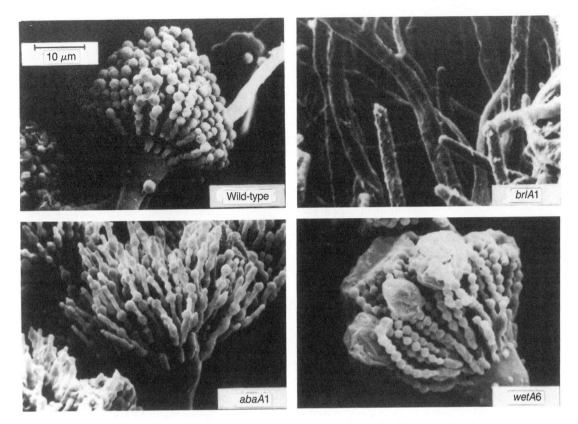

Figure 17.3 Morphology of wild-type and mutant conidiophores as visualized by scanning electron microscopy (Reproduced with permission from Boylan *et al.*, 1987).

Mazur *et al.*, 1990). It has been suggested that the mutational defect in these preinduction mutants deregulates an enzymatic activity that is common to both fatty acid and polyketide biosynthetic pathways (Champe and Simon, 1992).

Several other genetic loci have been defined that block the growth/development transition. Dorn (1970) analysed nine fluffy variants and showed that mutations in a minimum of five distinct loci can result in the fluffy phenotype. All of these fluffy mutants had simultaneously lost the ability to undergo development and taken on invasive properties. The term 'invasive' refers to the unusual ability of certain fluffy mutants to infiltrate other colonies by growing through the sharp zone of demarcation usually observed with wild-type *Aspergillus*. Additional fluffy mutants have been characterized by a number of other investigators (Yager *et al.*, 1982; Tamame *et al.*, 1983, 1988; Adams *et al.*, 1992).

Mutations that affect execution of the developmental pathway

One striking feature of mutational analysis of *Aspergillus* conidiation is the relative scarcity of loci that are required for specific steps in development following initiation (Clutterbuck, 1969; Martinelli and Clutterbuck, 1971; Clutterbuck, 1977). These include the potential regulatory genes *brlA*, *abaA*, *wetA*, *stuA* and *medA* as well as genes that are required for secondary characteristics of the conidiophore such as spore colour (*wA*, *yA*, *fwA*), spore wall structure (*rodA*), or conidiophore pigmentation (*ivoA*, *ivoB*). Given that many independent mutations have been isolated in most of these loci, the number of genes specifically involved in this process is apparently fairly small.

Only mutations in *brlA* (bristle mutants, Figure 17.3) and *abaA* (abacus mutants, Figure 3) completely block sporulation without affecting vegetative growth (Clutterbuck, 1969, 1977; Timberlake, 1990).

brlA⁻ mutants differentiate conidiophore stalks that grow to 10–20 times the height of wild-type conidiophores and never form apical vesicles (Clutterbuck, 1969). Thus, *brlA* is apparently required first during the transition from apical growth to bud formation. *abaA⁻* mutants are blocked in phialide differentiation (Clutterbuck, 1969; Sewall *et al.*, 1990a). Abacus mutants produce conidiophore stalks and vesicles but the metulae do not give rise to sporogenous phialides but instead form branching sterigmata that are reiterated resulting in cell chains with swellings at fairly specific intervals. Temperature sensitive *brlA* and *abaA* mutants were used to demonstrate that each gene is required not only for the developmental transitions described above but throughout development (Mirabito *et al.*, 1989).

wetA (wet-white, Figure 17.3), *stuA* (stunted) and *medA* (medusa) mutants all produce some spores and are therefore categorized as oligosporogenous (Clutterbuck, 1969; Martinelli and Clutterbuck, 1971). Each of these mutants has pleiotropic effects on conidiophore development. *wetA⁻* mutants do not synthesize crucial cell wall components. This leads to formation of conidia that either fail to become dormant or lyse due to structural defects (Sewall et al., 1990b). *stuA⁻* null mutants produce shortened, or stunted, conidiophores that do not form normal metulae and phialides (Clutterbuck, 1969; Martinelli, 1979; Miller *et al.*, 1991, 1992). In fact, conidia produced by *stuA⁻* mutants often appear to bud directly from the conidiophore vesicle surface. *medA⁻* mutants, in contrast, make multiple layers of metulae and phialides before producing spores and frequently produce secondary conidiophore structures (Clutterbuck, 1969). Both *stuA⁻* and *medA⁻* mutants are also self-sterile as they fail to produce cleistothecia, the sexual fruiting bodies.

(b) *Molecular genetic characterization of developmental regulators*

The ability to isolate wild-type copies of genes identified through mutational analysis using DNA-mediated transformation technology has provided novel means for examining development (Chapter 12). For instance, null mutations can be constructed by using gene flanking sequences attached to a selective marker to direct homologous gene replacement events (Timberlake and Marshall, 1989). Gain of function mutations can be constructed by increasing gene copy number or by placing coding sequences under the control of a strong, inducible promoter for a different gene (Adams *et al.*, 1988; Mirabito *et al.*, 1989; Marshall and Timberlake, 1991). Finally, *cis*-acting regulatory regions can be examined by investigating regulation of gene fusions to convenient reporters such as the *Escherichia coli lacZ* gene (Hamer and Timberlake, 1987; Adams and Timberlake, 1990; Aguirre *et al.*, 1990).

The central regulatory pathway

Genetic and biochemical analyses of *A. nidulans* conidiation have implicated *brlA*, *abaA* and *wetA* as the major post-initiation pathway specific regulators of conidiophore development and conidium maturation. Mutations in each gene specifically affect asexual development without altering vegetative growth or sexual reproduction and lead to formation of abnormal conidiophores that fail to produce viable, mature conidia (Figure 17.4). In addition, these mutations result in greatly reduced levels for most mRNAs encoded by developmentally regulated genes without affecting accumulation of mRNAs corresponding to non-regulated genes (Boylan *et al.*, 1987).

The isolation of the wild type *brlA*, *abaA* and *wetA* genes by complementation of mutant strains using a genomic library has allowed their further characterization (Boylan *et al.*, 1987). All three genes encode developmentally regulated transcripts that are not present (or are present at very low levels) in vegetative cells and appear sequentially during development; expression of each gene is required for expression of the next. These results confirm the conclusion of morphological epistasis tests (Martinelli, 1979) indicating that the genes define a linear dependent pathway: *brlA→ abaA→ wetA*.

The best characterized of these developmental regulatory genes is *brlA*. The presumed BrlA polypeptide contains a directly repeated sequence that closely resembles the CC/HH Zn(II) coordination sites that are typical of 'zinc finger' DNA binding motifs, indicating that *brlA* likely encodes a nucleic acid binding protein (Adams *et al.*, 1988). This possibility is supported by the fact that mutations altering the *brlA* CC/HH motif eliminate activity *in vivo* (Adams *et al.*, 1990). In addition, *A. nidulans brlA* expression in *Saccharomyces cerevisiae*

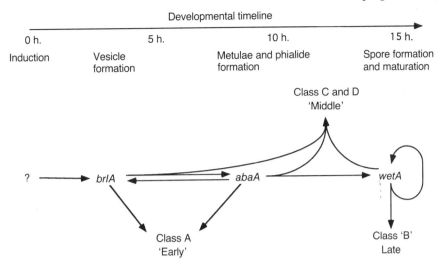

Figure 17.4 Developmental timeline for *Aspergillus* conidiophore development showing interactions of genes in the central regulatory pathway. Solid arrows represent positive interactions inferred from genetic analyses.

has been demonstrated to result in *brlA*-dependent activation of *Aspergillus* genes in *S. cerevisiae* providing that they contain proposed sites for BrlA protein interaction (BREs, BrlA Response Elements) (Chang and Timberlake, 1992). These results support the idea that *brlA* encodes a primary transcriptional regulator of the central regulatory pathway for development.

The *brlA* locus is complex in that it consists of two overlapping transcription units designated *brlAα* and *brlAβ*. *brlAβ* transcription initiates about 1 kb upstream of *brlAα* transcription which begins within *brlAβ* intronic sequences (Prade and Timberlake, 1993, Figure 17.5). The *brlAβ* transcript encodes two open reading frames (ORFs) that begin with AUGs, a short upstream ORF (μORF) and a downstream reading frame that encodes the same polypeptide as *brlAα* except that it includes 23 additional amino acids at the NH-terminus. Mutations that block expression of either transcript alone cause abnormal development. However, multiple copies of either *brlAα* or *brlAβ* can compensate for loss of the other gene. These results are consistent with the hypothesis that *brlAα* and *brlAβ* have evolved to provide a mechanism to separate responses to the multiple regulatory inputs activating *brlA* expression (Adams *et al.*, 1992).

The pivotal role of *brlA* in controlling this pathway is further supported by results of experiments involving misscheduled activation of *brlA* in inappropriate cell types. Adams *et al.* (1988) constructed an *A. nidulans* strain that allowed controlled transcription of *brlA* in vegetative cells by fusing the gene to the promoter for the *A. nidulans* catabolic alcohol dehydrogenase gene, *alcA(p)* (Patemen *et al.*, 1983; Gwynne *et al.*, 1987). Because *alcA* transcription is induced by threonine or ethanol and repressed in the presence of glucose, *brlA* could be induced or suppressed in vegetative cells by simply changing the medium. When the *alcA(p)::brlA* fusion strain was transferred from medium with glucose as a carbon source to medium with threonine, hyphal tips immediately stopped growing and differentiated into reduced conidiophores that produced viable conidia (Adams *et al.*, 1988; Adams and Timberlake, 1990). In addition, this forced activation of *brlA* resulted in activation of the *abaA* and *wetA* regulatory genes, as well as additional developmentally specific transcripts with known and unknown functions.

The predicted AbaA polypeptide contains a domain that is very closely related to the DNA binding domain of the SV40 enhancer factor TEF-1 (Andrianopoulos and Timberlake, 1991; Bürglin, 1991). This domain is also closely related to the yeast Tyl enhancer binding protein TEC1 and has been called the TEA (TEF-1 and TEC1, AbaA) or ATTS (AbaA, TEC1p, TEF-1 sequence) domain. AbaA also contains a potential leucine zipper for dimerization and like *brlA*, *abaA* is required for

Hyphae

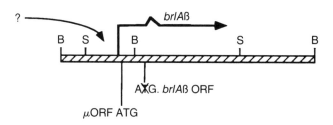

Induction

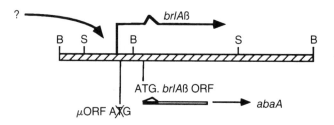

Figure 17.5   Model for differential control of *brlAα and brlAβ* during conidiation. *brlAβ* transcribed in hyphae before developmental induction but translation of the μORF represses translation of BrlA. Following induction, unknown regulators result in activation of BrlA translation from *brlAβ* by removing the translational block imposed by the mORF, increasing *brlAβ* transcription, or both. Activation of BrlA expression leads to transcription of *abaA* and other downstream regulatory genes which in turns activates a positive feedback regulatory loop to cause high levels of *brlAα* expression and further developmental changes.

Development

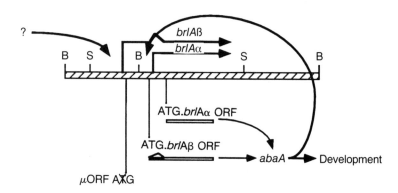

transcriptional activation of numerous sporulation specific genes (Boylan *et al.*, 1987; Mirabito *et al.*, 1989). Although AbaA has not yet been demonstrated to bind DNA directly, these results support the idea that the major role of AbaA in development is as a transcriptional regulator.

Forced activation of *abaA* from the *alcA(p)* in vegetative hyphae resulted in growth cessation and accentuated cellular vacuolization, but not in conidial differentiation (Mirabito *et al.*, 1989; Adams and Timberlake, 1990). *abaA* induction also led to activation of several developmentally specific genes with known and unknown functions including *wetA*, and perhaps surprisingly, *brlA*. Thus, *brlA* and *abaA* are reciprocal inducers but *brlA* expression must occur before *abaA* expression for productive conidiophore development.

wetA is predicted to encode a polypeptide that is rich in serine (14%), threonine (7%), and proline (10%) (Marshall and Timberlake, 1991). Analysis of the WetA sequence has given no clear indication of WetA function due to a lack of sequence similarities with other genes in databases. Forced activation of *wetA* in vegetative cells caused growth inhibition and excessive branching and resulted in accumulation of transcripts from several genes that are normally expressed only during spore formation and whose mRNAs are found in mature spores. *wetA* activation in hyphae did not result in *brlA* or *abaA* activation and never led to premature conidiation.

By examining patterns of RNA accumulation in a series of *brlA⁻*, *abaA⁻* and *wetA⁻* mutant strains in which *alcA(p):: brlA*, *alcA(p):: abaA* or *alcA(p):: wetA* was induced, the non-regulatory developmentally activated genes were divided into four categories (Mirabito *et al.*, 1989; Timberlake, 1990; Marshall and Timberlake, 1991). Class A genes are activated by either *brlA* or *abaA* or both, independent of *wetA* (Figure 17.4). These Class A genes are likely to be involved in early development events. *abaA* alone activates *wetA* which in turn activates Class B genes, independent of *brlA* and *abaA*. These Class B genes are likely to encode spore-specific functions. Class C and D genes require the combined activities of *brlA*, *abaA* and *wetA* for their expression and have been proposed to encode phialide-specific functions. Class C and D genes are distinguished from one another by their expression patterns during normally induced development in wild-type and mutant strains.

Accumulation of *wetA* mRNA requires *wetA⁺* activity both during normal conidiophore development and in forced expression experiments bringing up the interesting possibility that *wetA* is autogenously regulated (Boylan *et al.*, 1987; Marshall and Timberlake, 1991).

These results have been incorporated into the model describing the genetic processes underlying the temporal and spatial control of gene expression during differentiation of conidiophores shown in Figure 17.4. In this model, activation of *brlA* expression initiates a cascade of events that are coordinated by the interactions of *brlA*, *abaA* and *wetA*. The timing and extent of expression of the regulatory and non-regulatory genes is determined as development continues by intrinsic changes in the relative concentrations of the regulatory gene products in the various conidiophore cell types. There remains a great deal to learn regarding the molecular genetic processes that lead to *brlA* expression as well as the precise mechanisms through which *brlA*, *abaA* and *wetA* interact in controlling their own expression and the expression of other developmentally regulated genes.

Modifiers of development

The *stuA* and *medA* genes have been classified as modifiers of conidiophore development (Clutterbuck, 1977; Martinelli, 1979). Mutations in these genes result in abnormal cell differentiation and altered spatial organization in the conidiophore without blocking the ability to produce spores. A wild-type copy of the *stuA* gene has been isolated and analysed in some detail (Miller *et al.*, 1991, 1992). A detailed analysis of *medA* has not yet been reported.

As with *brlA*, *stuA* transcription is complex in that it consists of two transcripts *stuAα* and *stuAβ*. *stuAβ* transcription initiates about 650 bp upstream of *stuAα* transcription which begins within the first *stuAβ* intron (Miller *et al.*, 1992). The two transcripts are barely detectable in young vegetative cells but levels increase at least 50-fold when cells become developmentally competent. There is an additional increase in transcript levels during the first few hours after induction. Both *stuAα* and *stuAβ* are predicted to encode the same 63.5 kDa StuA polypeptide but they differ in that three mini-ORFs beginning with AUGs are found in the 5' end of the *stuAα* transcript and only two in the *stuA*β transcript. The functional significance of these upstream mini-ORFs has not been determined but there is some evidence for translational regulation of *stuA* expression.

A precise role for *stuA* in development is not yet clear. The predicted StuA polypeptide has no significant similarities to protein sequences found in databases but there is a highly basic (24%) amino-terminal domain and a potential bipartite nuclear localization in the carboxy-terminus (KRMR-18 a.a.-KRRK). Mutations in *stuA* lead to delocalized expression of *abaA(p)::lacZ* and *brlA(p)::lacZ* fusions indicating that one role for StuA may be to establish the proper cellular controls on expression of these genes needed for normal differentiation (Miller *et al.*, 1992). Understanding the molecular mechanism of *stuA* action awaits more detailed analyses.

Regulation of developmental induction: activation of *brlA*

When mutants of *A. nidulans* defective in asexual reproduction but having normal vegetative growth rates were classified by Martinelli and Clutterbuck (1971) according to the stage of development affected, 83% were found to be altered in their ability to initiate development. Thus, if mutation frequency accurately reflects the number of genes involved in a process, initiation is the most genetically complex step of conidiophore development. The complexity of this genetic step has made it difficult to approach experimentally. However, the demonstration that expression of *brlA* is sufficient to activate the remainder of the developmental pathway and results in spore formation helps to refocus the question of what activates the conidiation pathway to what activates *brlA* (Adams *et al.*, 1988; Adams and Timberlake, 1990).

The question of what activates *brlA* is complicated by the finding that the *brlA* locus encodes overlapping transcripts that could potentially be controlled by independent mechanisms (Prade and Timberlake, 1993, Figure 17.5). Han *et al.* (1993) used translational fusions between each of the *brlA*α and *brlA*β ORFs and the *Escherichia coli lacZ* gene to examine the possibility that the two *brlA* transcription units provide different mechanisms for controlling *brlA* expression during development. They found that BrlA expression from *brlA*β can be controlled by altering the choice of translational initiation codons. *brlA*α, on the other hand, is primarily controlled at the transcriptional level and is regulated by positive feedback controls that probably involve *abaA*.

A proposed model accounting for coordinated expression of *brlA*α and *brlA*β during conidiophore development is shown in Figure 17.5. *brlA*β message is transcribed in vegetative cells even before development is induced but no BrlA polypeptide is made because the μORF present at the 5′ end of the message represses translation from the internal AUG. During induction, unknown regulatory factors relieve the translational block imposed by the μORF on BrlA expression. This might be accomplished through altering the translational machinery to allow more efficient translation from internal AUGs, through increased transcription of *brlA*β, or by both mechanisms. Activation of BrlA translation leads to transcription of *abaA* and other downstream regulatory genes. This in turn activates a positive feedback regulatory loop leading to high levels of *brlA*α expression and causing further developmental changes. This model in no way precludes the possibility that initiation factors could also function through direct activation of *brlA*α transcription.

The mechanism of translational control by the *brlA*β μORF remains unclear. Small open reading frames like that present in the 5′ end of *brlA*β have been shown to regulate internal translational initiation for several eukaryotic mRNAs (see Kozak, 1991 for review). In some cases, this regulation requires a specific amino acid sequence in the regulatory peptide whereas in others the presence of an AUG alone is sufficient to regulate downstream translation (Hinnebusch, 1984; Mueller and Hinnebusch, 1986; Werner *et al.*, 1987; Abastado *et al.*, 1991). Frameshift mutations in the *brlA*β μORF had no measurable effect on *brlA* control but removing the AUG allowed derepressed expression. These results are consistent with the idea that specific amino acid sequences in the *brlA*β μORF are not required for imposition of the translational block but that a general change in the translational machinery occurs. It is noteworthy that at least one other *Aspergillus* developmental regulatory gene, *stuA*, has small open reading frames present upstream of the predicted initiation codon and there is some evidence for post-transcriptional regulation (Miller

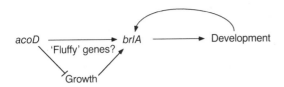

Figure 17.6 Model describing control of growth and *brlA* expression by *acoD*.

et al., 1992). Thus, translational changes may have general importance in controlling *Aspergillus* development.

Although many mutations have been described that result in morphological abnormalities indicating that wild type genes might exert their effects before *brlA* activation, only one gene, *acoD*, has been characterized further (Yager *et al.*, 1982; Adams *et al.*, 1992). *acoD* mutants are examples of a large class of putative developmental switching mutants known as 'fluffy'. *A. nidulans* fluffy mutants, like fluffy mutants in other fungi, characteristically proliferate as undifferentiated masses of vegetative cells that appear to grow faster than wild-type colonies and form large cotton-like masses. As described above, some of these mutants also overcome the barrier effect exhibited by wild type colonies in that they are able to grow into neighbouring colonies. Such fluffy mutants have therefore simultaneously lost the ability to respond to developmental signals and to the growth inhibitory signals that prevent wild-type colonies from merging.

Adams *et al.* (1992) isolated the wild type *acoD* gene through complementation of an *acoD* mutant strain and examined regulation of *brlA* expression in an *acoD* deletion strain. They demonstrated that whereas *acoD* is required for activation of *brlA* under optimal growth conditions, this requirement is partially overcome when growth is limited nutritionally (Figure 17.6). No *brlA* mRNA was detected in *acoD* mutant cultures even 48 h after developmental induction on complete medium but when the same mutant strain was grown on minimal medium *brlA* mRNA was detected within 9 h of induction and conidiophores appeared several hours later. *brlA* mRNA was detected within 4 h of inducing development in wild-type strains grown under both conditions. These results raise the possibility that *brlA* may be activated through more than one mechanism. In wild type *A. nidulans*, the major pathway for

conidiation requires *acoD* and results in programmed activation of development independent of nutrient status. In the absence of the *acoD* pathway, development can still be activated by a mechanism that senses growth rate or nutritional status directly. Whether *acoD* activates *brlA* expression through a direct mechanism or through indirect effects on growth remains to be understood (Figure 17.6).

One explanation for the results described above is that *acoD*-independent activation of *brlA* could be through effects of nutritional limitation on translational regulation of *brlAβ*. An analogy can be made to the translationally regulated expression of the *Saccharomyces cerevisiae* GCN4 gene. GCN4 encodes the transcriptional activator for general control of amino acid biosynthesis (Hinnebusch, 1984; Mueller and Hinnebusch, 1986). Small open reading frames present in a long transcribed leader repress GCN4 translation during growth in rich media. During amino acid starvation, translational repression is overcome through a mechanism that involves phosphorylation of the α subunit of eIF2 (Dever *et al.*, 1992). Programmed initiation of development directed by *acoD* would in some way override the need for nutritional signals. A more detailed analysis of *acoD* and of other genes identified by mutations resulting in fluffy morphologies should provide insights into the mechanisms controlling developmental activation.

17.3 CONIDIATION IN *NEUROSPORA CRASSA*

17.3.1 The developmental pathway

(a) Growth versus development

Asexual sporulation in *N. crassa* involves production of either of two morphologically and developmentally distinct conidial types called macroconidia and microconidia respectively (Turian and Bianchi, 1972). The most common conidial type is the macroconidium. These are large multinucleate spores that are produced from aerial hyphae or conidiophores. Microconidia are smaller, as the name implies, uninucleate spores that are typically made directly from hyphae in older cultures. Much more work has been devoted to understanding mechanisms underlying macroconidiation

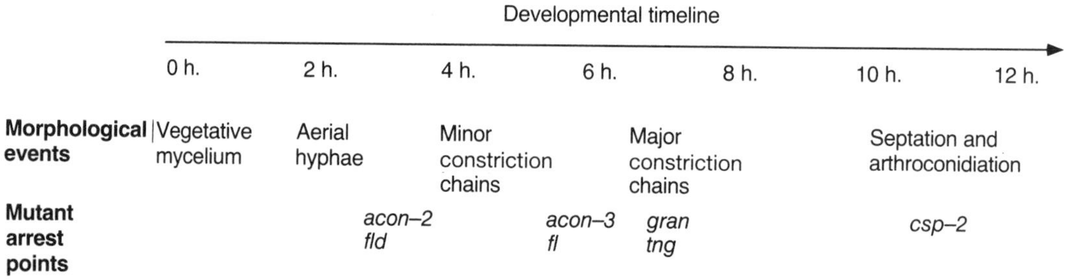

Developmental timeline

	0 h.	2 h.	4 h.	6 h.	8 h.	10 h.	12 h.
Morphological events	Vegetative mycelium	Aerial hyphae	Minor constriction chains		Major constriction chains	Septation and arthroconidiation	
Mutant arrest points		acon–2 fld		acon–3 fl	gran tng	csp–2	

Figure 17.7 Developmental timeline describing morphological changes that take place during *N. crassa* conidiation. Apparent positions of developmental arrest for several well-characterized morphologically aberrant mutants are indicated.

than microconidiation and it is this process that will be discussed here.

As with *A. nidulans*, the asexual lifecycle of *N. crassa* can be divided into growth and reproductive phases. These organisms differ in that the vegetative mycelium of *N. crassa* ramifies throughout the substrate before the reproductive phase begins so that the processes of growth and development for a *Neurospora* colony are separated in time (Turian and Bianchi, 1972; Matsuyama *et al.*, 1974; Springer and Yanofsky, 1989). In *Aspergillus*, both events occur in the same colony at the same time and are only separated spatially. In addition, there is no evidence for a requirement for developmental competence in *Neurospora* like that observed in *Aspergillus* (see above).

Initiation of conidial development by *N. crassa* appears to be strictly controlled by environmental conditions (Sargent and Kaltenborn, 1972; Turian and Bianchi, 1972). Conditions that favour conidiation include nutritional deprivation, desiccation, aeration, low carbon dioxide levels, and exposure to blue light. In addition, there is circadian control to *Neurospora* conidiation (Sargent and Briggs, 1967; Sargent and Woodward, 1969). As with *A. nidulans*, the cellular mechanisms required for interpreting these signals are not well defined.

(b) Morphological changes during Neurospora *macroconidiation*

Several investigators have described the morphological changes leading to formation of macroconidia in *N. crassa* (Turian and Bianchi, 1972; Seale, 1973; Matsuyama *et al.*, 1974; Springer and Yanofsky, 1989). Macroconidiation initiates after the vegetative hyphae have overtaken the

substrate and receive the proper environmental signals. Aerial hyphae are then produced that grow upwards and quickly begin to form lateral branches to result in construction of a large aerial mass (Figure 17.7). Within about 2 h of induction, some of the lateral branches change their growth mode from hyphal tip elongation to a process resembling the apical budding observed in *S. cerevisiae*. During the initial stages of budding, proconidial chains are produced in which the bud junctions are nearly as large in diameter as the proconidia. The structures produced are called 'minor constriction chains' and represent the first defined stage in conidiophore production. The majority of the aerial mycelium is composed of minor constriction chains by 4 h after developmental induction. As budding continues, bud junctions become narrower leading to formation of 'major constriction chains' which represent most of the proconidial chains by around 8 h after development is induced. The final step in conidium formation, septation, begins at this point. Septation occurs randomly along the proconidial chain and is first visible in SEM as ridges between adjacent proconidia. Subsequent septal thickening and furrow formation leads to formation of conidia held together by a structure termed a connective. Because macroconidia are produced through formation of a series of constrictions (or buds) followed by septation, *N. crassa* conidiation has both blastic and thallic properties and has been described as blastoarthro-conidiation.

17.3.2 Genetics

(a) Developmental mutants

A large number of mutants blocked in specific stages of conidiation have been identified and characterized genetically (Figure 17.7). Unfortunately, molecular genetic analysis of loci affected has lagged behind and for the most part there is little understanding of the nature of the defects. Several studies have relied upon morphological characterizations of single and double conidiation mutants. Most recently, Springer and Yanofsky (1989) examined the morphological defects in conidiation mutants to propose steps blocked in the developmental pathway. A summary of the model they developed to describe genes required for macroconidiation is shown in Figure 17.7.

The earliest acting mutations described are *acon-2* (aconidiate-2) and *fld* (fluffyoid). *acon-2* and *fld* mutants produce aerial hyphae but are specifically blocked before formation of minor constriction chains indicating that they fail to make the transition from hyphal tip elongation to bud formation (Matsuyama *et al.*, 1974). Interestingly, both mutations can be partially remediated by severe carbon starvation. These mutants are thus somewhat analogous to the *A. nidulans acoD* mutant which only conidiates when growth is nutritionally limited (see above).

acon-3 (aconidiate-3) and *fl* (fluffy) mutants produce minor constriction chains but fail to produce major constrictions (Matsuyama *et al.*, 1974). The production of major constriction chains has been postulated to be a commitment step in conidiation because minor constriction chains frequently revert to hyphal tip elongation. *acon-3* mutants make minor constriction chains that continue to grow and occasionally revert to hyphal growth whereas *fl* mutants produce short minor constriction chains that stop growing. As one might predict from the morphological blocks observed, *acon-2* and *fld* are epistatic to *acon-3* and *fl* in that double mutants do not develop beyond aerial hyphae formation.

gran (granular), *tng* (tangerine), *csp-1* (conidial separation-1), *csp-2* (conidial separation-2), *cy* (curly), and *eas* (easily wettable) are all affected specifically in aspects of macroconidium formation (Selitrennikoff *et al.*, 1974; Springer and Yanofsky, 1989). *gran* mutants produce amorphous interconidial junctions and *tng* produce proconidia in major constriction chains that frequently produce multiple (5–10) buds. Some of the proconidia produced by *tng* mutants continue to swell to near macroscopic size. Conidia produced by *csp-1*, *csp-2* and *cy* mutants are initially morphologically normal but

fail to separate from one another. Finally, *eas* mutant conidia lack the hydrophobic protein coat termed the rodlet layer and are therefore analogous to *Aspergillus rodA* mutants (Beever and Dempsey, 1978; Stringer *et al.*, 1991).

(b) Analysis of gene expression

Berlin and Yanofsky (1985a,b) characterized changes in protein levels during macroconidiation and isolated a series of genes based on their developmental regulation that they called *CON* (conidiation) genes. Transcripts for each of these genes begin to accumulate at specific times after development initiates. Some of these genes have been analysed with respect to their expression in developmental mutants. In general, *con* genes are either not expressed or expressed at low levels in the early developmental mutants *fl*, *acon-2* and *acon-3* but are expressed at relatively normal levels in mutants that are affected in later stages in development (Roberts *et al.*, 1988; Roberts and Yanofsky, 1989). Transcripts and protein for at least one *con* gene, *CON10*, are present in microconidia and ascospores of strains that do not produce microconidia (Springer *et al.*, 1992). The functions for *con* genes and the mechanisms controlling their expression remain unknown.

17.4 CONCLUSIONS

The ability to isolate wild type copies of fungal genes and manipulate them in sophisticated ways has allowed rapid progress towards understanding the molecular genetic controls for asexual sporulation in higher fungi, but many questions remain unanswered. Although several important regulatory genes for controlling morphogenic events in *Aspergilus* conidiation have been isolated, their biochemical activities have merely been inferred from sequence comparisons. Models like the one proposed in this chapter only hint at the complex genetic interactions that are likely to occur during conidiation. In addition, very few of the genes that contribute to conidiophore form and function in leading to production of dormant spores have been characterized in detail. Furthermore, the activities that result in developmental competence and the signals that allow activation of the conidiation pathway are not yet understood. The molecular details of these events need to be addressed if we are to achieve a complete understanding of the conidiation process in *Aspergillus*.

To date, none of the regulators of *Neurospora* conidiation have been isolated. Molecular characterization of development has mainly relied on examining genes that are specifically activated during macroconidiation. Nothing is known about the transcriptional regulators that control the sequential activation of these genes during spore differentiation. A number of developmental mutants have been described that are blocked at specific morphological stages. In some cases it has been shown that these mutants also fail to activate suites of conidiation specific genes, but it is not yet clear why transcription is blocked. It is essential that wild-type copies of genes required for conidiation and for activation of conidiation specific genes be isolated and characterized in order to understand *Neurospora* macroconidiation.

One of the most interesting challenges regarding sporulation in higher fungi is to understand the way in which so much diversity in mechanisms controlling conidiation and conidial form has arisen. *N. crassa* and *A. nidulans* are just two examples of the greater than 15 000 species of conidial fungi that have been described. The strategies undertaken by *Neurospora* and *Aspergillus* in regulating initiation of the conidiation pathway and in controlling morphological changes during development appear to be very different. *A. nidulans* conidiation begins as a programmed event at a relatively early stage in the life cycle, before growth has overtaken the substrate. By contrast, *N. crassa* conidiation is apparently strictly controlled by environmental parameters and typically does not initiate until the mycelium has ramified throughout the substrate. The degree of cell specialization in an *A. nidulans* conidiophore is much greater than that observed during macroconidiation in *N. crassa*. As the molecular genetic controls for these two pathways are better characterized, it will be interesting to learn how these apparent strategic differences are reflected in the developmental programme for each organism.

REFERENCES

Abastado, J.P., Miller, P.F., Jackson, B.M. and Hinnebusch, A.G. (1991) Suppression of ribosomal reinitiation at upstream open reading frames in amino acid-starved cells forms the basis for *GCN4* translational control. *Molecular and Cellular Biology*, **11**, 486–96.

Adams, T.H., Boylan, M.T. and Timberlake, W.E. (1988) *brlA* is necessary and sufficient to direct conidiophore development in *Aspergillus nidulans*. *Cell*, **54**, 353–62.

Adams, T.H., Deising, H. and Timberlake, W.E. (1990) *brlA* requires both zinc fingers to induce development. *Molecular and Cellular Biology*, **10**, 1815–17.

Adams, T.H., Hide, W.A., Yager, L.N. and Lee, B.N. (1992) Isolation of a gene required for programmed initiation of development by *Aspergillus nidulans*. *Molecular and Cellular Biology*, **12**, 3827–33.

Adams, T.H. and Timberlake, W.E. (1990) Developmental repression of growth and gene expression in *Aspergillus*. *Proceedings of the National Academy of Sciences of the USA*, **87**, 5405–9.

Aguirre, J., Adams, T.H. and Timberlake, W.E. (1990) Spatial control of developmental regulatory genes in *Aspergillus nidulans*. *Experimental Mycology*, **14**, 290–3.

Alexopoulos, C.J. and Mims, C.W. (1979) *Introductory Mycology*. Wiley, New York.

Andrianopoulos, A. and Timberlake, W.E. (1991) ATTS, a new and conserved DNA binding domain. *Plant Cell*, **3**, 747–8.

Aramayo, R., Adams, T.H. and Timberlake, W.E. (1989) A large cluster of highly expressed genes is dispensable for growth and development in *Aspergillus nidulans*. *Genetics*, **122**, 65–71.

Axelrod, D.E., Gealt, M. and Pastushok, M. (1973) Gene control of developmental competence in *Aspergillus nidulans*. *Developmental Biology*, **34**, 9–15.

Beever, R.E. and Dempsey, G.P. (1978) Function of rodlets on the surface of fungal spores. *Nature*, **272**, 608–10.

Berlin, V. and Yanofsky, C. (1985a) Isolation and characterization of genes differentially expressed during conidiation of *Neurospora crassa*. *Molecular and Cellular Biology*, **5**, 849–55.

Berlin, V. and Yanofsky, C. (1985b) Protein changes during the asexual cycle of *Neurospora crassa*. *Molecular and Cellular Biology*, **5**, 839–848.

Boylan, M.T., Mirabito, P.M., Willett, C.E. *et al.* (1987) Isolation and physical characterization of three essential conidiation genes from *Aspergillus nidulans*. *Molecular and Cellular Biology*, **7**, 3113–18.

Bürglin, T.R. (1991) The TEA domain: a novel, highly conserved DNA-binding motif. *Cell*, **66**, 11–12.

Butnick, N.Z., Yager, L.N., Hermann, T.E. *et al.* (1984a) Mutants of *Aspergillus nidulans* blocked at an early stage of sporulation secrete an unusual metabolite. *Journal of Bacteriology*, **160**, 533–40.

Butnick, N.Z., Yager, L.N., Kurtz, M.B. and Champe, S.P. (1984b) Genetic analysis of mutants of *Aspergillus nidulans* blocked at an early stage of sporulation. *Journal of Bacteriology*, **160**, 541–5.

Champe, S.P. and, El-Zayat, A.E. (1989) Isolation of a sexual sporulation hormone from *Aspergillus nidulans*. *Journal of Bacteriology*, **171**, 3982–8.

Champe, S.P. and Simon, L.D. (1992) in *Morphogenesis: An Analysis of the Development of Biological Form*, (eds E.F. Rossomando and S. Alexander), Marcel Dekker, New York, pp. 63–91.

Champe, S.P., Kurtz, M.B., Yager, L.N. *et al.* (1981) in *The Fungal Spore: Morphogenetic Controls*, (eds G. Turian and H.R. Hohl), Academic Press, London, pp. 255–76.

Champe, S.P., Rao, P. and Chang, A. (1987) An endogenous inducer of sexual development in *Aspergillus nidulans*. *Journal of General Microbiology*, **133**, 1383–7.

Chang, Y.C. and Timberlake, W.E. (1992) Identification of *Aspergillus* brlA response elements (BREs) by genetic selection in yeast. *Genetics*, **133**, 29–38.

Clutterbuck, A.J. (1969) A mutational analysis of conidial development in *Aspergillus nidulans*. *Genetics*, **63**, 317–27.

Clutterbuck, A.J. (1974) in *Handbook of Genetics*, (ed. R.C. King), Academic Press, London, pp. 447–510.

Clutterbuck, A.J. (1977) in *Genetics and Physiology of Aspergillus*, (eds J.E. Smith and J.A. Pateman), Academic Press, London, pp. 305–17.

Cole, G. (1986) Models of Cell Differentiation in Conidial Fungi. *Microbiology Reviews*, **50**, 95–132.

Dever, T.E., Feng, L., Wek, R.C. *et al.* (1992) Phosphorylation of Initiation factor 2α by protein kinase GCN2 mediates gene-specific translational control of GCN4 in yeast. *Cell*, **68**, 585–96.

Dorn, G.L. (1970) Genetic and morphological properties of undifferentiated and invasive variants of *Aspergillus nidulans*. *Genetics*, **66**, 267–79.

Gwynne, D.I., Buxton, F.P., Sibley, S. *et al.* (1987) Comparison of the *cis*-acting control regions of two coordinately controlled genes involved in ethanol utilization in *Aspergillus nidulans*. *Gene*, **51**, 205–16.

Hamer, J.E. and Timberlake, W.E. (1987) Functional organization of the *Aspergillus nidulans* trpC promoter. *Molecular and Cellular Biology*, **7**, 2352–9.

Han, S., Navarro, J., Greve, R.A. and Adams, T.H. (1993) Translational repression of brlA expression prevents premature development in *Aspergillus*.. *EMBO Journal* **12**, 2449–57.

Hinnebusch, A.G. (1984) Evidence for translational regulation of the activator of general amino acid control in yeast. *Proceedings of the National Academy of Sciences of the USA*, **81**, 6442–6.

Kozak, M. (1991) An analysis of vertebrate mRNA sequences: intimations of translational control. *Journal of Cellular Biology*, **115**, 887–903.

Marshall, M.A. and Timberlake, W.E. (1991) *Aspergillus nidulans* wetA activates spore-specific gene expression. *Molecular and Cellular Biology*, **11**, 55–62.

Martinelli, S.D. (1979) Phenotypes of double conidiation mutants of *Aspergillus nidulans*. *Journal of General Microbiology*, **114**, 277–287.

Martinelli, S.D. and Clutterbuck, A.J. (1971) A quantitative survey of conidiation mutants in *Aspergillus nidulans*. *Journal of General Microbiology*, **69**, 261–8.

Matsuyama, S.S., Nelson, R.E. and Siegel, R.W. (1974) Mutations specifically blocking differentiation of macroconidia in *Neurospora crassa*. *Developmental Biology*, **41**, 278–87.

Mazur, P., Meyers, H.V., Nakanishi, K *et al.*, (1990) Structural elucidation of sporogenic fatty acid metabolites in *Aspergillus nidulans*. *Tetrahedron Letters*, **31**, 3837.

Miller, K.Y., Toennis, T.M., Adams, T.H. and Miller, B.L. (1991) Isolation and transcriptional characterization of a morphological modifier: the *Aspergillus nidulans* stunted (*stuA*) gene. *Molecular and General Genetics*, **227**, 285–92.

Miller, K.Y., Wu, J. and Miller, B.L. (1992) stuA is required for cell pattern formation in *Aspergillus*. *Genes and Development*, **6**, 1770–82.

Mims, C.W., Richardson, E.A. and Timberlake, W.E. (1988) Ultrastructural analysis of conidiophore development in the fungus *Aspergillus nidulans* using freeze-substitution. *Protoplasma*, **44**, 132–41.

Mirabito, P.M., Adams, T.H. and Timberlake, W.E. (1989) Interactions of three sequentially expressed genes control temporal and spatial specificity in *Aspergillus* development. *Cell*, **67**, 859–68.

Mooney, J.L. and Yager, L.N. (1990). Light is required for conidiation in *Aspergillus nidulans*. *Genes and Development*, **4**, 1473–82.

Mooney, J.L., Hassett, D.E. and Yager, L.N. (1990) Genetic analysis of suppressors of the veA1 mutation in *Aspergillus nidulans*. *Genetics*, **126**, 869–74.

Morton, A.G. (1961) The induction of sporulation in mould fungi. *Proceedings of the Royal Society* London series B **153**, 548–69.

Mueller, P.P. and Hinnebusch, A.G. (1986) Multiple upstream AUG codons mediate translational control of GCN4. *Cell*, **45**, 201–7.

Pastushok, M. and Axelrod, D.E. (1976) Effect of glucose, ammonium and media maintenance on the time of conidiophore initiation by surface colonies of *Aspergillus nidulans*. *Journal of General Microbiology*, **94**, 221–4.

Patemen, J.A., Doy, C.H., Olsen, J.E. *et al.* (1983) Regulation of alcohol dehydrogenase (ADH) and aldehyde dehydrogenase (AldDH) in *Aspergillus nidulans*. *Proceedings of the Royal Society of London, Series B*, **217**, 243–64.

Perkins, D.D., Radford, A., Newmeyer, D. and Bjorkman, M. (1982) Chromosomal loci of *Neurospora crassa*. *Microbiology Reviews*, **46**, 426–570.

Pontecorvo, G., Roper, J.A., Hemmons, L.M. *et al.* (1953). The Genetics of *Aspergillus nidulans*. *Advances in Genetics*, **5**, 141–238.

Prade, R.A. and Timberlake, W.E. (1993) The *Aspergillus nidulans* brlA regulatory locus encodes two functionally redundant polypeptides that are individually required for conidiophore development. *EMBO Journal*, **12**, 2439–47.

Pratt, L.H. (1982) Phytochrome: the protein moiety. *Annual Review of Plant Physiology*, **33**, 557–82.

Roberts, A.N. and Yanofsky, C. (1989) Genes expressed during conidiation in *Neurospora crassa*: characterization of con-8. *Nucleic Acids Research*, **17**, 197–14.

Roberts, A.N., Berlin, V., Hager, K.M. and Yanofsky, C. (1988) Molecular analysis of a *Neurospora crassa* gene expressed during conidiation. *Molecular and Cellular Biology*, **8**, 2411–18.

Sargent, M.L. and Briggs, W.R. (1967) The effects of light on a circadian rhythm of conidiation in *Neurospora*. *Plant Physiology*, **42**, 1504–10.

Sargent, M.L. and Kaltenborn, S.H. (1972) Effects of medium composition and carbon dioxide on circadian conidiation in *Neurospora*. *Plant Physiology*, **50**, 171–5.

Sargent, M.L. and Woodward, D.O. (1969) Genetic determinants of circadian rhythmicity in *Neurospora*. *Journal of Bacteriology*, **97**, 861–6.

Seale, T. (1973) Life cycle of *Neurospora crassa* viewed by scanning electron microscopy. *Journal of Bacteriology*, **113**, 1015–25.

Selitrennikoff, C.P., Nelson, R.E. and Siegel, R.W. (1974) Phase-specific genes for macroconidiation in *Neurospora crassa*. *Genetics*, **78**, 679–90.

Sewall, T.C., Mims, C.W. and Timberlake, W.E. (1990a) *abaA* controls phialide differentiation in *Aspergillus nidulans*. *Plant Cell*, **2**, 731–9.

Sewall, T.C., Mims, C.W. and Timberlake, W.E. (1990b) Conidium differentiation in *Aspergillus nidulans* wild-type and wet-white (*wetA*) mutant strains. *Developmental Biology*, **138**, 499–508.

Springer, M.L. and Yanofsky, C. (1989) A morphological and genetic analysis of conidiophore development in *Neurospora crassa*. *Genes and Development*, **3**, 559–71.

Springer, M.L., Hager, K.M., Garrett, E.C. and Yanofsky, C. (1992) Timing of synthesis and cellular localization of two conidiation-specific proteins of *Neurospora crassa*. *Developmental Biology*, **152**, 255–62.

Stringer, M.A., Dean, R.A., Sewall, T.C. and Timberlake, W.E. (1991) Rodletless, a new *Aspergillus* developmental mutant induced by directed gene inactivation. *Genes and Development*, **5**, 1161–71.

Tamame, M., Antequera, F., Villanueva, J.R. and Santos, T. (1983) High-frequency conversion of a 'fluffy' developmental phenotype in *Aspergillus* spp. by 5-azacytidine treatment: evidence for involvement of a single nuclear gene. *Molecular and Cellular Biology*, **3**, 2287–97.

Tamame, M., Antequera, F. and Santos, E. (1988) Developmental characterization and chromosomal mapping of the 5-azacytidine-sensitive *fluF* locus of *Aspergillus nidulans*. *Molecular and Cellular Biology*, **8**, 3043–50.

Timberlake, W.E. (1980) Developmental gene regulation in *Aspergillus nidulans*. *Developmental Biology*, **78**, 497–510.

Timberlake, W.E. (1990) Molecular genetics of *Aspergillus* development. *Annual Review of Genetics*, **24**, 5–36.

Timberlake, W.E. and Marshall, M.A. (1988) Genetic regulation of development in *Aspergillus nidulans*. *Trends in Genetics*, **4**, 162–9.

Timberlake, W.E. and Marshall, M.A. (1989) Genetic engineering of filamentous fungi. *Science*, **244**, 1313–17.

Turian, G. and Bianchi, D.E. (1972) Conidiation in *Neurospora*. *Botanical Review*, **38**, 119–54.

Werner, M., Feller, A., Messenguy, F. and Pierard, A. (1987) The leader peptide of yeast gene *CPA1* is essential for the translational repression of its expression. *Cell*, **49**, 805–13.

Yager, L.N., Kurtz, M.B. and Champe, S.P. (1982) Temperature-shift analysis of conidial development in *Aspergillus nidulans*. *Developmental Biology*, **93**, 92–103.

SEXUAL REPRODUCTION IN HIGHER FUNGI

18

3

C. Staben

3

1

T.H. Morgan School of Biological Sciences, University of Kentucky, Lexington, KY, USA

18.1 INTRODUCTION

Sexual reproduction has two critical effects upon fungi: it provides the developmental mechanism to generate sexual spores and the potential during meiosis for changes (repair, recombination, and reassortment) to the genome and gene pool. Many other events are associated with sexual reproduction, as well. For mycologists, sexual reproduction provides a major traditional taxonomic character and a useful mechanism for studying the genome.

The sexual cycle in fungi is an alternation between haploid and diploid phases. Sexual reproduction involves five basic steps: sexual differentiation, plasmogamy, karyogamy, meiosis, and sporogenesis. Sexual differentiation is the formation of specialized mating structures (usually described as male or female structures: although male and female may seem to have obvious meanings, we will define male as the mating partner that does not donate mitochondrial DNA). Plasmogamy is the fusion of two distinct organisms at the cytoplasmic level, to form a dikaryotic or heterokaryotic and potentially heteroplasmic organism. Karyogamy is the fusion of two nuclei prior to meiosis. Meiosis is the complex set of DNA replication, recombination, and chromosome reassortments resulting in reduction of the diploid genome to haploid. Finally, sporogenesis is the packaging of the these haploid genomes into spores. Variations in the timing, mechanism, and regulation of these basic steps yields the incredibly varied life cycles encountered within the fungi. Life cycles are also classified as homothallic or heterothallic. Homothallic organisms are self-fertile, whereas heterothallic organisms are able to have sex only with an individual different from themselves.

The three primary regulators of sexual reproduction are environmental (e.g. temperature, light), nutritional (e.g. nitrogen, carbon) and genetic (e.g. mating type or fertility factors). Environment and nutrition have particularly strong effects on development; they determine whether sexual differentiation will occur or whether diploids will propagate or sporulate, for example. The most obvious effect of genetic regulation, particularly mating type, is to determine which individuals will have sex. By understanding the mechanics of sexual reproduction and its regulation, we can begin to understand reproduction at the organismal, molecular, and perhaps evolutionary levels.

We will confine our attention to the Ascomycete and Basidiomycete classes. Life cycles of lower fungi are described elsewhere in this volume, and by definition, Deuteromycetes do not undergo sexual reproduction. Primarily, this chapter will outline sexual reproduction in representative well-studied families. Attention will be concentrated on genetic control, primarily mating type control, of sexual reproduction. Development in the various families will be contrasted and compared to derive general models for the control of sexual reproduction and to evaluate evolutionary implications of current molecular findings.

18.2 SEXUAL SPORULATION IN ASCOMYCETES

The Ascomycetes are traditionally divided into two subclasses by their modes of ascus production. Hemiascomycete asci develop singly and not within a fruiting body. Euascomycete asci develop from ascogenous hyphae normally within a fruiting body. With some exceptions, the predominant

The Growing Fungus. Edited by Neil A.R. Gow and Geoffrey M. Gadd. Published in 1994 by Chapman & Hall, London. ISBN 0 412 46600 7

growth forms of Hemiascomycetes are single cells (yeasts) and those of Euascomycetes are filamentous (mycelia).

Sexual reproduction has been intensively studied, particularly at the molecular level, in two Hemiascomycetes: *Saccharomyces cerevisiae* and *Schizosaccharomyces pombe*. These systems have been intensively studied and extensively reviewed (Egel *et al.*, 1990; Herskowitz *et al.*, 1992). The yeast systems provide the seminal models for understanding mechanisms of cell-type control, mating type interconversion, signal transduction, cell–cell interaction, and nutritional regulation. An understanding of these yeast models is essential to an appreciation of molecular characterization of the life cycles of filamentous fungi.

18.2.1 Hemiascomycete sexual reproduction

(a) Life cycle of Saccharomyces cerevisiae

Essential features of the life cycle of a typical yeast (*S. cerevisiae*) are shown in Figure 18.1a. Yeasts, unlike filamentous fungi, do not differentiate sexes, though they undergo physiological and morphological changes before mating. Mating, and many other activities, are regulated by the single mating type locus. Plasmogamy is normally immediately followed by karyogamy. *S. cerevisiae* diploids can proliferate indefinitely; meiosis and sporulation are induced by specific nutritional conditions (carbon and nitrogen starvation). Spores are metabolically unlike vegetative cells and more resistant to environmental stresses such as solvents, extreme temperatures, and desiccation. Spores germinate spontaneously when placed on growth medium.

(b) Genetic regulation of the life cycle

The mating type locus specifies cell type in *S. cerevisiae*. Each haploid mating type has specific properties. For example, haploid **a** cells express *a*-pheromone, respond to α-pheromone, can mate with α cells, and cannot sporulate. Haploid α cells express α-pheromone, respond to **a**-pheromone, can mate with α cells, and cannot sporulate. Under certain conditions, haploids can switch mating type, facilitating mating. The diploid state is signalled by the presence of both mating type alleles in a single nucleus. Diploids do not express or respond to pheromones, are unable to mate,

and are able to sporulate. Diploids also have other physiological properties, including the ability to grow pseudohyphally (Gimeno *et al.*, 1992). The properties of haploid and diploid yeast can be explained by a molecular level understanding of the mating type locus and its regulatory targets, as shown in Figure 18.1b.

Haploid α cells have an α-specific 747 bp DNA segment (the *MAT*α allele) that encodes two genes α1 and α2. The α1 gene product binds to DNA and cooperates with a non-mating specific factor, *MCM1*, to promote expression of α-specific genes (encoding α-pheromone, **a**-pheromone receptor, for example). The product of the α2 gene binds to and represses expression of **a**-specific genes.

Haploid **a** cells have an **a**-specific 642 bp DNA segment (the *MAT***a** allele) at the *MAT* locus. Deletion of the mating type locus results in an **a** mating phenotype, but the inability to undergo meiosis. This indicates that the **a** allele has no essential function in haploid **a** cells (**a** cells express **a**-specific functions because they lack the α2 repressor), but is essential in diploids. This segment encodes two genes, **a**1 and **a**2.

Diploid (α/**a**) cells contain both alleles of mating type, but have novel properties because of the interaction between the **a**1 and α2 polypeptides. The **a**1/α2 dimer represses a set of haploid-specific genes, including α1 and *RME1*, a repressor of meiosis. Because these cells express α2, the **a**-specific genes are repressed. Because they do not express α1, the α-specific genes are not induced. Therefore, diploids do not produce or respond to mating pheromones, and are unable to mate. However, because *RME1* is repressed, diploids are able to undergo meiosis and sporulation. *RME1* probably regulates a series of other factors, including several genes that induce meiosis (*IME1*, *IME2*) and sporulation (*SPO* genes). This pathway depends upon several additional regulatory factors, including nutritional regulation (Herskowitz *et al.*, 1992).

In *S. pombe*, four mating type genes are essential for mating and sporulation (Kelly *et al.*, 1988). One mating type product, *Pi*, is a homeodomain polypeptide similar to the α2 polypeptide of *S. cerevisiae*. A second polypeptide, *Mc*, contains a DNA binding motif, the HMG box, that is similar to the *N. crassa mt a-1* polypeptide. The four mating type genes act in concert to activate the *mei3*[+] gene, which inactivates a protein kinase, the

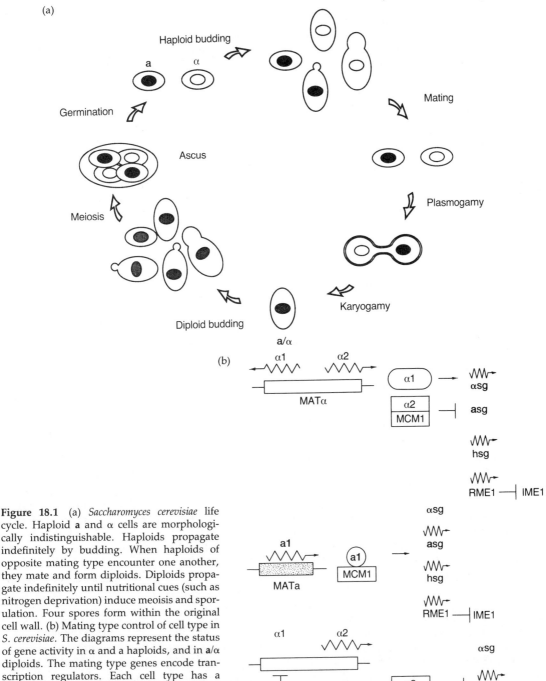

Figure 18.1 (a) *Saccharomyces cerevisiae* life cycle. Haploid **a** and α cells are morphologically indistinguishable. Haploids propagate indefinitely by budding. When haploids of opposite mating type encounter one another, they mate and form diploids. Diploids propagate indefinitely until nutritional cues (such as nitrogen deprivation) induce meiosis and sporulation. Four spores form within the original cell wall. (b) Mating type control of cell type in *S. cerevisiae*. The diagrams represent the status of gene activity in α and **a** haploids, and in **a**/α diploids. The mating type genes encode transcription regulators. Each cell type has a unique set of regulators, which causes specific patterns of gene expression. *MCM1* encodes a factor common to all cell types that has a role in repression and in activation. αsg, **a**sg and hsg stand for α-specific gene, **a**-specific gene and haploid-specific gene, respectively.

product of the *ran1*[+] gene (Beach and McLeod, 1988). The substrates of the protein kinase have not been identified. Nutritional regulation of sporulation is dependent on a pathway involving cAMP regulated transcription of another HMG box polypeptide, *STE11*[+] (Sugimoto *et al.*, 1991; Mochizuki and Yamamoto, 1992).

(c) Mating type interconversion

Mating type interconversion has been extensively studied in both *S. cerevisiae* and *S. pombe* (Klar, 1989, 1992). Both have intricate mechanisms that allow them to switch mating type in a regulated fashion, which expedites diploid formation. Silent copies of both mating type genes are present at other loci, but can be transposed to the *MAT* locus, where they become active. These silent copies are normally repressed by specific regulators. Transposition is regulated by many different factors. The details of mating type switching differ in the two systems, but the common features are that silent copies of the alternate mating type DNAs are transposed to an active locus. This requires both silencing and transposition mechanisms, which involve multiple additional genes.

(d) Signal transduction and cell–cell interaction

Haploid yeast cells secrete mating-type specific pheromones that mediate formation of diploids. Many features of the pheromone response in *S. cerevisiae* have been reviewed recently (Kurjan, 1992). The *S. cerevisiae* α-factor is a short peptide, the **a**-factor is a short peptide that is cysteine-isoprenylated and carboxy-methylated. *S. pombe* M-factor is also a short peptide modified like **a**-factor (Davey, 1992). Many proteins are involved in the processing and secretion of the yeast mating pheromones.

In both *S. cerevisiae* and *S. pombe*, mating factors are bound by pheromone-specific transmembrane receptors that interact with a common response pathway. The extracellular signal results in alteration of G-protein activity, which initiates a cascade of intracellular events (Kurjan, 1992). Changes induced by pheromone arrest cells in the G1 phase, change cell morphology, and induce agglutinability. These changes expedite mating and the formation of the diploid. Pheromone cannot be added exogenously; mating partners apparently respond to local production of mating pheromone to insure conjugation with the opposite mating type (Caplan and Kurjan, 1991).

(e) Nutritional regulation

In *S. cerevisiae*, mating generates a diploid that can proliferate and sporulation depends on both appropriate mating type gene expression and carbon/nitrogen starvation. Starvation is apparently sensed by changes in cAMP levels, which decrease during starvation. The mechanisms that relate this nutritional signal to the action of the *RME1* gene product are not clear.

In *S. pombe*, diploids are transient and mating depends upon starvation. Mating and sporulation are regulated by parallel pathways, one that responds to cAMP levels and one that responds to mating type (DeVoti *et al.*, 1991). Ultimately, induction of sporulation requires activation of both pathways, which results in activation of the *mei2*[+] gene and a cascade of other genes involved in induction of meiosis.

(f) Other roles of the mating type locus

S. cerevisiae mating type genes regulate many other properties besides those specifically required for mating. For example, transcription and transposition of yeast Ty elements is 20-fold lower in **a**/α diploids than in haploids or single mating-type diploids (Boeke and Sandmeyer, 1991). In addition, only **a**/α diploids grow pseudohyphally (Gimeno *et al.*, 1992). The mating pheromones themselves also have many effects in addition to those directly involved in mating. Either pheromone inhibits Ty1 transposition in a responsive cell (Xu and Boeke, 1991); α factor inhibits glycogen accumulation by inhibiting glycogen synthase (Francois *et al.*, 1991).

18.2.2 Euascomycete sexual reproduction

The Euascomycetes are normally divided into three classes based on the morphology of their fruiting bodies. The Plectomycete and Pyrenomycete classes are also consistent with analyses of the ribosomal RNA sequences (Berbee and Taylor, 1992; Bruns *et al.*, 1992). (No Discomycetes have been classified within RNA-based phylogenies.) The fruiting body of the Plectomycetes is a

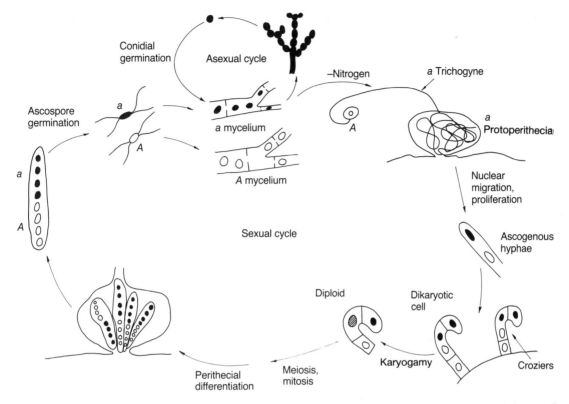

Figure 18.2 *Neurospora crassa* life cycle. The diagram represents the major asexual life cycle (upper) and the sexual life cycle. Protoperithecia (females) differentiate from either mating type in response to nitrogen deprivation. The diagram shows interaction of an *a* female with an *A* male, and detailed events within the ascogenous hyphae.

cleistothecium from which ascospores are released by rupture. Ascospores of the Pyrenomycetes develop within a flask-shaped perithecium and are discharged from a hole in the neck. The apothecium of Discomycetes is a flat disc bearing asci on its surface. Essential details of Euascomycete life cycles are very similar. Here, details of the *Neurospora crassa* sexual life cycle will be discussed then compared and contrasted with that of its relatives, both yeast and Euascomycetes.

(a) Neurospora crassa life cycle

The multinucleate (haploid) mycelium is the vegetative form of *N. crassa*. Two types of asexual spores differentiate from the mycelium: macroconidia and microconidia. Any form can serve as a male in a sexual cross. Under appropriate environmental conditions (temperature, nitrogen limitation and light), strains of either mating type develop female structures: protoperithecia and

trichogynes. Trichogynes (female hyphae specialized for mating) of one mating type respond specifically to pheromone secreted by the opposite mating type by orienting growth towards the male (Bistis, 1983). After fusion, a single male nucleus enters the ascogonium within the perithecium and associates with a female nucleus of opposite mating type. These nuclei divide synchronously and form dikaryotic ascogenous hyphae within the perithecium, which is enlarging and darkening. Karyogamy occurs in the penultimate cells of hook-shaped cells called croziers, and is immediately followed by meiosis and ascosporogenesis. The life cycle is shown in Figure 18.2. Microscopic observations of meiosis and ascosporogenesis have been described in detail (Raju, 1980). Life cycle and mating type of *Neurospora* have been recently reviewed (Glass and Staben, 1990; Glass and Kuldau, 1992).

The *N. crassa* and *S. cerevisiae* life cycles have

many similarities and many differences. Each fungus has a vegetative haplophase, but the diplophase of *S. cerevisiae* can also propagate. Each fungus has two mating types controlled by a single genetic locus containing mating type-specific (idiomorphic) DNA. These loci regulate many phenomena, including production and response to pheromones. Unlike *S. cerevisiae*, *N. crassa* has morphologically distinct male and female thalli. In *S. cerevisiae*, mitochondrial DNA is inherited from either parent or formed by recombination between parental mitochondrial DNAs. In *N. crassa*, mitochondrial DNA is inherited from the female (Taylor, 1986). *Neurospora* ascospores are formed from ascogenous hyphae within a differentiated fruiting body, unlike yeast asci that form singly.

(b)   Genetic control of the sexual cycle

The actions of many genes are essential to sexual reproduction. Genes important for normal development of females, fertility of males, or normal development of ascospores and perithecia have been identified by extensive classical mutation and limited molecular approaches (Perkins *et al.*, 1982; Nelson and Metzenberg, 1992; Raju, 1992). Some genes expressed during asexual sporulation are also expressed during sexual sporulation (Springer and Yanofsky, 1992). This chapter will concentrate on the mating type genes, which are the primary regulators of mate choice. I will describe the roles of mating type, the structure of the locus, and the genes encoded at the locus.

Roles of the *N. crassa* mating type locus

Mating occurs between males and differentiated females only of opposite mating type. The mating type genes control several individual steps within the sexual reproduction pathway. First, males produce mating-type specific pheromones. These pheromones orient the growth of female hyphae of opposite mating type towards the male (Bistis, 1983), *a* mating-type specific response. After the female hypha fuses with the male, the incipient fruiting body enlarges and darkens; morphological development is also controlled by mating type genes. Because asci always contain four *a* and four *A* spores, we know that a mating type-dependent mechanism must assure that one *a* and one *A* nucleus are partitioned into dikaryons that produce ascospores. Either nuclei are identified by

mating type, or only dikaryons containing one *a* and one *A* nucleus develop into asci.

The mating type locus of *N. crassa* also acts as a heterokaryon incompatibility locus during vegetative growth (Beadle and Coonradt, 1944). This function prohibits the formation of viable *a+A* heterokaryons during vegetative growth. Mating type is one of ten identified *het* loci in *N. crassa* (Perkins *et al.*, 1982). The heterokaryon incompatibility due to mating type, but not that due to several other *het* loci, can be suppressed by an extragenic suppressor, *tol*(erant) (Newmeyer, 1970). The function of the *N. crassa* mating type locus as a vegetative incompatibility locus is unusual; it is not a feature of the *Podospora* system, for example. Vegetative incompatibility is associated with mating type in *Aspergillus heterothallicus* (Kwon and Raper, 1967a).

Structure and function of the mating type locus

Mating type is determined by the DNA sequence present at the mating type locus. The *a* mating type is encoded in 3235 bp of DNA (Staben and Yanofsky, 1990) unique to *a* strains; *A* mating type is encoded by 5301 bp of DNA unique to *A* strains (Glass *et al.*, 1990). As noted earlier, these unique DNA segments, which may contain multiple genes, are termed idiomorphs (Metzenberg and Glass, 1990). DNA outside the idiomorphs is nearly identical in the regions directly flanking mating type that have been sequenced. The structure of the mating type locus is illustrated in Figure 18.3a. Classical genetics suggested that otherwise isogenic strains will mate provided only that they have different alleles (idiomorphs) at the mating type locus. Molecular genetics shows that these unique DNAs determine all mating type functions. Transformation of one mating type with idiomorphic DNA of the opposite strain can confer ability to mate as either mating type. Mutation of sequences within the idiomorphs (Glass *et al.*, 1990; Staben and Yanofsky, 1990; Glass and Lee, 1992) causes loss of mating type function. Substitution of one sequence (with accompanying flanking DNA) for the other by direct gene replacement changes mating type, allowing otherwise isogenic strains to mate (Chang and Staben unpublished).

The *a* idiomorph   A 1.6 kbp DNA segment within the *a* idiomorph confers both *a* mating type and vegetative incompatibility. This *a-1* region encodes

Figure 18.3 (a) Structure of idiomorph DNAs, *Neurospora crassa*. The mating type idiomorphs are flanked by homologous DNA. The *A* idiomorph contains at least two functional regions; the *a* idiomorph contains only one essential functional region. The approximate extents of the transcripts are indicated by the solid arrows. Location of the centromere is indicated. (b) Mechanism of *N. crassa* mating type control. During vegetative growth, the *a-1* gene product is presumed to induce expression of *a*-specific genes (asg). The *A-1* gene product is presumed to induce expression of *A*-specific genes. Interaction of the *a* and *A* gene products, in a pathway dependent on the *tol* gene, induces a vegetative incompatibility response (VCG). Under nitrogen limitation, females express a new set of genes (fsg), which cause them to express different functions in both *a* and *A* mating types and to repress VCG responses. The interaction of *a-1* and *A-1* in perithecia induces perithecial differentiation and ascosporogenesis, which are dependent on the *A-2* region. The nomenclature of *N. crassa* mating type, the genes that specify mating type and their encoded polypeptides have recently been standardized to the following convention. The two idiomorphs, *mt a* and *mt A*, contain multiple genes. These are designated *mt a-1*, *mt A-1* etc. The polypeptides encoded by these genes will be designated MT a-1, MT A-1 and so on.

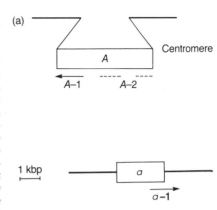

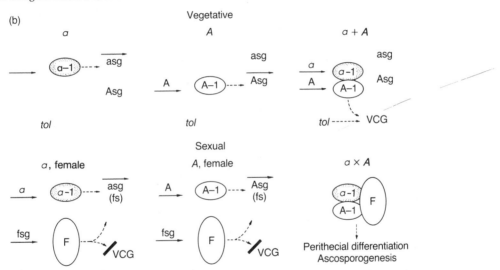

a 382 amino acid polypeptide (mt *a-1*) that mediates vegetative incompatibility and mating. Mt *a-1* is encoded by a spliced mRNA. Strains selected for escape from mating type vegetative incompatibility (Griffiths and DeLange, 1978) have mutations in *a-1* responsible for their mating type defects (Griffiths and DeLange, 1978; Staben and Yanofsky, 1990). Two characterized mutants with sequence changes in *a-1* have lost both vegetative incompatibility and mating activities. Another mutant, a^{m33}, can mate but has no vegetative incompatibility function. This mutant has a single base change in *a-1* that changes residue 258 of mt *a-1* from Arg to Ser.

The structure of mt *a-1* and its similarity to other proteins suggest that it is a DNA binding protein. The *a-1* polypeptide sequence is similar to the *S. pombe* mat-M_c (Kelly *et al.*, 1988), *Podospora anserina* FPR1 (Debuchy and Coppin, 1992) and *Cochliobolus heterostrophus* MAT-2 (Turgeon *et al.*, 1993) polypeptides. The similar sequences and mating functions suggest that the polypeptides are homologous. These polypeptides contain an essentially identical 12 amino acid segment presumed to play a critical role in mating. This segment is part of an 'HMG-box' DNA binding motif shared by high mobility group I (HMG1) proteins and other regulatory gene products, including sequence-specific DNA binding proteins and transcription activators: an RNA polymerase I factor (hUBF) (Jantzen *et al.*, 1990) male-determining factors (SRY proteins) (Nasrin *et al.*, 1991; Harley *et al.*, 1992), and T-cell specific transcription factor (Giese *et al.*, 1991). The HMG box in mt *a-1* and related polypeptides suggests that they are DNA binding proteins. The *N. crassa*, *P. anserina*, and *C. heterostrophus* polypeptides have acidic, proline-rich carboxy-terminal domains, similar to those of some transcription activator proteins. These structures suggest the hypothesis that mt *a-1* regulates target genes by binding to specific DNA sequences in those genes and enhancing transcription.

The 1.4 kbp of *a*-specific DNA outside the *a-1* region has no apparent function; deletion of this region does not interfere with fertility. Similarly, only *FPR1* is required for mating and fertility in *P. anserina* (Picard *et al.*, 1991; Debuchy and Coppin, 1992), and the *Cochliobolus heterostrophus* idiomorph is very short, containing only the *mt a-1* homologue coding region (Turgeon *et al.*, 1993).

The *N. crassa* A idiomorph Though dissimilar in sequence, the *A* and *a* idiomorphs resemble one another in organization (Glass and Staben, 1990; Glass *et al.*, 1990; Staben and Yanofsky, 1990). *A* has at least two functional regions (Glass and Lee, 1992). Mt *A-1* encodes vegetative incompatibility and mating functions. Both activities reside in a 288 amino acid polypeptide, mt *A-1*, produced from a spliced mRNA. Frameshift mutations (Glass *et al.*, 1990) within mt *A-1* coding region eliminate both *A* activities. Mt *A-1* is homologous to the *S. cerevisiae* α1 mating type gene (Astell *et al.*, 1981; Glass *et al.*, 1990). The α1 polypeptide combines with another polypeptide, PRTF (*MCM1*), to activate α-specific genes (Jarvis *et al.*, 1989; Passmore *et al.*, 1989). The other portion of *A* (*A-2*) is required for development of fertile perithecia (Glass and Lee, 1992). However, *A-2* can be inactivated by RIP (a premeiotic process that induces point mutations in all copies of duplicated DNA segments, section 11.4.1 and 18.2.6) Therefore, *A-2* DNA must not be required after RIP induces mutations. No *A-2* product has been identified, but the region can encode a protein with HMG box similarity (Glass and Lee, 1992).

A model for mating type action in *N. crassa* A model similar to the yeast model can explain *Neurospora* mating type gene action (Glass and Staben, 1990), though many details must be tested and clarified. Such a model must account for the roles of the mating type genes during vegetative growth and during each step in sexual development. Genes at the yeast MAT locus encode regulatory proteins that determine the specialized properties of each haploid mating type by causing mating-type specific gene expression. Fusion of cells of opposite mating type generates a new cell type, the diploid, in which products from the mating type genes from novel regulatory species that activate sets of target genes distinct from those active in either haploid (Herskowitz, 1988). During vegetative growth, *a-1* is expressed (Staben and Yanofsky, 1990) in *a Neurospora*, and *A-1* is expressed in *A Neurospora* (Glass *et al.*, 1990). These gene products cause expression of mating type specific genes, including those controlling pheromone production. In *a*, expression may result from DNA binding and transcription enhancing properties of the mt *a-1* polypeptide. If *a* and *A* mycelia fuse, stable heterokaryon formation is prevented unless *a-1*, *A-1*, or *tol* has been mutated. The vegetative incompatibility reaction is apparently due to the interaction of mt *a-1* and mt *A-1*, which are now present in the same cytoplasm. Interaction between mt *a-1* and mt *A-1* may be direct or indirect. However, induction of incompatibility is dependent on other genes, such as *tol* (Newmeyer, 1970).

The *a-1* and *A-1* genes are also expressed under nitrogen-limiting conditions that induce sexual differentiation (Staben and Yanofsky, 1990), and they are required for responsiveness to pheromone (Bistis, 1983). In addition to this role in mating-type specific gene expression, the idiomorphs are required for perithecial development and ascosporogenesis, as described earlier. I propose that the interaction of the *a* and *A* idiomorphs or their products after fertilization is required for gene expression necessary for sexual development. The result of this interaction is different from that which occurs during vegetative growth, so either the mating type gene products or the state of the organism are modified during female differentiation. Disruption indicates that the *A–2* region is required for events occurring after mating that are required for formation of fertile perithecia, perhaps in differentiation of *a*/*A* zygotes. It is not clear whether a second function resides in the *a* idiomorph, though experiments with *P. anserina* suggest that only a short region encoding *Mat-P*, the mt *a-1* homologue, is essential.

Many features of this model are testable by direct genetic and biochemical experiments. For example, one can determine with mating type polypeptides expressed in *E. coli* whether these polypeptides bind DNA and whether the proteins interact directly. Tests of mating type activities of site-directed mutants, both *in vivo* and *in vitro*, can correlate observed biochemical activities with the biological activities. For example, one can determine whether the presumed DNA binding activity of mt *a-1* is essential for vegetative incompatibility. Other tests of the model will require identifying the network of target genes controlled by the mating type genes.

18.2.3 Mating type switching in Euascomycetes

One of the most prominent and intensively studied mating type phenomena in *S. cerevisiae* and *S. pombe* is their ability to switch mating type. In each case, switching is accomplished by transposition of silent information of opposite mating type to the active mating type locus. The precise molecular mechanisms for this event and for the control of silent information appear to be different. Heterothallic *Neurospora* do not switch mating type and do not contain silent mating type information.

Mating type switching has not been widely observed among Euascomycetes, although it clearly occurs in certain genera (*Chromocrea, Sclerotinea* and *Glomerella*) (Perkins, 1987). In each case, switching can be explained if there are two mating types, one of which is stable and one of which is switchable. The mechanism(s) of switching are not known. They could be explained by transposition events similar to the Hemiascomycete switching in which only one type of silent information is available, but they may be different from those in the Hemiascomycetes. Molecular characterization of mating type loci should resolve basic questions about switching mechanisms.

18.2.4 Signal transduction and cell–cell interaction

Very little is known about signal transduction in *N. crassa*. It is clear that each mating type produces a specific pheromone that attracts trichogynes of the opposite mating type (Bistis, 1983). Production of pheromone and response to pheromone is regulated by the mating type loci. The pheromone response pathway is anticipated to be G-protein mediated, but these components have not been identified. Conservation of protein sequence within the G-protein family suggests that a direct molecular identification of the signalling components is possible (Chapter 9).

18.2.5 Environmental and nutritional regulation of sexual development

Nutritional and environmental factors regulate *Neurospora* sexual development primarily by their effect upon development of females. Protoperithecia develop under conditions of nitrogen limitation at temperatures below 30°C. The cellular mechanisms that sense nitrogen deprivation and temperature are not known, though they may be related to those that control general nitrogen metabolism (Marzluf, 1981). Protoperithecial development is enhanced significantly by blue light (Degli-Innocenti and Russo, 1984).

18.2.6 Other functions associated with sexual reproduction

Genetic events other than karyogamy and meiosis are triggered in *a*/*A* dikaryons and diploids. Transposition of the Tad element is associated with sexual reproduction (Kinsey and Helber, 1989). *N. crassa* also modifies repeated segments of DNA premeiotically, introducing point mutations in the repeated segments (the RIP process) (Selker *et al.*, 1987; Cambareri *et al.*, 1989). A potentially similar process involving premeiotic DNA methylation of repeated sequences occurs in *Ascobolus immersus* (Faugeron *et al.*, 1990). Unusual segregation of repeated DNA has also been observed in *Gibberella fujikuroi* (Leslie and Dickman, 1991). Other functions, such as the meiotic drive phenomenon, Spore killer (Sk), are also active during ascosporogenesis (Raju and Perkins, 1991).

18.2.7   Variants of the *N. crassa* life cycle within the genus *Neurospora*

The genus *Neurospora* has four heterothallic species classically defined by their inability to yield fertile offspring in interspecific crosses (Perkins and Turner, 1988). The life cycles of these species are indistinguishable from the *Neurospora crassa* life cycle. In addition, the genus contains one pseudo-homothallic species, *N. tetrasperma* and several homothallic isolates (Perkins and Turner, 1988). Unlike either *S. cerevisiae* or *S. pombe*, homothallism does not derive from mating type switching. Analysis of mating type in different *Neurospora* species is essential to understanding the origin of the variant life cycles and the relationship of mating type to species barriers.

(a)   Pseudohomothallic life cycle

Neurospora tetrasperma is a pseudohomothallic relative of *N. crassa*. Pseudohomothallism occurs because single ascospores normally derive from one *a* and one *A* nucleus and are therefore self-fertile. The arrangement of nuclei in this pattern depends on a specific arrangement of nuclei and spindle during meiosis, and occasional single mating type ascospores are observed. Obviously, the normal vegetative incompatibility associated with mating type is not operative in *N. tetrasperma*, although introgression of its *A* mating type allele into *N. crassa* indicates that its *A-1* gene retains this function (Metzenberg and Ahlgren, 1973). We can hypothesize that the pseudohomothallic life cycle is a variant that arose from a *N. crassa*-like cycle by loss of vegetative incompatibility 'sensor' function (such as a *tol* mutation) and the acquisition of a particular chromosomal segregation pattern. Alternatively, the *N. crassa*-like life style evolved from a pseudohomothallic life style by loss of the segregation mechanism and gain of incompatibility sensor. In either case, one would expect *N. tetrasperma* to be a monophyletic group distinct from the heterothallic *Neurospora*, as indicated by random DNA fragment analyses (Natvig *et al.*, 1987), rather than that this life style arose independently in several different lines.

(b)   Homothallic life cycle in *Neurospora*

Six members of the genus *Neurospora*, and close relatives in the genera *Gelasinospora* and *Annexiella*, have a homothallic life cycle. In these species, a single haploid ascospore can germinate, and replicate, and undergo sexual reproduction indistinguishable from *N. crassa* (Raju, 1978). However, the two nuclei that form the dikaryon, and all resulting progeny, appear to be genetically identical. Either sexual development in these organisms does not respond to a mating-type trigger, or this trigger has been encoded within a single haploid genome.

Initial molecular study of the mating type genes suggests that homothallism may have arisen twice within the genus, and possibly by both of these mechanisms (Glass *et al.*, 1988). DNA hybridization demonstrates that *Neurospora terricola* has DNA sequences extremely similar to both *a* and *A* idiomorphic DNAs. However, the other five homothallic species contain clear *A* homologues, but no *a* sequences detectable by low stringency DNA hybridization (Glass *et al.*, 1988).

Presumably, *N. terricola* has, like *N. tetrasperma*, obviated mating type associated vegetative incompatibility. *N. terricola*, however, has incorporated both mating type genes into a single haploid genome. The arrangement and activities of these genes is not yet clear. One model would be that the *a* genes are active in one nucleus, and *A* genes active in a neighbouring nucleus, and sexual development is similar to that in *N. crassa*. This model should be testable by hybridization *in situ* with cloned *a* and *A* probes.

Homothallic strains with only an *A* homologue may have arisen quite differently. Either sexual development is independent of an *a* gene or the *a* gene that these strains possess is quite different in DNA sequence from that of *N. crassa*. The conservation of *a-1*-like polypeptides in *S. pombe*, *P. anserina* and *C. heterostrophus* indicates the essential role of this factor in mating. The finding that the *P. anserina* FPR1 region encodes a polypeptide homologous to mt *a-1*, but that its DNA sequence is very divergent from *N. crassa a* (Debuchy and Coppin, 1992), suggests that these homothallic isolates could bear an *a* homologue. In the latter case the question becomes why the DNA sequence constraints upon the family of *A*-like genes are tighter than those upon *a*-like genes.

18.2.8 Other filamentous ascomycetes

P. anserina, like *N. tetrasperma*, is secondarily homothallic because two nuclei of opposite mating type are enclosed within a single spore. However, the two mating types, called (+) and (−) are occasionally found in asci containing five spores. Mating type is determined by idiomorphic DNAs similar in size to the *N. crassa* idiomorphs. The (−) idiomorph has two functional regions, the (+) idiomorph just one. The *FMR1* gene (− idiomorph) is a clear homologue of the *N. crassa A-1* gene; the *FPR1* gene is a homologue of *N. crassa a-1* (Picard *et al.*, 1991; Debuchy and Coppin, 1992). The *Podospora* system does not exhibit mating-type related vegetative incompatibility, and these genes do not confer vegetative incompatibility when transformed into *N. crassa* (Staben, unpublished).

Many other filamentous Ascomycetes have bipolar, heterothallic mating systems. For example, the mating type idiomorphs of the Pyrenomycete *C. heterostrophus* (Turgeon *et al.*, 1993) encode proteins similar to the *N. crassa* mt *a-1* and *mt A-1*. The Discomycete *Ascobolus immersus* has two mating types, which are associated with vegetative incompatibility. The typical Plectomycete, *Aspergillus nidulans* is homothallic, but a related species, *A. heterothallicus* is heterothallic (Kwon and Raper, 1967b).

Sexual development is sometimes associated with other phenomena. For example, the sexual spores of ascomycete plant pathogens are frequently the overwintering forms and sometimes actually the infectious forms. Mating type itself is associated with other phenomena. For example, in *Sclerotinia trifoliorum* (Fujii and Uhm, 1988) and *Chromocrea spinulosa* (Mathieson, 1952), ascospores of different mating types are also different sizes (ascospore dimorphism). The (+)-mating type of *Histoplasma capsulatum* is more prevalent among clinical isolates (Kwon-Chung *et al.*, 1974), though the relationship of virulence to mating type is not clear. The α mating type of the basidiomycete *Cryptococcus neoformans* has been proven by analysis of congenic strains to be more virulent than the *a* mating type (Kwon-Chung *et al.*, 1992).

18.3 SEXUAL REPRODUCTION IN BASIDIOMYCETES

Meiosis within the Basidiomycetes occurs in special cells called basidia. In the Heterobasidiomycetes, the basidium is septate; in the Homobasidiomycetes, the basidium is unicellular. The Heterobasidiomycetes include rusts and smuts, whereas Homobasidiomycetes are the typical 'mushrooms' and bracket fungi. Because the rusts (Uredinales) are largely obligate parasites with long life cycles, their sexual cycles are less well characterized than those of smuts (Ustilaginales) or mushrooms (Hymenomycetes). Representative Ustilaginales and Hymenomycetes will be described here, concentrating on the well-characterized *Ustilago maydis* in one order and *Schizophyllum commune* and *Coprinus cinereus* in the other. Other fungi will be added to the discussion for comparison, contrast, or to highlight additional features associated with sexual development in some organisms.

18.3.1 Sexual reproduction of *Ustilago maydis*

U. maydis is the most intensively studied basidiomycete. It is an important pathogen of maize. The normal life cycle is depicted in Figure 18.4. Recent reviews compare the *U. maydis* life cycle to that of other fungi, especially *N. crassa* and *S. cerevisiae* (Banuett and Herskowitz, 1988; Froeliger and Kronstad, 1990; Banuett, 1992).

Haploid, homokaryotic sporidia germinate to form yeasts. These yeasts can be propagated on artificial media, and readily manipulated by many of the techniques used in *S. cerevisiae* or *S. pombe*. Dikaryons form when yeast that differ at the *a* locus mate. Mating in many species of *Ustilago*, but not *U. maydis*, involves formation of an elongated conjugation tube, which may be preceded by formation of thin threads, fimbriae, connecting the mating partners (Day, 1976). Filamentous growth in the dikaryotic state normally requires difference at both *a* and *b* loci. Dikaryons propagate and differentiate only within the plant host, though mating tests on charcoal agar yield limited dikaryotic hyphae. Nutritional and environmental factors determined by the plant are suspected to have a crucial role in controlling sexual development. Factors stimulating growth of *U. violacea* hyphae have been isolated from susceptible plant tissue (Ruddat *et al.*, 1991). The dikaryotic mycelium does not appear to have clamp connections characteristic of the Hymenomycete dikaryotic mycelium. During dikaryotic growth, the fungus induces hypertrophy of plant tissue. Within this gall, diploid brandspores form

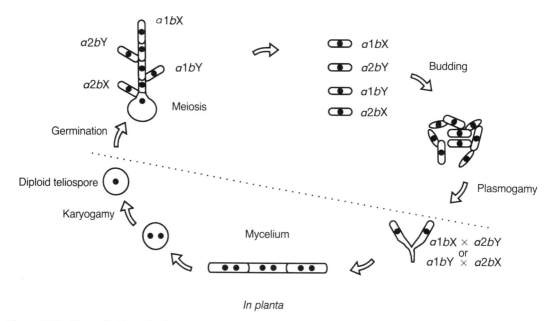

Figure 18.4   *U. maydis* life cycle. Portions of the life cycle that can occur outside the plant are represented above the dotted line, those that occur *in planta* are below the line. The *b*X and *b*Y alleles indicate any pair in which X does not equal Y.

from the mycelium and germinate to form the basidium in which meiosis occurs. Haploid basidiospores are released from galls and germinate to form sporidia that replicate by budding.

(a)   Mating and dikaryon formation in U. maydis

Considerable progress has been made in understanding the molecular details of mating in *U. maydis*. The haploid forms are manipulable by classical genetic techniques (relying on growth within the plant for meiosis) and molecular techniques such as transformation. Dikaryon formation is scored on charcoal agar, which has been critical to characterization of mating type and other genes essential for dikaryon formation.

Unlike the Ascomycetes, which have single mating type loci (some of which encode multiple gene products), *U. maydis* and many other Basidiomycetes have two distinct loci. In *U. maydis*, the *a* locus has two alleles (*a1* and *a2*), but the *b* locus is multiallelic, having at least 33 alleles. Some *Ustilago* species, e.g. *U. violacae* or *U. hordei*, have a single mating type locus, *a*. These *a* loci are presumed to have homologous functions to the *U.*

maydis a locus. Either there is no *b* locus, or it is very tightly linked to the *a* locus.

a locus

The *U. maydis a* locus contains idiomorphic DNA whose alternative forms are *a1* and *a2* (Figure 18.5a). Each idiomorph appears to encode two genes: a pheromone precursor and a pheromone receptor gene (Bolker *et al.*, 1992). The putative pheromone genes (*mfa1* and *mfa2*) encode short polypeptides whose C-terminus (C-aliphatic-aliphatic-X) is a consensus sequence for the addition of an isoprenyl group that is found on the *S. cerevisiae* a-factor (Kurjan, 1992), the *S. pombe* M-factor (Davey, 1992) and the *Rhodosporidium torulopsis* (Kamiya *et al.*, 1978) and *Tremella* (Sakagami *et al.*, 1979; Ishibashi *et al.*, 1984) pheromones. The putative receptor genes (*pra1* and *pra2*) encode homologues of the *S. cerevisiae* pheromone receptors (Bolker *et al.*, 1992). Introduction of the *mfa1* gene into a *a2b1* and *a2b2* strain allowed these strains to form dikaryons. Therefore, maintenance of the dikaryotic state apparently depends on an interactive pheromone-receptor pair, but does not require that

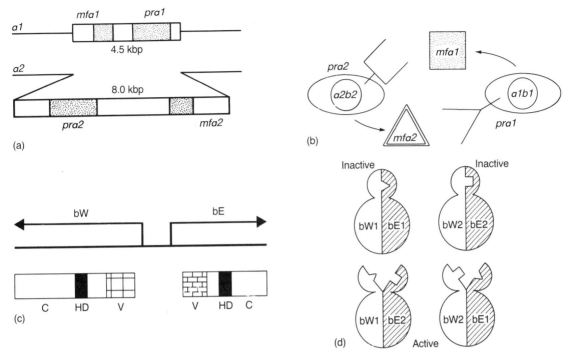

Figure 18.5 (a) Structures of the *U. maydis a* idiomorphs. The *mfa* genes encode putative pheromone precursors. The *pra* genes encode putative pheromone receptors. (b) Pheromone-receptor interactions in *U. maydis*. Normal interactions are diagrammed. Mating can occur with a single pheromone-receptor pair that can interact productively (*alb1* × *alb2*) (transformed with *pra2*). (c) Structure of a typical *b* allele of *U. maydis*. Both bW and bE transcripts encode a polypeptide with constant (C), homeodomain (HD), and variable (V) regions. The bE and bW polypeptides differ in size and sequence from one another. (d) Interaction of bE and bW products. This model of bE and bW function suggests that bE–bW dimers formed from the products of the same allele are inactive, but that dimers formed from heteroalleles are active, perhaps because a particular region of the polypeptides essential for functions remains exposed.

all *a*-idiomorph functions be complete (Bolker *et al.*, 1992) (Figure 18.5b).

b locus

The *U. maydis b* locus also contains idiomorphic DNA, though many different forms have been characterized (Kronstad and Leong, 1990; Schulz *et al.*, 1990). The DNA at the *b* locus is complex. Four characterized *b* idiomorphs encode two regulatory polypeptides, termed *bE* and *bW* (Figure 18.5c). Mating specificity of a given *b* locus can be conferred by either the *bE* or *bW* gene. A pair of *bE* and *bW* alleles from different mating types is both necessary and sufficient for development (Gillissen *et al.*, 1992). For example, a haploid that contains *bW1* and *bE2* is tumorigenic in plants, though it is not filamentous on charcoal agar because it has only a single *a* idiomorph (Banuett

and Herskowitz, 1989). A *a1/a2* diploid containing only *bW1* and *bE2* is both filamentous and pathogenic (see section 19.4.3).

The *bE* and *bW* polypeptides are presumed to be DNA binding proteins that regulate many target genes, such as those necessary for meiosis and pathogenesis. The polypeptide products of all characterized *bE* and *bW* genes contain DNA binding motifs related to homeodomain proteins (Kronstad and Leong, 1990; Schulz *et al.*, 1990; Gillissen *et al.*, 1992). The amino termini of these polypeptides show 40–60% similarity, the carboxy termini are 90% similar. The homeodomains are within the conserved regions. The bE and bW polypeptides share little similarity with one another apart from the homeodomain (Gillissen *et al.*, 1992).

Any model describing the actions of the *b*

polypeptides must account for the fact that all interactions between the *bE* and *bW* alleles of a single mating type are inactive, but that interaction between any different pair is active. One current model (Figure 18.5d) suggests that any different *b* pair forms a heterodimer able to bind DNA and that any single mating type pair forms hetero-dimers that are inactive (Gillissen *et al.*, 1992). Dimer formation and DNA binding are properties of the conserved regions. Inactive heterodimers would form when these activities were silenced by interactions between the subunits. Novel specifici-ties could be generated by mutations or recombi-nation events that conserved the basic functions, but prohibited silencing, allowing considerable flexibility in formation of new mating types.

18.3.2   Genes essential for dikaryon formation and growth *in planta*

The processes of dikaryon formation, mainten-ance, differentiation and gall formation involve many genes in addition to the mating type genes. Genes essential for dikaryon formation *in vitro* and tumour formation *in planta* have been found by screening for mutants that fail to give the typical 'fuzz' reaction signalling dikaryon formation on charcoal agar (Banuett, 1991). One of these genes, *rtf1*, appears to encode a negative regulator of tumour induction. Diploids formed from mutants in this gene bypass the normal requirement for different *b* alleles for pathogenesis. *rtf1* maps near the *b* locus (Gillissen *et al.*, 1992), so it could be a mutant *b* allele. Alternatively, *rtf1* may regulate tumorigenicity itself or perhaps activate the *b* alleles that are normally silent.

18.3.3   Homobasidiomycete sexual development

Sexual reproduction in two Hymenomycete fungi, *Coprinus cinereus* and *Schizophyllum commune* is intensively investigated. These organisms have similar life cycles though different habitats.

(a)   *The life cycle of* Schizophyllum commune

The life cycle of *Schizophyllum commune* is pre-sented in Figure 18.6. This description also applies in general to *C. cinereus*. Briefly, haploid spores germinate to form monokaryotic mycelia consis-ting of uninucleate haploid cells. Each homokaryon has two incompatibility factors, *A* and *B*. If two homokaryons with unlike factors encounter one another, they mate. After fusion, nuclei from each homokaryon migrate rapidly throughout both hyphae to form dikaryons. Dikaryotic nuclei divide synchronously by a special mechanism involving hook cell formation. Hook cells and their clamp connections are diagnostic of dikaryotic mycelium in *S. commune*. Dikaryotic mycelium can propagate indefinitely, unlike the transient di-karyons formed in filamentous Ascomycetes like *N. crassa*. Fruiting body formation is induced by environmental cues such as light exposure. Features of the life cycle, genetics and molecular genetics have been reviewed by Stankis *et al.* (1990).

(b)   *Genetic regulation of the life cycle*

Mating type regulation

The mating types of Homobasidiomycetes are ex-tremely complex multilocus, multiallelic systems. Mating type is determined by four loci: $A\alpha$, $A\beta$, $B\alpha$ and $B\beta$. The $A\alpha$ and $A\beta$ loci are linked, as are the B loci. However, the A and B loci are on different chromosomes. The requirement for completion of the life cycle is that two haploids must differ at at least one A and at least one B locus; the α and β loci are functionally redundant. Each locus has multiple alleles: 9 $A\alpha$, 32 $A\beta$, 9 $B\alpha$, and 9 $B\beta$ alleles have been found in *S. commune* worldwide. Many specificities have also been found in *C. cinereus*.

Each locus appears to govern particular aspects of the life cycle. In *S. commune*, heterokaryons alike at A and different at B undergo nuclear migration, but no further development. Hetero-karyons alike at B and different at A undergo nuclear pairing, hook cell formation, conjugate division, and hook cell septation. Therefore, the B locus is thought to regulate nuclear migration and hook cell fusion; the A locus regulates nuclear pairing, hook cell formation, conjugate division, and hook cell septation (Figure 18.6). The sum of these complementary pathways is required to form and maintain dikaryotic mycelium. Fruiting bodies are normally formed only from dikaryotic mycelium in response to environmental cues. Presumably, formation of fruiting bodies is regu-lated by mating type or requires features of the

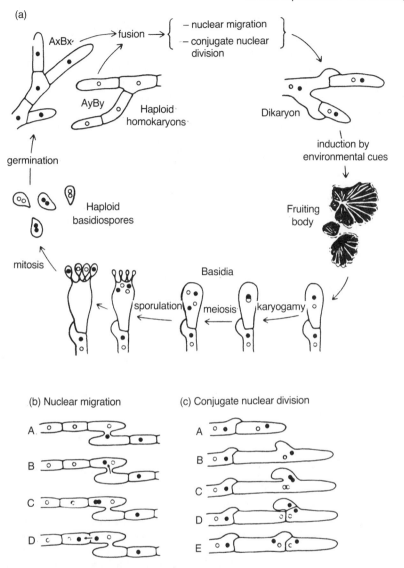

Figure 18.6 *S. commune* life cycle. (a) The steps in the sexual life cycle of a typical Hymenomycete. Haploid homokaryons mate and undergo rapid nuclear migration (b) to generate dikaryons. Dikaryotic mycelia proliferate and maintain the dikaryotic state via conjugate nuclear divisions involving hook cell formation (c). Environmental cues induce fruiting body formation. Within the fruiting body, karyogamy, meiosis, and sporulation generate haploid basidiospores. Redrawn from Stankis *et al.* (1990).

dikaryotic state that are normally induced by mating type. Formation of fruiting bodies and basidiospores undoubtedly requires the action of many additional genes. Estimates from mRNA hybridization analyses suggest that approximately 5% of the genome encodes dikaryon or fruiting-specific mRNAs.

A loci Idiomorphic DNA segments from *A* loci from both *S. commune* and *C. cinereus* have recently been cloned and characterized (May *et al.*, 1991; Kües *et al.*, 1992; Specht *et al.*, 1992; Stankis *et al.*, 1992; Tymon *et al.*, 1992). The *A* loci all appear to be complex and to contain several different genes. The sequenced regions encode multiple homeo-domain-containing polypeptides, suggesting that the polypeptides are DNA-binding transcription regulators. The homeodomains of these polypeptides are most similar to those encoded by the *U. maydis b* alleles and the *S. cerevisiae* α2 and α1 polypeptides. Other features of these polypeptides include acidic, proline-rich domains, similarity to POU-transcription factors, and nuclear localization signals (Specht *et al.*, 1992; Stankis *et al.*, 1992; Tymon *et al.*, 1992). A low level of sequence similarity of the amino termini of *A*α*Z* and *A*α*Y* to *U. maydis bE* and *bW* polypeptides suggests a functional similarity and possibly an evolutionary relationship between these fungal regulatory proteins (Stankis *et al.*, 1992).

The model for interaction of multiallelic DNA binding proteins described for *U. maydis* may also apply to these Hymenomycetes, though the diversity and complexity of their mating systems is even greater. An alternative model, based on independent DNA binding of factors has also been presented. In this model, transactivation domains of the polypeptides are active only when compatible factors are jointly bound to DNA (Kües and Casselton, 1992b, 1993).

B loci No molecular analysis of the *B* locus has been published. It will be very interesting to test by transformation whether the *B* locus controls steps involved in the early step of nuclear migration and the late step of hook cell fusion as suggested by the behavior of B-identical heterokaryons. Similarities between *U. maydis* and these Hymenomycetes suggest that the *B* loci could encode pheromones and receptors, though imagining the mechanisms for the necessary multiallelic interactions is difficult.

Genes involved in fruiting body formation
Genes associated with fruiting body formation in *S. commune* have been identified by classical and molecular techniques. Several homokaryotic fruiting genes have been identified in *S. commune*. First, constitutive mating type mutants are able to fruit in the homokaryotic state (Raper *et al.*, 1965). Other stimuli such as injury or a fruit-inducing substance can also cause homokaryons to fruit. Classical genetics indicated a minimum of six genes are required for homokaryotic fruiting (Leslie and Leonard, 1979). *FRT1* (Horton and Raper, 1991) is a DNA segment that conferred

homokaryotic fruiting when transformed into *S. commune*. *FRT1* may normally respond to mating type control, but the polypeptide encoded may have been separated from a *cis*-acting repressor of fruiting during cloning. Finally, several genes activated during fruiting have been identified (Mulder and Wessels, 1986). Expression of these *Sc* genes is regulated by environmental factors, mating type, and other genes (Ruiters *et al.*, 1988a; Springer and Wessels, 1989; Yli-Mattila *et al.*, 1989a, b). The *Sc* genes encode hydrophobic cysteine-rich polypeptides, related to the *Aspergillus nidulans rodA* polypeptide (Stringer *et al.*, 1991), which has a structural role in conidia. The *Sc* gene products are likely to have structural roles in the fruiting body.

18.3.4   Mating type in *Cryptococcus neoformans*

The single mating type locus in *Cryptococcus neoformans* regulates mating type and virulence (Kwon-Chung *et al.*, 1992). Initial results suggest that the mating type locus is over 60 kbp of DNA bearing interspersed unique and repetitive elements. A portion of this DNA can encode a pheromone precursor similar to those of other basidiomycetes such as *U. maydis, Tremella brasiliensis, Tremella mesenterica* or *Rhodosporidium toruloides* (Moore and Edman, 1993).

18.4   CONCLUSIONS

The fungi exhibit diverse modes of sexual reproduction regulated in a wide variety of ways (Kües and Casselton, 1992a). However, themes common to all fungal life cycles emerge. Fungi must identify an appropriate mating partner. In many fungi (but apparently not in many mushrooms such as *Coprinus*), the initial steps of mating depend on pheromones secreted by at least one mating partner and responded to by the other partner. Biosynthesis and response to the appropriate pheromone is regulated by mating type, either directly (as in *U. maydis*) or indirectly. At least one component of each characterized mating type system encodes proteins containing DNA binding and transcription-regulatory motifs. The properties of haploids, heterokaryons, and diploids are almost certainly determined by the interactions of these regulatory factors and parallel pathways that sense environmental cues.

Continued characterization of the mating type and environmental regulators will allow a molecular description of the actions and interactions of these factors.

Perhaps the most attractive aspect of studying fungal sexual reproduction is the possibility of understanding the evolution of life cycles and species in precise molecular terms. One interesting theme that is emerging is that different fungi use similar life cycle regulators in different combinations. The same cast of characters reappears in many fungi: peptide pheromones, transmembrane receptors coupled to G-proteins, and regulatory DNA binding proteins (homeodomain or HMG box). As the phylogeny of fungi with variant life histories (such as those within the genus *Neurospora*) is determined, it may be possible to determine what the barriers between species are and how they arose. It is exciting to realize that genes that control exchange within the gene pool and that probably shape fungal evolution itself are already in freezers, lab benches, and transformation vectors around the world. The startling increase in our understanding of sexual reproduction made possible by application of molecular genetic techniques has obviously just begun.

REFERENCES

Astell, C., Ahlstrom-Jonasson, L., Smith, M. *et al.* (1981) The sequence of DNAs coding for the mating-type loci of *Saccharomyces cerevisiae*. *Cell*, **27**, 15–23.

Banuett, F. (1991) Identification of genes governing filamentous growth and tumor induction by the plant pathogen *Ustilago maydis*. *Proceedings of the National Academy of Sciences of the USA*, **88**, 3922–6.

Banuett, F. (1992) *Ustilago maydis*, the delightful blight. *Trends in Genetics*, **8**, 174–80.

Banuett, F. and Herskowitz, I. (1988) *Ustilago maydis*, smut of maize, in *Advances in Plant Pathology*, Academic Press, New York, pp. 427–55.

Banuett, F. and Herskowitz, I. (1989) Different *a* alleles of *Ustilago maydis* are necessary for maintenance of filamentous growth but not for meiosis. *Proceedings of the National Academy of Sciences of the USA*, **86**, 5878–82.

Beach, D. and McLeod, M. (1988) A specific inhibitor of the *ran1*⁺ protein kinase regulates entry into meiosis in *Schizosaccharomyces pombe*. *Nature*, **332**, 509–14.

Beadle, G.W. and Coonradt, V.L. (1944) Heterocaryosis in *Neurospora crassa*. *Genetics*, **29**, 291–308.

Berbee, M.L. and Taylor, J.T. (1992) Two Ascomycete classes based on fruiting-body characters and ribosomal DNA sequence. *Molecular Biology and Evolution*, **9**, 278–84.

Bistis, G.N. (1983) Evidence for diffusible mating-type specific trichogyne attractants in *Neurospora crassa*. *Experimental Mycology*, **7**, 292–5.

Boeke, J.D. and Sandmeyer, S.B. (1991) Yeast transposable elements, in *The Molecular and Cellular Biology of the Yeast Saccharomyces: Genome Dynamics, Protein Synthesis, and Energetics*, (eds J.R. Broach, J.R. Pringle and E.W. Jones, Cold Spring Harbor Laboratory Press, USA, pp. 193–262.

Bolker, M., Urban, M. and Kahmann, R. (1992) The *a* mating type locus of *U. maydis* specifies cell signaling components. *Cell*, **68**, 441–50.

Bruns, T.D., Vilgalys, R., Barns, S.M. *et al.* (1992) Evolutonary relationships within the fungi: analyses of nuclear small subunit rRNA sequences. *Molecular Phylogenetics and Evolution*, **1**, 231–41.

Cambareri, E.B., Jensen, B.C., Schabtach, E. and Selker, E.U. (1989) Repeat-induced G-C to A-T mutations in *Neurospora*. *Science*, **244**, 1571–5.

Caplan, S. and Kurjan, J. (1991) Role of α-factor and the *MFα1* α-factor precursor in mating yeast. *Genetics*, **127**, 299–307.

Davey, J. (1992) Mating pheromones of the fission yeast *Schizosaccharomyces pombe*: purification and structural analysis of two genes encoding the pheromone. *EMBO Journal*, **11**, 951–60.

Day, A. (1976) Communication through fimbriae during conjugation in a fungus. *Nature*, **262**, 583–4.

Debuchy, R. and Coppin, E. (1992) The mating types of *Podospora anserina*: functional analysis and sequence of the fertilization domains. *Molecular and General Genetics*, **33**, 113–21.

Degli-Innocenti, F. and Russo, V.E.A. (1984) Isolation of new white collar mutants of *Neurospora crassa* and studies on their behaviour in the blue light-induced formation of protoperithecia. *Journal of Bacteriology*, **159**, 757–61.

DeVoti, J., Seydoux, G., Beach, D. and McLeod, M. (1991) Interaction between *ran1*⁺ protein kinase and cAMP dependent protein kinase as negative regulators of fission yeast meiosis. *EMBO Journal*, **10**, 3759–68.

Egel, R., Nielsen, O. and Weilguny, D. (1990) Sexual differentiation in fission yeast. *Trends in Genetics*, **6**, 369–73.

Faugeron, G., Rhounim, L. and Rossignol, J.-L. (1990) How does the cell count the number of ectopic copies of a gene in the premeiotic inactivation process acting in *Ascobolus immersus*? *Genetics*, **124**, 585–91.

Francois, J., Higgins, D.L., Chang, F. and Tatchell, K. (1991) Inhibition of glycogen synthesis in *Saccharomyces cerevisiae* by the mating pheromone α-factor. *Journal of Biological Chemistry*, **266**, 6174–80.

Froeliger, E.H. and Kronstad, J.W. (1990) Mating and pathogenesis in *Ustilago maydis*. *Seminars in Developmental Biology*, **1**, 185–93.

Fujii, H. and Uhm, J.Y. (1988) *Sclerotinia trifoliorum* cause of rots of *Trifolium* spp. *Advances Plant Pathology*, **6**, 233–47.

Giese, K., Amsterdam, A. and Grosschedl, R. (1991) DNA-binding properties of the HMG domain of the lymphoid-specific transcriptional regulator LEF-1. *Genes and Development*, **5**, 2567–78.

Gillissen, B., Bergemann, J., Sandmann, C. *et al.* (1992) A two-component regulatory system for self/non-self recognition in *Ustilago maydis*. *Cell*, **68**, 647–57.

Gimeno, C.J., Ljungdahl, P.O., Styles, C.A. and Fink, G.R. (1992) Unipolar cell divisions in the yeast *S. cerevisiae* lead to filamentous growth: regulation by starvation and *ras*. *Cell*, **68**, 1077–90.

Glass, N.L. and Kuldau, G.A. (1992) Mating type and vegetative incompatibility in filamentous Ascomycetes. *Annual Review of Phytopathology*, **30**, 201–24.

Glass, N.L. and Lee, L. (1992) Isolation of *Neurospora crassa A* mating type mutants by repeat induced point (RIP) mutation. *Genetics*, **132**, 125–33.

Glass, N.L. and Staben, C. (1990) Genetic control of mating in *Neurospora crassa*. *Seminars in Developmental Biology*, **1**, 177–84.

Glass, N.L., Vollmer, S.J., Staben, C. *et al* (1988) DNAs of the two mating-type alleles of *Neurospora crassa* are highly dissimilar. *Science*, **241**, 570–3.

Glass, N.L., Grotelueschen, J. and Metzenberg, R.L. (1990) *Neurospora crassa A* mating-type region. *Proceedings of the National Academy of Services of the USA*, **87**, 4912–16.

Griffiths, A.J.F. and DeLange, A.M. (1978) Mutations of the a mating-type gene in *Neurospora crassa*. *Genetics*, **88**, 239–54.

Harley, V.R., Jackson, D.I., Hextall, P.J. *et al.* (1992) DNA binding activity of recombinant SRY from normal males and XY females. *Science*, **255**, 453–6.

Herskowitz, I. (1988) Life cycle of the budding yeast *Saccharomyces cerevisiae*. *Microbiology Reviews*, **52**, 536–53.

Herskowitz, I., Rine, J. and Strathern, J. (1992) Mating-type determination and mating-type interconversion in *Saccharomyces cerevisiae*, in *The Molecular and Cellular Biology of the Yeast Saccharomyces: Gene Expression*, 1st edn, (eds J.R. Broach, J.R. Pringle and E.W.J. Jones), Cold Spring Harbor Laboratory Press, Plainview, NY, pp. 583–656.

Horton, J.S. and Raper, C.A. (1991) A mushroom-inducing DNA sequence isolated from the Basidiomycete, *Schizophyllum commune*. *Genetics*, **129**, 707–16.

Ishibashi, Y., Sakagami, Y., Isogai, A. and Suzuki, A. (1984) Structures of tremerogens A-9291-I and A-9291-VIII: peptidal sex hormones of *Tremella brasiliensis*. *Biochemistry*, **23**, 1399–404.

Jantzen, H.-M., Admon, A., Bell, S.P. and Tjian, R. (1990) Nucleolar transcription factor hUBF contains a DNA-binding motif with homology to HMG proteins. *Nature*, **344**, 830–6.

Jarvis, E.E., Clark, K.L. and Sprague, G.F. (1989) The yeast transcription factor, PRTF, a homolog of the mammalian serum response factor, is encoded by the *MCM1* gene. *Genes and Development*, **3**, 936–45.

Kamiya, Y., Sakurai, A., Tamura, S. and Takahashi, N. (1978) Structure of rhodotorucine A, a novel lipopeptide, inducing mating tube formation in *Rhodosporidium toruloides*. *Biochemical and Biophysical Research Communications*, **83**, 1077–83.

Kelly, M., Burke, J. Smith, M. *et al.* (1988) Four mating-type genes control sexual differentiation in the fission yeast. *EMBO Journal*, **7**, 1537–47.

Kinsey, J.A. and Helber, J. (1989) Isolation of a transposable element from *Neurospora crassa*. *Proceedings of the National Academy of Sciences of the USA*, **86**, 1929–33.

Klar, A.J.S. (1989) The interconversion of yeast mating type: *Saccharomyces cerevisiae* and *Schizosaccharomyces pombe*, in *Mobile DNA*, American Society of Microbiology, Washington, DC, pp. 671–91.

Klar, A.J.S. (1992) Developmental choices in mating-type interconversion in fission yeast. *Trends in Genetics*, **8**, 208–13.

Kronstad, J.W. and Leong, S.A. (1990) The b mating-type locus of *Ustilago maydis* contains variable and constant regions. *Genes and Development*, **4**, 1384–95.

Kües, U. and Casselton, L.A. (1992a) Fungal mating type genes – regulators of sexual development. *Mycological Research*, **96**, 993–1006.

Kües, U. and Casselton, L. (1992b) Homeodomains and regulation of sexual development in basidiomycetes. *Trends in Genetics*, **8**, 154–5.

Kües, U. and Casselton, L.A. (1993) The origin of multiple mating types in mushrooms. *Journal of Cell Science*, **104**, 227–30.

Kües, U., Richardson, W.V.J., Tymon, A.M. *et al.* (1992) The combination of dissimilar alleles of the Aα and Aβ gene complexes, whose proteins contain homeodomain motifs, determines sexual development in the mushroom *Coprinus cinereus*. *Genes and Development*, **6**, 568–77.

Kurjan, J. (1992) Pheromone response in yeast. *Annual Reviews of Biochemistry*, **61**, 1097–129.

Kwon, K.-J. and Raper, K.B. (1967a) Heterokaryon formation and genetic analysis of color mutants in *Aspergillus heterothallicus*. *American Journal of Botany*, **54**, 49–60.

Kwon, K.-J. and Raper, K.B. (1967b) Sexuality and cultural characteristics of *Aspergillus heterothallicus*. *American Journal of Botany*, **54**, 36–48.

Kwon-Chung, K.J., Weeks, R.J. and Larsh, H.W. (1974) Studies on *Emmonsiella capsulatum* (*Histoplasma capsulatum*) II. Distribution of two mating types in 13 endemic states of the US. *American Journal of Epidemiology*, **99**, 44–9.

Kwon-Chung, K.J., Edman, J.C. and Wickes, B.L. (1992) Genetic Association of mating types and virulence in *Cryptococcus neoformans*. *Infection and Immunity*, **60**, 602–5.

Leslie, J.F. and Dickman, M.B. (1991) Fate of DNA encoding hygromycin resistance after meiosis in transformed strains of *Gibberella fujikuroi* (*Fusarium moniliforme*). *Applied and Environmental Microbiology*, **57**, 1423–9.

Leslie, J.F. and Leonard, T.J. (1979) Three independent genetic systems that control initiation of a fungal fruiting body. *Molecular and General Genetics*, **171**, 257–60.

Marzluf, G.A. (1981) Regulation of nitrogen metabolism and gene expression in fungi. *Microbiology Reviews*, **45**, 437–61.

Mathieson, M.J. (1952) Ascospore dimorphism and mating type in *Chromocrea spinulosa* (Fuckel) Petch n. comb. *Annals of Botany*, **16**, 449–67.

May, G., LeChevanton, L. and Pukkila, P.J. (1991) Molecular analysis of the *Coprinus cinereus* mating type A factor demonstrates an unexpectedly complex structure. *Genetics*, **128**, 529–38.

Metzenberg, R.L. and Ahlgren, S.K. (1973) Behaviour of *Neurospora tetrasperma* mating type genes introgressed into *N. crassa*. *Canadian Journal of General Cytology*, **15**, 571–6.

Metzenberg, R.L. and Glass, N.L. (1990) Mating type and mating strategies in *Neurospora*. *Bioessays*, **12**, 53–9.

Mochizuki, N. and Yamamoto, M. (1992) Reduction in the intracellular cAMP level triggers initiation of sexual development in fission yeast. *Molecular and General Genetics*, **233**, 17–24.

Moore, T.D. and Edman, J.C. (1993) The α-mating type locus of *Cryptococcus neoformans* contains a peptide pheromone gene. *Molecular and Cellular Biology*, **13**, 1962–70.

Mulder, G.H. and Wessels, J.G. (1986) Molecular cloning of RNAs differentially expressed in monokaryons and dikaryons of *Schizophyllum commune* in relation to fruiting. *Experimental Mycology*, **10**, 214.

Nasrin, N., Buggs, C., Kong, X.F. *et al.* (1991) DNA-binding properties of the product of the testis-determining gene and a related protein. *Nature*, **354**, 317–20.

Natvig, D.O., Jackson, D.A. and Taylor, J.W. (1987) Random-fragment hybridization analysis of evolution in the genus *Neurospora*: the status of four-spored strains. *Evolution*, **41**, 1003–21.

Nelson, M.A. and Metzenberg, R.L. (1992) Sexual development genes of *Neurospora crassa*. *Genetics*, **132**, 149–62.

Newmeyer, D. (1970) A suppressor of the heterokaryon-incompatibility associated with mating type in *Neurospora crassa*. *Canadian Journal of General Cytology*, **12**, 914–26.

Passmore, S., Elble, R. and Tye, B.-K. (1989) A protein involved in minichromosome maintenance in yeast binds a transcriptional enhancer conserved in eukaryotes. *Genes and Development*, **3**, 921–35.

Perkins, D.D. (1987) Mating-type switching in filamentous Ascomycetes. *Genetics*, **115**, 215–6.

Perkins, D.D. and Turner, B.C. (1988) *Neurospora* from natural populations: toward the population biology of a haploid eukaryote. *Experimental Mycology*, **12**, 91–131.

Perkins, D.D., Radford, A., Newmeyer, D. and Bjorkman, M. (1982) Chromosomal loci of *Neurospora crassa*. *Microbiological Reviews*, **46**, 462–570.

Picard, M., Debuchy, R. and Coppin, E. (1991) Cloning the mating types of the heterothallic fungus *Podospora anserina*: developmental features of haploid transformants carrying both mating types. *Genetics*, **128**, 539–47.

Raju, N.B. (1978) Meiotic nuclear behavior and ascospore formation in five homothallic species of *Neurospora*. *Canadian Journal of Botany*, **56**, 754–63.

Raju, N.B. (1980) Meiosis and ascospore genesis in *Neurospora*. *European Journal of Cell Biology*, **23**, 208–23.

Raju, N.B. (1992) Genetic control of the sexual cycle in *Neurospora*. *Mycological Research*, **96**, 241–62.

Raju, N.B. and Perkins, D.D. (1991) Expression of meiotic drive elements *Spore killer*-2 and *Spore killer*-3 in asci of *Neurospora tetrasperma*. *Genetics*, **129**, 25–37.

Raper, J.R., Boyd, D.H. and Raper, C.A. (1965) Primary and secondary mutations at the incompatibility loci in *Schizophyllum*. *Proceedings of the National Academy of Sciences of the USA*, **53**, 1324–32.

Ruddat, M., Kokontis, J., Birch, L. *et al.* (1991) Interactions of *Microbotryum* (*Ustilago violacea*) with its host *Silene alba*. *Plant Science* (*Limerick*), **80**, 157–66.

Ruiters, M.H.J., Sietsma, J.H. and Wessels, J.G.H. (1988) Expression of dikaryon-specific mRNAs of *Schizophyllum commune* in relation to incompatibility genes, light, and fruiting. *Experimental Mycology*, **12**, 60–9.

Sakagami, Y., Isogai, A., Suzuki, A. *et al.* (1979) Structure of tremerogen A-10, a peptidal hormone inducing conjugation tube formation in compatible mating-type cells of *Tremella mesenterica*. *Agricultural Biology and Chemistry*, **43**, 2643–5.

Schulz, B., Banuett, F., Dahl, M. *et al.* (1990) The *b* alleles of *U. maydis*, whose combinations program pathogenic development, code for polypeptides containing a homeodomain-related motif. *Cell*, **60**, 295–306.

Selker, E.U., Cambareri, E.B., Jensen, B.C. and Haack, K.R. (1987) Rearrangement of duplicated DNA in specialized cells of *Neurospora*. *Cell*, **51**, 741–52.

Specht, C.A., Stankis, M.M., Giasson, L. *et al.* (1992) Functional analysis of the homeodomain-related proteins of the *Aα* locus of *Schizophyllum commune*. *Proceedings of the National Academy of Sciences of the USA*, **89**, 7174–8.

Springer, J. and Wessels, J.G.H. (1989) A frequently occurring mutation that blocks the expression of fruiting genes in *Schizophyllum commune*. *Molecular and General Genetics*, **219**, 486–8.

Springer, M.L. and Yanofsky, C. (1992) Expression of *con* genes along the three sporulation pathways of *Neurospora crassa*. *Genes and Development*, **6**, 1052–7.

Staben, C. and Yanofsky, C. (1990) *Neurospora crassa a* mating type region. *Proceedings of the National Academy of Sciences of the USA*, **87**, 4917–21.

Stankis, M.M., Specht, C.A. and Giasson, L. (1990) Sexual incompatibility in *Schizophyllum commune*: from classical genetics to a molecular view. *Seminars in Developmental Biology*, **1**, 195–206.

Stankis, M.M., Specht, C.A., Yang, H. *et al.* (1992) The *A*α mating locus of *Schizophyllum commune* encodes two dissimilar multiallelic proteins. *Proceedings of the National Academy Sciences of the USA*, **89**, 7169–73.

Stringer, M.A., Dean, R.A., Sewall, T.C. and Timberlake, W.E. (1991) *Rodletless*, a new *Aspergillus* developmental mutant induced by directed gene inactivation. *Genes and Development*, **5**, 1161–71.

Sugimoto, A., Iino, Y., Maeda, T. *et al.* (1991) *Schizosaccharomyces pombe stel1*+ encodes a transcription factor with an HMG motif that is a critical regulator of sexual development. *Genes and Development*, **5**, 1990–9.

Taylor, J.W. (1986) Fungal evolutionary biology and mitochondrial DNA. *Experimental Mycology*, **10**, 259–69.

Turgeon, B.G., Bohlmann, H., Ciuffetti, L.M. *et al.* (1993) Cloning and analysis of the mating type genes from *Cochliobolus heterostrophus*. *Molecular and General Genetics* **238**, 270–84.

Tymon, A.M., Kües, U., Richardson, W.V.J. and Casselton, L.A. (1992) A fungal mating type protein that regulates sexual and asexual development contains a POU-related domain. *EMBO Journal*, **11**, 1805–13.

Xu, H. and Boeke, J.D. (1991) Inhibition of Ty1 transposition by mating pheromones in *Saccharomyces cerevisiae*. *Molecular and Cellular Biology*, **11**, 2736–43.

Yli-Mattila, T., Ruiters, M.H.J. and Wessels, J.G.H. (1989a) Photoregulation of dikaryon-specific mRNAs and proteins by UV-A light in *Schizophyllum commune*. *Current Microbiology*, **18**, 289–95.

Yli-Mattila, T., Ruiters, M.H.J., Wessels, J.G.H. and Raudaskoski, M. (1989b) Effect of inbreeding and light on monokaryotic and dikaryotic fruiting in the homobasidiomycete *Schizophyllum commune*. *Mycological Research*, **93**, 535–42.

N.A.R. Gow
Department of Molecular and Cell Biology, Marischal College, University of Aberdeen, Aberdeen, UK

19.1 INTRODUCTION

All fungi have some capacity to grow in two basic morphological forms – spheres and tubes – therefore it could be argued that they are all, to some extent, dimorphic. For many filamentous fungi spherical growth may only be expressed during the formation of spores and many yeast-like fungi have only the remnants of a true filamentous growth habit. However, the many shapes and forms found among the 64 000 recognized species of fungi are by and large generated by employing these two basic patterns of cell wall expansion. The dimorphic fungi are taken to represent those organisms in which the equilibrium between polarized and spherical growth is such that vegetative growth can occur in either a hyphal or budding mode according to environmental conditions. The term dimorphism is a misnomer since many of the so-called dimorphic fungi produce a variety of cell forms and therefore are really polymorphic. These fungi do not represent in any way a taxonomic grouping and the dimorphic fungi include certain Zygomycetes, Ascomycetes, Basidiomycetes and Deuteromycetes. Some of these are important pathogens of humans or plants, others are saprophytes. The pathogenic forms have been studied extensively because of their medical or agricultural impact and also because they are fascinating systems in which questions of cell shape regulation and cell polarity can be explored. Because dimorphism is a very common phenomenon it is not possible to present a comprehensive review of all dimorphic fungi that have been studied but the reader is referred to two specialist texts (Szaniszlo, 1985; Vanden Bossche *et al.*, 1993). The examples selected here are intended to illustrate some of the systems in which detailed analyses have been made.

Unfortunately for the experimental mycologist, many of the most interesting species are asexual and may be diploid so that genetic studies are difficult. This has significantly impeded progress in understanding the basis of the dimorphic response of these fungi. The dimorphic behaviour of fungi with well-characterized sexual genetics such as *Saccharomyces cerevisiae* and *Ustilago maydis* (discussed later) have advantages in this regard and can serve as model systems to explore basic regulatory mechanisms that may also operate in genetically recalcitrant species.

In some cases, such as *Mucor rouxii* and *Histoplasma capsulatum*, the filamentous form may comprise unconstricted branching hyphae that are true branching mycelia. With other fungi such as *S. cerevisiae* (Gimeno *et al.*, 1992) true, unconstricted hyphae are not produced but pseudohyphae consisting of chains of elongated yeast cells joined end to end are formed. Hyphae or pseudohyphae may branch or give rise to lateral buds (sometimes called blastospores) at the junctions between the daughter cells. With many dimorphic fungi, such as *Candida albicans*, a range of cell types can exist that represent a continuum between yeast cells, pseudohyphae (that can vary markedly in the extent of bud elongation) and true unconstricted hyphae. The precise morphology can be important in modelling hyphal development and morphology indices have recently been developed that describe numerically the extent of cell elongation (Merson-Davies and Odds, 1992).

19.2 PATTERNS OF CELL WALL GROWTH

The dimorphic process is an example of spatial differentiation of cell wall biosynthesis since the cell wall ultimately determines the cell shape of the fungus. The pattern of cell wall synthesis is in turn related to the activities of the cytoskeleton which directs the vesicles that carry the cell wall

The Growing Fungus. Edited by Neil A.R. Gow and Geoffrey M. Gadd. Published in 1994 by Chapman & Hall, London. ISBN 0 412 46600 7

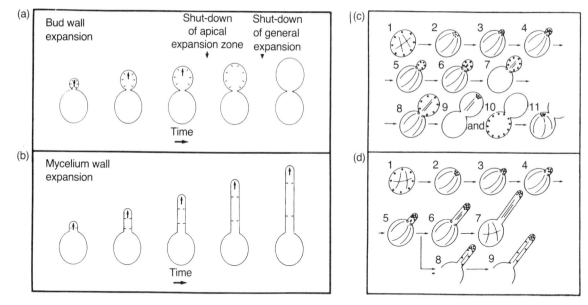

Figure 19.1   Patterns of cell wall growth in the yeast form (a) and hyphal form (b) of *C. albicans* as determined from microbead labelling of the cell surface. The small arrows indicate general expansion and the large arrows apical expansion. The pattern of actin, revealed by staining with rhodamine phalloidin, is shown during (c) budding and (d) hyphal growth. Modified from Staebell and Soll (1985) (a,b) and Anderson and Soll (1986) (c,d).

biosynthetic enzymes to the cell surface (Chapters 6 and 13). In general, the transition from yeast to hyphal cell represents an increasing polarization of the pattern of cell growth, with pseudohyphae representing cells of intermediate polarity.

19.2.1   Patterns of budding

In yeast cells in general, the pattern of cell wall growth varies between different species. The yeast cells of *Mucor* species and *Paracoccidioides brasiliensis* exhibit multipolar budding where buds may be formed anywhere on the surface. Some yeast cells form buds alternately at each pole (bipolar budding). In the haploid cells of *S. cerevisiae* the pattern is axial, with mother and daughter cells forming buds at the pole that is adjacent to the previous mother–daughter cell junction. Diploid cells, such as *C. albicans* and the diploid form of *S. cerevisiae* exhibit polar budding where a virgin daughter cell forms its first several buds at the cell pole that is opposite the birth scar, where it separated from the mother cell. This polar form of budding is essential for pseudohyphal growth of *S. cerevisiae* (Gimeno *et al.*, 1992) and the

filamentous form of other dimorphic fungi such as *C. albicans* (Chaffin, 1984), because the mycelium and pseudomycelium are polarized growth forms requiring the maintenance of polar development between successive cell or duplication cycles. In *S. cerevisiae* the importance of the polar budding pattern was demonstrated clearly, by showing that mutations in the *RSR1/BUD1* gene, which causes a random pattern of budding in diploid cells, suppressed pseudohyphal growth (Gimeno *et al.*, 1992).

In *C. albicans* the precise pattern of cell wall expansion in the yeast and hyphal forms has been studied by decorating the surface of cells with polylysine coated microbeads and observing their mutual displacement during cell expansion (Staebell and Soll, 1985; Merson-Davies and Odds, 1992). The results showed that during the first two-thirds of bud expansion, growth was localized to a small apical zone, while in the final one third of the bud growth cycle growth was due to general, isotropic expansion (Figure 19.1). The zones of cell growth are also reflected in the pattern of actin staining in growing buds (Figure 19.1) (Anderson and Soll, 1986). Thus polar growth is an important aspect of the growth of buds as well as germ tubes and hyphae (Chapter 13).

19.2.2 Patterns of hyphal growth

Hyphae of *C. albicans* and other dimorphic and mycelial fungi expand apically. This has been demonstrated using microbead-labelling experiments (Figure 19.1) (Staebell and Soll, 1985; Merson-Davies and Odds, 1992), by autoradiography to detect areas of *de novo* chitin synthesis (Braun and Calderone, 1978) and from kinetic measurements (Gow and Gooday, 1982). The hyphal tip of *C. albicans* has been shown to be actin-rich by rhodamine–phalloidin staining (Anderson and Soll, 1986) and microtubules have also been shown to be important for germ tube formation and growth (Barton and Gull, 1988; Yokoyama *et al.*, 1990) (Chapter 13). The septa of *C. albicans* have a micropore 25 nm in diameter (Gow *et al.*, 1980) which is too small to allow effective intercompartment communication and the passage of organelles. Indeed, the mycelium can be broken apart by ultrasound, which shears the hyphae at septal plates into viable cell compartments (Gooday and Gow, 1983). Thus the mycelium of *C. albicans* behaves as a series of more or less autonomous cell compartments, like elongated yeast cells joined end to end. In contrast, hyphae of *M. rouxii* and *H. capsulatum* resemble closely the hyphae of true mycelial fungi. Hyphae of *H. capsulatum* have Woronin bodies (Chapter 5) to occlude the septal pores (Maresca and Kobayashi, 1989) implying that there is cytoplasmic continuity across septa and a true coenocytic nature to the mycelium.

Hyphae normally have vacuoles situated behind the apex. In *C. albicans* the apical cell compartment of the hypha may have a few vacuoles but the subapical compartments are mostly highly vacuolated (Gow and Gooday, 1984). This may explain why branch formation is sparse in *C. albicans* hyphae. During germ tube formation the cytoplasm from the parent yeast cell migrates into the growing germ tube leaving behind a vacuolated mother cell with a single nucleus (Figure 19.2). The cytoplasmic space is then regenerated and the vacuolar space correspondingly reduced prior to the formation of a second germ tube (Gow and Gooday, 1984). This process can be seen to be economical for cells forming hyphae in response to nitrogen starvation since germ tube formation occurs with a minimal demand on protein synthesis. This mode of growth is common in other types of germ tubes such as those of certain rust and smut pathogens of plants.

C. albicans hyphae grow on surfaces and within tissues of epithelial cells of the human host. Their hyphae are apparently imbued with touch sensitivity so that they can respond to topographical features of the substrate, i.e. they are thigmotropic (Sherwood *et al.*, 1992). On artificial substrates they have been shown to grow along grooves and ridges and to invade the pores of polycarbonate Nuclepore membranes (Sherwood *et al.*, 1992; Gow, 1993). This may facilitate the invasion of the host epithelium by seeking out microsites that are wounded or have a weakened integrity.

19.2.3 Cell and duplication cycles

The transition between yeast-like and mycelial growth represents an intriguing problem for cell cycle regulation. The cell cycle of filamentous cells is called the duplication cycle (Chapter 14). Yeast cells of *C. albicans*, *S. cerevisiae* and *H. capsulatum* are uninucleate whereas those of *Blastomyces dermatiditis* and *P. brasiliensis* are multinucleate. Those fungi with multinucleated yeast cells also tend to have several nuclei in each hyphal compartment. Yokoyama and Takeo (1983) showed that after accounting for the substantial volume of the vacuolar space of pseudohyphal cells of *C. albicans*, a cell size-mediated control process of the cell and duplication cycles could be inferred. It is not yet clear whether the cell cycle regulatory genes encoding cyclins, kinases and phosphatases are the same in yeast and hyphal cells or whether there is a specific cross-roads in the cycle where the decision to form a yeast cell or a hypha must be made. This is, however, an interesting area that would seem particularly well suited for analysis in *S. cerevisiae* where well-defined cell division cycle (*cdc*) mutants are available.

19.3 DIMORPHIC FUNGAL SYSTEMS

The availability of cells growing in two morphological forms suggests immediately comparative experiments to describe biochemical differences

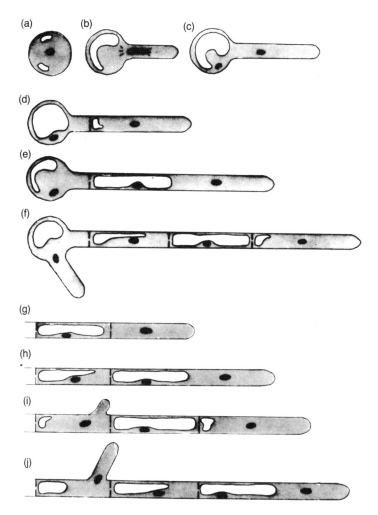

Figure 19.2 Vacuolation and growth of germ tubes of C. *albicans*. Cytoplasm moves into the germ tube and the germ tube extends leaving behind an extensively vacuolated uninucleated parent yeast cell. Hyphal compartments with the exception of the apical cell are highly vacuolated. Vacuolated compartments, including the parental mother cell, regenerate cytoplasm prior to forming secondary germ tubes or branches. From Gow and Gooday (1984), with permission.

between the two cell types that might help to explain the differences in morphology. Another type of obvious experiment is to alter environmental conditions and monitor the effect on cell shape. These approaches have facilitated understanding of the basic physiology of dimorphism but they have failed, for the most part, to illuminate the precise nature of the regulatory mechanisms. These may be more profitably investigated using genetic approaches, as described later.

19.3.1 *Candida albicans*

There a voluminous literature describing a multitude of environmental factors capable of influencing the equilibrium between yeast and hyphal growth of C. *albicans*. This suggests the presence of several signalling systems that are probably regulated independently (Gow, 1988). For example media containing N-acetylglucosamine or serum are both able to stimulate hyphal growth (Odds, 1988), but strains have been described that fail to produce germ tubes in *N*-acetylglucosamine but retain the ability to form germ tubes in media containing serum (Corner *et al.*, 1986). Clearly serum and *N*-acetylglucosamine-mediated hyphal induction operate independently. Also, 2-deoxyglucose inhibits *N*-acetylglucosamine-induced germ tube formation of *C. albicans* but does not affect serum-induced dimorphism (Torosantucci *et al.*, 1984). Clearly the diversity of possible external signals known to influence dimorphism in *C. albicans* (Odds, 1988) in itself means that it is not helpful to think of dimorphic regulation in this organsim in terms of the consequence of any single environmental parameter or perturbation. An elevated temperature, neutral pH and relatively nutrient poor growth medium are general parameters that selectively encourage the growth of hyphae over yeasts, but reports exist in which each of these conditions has been shown to be non-essential for filamentous growth. This situation is in contrast to several other dimorphic fungi where a single environmental parameter (for example, temperature for *H. capsulatum* or anaerobiosis for *Mucor*), can be demonstrated to be critical for the dimorphic switch.

Several second messenger systems have been implicated in dimorphic regulation including those based on cAMP (Chattaway *et al.*, 1981; Sabie and Gadd, 1992), Ca^{2+} and calmodulin or inositol phosphates (Roy and Datta, 1987; Sabie and Gadd, 1989) or intracellular pH (Stewart *et al.*, 1988; Kaur *et al.*, 1988) (Chapter 9). These studies have demonstrated that addition of exogenous second messenger, such as cAMP or its precursors or analogues, can elicit a dimorphic response, or that antagonism of second messenger systems (such as addition of calmodulin inhibitors (Sabie and Gadd, (1989)) may retard it. Finally, changes in the intracellular pH or second messenger concentration have been measured during the morphological transition (Niimi *et al.*, 1980; Kaur *et al.*, 1988; Stewart *et al.*, 1988; Egidy *et al.*, 1989). These results suggest that several second messenger systems may be employed to transduce environmental changes into changes in gene expression, metabolism and cellular organization. An implication of this is that these second messengers may in turn influence the expression of some gene or genes that function as a master switch leading to a change in the morphogenetic programme. Such a master switch is as yet entirely hypothetical.

19.3.2 *Histoplasma capsulatum*

H. capsulatum is a pathogen that is found worldwide but is particularly widespread in the Ohio river valley in the USA. The mycelium is found in soil and guano of bats and birds. The yeast form is the parasitic form which is found multiplying in the lungs and within the reticuloendothelial cells of infected humans and animals. This fungus may have accounted for the death of the archaeologists of Egypt such as Lord Caernarvon who sponsored the excavation of Tutankhamun's tomb and fell victim to the 'mummy's curse' after inhaling the stagnant air in the recently opened vault that was full of the stench of bats' droppings. *Blastomyces dermatitidis* is a closely related dimorphic pathogen with a similar geographical range and pattern of disease. The sexual phases of *H. capsulatum* and *B. dermatitidis* are classified in the same genus, *Ajellomyces*.

There are many contrasts between the hyphae and yeasts of *C. albicans* and those of other dimorphic human pathogens such as *H. capsulatum*, *P. brasiliensis* and *B. dermatitidis*. With the exception of *C. albicans* these fungi produce yeast cells at body temperature and hyphae at lower temperatures. Only *C. albicans* has more chitin and cAMP in the hyphal form compared to the yeast form. Thus in *H. capsulatum*, the dimorphic transition is normally studied in the mycelium to yeast direction and is triggered by elevating the temperature from 25°C to 37°C.

Important biochemical events associated with the mycelium to yeast transition in *H. capsulatum* include an initial uncoupling of oxidative phosphorylation followed by a drop in ATP levels and a decline in the rate of respiration (Maresca and Kobayashi, 1989). Following this, the respiration rate recovers via respiratory bypass pathways that require exogenous cysteine or other sulphydryl components (Figure 19.3). Sulphydryl inhibitors such as *p*-chloromercuriphenylsulphonic acid (PCMS) irreversibly block the transition by interfering with this period of respiratory recovery (Medoff *et al.*, 1986, 1987). Because an increase in temperature is critical for morphogenesis in this system there has been considerable interest in the possible involvement of heat-shock genes and

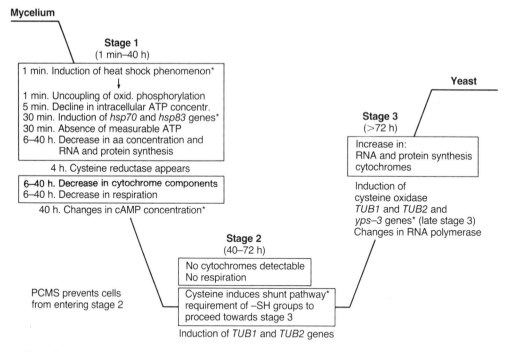

Figure 19.3 Respiration and other events associated with the hyphal to yeast conversion of *H. capsulatum*. The respiration rate is indicated by the bold line and the boxes indicate related events. Other events (*) whose activation does not influence the phase transition are also shown. From Maresca and Kobayashi (1989), with permission.

proteins in cellular regulation (Maresca and Kobayashi, 1989; Minchiotti *et al.*, 1991; Patriaca *et al.*, 1992).

19.3.3 *Paracoccidioides brasiliensis*

P. brasiliensis causes primary lung disease and secondary systemic disease and is endemic in Central and South America (San-Blas, 1985). The disease is an order of magnitude more common in males rather than females because of the effect of human sex hormones (Gooday and Adams, 1993). As with *H. capsulatum*, temperature is the critical environmental parameter and it is the yeast form that is produced at 37°C. Medium composition seems to be irrelevant to dimorphism in this case (San-Blas, 1986). The main focus for research has been in the relative composition of the yeast and hyphal cell walls which consist of α(1–3)- and β(1–3)-glucan, chitin, galactomannan and protein. The yeast cells are rich in α(1–3)-glucan which is completely absent from the hyphal cells. Mutants that were avirulent due to serial passage *in vitro* were found to be also almost devoid of this

polysaccharide (Kanetsuna *et al.*, 1969). The α (1–3)-glucan may make the cells difficult to phagocytose and hence increase the virulence of this form of the fungus. This molecule is one of the most unequivocal examples of a single virulence factor for a fungal pathogen of humans.

19.3.4 *Coccidioides immitis*

C. immitis is a lung pathogen that grows saprophytically in the soil in the southwest USA and in Central and South America. The parasitic phase involves the development of large 'spherules' from arthroconidia that are formed by the saprophytic mycelium and which can be breathed into the lungs (Cole and Sun, 1985). The cylindrical arthroconidia develop into spherules at 37°C by a process of cell expansion, nuclear division and finally segmentation that subdivides the cytoplasm into uninucleate, walled endospores which are disseminated in the body when the spherule ruptures. In the body this process is repeated as the infection deepens. Thus although this fungus is dimorphic, it is the morphogenesis of the spherule that is most important for pathogenicity. Biochemical activities and genes that may be important for pathogenicity, such as spherule-associated protease (Cole *et al.*, 1992), β(1–3)-glucanase (Kruse and Cole, 1992) and surface antigens have been studied and characterised (Cole *et al.*, 1993).

19.3.5 *Sporothrix schenckii*

S. schenckii is another widely distributed dimorphic pathogenic fungus that is parasitic in the yeast phase and saprophytic in the mycelial phase. As with *H. capsulatum* and *P. brasiliensis* the yeast form is produced at body temperatures and mycelial development is favoured at lower temperatures (Rodrìguez-Del Valle *et al.*, 1983). There is evidence for cAMP acting as a signal transducer since both dibutyrl cAMP and dibutyrl-cGMP induced yeast-like growth when added to the medium (Rodrìguez-Del Valle *et al.*, 1984).

19.3.6 *Wangiella dermatitidis*

This pathogenic fungus is darkly pigmented due to the production of melanin and normally forms clinical lesions in the epidermal and dermal regions of the skin. Vegetative growth can occur as budding yeast cells, hyphae or pseudohyphae and as a thick-walled segmented multicellular (Mc) form that resembles sclerotic bodies of other fungal pathogens (Oujezdsky *et al.*, 1973; Geis and Jacobs, 1985). This organism is asexual but parasexual genetics has been developed and then used to examine mutants in pigment production, and in cell cycle regulation (Cooper, 1993). Temperature-sensitive mutants in Mc formation have also been isolated and characterized (Jacobs *et al.*, 1985; Cooper, 1993).

19.3.7 *Mucor* species

Dimorphism is a rare phenomenon among the Zygomycetes but several *Mucor* species including *M. racemosus* and *M. rouxii* as well as members of the genera *Mycotypha* and *Cokoromyces* are dimorphic (Orlowski, 1991). Dimorphism in *Mucor* was probably first described by Robert Hooke in 1665 and later by Louis Pasteur who observed it as a contaminant in beer. This dimorphic process is different from those described above where temperature or nutrients are the critical environmental parameters. In *Mucor*, anaerobiosis favours the growth of the yeast form and even low concentrations of oxygen promote mycelial development (Bartnicki-Garcia, 1963; Sypherd *et al.*, 1978). Hexose sugars are also required for anaerobic (or perhaps more correctly microaerophilic), fermentative growth and hence for development of yeast cells (Sypherd *et al.*, 1978). Inhibition of oxidative phosphorylation and electron transport with compounds such as potassium cyanide, antimycin A and oligomycin will allow growth by budding under aerobic conditions (Freidenthal *et al.*, 1974; Orlowski, 1991) and phenethyl alcohol, which includes aerobic alcohol fermentation, also induces yeast cell growth. The regulation of morphogenesis again seems to be regulated in some way by cAMP (Orlowski and Ross, 1981; Pereyra *et al.*, 1992). Genetic regulation of *Mucor* dimorphism is discussed below.

19.3.8 *Ophiostoma ulmi*

O. ulmi (formerly *Ceratocystis ulmi*) is the causative agent of Dutch elm disease. The dimorphic transition can be regulated in the laboratory by controlling the supply of nitrogen in the growth medium (Kulkarni and Nickerson, 1981). Ammonium ions, yeast extract and certain amino acids promote hyphal development which may be regulated by a calcium-calmodulin based second messenger system (Muthukumar and Nickerson, 1984; Brunton and Gadd, 1989; Gadd and Brunton, 1992; Chapter 9). Other evidence suggests that inositol phosphate signalling may be involved in dimorphic regulation (Brunton and Gadd, 1991).

19.4 GENETIC STUDIES

Few of the dimorphic fungi mentioned above have well-defined genetic systems so progress towards understanding the regulation of dimorphism at the genetic level has, until recently, been slow. Nonetheless, the importance of many of these organisms as animal and plant pathogens has provided a need for the development of molecular approaches which are now making a significant impact on our understanding of the underlying mechanisms and the relationship between morphogenesis and pathogenesis.

19.4.1 *Candida albicans*

Genetic studies in *C. albicans* have been limited by the organism's asexual diploid nature. A desirable starting point for the analysis of any developmental system such as dimorphism is to examine mutants that are unable to grow in one vegetative growth form or the other. However, morphological mutants of diploid organisms such as *C. albicans* are difficult to obtain because there are two copies of each gene at each locus, and for most diploid organisms, both alleles must be mutated in order to observe a phenotype. The likelihood of discovering a haploid phase of this fungus is slight since the genome apparently contains many recessive-mutations that would be lethal to the fungus if reduction division to a haplotype were possible (Scherer and Magee, 1990). Mutants obtained by chemical or UV mutagenesis have often been subjected to heavy mutagenic damage and so the phenotypes of such strains may be complicated by the presence of unrecognized mutations at several loci. Recently however, *ura3* auxotrophs of *C. albicans*, which have a strict requirement for exogenous uridine, have been created which are otherwise isogenic with the parental clinical isolate (Fonzi and Irwin, 1993). These serve as good strains in which to perform genetic analyses. Systems of reverse genetics have also been devised (Chapter 12) in which disrupted genes can be generated by integrative transformation without affecting other alleles (Alani *et al.*, 1987; Birse *et al.*, 1993). These methods require knowledge of the structure of the target gene and so assumptions have to be made of the types of genes that may play roles in dimorphic regulation. However, the availability of these new genetically marked strains and systems of reverse genetics opens the way for an analysis of genes encoding putative virulence factors through the creation of specific null mutations in these genes then assessing the result of such mutations on cell morphology or invasiveness in animal models.

Another methodological obstacle in work on the molecular biology of *C. albicans* relates to the recent discovery that this fungus has a non-standard usage of CUG codons which in *C. albicans* is unusual in being recognized by a tRNA that is charged with serine rather than leucine (Santos *et al.*, 1993). This may explain the general observation that *C. albicans* will only express heterologous genes poorly and has in turn so far prevented the use of reporter genes in the analysis of gene expression in this fungus.

Northern analyses indicate that the expression of many genes of *C. albicans* varies during the conversion of yeasts into hyphae, but many of these changes relate to changes in the medium composition or cultural conditions, such as elevation of temperature, that are used to induce the morphogenetic transition (Swoboda *et al.*, 1993). Similarly, several studies have used two-dimensional electrophoresis to characterize differences in the proteins of yeast and hyphal cells (Manning and Mitchell, 1980; Brown and Chaffin, 1981; Ahrens *et al.*, 1983; Finney *et al.*, 1985). The results show that there are few major proteins that are unique to one form or the other but that several proteins are modified or represented at different levels in the two cell types. Gene knockout experiments can now be performed in order to obtain firm evidence that a given gene is actively involved in regulating hyphal growth. Disruption of *ECE1*, a gene which is expressed in relation to the e̲xtent of c̲ell e̲longation, and *CHS2*, a chitin synthase gene which is expressed preferentially in the hyphal form, did not yield hyphal-minus phenotypes (Birse *et al.*, 1993; Gow *et al.*, 1993). Thus, even genes whose expression is apparently linked to morphogenesis may ultimately be found, through gene disruption experiments, to be dispensable as far as morphogenesis is concerned. Thus far there has been no gene identified whose regulation has been demonstrated to be linked constitutively to the dimorphic transition of *C. albicans*, but several groups have created cDNA libraries from one cell type or another and used these to screen for dimorphic-specific signals. The results of these types of studies are likely eventually to identify genes that play a direct role in dimorphism.

(a) *Cell wall adhesins and ligands*

Molecular analysis of components of the *C. albicans* cell wall that are important for interactions between the fungus and its host are as yet in their infancy. A great deal of immunocytochemical work with polyclonal and monoclonal antibodies has demonstrated that there are many antigenic epitopes that are expressed only on the surface of the walls of hyphae or yeast cells (Brawner and Cutler, 1986; Ponton and Jones, 1986; Sundstrom *et al.*, 1988; Cassanova *et al.*, 1989; Bouchara *et al.*, 1990; Brawner *et al.*, 1990; Marot-Leblond *et al.*, 1993). The most common epitopes seem to be mannoproteins which must therefore be seen as being crucial to the host–pathogen interaction. It is clear that the mannoproteins of the cell wall change dramatically during the yeast–hyphal conversion, and that some of these molecules are capable of recognizing specific receptors on epithelial or endothelial membranes as well as complement proteins and components of macrophage membranes (Calderone, 1993). Also, the adhesiveness of the cell wall has been shown to be related to the mannoprotein composition which can vary according to the substrate for growth (Douglas, 1992). Interactions with several host proteins including iC3b, fibronectin, laminin, types I and IV collagen, fibrin and fibrinogen have been described (Cutler, 1991). The molecular interaction may be via oligosaccharide–protein interactions (lectin-like), protein–protein interactions or may involve polysaccharides such as chitin in the cell wall (Douglas, 1992; Calderone, 1993). However, it is not yet known what importance these recognition systems have in pathogenesis and the genes that encode the proteins and cause protein mannosylation have yet to be characterized. One recent advance has been to devise a screen for *Candida* genes which, when transformed into *S. cerevisiae*, render the transformants sticky to mammalian tissues (Barki *et al.*, 1993). Such a screen could be modified to pick up genes concerned with specific interactions with particular human surfaces.

(b) Proteases

C. albicans secretes protease enzymes that may digest host tissues and thereby contribute in a direct way to pathogenesis. There are at least seven genes (*SAP*) encoding secretory aspartyl proteinase enzymes (Hube *et al.*, 1991; Wright *et al.*, 1992; White *et al.*, 1993; Monod, unpublished). These genes are all regulated differentially and at least three are synthesized specifically during hyphal growth (Hube, Monod, Schofield, Brown and Gow, unpublished). However, the precise role of these enzymes in pathogenesis again requires an analysis of the virulence of null mutants in the *SAP* structural genes.

(c) Phenotypic switching in C. albicans

C. albicans is asexual and therefore cannot create genetic variation through meiosis. However, most strains of this organism can switch reversibly at high frequency between a repertoire of colony types (Figure 19.4) (Soll, 1992). The colony morphologies can vary so much as to give the impression that they could be entirely different strains or even different species. Yet it has been demonstrated that an individual strain may generate eight or more colony types which are heritable yet interconvertible through a switching event (Slutsky *et al.*, 1985, 1987). The colony morphology apparently reflects spatial changes in the proportion of cells in the yeast, hyphal and pseudohyphal forms in different parts of the colony. These spatial patterns of cells with different shapes generate colonies that may be smooth or wrinkled, fuzzy, stippled, ringed, star or fried-egg shaped (Figure 19.4). In the well-studied white to opaque transition of strain WO-1 the opaque cells are larger, elongated and have pimples on the cell wall whereas the white cells are normally shaped spherical yeast cells. They also vary in a wide range of other biochemical properties (Soll, 1992). Sectoring of individual colonies can be observed in which a hybrid of two or more colony types can be seen (Figure 19.4).

The frequencies at which the colonies switch from one form to another is very rapid (1 in 10 to 1 in 10^5) and the frequency is stimulated by UV light and repressed in media containing high zinc concentrations. Clearly the high frequency and reversible nature of phenotypic switching is incompatible with a mutational event. The mechanism underlying the transition is not yet understood but may involve some reversible transposition event or genomic rearrangement. *C. albicans* was not found to have any homologous sequences to the switchable mating locus *MAT*a and the transposable element Ty of *S. cerevisiae* which have properties that may be appropiate for such a regulatory mechanism (Soll, 1992). A transposon-like element called Tca1 has been isolated from *C. albicans* and was found to be dispersed in the genome but as yet this sequence has not been

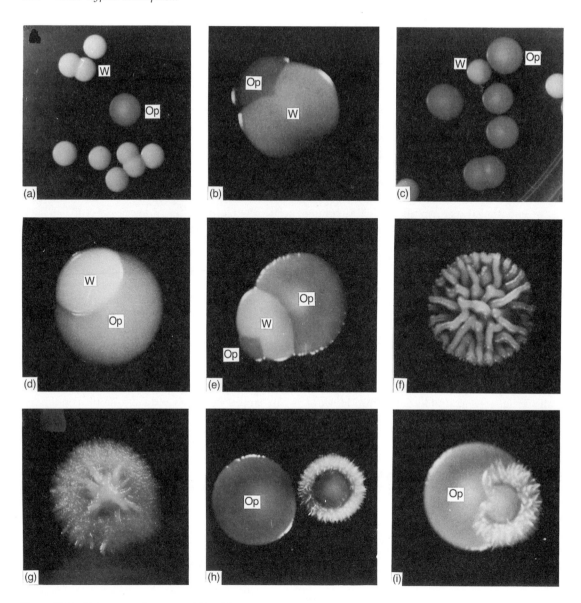

Figure 19.4   Colony morphotypes exhibited by *C. albicans* WO-1 due to phenotypic switching. (a) An opaque colony (Op) arising from cells from a white colony (W) and reciprocally (e) a white colony that arose from cells of an opaque colony. Sectoring of colonies is seen in (a,c,f and i). A wrinkled colony (d) and two fuzzy variants (g,h and i) that can also be produced by the strain are also shown. From Slutsky *et al.* (1987), with permission.

observed to move during switching (Chen and Fonzi, 1992). Several moderately repeated sequences have also been described for *C. albicans* including 27A (Scherer and Stevens, 1988), *Msp*I (Cutler *et al.*, 1988), Ca3 (Sadhu *et al.*, 1991), the telomere associated sequence Ca7 (Sadhu *et al.*,

1991), and CARE-1 and CARE 2 (Lasker *et al.*, 1991, 1992) and RPS1 (Iwaguchi *et al.*, 1992). Some of these repeated sequences were found to be highly recombinogenic or promoted reorganization of the DNA at reasonable frequencies. However, in no case to date have these events been shown to correlate well with phenotypic switching (Scherer and Stevens, 1988; Sadhu *et al.*, 1991; Soll, 1992) and it must be concluded that recombination between these sequences is unlikely to represent the mechanism behind this phenomenon.

A controversial suggestion for the mechanism of phenotypic switching relates to observations of dramatic changes that can occur in the arrangements of *C. albicans* chromosomes. It is now clear from pulsed-field gel electrophoresis experiments that the karyotype of *C. albicans* is highly variable (Suzuki *et al.*, 1989; Scherer and Magee, 1990; Rustchenko-Bulgac and Howard, 1993). This variability apparently relates to both chromosome length polymorphisms and chromosome instability which may result from recombination events between repetitious DNA sequences. Several reports have been published where changes in karyotype correlate with changes in the ability to form germ tubes, colony morphologies, pigmentation and other traits (Suzuki *et al.*, 1989; Rustchenko-Bulgac *et al.*, 1990; Rustchenko-Bulgac and Howard, 1993). In other cases, such as in the white/opaque transition no obvious chromosomal changes were seen during switching between the white and opaque colony types (Soll, 1992). Thus it is difficult to ascribe a functional role for chromosome instability in the regulation of phenotypic switching.

An alternative suggestion for the mechanism of phenotypic switching has been suggested by Soll (1992), which stems from an observation by Aparicho *et al.* (1991) of how *SIR* (silent information regulation) genes in *S. cerevisiae* may regulate gene expression by modulating heterochromatin formation near the telomeres of chromosomes. Condensed heterochromatin is transcriptionally silent. Genes that were cloned into chromosomes near the telomeric regions could switch between repressed and expressed states. These states were heritable, reversible, occurred at high frequency and depended on functional *SIR2*, *SIR3* and *SIR4* genes. Thus Soll (1992) has suggested that in *C. albicans* a regulatory gene(s) for dimorphism or phenotypic switching might be positioned near a heterochromatic region of the telomeres of one or more chromosomes and that the expression of this hypothetical regulatory gene may in turn be determined by the silencing function of the *SIR* genes that regulate condensation of the chromatin near the telomeric regions.

Parasexual analysis of phenotypic switching has been performed by fusing spheroplasts of switching and non-switching strains. The results showed that the progeny were non-switching, suggesting that the white–opaque switching phenotype is recessive due to the presence of a repressor in the wild-type cell (Chu *et al.*, 1992). Moreover, the location of the repressor and the switching genes could be inferred from experiments in which heat was used to induce chromosome loss and this suggested that the repressor was located on chromosomes 2,5 and/or 6 and that the switching genes (*SW1*) that were essential for the transition were on chromosome 3 (Chu *et al.*, 1992).

The significance of phenotypic switching to the pathogenicity of *C. albicans* is related to the fact that many different aspects of the phenotype are under the regulatory influence of this mechanism. These include cell morphology, cell size, mannoprotein biosynthesis, antigenicity, protease production, sensitivity to antifungal drugs and to phagocytic cells, sterol content and the assimilation profile for various solutes (Soll, 1992). Several genes that are expressed specifically in the white or opaque cells have recently been cloned from libraries of genes generated from cDNAs from the white or opaque cells of WO-1 (Morrow *et al.*, 1993). One of the opaque-specific genes is *SAP1* (originally called *PEP1*) which is a structural gene for a secretory aspartyl protease. A second opaque-specific gene Op4, which has no known sequence homology to a known gene, was regulated coordinately with *SAP1*. No rearrangement of the DNA was found in the vicinity of *SAP1* or Op4 during the switching transition (Morrow *et al.*, 1993).

Other pathogenic microorganisms have switching systems for virulence determinants. One of the best known is the antigenic variation phenomenon in *Trypanosoma brucei*. However, the switching system of trypanosomes and bacterial pathogens are normally restricted to alterations in antigenicity alone and thus phenotypic switching in *C. albicans*, which coordinately regulates a wide range of putative virulence functions including cellular dimorphism, is of considerable importance and interest. Moreover, it has also been shown that when *C. albicans* was isolated from patients suffering from recurrent vulvovaginal candidosis, an individual strain had undergone phenotypic switching between each episode of vaginitis

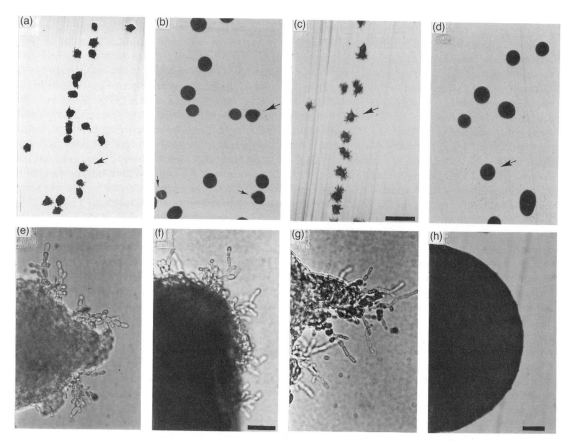

Figure 19.5 Genetic and physiological regulation of pseudohyphal growth in *S. cerevisiae*. The photographs in e–f are of the colonies indicated in the above panel with large arrows (a–d). Panels a,b,e and f show a parental diploid strain growing on a synthetic medium, deficient in nitrogen and containing low concentrations of ammonia (a,e) or proline as the sole nitrogen source (b,f). Proline is a poor nitrogen source for *S.cerevisiae*. Panels c,d,g and h are the same diploid strain but with a homozygous *shr3* mutation (described in the text) which reduces the ability of cells to take up amino acids. In (c) and (g) the mutant is grown in medium with proline as the only source of nitrogen. The pseudohyphal growth is exaggerated in this strain in this medium. In (d) and (h) pseudohyphal growth is repressed by the presence of standard amounts of ammonium in the medium. The scale bar in (c) is 0.5 mm and in (f) and (h) 30 μm. From Gimeno *et al.* (1992), with permission.

(Soll *et al.*, 1989). This suggests that the switching mechanism may have evolved to allow the fungus to create the variability necessary to avoid capture by the immune system.

19.4.2 *Saccharomyces cerevisiae*

Perhaps the best developed and most readily exploited system of genetic analysis in biology is that in the brewer's yeast, *S. cerevisiae* (Guthrie and Fink, 1991). Many wild-type strains of this yeast

exhibit some dimorphic capacity in nature (Lodder, 1970) but filamentous forms have been outbred in the many laboratory strains that have been the focus of interest for biochemical, molecular and genetic experiments. However, the dimorphic response was rediscovered recently in diploid cells which, upon nitrogen starvation, formed pseudo-hyphal cells that enabled the fungus to move away from zones of nutrient depletion (Figure 19.5) (Gimeno *et al.*, 1992). The foraging hyphae were observed to invade the agar, unlike the parental

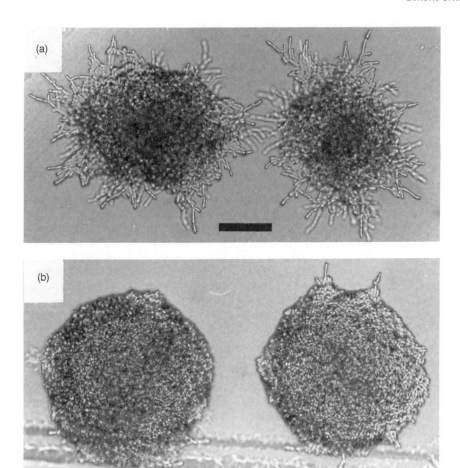

Figure 19.6   Pseudohyphal growth in *S.cerevisiae* is induced by the transforming cells with a vector carrying the dominant RAS^{val19} mutation leading to elevated cytoplasmic cAMP levels. In (a) the transformant with the RAS^{val19} mutation exhibits pseudohyphal growth. In (b) the control is transformed with vector sequence alone. Scale bar = 60 μm. From Gimeno *et al.* (1992).

haploid cells or non-starved diploid cells that formed normal colonies, on top of the agar surface. The pattern of cell growth of the diploid pseudo-hyphal cells is described above. Dimorphic mutants have also been described in *S. cerevisiae* (Gimeno *et al.*, 1992; Blacketer *et al.*, 1993) as well as in the fission yeast *Schizosaccharomyces pombe* (Sipiczki *et al.*, 1993) which also help us to understand the mechanism of dimorphism in other fungi.

The power of the well-defined *Saccharomyces* genetic system will enable rapid progress in the understanding of many aspects of the dimorphic process that is likely to be similar to that in

genetically intransigent species such as *C. albicans*. For example, it was shown in *S. cerevisiae* that constitutive activation of the *RAS2* gene led to stimulation of the pseudohyphal growth form suggesting that elevated cAMP levels may be involved in the signal transduction pathway for the morphogenetic process (Figure 19.6) (Gimeno *et al.*, 1992). This is of particular significance in the light of studies in *C. albicans* that showed that hyphal cells had higher levels of cAMP than yeast cells (Niimi *et al.*, 1980; Egidy *et al.*, 1989; Sabie and Gadd, 1992) and that exogenous cAMP or dibutyrl cAMP stimulated a yeast to hyphal transition

(Chattaway *et al.*, 1981; Sabie and Gadd, 1992). In addition, mutations in *SHR3* (a gene which is required for amino acid uptake in *S. cerevisiae*) also increased pseudohyphal growth. Interestingly, nitrogen starvation is also one environmental factor that stimulates germ tube formation in *C. albicans* (Odds, 1988). The signal that induces pseudohyphal growth in *S. cerevisiae* is apparently transmitted via a set of protein kinases that act as a kinase cascade (Liu *et al.*, 1993). This cascade is the same one that mediates the activation of mating functions in haploid cells that are exposed to mating pheromones. This pathway ultimately affects the transcription of *STE12*, which acts as a transcription factor for mating-specific genes. These observations provide an elegant explanation for the observation that *STE12* and several other *STE* genes (*STE20*, *STE11* and *STE7*) that encode other protein kinases in the cascade are transcribed in diploid cells that do not mate but which form pseudohyphae (Liu *et al.*, 1993). Further support for the involvement of mating pathway genes in dimorphic regulation was obtained by the analysis of null mutations in *STE12*, *STE20*, *STE11* and *STE7* which all suppressed pseudohyphal growth (Liu *et al.*, 1993).

Other mutations in *S. cerevisiae* lead to cell elongation and pseudohyphal growth. For example *elm1*, *elm2* and *elm3* mutations cause formation of pseudohyphal cells which again 'invade' the agar (Blacketer *et al.*, 1993). Strains that were heterozygous for *ELM2* and *ELM3* were more prone to undergo a dimorphic transition as a consequence of nitrogen starvation suggesting a common underlying regulatory pathway to *shr3*-mediated dimorphism. Dimorphism in *S. cerevisiae* is also apparently regulated by protein phosphorylation. The *ELM1* gene encodes a novel protein kinase and pseudohyphal growth induced by nitrogen starvation depends on the activity of the *CDC55* gene product which may be a regulatory subunit of protein phosphatase 2A (Blacketer *et al.*, 1993).

Many genes in *S. cerevisiae* have been shown to have homologues in *C. albicans* (Scherer and Magee, 1990), and many *Candida* genes have been cloned using *Saccharomyces* genes as probes or by functional complementation in *Saccharomyces*. Because there are many physiological similarities between these two organisms progress in our understanding of the dimorphic response of *C. albicans* will be catalysed by detailed analysis of the dimorphic behaviour of *S. cerevisiae*.

19.4.3 *Ustilago maydis*

U. maydis is a pathogen of maize and only one other close relative, teosinte. The disease, corn smut, is a systemic infection of the cereal by the mycelial form of the fungus. Infection leads to the formation of tumours or galls on the green parts of the plants. The fungus is a Basidiomycete and has a well-characterized mating pathway that is central to the regulation of mycelium formation and pathogenicity (Chapter 18). The haploid cells are unicellular, elongated yeast cells that can be grown saprophytically in laboratory media, but cannot infect maize plants. Fusion of compatible haploid cells leads to the formation of a dikaryotic filamentous form that is difficult to grow on agar, but can be cultivated on media containing charcoal (Day and Anagnostakis, 1974) and can readily infect the host maize plant. The mating type loci, or incompatibility loci therefore also regulate dimorphism in this fungus. Because this fungus has an exploitable genetic system with haploid forms where mutants can be selected readily, the dimorphic process of this fungus has been subjected to a detailed genetic and molecular analysis (Schultz *et al.*, 1990; Banuett, 1992; Bölker *et al.*, 1992; Gillissen *et al.*, 1992).

Dimorphism and dikaryon formation are regulated by two unlinked mating type loci, *a* and *b*. These loci contain genes encoding elements required for the recognition of compatible partners and the regulation of other genes concerned with the stimulation of the dimorphic response. The *a* locus is concerned with cell recognition that is mediated by lipopeptide mating pheromones. There are two *a* alleles *a1* and *a2* which are flanked by identical sequences of DNA. Within each *a* allele are regions specifying genes for lipopeptide mating factors and a receptor for the mating factor from the other allele (Bölker *et al.*, 1992). The *b* locus is multiallelic (at least 25 alleles exist) and controls sexual and pathogenic development (Gillissen *et al.*, 1992). Each allele consists of two genes *bE* and *bW* which have a similar organization in having a constant and a variable domain, but which are not homologous at the amino acid level. The variable regions of these genes probably determine self/non-self interactions whereas the constant regions interact with DNA thereby regulating the expression of genes concerned with filamentous growth and tumour formation. Within the constant regions of these genes are sequences that are similar to the homeodomains of higher eukaryotes and which are involved in sequence-specific DNA binding (Schultz *et al.*, 1990). Mutational analysis suggests that *bE* and *bW* polypeptides from different alleles come together to form a multimeric complex that is a positive regulator of genes concerned with filamentous growth and tumorigenesis. A similar interaction occurs between the **a**1 and α2 mating factor polypeptides of *S. cerevisiae* (Dranginis, 1990). The genes that function downstream from *a* and *b* are therefore likely to be directly involved in the regulation of dimorphism and filamentous growth.

19.4.4 *Mucor* species

It seems likely that DNA methylation or demethylation plays an important role in dimorphic regulation in *Mucor* (Cano-Canchola *et al.*, 1992). The DNA from sporangiospores is highly methylated and becomes demethylated when the spores germinate to form hyphae (Cano *et al.*, 1987). Demethylation requires DNA replication and may regulate gene expression. This might in turn be related to the activity of the ornithine decarboxylase enzyme, which plays an important role in the regulation of polyamine biosynthesis. The ornithine decarboxylase inhibitor 1,4-diamino-2-butanone (DAB) prevented the formation of hyphae from germinating sporangiospores, blocked demethylation of DNA and resulted in a decrease of polyamine levels (Ruiz-Herrera and Calvo-Mendez, 1987). The level of methylation of *CUP* genes was found to decrease at the onset of polarized hyphal growth. At the same time the gene became transcriptionally active suggesting that demethylation led to the activation of this gene and other genes that may play a role in the dimorphic transition (Cano-Canchola *et al.*, 1992).

19.5 DIMORPHISM AND PATHOGENICITY

Many dimorphic fungi including *C. albicans*, *H. capsulatum*, *P. brasiliensis*, *B. dermatitidis*, *C. immitis*, *W. dermatitidis* and *S. schenckii* are human pathogens and some such as *O. ulmi* and *U. maydis* are plant pathogens. For *C. albicans* and *U. maydis* it is thought that mycelial cells are invasive and more pathogenic than yeast cells. However, it should be noted that the human pathogens *H. capsulatum*, *P. brasiliensis* and *B. dermatidis* are pathogenic in the yeast form and saprophytic in the mycelial form. This has led to confusion and controversy about ascribing any functional relationship between the vegetative growth form and pathogenicity. In *H. capsulatum* the transition from hyphal to yeast is thought to be obligate for pathogenicity since the sulphydryl blocking agent *p*-chloromercuriphenylsulphonic acid (PCMS) can block the transition from hypha to yeast at 37°C and this suppressed infection in mice (Medoff *et al.*, 1987). Moreover, hyphal cells of *H. capsulatum*, *P. brasiliensis*, *C. immitis* and *B. dermatitidis* are not found in diseased tissue in the body.

Clearly, hyphal growth allows cells to move and penetrate insoluble barriers such as human tissues. The pseudohyphal cells of *S. cerevisiae* have been reported to be able to penetrate into agar media, whereas colonies of yeast cells always remained on top of the agar surface (Gimeno *et al.*, 1992). Thus, invasive growth is likely to be potentiated by the capacity to form hyphae.

In *C. albicans* the debate over the relative pathogenicity of the yeast and hyphal forms has also been fuelled by observations in which hyphal elements were found to be rare or even absent in histological examinations of diseased lesions (Odds, 1988) and reports where the yeast phase was reported as being the more pathogenic (Simonetti and Strippoli, 1973). Unfortunately, there have been no clean, stable hyphal-minus mutants with which the problem can be addressed experimentally, although natural isolates of strains that have a decreased capacity to form hyphae have been compared with hyphal-positive strains for their ability to cause infections (Martin *et al.*, 1984). In addition heavily mutagenized strains that are impaired in germ tube formation have also been examined (Shepherd, 1985; Ryley and Ryley, 1990) in this regard. These experiments must be interpreted with caution since such strains are almost certain to have become altered at a number of genetic loci, including those concerned with other virulence determinants. These experiments have not clarified the issue of whether hyphal formation should be regarded as a virulence factor. Hyphal-minus strains were shown to be able to kill mice, but the lesions did not spread as deeply into organs such as kidneys as did hyphal-positive strains (Ryley and Ryley, 1990). Also it should be borne in mind that hyphal penetration is more likely to be advantageous in the *initiation* of infections where the fungus must cross some epithelial barrier, such as the mucosa. If the route of experimentally-induced infection is via intravenous injection the advantages conferred by the penetrating hyphal growth habit may be bypassed and thus infections can be initiated by yeast cells. Interestingly, it was demonstrated that one strain of *C. albicans* that was hyphal-minus in *in vitro* experiments formed hyphae in the vaginal cavity when used to challenge animals in virulence studies (De Bernardis *et al.*, 1993).

Dimorphic human pathogens, such as *H. capsulatum*, *P. brasiliensis* and *C. immitis*, are mostly pathogenic in the yeast form. In these cases, however, the primary route of infection is not across the mucous epithelium but via the lungs. Spores are inhaled and infections are established in the lungs, but may spread systemically, especially if the host's immune system is impaired, for example in patients undergoing chemotherapy. The yeast cells of *H. capsulatum* and *B. dermatitidis* are harboured intracellularly and thus are able to avoid the antagonism of the immune system. In these cases the yeast cells only are responsible for dissemination of the disease. In *P. brasiliensis* this has been shown to relate to the composition of the yeast cell wall which, unlike walls of hyphae, and avirulent strains of the yeast form, has a high α(1–3)-glucan content that renders the wall indigestible to macrophages and other phagocytes (San-Blas, 1986). This emphasizes the point that the cell morphology may not be a primary virulence factor, but may be associated with other changes in cell physiology that are important in pathogenesis.

Fungal virulence in humans is therefore multifactorial depending on a variety of features including adherence, the production of proteases and other hydrolytic enzymes and the ability to escape the cellular, humoral and chemical defences of the host (Cutler, 1991). In *C. albicans*, where the most extensive analysis of virulence and pathogenicity has been undertaken, it is still not clear which of these factors have a dominant effect during infections. Future experiments in which specific mutations are introduced into the genes of these putative virulence factors will improve

greatly our understanding of the essential processes. However, it must be borne in mind that many human pathogenic fungi are only weakly pathogenic, or opportunistic and cannot establish serious infections in healthy individuals. Thus, the physiological condition of the host is paramount when considering fungus–host interactions.

For plant pathogens such as *O. ulmi* and *U. maydis* morphological plasticity may confer similar advantages to those described above. Primary invasion and penetration may be facilitated by tip growth of hyphae. Systemic spread in the vascular tissues may, however, be more rapid for smaller unicellular yeast cells. With *U. maydis* the budding, haploid yeast cells are unable to invade the maize plant and here the penetrative properties of the dikaryotic hyphae seem important to the parasitic properties of the fungus. In this case the formation of the filamentous form is part of the sexual cycle of the fungus. Crossing experiments with compatible and non-compatible partners have shown that filamentous growth is required absolutely for invasion and tumour production in the host (Banuett, 1992).

19.6   CONCLUSIONS

Dimorphism is a common property of fungal cells. Some aspects of the regulation of the transition between yeast and hyphal development or hyphal to yeast development may be similar, even between taxonomically unrelated species. However, other features, particularly with regard to the response to environmental factors, may be very different between different species. This is suggested in the diversity of environmental signals discussed above that are capable of influencing morphogenesis. It is clear that future research will be influenced strongly by rapid advances that can be made in model systems such as in *S. cerevisiae*. However, the use of brewer's yeast may be limited as a model for understanding the pathogenicity of dimorphic fungi, since there are important differences between organisms, for example in the repertoire of proteases, adhesins, antigens etc., that are crucial in defining the virulence properties of pathogenic species (Cutler, 1991). Recent advances in the creation of molecular techniques that can be applied to asexual, genetically intractable fungi will enable the requisite analyses of specific virulence factors or signalling mechanisms to be made. These concerted activities should work synergistically in furthering our knowledge of these interesting organisms.

REFERENCES

Ahrens, J.C., Daneo-Moore, L. and Buckley, H.R. (1983) Differential protein synthesis in *Candida albicans* during blastospore formation at 24.5°C and during germ tube formation at 37°C. *Journal of General Microbiology*, **129**, 1133–9.

Alani, E., Cao, L and Kleckner, N. (1987) A method for gene disruption that allows repeated use of *URA3* selection in the construction of multiply disrupted yeast strains. *Genetics*, **116**, 541–45.

Anderson, J. and Soll, D.R. (1986) Differences in actin localization during bud and hypha formation in the yeast *Candida albicans*. *Journal of General Microbiology*, **132**, 2035–47.

Aparicho, O.M., Billington, B.C. and Gottschling, D.E. (1991) Modifiers of position effects are shared between telomeric and silent mating-type loci in *S. cerevisiae*. *Cell*, **66**, 1279–87.

Barki, M., Koltin, Y., Yanko, M. *et al.* (1993) Isolation of a *Candida albicans* sequence conferring adhesion and aggregation on *Saccharomyces cerevisiae*. *Journal of Bacteriology*, **175**, 5683–9.

Bartnicki-Garcia, S. (1963) Symposium on biochemical bases of morphogenesis in fungi III. Mold-yeast dimorphism of *Mucor*. *Bacteriological Reviews*, **27**, 293–304.

Barton, R. and Gull, K. (1988) Variation in cytoplasmic microtubule organization and spindle length in the two forms of the dimorphic fungus *Candida albicans*. *Journal of Cell Science.*, **91**, 211–20.

Banuett, F. (1992) *Ustilago maydis*, the delightful blight. *Trends in Genetics*, **8**, 174–80.

Birse, C.E., Irwin, M.Y., Fonzi, W.A. and Sypherd, P.S. (1993) Cloning and characterization of *ECE1*, a gene expressed in association with cell elongation of the dimorphic pathogen *Candida albicans*. *Infection and Immunity*, **61**, 3648–55.

Blacketer, M.J., Koehler, C.M., Coats, S.G. *et al.* (1993) Regulation of dimorphism in *Saccharomyces cerevisiae*: involvement of the novel protein kinase homolog Elm1p and protein phosphatase 2A. *Molecular and Cellular Biology*, **13**, 5567–81.

Bölker, M., Urban, M. and Kahmann, R. (1992) The *a* mating type locus of *U. maydis* specifies cell signalling components. *Cell*, **68**, 441–50.

Bouchara, J-P., Tronchin, G., Annaix, V. (1990) Laminin receptors on *Candida albicans* germ tubes. *Infection and Immunity*, **58**, 48–54.

Braun, P.C. and Calderone, R.A. (1978) Chitin synthesis in *Candida albicans*: comparison of yeast and hyphal forms. *Journal of Bacteriology*, **135**, 1472–7.

Brawner, D.L. and Cutler, J.E. (1986) Ultrastructural and biochemical studies of two dynamically expressed cell surface determinants on *Candida albicans. Infection and Immunity*, **51**, 327–36.

Brawner, D.L., Cutler, J.E. and Beatty, W.L. (1990) Caveats in the investigation of form-specific molecules of *Candida albicans. Infection and Immunity*, **58**, 378–83.

Brown, L.A. and Chaffin, W.L. (1981) Differential expression of cytoplasmic proteins during yeast bud and germ tube formation in *Candida albicans. Canadian Journal of Microbiology*, **27**, 580–5.

Brunton, A.H. and Gadd, G.M. (1991). The effect of exogenously-supplied nucleosides and nucleotides of adenosine 3′:5′-cyclic monophosphate (cyclic AMP) in the yeast mycelium transition of *Ceratocystis* (= *Ophiostoma*) *ulmi*. *FEMS Microbiology Letters*, **60**, 49–54.

Brunton, A.H. and Gadd, G.M. (1991) Evidence for an inositol lipid signal pathway in the yeast-mycelium transition of *Ophiostoma ulmi*, the Dutch elm disease fungus. *Mycological Research*, **95**, 484–91.

Calderone, R.A. (1993) Recognition between *Candida albicans* and host cells. *Trends in Microbiology*, **1**, 55–8.

Cano, C., Herrera-Estrella, L. and Ruiz-Herrera, J. (1987) DNA methylation and polyamines in regulation of development of the fungus *Mucor rouxii*. *Journal of Bacteriology*, **170**, 5946–8.

Cano-Canchola, C., Sosa, L., Fonzi, W.A. *et al.* (1992) Developmental regulation of *CUP* gene expression through DNA methylation in *Mucor* spp. *Journal of Bacteriology*, **174**, 362–6.

Cassanova, M., Gil, M.L., Cardenoso, L. *et al.* (1989) Identification of wall-specific antigens synthesized during germ tube formation by *Candida albicans*. *Infection and Immunity*, **57**, 262–71.

Chaffin, W.L. (1984) Site selection for bud and germ tube emergence in *Candida albicans. Journal of General Microbiology*, **130**, 431–40.

Chattaway, F.C., Wheeler, P.R. and O'Reilly, J. (1981) Involvement of adenosine 3′:5′-cyclic monophosphate in the germination of blastospores of *Candida albicans*. *Journal of General Microbiology*, **123**, 233–40.

Chen, J-Y. and Fonzi, W.A. (1992) A temperature-regulated, retrotransposon-like element from *Candida albicans*. *Journal of Bacteriology*, **174**, 5624–32.

Chu, W-S., Rikkerink, E.H.A. and Magee, P.T. (1992) Genetics of white–opaque transition in *Candida albicans*: demonstration of switching recessivity and mapping of switching genes. *Journal of Bacteriology*, **174**, 2951–7.

Cole, G.T. and Sun, S.H. (1985) Arthroconidium–spherule–endospore transformation in *Coccidioides immitis*, in *Fungal Dimorphism*, (ed. P.J. Szaniszlo), Plenum Press, New York, pp. 281–333.

Cole, G.T., Zhu, S., Hsu, L. *et al.* (1992) Isolation and expression of a gene which encodes a wall-associated proteinase of *Coccidioides immitis. Infection and Immunity*, **60**, 416–27.

Cole, G.T., Kruse, D., Seshan, K.R. *et al* (1993) Factors regulating morphogenesis in *Coccidioides immitis*, in *Dimorphic Fungi in Biology and Medicine*, (eds H. Vanden

Bossche, F.C. Odds and D. Kerridge), Plenum Press, New York, pp. 191–212.

Cooper, C.R. (1993) Phase transition in *Wangiella dermatitidis*: identification of cell-division-cycle genes involved in yeast bud emergence, in *Dimorphic Fungi in Biology and Medicine*, (eds H. Vanden Bossche, F.C. Odds and D.Kerridge), Plenum Press, New York, pp. 105–19.

Corner, B.E., Poulter, R.T.M., Shepherd, M.G. and Sullivan, P.A. (1986) A *Candida albicans* mutant impaired in the utilization of *N*-acetyl glucosamine. *Journal of General Microbiology*, **132**, 15–19.

Cutler, J.E. (1991) Putative virulence factors of *Candida albicans*. *Annual Review of Microbiology*, **45**, 187–218.

Cutler, J.E., Glee, P.M. and Horn, H.L. (1988) *Candida albicans* and *Candida stellatoidea*-specific DNA fragment. *Journal of Clinical Microbiology*, **26**, 1720–4.

De Bernardis, F., Adriani, D., Lorenzini, R. *et al* (1993) Filamentous growth and elevated vaginopathic potential of a nongerminative variant of *Candida albicans* expressing low virulence in systemic infection. *Infection and Immunity*, **61**, 1500–8.

Douglas, L.J. (1992) Mannoprotein adhesins of *Candida albicans*, in *New Strategies in Fungal Disease*, (eds J.E. Bennett, R.H. Hay and P.K. Peterson), Churchill Livingstone, Edinburgh, pp. 34–50.

Dranginis, A.M. (1990) Binding of yeast a1 and α2 as a heterodimer to the operator DNA of a haploid-specific gene. *Nature*, **347**, 682–5.

Egidy, G., Paveto, M.C. Passeron, S. and Galvagno, M.A. (1989) Relationship between cyclic adenosine 3':5'-monophosphate and germination in *Candida albicans*. *Experimental Mycology*, **13**, 428–32.

Finney, R., Langtimm, C.J. and Soll, D.R. (1985) The programs of protein synthesis accompanying the establishment of alternative phenotypes in *Candida albicans*. *Mycopathologia*, **91**, 3–15.

Fonzi, W.A. and Irwin, M.Y. (1993) Isogenic strain construction and gene mapping in *Candida albicans*. *Genetics*, **134**, 717–28.

Friedenthal, M., Epstein, A. and Passeron, S. (1974) Effect of potassium cyanide, glucose and anaerobiosis on morphogenesis of *Mucor rouxii*. *Journal of General Microbiology*, **82**, 15–24.

Gadd, G.M. and Brunton, A.H. (1992) Calcium involvement in dimorphism of *Ophiostoma ulmi*, the Dutch elm disease fungus, and characterization of calcium uptake by yeast cells and germ tubes. *Journal of General Microbiology*, **138**, 1561–71.

Geis, P.A. and Jacobs, C.W. (1985) Polymorphism of *Wangiella dermatitidis*, in *Fungal Dimorphism*, (ed P.J. Szaniszlo), Plenum Press, New York, pp. 205–33.

Gillissen, B., Bergemann, J., Sandmann, C. *et al.* (1992) A two-component regulatory system for self/non-self recognition in *Ustilago maydis*. *Cell*, **68**, 647–57.

Gimeno, C.J., Ljungdahl, P.O. and Fink, G.R. (1992) Unipolar cell divisions in the yeast *S. cerevisiae* lead to filamentous growth: regulation by starvation and *RAS*. *Cell*, **68**, 1077–90.

Gooday, G.W. and Adams, D.J. (1993) Sex hormones and fungi. *Advances in Microbial Physiology*, **34**, 69–145.

Gooday, G.W. and Gow, N.A.R. (1983) A model of the hyphal septum of *Candida albicans*. *Experimental Mycology*, **7**, 370–3.

Gow, N.A.R. (1988) Biochemical and biophysical aspects of dimorphism in *Candida albicans*, in *Congress of the X International Society for Human and Animal Mycology – ISHAM*, (ed. J.M. Torres-Rodriguez), J.R. Prous Science, Barcelona, pp. 73–7.

Gow, N.A.R. (1993) Non-chemical signals used for host location and invasion by fungal pathogens. *Trends in Microbiology*, **1**, 45–50.

Gow, N.A.R. and Gooday, G.W. (1982) Growth kinetics and morphology of colonies of the filamentous form of *Candida albicans*. *Journal of General Microbiology*, **128**, 2187–98.

Gow, N.A.R. and Gooday, G.W. (1984) A model for the germination and mycelial growth form of *Candida albicans*. *Sabouraudia*, **22**, 137–42.

Gow, N.A.R., Gooday, G.W., Newsam, R. and Gull, K. (1980) Ultrastructure of the septum of *Candida albicans*. *Current Microbiology*, **4**, 357–9.

Gow, N.A.R., Swoboda, R., Bertram, G. *et al.* (1993) Key genes in the regulation of dimorphism of *Candida albicans*, in *Dimorphic Fungi in Biology and Medicine* (eds H. Vanden Bossche, F.C. Odds and D. Kerridge), Plenum Press, New York, pp. 61–71.

Guthrie, C.G. and Fink, G.R. (eds) (1991) Guide to yeast genetics and molecular biology. *Methods in Enzymology*, **194**.

Hube, B., Turver, C.J., Odds, F.C. *et al.* (1991) Sequence of the *Candida albicans* gene encoding the secretory aspartate proteinase. *Journal of Medical and Veterinary Mycology*, **29**, 129–132.

Iwaguchi, S-I., Homma, M., Chibana, H. and Tanaka, K. (1992) Isolation and characterization of a repeated sequence (RSP1) of *Candida albicans*. *Journal of General Microbiology*, **138**, 1893–900.

Jacobs, C.W., Roberts, R.L. and Szaniszlo, P.J. (1985) Reversal of multicellular-form development in a conditional morphological mutant of the fungus *Wangiella dermatitidis*. *Journal of General Microbiology*, **131**, 1719–28.

Kanetsuna, F., Carbonell, L.M., Moreno, R.E. and Rodriguez, J. (1969) Cell wall composition of the yeast and mycelial forms of *Paracoccidioides brasiliensis*. *Journal of Bacteriology*, **97**, 1046–1.

Kaur, S., Mishra, P. and Prasad, R. (1988), Dimorphism-associated changes in internal pH of *Candida albicans*. *Biochimica et Biophysica Acta*, **972**, 227–82.

Kruse, D. and Cole, G.T. (1992) A seroactive 12-kilodalton β-1,3-glucanase of *Coccidioides immitis* which may participate in spherule morphogenesis. *Infection and Immunity*, **60**, 4350–63.

Kulkarni, R.K. and Nickerson, K.W. (1981) Nutritional control of dimorphism in *Ceratocystis ulmi*. *Experimental Mycology*, **5**, 148–54.

Lasker, B.A., Page, L.S., Lott, T.J. *et al.* (1991). Characterization of CARE-1: *Candida albicans* repetitive element-1. *Gene* **102**, 45–50.

Lasker, B.A., Page, L.S., Lott, T.J. and Kobayashi, G.S. (1992) Isolation, characterization, and sequencing of *Candida albicans* repetetive element 2. *Gene*, **116**, 51–7.

Liu, H., Styles, C.A. and Fink, G.R. (1993) Elements of the yeast pheromone response pathway required for filamentous growth of diploids. *Science*, **262**, 1741–4.

Lodder, J. (ed.) (1970) *The Yeasts: a Taxonomic Study*. Elsevier, North Holland, Amsterdam.

Manning, M. and Mitchell, T.G. (1980) Morphogenesis of *Candida albicans* and cytoplasmic proteins associated with differences in morphology, strain or temperature. *Journal of Bacteriology*, **144**, 258–73.

Maresca, B. and Kobayashi, G.S. (1989) Dimorphism in *Histoplasma capsulatum*: a model for the study of cell differentiation in pathogenic fungi. *Microbiological Reviews*, **53**, 186–209.

Marot-Leblond, A., Robert, R., Aubry, J. Ezcurra, P. and Senet, J-M. (1993) Identification and immunochemical charaterization of a germ tube specific antigen of *Candida albicans*. *FEMS Immunology and Medical Microbiology*, **7**, 175–86.

Martin, M.V., Craig, G.T. and Lamb, D.J. (1984) An investigation of the role of true hypha production in the pathogenesis of experimental oral candidosis. *Journal of Veterinary and Medical Mycology*, **22**, 471–6.

Medoff, G., Sacco, M., Maresca, B. *et al.* (1986) Irreversible block of the mycelial-to-yeast phase transition of *Histoplasma capsulatum*. *Science*, **231**, 476–9.

Medoff, G., Kobayashi, G.S., Painter, A and Travis, S. (1987) Morphogenesis and pathogenicity of *Histoplasma capsulatum*. *Infection and Immunity*, **55**, 1355–88.

Merson-Davies, L.A. and Odds, F.C. (1989) A morphology index for characterization of cell shape in *Candida albicans*. *Journal of General Microbiology*, **135**, 3143–52.

Merson-Davies, L.A. and Odds, F.C. (1992) Expansion of the *Candida albicans* cell envelope in different morphological forms of the fungus. *Journal of General Microbiology*, **138**, 461–6.

Minchiotti, G., Gargano, S. and Maresca, B. (1991) The intron-containing *hsp80* gene of the dimorphic pathogenic fungus *Histoplasma capsulatum* is properly spliced in severe heat shock conditions. *Molecular and Cellular Biology*, **11**, 5624–30.

Morrow, B., Srikantha, T., Anderson, J. and Soll, D.R. (1993) Coordinate regulation of two opaque-phase-specific genes during white-opaque switching in *Candida albicans*. *Infection and Immunity*, **61**, 1823–28.

Muthukumar, G. and Nickerson, K.W. (1984) Ca(II)-calmodulin regulation of morphological commitment in *Ceratocystis ulmi*. *FEMS Microbiology Letters*, **27**, 199–202.

Niimi, M., Niimi, K., Tokunaga, J. and Nakayama, H. (1980) Changes in the cyclic nucleotide levels and dimorphic transition in *Candida albicans*. *Journal of Bacteriology*, **142**, 1010–14.

Odds, F.C. (1988) *Candida and Candidosis*. Ballière Tindall, London.

Orlowski, M. (1991) *Mucor* dimorphism. *Microbiological Reviews*, **55**, 234–58.

Orlowski, M. and Ross, J.F. (1981) Relationship between internal cyclic AMP levels, rates of protein synthesis and *Mucor* dimorphism. *Archives of Microbiology*, **129**, 353–6.

Oujezdsky, K.B., Grove, S.N. and Szaniszlo, P.J. (1973) Morphological and structural changes during yeast-to-mould conversion of *Phialiphora dermatitidis*. *Journal of Bacteriology*, **113**, 468–77.

Patriaca, E.J., Kobayashi, G.S. and Maresca, B. (1992) Mitochondrial activity and heat-shock response during morphogenesis in the pathogenic fungus *Histoplasma capsulatum*. *Biochemistry and Cell Biology*, **70**, 207–14.

Peryra, E., Zaremberg, V. and Moreno, S. (1992) Effect of dibutyrl-cAMP on growth and morphology of germinating *Mucor rouxii* sporangiospores. *Experimental Mycology*, **16**, 93–101.

Ponton, J. and Jones, J.M. (1986) Identification of two germ-tube specific cell wall antigens of *Candida albicans*. *Infection and Immunity*, **54**, 864–8.

Rodrìguez-Del Valle, N, Rosario, M. and Torres-Blasini, G. (1983) Effects of pH, temperature, aeration and carbon source on the development of the mycelial and yeast forms of *Sporothrix schenckii* from conidia. *Mycopathologia*, **82**, 83–8.

Rodrìguez-Del Valle, N., Debs-Elías, N. and Alsina, A. (1984) Effects of caffeine, cyclic 3'; 5' adenosine monophosphate and cyclic 3', 5' guanosine monophosphate in the development of the mycelial form of *Sporothrix schenckii*. *Mycopathologia*, **86**, 29–33.

Roy, B.G. and Datta, A. (1987) A calmodulin inhibitor blocks morphogenesis in *Candida albicans*. *FEMS Microbiology Letters*, **41**, 327–9.

Ruiz-Herrera, J. and Calvo-Mendez, C. (1987) Effect of ornithine decarboxylase inhibitors on the germination of sporangiospores of mucorales. *Experimental Mycology*, **11**, 287–96.

Rustchenko-Bugac, E.P. and Howard, D.H. (1993) Multiple chromosomal and phenotypic changes in spontaneous mutants of *Candida albicans*. *Journal of General Microbiology*, **139**, 1195–207.

Rustchenko-Bulgac, E.P., Sherman, F. and Hicks, J.B. (1990) Chromosomal rearrangments associated with morphological mutants provide a means for genetic variation of *Candida albicans*. *Journal of Bacteriology*, **172**, 1276–83.

Ryley, J.F. and Ryley, N.G. (1990) *Candida albicans* – do mycelia matter? *Journal of Medical and Veterinary Mycology*, **28**, 225–39.

Sabie, F.T. and Gadd, G.M. (1989) Involvement of a Ca^{2+}-calmodulin interaction in the yeast–mycelial transition of *Candida albicans*. *Mycopathologia*, **108**, 47–54.

Sabie, F.T. and Gadd, G.M. (1992) Effect of nucleosides and nucleotides and the relationship between cellular adenosine 3': 5'-cyclic monophosphate (cyclic AMP) and germ tube formation in *Candida albicans*. *Mycopathologia*, **119**, 147–56.

Sadhu, C., McEachern, M., Rustchenko-Bulgac, E.P. *et al.* (1991) Telomeric repeated sequences in *Candida* yeasts and their use in strain identification. *Journal of Bacteriology*, **173**, 842–50.

San-Blas, G. (1985) *Paracoccidioides brasliensis*: cell wall glucans, pathogenicity and dimorphism. *Current Topics in Medical Mycology*, **1**, 235–57.

Santos, M.A.S., Keith, G. and Tuite, M.F. (1993) Non-standard translational events in *Candida albicans* mediated by an unusual seryl-tRNA with a 5'-CAG-3' (leucine) anticodon. *EMBO Journal*, **12**, 607–16.

Scherer, S. and Magee, P.T. (1990) Genetics of *Candida albicans*. *Microbiological Reviews*, **54**, 226–41.

Scherer, S. and Stevens, D.A. (1988) A *Candida albicans* dispersed, repeated gene family and its epidemiologic applications. *Proceedings of the National Academy of Sciences of the USA*, **85**, 1452–6.

Schultz, B., Banuett, F., Dahl, M. *et al.* (1990) The *b* alleles of *U. maydis* whose combinations program pathogenic development, code for polypeptides containing a homeodomain-related motif. *Cell*, **60**, 295–306.

Shepherd, M.G. (1985) Pathogenicity of morphological and auxotrophic mutants of *Candida albicans* in experimental infections. *Infection and Immunity*, **50**, 541–4.

Sherwood, J., Gow, N.A.R., Gooday, G.W. *et al.*, (1992) Contact sensing in *Candida albicans*: a possible aid to epithelial penetration. *Journal of Medical and Veterinary Mycology*, **30**, 461–9.

Simonetti, N. and Strippoli, V. (1973) Pathogenicity of the Y form as compared to M form in experimentally induced *Candida albicans* infections. *Mycopathologia et Mycologia Applicata*, **51**, 19–28.

Sipicki, M., Grallert, B. and Miklos, I. (1993) Mycelial and syncytial growth in *Schizosaccharomyces pombe* induced by novel septation mutants. *Journal of Cell Science*, **104**, 485–93.

Slutsky, B., Buffo, J. and Soll, D.R. (1985) High-frequency switching of colony morphology in *Candida albicans*. *Science*, **230**, 666–9.

Slutsky, B., Staebell, M., Anderson, J. *et al.*, (1987) 'White–opaque transition': a second high-frequency switching system in *Candida albicans*. *Journal of Bacteriology*, **169**, 189–97.

Soll, D.R. (1992) High-frequency switching in *Candida albicans*. *Clinical Microbiology Reviews*, **5**, 183–203.

Soll, D.R., Galask, R., Isley, S. *et al.* (1989) Switching of *Candida albicans* during recurrent episodes of recurrent vaginitis. *Journal of Clinical Microbiology*, **27**, 681–90.

Staebell, M. and Soll, D.R. (1985) Temporal and spatial differences in cell wall expansion during bud and mycelium formation in *Candida albicans*. *Journal of General Microbiology*, **131**, 1467–80.

Stewart, E., Gow, N.A.R. and Bowen, D.V. (1988) Cytoplasmic alkalinization during germ tube formation in *Candida albicans*. *Journal of General Microbiology*, **134**, 1079–87.

Sundstrom, P.M., Tam, M.R., Nicholls, E.J. and Kenny, G.E. (1988) Antigenic differences in the surface mannoproteins of *Candida albicans* as revealed by monoclonal antibodies. *Infection and Immunity*, **56**, 6011–606.

Suzuki, T., Kobayashi, I., Kanbe, T. and Tanaka, K. (1989) High frequency variation of colony morphology and chromosome reorganization in the pathogenic yeast *Candida albicans*. *Journal of General Microbiology*, **135**, 425–34.

Swoboda, R.K., Bertram, G., Hollander, D. *et al.* (1993) Glycolytic enzymes of *Candida albicans* are nonubiquitous immunogens during candidiasis. *Infection and Immunity*, **61**, 4263–71.

Sypherd, P.S., Borgia, P.T and Paznokas, J.L. (1978) The biochemistry of morphogenesis in the fungus *Mucor*. *Advances in Microbial Physiology*, **18**, 67–104.

Szaniszlo, P.J. (ed.) (1985) *Fungal Dimorphism*. Plenum Press, New York.

Torosantucci, A., Angiolella, L. and Cassone, A. (1984) Antimorphogenetic effects of 2-deoxy-D-glucose in *Candida albicans*. *FEMS Microbiology Letters*, **24**, 335–9.

Vanden Bossche, H., Odds, F.C. and Kerridge, D. (eds) (1993) *Dimorphic Fungi in Biology and Medicine*. Plenum Press, New York.

White, T., Miyasaki, S.H. and Agabian, N. (1993) Three distinct secreted aspartyl proteinases in *Candida albicans*. *Journal of Bacteriology*, **175**, 6126–6123.

Wright, R.J., Carne, A., Hieber, A.D. *et al.* (1992) Two genes for secreted aspartate proteinase in *Candida albicans*. *Journal of Bacteriology*, **174**, 7848–53.

Yokoyama, K. and Takeo, K. (1983) Differences of assymetrical division between the pseudomycelial and yeast forms of *Candida albicans* and their effect on multiplication. *Archives of Microbiology*, **134**, 251–3.

Yokoyama, K., Kaji, H., Nishimura, K. and Miyaji, M. (1990) The role of microfilaments and microtubules in apical growth and dimorphism of *Candida albicans*. *Journal of General Microbiology*, **136**, 1067–75.

TISSUE FORMATION

D. Moore
*Microbiology Research Group, Department of Cell and Structural Biology,
School of Biological Sciences, The University of Manchester, Manchester, UK*

20.1 INTRODUCTION

During particular stages in the life history of many fungi, hyphae become differentiated and aggregated to form tissues distinct from the vegetative hyphae which ordinarily compose a mycelium. The mycelium, of course, is a diverse, dynamic population of hyphae which is a fascinating study in its own right (Boddy and Rayner, 1983a,b; Gregory, 1984; Rayner and Webber, 1984; Rayner, 1993; Chapter 2), but this chapter will deal specifically with the patterns which result in formation of defined tissues in multihyphal fungal structures.

20.2 MULTIHYPHAL STRUCTURES

The majority of the macroscopic fungal structures are formed by hyphal aggregation into either linear organs – strands, rhizomorphs and fruit body stipes – or globose masses – sclerotia and the familiar fruit bodies, as well as less-familiar sporulating structures, of the larger Ascomycotina, Deuteromycotina and Basidiomycotina.

20.2.1 Basic vocabulary and basic principles

The general term plectenchyma (Greek *plekein*, to weave, with *enchyma*=infusion, meaning an intimately woven tissue) is used to describe organized fungal tissues. There are two types of plectenchyma: prosenchyma (Greek *pros*, toward and *enchyma*; i.e. approaching or almost a tissue) is, visually, a rather loosely organized tissue in which the components can be seen to be hyphae; and pseudoparenchyma (Greek *pseudo*=false with *parenchyma*= a type of plant tissue) which, as seen in microscope sections, is composed of tightly packed cells resembling plant tissue. In pseudoparenchyma, the hyphae are not immediately obvious as such, though the hyphal nature of the

components can be demonstrated by reconstruction from serial sections or by scanning electron microscopy.

The majority of the lower fungi have coenocytic hyphae, but lower fungi do not form multicellular (multihyphal) structures. Read (1983, 1993) and Read and Becket (1985) have argued for a simple classification of fungal tissues and suggested that the term cellular element should be used in preference to 'cell' because fungal cells are always hyphal compartments and consequently different from the concept of the cell which emerges from elementary biological education. However, a case can be made for using the term 'cell'.

Primary septa in fungal hyphae are formed by a constriction process in which a belt of microfilaments around the hyphal periphery interacts with microvesicles and other membranous cell organelles (Girbardt, 1979). Girbardt (1979) emphasized the correspondence between fungal septation and animal cell cleavage. The completed septum has a pore which may be elaborated with the parenthesome apparatus in most Basidiomycetes or be associated with Woronin bodies in Ascomycetes; in either case the movement or migration of cytoplasmic components between neighbouring compartments is under effective control. So, the cellular nature of the hypha extends, at least, to its being separated into compartments whose interactions are carefully regulated and which can exhibit contrasting patterns of differentiation. Further, Griffin *et al.* (1974) pointed out that, by increasing the number of growing tips, mycelial branching is the equivalent of cell division, and kinetic analyses (Trinci, 1974, 1984) show clearly that fungal filamentous growth can be interpreted on the basis of a regular cell cycle. Thus, fungi can quite reasonably be considered to be cellular organisms producing differentiated tissues

The Growing Fungus. Edited by Neil A.R. Gow and Geoffrey M. Gadd. Published in 1994 by Chapman & Hall, London. ISBN 0 412 46600 7

composed of cells which are the progeny of an initial cell or cell population which is induced to start multiplication and differentiation. But there are fundamental differences between this cell concept and one which might be applied to plants; one is the way in which proliferation occurs and the other is the nuclear ploidy-cytoplasm relationship.

Plants, animals and fungi are distinct eukaryotic Kingdoms (Whittaker, 1969; Margulis, 1974; Cavalier-Smith, 1981) which are thought to have separated at some protistan level prior to the establishment of the multicellular grade of organization in any of them. These three Kingdoms are very different from one another in ways that are critical to determining the morphology of multicellular structures as well as in their nutrition (animals engulf, plants use radiant energy, fungi absorb) which was part of the original definition of the Kingdoms (Whittaker, 1969). Among the crucial evolutionary steps leading to organized multicellularity were probably the development of mechanisms for dividing a cell, together with a mechanism for controlling the placement of the plane of cell division in particular relation to the orientation of nuclear division. A key characteristic of embryology throughout the animal Kingdom is the movement of cells and cell populations. Plant cells have little scope for movement and their morphogenesis depends upon control of the orientation and position of the daughter cell wall which forms at the equator of the mitotic division spindle. Fungi are also encased in walls but their basic structural unit, the hypha, exhibits two features which cause fungal morphogenesis to be totally different from plant morphogenesis. These are that hyphae grow only at their tips and that cross walls form only at right angles to the long axis of the hypha. The consequence is that fungal morphogenesis depends on the placement of hyphal branches. To proliferate a hypha must branch, and to form a structure the position of branch emergence and its direction of growth must be controlled.

A relevant contrast is the development of the protonema following fern spore germination. Germination produces a uniseriate filament of cells – very similar to outgrowth of a hyphal germ tube from a fungal spore. Eventually, and usually in the apical cell, the division plane becomes reoriented so that the new cell walls are formed obliquely or parallel to the long axis of the filament and a flat plate of cells (the gametophyte prothallus) is formed (Miller, 1980; and Figure 20.1). This transition epitomizes the importance of the mitotic orientation in plant morphogenesis and serves to emphasize how totally different is the fungal approach to solution of the same problem. Crosswalls in fungal hyphae are formed at right angles to the long axis of the hypha; the only exceptions are in cases of injury or in hyphal tips already differentiated to form sporing structures. Hyphal tip cells are not subdivided by oblique cross-walls, nor by longitudinally oriented ones. Even in fission yeasts producing irregular septation patterns under experimental manipulation, the plane of the septum is always perpendicular to the longest axis of the cell (Miyata *et al.*, 1986). In general, then, the characteristic fungal response to the need to convert the one-dimensional hypha into a two-dimensional plate or three-dimensional block is the formation of lateral branches (Figure 20.1).

Thus, the regulatory target in fungal morphogenesis is the machinery, presently completely unknown, which is involved in generating a new apical growth centre (to become the hyphal tip of the new lateral branch) and the determination of its position, orientation and direction of outgrowth from the parent hypha. Origin of the branch seems to be the formal equivalent of determination of morphogenetic growth by orienting the plane of division and the new crosswall, as occurs in plants, and directional growth of the new hyphal apex has much in common with the morphogenetic cell migrations that contribute to development of body form and structure in animals. Septation in the main hypha is in some way defined by the position of the dividing nucleus (Talbot, 1968; Girbardt, 1979), but, for branch formation, there does not seem to be any dependence on orientation of the nuclear division spindle. Cytoplasmic vesicles are crucial to the growth of hyphal apices (Grove, 1978; Bartnicki-Garcia, 1990) and distribution of microvesicles may be closely connected with branch initiation. During initial outgrowth of clamp connections of *Schizophyllum commune* vesicles are displaced in the direction of the curvature of the clamp cell soon after its emergence (Todd and Aylmore, 1985) and localized accumulation of microvesicles may be a cause of branch intiation (Trinci, 1978). Differential ion fluxes could direct this. Applied electrical

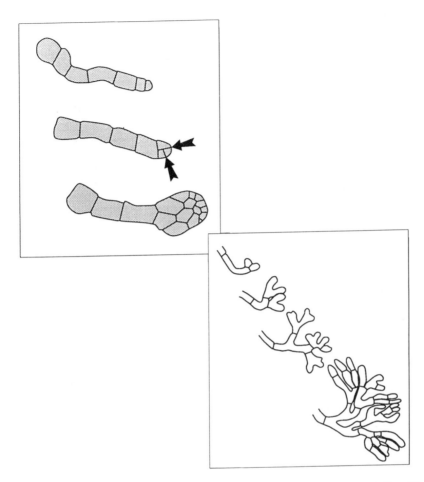

Figure 20.1 Transforming a filament into a two-dimensional plate. A comparison of the cell proliferation strategy employed by a fern (top) and a filamentous fungus (bottom). Top panel: change in the orientation of the plane of division in the apical cell (arrows) of the protonema of the fern *Onoclea sensibilis* converts it into a meristematic cell which can give rise to the planar gametophyte by further regulation of the mitotic division plane (drawings made from photographic illustrations in Miller, 1980). Bottom panel: orientation of mitotic division spindles is irrelevant in the fungus *Botrytis allii* as cross-walls are always formed at right angles to the long axis of the hypha. A two-dimensional plate can only be formed by controlling the position, orientation and direction of growth of new hyphal tips formed as lateral branches (redrawn from Townsend and Willetts, 1954).

fields affect the site of branch formation and the direction of hyphal growth in young mycelia of several fungi (McGillivray and Gow, 1986). Studies with *Achlya* (Kropf *et al.*, 1983, 1984) revealed an ion current caused by influx of protons at the hyphal tip (as an amino acid symport in this organism) and that a new zone of proton influx often preceded and predicted the emergence of a branch. However, the causality of this process in branching is doubtful as it seems to be more related to nutrient uptake than tip growth (Harold and Caldwell, 1990; de Silva *et al.*, 1992).

The information we have is sparse, and derives from experiments with vegetative hyphae where the connections between nuclear division, cytokinesis and branch formation may well be relaxed. In the pseudoparenchymatous (or prosenchymatous) 'generative' tissues which precede final differentiation in developing fungal multicellular structures the constituent cells are generally smaller and less vacuolated than typical hyphal cells. Small size and dense cytoplasmic content are also often associated with rapidly dividing cells in animals and plants. In the fungi, a consequence of rapid karyogamy and frequent branching might be that a much closer correlation is maintained between nuclear/cell division and branch formation. However, the higher fungi do seem to have looser connection between cell differentiation and nuclear number and ploidy than is usual in plants and animals. The cells of *Agaricus* mycelia have 6–20 nuclei per cell (Colson, 1935; Kligman, 1943) and cells of the mushroom fruit body have an average of six nuclei (Evans, 1959). Conversely, in a basidiomycete with the most classically regular vegetative dikaryon, *Coprinus cinereus*, cells of the fruit body stipe can become multinucleate by a series of consecutive conjugate divisions, a peculiarity exhibited by other agarics (Stephenson and Gooday, 1984; Gooday, 1985). As *Armillaria* species have diploid tissues in the fruit body, it is clear that the ploidy level and the number of nuclei are both variable, being controlled by factors other than those imposed by the need to assemble multicellular structures.

Although the orientation of the branch tip decides the initial direction of the new growth, the hyphal tip is an invasive, migratory structure. Its direction of growth after initial emergence must be under precise control as it determines the nature and relationships of the cells the hypha will form. This is clearly seen in hymenial layers which are constructed from branches of determinate growth in a precise spatial array (Figure 20.9), in the behaviour of binding hyphae in fruit body stipes (Williams *et al.*, 1985), tendril hyphae in mycelial strands (Butler, 1958; and Figure 20.2) and generally in structures constructed from closely appressed axially arranged hyphae such as strands (Figure 20.2), rhizomorphs (Figure 20.3) synnemata (Figure 20.4), and necks of ascomata (Rayner *et al.*, 1985; Read, 1993). Very little is known about the mechanisms involved but since even the most open fungal tissues appeaar to be filled with an extracellular mucilaginous material, reactions akin to the cell-matrix interactions of animal tissues may guide and coordinate hyphal growth in tissues. A highly hydrated extracellular matrix, composed predominantly of glucans, fills the interhyphal spaces within sclerotia (Willetts and Bullock, 1992), similar to that found in fruit

bodies (Williams *et al.*, 1985) and rhizomorphs (Rayner *et al.*, 1985) and may be important in providing an environment in which morphogenetic control agents can interact. Different tissues may synthesize and secrete specific polysaccharides and/or glycoproteins to provide a local environment within which hyphae and growth control factors interact with the specificity associated with the notion of 'control by context' where the response of the cell to a growth factor is influenced by the extracellular matrix within which the interaction takes place (Nathan and Sporn, 1991). In animal systems it is becoming increasingly apparent that the extracellular matrix regulates transcription directly, probably through integrin-mediated signalling (Damsky and Werb, 1992; Hynes, 1992; Streuli, 1993). If applicable to fungi, this raises the possibility of a hypha directly influencing the gene expression of its neighbours through the extracellular matrix molecules it secretes.

20.2.2 Linear organs: strands, cords, rhizomorphs and stipes

Formation of parallel aggregates of morphologically similar hyphae is common among Basidiomycotina, Ascomycotina and Deuteromycotina. Mycelial strands and cords provide the main translocation routes of the mycelium, developing under circumstances which require large-scale movement of nutrients (including water) to and from particular sites. They are formed in mushroom cultures to channel nutrients towards developing fruit bodies; Mathew (1961) described their development in the cultivated mushroom *Agaricus bisporus*. For a similar function, they are also formed by mycorrhizal fungi to radiate into the soil, greatly supplementing the host plant's root system and gathering nutrients for the host (Read *et al.*, 1989; Read, 1991). Although mycelial strands extend into the soil, the mycelium does not aggregate into the pseudoparenchymatous tissue of the mycorrhizal sheath either in the soil or *in vitro*, this tissue development requires a surface, oxygen and a supply of nutrients (Read and Armstrong, 1972).

In saprotrophic phases, strands are also migratory organs, extending from an existing food base to explore nutrient-poor surrounding for new nutrient sources. Strands of *Serpula lacrimans*, the

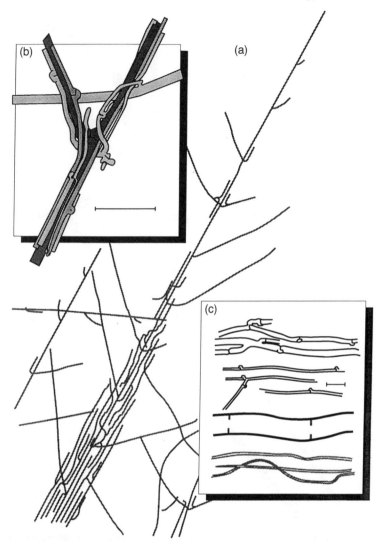

Figure 20.2 Hyphal strands of *Serpula lacrymans*. Strands originate when branches of a leading hypha form at an acute angle to grow parallel to the parent hypha which also tends to grow alongside other hyphae it may encounter. Anastomoses between the hyphae of the strands consolidate them and narrow hyphal branches ('tendril' hyphae) from the older regions of the main hyphae grow around the main hyphae and ensheath them. (a) Shows the strand in a general habit sketch and (b) shows tendril hyphae intertwined around main hyphae (redrawn after Butler, 1958, 1966). (c) Shows some of the cell types encountered in strands, with undifferentiated hyphae at the top then tendril hyphae, a vessel hypha and fibre hyphae (redrawn after Jennings and Bravery, 1991), scale bar = 20 μm.

dry-rot fungus, are able to penetrate several metres of brick-work from a food base in decaying wood (Butler, 1957, 1958; Watkinson, 1971; Jennings and Watkinson, 1982) and to overgrow Perspex and many building materials (Jennings, 1991). The strands hasten capture of new substrate by increasing the inoculum potential of the fungus at the point of contact with it (Garrett, 1954, 1956, 1960, 1970) but they also facilitate concentration of mycelial resources on capture and consolidation of the new food base by providing translocation routes in both directions. The distribution of strands around a food base changes with time (Thompson, 1984). By resorption of hyphae and

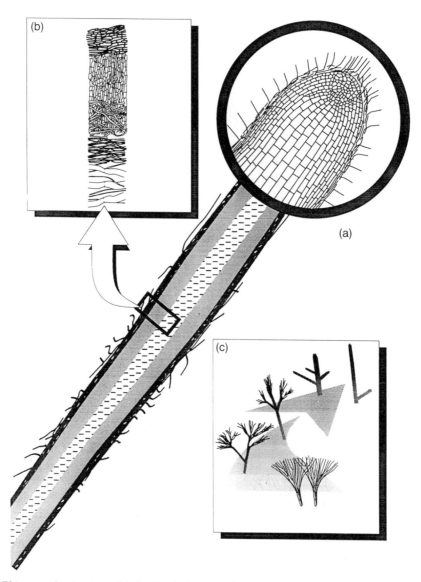

Figure 20.3 Rhizomorph structure. (a) Sectional drawing showing general structure, with the apical region magnified to show the appearance of a growing point of tightly packed cells (redrawn after de Bary, 1887). Behind the tip is a medullary zone containing swollen, vacuolated and often multinucleate cells surrounded by copious air- or mucilage-filled spaces. The medullary region forms a central channel through the rhizomorph and, in mature tissues, is traversed by narrow fibre hyphae and wide-diameter vessel hyphae, the microscopic appearance being indicated in (b) (redrawn after Webster, 1980). (c) Shows mycelial fans, strands, cords and rhizomorphs as a series showing increasing apical dominance (redrawn after Rayner *et al.*, 1985a).

redistribution of the nutrients so recovered the strands enable migration of the colony from place to place (Rayner *et al.*, 1985a,b; Boddy, 1993; Rayner, 1994; Chapter 2).

Although mycelial strands contain morphologically differentiated hyphae (see below), their constituent hyphae are relatively loosely aggregated. Certain fungi produce highly differentiated

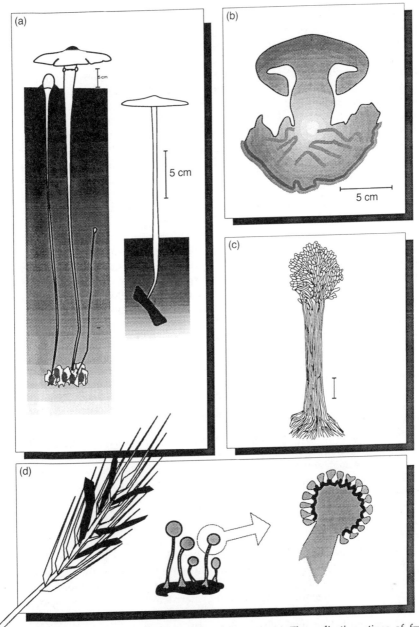

Figure 20.4 Relationships between fungal multicellular structures. (a) The radicating stipes of fruit bodies of *Termitomyces* spp. (left), connected by their pseudorhizas to the fungal galleries of an abandoned termitarium many feet below, and of *Oudemansiella* (= *Xerula*) *radicata*, where the pseudorhiza extends from tree roots or buried wood (scale bar = 50 mm; redrawn from Ingold, 1979). (b) Diagram showing the giant sclerotium of *Polyporus mylittae* germinating to form the fruit body (scale bar = 50 mm; redrawn from photographs in Macfarlane *et al.*, 1978). (c) The synnemata (bunched conidiophores) of *Podosporium elongatum* (scale bar = 50 μm; redrawn after Chen and Tzean, 1993). (d) *Claviceps purpurea* which transforms the ovaries of its host (grasses and cereals, especially rye) into a hard, blackish, banana-shaped sclerotium (the ergot) which overwinters on the ground and the following year gives rise to pinkish, drumstick-like perithecial stromata which contain the flask-shaped perithecia, within which the asci are formed (redrawn after Burnett, 1968).

aggregations of hyphae with well-developed tissues (Figure 20.3). These structures are very root-like in appearance and are called rhizomorphs. A prime example is *Armillaria mellea*, a pathogen of trees and shrubs, which spreads from one root system to another by means of its rhizomorphs (Rishbeth, 1985). Here, again, the structure serves translocatory and migratory functions and, as with strands, translocation is bidirectional, glucose being translocated towards and away from the apex simultaneously (Granlund *et al.*, 1985). In moist tropical forests aerial rhizomorphs, mainly of *Marasmius* spp., form a network which intercepts and traps freshly fallen leaves, forming a suspended litter layer (Hedger, 1985; Hedger *et al.*, 1993).

Mycelial strands originate when young branches adhere to, and grow over, an older leading hypha (Figure 20.2). Further localized growth and incorporation of other hyphae it may meet leads to increase in size of the strand (Nuss *et al.*, 1991). Anastomoses between the hyphae of the strands consolidates them and narrow hyphal branches ('tendril' hyphae) from the older regions of the main hyphae intertwine around the other hyphae (Figure 20.2). From the beginning, some of the central hyphae may be wide-diameter, thin walled so-called vessel hyphae and in older strands narrow, but thick-walled, 'fibre' hyphae appear, running longitudinally through the mature strands. Strand formation occurs in ageing mycelium on an exhausted substrate when the hyphae are likely to be the main repositories of nutrients (especially nitrogen) and it has been argued that stranding results from the limitation of new growth to the immediate vicinity of the remaining nutrient (Watkinson, 1975, 1979). As long as the strand is the main supplier of nutrient the integrity of the strand will be reinforced, but when the strand encounters an external source greater than its own endogenous supply the stimulus to cohesive growth will be lost and spreading, invasive, hyphal growth will envelop the new substrate.

Rhizomorphs differ from strands fundamentally by having a highly organized apical growing point and extreme apical dominance. The apical region of the rhizomorph contains a compact growing point of tightly packed cells, protected by a cap of intertwined hyphae in (and producing) a mucilaginous matrix. Behind is a medullary zone containing vessel-hyphae composed of swollen, vacuolated and often multinucleate cells surrounded by copious air- or mucilage-filled spaces. The medullary region forms a central channel through the rhizomorph and, in mature tissues, is traversed by narrow-diameter, thick-walled fibre hyphae (Figure 20.3) (Townsend, 1954; Motta, 1969, 1971; Botton and Dexheimer, 1977; Motta and Peabody, 1982; Powell and Rayner, 1983; Cairney *et al.*, 1989; Cairney and Clipson, 1991). Towards the periphery of the rhizomorph, the cells are smaller, darker, and thicker-walled, and there is a fringing mycelium extending outwards between the outer layers of the rhizomorph, resembling the root-hair zone in a plant root. The similarity, at least in microscope sections, with the plant root has prompted the suggestion that rhizomorph extension results from meristematic activity (Motta, 1967; 1969, 1971; Motta and Peabody, 1982). However, ultrastructural (especially scanning electron microscopical) observations reveal the hyphal structure of the rhizomorph tip (Botton and Dexheimer, 1977; Powell and Rayner, 1983; Rayner *et al.*, 1985a). A meristem-like structure would be totally alien to the growth strategy of the fungal hypha (Figure 20.1) and, as suggested by Rayner *et al.* (1985a) the impression of central apical initials giving rise to axially arranged tissues is undoubtedly an artefact caused by sectioning compact aggregations of parallel hyphae.

The rhizomorph apex is not unique in this: tissue layers involved in rapid cell formation in which the hyphae run parallel to one another have been recognized in agaric fruit bodies. They frequently demarcate the major tissue layers of the fruit body and were called meristemoids by Reijnders (1977). However, Reijnders was careful to emphasize how they differed from true meristems: 'These meristemoids closely resemble the meristems of the phanerogams; sometimes they are referred to as such. This is not correct because the meristemoids initiate from hyphae which together form a simple tissue; cell division, therefore, can only take place in one direction, and the cell walls between the cells of different hyphae are double' (Reijnders, 1977). Some of the Deuteromycotina are said to have 'meristem arthrospores' or 'meristem blastospores' (Hughes, 1953), but this is, again, an unfortunate misuse of the word. Hughes (1971) compared cell proliferation in fungi with some lower plants and concluded that '. . . septation of fungal conidia results, not from the activity of a single [i.e. meristematic] cell, but by division of any or all of the cells. An apparent dictyoseptate condition of some conidia or other reproductive units in fungi may arise from the compacting of a coiled septate hypha or of hyphae which may branch repeatedly to form a more or less solid mass of cells.' It is quite clear that meristems do not occur in fungi.

The development of rhizomorphs *in vitro* was described by Garrett (1953, 1970) and Snider (1959) but details of their inception are sparse. They have arisen as protuberances on mounded mycelial aggregates of strains of *Stereum hirsutum*, a fungus which does not normally produce rhizomorphs (Rayner *et al.*, 1985a). Usually, rhizomorphs are initiated as compact masses of aggregated cells the ultimate origin being ascribed to locally enhanced acute-angled branching of some marginal hyphae in a mycelium; a phenomenon described as 'point-growth' (Coggins *et al.*, 1980). Thus it seems likely that these linear organs originate from originally unpolarized hyphal aggregations which somehow become apically polarized.

Mycelial strands and rhizomorphs are extremes in a range of hyphal linear aggregations. Many intergrading forms can be recognized and have been given particular names (Townsend, 1954; Butler, 1966; Garrett, 1970) which have value in a descriptive sense, enabling distinctions to be made between species and their life style strategies, but different names should not obscure the close developmental relationships which exist between the structures. By all means call a hand a hand and a flipper a flipper; but recognize that a human hand and a dolphin flipper are variations on the same theme. Rayner *et al.* (1985a) suggested that all linear hyphal aggregations could be related together in a hierachy depending on apical dominance (Figure 20.3).

Fruit bodies should also be included in this arrangement. In describing the structure of litter-trapping rhizomorph networks in moist tropical forests, Hedger *et al.* (1993) showed that the rhizomorphs have a reduced fruit body pileus at their tips which may protect the apex against desiccation. Overall, these linear organs are functionally analogous to soil rhizomorphs, but developmentally analogous to indefinitely extending fruit body stipes (Jacques-Félix, 1967). Many fruit

bodies are served by radiating strands which convey nutrients towards the fruiting structure. In cases where the fruit body is stipitate (i.e. it has a stipe (=stem)) these can be so highly developed that the junction between strand and stipe is obscure. The term 'radicating' is used to describe fruit bodies whose stipes are elongated into root-like pseudorhizas which extend to the surface from some buried substrate (Figure 20.4). Even in species which do not normally produce pseudo-rhizas, they can be induced by keeping fruiting cultures in darkness (Buller, 1924) whereupon the stipe base can extend for many centimetres, driving the fruit body primordium on its tip towards any source of light. Rhizomorphs? Pseudorhizas? Extending stipes? What they are called is less important than the implication that a close morphogenetic relationship underlies all fungal linear hyphal aggregations.

20.2.3 Globose structures: sclerotia, stromata and fruit bodies

Sclerotia are pseudoparenchymatous hyphal aggregations in which concentric zones of tissue form an outer rind and inner medulla, with a cortex sometimes distinguishable between them. Sclerotia are tuber-like and detach from their parental mycelium at maturity. Some sclerotia consist of very few cells and are therefore of microscopic dimensions. At the other extreme, the sclerotium of *Polyporus mylittae*, found in the deserts of Australia, can reach 20–35 cm in diameter and is known as native or blackfellow's bread.

Sclerotia are resistant survival structures which pass through a period of dormancy before utilizing accumulated reserves to 'germinate', often by producing fruiting structures (Figure 20.4). Dormant sclerotia may survive for several years (Sussman, 1968; Coley-Smith and Cooke, 1971; Willets, 1971), owing their resistance to the rind being composed of tightly-packed hyphal tips which become thick-walled and pigmented (mela-nized) to form an impervious surface layer. The medulla forms the bulk of the sclerotium, and its cells (and those of the cortex where present) may accumulate reserves of glycogen, polyphosphate, protein and lipid.

Sclerotium ontogeny comprises initiation, when the hyphae begin to aggregate to form small, distinct initials; development, when the initials expand and grow to full size, accumulating nutritional reserves from the parent mycelium; and **maturation**, which is most obviously characterized by clear demarcation of the surface and pigmentation of its constituent cell walls, but which also involves conversion of the reserve nutrients to forms suitable for long-term storage (Chet and Henis, 1975). Townsend and Willets (1954) and Willetts and Wong (1971) distinguished several kinds of development in sclerotia. In the loose type (as in *Rhizoctonia solani*) sclerotial initials arise by branching and septation of hyphae; the cells become inflated and fill with dense contents and numerous vacuoles. The mature sclerotium is pseudoparenchymatous, but the tissue has an open structure and its hyphal nature is readily seen. At the periphery of this type of sclerotium the hyphae are more loosely arranged and generally lack thickened walls (Willetts, 1969). The terminal type of development is characterized by repeated dichotomous branching and cross-wall formation. It is exemplified by *Botrytis cinerea* and *B. allii* and an illustration of it is used in Figure 20.1 as an example of the fungal space-filling strategy. Eventually the hyphal branches cohere to give the appearance of a solid tissue. In *Botrytis*, a (usually flattened) mature sclerotium may be about 10 mm long, 3–5 mm wide and 1–3 mm thick. It is differentiated into a rind composed of several layers of round cells with thickened, pigmented walls, a narrow cortex of thin-walled pseudoparenchymatous cells with dense contents, and a medulla of loosely arranged filaments. The lateral type is illustrated by *Sclerotinia gladioli* (which causes dry rot of corms of *Gladiolus*, *Crocus* and other plants). Sclerotial initials arise by formation of numerous side branches from one or more main hypha or strands of several parallel hyphae. The mature sclerotium, about 0.1–0.3 mm in diameter, is differentiated into a rind of small thick-walled cells and a medulla of large thin-walled cells.

More complex sclerotia and other types of sclerotial development have been found (Butler, 1966; Chet *et al.*, 1969). Many fungi enclose portions of the substrate and/or substratum (which may include host cells if the fungus is a pathogen) within a layer of pigmented, thick-walled cells; the whole structure may be regarded as a kind of sclerotium. Such a layer of impervious tissue, which is especially protective against desication,

may be developed in other circumstances to form what has been called a pseudosclerotial plate, for example, over the surface of hyphal structures of the fungus *Hymenochaete corrugata* binding hazel branches together (Ainsworth and Rayner, 1990) and in similar adhesions between rhizomorphs and leaf litter in tropical forests (Hedger *et al.*, 1993). When such plates completely enclose the mass on which they are formed, the structure which results is called a sclerotium. The term sclerotium is, therefore, a functional one. The structure is defined as a multihyphal aggregate which can remain dormant or quiescent when the environment is adverse and then, when conditions improve, germinate to reproduce the fungus (Willetts and Bullock, 1992). Since a number of dissimilar structures are encompassed by this definition it is thought that the different forms arose by convergent evolution, most probably evolving, in Ascomycotina, from aborted spore-forming organs like perithecia, cleistothecia and conidial masses (Willetts, 1972; Cooke, 1983; Willetts and Bullock, 1992). Their close relationship with sporulating organs is shown in many ways. Inhibition of conidial differentiation in *Monilinia fructicola* by high humidity occasionally leads to formation of sclerotium-like stromata (Willetts, 1968; Willetts and Calonge, 1969). This is an example of the environment directly influencing the pathway of development, but another example provides evidence for the same genes being involved in both sclerotium and fruit body initiation in the basidiomycete *Coprinus cinereus* (Moore, 1981; and section 20.4.1 and Figure 20.17). Sclerotia of *Coprinus cinereus* are polymorphic (Hereward and Moore, 1979) and strains which were either unable to make sclerotia or made abnormal sclerotia were shown to result from single gene recessive defects. In crosses with wild type strains the 'heterozygous' dikaryons were able to make both sclerotia and fruit bodies, but dikaryons 'homozygous' for sclerotium-defective genes were unable to make fruit bodies. The conclusion was that a common initiation pathway gave rise to hyphal aggregations which, under permissive environmental conditions (22–26°C plus illumination), developed axial symmetry and became fruit body initials, but under non-permissive conditions (30–37°C plus continuous darkness) developed radial symmetry and became sclerotia.

Sclerotia may 'germinate' to form mycelium, conidia, ascomata or basidiomata. Mode of germination seems in some cases to be a matter of size. The small sclerotia of *Coprinus cinereus* germinate to produce a mycelium but many other species produce spores and fruit bodies (Figure 20.4). The giant sclerotium of *Polyporus mylittae* can form a basidiodome (Figure 20.4) without being supplied with water as the flesh is honeycombed with blocks of translucent tissue where the hyphae form copious amounts of an extrahyphal gel which is thought to serve as both nutrient and water store (Macfarlane *et al.*, 1978).

In many Ascomycotina, Basidiomycotina and Deuteromycotina hyphae may aggregate to form fruiting structures which are responsible for producing and, just as important, distributing spores of various kinds. In Ascomycotina, the sexually produced ascospores are contained in asci (singular: ascus) enclosed in an aggregation of hyphae termed an ascoma. Ascomata are formed from non-dikaryotic sterile hyphae surrounding the ascogonical hyphae of the centrum; a number of distinct types can be recognized (Booth, 1966; Turian, 1978; Reynolds, 1981; Chadefaud, 1982 a,b,c; Read, 1993) (Figure 20.5). The fruit-bodies of Basidiomycotina, the mushrooms, toadstools, bracket fungi, puff-balls, stinkhorns, bird's nest fungi, etc., are all examples of basidiomata which bear the sexually produced basidiospores on basidia. Simplified diagrammatic drawings of some of the different types of ascomata and basidiomata are shown in Figures 20.5 and 20.6 and the rest of this chapter will concentrate on the establishment of the patterns which result in formation of defined tissues in multicellular fungal structures. Many more details could be added but, in most cases, it is impossible to give much further information about possible mechanisms involved in defining morphogenesis of the structures, because of the lack of appropriately stuctured research. The great bulk of the published research has been done with taxonomic intentions. It has great value for its descriptive and comparative content, but precise developmental accounts are extremely rare and experimental approaches rarer still. Most work has been done on development of mushroom fruit bodies, particularly with *Coprinus cinereus* and other *Coprinus* spp., so the main theme here will be morphogenesis of the basidiome of *C. cinereus* with emphasis on how tissue patterns may

arise, how the patterns contribute to morphogenesis and how they are expressed in morphological and biochemical differentiation.

20.3   MORPHOGENETIC PATTERNS

Development of any multicellular structure in fungi requires modification of the normal invasive growth of vegetative mycelium so that hyphae no longer characteristically diverge, but grow towards one another to cooperate in forming the differentiating organ. Although we are beginning to understand the kinetics of growth and branching patterns (Prosser, 1993) we remain ignorant of the control processes responsible for changing the fundamental growth pattern of the hyphae. The sex hormone of lower aquatic and terrestrial fungi and molecules determining mating-type specific agglutination in yeast (Beavan *et al.*, 1979; Rahary *et al.*, 1985; Kihn *et al.*, 1988) are about the limit of current knowledge about the control of hyphal interactions (Moore, 1984a). As pointed out before (Reijnders and Moore, 1985), fungi offer a unique system for study of cell-to-cell tropisms and specific cell-to-cell adhesion since the change from one state to another is part of their normal development. Once the prosenchymal mycelial tuft (constituting the initial of the developing structure) is established, major tissue domains are demarcated very quickly. In *Coprinus cinereus*, fruit body initials only 800 μm tall are clearly differentiated into pileus and stipe (Moore *et al.*, 1979) though this is only 1% of the size of a mature fruit body (Figure 20.7). This sort of example brings to mind establishment of the basic body plan very early in development of the animal embryo and this is worth exploration. The components in the overall process of animal embryo development have been characterized as: formation of inhomogeneous cell populations from homogeneous ones; regional specification of tissues (pattern formation) directed by organizers producing morphogens; specification and commitment of particular cells to particular fates; cell differentiation; and regulation of gene activity in ways specifically geared to morphogenesis (Slack, 1983). These statements highlight the major events contributing to morphogenesis in both animals and plants. The challenge is whether evidence exists for such mechanisms in the development of fungal structures.

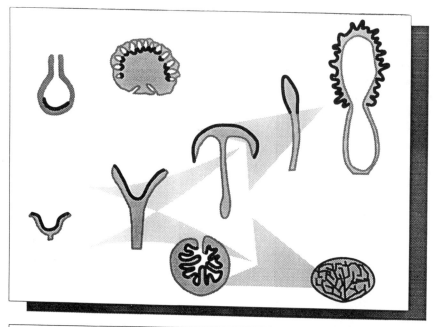

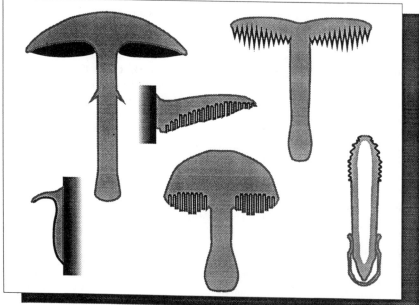

Figure 20.5   The variety of multicellular fruiting bodies of fungi in the form of simplified diagrammatic sectional drawings (redrawn after Burnett, 1968); in each case the hymenial tissue is represented by the black line. The upper panel shows a range of ascomata with, at top left a basic flask-like perithecium (e.g. *Sordaria*) alongside the perithecial stroma of *Daldinia*. The rest of the drawings are arranged to show how the simple cup-like ascome of *Peziza* might, hypothetically, have given rise, on the one hand, to the morel (*Morchella*) via the fruit body forms of *Sarcoscypha*, *Helvella* and *Mitrula*, and on the other hand to the subterranean fruit body of *Tuber* via a form like *Genea*. The lower panel shows a number of basidiomata, including the agaric (mushroom) form of fruit body, the poroid bracket and toadstool polypore, and toothed (hydnoid) form. At bottom left is an encrusting fruit body (as in *Stereum* and *Phellinus*) and at bottom right the mature stink-horn (*Phallus*).

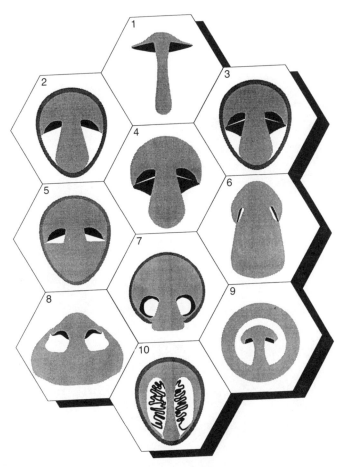

Figure 20.6   Ten ways to make a mushroom. A montage of diagrammatic sections illustrating the various primordial tissue patterns which eventually mature to form mushroom-like basidiomata; hymenial tissues are shown in black (redrawn after Watling and Moore, 1993). 1, gymnocarpic, where the hymenium is naked at first appearance and develops to maturity on the fruit body surface; 2, monovelangiocarpic, with a single (universal) veil enveloping the whole primordium; 3, bivelangiocarpic, in which an inner (partial) veil provides additional protection to the hymenium; 4, paravelangiocarpic, where the veil is reduced and often lost at maturity; 5, metavelangiocarpic, where a union of secondary tissues emerging from the pileus and/or stipe forms an analogue of the universal veil; 6, gymnovelangiocarpic, in which the hymenium is protected by a very reduced veil, seen only at adolescence, formed between the stipe and the closely applied pileus; 7 pilangiocarpic, the hymenium is protected by tissue extending downwards from the margin of the pileus; 8, stipitoangiocarpic, the hymenium is protected by tissue extending upwards from the stipe base, but this does not enclose the primordium; 9, bulbangiocarpic, where the tissue protecting the hymenium is largely derived from the basal bulb of the stipe and initially completely encloses the primordium; 10, endocarpic, where the mature hymenium is enclosed or covered over, just one (the pileate type) of a number of patterns of this gasteromycetous form of fruit body is shown.

20.3.1   Regional specification of tissue domains

Initials of multihyphal structures are composed of a mass of prosenchymal tissue which originates from communities of hyphae and their branches which grow together. This is clearly the case for strand formation (Butler, 1958; Nuss *et al.*, 1991) and initiation of fruit bodies and sclerotia involves aggregation by hyphal congregation (Matthews and Niederpruem, 1972; Waters *et al.*, 1975a; Van der Valk and Marchant, 1978). A range of observations show that fungal fruit bodies consist of a

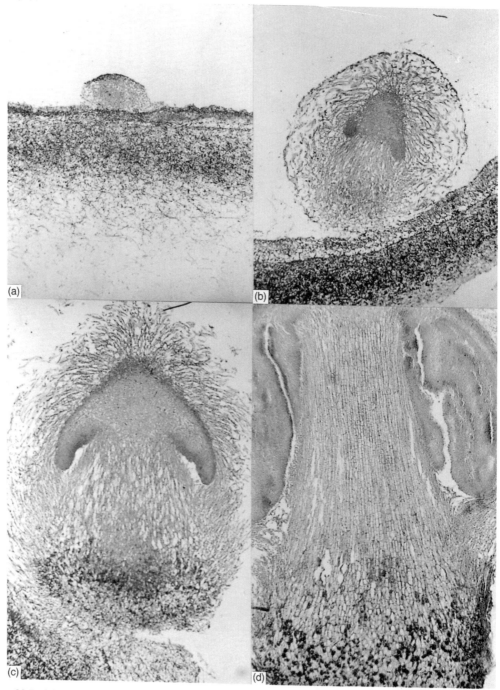

Figure 20.7   Micrographs of median vertical sections of basidiome initials and primordia of *Coprinus cinereus*. (a) The initial is approx. 0.2 mm tall, showing hyphal aggregation but no internal pattern.(b) Shows a 0.8 mm tall initial which is clearly demarcated into veil, pileus and stipe tissues. (c) Shows a 1.2 mm basidiome initial and (d) a 3 mm tall primordium.

population of cells assembled from contributions of a number of cooperating hyphal systems. Hyphal congregation is so fundamental that it can lead to the formation of chimeric fruit bodies. Kemp (1977) described fruit bodies of *Coprinus* consisting of two different species, *C. miser* and *C. pellucidus*. The hymenium comprised a mixed population of basidia bearing the distinctive spores of the two species but the chimera extended throughout the fruit body as both species could be recovered by outgrowth from stipe segments incubated on nutrient medium.

Development of the basidiome has been a rich source of variety in terminology. Some of this can be traced back to the last century (Brefeld, 1877; de Bary, 1884), but in recent years Reijnders (1948, 1963, 1979) has been instrumental in formalizing the descriptive terminology on the basis of extremely extensive observation. He stresses the importance of three sets of features: (a) development and nature of the veil and pileipellis (the 'epidermis' of the pileus) in relation to covering the developing hymenophore (the hymenophore carries the hymenium, a cell layer responsible for eventually producing the basidiospores); (b) the sequence of development of the stipe, pileus and hymenophore, which are the major functional zones of the basidiome; (c) the mode of development of the hymenophore. The terminology is discussed by Watling (1985) and illustrated in Figure 20.6.

Microscope sections of even extremely small fruit body initials can be resolved into regions of recognizable pileus and stipe (Figure 20.7), and provide *prima facie* evidence for regional specification, since the creation of such histologically distinct regions requires that some organization is imposed on the homogeneous prosenchyma. The most highly differentiated cellular elements seem to occur at the boundaries of tissue regions (Williams, 1986). In the youngest specimens, before this cell differentiation has occurred, the boundaries are frequently denoted by closely appressed parallel hyphae (called meristemoids by Reijnders, 1977; section 20.2.2) which seem to be involved in rapid cell formation in the sense that the distance between successive hyphal cross walls is minimized.

20.3.2 Mushrooms make gills

The gills of agaric fungi are plates suspended from the fruit body pileus tissue. Intuitively one might expect such plates to develop and extend by 'downward' growth of the distal edge of the gill (that is, the edge which is eventually exposed) but this is not the case. The direction of gill development has been a matter of controversy for many years but experimental proof has recently become available from work with both *Coprinus cinereus* and *Volvariella bombycina*. A crucial aspect of understanding how the final structure of the fruit body is attained is appreciation of the geometrical consequences of the differential growth of the primordium (Figure 20.7). As a typical fruit body of *C. cinereus* grows from 1 to 34 mm in height, the circumference of the stipe increases 9-fold and the circumference of the outer surface of the pileus increases 15-fold; this latter corresponds to more than a 3000-fold increase in volume (Figure 20.7). The implications of primordium enlargement for tissue relationships must be kept in mind as we turn to the first question, which is the origin of the space between gills.

(a) Cavitation

The terminology describing hymenophore development was originally defined by Locquin (1953). Fruit body primordia may be categorized as being levhymenial (an initially continuous hymenium becomes folded) or rupthymenial (the hymenium originates in fragmented form). To these categories was later added schizohymenial (where gills and gill cavities differentiate together from the background tissue). Since originally coined, the definitions of these terms have become vectorized: rupthymenial gills are assumed to develop away from the stipe whereas levhymenial gills are alleged to push down into a preformed gill cavity (see definitions in Watling, 1985). Unfortunately, this latter definition is wrong, all gills grow at their roots – the developmental vector moves away from the stipe (see below). However, as defined in the sense of contrasting preformed or simultaneously formed gill cavities, the terms levhymenial and schizohymenial are useful alternatives.

The internal structure of the *Coprinus* primordium is uniformly solid at the time that gills begin to arise (Figure 20.7) so gills and gill space arise together. Lu (1991) has claimed that the gill cavities arise as a result of 'programmed cell death' which plays a part in development in many plants and animals. However, no such suggestions have arisen in previous work on gill formation (e.g. Reijnders, 1963, 1979) and Lu's identification of cell degeneration depends on observation of precipitates in cell vacuoles and in gill cavities which have been dismissed previously as artefacts produced by precipitation of vacuolar contents during fixation (Waters, 1972; Waters *et al.*, 1975b). Because of the enormous increase in size of the fruit body primordium, however, programmed cell death is not necessary to form a gill cavity. When two groups of hyphal tips are formed opposing one another as a pair of palisaded cell plates, like the opposing hymenia of neighbouring gills, they form an incipient fracture plane which can be opened out into a cavity when the expansion of the underlying tissue puts tension across the 'fracture' and pulls the palisades apart (Figure 20.8). If the 'fracture planes' form an annulus around the top of the stipe (one palisade might be the stipe apical meristemoid, the other might be the hymenophore meristemoid), then an annular cavity may arise before gill formation (a mode of development which would be described as levhymenial). On the other hand, where the fracture planes are defined by the hymenia of neighbouring gills, cavities would be isolated, no annular cavity would be apparent, and the process would be described as schizohymenial. Though applied here to agaric primordia, this argument applies to cavitation in all differentially expanding cellular structures. It is a mechanical consequence of developmental change in their geometry like the faceting of cells in compressed tissues discussed by Dormer (1980).

(b) Coprinus *gills*

In *Coprinus*, the pileus of the fruit body primordium encloses the top of the stipe and gills are formed as essentially vertical plates arranged radially around the stipe. Transverse sections show the pileus as an annulus concentric with the stipe, the inner circumference of the annulus being the surface of the stipe and the outer the surface of the pileus tissue (Figure 20.9). There are two types of gill: primary gills which, from formation, have their inner, tramal tissue in continuity with the outer layers of the stipe (Figure 20.9) and secondary (and lesser ranked) gills in which the hymenium is continuous over the gill edge (Figure 20.10; and Reijnders, 1979; Rosin and Moore, 1985a; Rosin *et al.*, 1985; Moore, 1987). Tramal tissues of primary gills remain connected to the stipe until being freed when expansion of the pileus detaches them from the stipe (Rosin and Moore, 1985a; Moore, 1987).

Primary gills are connected with pileus tissue at their outer edge and with the stipe at their inner edge; since the circumference of the stipe increases so much during maturation why doesn't the gill thickness increase by the same extent? The answer is that the tendency to widen as the stipe circumference increases is compensated by gill replication, and specifically by formation of a new gill cavity and its bounding pair of hymenia within the trama of a pre-existing gill. This forms a Y-shaped structure (Figure 20.10). Observation of fruit body sections shows that these Y-shaped gill structures are oriented exclusively as though the new gill organizer originates at the stipe circumference (Moore, 1987) so that the crotch of the Y-shape moves outwards towards the pileus (Figure 20.10). This clearly sets the direction of development as outwards from the stipe.

These observations indicate that gills in the *C. cinereus* fruit body grow radially outwards, their roots extending into the undifferentiated tissue of the pileus context. The formative element appears to be an organizer in the tissue at the extreme end of the gill cavity where the change in structure occurs from the randomly intertwined prosenchymatous tissue of the pileus context to the highly compacted hymenial plates separated by the gill cavity. The gill organizer is responsible for the progression of the gill cavity radially outwards, away from the stipe. The prosenchyma/protohymenium transition need be no more complex than an increase in branch frequency to produce branches of determinate growth which are mutually 'attracted' so that they form the opposing palisades of a fracture plane allowing pileus expansion to separate the two protohymenia (Figure 20.8). The organizer responsible for this is moving radially outwards penetrating, as the primordium grows, successive generations of undifferentiated prosenchymatous tissue in the context which lies between the gills and the 'epidermis' (pileipellis) of the pileus (Rosin and Moore, 1985b). Since it is a radial progression, neighbouring organizers become further and

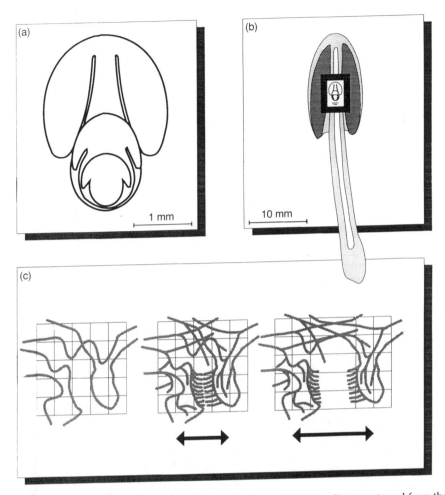

Figure 20.8   Dynamics of basidiome expansion in *Coprinus cinereus*. (a) Outline diagrams traced from the basidiome initials shown in Figure 20.7 are nested together to illustrate the steady outward expansion of the tissue layers. These diagrams are superimposed (to scale) onto a median diagrammatic section of a mature basidiome in (b) to demonstrate the extent of the outward movement of tissue boundaries. (c) Shows how this expansion can generate cavities by putting tension stress across an incipient fracture plane. In these diagrams, a small region of prosenchyma is shown first against a reference grid. In the central diagram a round of branching is assumed to have taken place forming branches of determinate growth arranged in two opposing palisades. This constitutes the incipient fracture plane and when tension is applied the palisades will be pulled apart (right hand figure).

further separated from one another as development proceeds (Figure 20.9). As the distance between neighbouring organizers increases a new one can arise between them (Rosin and Moore, 1985a; Figure 20.10); when a new gill organizer emerges, the margin of a new (but 'secondary') gill is formed. It is extended not by growth of its margin, but by continued radial outward progression of the two gill organizers which bracket it into

the undifferentiated prosenchyma of the pileus context.

The observations summarized in Figure 20.10 and the explanation given above provide *prima facie* evidence for two classic components of theoretical morphogenesis – activation and inhibition by diffusing morphogens. First, we can suggest that diffusion of an activating signal along the fruit body radius assures progression of the gill

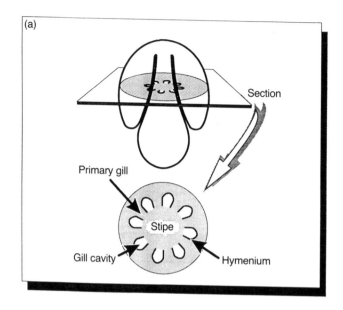

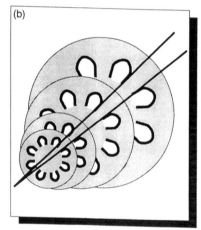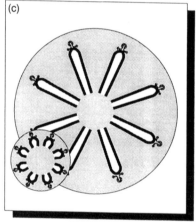

Figure 20.9   Primary gills of *Coprinus* are connected to the stipe. (a) Is intended to orient the reader and to illustrate the basic layout of a transverse section of the primordial pileus. If these geometrical relationships remained unchanged during expansion, gill thickness would increase greatly for two reasons: (1) because the primary gills are connected to the stipe circumference, increase in stipe circumference would be accompanied by increase in gill thickness (b) and (2) because gill organizers migrate radially outwards (generating the branching pattern which forms the protohymenia as palisades bounding an incipient fracture plane as illustrated in Figure 20.8) the gills would inevitably become thicker as their radial paths diverge (c). The solutions to these problems are illustrated in Figure 20.10.

organizer along its radial path. Second, each organizer can be assumed to produce an inhibitor which prevents formation of a new organizer within its diffusion range (i.e. the gill organizer uses this inhibitor to control its morphogenetic field). As radial progression into the extending pileus context causes neighbouring organizers to diverge, a region appears between them which is beyond the range of their inhibitors – at this point a new organizer can arise in response to the radial activating signal. Interaction between the diffusion characteristics of the activator and the inhibitor

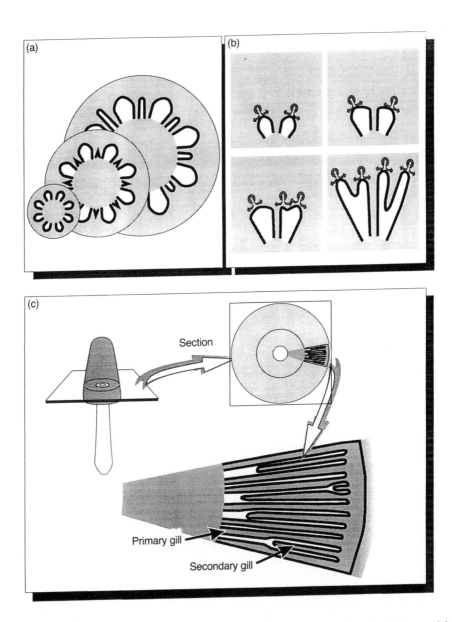

Figure 20.10   Generation of new primary and secondary gills in *Coprinus*. (a) Increase in the thickness of the primary gill at its junction with the expanding stipe is compensated by the appearance of new gill cavities within the trama of the original primary gills. (b) The formative element in outward extension of the gill is the gill organizer located in the tissue which borders the outermost part of the gill cavity. As their radial paths diverge, this part of the gill cavity expands tangentially until sufficient space exists between neighbouring gill organizers for a new organizer to appear between them. Continued outward migration of parent and daughter gill organizers creates a secondary gill between them. These two mechanisms are not alternatives; they occur together to generate the community of radial, narrow gills which characterize the mature *Coprinus* basidiome (c).

is all that is necessary to control gill spacing, gill number, gill thickness, and the radial orientation of the gill field.

(c) Volvariella *gills*

In fruit body development of *Volvariella bombycina*, primary gills arise as ridges on the lower surface of the pileus, projecting into a preformed annular cavity – a levhymenial mode of development. The gills clearly project into the annular cavity, but the question is, do they grow into it? I believe that the answer to this question is no, and see gill development as being exactly homologous with the process in *Coprinus*, i.e. growth of any one gill occurs by outward progression into outwardly expanding pileus context, of gill organizers either side of the foot of the gill. Effectively, therefore, the gill margins (the edges which arise as projections into the cavity when the under-surface of the pileus first becomes folded) remain positionally fixed in space while the gill cavities enfold and extend around them. Research on *Coprinus* development has been aided by the geometrical structure of the primordial fruit body which provides spatial reference points (e.g. the initial attachment of primary gills to the stipe). The *Volvariella* primordium is not so 'user-friendly'.

Volvariella bombycina is bulbangiocarpic (Figure 20.6) but occasionally primordia with exposed gills arise. Young fruit bodies of this sort were used to trace the relative growth rates of the different parts of the hymenophore by painting black ink marks on the tissues (Chiu and Moore, 1990a). During further fruit body development, ink marks placed on the pileus margin and those placed on the edges of the gills remained at the margin or the gill edges respectively. The growth increment here is quite considerable, the radius of the pileus increasing from 0.5 to 2.5 cm and the depth of the gills from 1.5 to 5 mm. If growth of the pileus and gill margins resulted from apical growth of the hyphal tips which occupied the margin, then ink particles placed on those hyphal tips would be left behind as the hyphal apices extended (Figure 20.11). Indeed, this approach has been used to study growth of sporangiophores of *Phycomyces* (Castle, 1942) and conidiophores of *Aspergillus giganteus* (Trinci and Banbury, 1967); both reports present time-lapse photographic sequences showing the sporangiophore/condiophore tip growing beyond

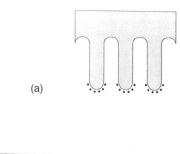
(a)

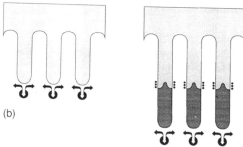
(b)

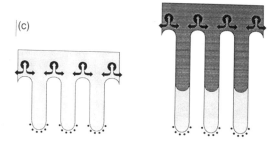
(c)

Figure 20.11 Gill formation in *Volvariella bombycina*. Line drawings illustrating the outcome of marking experiments (Chiu and Moore, 1990a) and their implications for development of gills in this organism. (a) Diagram of ink particles on the primordial gills, shown in section. (b and c) Alternative strategies for gill growth and their predicted outcome for this simple experiment. If the gill organizer is located at the gill margin (b) growth at the margin will extend beyond the initial ink marks burying the ink deep within the gill cavities. If the gill organizer is located at the foot of the gill (c) originating the gill cavity as illustrated in Figure 20.10, then growth of the gill will simply push the already-formed, ink-marked, gill margins further from the cap context but will leave the ink marks in full view. In the experiments reported by Chiu and Moore (1990a), ink marks were painted on the gill margins of primordial fruit bodies and were still clearly visible on the gill margins of mature fruit bodies. It is concluded that agaric gills grow by extension at their roots, and not by extension from the free margin.

externally applied markers (*Lycopodium* spores in Castle (1942), starch grains in Trinci and Banbury (1967)). In the *Volvariella* experiment, extension at the margin would consequently have resulted in the ink marks being left at their original absolute positions, being buried beneath 4–20 mm of newly formed tissue by the end of the experiment (Figure 20.11). It follows, therefore, that gills of *V. bombycina* extend in depth by growth of their roots into the pileus context and by insertion of hymenial elements into their central and root regions. The developmental vector is directed away from the stipe, as it is in *Coprinus*, and in both the schizohymenial and levhymenial modes of gill development the gill grows at its root and not at its margin (Moore, 1987). Similarly, the hyphal tips which form the pileus margin when it is established at the very earliest stage of development remain at the margin. They do not continue to grow apically to extend the margin radially, nor are they overtaken by other hyphae; instead they are 'pushed' radially outwards by the press of fresh growth behind, and they are joined by fresh branches appearing alongside as the circumference of the margin is increased.

Because so much stress is placed on apical wall growth, there is a popular misconception that growth occurs only at the hyphal tip, so the above interpretations are judged as radical by many people. Yet the hyphal tips at pileus and gill margins are growth-limited, differentiated cells, so it should come as no surprise that growth of the structure they comprise is concentrated behind them. This fact has been appreciated for over 30 years, as the *Dictionary of the Fungi* defines 'inflated hypha' as 'one in which cells behind the growing apex enlarge and cause the apparent rapid rate of growth characteristic of most agaric and gasteromycete fruit bodies' (Ainsworth, 1961). If a parallel to the paradigm that the agaric gill grows at its base is required, then consider the lilies of the field, how they grow; they toil not, neither do they spin, but their leaves grow at the base.

(d) Mushroom mechanics

In both *C. cinereus* and *V. bombycina* the first-formed gills were radially arranged. As the pileus expanded more gills were formed. In *V. bombycina*, new gills were formed in two ways (Chiu and Moore, 1990a). First, by bifurcation of an existing gill near its free edge. Initiation of the folding which produced bifurcations on existing gills was a localized and irregular event, resulting in sinuous, contorted gills. The formation of two daughter gills depended on completion of the bifurcation along the entire edge of the parental gill. Second, new generations of gills appeared as ridges in the region between existing gill roots, creating new folds on the pileus context representing the free edges of new secondary or tertiary gills, the gill spaces on either side extending into the pileus context as the gill grew by its root differentiating from the context. In *C. cinereus*, more gills are added as the basidiome enlarges by bifurcation of existing gills either on one side or at the stipe-gill junction, and by division of gill organizers at the roots of existing gills (section 20.3.22 and Figure 20.10). Consequently, *Coprinus* gills are also formed as convoluted plates (Chiu and Moore, 1990b). Thus, the summary description is the same in each case and a sinuous, labyrinthiform hymenophore is a normal 'embryonic' stage in basidiome development in agarics, yet a regular radial arrangement of the gills is characteristic of the mature basidiome. How this is achieved seems, again, to be a function of the expansion of the maturing primordium but in this case the tensions generated by differential growth between tissue layers stretch the convoluted gills into strict radii.

In *C. cinereus*, tension stresses generated by growth of other parts of the basidiome place geometrical stress on the 'embryonic' gills – like a folded cloth being straightened by stretching. Such a mechanism requires that the folded elements (in this case the gills) are anchored. The connection of primary gills to the stipe provides the initial anchorage; subsequently cystidium–cystesium pairs interconnect gill plates around the stipe. Tensions generated by expansion of the pileus will then be communicated and balanced throughout the structure. Cystidium–cystesium pairs act, therefore, not as buttresses to keep hymenia apart (the conventional view; Buller, 1924) but as tension elements whose function is to hold adjacent hymenia together as pileus expansion pulls the gills into shape.

For *Volvariella bombycina* an alternative mechanism operates because of the lack of cystidium–cystesium pairs. The hymenium of *V. bombycina* is a layer of tightly appressed cells, and the trama of the gill becomes filled with greatly inflated cells as maturation proceeds. These features suggest that expansion of tramal cells in gills enclosed by the hymenial 'epidermis' will generate compression forces which will effectively inflate, and so stretch, the embryonic gills to form the regularly radial pattern of the mature pileus.

20.3.3 Cell distribution patterns in hymenia and fruit body stipes

(a) *Cellular elements in the hymenium of* Coprinus

What the gill organizer leaves behind in the developing hymenophore of *Coprinus* is a proto-hymenium. A hyphal tip in this protohymenium has a probability of about 40% of becoming a cystidium, but when a cystidium does arise, it inhibits formation of further cystidia in the same hymenium within a radius of about 30 μm (section 20.3.3c). As a result, only about 8% of these tramal hyphal branches become cystidia; the rest become probasidia (Horner and Moore, 1987) which proceed to karyogamy and initiate the meiotic cycle ending with sporulation. Paraphyses arise as branches of sub-basidial cells and insert into the hymenium. About 75% of the paraphysis population is inserted before the end of meiosis, the rest insert at later stages of development (Rosin and Moore, 1985b). There is, therefore, a defined temporal sequence: probasidia appear first and then paraphyses arise as branches from sub-basidial cells. Another cell type in the hymenium of *C. cinereus* (called the cystesium; Horner and Moore, 1987) illustrates how a contact stimulus can set in train a pathway of differentiation leading to an adhesive cell type. At early stages in growth of the cystidium across the gill cavity the cell(s) with which it will collide in the opposing hymenium are indistinguishable from their fellow probasidia. However, when the cystidium contacts the opposing hymenium, the cells with which it collides develop a granular, vacuolated cytoplasm, more similar to that of the cystidium than to their neighbouring probasidia. The contact triggers the differentiation – a phenomenon which is met in other organisms.

(b) *Cellular elements in the stipe of* Coprinus

Until recently there was surprisingly little information concerning the structure of the stipe of *Coprinus* basidiomata. Recently, though, Hammad *et al.* (1993a) have demonstrated that the stipe contains both narrow and inflated hyphae. Narrow hyphae (cross-sectional area < 20 μm²) always comprise a significant numerical proportion (23–54%) of the cells seen in microscope sections of stipe tissue, although they only contribute 1–4% to the overall cross-sectional area of the stipe (Figure 20.12). Narrow hyphae tend to be concentrated on the outside of the stipe and fringe the central lumen (they fill the lumen at the extreme bases and apices of elongating stipes). Elsewhere, narrow hyphae were interspersed between the inflated hyphae (section 20.3.3c below). Narrow hyphae stain differentially as well as having varied spatial arrangements, suggesting that although morphologically alike, they may serve distinct functions. The narrow hyphae form interconnections independent of inflated hyphae; being branched and fused laterally with other narrow hyphae, whilst there is no evidence that inflated hyphae are either branched or associated in networks of this sort. Since nutrients can be translocated through the stipe, avoiding barriers inserted into the tissue (Ji and Moore, 1993), it is likely that the network of narrow hyphae is important in nutrient translocation.

During normal stipe growth the greatest cell expansion is seen in the inflated cells situated between the mid-cortex and the lumen rather than at the periphery of the stipe (Figure 20.12). Such a distribution of expansion would magnify the tensions (especially compression forces) generated by the expansion process and would obviously contribute greatly to the stretching mechanisms (remember that the barrel-like pileus surrounds the stipe) discussed in section 20.3.2d.

(c) *Patterns of distribution*

Nearest neighbour analysis of cell distributions in *Coprinus* stipes shows that inflated hyphae are evenly rather than randomly distributed regardless of the age of the basidiome or position within the stipe (Hammad *et al.*, 1993a). The proportion of narrow hyphae decreases with time, presumably due to some becoming inflated, and the even distribution of inflated hyphae could be due to some sort of control over the pattern of inflation.

Other evidence for local control of morphogenesis has been obtained from a comparison of the

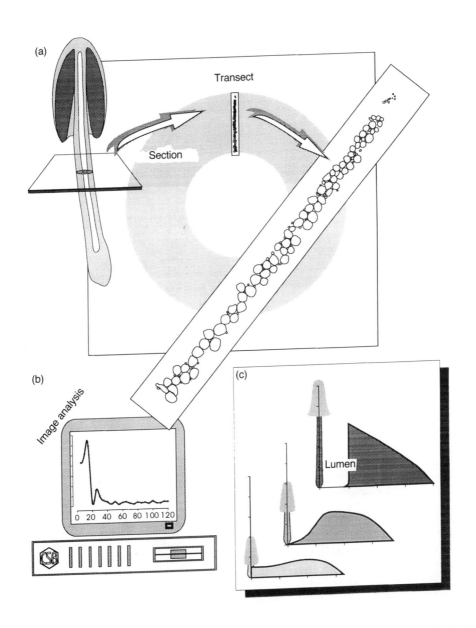

Figure 20.12 Stipe structure and development in *Coprinus cinereus*. (a) Shows how cell-size distribution data were obtained from microscope sections of basidiome stipes. The cell size distribution plot on the computer monitor (b) illustrates the very distinct population of narrow hyphae and the very disperse population of inflated hyphae. (c) Shows average cell area along a radius extending outwards from the centre of the stipe for basidiomes which were 27, 45 and 70 mm tall. The changing distribution of cell size shows that stipe growth is accompanied by inflation of cells in an annulus deep within the original stipe tissue. The outcome of this is that the inner region is torn apart (forming the central lumen) and the outer region is stretched.

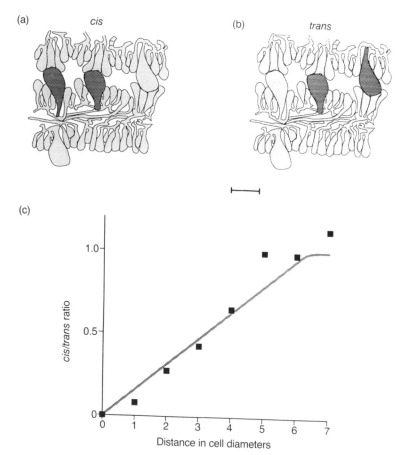

Figure 20.13 Cystidium distribution in the *Coprinus cinereus* hymenium. (a) and (b) show the categorization of neighbouring pairs of cystidia in micrographs as either *cis* (both emerge from the same hymenium) or *trans* (emerging from opposite hymenia). (c) Compares the frequencies of these two types over various distances of separation and shows that closely spaced *cis* neighbours are less frequent than closely spaced *trans* neighbours, implying some inhibitory influence over the patterning of cystidia emerging from the same hymenium.

distributions of cystidia on adjacent hymenia in the *Coprinus* fruit body (Horner and Moore, 1987). Cystidia are large, inflated cells which are readily seen in microscope sections so their relationships are open to numerical analysis. Thus, cystidia spanning the gill cavity may be 'distant', having other cells separating them, or 'adjacent', with no intervening cells; and, in either case, both cystidia may emerge from the same hymenium (described as *cis*) or from opposite hymenia (*trans*) (Figure 20.13). If the distribution of cystidia is entirely randomized there should be an equal number of *cis* and *trans* in both the distant and adjacent categories. However, quantitative data showed a

distinct shortage of adjacent-*cis* cystidia (Figure 20.13), suggesting that formation of a cystidium lowers the probability of another being formed in the immediate vicinity (Horner and Moore, 1987). The inhibition extends over a radius of about 30 μm and is limited to the hymenium of origin. The region around a cystidium is a morphogenetic field controlled by the cystidium at its centre.

The distribution pattern of cystidia might, therefore, be dependent on interplay between activating and inhibiting factors. In this instance (as with the determination of gill development mentioned in section 20.3.2b) the patterning process is open to interpretation using the activator–inhibitor model (Meinhardt and Gierer, 1974; Meinhardt, 1984) which suggests that a morphogenetic pattern results from interaction with an activator which autocatalyses its own synthesis, and an inhibitor which inhibits synthesis of the activator. Both diffuse from the region where they are synthesized, the inhibitor diffusing more rapidly and consequently preventing activator production in the surrounding cells. A wide variety of patterns can be generated in computer simulations by varying diffusion coefficients, decay rates and other parameters (Meinhardt, 1984). The model readily accounts for stomatal, cilial, hair and bristle distributions, and has been applied successfully to simulate leaf venation and phylotaxy (Meinhardt, 1984). For animal systems, Green and Smith (1991) conclude that the answer is 'probably yes' to the question 'do gradients and thresholds of growth factors acting as morphogens establish body plan?' A belief that the same may be true for such fungal phenomena as have been outlined in this chapter is implicit in the way in which they are described here. Unfortunately, there are no clues to the nature of the morphogens which might be the activating and/or inhibiting growth factors in these phenomena. Another problem is that lateral contacts between fungal hyphae are extremely rare, being represented only by lateral hyphal fusions. The constituent cells of plant and animal tissues are interconnected laterally by frequent plasmodesmata, gap junctions and cell processes. The absence of similar structures connecting adjacent hyphae suggests that any morphogens which do exist are likely to be communicated exclusively through the extracellular environment (Reijnders and Moore, 1985; and section 20.2.1). Although it is abundantly clear that coordination of developmental processes is successfully achieved in fungal multicellular structures, the evidence for chemicals able to perform the signal communication involved is sparse and disappointingly unconvincing (discussed in Moore, 1991).

20.3.4 Reijnders' hyphal knots: a basic building block?

Although the work of Hammad *et al.* (1993a,b) is the first quantitative hyphal analysis, hyphal analysis of basidiomata was introduced into taxonomy by Corner (1932), coining the terms monomitic, dimitic and trimitic to describe tissues consisting of one, two or three kinds of hyphae. Later, sarcodimitic and sarcotrimitic were used to describe basidiomata where there are two or three types of hyphae of which one is inflated and thickened (Corner, 1966), and Redhead (1987) recognized a group of closely related agarics with such structures. Some specialized gill trama structures have been highly rated as taxonomic criteria: the heteromerous trama of the Russulaceae, acrophysalidic trama of the Amanitaceae and Pluteaceae, and the sarcodimitic trama of Trogia (Corner, 1991) and the Xerulaceae (Redhead, 1987). The heteromerous trama in Russulaceae (see below) was described by Fayod (1889) who was also aware of the presence of narrow hyphae among the more easily seen cells of the basidiome tissues he examined, but although differentiated hyphae and cells have been recognized as taxonomically important (Lentz, 1971), they have not been used in identification of agarics to the extent that the mitic system has in polypores.

Recently, these have all been examined in detail by Reijnders (1993) who has come to the general conclusion that cell structures which are considered peculiar to each of the specialized trama types can be found in some form (either less well developed or restricted to a particular developmental stage) in many other, unrelated taxa. Of special interest is the heteromerous trama, to which has been attached particular taxonomic importance. The heteromerous trama is characterized by sphaerocysts, first depicted by Corda (1839), which are inflated cells situated in a ring surrounding a central ('induction') hypha (Reijnders, 1976; Watling and Nicoll, 1980). Very similar aggregations of hyphae, termed hyphal knots (Reijnders, 1977) have been observed in a wide range of species (Reijnders, 1993). The common features of Reijnders' hyphal knots seem to be a central hypha (which remains hyphal) and an immediately surrounding family of hyphae which differentiate in concert. Some hyphal knots do not show conspicuous differentiation, remaining as systems of tightly interwoven hyphae of uniform structure, but at the extreme, swollen cells in a ring or cylinder around a central hypha may be formed in species taxonomically far removed from the Russulaceae.

Hyphal knots are found particularly frequently in plectenchymatous tissue and occur in bulb, stipe, veil and pileus as well as tramal tissues.

Reijnders (1993) discusses the impact of their widespread occurrence on taxonomic and phylogenetic arguments about the Agaricales. It should be noted that their analogues (perhaps, even, homologues) might occur much more widely than that because elements of Reijnders' description of hyphal knots are detectable in descriptions of many fungal multihyphal structures, from strand formation (Butler, 1958) through development of sclerotia (Townsend and Willetts, 1954), and on to the descriptions given above of paraphysis distributions around basidia as well as the influence exerted by cystidia on their surroundings in the *Coprinus* hymenium, and the possible relationships between narrow and inflated hyphae in the *Coprinus* stipe.

Perhaps, in all multihyphal fungal structures, the ultimate morphogenetic regulatory structure is the Reijnders hyphal knot – a little community comprising an induction hypha (or hyphal tip, or hyphal compartment) and the immediately surrounding hyphae (or tips, or compartments) which can be brought under its influence. Larger-scale morphogenesis could be coordinated by 'knot-to-knot' interactions.

20.4 CONTROLLING CELL DIFFERENTIATION

It is important to appreciate that all the cell types so far mentioned arise as branches from some other hypha-like elements. At some point sister branches of the same hyphal system, cellular elements separated only by a dolipore septum in Basidiomycotina or a much simpler septum in Ascomycotina, will follow totally different pathways of differentiation (Figure 20.14). Although evidence for different dolipore ultrastructures has been found in the basidiomycete hymenium (Gull, 1976, 1978; Moore, 1985), even in this extreme example there is no indication that the septa are physically sealed, so it is not at all clear how alternate pathways of differentiation are regulated on the two sides of a septum which appear to be in physiological contact.

20.4.1 Genetic regulation

Despite a generally smaller genome size, fungi share with other organisms the fact that only a small proportion of the genome is associated with any particular morphogenetic process. This has

been demonstrated in *Schizophyllum* (Zantinge *et al.*, 1979; de Vries *et al.*, 1980; de Vries and Wessels, 1984; Wessels *et al.*, 1987), *Coprinus* (Yashar and Pukkila, 1985; Pukkila and Casselton, 1991), *Neurospora* (Nasrallah and Srb, 1973, 1977, 1978), *Sclerotinia* (Russo *et al.*, 1982), *Sordaria* (Broxholme *et al.*, 1991), and *Saccharomyces*. The yeast example is especially interesting because of its small genome (haploid genome = about 1.4×10^4 kilobase pairs, which is less than four times the size of the genome of the bacterium *Escherichia coli*), yet only 21–75 of the estimated 12 000 genes in yeast are specific to sporulation (i.e. meiosis and ascospore formation)(Esposito *et al.*, 1972). Thus, the emphasis in morphogenetic gene regulation is on differential integration of activity rather than on large-scale replacement of one set of gene products by another. This could be seen as conflicting with Timberlake's (1980) estimates that 11–18% of mRNA sequences accumulated in conidiating cultures of *Aspergillus nidulans* are not detectable in vegetative hyphae and that 6% of the unique sequences are expressed during this aspect of development (Timberlake and Marshall, 1988; Chapter 17). But this estimation is for the entire portfolio of differentiation events rather than morphogenetic events *per se*: vegetative material comprised spores germinated for only 16 h, conidiating cultures were grown for 40 h (Timberlake, 1980). The other comparisons between fruiting and non-fruiting cultures referred to above involve cultures of similar age which, for environmental or genetic reasons, differ in their ability to fruit. As mycelia undergo vegetative morphogenesis, many aspects of cell differentiation expected of the fruit body (cell inflation, wall thickening, accumulation of metabolites, etc.) will be found in non-fruiting mycelia, so comparisons made in these cases emphasize differences ascribable to the morphogenetic events contributing to fruit body formation rather than the differentiation of its component cells. It is these morphogenetic processes which we might hope to identify by studying the genetic control of fruiting.

Classical genetic studies, namely identification of variant strains, application of complementation tests to establish functional cistrons, construction of heterokaryons to determine dominance/recessive and epistatic relationships (to indicate the sequence of gene expression) have revealed a 'developmental pathway' for perithecium formation in

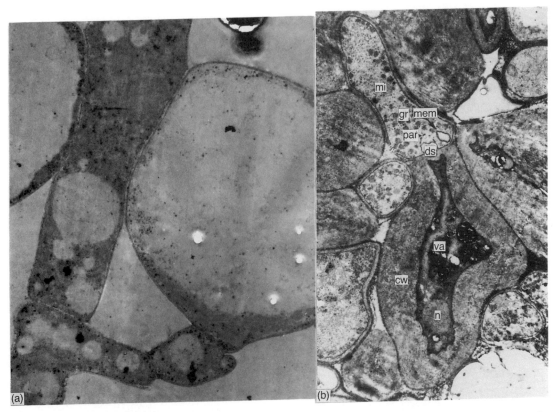

Figure 20.14 Differentiation across the dolipore septum in *Coprinus cinereus*. The transmission electron micrographs illustrate neighbouring cells showing extremes of differentiation in the hymenium and sclerotium. In the hymenium (a) a greatly inflated paraphysis is developing at the apex of a branch which emerged from the cell immediately beneath a basidium. (b) Shows adjacent compartments of a hypha in a section of the medulla of an aerial sclerotium. The upper compartment is thin-walled and has an electron-translucent cytoplasm with scattered glycogen 'rosettes' (gr) whereas the lower compartment is thick-walled, has a more electron-dense cytoplasm and vacuoles (va) with extremely electron-dense contents. Cw = cell wall; ds = dolipore septum; mi = mitochondrion; n = nucleus; par = parenthosome of dolipore septum. Scale bars = 5 μm.

Sordaria (Esser and Straub, 1958; Figure 20.15), and 29 complementation groups involved in perithecium development have been identified in *Neurospora* (Johnson, 1978). Johnson (1976) used genetic mosaics (heterokaryons in which one nucleus carried a recessive colour mutant) to show that perithecia of *Neurospora* arise from an initiating population of 100–300 nuclei, and that the perithecium wall is composed of three developmentally distinct layers. Ashby and Johnstone (1993) have used the *E. coli* β-glucuronidase gene as a reporter gene to study development of ascomata in *Pyrenopeziza brassicae* and have also revealed three tissue layers. One to which both mating types contribute and two to which the two mating types contribute separately. The significance of extensive tissue layers in which only one mating type is expressed is unknown.

In Basidiomycetes, the picture revealed by classical genetic approaches is less clear. One reason is that fruit bodies in Basidiomycetes are normally formed by heterokaryotic mycelia and the presence of two (or more) nuclei (and, hence, two or more genotypes) makes study of the genetics of development by conventional means very difficult. However, fruiting on monokaryotic mycelia has been reported for a diverse selection of agarics and polypores (Stahl and Esser, 1976; Elliott, 1985) and this allowed a start to be made on the genetic control of fruit body development.

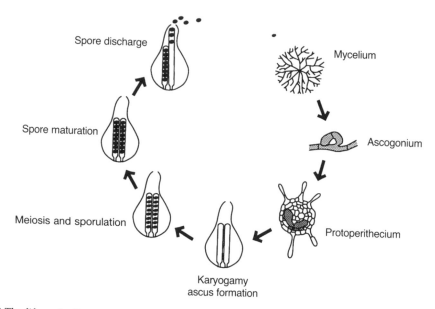

Figure 20.15 The life cycle diagram and perithecium developmental pathway of *Sordaria macrospora* (after Esser and Straub, 1958). A variety of mutants are known which block the pathway at each of the stages represented by arrows, so the whole pathway is interpreted as being essentially a single sequence. Contrast this with the multiple parallel 'subroutines' which seem to characterize basidiome development (Figures 20.16 and 20.17).

(a)   Monokaryotic fruiting

The frequency of monokaryons able to fruit differs widely between genera: 27% of *Sistotrema* isolates formed monokaryotic fruit bodies (Ullrich, 1973), 7% of *Schizophyllum* strains did so (Raper and Krongelb, 1958), but only one of 16 monokaryons of *Coprinus cinereus* tested by Uno and Ishikawa (1971). Horton and Raper (1991) identified a DNA sequence which induced monokaryotic fruiting in strains of *Schizophyllum commune* into which it was introduced by transformation, but there is no indication yet as to how it operates.

Stahl and Esser (1976) carried out genetic crosses between various monokaryotic fruiting strains of *Polyporus ciliatus* and identified three unlinked genes involved in monokaryotic fruiting. The way in which the genes function is unknown but fi^+ is interpreted as initiating monokaryotic fruiting whereas fb^+ is seen as being responsible for 'moulding' the structure initiated by fi^+ into a fruit body, whereas mod^+ directs development into a futile pathway which leads to formation of compact mycelial masses called stromata. In the dikaryon mod^+ inhibited fruiting, but neither fi^+ nor fb^+ showed any expression even when homozygous.

A broadly similar genetic system was revealed in analogous experiments with the agaric *Agrocybe aegerita* (Esser and Meinhardt, 1977). Here, again, one gene, fi^+, was identified as being responsible for initiation of monokaryotic fruiting, whereas a second, fb^+, was thought to be responsible for modelling the initiated structures into fruit bodies. Unlike *Polyporus*, these genes were found to be concerned with fruiting in the dikaryon of *A. aegerita*. Only dikaryons carrying at least one allele of both fi^+ and fb^+ were able to produce fertile fruit bodies.

In *Schizophyllum commune*, Raper and Krongelb (1958) examined some monokaryotic fruiting strains (called *hap*) and were able to show that there was no correlation between monokaryotic and dikaryotic fruiting, and the former was probably under polygenic control. The polygene complex involved may have been identified by Esser *et al.* (1979), who found monokaryotic fruiting to be controlled by at least four genes in *S. commune*. Two 'fruiting initiation genes' (fi-1^+ and fi-2^+, either of which alone allowed differentiation into 2–3 mm 'initials', whereas when both were

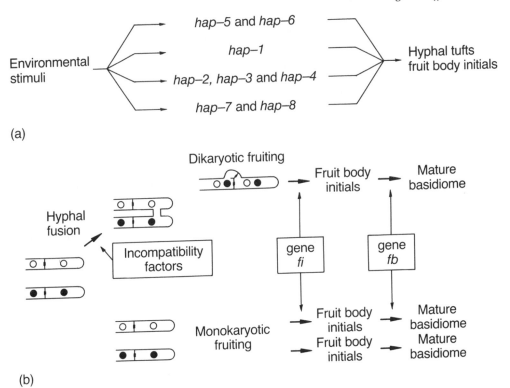

(a)

(b)

Figure 20.16 Models for the genetic control of basidiome development. (a) Shows the genes involved in monokaryotic fruiting in *Schizophyllum commune* (after Leslie and Leonard, 1979b) in which *hap-5* and *hap-6* control the spontaneous initiation of basidiomes, *hap-1* alone or *hap-2, -3* and *-4* acting together control fruiting as a response to injury, and *hap-7* and *hap-8* determine fruiting in response to applied chemicals. (b) A proposed model for the action of major genes controlling basidiome formation in *Agrocybe aegerita* (after Esser and Meinhardt, 1977).

present stipes 6–8 mm long were formed), a third gene (*fb*⁺) was required for formation of complete monokaryotic fruit bodies. The fourth gene (*st*⁺) prevented expression of the others; any mycelium carrying *st*⁺ produced only stromata. Although *st*⁺ also blocked dikaryotic fruiting when homozygous, the other three genes had no effect on differentiation of fruit bodies in the dikaryon though they did influence the time of fruiting. The most rapid fruiting occurred on dikaryons which were homozygous for all three monokaryotic fruiter genes. The slowest fruiting occurred when the dikaryon did not carry any of the monokaryotic fruiter alleles, but the fact that fruiting did eventually occur in this latter case implies a major difference between the genetic control of monokaryotic and dikaryotic fruiting. Barnett and Lilly (1949) also reported an increased frequency of fruiting in dikaryons made from monokaryotic

fruiters in *Lenzites trabea*, but they concluded that the same factors might be involved in the genetic control of monokaryotic and dikaryotic fruiting.

Leslie and Leonard (1979a,b) identified eight genes in *Schizophyllum commune* involved in four distinct pathways enabling monokaryons to initiate fruiting bodies in response to mechanical and chemical treatments (Figure 20.16). As Leslie and Leonard (1979b) place the operation of these genes prior to fruiting stage I, which corresponds to masses of aggregated cells which have no defined shape (Leonard and Dick, 1968), these systems are probably distinct from and perhaps operate at stages prior to those governed by the genes identified by Esser *et al.* (1979), which produce stipe-like structures. However, descriptive comparisons are the only basis for speculation about relationships between these systems.

The multiplicity of genetic factors involved in

monokaryotic fruiting mirrors the multiplicity of physiological conditions which are able to promote such fruiting. In the overwhelming majority of cases 'monokaryotic fruits' are abnormal structures, being incomplete, sterile or both. An essential question, then, is whether genes which influence fruiting in monokaryons are actually relevant to the normal process of dikaryotic fruiting. Stahl and Esser (1976) report that neither *fi*$^+$ nor *fb*$^+$ showed expression in the dikaryon of *Polyporus ciliatus* even when homozygous, and Esser *et al.* (1979) came to a similar conclusion for the genes to which they gave the same names in *Schizophyllum commune*, but in both species there are examples of genes with some expression in both mycelial states: the gene *mod*$^+$ in the former and *st*$^+$ in the latter each blocked fruiting in the dikaryon. In *Agrocybe aegerita*, both *fi*$^+$ and *fb*$^+$ were essential for the dikaryon to produce fertile basidiomata (Esser and Meinhardt, 1977; Esser *et al.*, 1979; Figure 20.16).

(b)   Dikaryotic fruiting

Uno and Ishikawa (1971) concluded that more than one gene is involved in controlling monokaryotic fruiting in *Coprinus cinereus* (= *macrorhizus*) but *C. cinereus* is the only organism in which any attempt has been made to study the genetic control of fruit body formation by the dikaryon. Dikaryons of *C. cinereus* can form sclerotia and basidiomata, monokaryons can also form sclerotia but normally do not form basidiomata. The initial stages in the development of both structures have been described separately and the descriptions are remarkably similar (Matthews and Niederpruem, 1972; Waters *et al.*, 1975a). For both structures, development from the mycelium must involve equivalent patterns of hyphal aggregation so the likeness observed may indicate a shared initial pathway of development or coincidentally analogous separate, but parallel, pathways. The opportunity to distinguish between these possibilities arose with the identification of monokaryons unable to form sclerotia.

This monokaryon phenotype segregated in crosses as though controlled by a single major gene. Four such *scl* (sclerotium-negative) genes were characterized; one, *scl*-4, caused abortion of developing fruit body primordia even when paired in the dikaryon with a wild-type nucleus

but the other three behaved as recessive genes in such heteroallelic dikaryons and were mapped to existing linkage groups (Waters *et al.*, 1975a). Subsequently, Moore (1981) showed that homoallelic dikaryons (i.e. dikaryons in which both nuclei carried the same *scl* allele) were unable to form either sclerotia or fruit bodies. Since these single genetic defects blocked development of both dikaryon structures it was concluded that in the initial stages sclerotia and basidiomata share a common developmental pathway which is governed by the *scl* genes (Figure 20.17). When they mutate they are usually recessive so the pathway can proceed only in the heteroallelic dikaryon where the missing *scl* function is provided by the wild-type nucleus from the other parent. The *scl* mutations remove the ability to make sclerotia – a normal aspect of the monokaryon phenotype. In this respect they seem to be the antithesis of the *fis* mutants, some of which cause monokaryotic fruiting (Uno and Ishikawa, 1971), and the *roc* gene, which causes stromatic proliferations (Nyunoya and Ishikawa, 1979) of *C. cinereus*, and the *hap*, *fi* and *fb* genes in *Schizophyllum* (discussed above) which confer on the monokaryon the ability to form a fruit body – a phenotype which is normally a character of the dikaryon. Thus, recessive mutations can lead both to loss and gain of the ability to form multicellular structures. Attempts have been made to simplify many of these observations into a single developmental pathway (Esser and Hoffman, 1977; Esser *et al.*, 1977; Meinhardt and Esser, 1983; Figure 20.16), yet much of the evidence points to there being a number of discrete partial pathways which can run in parallel.

Basic characterization of the genetic control of dikaryon fruit body development has only been attempted in *Coprinus cinereus* (under the name *C. macrorhizus*) by Takemaru and Kamada (1971, 1972). These workers treated macerated dikaryon fragments with mutagens and then searched for developmental abnormalities among the survivors. Including spontaneous mutations, a total of 1594 were identified out of 10 641 dikaryotic survivors tested. They were classified into categories on the basis of the phenotype of the fruit body produced:

1. 'knotless', no hyphal aggregations are formed;

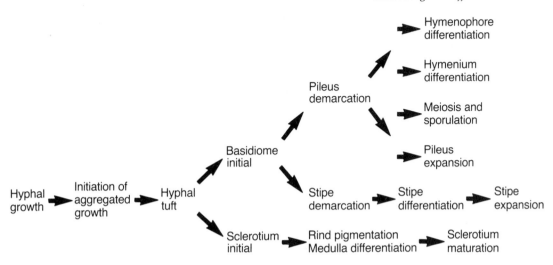

Figure 20.17 Genetically distinct pathways involved in sclerotium and basidiome development in *Coprinus cinereus* (revised after Moore, 1981).

2. 'primordiumless', aggregations are formed but they do not develop further;
3. 'maturationless', primordia are produced which fail to mature;
4. 'elongationless', stipe fails to elongate but pileus development is normal;
5. 'expansionless', stipe elongation normal but pileus fails to open;
6. 'sporeless', few or no spores are formed in what may otherwise be a normal fruit body.

Since dikaryotic mutagen survivors were isolated, the genetic defects identified are all dominant. Elongationless mutants have been exploited to study stipe elongation (Kamada and Takemaru, 1977a,b, 1983), and sporeless mutants have been used to study sporulation (Miyake *et al.*, 1980a,b).

These mutants suggest that different aspects of basidiome development are genetically separate. Prevention of meiosis still permits the fruit body to develop normally, demonstrating, as do mono-karyotic fruit bodies, that meiosis and spore formation are entirely separable from construction of the spore-bearing structure. Perhaps more interesting is the fact that mutants were obtained with defects in either pileus expansion or stipe elongation. Both processes depend on enormous cell inflation, and the fact that they can be separated by mutation indicates that the same result (increase in cell volume) is achieved by different means (Moore *et al.*, 1979).

However, there is a profound problem in accounting for the induction of dominant mutations at the high frequency observed by Takemaru and Kamada (1972). Among the mutagens used, ultraviolet light and *N*-methyl-*N*'-nitro-*N*-nitrosoguanidine both increased the mutation frequency, together producing a total of 1582 mutants among 8547 survivors examined. This 18.5% mutation rate is extraordinarily high. Another peculiarity is that over 72% of the mutants belong to just two phenotypes; there being 595 maturationless and 582 sporeless isolates. Takemaru and Kamada (1972) account for these frequencies with the suggestion that genes involved in development may be easy to mutate. The absence of reports of such mutations in other populations of *C. cinereus* dikaryons argues against this proposition. An alternative interpretation (Moore, 1981) is that the genes which were being caused to mutate were not those involved directly in development, but rather genes which modify the dominance of pre-existing developmental variants; i.e. the dikaryon subjected to mutagenesis carried genetic defects affecting maturation or spore formation, but that their recessiveness depended on modifying genes. Exposure to mutagen caused mutations in one or other of these modifiers and consequent change in the balance dominance modification resulted in the recessive phenotype becoming a dominant one. That dominance (or penetrance) is dependent on the modify ing action of other genes is a well established and perfectly respectable idea in genetical theory (Fisher, 1928, 1931; Sheppard, 1967; Manning, 1976, 1977), indeed one can imagine considerable selective advantage in a system which imposed recessiveness on variants in genes concerned with development. This interpretation was arrived at following work showing that the penetrance of *scl* genes in heteroallelic dikaryons depended on the segregation of modifiers (Moore, 1981). Nyunoya and Ishikawa (1979) also showed that *roc* and *fis*[c] segregated in ratios suggestive of multiple gene control and dominance modification has also been invoked to explain segregation patterns of a gene conferring resistance to *p*-fluorophenylalanine in *C. cinereus* (Senathirajah and Lewis, 1975; Lewis and Vakeria, 1977). As differentiation in basidiomycetes involves extensive protein processing (Zantinge *et al.*, 1979; de Vries *et al.*, 1980; Moore and Jirjis, 1981), modifiers might be involved in processing signal sequences of structural proteins. In the presence of particular modifier alleles (those which cause the change in penetrance), signal processing might lead to normal structural proteins failing to reach their correct destination, or abnormal proteins being partially corrected so that they do reach the target site, despite being defective.

Isolation of strains of *C. cinereus* which have mutations in both mating type factors (*Amut Bmut* strains) has opened up new possibilities for genetic analysis of morphogenesis in this organism. These strains are homokaryotic dikaryon phenocopies; i.e. they emulate the dikaryon in that their hyphae have binucleate compartments and extend by conjugate nuclear division with formation of clamp connections, and the cultures can produce apparently normal basidiomata. On the other hand they are homokaryons, being able to produce asexual spores (usually called oidia) and, most importantly, containing only one (haploid) genetic complement (Swamy *et al.*, 1984). This last feature allows expression of recessive developmental mutants and these strains have been used to study a number of developmental mutants (Kanda and Ishikawa, 1986) especially in meiosis and spore formation (Zolan *et al.*, 1988; Kanda *et al.*, 1989a, 1990; Kamada *et al.*, 1989) and in the formation of basidiome primordia (Kanda *et*

al., 1989b), but no overall basidiome developmental pathway has yet emerged.

(c) Expression of fruiting genes

The work discussed so far gives no guidance about the way in which genes causing developmental variants exercise their effects. As stated at the beginning of this section, only a small fraction of the genome is specific to morphogenesis, and correspondingly few morphogenesis-specific polypeptides have been identified. A development specific protein has been identified in sclerotia of *Sclerotinia sclerotiorum* (Russo *et al.*, 1982) and a polypeptide specific to ascomatal development has been detected in *Neurospora tetrasperma* (Nasrallah and Srb, 1973, 1977) and localized to the mucilaginous matrix surrounding the asci and paraphyses (Nasrallah and Srb, 1978). In *Sordaria brevicollis*, 17 out of over 200 polypeptides detected after pulse-labelling were found in perithecia after crossing (Broxholme *et al.*, 1991). De Vries and Wessels (1984) found only 15 polypeptides specifically expressed in fruit body primordia of *Schizophyllum commune*. Other techniques also suggest that expression of only a small proportion of the genome is devoted to morphogenesis in both *S. commune* (Zantinge *et al.*, 1979; de Vries *et al.*, 1980) and *Coprinus cinereus* (Yashar and Pukkila; 1985; Pukkila and Casselton, 1991).

Reallocation of ribosomal-RNA between fruit bodies and their parental vegetative mycelium was demonstrated by *in situ* hybridization in *S. commune* (Ruiters and Wessels, 1989a) and concentration of fruiting-specific RNAs in the basidiomata has also been demonstrated (Mulder and Wessels, 1986; Ruiters and Wessels, 1989b). From among the fruiting-specific sequences, Dons *et al.* (1984) cloned a gene belonging to a family of sequences encoding hydrophobins, cysteine-rich polypeptides which are excreted into the culture medium but polymerize on the wall of hyphae which emerge into the air (to form fruit body initials, for example) and invest them with a hydrophobic coating (Wessels, 1992; Wasten *et al.*, 1993, 1994; and Chapter 3).

In *Coprinus cinereus* some enzymes are specifically derepressed in the fruit body pileus, being absent from its stipe; among these is the NADP-linked glutamate dehydrogenase (NADP-GDH) and glutamine synthetase (GS) (Moore, 1984b). Cytochemical examination shows that the NADP-GDH appears first in isolated islands of cells in the very young primordium (Elhiti *et al.*, 1979) where the enzyme is localized to cytoplasmic microvesicles in the peripheral regions of specific cells (Elhiti *et al.*, 1987). Other analyses associated initial derepression of the enzyme with karyogamy and the progress of meiosis (Moore *et al.*, 1987b). It was suggested that the enzymes functioned in ammonium detoxification, rather than being primarily ammonium assimilators, to protect meiosis and sporulation (Moore *et al.*, 1987a). Subsequently, it was demonstrated that supplementation of the transplantation medium with ammonium salts abolished the commitment to sporulation that basidia otherwise show *in vitro*, causing them to form vegetative hyphae (Chiu and Moore, 1988b). Another remarkable feature of *C. cinereus* is that the helical arrangement of chitin microfibrils, which has been related to stipe extension growth in both *C. cinereus* (Kamada *et al.*, 1991) and *Agaricus bisporus* (Mol *et al.*, 1990), is established in the hyphal tufts which represent the earliest discernible fruit body initials, the walls of the latter being distinguishable from vegetative hyphal walls both in having less chitin and in having the helical microfibril arrangement (Kamada and Tsuru, 1993).

Schizophyllum hydrophobins, and *Coprinus* NADP-GDH and wall structure are disparate examples of two generalizations. First, the roles of morphogenesis-specific proteins provide surprises; they do not promote the sorts of function which might be postulated in speculative models. Second, they operate at, or in the vicinity of, the cell surface so as to modify the immediate environment of the differentiating hyphae. Hydrophobicity is presumably essential for aerial hyphae (and perhaps aggregating hyphae?); removing ammonium ions from the vicinity is certainly crucial for even a committed basidium to remain committed. Could this be the biochemical/molecular basis of the 'tenuous grasp on differentiation' and 'need for continual reinforcement' discussed in section 20.4.2? Perhaps fungal morphogenesis depends on the differentiating hypha controlling its immediate environment. The degree and complexity of differentiation which can be achieved may be a function of the extent and potency of that control so that when control is diluted (by experimental

explantation, for example) the state of differentiation can no longer be maintained.

Whatever genes are directly involved in morphogenesis, they are presumably ultimately controlled by the mating type factors (Kües and Casselton, 1992; and Chapter 18) and most seem to be transcriptionally regulated (Schuren *et al.*, 1993). Since variation in fruit body morphology is common in higher fungi (Watling, 1971; Chiu *et al.*, 1989) and can span generic (Bougher *et al.*, 1993) and even wider taxonomic boundaries (Watling and Moore, 1993), it has been suggested that normal morphogenesis may be an assemblage of distinct developmental subroutines (Chiu *et al.*, 1989); in other words that the genetic control of overall morphogenesis is compartmentalized into distinct segments which can be put into operation independently of one another. Thus, this model postulates subroutines for hymenophore, hymenium, stipe, pileus, etc., which in normal development appear to be under separate genetic control (Figure 20.17). In any one species they are thought to be invoked in a specific sequence which generates the particular ontogeny and morphology of that species but the same subroutines may be invoked in a different sequence as an abnormality in that same species or as the norm in a morphologically different species. The model provides a unifying theme for categorizing fruit body ontogeny and for clarifying phylogenetic and taxonomic relationships (Watling and Moore, 1994).

20.4.2 Concepts of commitment

There is, then, a wealth of evidence for highly specific differentiation of individual cells in fungi, but very little direct evidence for the developmentally important concept of commitment. This is the process whereby a cell becomes firmly committed to one of the developmental pathways open to it before expressing the phenotype of the differentiated cell type. The classic demonstration of embryological commitment involves transplantation of the cell into a new environment; if the transplanted cell continues along the developmental pathway characteristic of its origin then it is said to have been committed prior to transplant. On the other hand, if the transplanted cell embarks on the pathway appropriate to its new environment then it was clearly not committed at the time of transplant. Most fungal structures produce vegetative hyphae very readily when disturbed and 'transplanted' to a new 'environment' or medium. This is an essential feature of collecting fungi 'from the wild'. Mycologists would expect to be able to recover cultures from the stipes of many fruit bodies collected from the field, incubate them with the minimum of fuss and obtain a fresh generation of fruit bodies *in vitro* within a matter of weeks. Try doing that with tissue pulled off a cow from the same field! The readiness with which fungi regenerate creates the impression that fungal cells express little commitment to their state of differentiation.

Unfortunately, very little transplantation experimentation has been reported with fungal multicellular structures. The clearest examples of commitment to a developmental pathway have been provided by Bastouill-Descollonges and Manachère (1984) and Chiu and Moore (1988a) who, respectively, demonstrated that basidia of isolated gills of *Coprinus congregatus* and *C. cinereus* continued development to spore production if removed to agar medium after initiation of meiosis, but immediately regressed to form vegetative mycelium if removed 12 h earlier. In *C. cinereus*, basidia in the prekaryogamy (dikaryotic) stage became arrested on transplantation, neither developing further nor reverting to hyphal growth. Cystidia, paraphyses and tramal hyphae readily reverted to hyphal growth at all times. Evidently, probasidia become committed to complete the sporulation programme during meiotic prophase I. Once initiated, the maturation of basidia is able to proceed *in vitro*, though it can be inhibited by application of ammonium salts and some other compounds (Chiu and Moore, 1988b, 1990c). Clearly, then, even if only to a limited extent (only cells entering meiosis?), commitment to a pathway of differentiation some time before realization of the differentiated phenotype can evidently occur in fungi. Butler (1988, 1992a) described development of the pore field in cultures of the resupinate polypore *Phellinus contiguus*, and has demonstrated that hyphal differentiation is autonomous in explants (Butler, 1992b). Differentiation in explants was similar to that which occurred in intact basidiomata at positions equivalent to those from which the explants were removed. This implies that the mode of differentiation of the explant was somehow specified, i.e. the cells were developmentally committed, prior to its excision.

20.4.3   Absolutes or probabilities?

The above discussion (section 20.4.2) of the explantation experiments of Chiu and Moore (1988b) concentrated on the commitment shown by basidia to the sporulation pathway. It is also important to appreciate that the other cells of the hymenium and hymenophore, which immediately reverted to hyphal growth on explantation, do not default to hyphal growth *in situ*. While the fruit body is intact these cells exhibit extremes of differentiation, yet as soon as they are separated from their fellows they produce hyphal outgrowths. It is as though they have an extremely tenuous grasp on their state of differentiation, and that, *in situ*, their state of differentiation is somehow continually reinforced by something in the environment of the tissue which they comprise.

A cell described as a basidium is quite clearly characterized by karyogamy, meiosis and the formation of basidiospores. In other words, the nomenclature encompasses a portfolio of features which cells of that sort are expected to express. However, on the one hand, the full portfolio of features may not always be expressed, and on the other hand, the portfolio of features that are expressed may not all be appropriate to the cell expressing them. Facial cystidia (those in the hymenium on the surface of the gill) in *C. cinereus* are established as components of the very first population of dikaryotic hyphal tips which form hymenial tissue (Rosin and Moore, 1985b; Horner and Moore, 1987) and are generally binucleate, reflecting their origin and the fact that they are sterile cells. Yet, occasional examples can be found of cystidia in which karyogamy has occurred (Chiu and Moore, 1993). This suggests that entry to the cystidial pathway of differentiation does not totally preclude expression of at least the start of the nuclear differentiation pathway characteristic of a different cell type.

Facets of developmental uncertainty are the facts that cells of different sorts can serve the same function, and cells of similar form can serve different functions and arise in different ways. To illustrate the first of these propositions: the 'epidermal pavement'-like structure of the *Coprinus* hymenium is constructed of highly differentiated cells – paraphyses – which arise as branches from beneath the basidia (Rosin and Moore, 1985b). At maturity, individual basidia are surrounded by about five paraphyses, so that about 80% of the hymenial cells serve a structural function. In the hymenium of *Agaricus bisporus*, the 'epidermal pavement' which provides the structural support for basidia is made up of immature basidia in an arrested meiotic state (Allen *et al.*, 1992). Even after many days existence, when the fruit body is close to senescence, 30–70% of the immature basidia are in meiotic prophase. Rather than this being wastage of reproductive potential, it is constructive use of one differentiation pathway to serve two distinct but essential functions. *Agaricus* and *Coprinus* hymenophore tissues reach essentially the same structural composition by radically different routes.

Marginal and facial cystidia of *C. cinereus* are very similar in shape and size, but whereas facial cystidia (those in the hymenium on the surface of the gill) arise from the hymenial population of hyphal tips, cystidia on the edges of primary gills (marginal cystidia) are the apical cells of branches from the multinucleate gill trama, which become swollen to repair the injury caused when primary gills are pulled away from the stipe. Marginal cystidia retain the multinucleate character of their parental tramal hyphae (Chiu and Moore, 1993), serve to heal an injury rather than act as tension ties, and yet, apart from position, are indistinguishable from their analogues, the facial cystidia. Basidia provide more examples of 'convergent development'. A hymenial cell initial being binucleate in *C. cinereus* but uninucleate in *V. bombycina* leads to inevitable differences in the course of meiotic division, which are magnified by the fact that DNA synthesis takes place before karyogamy in *C. cinereus* but after karyogamy in *V. bombycina*. Yet the two basidia exhibit remarkably similar morphologies and clearly serve the same function.

Discussion of fungal cell differentiation often involves use of the word 'switch' in phrases which imply wholesale diversion between alternative developmental pathways at some point – like the 'mode switches' in vegetative mycelial development discussed by Gregory (1984). But the examples given above suggest that fungal cells can assume a phenotype even when all conditions for it have not been met. Rather than rigidly following

a prescribed sequence of steps, such differentiation pathways seem to be based on allowing latitude in interpreting the rules on which the pathways are based. From his work with genetic mosaics, Johnson (1976) concluded:

> My observations suggest a very different mechanism of pattern formation for *Neurospora* perithecia than is observed in *Drosophila* development. In contrast to *Drosophila*, no rigid determination of the number of nuclear divisions and the pattern of cellular growth can be detected. Rather, there seems to be control on the average amount of growth which an entire tissue or perithecium undergoes.

Perhaps choice between, and progress of, fungal differentiation pathways has more to do with the balance of probabilities than with switching between absolute alternatives.

20.5 CONCLUSIONS

The existing information about developmental mycology is fragmentary and relies too much on inelegant, long-superseded techniques which inevitably give rise to inelegant interpretations. Even at the basic observational level, inappropriate techniques are still employed. Observations made with the light-microscope using sections 60, 40 or even 20 μm thick are useless. The ready availability of resins which can be sectioned routinely at 5–10 μm thick leaves no excuse for continued use of such material. Similarly, the accessibility of electron microscopy, especially cryo-SEM, should abolish all attempts to interpret, often with more imagination than accuracy, light microscope images at, and even beyond, the limits of resolution and magnification of the equipment. The value of recent technical innovations should also be more rapidly appreciated. The personal computer can be applied easily to gain new insight into the dynamic communities of hyphae in developing fruit bodies. Capture of video images for computerized image analysis takes some of the pain out of the sort of painstaking analysis which is essential to obtain the precise and detailed quantitative descriptions of temporal and spatial relationships between hyphae in developing structures.

These types of investigation can be done relatively easily. Going up a step we need to know more about the physiology and biochemistry of multihyphal structures. In many cases even the most simple questions cannot be answered at the moment. For example, how is oxygen supplied to the centre of a 20 cm diameter sclerotium of *Polyporus mylittae*? Oxygen tension has been shown to fall to zero within 150 µm of the surface of pellets of *Penicillium chrysogenum* formed in liquid cultures (Wittler *et al.*, 1986) but no similar information is available for more normal hyphal aggregates. It goes without saying that the study of fungal morphogenesis would be greatly advanced by deeper understanding of the molecular events involved. Involved in what? Involved in defining hyphal branch initiation, in directional growth of branches, in the nature of the hyphal surface, in the nature of the extracellular matrix, in hypha-to-hypha and hypha-to-extracellular matrix interactions, in the signalling processes and molecules which discriminate between tissues, in the expression of developmental specific genes and the nature of their polypeptides.

REFERENCES

Ainsworth, A.M. and Rayner, A.D.M. (1990) Aerial mycelial transfer by *Hymenochaete corrugata* between stems of hazel and other trees. *Mycological Research*, **92**, 263–6.

Ainsworth, G.C. (1961) *Ainsworth and Bisby's Dictionary of the Fungi*, 5th edn. Commonwealth Mycological Institute, Kew, Surrey, UK.

Allen, J.J., Moore, D. and Elliott, T.J. (1992) Persistent meiotic arrest in basidia of *Agaricus bisporus*. *Mycological Research*, **96**, 125–7.

Ashby, A.M. and Johnstone, K. (1993) Expression of the *E. coli* β-glucuronidase gene in the light of leaf spot pathogen *Pyrenopeziza brassicae* and its use as a reporter gene to study developmental interactions in fungi. *Mycological Research*, **97**, 575–81.

Barnett, H.L. and Lilly, V.G. (1949) Production of haploid and diploid fruit bodies of *Lenzites trabea* in culture. *Proceedings of the West Virginia Academy of Science*, **19**, 34–9.

Bartnicki-Garcia, S. (1990) Role of vesicles in apical growth and a new mathematical model of hyphal morphogenesis, in *Tip Growth in Plant and Fungal Cells*, (ed. I.B. Heath), Academic Press, San Diego and London, pp. 211–32.

Bastouill-Descollonges, Y. and Manachère, G. (1984) Photosporogenesis of *Coprinus congregatus*: correlations between the physiological age of lamellae and the development of their potential for renewed fruiting. *Physiologia Plantarum*, **61**, 607–10.

Beavan, M.J., Belk, D.M., Stewart, G.G. and Rose, A.H.

(1979) Changes in electrophoretic mobility and lytic enzyme activity associated with the development of flocculating ability in *Saccharomyces cerevisiae*. *Canadian Journal of Microbiology*, **25**, 888–95.

Boddy, L. (1993) Saprotrophic cord-forming fungi: warfare strategies and other ecological aspects. *Mycological Research* **97**, 641–55.

Boddy, L. and Rayner, A.D.M. (1983a) Mycelial interactions, morphogenesis and ecology of *Phlebia radiata* and *P. rufa* from oak. *Transactions of the British Mycological Society*, **80**, 437–48.

Boddy, L. and Rayner, A.D.M. (1983b) Ecological roles of basidiomycetes forming decay columns in attached oak branches. *New Phytologist*, **93**, 77–88.

Booth, C. (1966) Fruit bodies in Ascomycetes, in *The Fungi: An Advanced Treatise*, vol. II, (eds G.C. Ainsworth and A.S. Sussman), Academic Press, New York and London, pp. 133–50.

Botton, B. and Dexheimer, J. (1977) The ultrastructure of the rhizomorphs of *Sphaerostilbe repens* B. & B. *Zeitschrift für Pflanzenphysiologie*, **85**, **42**, 429–43.

Bougher, N.L., Tommerup, I.C. and Malajczuk, N. (1993) Broad variation in developmental and mature basidiome morphology of the ectomycorrhizal fungus *Hydnangium sublamellatum* sp. nov. bridges morphologically-based generic concepts of *Hydnangium*, *Podohydnangium* and *Laccaria*. *Mycological Research*, **97**, 613–19.

Brefeld, O. (1877) *Botanische Untersuchungen über Schimmelpilze*. III Heft. *Basidiomyceten I*. Arthur Felix, Leipzig.

Broxholme, S.J., Read, N.D. and Bond, D.J. (1991) Developmental regulation of proteins during fruit-body morphogenesis in *Sordaria brevicollis*. *Mycological Research*, **95**, 958–69.

Buller, A.H.R. (1924) *Researches on Fungi*, vol. 3. Longman Green, London.

Burnett, J.H. (1968) *Fundamentals of Mycology*. Edward Arnold, London.

Butler, G.M. (1957) The development and behaviour of mycelial strands in *Merulius lacrymans* (Wulf.) Fr. I. Strand development during growth from a food-base through a non-nutrient medium. *Annals of Botany*, **21**, 523–37.

Butler, G.M. (1958) The development and behaviour of mycelial strands in *Merulius lacrymans* (Wulf.) Fr. II. Hyphal behaviour during strand formation. *Annals of Botany*, **22**, 219–36.

Butler, G.M. (1966) Vegetative structure, in *The Fungi: An Advanced Treatise*, **vol. II**, (eds G.C. Ainsworth and A.S. Sussman), Academic Press, New York and London, pp. 83–112.

Butler, G.M. (1988) Pattern of pore morphogenesis in the resupinate basidiome of *Phellinus contiguus*. *Transactions of the British Mycological Society*, **91**, 677–86.

Butler, G.M. (1992a) Location of hyphal differentiation in the agar pore field of the basidiome of *Phellinus contiguus*. *Mycological Research*, **96**, 313–17.

Butler, G.M. (1992b) Capacity for differentiation of setae and other hyphal types of the basidiome in explants from cultures of the polypore *Phellinus contiguus*. *Mycological Research*, **96**, 949–55.

Cairney, J.W.G. and Clipson, N.J.W. (1991) Internal structure of rhizomorphs of *Trechispora vaga*, **95**, 764–7.

Cairney, J.W.G., Jennings, D.H. and Veltkamp, C.J. (1989) A scanning electron microscope study of the internal structure of mature linear mycelial organs of four basidiomycete species. *Canadian Journal of Botany*, **67**, 2266–71.

Castle, E.S. (1942) Spiral growth and reversal of spiralling in *Phycomyces*, and their bearing on primary wall structure. *American Journal of Botany*, **29**, 664–72.

Cavalier-Smith, T. (1981) Eukaryote kingdoms: seven or nine? *BioSystems*, **14**, 461–81.

Chadefaud, M. (1982a) Les principaux types d'ascocarpes: leur organisation et leur évolution. *Cryptogamie Mycologie*, **3**, 1–9.

Chadefaud, M. (1982b) Les principaux types d'ascocarpes: leur organisation et leur évolution. Deuxième partie: les discocarpes. *Cryptogamie Mycologie*, **3**, 103–44.

Chadefaud, M. (1982c) Les principaux types d'ascocarpes: leur organisation et leur evolution. Troisieme partie: les pyrenocarpes. *Cryptogamie Mycologie*, **3**, 199–235.

Chen, J.L. and Tzean, S.S. (1993) *Podosporium elongatum*, a new synnematous hyphomycete from Taiwan. *Mycological Research*, **97**, 637–40.

Chet, I. and Henis, Y. (1975) Sclerotial morphogenesis in fungi. *Annual Review of Phytopathology*, **13**, 169–92.

Chet, I., Henis, Y. and Kislev, N. (1969) Ultrastructure of sclerotia and hyphae of *Sclerotium rolfsii* Sacc. *Journal of General Microbiology*, **57**, 143–7.

Chiu, S.W. and Moore, D. (1988a) Evidence for developmental commitment in the differentiating fruit body of *Coprinus cinereus*. *Transactions of the British Mycological Society*, **90**, 247–53.

Chiu, S.W. and Moore, D. (1988b) Ammonium ions and glutamine inhibit sporulation of *Coprinus cinereus* basidia assayed *in vitro*. *Cell Biology International Reports*, **12**, 519–26.

Chiu, S.W. and Moore, D. (1990a) Development of the basidiome of *Volvariella bombycina*. *Mycological Research*, **94**, 327–37.

Chiu, S.W. and Moore, D. (1990b) A mechanism for gill pattern formation in *Coprinus cinereus*. *Mycological Research*, **94**, 320–6.

Chiu, S.W. and Moore, D. (1990c) Sporulation in *Coprinus cinereus*: use of an *in vitro* assay to establish the major landmarks in differentiation. *Mycological Research*, **94**, 249–53.

Chiu, S.W. and Moore, D. (1993) Cell form, function and lineage in the hymenia of *Coprinus cinereus* and *Volvariella bombycina*. *Mycological Research*, **97**, 221–6.

Chiu, S.W., Moore, D. and Chang, S.T. (1989) Basidiome polymorphism in *Volvariella bombycina*. *Mycological Research*, **92**, 69–77.

Coggins, C.R., Hornung, U., Jennings, D.H. and Veltkamp, C.J. (1980) The phenomenon of 'point-growth' and its relation to flushing and strand formation in mycelium of *Serpula lacrimans*. *Transactions of the British Mycological Society*, **75**, 69–76.

Coley-Smith, J.R. and Cooke, R.C. (1971) Survival and germination of fungal sclerotia. *Annual Review of Phytopathology*, **9**, 65–92.

Colson, B. (1935) The cytology of the mushroom *Psalliota campestris* Quél. *Annals of Botany*, **49**, 1–17.

Cooke, R.C. (1983) Morphogenesis of sclerotia, in *Fungal Differentiation, a Contemporary Synthesis*, (ed. J.E. Smith), Marcel Dekker: New York, pp. 397–418.

Corda, A.C.J. (1839) *Icones Fungorum Hucusque Cognitorum* III. Prague.

Corner, E.J.H. (1932) A *Fomes* with two systems of hyphae. *Transactions of the British Mycological Society*, **17**, 51–81.

Corner, E.J.H. (1966) *A Monograph of Cantharelloid Fungi*. Annals of Botany Memoirs no. 2. Oxford University Press, London.

Corner, E.J.H. (1991) *Trogia* (Basidiomycetes). *The Garden's Bulletin, Singapore*, supplement **2**, 1–100.

Damsky, C.H. and Werb, Z. (1992) Signal transduction by integrin receptors for extracellular matrix: cooperative processing of extracellular information. *Current Opinions in Cell Biology*, **4**, 772–81.

de Bary, A. (1884) *Vergleichende Morphologie und Biologie der Pilze*. USW Leipzig.

de Bary, A. (1887) *Comparative Morphology and Biology of the Fungi, Mycetozoa and Bacteria*. Oxford University (Clarendon) Press, London and New York.

de Silva, L.R., Youatt, J., Gooday, G.W. and Gow, N.A.R. (1992) Inwardly directed ionic currents of *Allomyces macrogynus* and other water moulds indicate sites of proton-driven nutrient transport but are incidental to tip growth. *Mycological Research*, **96**, 925–31.

de Vries, O.M.H. and Wessels, J.G.H. (1984) Patterns of polypeptide synthesis in non-fruiting monokaryons and a fruiting dikaryon of *Schizophyllum commune*. *Journal of General Microbiology*, **133**, 145–54.

de Vries, O.M.H., Hoge, J.H.C. and Wessels, J.G.H. (1980) Translation of RNA from *Schizophyllum commune* in a wheat germ and rabbit reticulocyte cell-free system: comparison of *in vitro* and *in vivo* products after two-dimensional gel electrophoresis. *Biochimica et Biophysica Acta*, **607**, 373–8.

Dons, J.J.M., Springer, J., de Vries, S.C. and Wessels, J.G.H. (1984) Molecular cloning of a gene abundantly expressed during fruiting body initiation in *Schizophyllum commune*. *Journal of Bacteriology*, **157**, 802–8.

Dormer, K.J. (1980) *Fundamental Tissue Geometry for Biologists*. Cambridge University Press, Cambridge, UK.

Elhiti, M.M.Y., Butler, R.D. and Moore, D. (1979) Cytochemical localization of glutamate dehydrogenase during carpophore development in *Coprinus cinereus*. *New Phytologist*, **82**, 153–7.

Elhiti, M.M.Y., Moore, D. and Butler, R.D. (1987) Ultrastructural distribution of glutamate dehydrogenases during fruit body development in *Coprinus cinereus*. *New Phytologist*, **107**, 531–9.

Elliott, T.J. (1985) Developmental genetics – from spore to sporophore, in *Developmental Biology of Higher Fungi*, (eds D. Moore, L.A. Casselton, D.A. Wood and J.C. Frankland), Cambridge University Press, Cambridge, pp. 451–65.

Esposito, R.E., Frink, N., Bernstein, P. and Esposito, M.S. (1972) The genetic control of sporulation in *Saccharomyces*. II. Dominance and complementation of mutants of meiosis and spore formation. *Molecular and General Genetics*, **114**, 241–8.

Esser, K. and Hoffman, F. (1977) Genetic basis for speciation in higher basidiomycetes with special reference to the genus *Polyporus*, in *The Species Concept in Hymenomycetes*, (ed. H. Clémençon), Cramer, Vaduz, pp. 189–214.

Esser, K. and Meinhardt, F. (1977) A common genetic control of dikaryotic and monokaryotic fruiting in the basidiomycete *Agrocybe aegerita*. *Molecular and General Genetics*, **155**, 113–15.

Esser, K. and Straub, J. (1958) Genetische Untersuchungen an *Sordaria macrospora* Auersw., Kompensation und induktion bei genbedingten Entwicklungsdefekten. *Zeitschrift für Vererbungslehre*, **89**, 729–46.

Esser, K., Stahl, U. and Meinhardt, F. (1977) Genetic aspects of differentiation in fungi, in *Biotechnology and Fungal Differentiation*, (eds J. Meyrath and J.D. Bu'-Lock), Academic Press, London, pp. 67–75.

Esser, K., Saleh, F. and Meinhardt, F. (1979) Genetics of fruit body production in higher basidiomycetes. II. Monokaryotic and dikaryotic fruiting in *Schizophyllum commune*. *Current Genetics*, **1**, 85–8.

Evans, H.J. (1959) Nuclear behaviour in the cultivated mushroom. *Chromosoma*, **10**, 115–35.

Fayod, V. (1889) Prodrome d'une histoire naturelle des Agaricinés. *Annales des Sciences Naturelles. Botanique Série*, **7–9**, 179–411.

Fisher, R.A. (1928) The possible modifications of the wild type to recurrent mutations. *American Naturalist*, **62**, 115–26.

Fisher, R.A. (1931) The evolution of dominance. *Biological Reviews*, **6**, 345–68.

Garrett, S.D. (1953) Rhizomorph behaviour in *Armillaria mellea* (Vahl) Quél. I. Factors controlling rhizomorph initiation by *Armillaria mellea* in pure culture. *Annals of Botany*, **17**, 63–79.

Garrett, S.D. (1954) Function of the mycelial strands in substrate colonization by the cultivated mushroom *Psalliota hortensis*. *Transactions of the British Mycological Society*, **37**, 51–7.

Garrett, S.D. (1956) *Biology of Root-Infecting Fungi*. Cambridge University Press, Cambridge.

Garrett, S.D. (1960) Inoculum potential, in *Plant Pathology: an Advanced Treatise*, vol. 3, (eds J.G. Horsfall and A.E. Dimond), Academic Press: New York and London, pp. 23–56.

Garrett, S.D. (1970) *Pathogenic Root-Infecting Fungi*. Cambridge University Press, Cambridge.

Girbardt, M. (1979) A microfilamentous septal belt (FSB) during induction of cytokinesis in *Trametes versicolor* (L. ex Fr.). *Experimental Mycology*, **3**, 215–28.

Gooday, G.W. (1985) Elongation of the stipe of *Coprinus cinereus*, in *Developmental Biology of Higher Fungi*, (eds D. Moore, L.A. Casselton, D.A. Wood and J.C. Frankland), Cambridge University Press, Cambridge, pp. 311–31.

Granlund, H.I., Jennings, D.H. and Thompson, W. (1985) Translocation of solute along rhizomorphs of *Armillaria mellea*. *Transactions of the British Mycological Society*, **84**, 111–19.

Green, J.B.A. and Smith, J.C. (1991) Growth factors as morphogens: do gradients and thresholds establish body plan? *Trends in Genetics*, **7**, 245–50.

Gregory, P.H. (1984) The fungal mycelium – an historical perspective, in *The Ecology and Physiology of the Fungal Mycelium*, (eds D.H. Jennings and A.D.M. Rayner), Cambridge University Press: Cambridge, pp. 383–417.

Griffin, D.H., Timberlake, W.E. and Cheney, J.C. (1974) Regulation of macromolecular synthesis, colony development and specific growth rate of *Achlya bisexualis* during balanced growth. *Journal of General Microbiology*, **80**, 381–8.

Grove, S.N. (1978) The cytology of hyphal tip growth, in *The Filamentous Fungi*, vol. 3, *Developmental Mycology*, (eds J.E. Smith and D.R. Berry), Edward Arnold, London, pp. 28–50.

Gull, K. (1976) Differentiation of septal ultrastructure according to cell type in the basidiomycete *Agrocybe praecox*. *Journal of Ultrastructure Research*, **54**, 89–94.

Gull, K. (1978) Form and function of septa in filamentous fungi, in *The Filamentous Fungi*, vol. 3, *Developmental Mycology*, (ed. J.E. Smith and D.R. Berry), Edward Arnold, London, pp. 78–93.

Hammad, F., Watling, R. and Moore, D. (1993a) Cell population dynamics in *Coprinus cinereus*: narrow and inflated hyphae in the basidiome stipe. *Mycological Research*, **97**, 269–74.

Hammad, F., Ji, J., Watling, R. and Moore, D. (1993b) Cell population dynamics in *Coprinus cinereus*: co-ordination of cell inflation throughout the maturing basidiome. *Mycological Research*, **97**, 275–82.

Harold, F.M. and Caldwell, J.H. (1990) Tips and currents: electrobiology of apical growth, in *Tip Growth in Plant and Fungal Cells* (ed. I.B. Heath), Academic Press, San Diego and London, pp. 59–90.

Hedger, J.N. (1985) Tropical agarics: resource relations and fruiting periodicity, in *Developmental Biology of Higher Fungi*, (eds D. Moore, L.A. Casselton, D.A. Wood and J.C. Frankland), Cambridge University Press, Cambridge, pp. 41–86.

Hedger, J.N., Lewis, P. and Gitay, H. (1993) Litter-trapping by fungi in moist tropical forest, in *Aspects of Tropical Mycology*, (eds S. Isaac, R. Watling, A.J.S. Whalley and J.C. Frankland), Cambridge University Press, Cambridge, pp. 15–35.

Hereward, F.V. and Moore, D. (1979) Polymorphic variation in the structure of aerial sclerotia of *Coprinus cinereus*. *Journal of General Microbiology*, **113**, 13–18.

Horner, J. and Moore, D. (1987) Cystidial morphogenetic field in the hymenium of *Coprinus cinereus*. *Transactions of the British Mycological Society*, **88**, 479–88.

Horton, J.S. and Raper, C.A. (1991) A mushroom-inducing DNA sequence isolated from the basidiomycete, *Schizophyllum commune*. *Genetics*, **129**, 707–16.

Hughes, S.J. (1953) Conidiophores, conidia, and classification. *Canadian Journal of Botany*, **31**, 577–659.

Hughes, S.J. (1971) On conidia of fungi, and gemmae of algae, bryophytes, and pteridophytes. *Canadian Journal of Botany*, **49**, 1319–39.

Hynes, R.O. (1992) Integrins: versatility, modulation, and signalling in cell adhesion. *Cell*, **69**, 11–25.

Ingold, C.T. (1979), *The Nature of Toadstools*. Studies in Biology Series, no. 113. Edward Arnold, London.

Jacques-Félix, M. (1967) Recherches morphologiques, anatomiques, morphogénétiques et physiologiques sur des rhizomorphes de champignons supérieurs et sur le déterminisme de leur formation. I. Observations sur les formations 'synnémiques' des champignons supérieurs dans le milieu naturel. *Bulletin Trimestrial de la Societé Mycologique de France*, **83**, 5–103.

Jennings, D.H. (1991) The physiology and biochemistry of the vegetative mycelium, in *Serpula lacrymans: Fundamental Biology and Control Strategies*, (eds D.H. Jennings and A.F. Bravery), Wiley, Chichester, pp. 55–79.

Jennings, D.H. and Bravery, A.F. (1991) *Serpula lacrymans: Fundamental Biology and Control Strategies*. Wiley, Chichester.

Jennings, D.H. and Watkinson, S.C. (1982). Structure and development of mycelial strands in *Serpula lacrimans*. *Transactions of the British Mycological Society*, **78**, 465–74.

Ji, J. and Moore, D. (1993) Glycogen metabolism in relation to fruit body maturation in *Coprinus cinereus*. *Mycological Research*, **97**, 283–9.

Johnson, T.E. (1976) Analysis of pattern formation in *Neurospora* perithecial development using genetic mosaics. *Developmental Biology*, **54**, 23–36.

Johnson, T.E. (1978) Isolation and characterisation of perithecial development mutants in *Neurospora*. *Genetics* **88**, 27–47.

Kamada, T. and Takemaru, T. (1977a) Stipe elongation during basidiocarp maturation in *Coprinus macrorhizus*: mechanical properties of stipe cell wall. *Plant and Cell Physiology*, **18**, 831–40.

Kamada, T. and Takemaru, T. (1977b) Stipe elongation during basidiocarp maturation in *Coprinus macrorhizus*: changes in polysaccharide composition of stipe cell wall during elongation. *Plant and Cell Physiology*, **18**, 1291–1300.

Kamada, T. and Takemaru, T. (1983) Modifications of cell wall polysaccharides during stipe elongation in the basidiomycete *Coprinus cinereus*. *Journal of General Microbiology*, **129**, 703–9.

Kamada, T., Sumiyoshi, T., Shindo, Y. and Takemaru, T. (1989) Isolation and genetic analysis of resistant mutants to the benzimidazole fungicide benomyl in *Coprinus cinereus*. *Current Microbiology*, **18**, 215–18.

Kamada, T., Takemaru, T., Prosser, J.I. and Gooday, G.W. (1991) Right and left handed helicity of chitin microfibrils in stipe cells in *Coprinus cinereus*. *Protoplasma*, **165**, 64–70.

Kamada, T. and Tsuru, M. (1993) The onset of the helical arrangement of chitin microfibrils in fruit body development of *Coprinus cinereus*. *Mycological Research*, **97**, 884–8.

Kanda, T. and Ishikawa, T. (1986) Isolation of recessive developmental mutants in *Coprinus cinereus*. *Journal of General and Applied Microbiology*, **32**, 541–3.

Kanda, T., Goto, A., Sawa, K. *et al.* (1989a). Isolation and characterization of recessive sporeless mutants in the basidiomycete *Coprinus cinereus*. *Molecular and General Genetics*, **216**, 526–9.

Kanda, T., Ishihara, H. and Takemaru, T. (1989b) Genetic analysis of recessive primordiumless mutants in the basidiomycete *Coprinus cinereus*. *Botanical Magazine, Tokyo*, **102**, 561–4.

Kanda, T., Arakawa, H., Yasuda, Y. and Takemaru, T. (1990), Basidiospore formation in a mutant of the incompatibility factors and mutants that arrest at meta-anaphase I in *Coprinus cinereus*. *Experimental Mycology*, **14**, 218–26.

Kemp, R.F.O. (1977) Oidial homing and the taxonomy and speciation of basidiomycetes with special reference to the genus *Coprinus*, in *The Species Concept in Hymenomycetes*, (ed. H. Clémencon), Cramer, Vaduz, pp. 259–73.

Kihn, J.C., Masy, C.L. and Mestdagh, M. M. (1988) Yeast flocculation: competition between nonspecific repulsion and specific bonding in cell adhesion. *Canadian Journal of Microbiology*, **34**, 773–8.

Kligman, A.M. (1943) Some cultural and genetic problems in the cultivation of the mushroom *Agaricus campestris*. *American Journal of Botany*, **30**, 745–63.

Kropf, D.L., Lupa, M.D.A., Caldwell, J.C. and Harold, F.M. (1983) Cell polarity: endogenous ion currents precede and predict branching in the water mould *Achlya*. *Science*, **220**, 1385–7.

Kropf, D.L., Caldwell, J.C., Gow, N.A.R. and Harold, F.M. (1984) Transcellular ion currents in the water mould *Achlya*. Amino acid proton symport as a mechanism of current entry. *Journal of Cell Biology*, **99**, 486–96.

Kües, U. and Casselton, L.A. (1992) Fungal mating type genes – regulators of sexual development. *Mycological Research*, **96**, 993–1006.

Lentz, P.L. (1971) Analysis of modified hyphae as a tool in taxonomic research in the higher Basidiomycetes, in *Evolution in the Higher Basidiomycetes*, (ed. R.H. Petersen), University of Tennessee Press, Knoxville, pp. 99–127.

Leonard, T.J. and Dick, S. (1968) Chemical induction of haploid fruiting bodies in *Schizophyllum commune*.

Proceedings of the National Academy of Sciences of the USA, **59**, 745–51.

Leslie, J.F. and Leonard, T.J. (1979a) Three independent genetic systems that control initiation of a fungal fruiting body. *Molecular and General Genetics*, **171**, 257–60.

Leslie, J.F. and Leonard, T.J. (1979b) Monokaryotic fruiting in *Schizophyllum commune*: genetic control of the response to mechanical injury. *Molecular and General Genetics*, **175**, 5–12.

Lewis, D. and Vakeria, D. (1977) Resistance to *p*-fluorophenylalanine in diploid/haploid dikaryons: dominance modifier gene explained as a controller of hybrid multimer formation. *Genetical Research*, **30**, 31–43.

Locquin, M. (1953) Recherches sur l'organisation et le développement des Agarics, des Bolets et des Clavulaires. *Bulletin de la Société Mycologique de France*, **69**, 389–402.

Lu, B.C. (1991) Cell degeneration and gill remodelling during basidiocarp development in the fungus *Coprinus cinereus*. *Canadian Journal of Botany*, **69**, 1161–9.

Macfarlane, T.D., Kuo, J. and Hilton, R.N. (1978) Structure of the giant sclerotium of *Polyporus mylittae*. *Transactions of the British Mycological Society*, **71**, 359–65.

Manning, J.T. (1976) Is sex maintained to facilitate or minimise mutational advance? *Heredity*, **36**, 351–7.

Manning, J.T. (1977) The evolution of dominance: Haldane v Fisher revisited. *Heredity*, **38**, 117–19.

Margulis, L. (1974) Five-Kingdom classification and the origin and evolution of cells. *Evolutionary Biology*, **7**, 45–78.

Mathew, K.T. (1961) Morphogenesis of mycelial strands in the cultivated mushroom *Agaricus bisporus*. *Transactions of the British Mycological Society*, **44**, 285–90.

Matthews, T.R. and Niederpruem, D.J. (1972) Differentiation in *Coprinus lagopus*. I. Control of fruiting and cytology of initial events: *Archives of Microbiology*, **87**, 257–68.

McGillivray, A.M. and Gow, N.A.R. (1986) Applied electrical fields polarize the growth of mycelial fungi. *Journal of General Microbiology*, **132**, 2515–25.

Meinhardt, F. and Esser, K. (1983) Genetic aspects of sexual differentiation in fungi, in *Fungal Differentiation*, (ed. J.E. Smith), Marcel Dekker, New York, pp. 537–57.

Meinhardt, H. (1984) Models of pattern formation and their application to plant development, in *Positional Controls in Plant Development*, (ed. P.W. Barlow and D.J. Carr), Cambridge University Press, Cambridge, pp. 1–32.

Meinhardt, H. and Gierer, A. (1974) Applications of a theory of biological pattern formation based on lateral inhibition. *Journal of Cell Science*, **15**, 321–46.

Miller, J.H. (1980) Orientation of the plane of cell division in fern gametophytes: the roles of cell shape and stress. *American Journal of Botany*, **67**, 534–42.

Miyake, H., Takemaru, T. and Ishikawa, T. (1980a) Sequential production of enzymes and basidiospore formation in fruiting bodies of *Coprinus macrorhizus*. *Archives of Microbiology*, **126**, 201–5.

Miyake, H., Tanaka, K. and Ishikawa, T. (1980b) Basidiospore formation in monokaryotic fruiting bodies of a mutant strain of *Coprinus macrorhizus*. *Archives of Microbiology*, **126**, 207–12.

Miyata, M., Miyata H. and Johnson, B.F. (1986) Establishment of septum orientation in a morphologically altered fission yeast, *Schizosaccharomyces pombe*. *Journal of General Microbiology*, **132**, 2535–40.

Mol, P.C., Vermeulen, C.A. and Wessels, J.G.H. (1990) Diffuse extension of hyphae in stipes of *Agaricus bisporus* may be based on a unique wall structure. *Mycological Research*, **94**, 480–8.

Moore, D. (1981) Developmental genetics of *Coprinus cinereus*: genetic evidence that carpophores and sclerotia share a common pathway of initiation. *Current Genetics*, **3**, 145–50.

Moore, D. (1984a) Positional control of development in fungi, in *Positional Controls in Plant Development*, (eds P.W. Barlow and D.J. Carr), Cambridge University Press, Cambridge, pp. 107–35.

Moore, D. (1984b) Developmental biology of the *Coprinus cinereus* carpophore: metabolic regulation in relation to cap morphogenesis. *Experimental Mycology*, **8**, 283–97.

Moore, D. (1987) The formation of agaric gills. *Transactions of the British Mycological Society*, **89**, 105–8.

Moore, D. (1991) Perception and response to gravity in higher fungi – a critical appraisal. *New Phytologist*, **117**, 3–23.

Moore, D. and Jirjis, R.I. (1981) Electrophoretic studies of carpophore development in the basidiomycete *Coprinus cinereus*. *New Phytologist*, **87**, 101–13.

Moore, D., Elhiti, M.M.Y. and Butler, R.D. (1979) Morphogenesis of the carpophore of *Coprinus cinereus*. *New Phytologist*, **83**, 695–722.

Moore, D., Horner, J. and Liu, M. (1987a) Co-ordinate control of ammonium-scavenging enzymes in the fruit body cap of *Coprinus cinereus* avoids inhibition of sporulation by ammonium. *FEMS Microbiology Letters*, **44**, 239–42.

Moore, D., Liu, M. and Kuhad, R.C. (1987b) Karyogamy-dependent enzyme derepression in the basidiomycete *Coprinus*. *Cell Biology International Reports*, **11**, 335–41.

Moore, R.T. (1985). The challenge of the dolipore/parenthesome septum, in *Developmental Biology of Higher Fungi*, (eds D. Moore, L. A. Casselton, D.A. Wood and J.C. Frankland), Cambridge University Press, Cambridge, pp. 175–212.

Motta, J.J. (1967) A note on the mitotic apparatus in the rhizomorph meristem of *Armillaria mellea*. *Mycologia*, **59**, 370–5.

Motta, J.J. (1969) Cytology and morphogenesis in the rhizomorph of *Armillaria mellea*. *American Journal of Botany*, **56**, 610–19.

Motta, J.J. (1971) Histochemistry of the rhizomorph meristem of *Armillaria mellea*. *American Journal of Botany*, **58**, 80–87.

Motta, J.J. and Peabody, D.C. (1982), Rhizomorph cytology and morphogenesis in *Armillaria tabescens*. *Mycologia*, **74**, 671–4.

Mulder, G.H. and Wessels, J.G.H. (1986) Molecular cloning of RNAs differentially expressed in mono-karyons and dikaryons of *Schizophyllum commune* in relation to fruiting. *Experimental Mycology*, **10**, 214–27.

Nasrallah, J.B. and Srb, A.M. (1973) Genetically related protein variants specifically associated with fruiting body maturation in *Neurospora*. *Proceedings of the National Academy of Sciences of the USA*, **70**, 1891–3.

Nasrallah, J.B. and Srb, A.M. (1977) Occurrence of a major protein associated with fruiting body development in *Neurospora* and related Ascomycetes. *Proceedings of the National Academy of Sciences of the USA*, **74**, 3831–4.

Nasrallah, J.B. and Srb, A.M. (1978) Immunofluorescent localization of a phase-specific protein in *Neurospora tetrasperma* perithecia. *Experimental Mycology*, **2**, 211–15.

Nathan, C. and Sporn, M. (1991) Cytokines in context. *Journal of Cell Biology*, **113**, 981–6.

Nuss, I., Jennings, D.H. and Veltkamp, C.J. (1991) Morphology of *Serpula lacrymans*, in *Serpula lacrymans: Fundamental Biology and Control Strategies*, (eds D.H. Jennings and A.F. Bravery), Wiley, Chichester, pp. 9–38.

Nyunoya, H. and Ishikawa, T. (1979) Control of unusual hyphal morphology in a mutant of *Coprinus macrorhizus*. *Japanese Journal of Genetics*, **54**, 11–20.

Powell, K.A. and Rayner, A.D.M. (1983) Ultrastructure of the rhizomorph apex in *Armillaria bulbosa* in relation to mucilage production. *Transactions of the British Mycological Society*, **81**, 529–34.

Prosser, J.I. (1993) Growth kinetics of mycelial colonies and hyphal aggregates of ascomycetes. *Mycological Research*, **97**.

Pukkila, P.J. and Casselton, L.A. (1991) Molecular genetics of the agaric *Coprinus cinereus*. *More Gene Manipulations in Fungi*, (ed. J.W. Bennett and L.A. Lasure), Academic Press, New York, pp. 126–50.

Rahary, L., Bonaly, R., Lematre, J. and Poulain, D. (1985) Aggregation and disaggregation of *Candida albicans* germ tubes. *FEMS Microbiology Letters*, **30**, 383–7.

Raper, J.R. and Krongelb, G.S. (1958) Genetic and environmental aspects of fruiting in *Schizophyllum commune* Fr. *Mycologia*, **50**, 707–40.

Rayner, A.D.M. (1994) Differential insulation and the generation of mycelial patterns, in *Shape and Form in Plants and Fungi*, (ed. D.S. Ingram), Academic Press, London, pp. 291–310.

Rayner, A.D.M. and Webber, J.F. (1984) Interspecific mycelial interactions – an overview, in *The Ecology and Physiology of the Fungal Mycelium*, (eds D.H. Jennings and A.D.M. Rayner), Cambridge University Press, Cambridge, pp. 383–417.

Rayner, A.D.M., Powell, K.A., Thompson, W. and Jennings, D.H. (1985a) Morphogenesis of vegetative organs, in *Developmental Biology of Higher Fungi*, (eds D. Moore, L. A. Casselton, D.A. Wood and J.C. Frankland), Cambridge University Press, Cambridge, pp. 249–79.

Rayner, A.D.M., Watling, R. and Frankland, J.C. (1985b). Resource relations – an overview, in *Developmental Biology of Higher Fungi*, (eds D. Moore, L.A. Casselton, D.A. Wood and J.C. Frankland), Cambridge University Press: Cambridge, pp. 1–40.

Read, D.J. (1991) Mycorrhizas in ecosystems - Nature's response to the 'Law of the Minimum', in *Frontiers of Mycology*, (ed. D.L. Hawksworth), CAB International, Wallingford, pp. 101–30.

Read, D.J. and Armstrong, W. (1972) A relationship between oxygen transport and the formation of the ectotrophic mycorrhizal sheath in conifer seedlings. *New Phytologist*, **71**, 49–53.

Read, D.J., Leake, J.R. and Langdale, A.R. (1989) The nitrogen nutrition of mycorrhizal fungi and their host plants, in *Nitrogen, Phosphorus and Sulphur Utilization by Fungi*, (eds L. Boddy, R. Marchant and D.J. Read), Cambridge University Press, Cambridge, pp. 181–204.

Read, N.D. (1983) A scanning electron microscopic study of the external features of perithecium development in *Sordaria humana*. *Canadian Journal of Botany*, **61**, 3217–29.

Read, N.D. (1993) Multicellular development in fungi, in *Shape and Form in Plants and Fungi*, (ed. D.S. Ingram), Academic Press, London.

Read, N.D. and Beckett, A. (1985) The anatomy of the mature perithecium in *Sordaria humana*; and its significance for fungal multicellular development. *Canadian Journal of Botany*, **63**, 281–96.

Redhead, S.A. (1987) The Xerulaceae (Basidiomycetes), a family with sarcodimitic tissues. *Canadian Journal of Botany*, **65**, 1551–62.

Reijnders, A.F.M. (1948) Études sur le développement et l'organisation histologique des carpophores dans les Agaricales. *Recuil des Travaux Botaniques Néerlandais*, **41**, 213–396.

Reijnders, A.F.M. (1963) *Les problèmes du Développement des Carpophores des Agaricales et de Quelques Groupes Voisins*. Dr W. Junk, The Hague.

Reijnders, A.F.M. (1976) Recherches sur le développement et l'histogénèse dans les Asterosporales. *Persoonia*, **9**, 65–83.

Reijnders, A.F.M. (1977) The histogenesis of bulb and trama tissue of the higher Basidiomycetes and its phylogenetic implications. *Persoonia*, **9**, 329–62.

Reijnders, A.F.M. (1979) Developmental anatomy of *Coprinus*. *Persoonia*, **10**, 383–424.

Reijnders, A.F.M. (1993) On the origin of specialised trama types in the Agaricales. *Mycological Research*, **97**, 257–68.

Reijnders, A.F.M. and Moore, D. (1985) Developmental biology of agarics – an overview, in *Developmental Biology of Higher Fungi*, (ed. D. Moore, L.A. Casselton, D.A. Wood and J.C. Frankland), Cambridge University Press, Cambridge, pp. 333–51.

Reynolds, D.R. (1981), *Ascomycete Systematics: The Luttrellian Concept*. Springer-Verlag, New York.

Rishbeth, J. (1985) *Armillaria*: resources and hosts, in *Developmental Biology of Higher Fungi*, (eds D. Moore, L.A. Casselton, D.A. Wood and J.C. Frankland), Cambridge University Press, Cambridge, pp. 87–101.

Rosin, I.V. and Moore, D. (1985a) Origin of the hymenophore and establishment of major tissue domains during fruit body development in *Coprinus cinereus*. *Transactions of the British Mycological Society*, **84**, 609–19.

Rosin, I.V. and Moore, D. (1985b) Differentiation of the hymenium in *Coprinus cinereus*. *Transactions of the British Mycological Society* **84**, 621–8.

Rosin, I.V., Horner, J. and Moore, D. (1985) Differentiation and pattern formation in the fruit body cap of *Coprinus cinereus*, in *Developmental Biology of Higher Fungi*, (eds D. Moore, L.A. Casselton, D.A. Wood and J.C. Frankland), Cambridge University Press, Cambridge, pp. 333–51.

Ruiters, M.H.J. and Wessels, J.G.H. (1989a) *In situ* localization of specific RNAs in whole fruiting colonies of *Schizophyllum commune*. *Journal of General Microbiology*, **135**, 1747–54.

Ruiters, M.H.J. and Wessels, J.G.H. (1989b) *In situ* localization of specific RNAs in developing fruit bodies of the basidiomycete *Schizophyllum commune*. *Experimental Mycology*, **13**, 212–22.

Russo, G.M., Dahlberg, K.R. and Van Etten, J.L. (1982) Identification of a development-specific protein in sclerotia of *Sclerotinia sclerotiorum*. *Experimental Mycology*, **6**, 259–67.

Schuren, F.H.J., van der Lende, T.R. and Wessels, J.G.H. (1993) Fruiting genes of *Schizophyllum commune* are transcriptionally regulated. *Mycological Research*, **97**, 538–42.

Senathirajah, S. and Lewis, D. (1975) Resistance to amino acid analogues in *Coprinus*: dominance modifier genes and dominance reversal in dikaryons and diploids. *Genetical Research*, **25**, 95–107.

Sheppard, P.M. (1967) *Natural Selection and Heredity*. Hutchinson, London.

Slack, J.M.W. (1983) *From Egg to Embryo: Determinative Events in Early Development*. Cambridge University Press, Cambridge.

Snider, P.J. (1959) Stages of development in rhizomorphic thalli of *Armillaria mellea*. *Mycologia*, **51**, 693–707.

Stahl, U. and Esser, K. (1976) Genetics of fruit body production in higher basidiomycetes. I. Monokaryotic fruiting and its correlation with dikaryotic fruiting in *Polyporus ciliatus*. *Molecular and General Genetics*, **148**, 183–97.

Stephenson, N.A. and Gooday, G.W. (1984) Nuclear numbers in the stipe cells of *Coprinus cinereus*. *Transactions of the British Mycological Society*, **82**, 531–4.

Streuli, C.H. (1993) Extracellular matrix and gene expression in mammary epithelium. *Seminars in Cell Biology*, **4**, 203–12.

Sussman, A.S. (1968) Longevity and survivability of fungi, in *The Fungi: An Advanced Treatise*, **vol. III**, (ed. G.C. Ainsworth and A.S. Sussman), Academic Press, New York and London, pp. 447–86.

Swamy, S., Uno, I. and Ishikawa, T. (1984) Morphogenetic effects of mutations at the *A* and *B* incompatibility factors in *Coprinus cinereus*. *Journal of General Microbiology*, **130**, 3219–24.

Takemaru, T. and Kamada, T. (1971) Gene control of basidiocarp development in *Coprinus macrorhizus*. *Reports of the Tottori Mycological Institute, Japan*, **9**, 21–35.

Takemaru, T. and Kamada, T. (1972) Basidiocarp development in *Coprinus macrorhizus*. I. Induction of developmental variations. *Botanical Magazine (Tokyo)*, **85**, 51–7.

Talbot, P.H.B. (1968) Fossilized pre-Patouillardian taxonomy? *Taxon*, **17**, 622–8.

Thompson, W. (1984) Distribution, development and functioning of mycelial cord systems of decomposer basidiomycetes of the deciduous woodland floor, in *The Ecology and Physiology of the Fungal Mycelium*, (eds D.H. Jennings and A.D.M. Rayner), Cambridge University Press, Cambridge, pp. 185–214.

Timberlake, W.E. (1980) Developmental gene regulation in *Aspergillus nidulans*. *Developmental Biology*, **78**, 497–510.

Timberlake, W.E. and Marshall, M.A. (1988) Genetic regulation of development in *Aspergillus nidulans*. *Trends in Genetics*, **4**, 162–9.

Todd, N.K. and Aylmore, R.C. (1985) Cytology of hyphal interactions and reactions in *Schizophyllum commune*, in *Developmental Biology of Higher Fungi*, (eds D. Moore, L.A. Casselton, D.A. Wood and J.C. Frankland), Cambridge University Press, Cambridge, pp. 231–48.

Townsend, B.B. (1954) Morphology and development of fungal rhizomorphs. *Transactions of the British Mycological Society*, **37**, 222–33.

Townsend, B.B. and Willetts, H.J. (1954) The development of sclerotia of certain fungi. *Transactions of the British Mycological Society*, **37**, 213–21.

Trinci, A.P.J. (1974) A study of the kinetics of hyphal extension and branch initiation of fungal mycelia. *Journal of General Microbiology*, **81**, 225–36.

Trinci, A.P.J. (1978) The duplication cycle and vegetative development in moulds, in *The Filamentous Fungi*, vol. 3, *Developmental Mycology*, (eds J.E. Smith and D.R. Berry), Edward Arnold, London, pp. 133–63.

Trinci, A.P.J. (1984) Regulation of hyphal branching and hyphal orientation, in *The Ecology and Physiology of the Fungal Mycelium*, (eds D.H. Jennings and A.D.M. Rayner), Cambridge University Press, Cambridge, pp. 23–52.

Trinci, A.P.J. and Banbury, G.H. (1967) A study of the tall conidiophores of *Aspergillus giganteus*. *Transactions of the British Mycological Society*, **50**, 525–38.

Turian, G. (1978) Sexual morphogenesis in the Ascomycetes, in *The Filamentous Fungi*, vol. 3, *Developmental*

Mycology, (eds J.E. Smith and D.R. Berry), 315–33. Edward Arnold, London.

Ullrich, R.C. (1973) Sexuality, incompatibility, and inter-sterility in the biology of the *Sistotrema brinkmannii* aggregate. *Mycologia*, **65**, 1234–49.

Uno, I. and Ishikawa, T. (1971) Chemical and genetical control of induction of monokaryotic fruiting bodies in *Coprinus macrorhizus*. *Molecular and General Genetics*, **113**, 228–39.

Van der Valk, P. and Marchant, R. (1978) Hyphal ultrastructure in fruit body primordia of the basidio-mycetes *Schizophyllum commune* and *Coprinus cinereus*. *Protoplasma*, **95**, 57–72.

Waters, H. (1972) *Aspects of sclerotium morphogenesis in Coprinus lagopus (sensu) Bull*. PhD Thesis, University of Manchester.

Waters, H., Moore, D. and Butler, R.D. (1975a) Morpho-genesis of aerial sclerotia of *Coprinus lagopus*. *New Phytologist*, **74**, 207–13.

Waters, H., Butler, R.D. and Moore, D. (1975b) Structure of aerial and submerged sclerotia of *Coprinus lagopus*. *New Phytologist*, **74**, 199–205.

Watkinson, S.C. (1971) The mechanism of mycelial strand induction in *Serpula lacrimans*: a possible effect of nutrient distribution. *New Phytologist*, **70**, 1079–88.

Watkinson, S.C. (1975) The relation between nitrogen nutrition and the formation of mycelial strands in *Serpula lacrimans*. *Transactions of the British Mycological Society*, **64**, 195–200.

Watkinson, S.C. (1979) Growth of rhizomorphs, mycelial strands coremia and sclerotia, in *Fungal Walls and Hyphal Growth*, (ed. J.H. Burnett and A.P.J. Trinci), Cambridge University Press, Cambridge, pp. 93–113.

Watling, R. (1971) Polymorphism in *Psilocybe merdaria*. *New Phytologist*, **70**, 307–26.

Watling, R. (1985) Developmental characters of agarics, in *Developmental Biology of Higher Fungi*, (ed. D. Moore, L.A. Casselton, D.A. Wood and J.C. Frankland), Cambridge University Press, Cambridge, pp. 281–310.

Watling, R. and Moore, D. (1994) Moulding moulds into mushrooms: shape and form in the higher fungi, in *Shape and Form in Plants and Fungi*, (ed. D.S. Ingram), Academic Press, London, pp. 274–92.

Watling, R. and Nicoll, H. (1980) Sphaerocysts in *Lactarius rufus*. *Transactions of the British Mycological Society* **75**, 331–3.

Webster, J. (1980) *Introduction to Fungi*, 2nd edn. Cambridge University Press, Cambridge.

Wessels, J.G.H. (1992) Gene expression during fruiting in *Schizophyllum commune*. *Mycological Research*, **96**, 609–20.

Wessels, J.G.H., Mulder, G.H. and Springer, J. (1987) Expression of dikaryon-specific and non-specific mRNAs of *Schizophyllum commune* in relation to environmental conditions and fruiting. *Journal of General Microbiology*, **133**, 2557–61.

Whittaker, R.H. (1969) New concepts of kingdoms of organisms. *Science*, **163**, 150–60.

Willetts, H.J. (1968) The development of stromata of *Sclerotinia fructicola* and related species. II. In fruits. *Transactions of the British Mycological Society*, **51**, 633–42.

Willetts, H.J. (1969) Structure of the outer surfaces of sclerotia of certain fungi. *Archiv für Mikrobiologie*, **69**, 48–53.

Willetts, H.J. (1971) The survival of fungal sclerotia under adverse environmental conditions. *Biological Reviews*, **46**, 387–407.

Willetts, H.J. (1972) The morphogenesis and possible evolutionary origins of fungal sclerotia. *Biological Reviews*, **47**, 515–36.

Willetts, H.J. and Bullock, S. (1992) Developmental biology of sclerotia. *Mycological Research*, **96**, 801–16.

Willetts, H.J. and Calonge, F.D. (1969) Spore develop-ment in the brown rot fungi (*Sclerotinia* spp.). *New Phytologist*, **68**, 123–31.

Willetts, H.J. and Wong, A.L. (1971) Ontogenetic diver-sity of sclerotia of *Sclerotinia sclerotiorum* and related species. *Transactions of the British Mycological Society*, **57**, 515–24.

Williams, M.A.J. (1986) *Studies on the structure and development of Flammulina velutipes (Curtis: Fries) Singer*. PhD Thesis, University of Bristol.

Williams, M.A.J., Beckett, A. and Read, N.D. (1985) Ultrastructural aspects of fruit body differentiation in *Flammulina velutipes*, in *Developmental Biology of Higher Fungi* (ed. D. Moore, L.A. Casselton, D.A. Wood and J.C. Frankland), Cambridge University Press, Cam-bridge, pp. 429–50.

Wittler, R., Baumgartl, H., Lubbers, D.W. and Shugerl, K. (1986) Investigations of oxygen transfer into *Penicil-lium chrysogenum* pellets by microprobe measure-ments. *Biotechnology and Bioengineering*, **28**, 1024–36.

Wosten, H.A.B., De Vries, O.M.H. and Wessels, J.G.H. (1993) Interfacial self-assembly of a fungal hydrophobin into a hydrophobic rootlet layer. *Plant Cell*, **5**, 1567–74.

Wosten, H.A.B., Asgeirsdottir, S.A., Krook, J.H., Drenth, J.H.H. and Wessels, J.G.H. (1994) The fungal hydrophobin Sc3p self-assembles at the surface of aerial hyphae as a protein membrane constituting the hydrophobic rootlet layer. *European Journal of Cell Biology*, **63**, 122–9.

Yashar, B.M. and Pukkila, P.J. (1985) Changes in polyadenylated RNA sequences associated with fruit-ing body morphogenesis in *Coprinus cinereus*. *Trans-actions of the British Mycological Society*, **84**, 215–26.

Zantinge, B., Dons, H. and Wessels, J.G.H. (1979) Comparison of poly(A)-containing RNAs in different cell types of the lower eukaryote *Schizophyllum com-mune*. *European Journal of Biochemistry*, **101**, 251–60.

Zolan, M.E., Tremel, C.J. and Pukkila, P.J. (1988) Produc-tion and characterization of radiation-sensitive meiotic mutants of *Coprinus cinereus*. *Genetics*, **120**, 379–87.

Index